Mathematical Methods in Interdisciplinary Sciences

Mathematical Methods in Interdisciplinary Sciences

Edited by

Snehashish Chakraverty

This edition first published 2020

Registered Office
John Wiley & Sons, Inc., 111 River Street, Hoboken, NJ 07030, USA

Editorial Office
111 River Street, Hoboken, NJ 07030, USA

For details of our global editorial offices, customer services, and more information about Wiley products visit us at www.wiley.com.

Wiley also publishes its books in a variety of electronic formats and by print-on-demand. Some content that appears in standard print versions of this book may not be available in other formats.

Library of Congress Cataloging-in-Publication Data

Names: Chakraverty, Snehashish, editor.
Title: Mathematical methods in interdisciplinary sciences / edited by Snehashish Chakraverty.
Description: Hoboken, NJ : Wiley, 2020. | Includes bibliographical references and index.
Identifiers: LCCN 2020001570 (print) | LCCN 2020001571 (ebook) | ISBN 9781119585503 (cloth) | ISBN 9781119585619 (adobe pdf) | ISBN 9781119585657 (epub)
Subjects: LCSH: Science–Mathematics. | Mathematical analysis.
Classification: LCC Q175.32.M38 M38 2020 (print) | LCC Q175.32.M38 (ebook) | DDC 501/.51–dc23
LC record available at https://lccn.loc.gov/2020001570
LC ebook record available at https://lccn.loc.gov/2020001571

Cover design by Wiley
Cover image: © Sharlotta/Shutterstock

Set in 9.5/12.5pt STIXTwoText by SPi Global, Chennai, India

10 9 8 7 6 5 4 3 2 1

Contents

Notes on Contributors *xv*
Preface *xxv*
Acknowledgments *xxvii*

1 Connectionist Learning Models for Application Problems Involving Differential and Integral Equations *1*
Susmita Mall, Sumit Kumar Jeswal, and Snehashish Chakraverty
1.1 Introduction *1*
1.1.1 Artificial Neural Network *1*
1.1.2 Types of Neural Networks *1*
1.1.3 Learning in Neural Network *2*
1.1.4 Activation Function *2*
1.1.4.1 Sigmoidal Function *3*
1.1.5 Advantages of Neural Network *3*
1.1.6 Functional Link Artificial Neural Network (FLANN) *3*
1.1.7 Differential Equations (DEs) *4*
1.1.8 Integral Equation *5*
1.1.8.1 Fredholm Integral Equation of First Kind *5*
1.1.8.2 Fredholm Integral Equation of Second Kind *5*
1.1.8.3 Volterra Integral Equation of First Kind *5*
1.1.8.4 Volterra Integral Equation of Second Kind *5*
1.1.8.5 Linear Fredholm Integral Equation System of Second Kind *6*
1.2 Methodology for Differential Equations *6*
1.2.1 FLANN-Based General Formulation of Differential Equations *6*
1.2.1.1 Second-Order Initial Value Problem *6*
1.2.1.2 Second-Order Boundary Value Problem *7*
1.2.2 Proposed Laguerre Neural Network (LgNN) for Differential Equations *7*
1.2.2.1 Architecture of Single-Layer LgNN Model *7*
1.2.2.2 Training Algorithm of Laguerre Neural Network (LgNN) *8*
1.2.2.3 Gradient Computation of LgNN *9*
1.3 Methodology for Solving a System of Fredholm Integral Equations of Second Kind *9*
1.3.1 Algorithm *10*
1.4 Numerical Examples and Discussion *11*
1.4.1 Differential Equations and Applications *11*
1.4.2 Integral Equations *16*
1.5 Conclusion *20*
References *20*

2 **Deep Learning in Population Genetics: Prediction and Explanation of Selection of a Population** *23*
Romila Ghosh and Satyakama Paul
2.1 Introduction *23*
2.2 Literature Review *23*
2.3 Dataset Description *25*
2.3.1 Selection and Its Importance *25*
2.4 Objective *26*
2.5 Relevant Theory, Results, and Discussions *27*
2.5.1 automl *27*
2.5.2 Hypertuning the Best Model *28*
2.6 Conclusion *30*
References *30*

3 **A Survey of Classification Techniques in Speech Emotion Recognition** *33*
Tanmoy Roy, Tshilidzi Marwala, and Snehashish Chakraverty
3.1 Introduction *33*
3.2 Emotional Speech Databases *33*
3.3 SER Features *34*
3.4 Classification Techniques *35*
3.4.1 Hidden Markov Model *36*
3.4.1.1 Difficulties in Using HMM for SER *37*
3.4.2 Gaussian Mixture Model *37*
3.4.2.1 Difficulties in Using GMM for SER *38*
3.4.3 Support Vector Machine *38*
3.4.3.1 Difficulties with SVM *39*
3.4.4 Deep Learning *39*
3.4.4.1 Drawbacks of Using Deep Learning for SER *41*
3.5 Difficulties in SER Studies *41*
3.6 Conclusion *41*
References *42*

4 **Mathematical Methods in Deep Learning** *49*
Srinivasa Manikant Upadhyayula and Kannan Venkataramanan
4.1 Deep Learning Using Neural Networks *49*
4.2 Introduction to Neural Networks *49*
4.2.1 Artificial Neural Network (ANN) *50*
4.2.1.1 Activation Function *52*
4.2.1.2 Logistic Sigmoid Activation Function *52*
4.2.1.3 tanh or Hyperbolic Tangent Activation Function *53*
4.2.1.4 ReLU (Rectified Linear Unit) Activation Function *54*
4.3 Other Activation Functions (Variant Forms of ReLU) *55*
4.3.1 Smooth ReLU *55*
4.3.2 Noisy ReLU *55*
4.3.3 Leaky ReLU *55*
4.3.4 Parametric ReLU *56*
4.3.5 Training and Optimizing a Neural Network Model *56*

4.4 Backpropagation Algorithm *56*
4.5 Performance and Accuracy *59*
4.6 Results and Observation *59*
References *61*

5 Multimodal Data Representation and Processing Based on Algebraic System of Aggregates *63*
Yevgeniya Sulema and Etienne Kerre
5.1 Introduction *63*
5.2 Basic Statements of ASA *64*
5.3 Operations on Aggregates and Multi-images *65*
5.4 Relations and Digital Intervals *72*
5.5 Data Synchronization *75*
5.6 Fuzzy Synchronization *92*
5.7 Conclusion *96*
References *96*

6 Nonprobabilistic Analysis of Thermal and Chemical Diffusion Problems with Uncertain Bounded Parameters *99*
Sukanta Nayak, Tharasi Dilleswar Rao, and Snehashish Chakraverty
6.1 Introduction *99*
6.2 Preliminaries *99*
6.2.1 Interval Arithmetic *99*
6.2.2 Fuzzy Number and Fuzzy Arithmetic *100*
6.2.3 Parametric Representation of Fuzzy Number *101*
6.2.4 Finite Difference Schemes for PDEs *102*
6.3 Finite Element Formulation for Tapered Fin *102*
6.4 Radon Diffusion and Its Mechanism *105*
6.5 Radon Diffusion Mechanism with TFN Parameters *107*
6.5.1 EFDM to Radon Diffusion Mechanism with TFN Parameters *108*
6.6 Conclusion *112*
References *112*

7 Arbitrary Order Differential Equations with Fuzzy Parameters *115*
Tofigh Allahviranloo and Soheil Salahshour
7.1 Introduction *115*
7.2 Preliminaries *115*
7.3 Arbitrary Order Integral and Derivative for Fuzzy-Valued Functions *116*
7.4 Generalized Fuzzy Laplace Transform with Respect to Another Function *118*
References *122*

8 Fluid Dynamics Problems in Uncertain Environment *125*
Perumandla Karunakar, Uddhaba Biswal, and Snehashish Chakraverty
8.1 Introduction *125*
8.2 Preliminaries *126*
8.2.1 Fuzzy Set *126*
8.2.2 Fuzzy Number *126*

8.2.3 δ-Cut 127
8.2.4 Parametric Approach 127
8.3 Problem Formulation 127
8.4 Methodology 129
8.4.1 Homotopy Perturbation Method 129
8.4.2 Homotopy Perturbation Transform Method 130
8.5 Application of HPM and HPTM 131
8.5.1 Application of HPM to Jeffery–Hamel Problem 131
8.5.2 Application of HPTM to Coupled Whitham–Broer–Kaup Equations 134
8.6 Results and Discussion 136
8.7 Conclusion 142
References 142

9 Fuzzy Rough Set Theory-Based Feature Selection: A Review *145*
Tanmoy Som, Shivam Shreevastava, Anoop Kumar Tiwari, and Shivani Singh
9.1 Introduction 145
9.2 Preliminaries 146
9.2.1 Rough Set Theory 146
9.2.1.1 Rough Set 146
9.2.1.2 Rough Set-Based Feature Selection 147
9.2.2 Fuzzy Set Theory 147
9.2.2.1 Fuzzy Tolerance Relation 148
9.2.2.2 Fuzzy Rough Set Theory 149
9.2.2.3 Degree of Dependency-Based Fuzzy Rough Attribute Reduction 149
9.2.2.4 Discernibility Matrix-Based Fuzzy Rough Attribute Reduction 149
9.3 Fuzzy Rough Set-Based Attribute Reduction 149
9.3.1 Degree of Dependency-Based Approaches 150
9.3.2 Discernibility Matrix-Based Approaches 154
9.4 Approaches for Semisupervised and Unsupervised Decision Systems 154
9.5 Decision Systems with Missing Values 158
9.6 Applications in Classification, Rule Extraction, and Other Application Areas 158
9.7 Limitations of Fuzzy Rough Set Theory 159
9.8 Conclusion 160
References 160

10 Universal Intervals: Towards a Dependency-Aware Interval Algebra *167*
Hend Dawood and Yasser Dawood
10.1 Introduction 167
10.2 The Need for Interval Computations 169
10.3 On Some Algebraic and Logical Fundamentals 170
10.4 Classical Intervals and the Dependency Problem 174
10.5 Interval Dependency: A Logical Treatment 176
10.5.1 Quantification Dependence and Skolemization 177
10.5.2 A Formalization of the Notion of Interval Dependency 179
10.6 Interval Enclosures Under Functional Dependence 184
10.7 Parametric Intervals: How Far They Can Go 186
10.7.1 Parametric Interval Operations: From Endpoints to Convex Subsets 186
10.7.2 On the Structure of Parametric Intervals: Are They Properly Founded? 188

10.8 Universal Intervals: An Interval Algebra with a Dependency Predicate *192*
10.8.1 Universal Intervals, Rational Functions, and Predicates *193*
10.8.2 The Arithmetic of Universal Intervals *196*
10.9 The S-Field Algebra of Universal Intervals *201*
10.10 Guaranteed Bounds or Best Approximation or Both? *209*
Supplementary Materials *210*
Acknowledgments *211*
References *211*

11 Affine-Contractor Approach to Handle Nonlinear Dynamical Problems in Uncertain Environment *215*
Nisha Rani Mahato, Saudamini Rout, and Snehashish Chakraverty
11.1 Introduction *215*
11.2 Classical Interval Arithmetic *217*
11.2.1 Intervals *217*
11.2.2 Set Operations of Interval System *217*
11.2.3 Standard Interval Computations *218*
11.2.4 Algebraic Properties of Interval *219*
11.3 Interval Dependency Problem *219*
11.4 Affine Arithmetic *220*
11.4.1 Conversion Between Interval and Affine Arithmetic *220*
11.4.2 Affine Operations *221*
11.5 Contractor *223*
11.5.1 SIVIA *223*
11.6 Proposed Methodology *225*
11.7 Numerical Examples *230*
11.7.1 Nonlinear Oscillators *230*
11.7.1.1 Unforced Nonlinear Differential Equation *230*
11.7.1.2 Forced Nonlinear Differential Equation *232*
11.7.2 Other Dynamic Problem *233*
11.7.2.1 Nonhomogeneous Lane–Emden Equation *233*
11.8 Conclusion *236*
References *236*

12 Dynamic Behavior of Nanobeam Using Strain Gradient Model *239*
Subrat Kumar Jena, Rajarama Mohan Jena, and Snehashish Chakraverty
12.1 Introduction *239*
12.2 Mathematical Formulation of the Proposed Model *240*
12.3 Review of the Differential Transform Method (DTM) *241*
12.4 Application of DTM on Dynamic Behavior Analysis *242*
12.5 Numerical Results and Discussion *244*
12.5.1 Validation and Convergence *244*
12.5.2 Effect of the Small-Scale Parameter *245*
12.5.3 Effect of Length-Scale Parameter *247*
12.6 Conclusion *248*
Acknowledgment *249*
References *250*

13 **Structural Static and Vibration Problems** *253*
M. Amin Changizi and Ion Stiharu
13.1 Introduction *253*
13.2 One-parameter Groups *254*
13.3 Infinitesimal Transformation *254*
13.4 Canonical Coordinates *254*
13.5 Algorithm for Lie Symmetry Point *255*
13.6 Reduction of the Order of the ODE *255*
13.7 Solution of First-Order ODE with Lie Symmetry *255*
13.8 Identification *256*
13.9 Vibration of a Microcantilever Beam Subjected to Uniform Electrostatic Field *258*
13.10 Contact Form for the Equation *259*
13.11 Reducing in the Order of the Nonlinear ODE Representing the Vibration of a Microcantilever Beam Under Electrostatic Field *260*
13.12 Nonlinear Pull-in Voltage *261*
13.13 Nonlinear Analysis of Pull-in Voltage of Twin Microcantilever Beams *266*
13.14 Nonlinear Analysis of Pull-in Voltage of Twin Microcantilever Beams of Different Thicknesses *268*
References *272*

14 **Generalized Differential and Integral Quadrature: Theory and Applications** *273*
Francesco Tornabene and Rossana Dimitri
14.1 Introduction *273*
14.2 Differential Quadrature *274*
14.2.1 Genesis of the Differential Quadrature Method *274*
14.2.2 Differential Quadrature Law *275*
14.3 General View on Differential Quadrature *277*
14.3.1 Basis Functions *278*
14.3.1.1 Lagrange Polynomials *281*
14.3.1.2 Trigonometric Lagrange Polynomials *282*
14.3.1.3 Classic Orthogonal Polynomials *282*
14.3.1.4 Monomial Functions *291*
14.3.1.5 Exponential Functions *291*
14.3.1.6 Bernstein Polynomials *291*
14.3.1.7 Fourier Functions *292*
14.3.1.8 Bessel Polynomials *292*
14.3.1.9 Boubaker Polynomials *292*
14.3.2 Grid Distributions *293*
14.3.2.1 Coordinate Transformation *293*
14.3.2.2 δ-Point Distribution *293*
14.3.2.3 Stretching Formulation *293*
14.3.2.4 Several Types of Discretization *293*
14.3.3 Numerical Applications: Differential Quadrature *297*
14.4 Generalized Integral Quadrature *310*
14.4.1 Generalized Taylor-Based Integral Quadrature *312*
14.4.2 Classic Integral Quadrature Methods *314*
14.4.2.1 Trapezoidal Rule with Uniform Discretization *314*
14.4.2.2 Simpson's Method (One-third Rule) with Uniform Discretization *314*

14.4.2.3 Chebyshev–Gauss Method (Chebyshev of the First Kind) *314*
14.4.2.4 Chebyshev–Gauss Method (Chebyshev of the Second Kind) *314*
14.4.2.5 Chebyshev–Gauss Method (Chebyshev of the Third Kind) *315*
14.4.2.6 Chebyshev–Gauss Method (Chebyshev of the Fourth Kind) *315*
14.4.2.7 Chebyshev–Gauss–Radau Method (Chebyshev of the First Kind) *315*
14.4.2.8 Chebyshev–Gauss–Lobatto Method (Chebyshev of the First Kind) *315*
14.4.2.9 Gauss–Legendre or Legendre–Gauss Method *315*
14.4.2.10 Gauss–Legendre–Radau or Legendre–Gauss–Radau Method *315*
14.4.2.11 Gauss–Legendre–Lobatto or Legendre–Gauss–Lobatto Method *316*
14.4.3 Numerical Applications: Integral Quadrature *316*
14.4.4 Numerical Applications: Taylor-Based Integral Quadrature *320*
14.5 General View: The Two-Dimensional Case *324*
References *340*

15 Brain Activity Reconstruction by Finding a Source Parameter in an Inverse Problem *343*
Amir H. Hadian-Rasanan and Jamal Amani Rad
15.1 Introduction *343*
15.1.1 Statement of the Problem *344*
15.1.2 Brief Review of Other Methods Existing in the Literature *345*
15.2 Methodology *346*
15.2.1 Weighted Residual Methods and Collocation Algorithm *346*
15.2.2 Function Approximation Using Chebyshev Polynomials *349*
15.3 Implementation *353*
15.4 Numerical Results and Discussion *354*
15.4.1 Test Problem 1 *355*
15.4.2 Test Problem 2 *357*
15.4.3 Test Problem 3 *358*
15.4.4 Test Problem 4 *359*
15.4.5 Test Problem 5 *362*
15.5 Conclusion *365*
References *365*

16 Optimal Resource Allocation in Controlling Infectious Diseases *369*
A.C. Mahasinghe, S.S.N. Perera, and K.K.W.H. Erandi
16.1 Introduction *369*
16.2 Mobility-Based Resource Distribution *370*
16.2.1 Distribution of National Resources *370*
16.2.2 Transmission Dynamics *371*
16.2.2.1 Compartment Models *371*
16.2.2.2 SI Model *371*
16.2.2.3 Exact Solution *371*
16.2.2.4 Transmission Rate and Potential *372*
16.2.3 Nonlinear Problem Formulation *373*
16.2.3.1 Piecewise Linear Reformulation *374*
16.2.3.2 Computational Experience *374*
16.3 Connection–Strength Minimization *376*
16.3.1 Network Model *376*

16.3.1.1 Disease Transmission Potential *376*
16.3.1.2 An Example *376*
16.3.2 Nonlinear Problem Formulation *377*
16.3.2.1 Connection Strength Measure *377*
16.3.2.2 Piecewise Linear Approximation *378*
16.3.2.3 Computational Experience *379*
16.4 Risk Minimization *379*
16.4.1 Novel Strategies for Individuals *379*
16.4.1.1 Epidemiological Isolation *380*
16.4.1.2 Identifying Objectives *380*
16.4.2 Minimizing the High-Risk Population *381*
16.4.2.1 An Example *381*
16.4.2.2 Model Formulation *382*
16.4.2.3 Linear Integer Program *383*
16.4.2.4 Computational Experience *383*
16.4.3 Minimizing the Total Risk *384*
16.4.4 Goal Programming Approach *384*
16.5 Conclusion *386*
References *387*

17 Artificial Intelligence and Autonomous Car *391*
Merve Antürk, Sırma Yavuz, and Tofigh Allahviranloo
17.1 Introduction *391*
17.2 What Is Artificial Intelligence? *391*
17.3 Natural Language Processing *391*
17.4 Robotics *393*
17.4.1 Classification by Axes *393*
17.4.1.1 Axis Concept in Robot Manipulators *393*
17.4.2 Classification of Robots by Coordinate Systems *394*
17.4.3 Other Robotic Classifications *394*
17.5 Image Processing *395*
17.5.1 Artificial Intelligence in Image Processing *395*
17.5.2 Image Processing Techniques *395*
17.5.2.1 Image Preprocessing and Enhancement *396*
17.5.2.2 Image Segmentation *396*
17.5.2.3 Feature Extraction *396*
17.5.2.4 Image Classification *396*
17.5.3 Artificial Intelligence Support in Digital Image Processing *397*
17.5.3.1 Creating a Cancer Treatment Plan *397*
17.5.3.2 Skin Cancer Diagnosis *397*
17.6 Problem Solving *397*
17.6.1 Problem-solving Process *397*
17.7 Optimization *399*
17.7.1 Optimization Techniques in Artificial Intelligence *399*
17.8 Autonomous Systems *400*
17.8.1 History of Autonomous System *400*
17.8.2 What Is an Autonomous Car? *401*

17.8.3 Literature of Autonomous Car *402*
17.8.4 How Does an Autonomous Car Work? *405*
17.8.5 Concept of Self-driving Car *406*
17.8.5.1 Image Classification *407*
17.8.5.2 Object Tracking *407*
17.8.5.3 Lane Detection *408*
17.8.5.4 Introduction to Deep Learning *408*
17.8.6 Evaluation *409*
17.9 Conclusion *410*
References *410*

18 Different Techniques to Solve Monotone Inclusion Problems *413*
Tanmoy Som, Pankaj Gautam, Avinash Dixit, and D. R. Sahu
18.1 Introduction *413*
18.2 Preliminaries *414*
18.3 Proximal Point Algorithm *415*
18.4 Splitting Algorithms *415*
18.4.1 Douglas–Rachford Splitting Algorithm *416*
18.4.2 Forward–Backward Algorithm *416*
18.5 Inertial Methods *418*
18.5.1 Inertial Proximal Point Algorithm *419*
18.5.2 Splitting Inertial Proximal Point Algorithm *421*
18.5.3 Inertial Douglas–Rachford Splitting Algorithm *421*
18.5.4 Pock and Lorenz's Variable Metric Forward–Backward Algorithm *422*
18.5.5 Numerical Example *428*
18.6 Numerical Experiments *429*
References *430*

Index *433*

Notes on Contributors

Snehashish Chakraverty has an experience of 29 years as a researcher and teacher. Presently, he is working in the Department of Mathematics (Applied Mathematics Group), National Institute of Technology Rourkela, Odisha, as a senior (HAG) professor. Before this, he was with CSIR-Central Building Research Institute, Roorkee, India. After completing graduation from St. Columba's College (Ranchi University), his career started from the University of Roorkee (Now, Indian Institute of Technology Roorkee) and did MSc (Mathematics) and MPhil (Computer Applications) from there securing the first position in the university. Dr. Chakraverty received his PhD from IIT Roorkee in 1992. Thereafter, he did his postdoctoral research at the Institute of Sound and Vibration Research (ISVR), University of Southampton, UK, and at the Faculty of Engineering and Computer Science, Concordia University, Canada. He was also a visiting professor at Concordia and McGill Universities, Canada, during 1997–1999 and a visiting professor of University of Johannesburg, South Africa, during 2011–2014. He has authored/coauthored 16 books, published 333 research papers (till date) in journals and conferences, 2 more books are in press, and 2 books are ongoing. He is in the editorial boards of various International Journals, Book Series, and Conferences. Prof. Chakraverty is the chief editor of the "*International Journal of Fuzzy Computation and Modelling*" (IJFCM), Inderscience Publisher, Switzerland (http://www.inderscience.com/ijfcm), associate editor of "Computational Methods in Structural Engineering, Frontiers in Built Environment," and happens to be the editorial board member of "Springer Nature Applied Sciences," "IGI Research Insights Books," "Springer Book Series of Modeling and Optimization in Science and Technologies," "**Coupled Systems Mechanics (Techno Press)**," "Curved and Layered Structures (De Gruyter)," "Journal of Composites Science (*MDPI*)," "Engineering Research Express (IOP)," *and* "Applications and Applied Mathematics: An International Journal." He is also the reviewer of around 50 national and international journals of repute, and he was the President of the Section of Mathematical Sciences (including Statistics) of "Indian Science Congress" (2015–2016) and was the Vice President – "Orissa Mathematical Society" (2011–2013). Prof. Chakraverty is a recipient of prestigious awards, viz. Indian National Science Academy (INSA) nomination under International Collaboration/Bilateral Exchange Program (with Czech Republic), Platinum Jubilee ISCA Lecture Award (2014), CSIR Young Scientist (1997), BOYSCAST (DST), UCOST Young Scientist (2007, 2008), Golden Jubilee Director's (CBRI) Award (2001), INSA International Bilateral Exchange Award (2010–2011 [selected but could not undertake], 2015 [selected]), Roorkee University Gold Medals (1987, 1988) for first positions in MSc and MPhil (Computer Applications), etc. He has already guided 15 PhD students and 9 are ongoing. Prof. Chakraverty has undertaken around 16 research projects as the principle investigator funded by international and national agencies totaling about ₹1.5 crores. A good number of international and national conferences, workshops, and training programs have also been organized by him. His present research area includes differential equations (ordinary, partial, and fractional), Numerical Analysis and Computational Methods, Structural Dynamics (FGM, Nano) and Fluid Dynamics, Mathematical Modeling and Uncertainty Modeling, Soft Computing and Machine Intelligence (Artificial Neural Network, Fuzzy, Interval, and Affine Computations).

Susmita Mall received her PhD degree in Mathematics from the National Institute of Technology Rourkela, Rourkela, Odisha, India, in 2016. Currently, she is a postdoctoral fellow in National Institute of Technology, Rourkela – 769 008, Odisha, India. She has been awarded Women Scientist Scheme-A (WOS-A) fellowship, under Department of Science and Technology (DST), Government of India, to undertake her PhD studies and Post Doc. Her current research interest includes mathematical modeling, artificial neural network, differential equations, and numerical analysis.

Sumit Kumar Jeswal received the undergraduation degree from G.M. College, Sambalpur, India, the MSc degree in mathematics from Utkal University, Bhubaneswar, India, and the MTech degree in computer science and data processing from Indian Institute of Technology Kharagpur, Kharagpur, India. He is currently pursuing the PhD degree with the Department of Mathematics, National Institute of Technology Rourkela, Rourkela, India. His current research interests include artificial neural network, fuzzy and interval analysis, and numerical analysis.

Romila Ghosh did her master's graduate in statistics from Amity University, India. At present, she is working with the Data Sciences team in one of the reputed Big 4 consulting MNCs. Her areas of interest include biostatistics, population genetics, and data science.

Satyakama Paul is a PhD and postdoctorate in computational intelligence from the University of Johannesburg, South Africa. At present, he is a senior data scientist of a consulting firm in Kolkata, India. His research interests lie in deep learning and its application in various domains such as finance, medical image processing, etc.

Tanmoy Roy is doing his doctoral studies from the University of Johannesburg. His primary research interest is speech emotion recognition (SER), where he is applying the latest machine learning and deep learning techniques to solve the SER problem. His publications include one journal paper in an Elsevier journal named "Communications in Nonlinear Science and Numerical Simulations" and two conference papers, one IEEE and another Elsevier. Apart from that, he has also submitted one book chapter with Springer. Before his doctoral studies, he was an experienced software programmer and worked as an associate consultant at SIEMENS.

Tshilidzi Marwala received his PhD from the University of Cambridge in Artificial Intelligence. He is presently the Vice Chancellor and Principal of the University of Johannesburg. University of Johannesburg is making remarkable progress under his energetic and dynamic leadership. After his PhD, he became a postdoctoral research associate at the University of London's Imperial College of Science, Technology, and Medicine, where he worked on intelligence software. He was a visiting fellow at Harvard University and Wolfson College, Cambridge. His research interests include the theory and application of artificial intelligence to engineering, computer science, finance, economics, social science, and medicine. He has made fundamental contributions to engineering science including the development of the concept of pseudo-modal energies, proposing the theory of rational counterfactual thinking, rational opportunity cost, and the theory of flexibly bounded rationality. He was a coinventor of the innovative methods of radiation imaging with Megan Jill Russell and invented the artificial larynx with David Rubin. His publication list is long, which includes important patents, books, and articles. He has published his work in reputed journals and conferences. His work has been cited by many organizations, including the International Standards Organizations (ISO) and NASA. He collaborates with researchers in South Africa, the United States of America, England, France, Sweden, and India. He regularly acts as a reviewer for international journals. He has received numerous awards and honors including winning the National Youth Science Olympiad, and because of this, he attended the 1989 London International Youth Science Forum, the Order of Mapungubwe, Case Western Reserve University Professional Achievement Award, The Champion of Research Capacity Development and Transformation at SA Higher Education Institutions, NSTF Award, Harvard-South Africa Fellowship, TWAS-AAS-Microsoft Award, and ACM Distinguished Member Award.

Srinivasa Manikant Upadhyayula is the lead analyst for machine learning and automation team at CRISIL Global Research & Analytics with an extensive experience of over a decade in financial services and technology across investment banking, private banking, risk, and finance. He is passionately pursuing an active innovation roadmap to enable artificial intelligence (AI)-based products in financial risk and analytics. He is pioneering initiatives that apply AI, ML, NLP, and big data platforms to automate business processes and control functions, validate model risk scenarios, and proactively manage operational and reputational risk for various large banks.

He holds an MBA in finance with honors from Great Lakes Institute of Management and Bachelor's in electronics and communication engineering from Sri Chandrasekharendra Saraswathi Viswa Mahavidyalaya (SCSVMV) University.

Kannan Venkataramanan is a senior quantitative analyst for risk and analytics at CRISIL Global Research & Analytics, which has a broad mandate to build and implement ML- and AI-based automation solutions across clients. He is involved in driving efforts to apply word embedding techniques and NLP algorithms to streamline market news analysis as well as automate financial insights for investments banks. Before CRISIL, he was a decision scientist at MuSigma Inc., where he was involved in analyzing big data, identifying anomalies and patterns, and designing business strategies to solve complex, real-time problems.

Earlier, he completed his bachelor in mechatronics engineering with a gold medal from Sastra University. Also, he has published a couple of research papers on Euler's totient function and Jacobsthal numbers.

Yevgeniya S. Sulema is an associated professor at the Computer Systems Software Department and a vice-dean at the Faculty of Applied Mathematics of the National Technical University of Ukraine "Igor Sikorsky Kyiv Polytechnic Institute." She received her PhD degree from the National Technical University of Ukraine "Kyiv Polytechnic Institute" in 1999. She is a member of the editorial advisory board of Systemics, Cybernetics and Informatics Journal and a member of the program committees at several international conferences. She is a member of the International Institute of Informatics and Systemics. The research in her laboratory of Multimedia, Mulsemedia, and Immersive Technologies covers multimodal data representation and processing, mulsemedia and immersive applications, image and audio processing, and multimedia data protection methods. She is an author of more than 130 scientific publications. She participates in numerous European and national research projects. She teaches the course on multimedia technologies and the course on digital signal and image processing to Master program students and the course on mathematical and algorithmic fundamentals of computer graphics to Bachelor program students at her university. She also gave lectures on multimedia, mulsemedia, and immersive technologies to students at several European universities as a guest lecturer within academic mobility projects.

Etienne E. Kerre was born in Zele, Belgium on 8 May 1945. He obtained his MSc degree in mathematics in 1967 and his PhD in mathematics in 1970 from Ghent University. Since 1984, he has been a lector and since 1991 a full-time professor at Ghent University. In 2010, he became a retired professor. He is a referee for more than 80 international scientific journals and also a member of the editorial board of international journals and conferences on fuzzy set theory. He was an honorary chairman at various international conferences. In 1976, he founded the Fuzziness and Uncertainty Modeling Research Unit (FUM), and since then, his research has been focused on the modeling of fuzziness and uncertainty and has resulted in a great number of contributions in fuzzy set theory and its various generalizations. Especially, the theories of fuzzy relational calculus and of fuzzy mathematical structures owe a very great deal of him. Over the years, he has also been a promoter of 30 PhDs on fuzzy set theory. His current research interests include fuzzy and intuitionistic fuzzy relations, fuzzy topology, and fuzzy image processing. He has authored or coauthored 25 books, and more than 500 papers appeared in international refereed journals and proceedings.

Sukanta Nayak is an assistant professor of mathematics at Amrita Vishwa Vidyapeetham, Coimbatore, India. Before coming to Amrita Vishwa Vidyapeetham, he was a postdoctoral research fellow at University of Johannesburg, South Africa. He received PhD in mathematics from the National Institute of Technology Rourkela in 2016. Dr. Nayak has received Global Excellence Stature Postdoctoral Fellowship in 2016 and postgraduate scholarship by Government of Odisha in 2008. In addition, he qualified all India Graduate Aptitude Test in Engineering (GATE) and awarded as best presentation of Department of Mathematics at Research Scholar Week (RSW-2015), NIT Rourkela in 2015. His research interests span uncertainty modeling and investigation of various diffusion problems, viz., heat transfer and neutron diffusion. He is also an author of the book entitled "Neutron Diffusion: Concepts and Uncertainty Analysis for Engineers and Scientists" of CRC Press (Taylor and Francis) and "Interval Finite Element Method with MATLAB" of Academic Press (Elsevier), as well as he has several book chapters and numerous articles in academic journals.

Tharasi Dilleswar Rao is presently a PhD scholar at the National Institute of Technology (NIT) Rourkela, Odisha, India. He received his MSc degree from Osmania University, Telangana, in 2013. Then, he worked as a faculty from 2014 to 2015. Then, he started his research work as a PhD student at NIT Rourkela in 2015. His research area includes numerical methods, fuzzy/interval uncertainty, and modeling radon transport mechanisms. He has four published/accepted papers in reputed journals, two book chapters (accepted) in Springer, and he is a coauthor of a book entitled "Advanced Numerical Methods and Semi-Analytical Methods for Differential Equations" in John Wiley and Sons till now. He also has two conference papers.

Tofigh Allahviranloo is a senior full-time professor in applied mathematics at Bahcesehir World International University (BAU) in Istanbul.

As a trained mathematician and computer scientist, Tofigh has developed a passion for multi- and interdisciplinary research. He is not only deeply involved in fundamental research in fuzzy applied mathematics, especially fuzzy differential equations, but he also aims at innovative applications in the applied biological sciences.

Soheil Salahshour is an Assistant Professor of Applied Mathematics at Bahcesehir University, Istanbul, Turkey. He not only pursues fundamental research in fuzzy applied mathematics, especially fuzzy differential equations of arbitrary order, but also develops innovative applications in the applied sciences like as mechanical, electrical and industrial engineering.

Karunakar Perumandla is working as an assistant professor in the Department of Mathematics, Amrita School of Engineering, Amrita Vishwa Vidyapeetham, Chennai Campus, India. He has submitted his PhD thesis on "Shallow Water Wave Equations with Uncertain Parameters" to National Institute of Technology Rourkela, Odisha in 2019. He received his MSc degree from NIT Warangal in 2009. His research area includes numerical methods, fuzzy/interval uncertainty, and shallow water wave equations. He has eight published/accepted papers in reputed journals, a book and two book chapters in Springer till now. Also, he has three conference papers. He is the editorial assistant of International Journal of Fuzzy Computation and Modelling (Inderscience publisher).

Uddhaba Biswal is presently a PhD scholar at National Institute of Technology (NIT) Rourkela, Odisha, India. He received his BSc degree from Rajendra (Auto.) college, Balangir, Odisha, India, in 2014 and MSc degree from Pondicherry University, Puducherry, India, in 2016. His research area includes fluid dynamics, fuzzy/interval uncertainty, and numerical analysis.

Tanmoy Som is serving as a professor in the Department of Mathematical Sciences, Indian Institute of Technology (Banaras Hindu University), Varanasi, India, since last 10 years and has more than 30 years of teaching and research experience. He has done PhD from Institute of Technology, Banaras Hindu University, Varanasi, India, in 1986. His main research areas are "fixed point theory, fuzzy and soft set theory, image processing and mathematical modeling." Around 125 research publications are there to his credit in reputed journals and edited book chapters of National and International repute. He has delivered several invited talks at National and International conferences/workshops/refresher courses, which includes talks at Texas A&M University, Kingsville, USA, Naresuan University, Thailand, and University of California, Berkeley. Thirteen scholars have completed PhD under his supervision. He is also a reviewer/editorial board member/guest editor of few reputed journals. He has organized quite a few national/international level conferences/workshops.

Shivam Shreevastava received his BSc (Hons) from Institute of Science, Banaras Hindu University, Varanasi, India, in 2011 and MSc in mathematics from Indian Institute of Technology, Kanpur, in 2013. He has completed his PhD from the Department of Mathematical Sciences, Indian Institute of Technology (BHU), Varanasi, India, in 2018. His research area focuses on applications of fuzzy set theory in decision making problems. He has several research publications in reputed journals in the last few years.

Anoop Kumar Tiwari has completed his BSc, MSc, and PhD degree from Department of Computer Science and Engineering, Dr. K. N. Modi University, Tank, Rajasthan, India. His research interests are machine learning, bioinformatics, and medical image processing. He has quite a few good research publications in the above areas to his credit.

Shivani Singh received her BSc in 2013 and MSc in mathematics in 2015 from PPN PG College, Kanpur, UP. She is currently pursuing her PhD in DST-Centre for Interdisciplinary Mathematical Sciences, Institute of Science, Banaras Hindu University, Varanasi, India. Her research area focuses on fuzzy rough set-based techniques for knowledge discovery process.

Hend Dawood is a senior lecturer of computational mathematics in the Department of Mathematics at Cairo University. Her current research interests include algebraic systems of interval mathematics, ordered algebraic structures and algebraic logic, computable analysis, nonclassical logics, formal fuzzy logics, automatic differentiation, and uncertainty quantification. Dr. Dawood is a member of the IEEE 1788 committee for standardizing interval arithmetic, the Cairo University Interval Arithmetic Research Group (CUIA), and the Egyptian Mathematical Society (EMS). She also serves in the editorial board of a number of international journals of repute in the field of computational mathematics.

Yasser Dawood is currently the head of the Department of Mathematical and Computational Urbanism, Go Green Architects, Giza, Egypt. He has over 20 years of experience in the fields of computational and applied mathematics. His current research interests include algebraic logic, nonclassical logics, formal fuzzy logics, reasoning under uncertainty, and formal representations of uncertainty, with emphasis on applying algebraic and logical methods to engineering and real-world problems. Dr. Dawood serves as a consultant for many Egyptian and international companies and institutes. He also serves in the editorial and reviewing boards of a number of international journals of repute in the field of applied and computational mathematics.

Nisha Rani Mahato is a PhD scholar at the Department of Mathematics, National Institute of Technology, Rourkela, Odisha. She completed her MSc degree in mathematics at the National Institute of Technology, in 2011. She was awarded Raman–Charpak Fellowship-2016 (RCF-2016) by CEFIPRA, New Delhi. Also, she was awarded the best paper at the 38th Annual Conference of Orissa Mathematical Society in 2011 and best poster in mathematics at Research Scholar Week 2018, NIT Rourkela. Further, she has participated in various conferences/workshops and published a few research papers, two book chapters, and two books. Her current research areas include interval computing, fuzzy set theory, interval/fuzzy eigenvalue problems, interval/fuzzy simultaneous equations, and interval/fuzzy differential equations.

Saudamini Rout is currently pursuing her PhD degree under the supervision of Prof. S. Chakraverty at the Department of Mathematics, National Institute of Technology Rourkela, Rourkela, Odisha, India. She completed her MPhil degree in Mathematics in 2016, MSc degree in Mathematics in 2015, and BSc degree in 2013 from Ravenshaw University, Cuttack, Odisha, India. She was awarded a gold medal for securing first position in the university for MSc degree. Her current research interests include fuzzy set theory, interval analysis, affine arithmetic, uncertain linear, and/or nonlinear static and dynamic problems and numerical analysis.

Subrat Kumar Jena is currently working as a research fellow at the Department of Mathematics (MA), National Institute of Technology Rourkela. He is also working in Defence Research and Development Organisation (DRDO) sponsored project entitled "Vibrations of Functionally Graded Nanostructural Members" in collaboration with Defence Metallurgical Research Laboratory (DMRL). Subrat does research in structural dynamics, nano vibration, applied mathematics, computational methods, and numerical analysis, differential equations, fractional differential equations, mathematical modeling, uncertainty modelling, soft computing, etc. He has published 13 research papers (till date) in journals, 4 conference papers, and 3 book chapters. He has also served as the reviewer for International Journal of Fuzzy Computation and Modelling (IJFCM), and Computational Methods in Structural Engineering. Subrat has attended one national conference and one GIAN course on "Structural Dynamics." He has also been continuing collaborative research works with renowned researchers from Italy, Estonia, Turkey, Iran, etc.

Rajarama Mohan Jena is currently working as a senior research fellow at the Department of Mathematics (MA), National Institute of Technology Rourkela. He is an INSPIRE (Innovation in Science Pursuit for Inspired Research) fellow of Department of Science of Technology, Ministry of Science and Technology, Govt. of India, and doing his research under this fellowship scheme. Rajarama does research in fractional dynamical systems, applied mathematics, computational methods, numerical analysis, partial differential equations, fractional differential equations, uncertainty modeling, and soft computing. He has published 12 research papers (till date) in journals, 1 conference paper, and 1 book chapter. Rajarama also served as the reviewer for International Journal of Fuzzy Computation and Modelling (IJFCM), Abstract and Applied Analysis, and SN Applied Sciences. He has attended one national conference. He has been continuing collaboration works with renowned researchers from Turkey, Canada, Iran, Egypt, Nigeria, etc.

M. Amin Changizi received the PhD degree from Concordia University, Montréal, Canada, in 2011. He is currently a senior mechanical engineer researcher in Mohawk Innovative Technology. He is working in turbomachine and bearing industry. Previously, he was a knowledge engineer with Intelliquip, USA. He was working on pump industry, from 2014 to 2018. Before Intelliquip, he was with Carrier Corporation, USA, as a staff engineer, from 2012 to 2013. Before Carrier Corporation, he was a part-time faculty member with Concordia University, Montréal, from 2008 to 2011. He has been a full-time faculty member with Tabriz Azad University, Tabriz, for seven years, and an engineer in several different positions in PETCO pump industry. His research interests are theoretical nonlinear mechanics, Lie symmetry method with applications on nonlinear differential equations and nonlinear behaviors of MEMS devices.

Ion Stiharu is graduated as a diplomat engineer of the Polytechnic University of Bucharest – Romania – in 1979 and PhD in 1989 from the same institution. In 1991, he joined Concordia University in Montreal in the Department of Mechanical, Industrial, and Aerospace Engineering. His research interests are focusing on microsystems performance, microfabrication, and applications. He has supervised and co-supervised 31 PhD students who successfully graduated and he authored and coauthored more than 100 journal publications and over 250 conference and presentations. He is one of the editors of "Micromachines" journal and he is part of the editorial board of few journals. He is a fellow of the ASME and CSME.

Francesco Tornabene was born in Bologna, 13 January 1978, and received a degree in mechanical engineering (Course of Studies in Structural Mechanics) on 23 July 2003. He submitted a thesis under the title (in Italian) Dynamic Behavior of Cylindrical Shells: Formulation and Solution. He obtained the first position in the competition for admission to the PhD in structural mechanics in December 2003. He is the winner of the scholarship Carlo Felice Jodi for a degree in structural mechanics in 2004. He did PhD in structural mechanics on 31 May 2007. He is the author of 11 books. Some of them entitled Meccanica delle Strutture a Guscio in Materiale Composito. Il metodo Generalizzato di Quadratura Differenziale, 2012; Mechanics of Laminated Composite Doubly-Curved Shell Structures. The Generalized Differential Quadrature Method and the Strong Formulation Finite Element Method, 2014; Laminated Composite Doubly-Curved Shell Structures I. Differential Geometry. Higher-Order Structural Theories, 2016; Laminated Composite Doubly-Curved Shell Structures II. Differential and Integral Quadrature. Strong Formulation Finite Element Method, 2016; and Anisotropic Doubly-Curved Shells. Higher-Order Strong and Weak Formulations for Arbitrarily Shaped Shell Structures, 2018. He is an author of more than 190 research papers since 2004. He is an assistant professor at the University of Bologna from 2012 to 2018 and from 2018 until now at the University of Salento. His research focuses on structural mechanics, mechanics of solids and structures, computational mechanics, composite and innovative materials.

Rossana Dimitri is an associate professor at the School of Engineering, Department of Innovation Engineering, University of Salento, Lecce, Italy. She received from the University of Salento, a MSc degree in "materials engineering" in 2004, a PhD degree in "materials and structural engineering" in 2009, and a PhD degree in "industrial and mechanical engineering" in 2013. In 2005, she received from the University of Salento the "Best M. Sc. Thesis Price 2003–2004" in memory of Eng. Gabriele De Angelis; in 2013, she was awarded by the Italian Group for Computational Mechanics (GIMC) for the Italian selection of the 2013 ECCOMAS PhD Award. Her current interests include structural mechanics, solid mechanics, damage and fracture mechanics, contact mechanics, isogeometric analysis, high-performance finite elements, consulting in applied technologies, and technology transfer. During 2011 and 2012, she was a visiting scientist with a fellowship at the Institut für Kontinuumsmechanik Gottfried Wilhelm Leibniz Universität Hannover to study interfacial problems with isogeometric approaches. From 2013 to 2019, she was a researcher at the University of Salento, working on the computational mechanical modeling of structural interfaces, and on the comparative assessment of some advanced numerical collocation methods with lower computational cost for fracturing problems and structural mechanics of composite plates and shells, with complex geometry and material. She also collaborates, as a reviewer, with different prestigious international journals.

Jamal Amani Rad received his PhD degree in numerical analysis (scientific computing) from Shahid Beheshti University (SBU) in 2015. His doctoral dissertation was focused on numerical algorithms to evaluate financial option models. After graduation, he joined the Department of Cognitive Modeling, Institute for Cognitive and Brain Sciences, Shahid Beheshti University, as a postdoctoral fellow and is currently an assistant professor focused on the development of mathematical methods as tools in cognitive modeling.

Amir Hosein Hadian-Rasanan received the BSc degree in computer sciences from Shahid Beheshti University in 2019 and he is currently an MSc student of cognitive sciences in the Institute for Cognitive and Brain Sciences at Shahid Beheshti University. His research interests are brain data analysis using tensor methods and modeling of human decision making using fractional dynamics.

S.S.N. Perera has 20 years of research and teaching experience. He received his PhD in mathematical modeling from the TU Kaiserslautern, Germany, and University of Colombo, Sri Lanka (2008), under the DAAD scholarship program, and his MSc from ICTP/SISSA, Trieste, Italy (2004). Presently, he is the head of Department of Mathematics, University of Colombo. His current research interests are on multidisciplinary areas that bridge applied mathematics, numerics and natural sciences together. He is working in the area of mathematical modeling on epidemiology, biology, ecology, and finance/actuarial science. He has published 57 research papers in leading national and international journals and over 100 publications local/international conferences and symposia through collaborative research work. He has published eight book chapters, three books, and has edited one book. He is a member of the editorial board of the Journal of National Science Foundation, which is the only Sri Lankan journal in the science citation index. He served as the keynote speaker/plenary speaker/invited speaker/resource person in more than 30 international conferences/symposia/workshops. He was awarded many national and international awards in recognition of research excellence.

A.C. Mahasinghe received BSc degree in mathematics with a first class honors from University of Colombo and completed a collaborative PhD program with the School of Physics, University of Western Australia in Perth. Later, he worked as a postdoctoral research fellow at the Department of Computer Science, University of Auckland. He is currently teaching at the Department of Mathematics, University of Colombo, where he is the deputy director of the Centre for Mathematical Modelling. His works are mainly on mathematical aspects of quantum walks, quantum algorithms, and computational mathematics. He has published several papers in theoretical physics and computer science journals, given talks at several international conferences, and won national awards for scientific publications.

K.K.W.H. Erandi received her BSc honors in mathematics from University of Colombo. She is currently pursuing her PhD at the same institution. She is a former visiting fellow on dynamical modelling at the Department of Mathematics, University of Koblenz.

Merve Arıtürk is a research assistant of Software Engineering Department at Bahcesehir University in Istanbul. She received her bachelor's degree from Baskent University and master's degree from Bahcesehir University. She is now a PhD candidate at Yıldız Technical University. Her research interests are artificial intelligence, autonomous systems and vehicles, deep learning, and natural language processing.

Sırma Yavuz is an associate professor of computer engineering at Yıldız Technical University. She received her BE, ME, and PhD from Yıldız Technical University. Her research interests include autonomous robots, real-time robot localization, mapping, exploration, and navigation. For the past five years, she has been teaching courses on deep learning and robotics, in addition to supervising the research of graduate students in these fields. She has also served in numerous capacities successfully designing, implementing, and leading deep learning-based projects especially in real-world robotic applications. She is a founder of Probabilistic Robotics Group at Yıldız Technical University. With her colleagues and students, she has developed teams of real and virtual rescue robots, which have participated in RoboCup championships and won a number of top awards. She is a member of IEEE and ACM.

Pankaj Gautam received his BSc and MSc degrees from the University of Allahabad, Allahabad, in 2013 and 2015, respectively. He is currently pursuing his PhD in the Department of Mathematical Sciences, Indian Institute of Technology (BHU), Varanasi, India, since July 2016. His research area focuses on "approximation of some nonlinear problems in Hilbert spaces and its applications." He has attended several workshops and presented his work in several international conferences with one held in the Technical University of Berlin, Germany. He is an active member of European Mathematical Society, Working Group on Generalized Convexity, Society for Industrial and Applied Mathematics, Institute of Mathematics and its Applications.

Avinash Dixit received his BSc degree from University of Allahabad, Allahabad, India, in 2013 and MSc degree in mathematics from University of Allahabad, Allahabad, in 2015. He is pursuing his PhD in the Department of Mathematical Sciences, Indian Institute of Technology (BHU), Varanasi, India. His research area focuses on "fixed point theory and its application to machine learning and image processing." He has few research publications in reputed journals in last few years.

D.R. Sahu is a professor at the Department of Mathematics, Institute of Science, Banaras Hindu University, Varanasi. He has more than 25 years of research and teaching experience. Professor Sahu has published more than 120 research papers in reputed journals and conferences and has more than 1900 citations. He is a coauthor of a book on "Fixed Point Theory for Lipschitzian-Type Mappings with Applications: Topological Fixed Point Theory and Its Applications, Springer, New York, 2009." He has been a referee and an editor of many mathematical communities including Applied Mathematics Letters, Nonlinear Analysis, Journal of Mathematical Analysis and Applications, Fixed Point Theory and Applications, International Journal of Mathematics and Mathematical Sciences, Journal of Inequalities and Applications, Computers and Mathematics with Applications, International Journal of Nonlinear Operators Theory and Applications, JOTA, AAA, etc. His fields of interest include fixed point theory, operator theory, best approximation theory, variational inequality theory, and Newton-like methods for solving generalized operator equations. Currently, he is working on numerical techniques for some engineering problems.

Preface

Interdisciplinary sciences include various challenging problems of engineering and computational science along with different physical, biological, and many other sciences. Computational, mathematical, and machine intelligence are the bridge to handle various interdisciplinary problems. In recent years, correspondingly increased dialog between the disciplines has led to this new book. The purpose of this book is to meet the present and future needs for the interaction between various science and technology/engineering areas on the one hand and different branches of mathematics on the other hand. It may be challenging to know the ways that mathematics may be applied in traditional areas and innovative steps of applications.

This book deals with recent developments on mathematical methods on theoretical as well as applied science and engineering. It focuses mainly on subjects based on mathematical methods with concepts of modeling, simulation, waves and dynamics, machine intelligence, uncertainty, and pure mathematics that are written clearly and well organized. The idea has been to bring together the leading-edge research on mathematics combining various fields of science and engineering to present the original work with high quality. This perspective acknowledges the inherent characteristic of current research on mathematics operating in parallel over different subject fields.

In view of the above, the present book consists of 18 chapters. Chapters 1–4 include different problems of connectionist learning methods. As such, Chapter 1 contains the solution of various types of differential and integral equations by using single-layer artificial neural network (ANN) and functional link artificial neural network (FLANN) models. Multilayer perceptron-based deep learning architecture has been used in population genetics – specifically selection in *Drosophila melanogaster* in Chapter 2. Chapter 3 provides an overview of different classification models used in speech emotion recognition (SER) research. Prominent classification techniques are discussed in the SER context with brief mathematical elaborations. Latest deep learning architecture and its implementation to solve the SER problem is also discussed here. Deep learning fundamentals and related applications are addressed in Chapter 4.

Chapters 5–9 incorporate data processing and fuzzy uncertainty problems. A new approach to time-wise complex representation and processing of multimodal data based on the algebraic system of aggregates have been investigated in Chapter 5. In Chapter 5, authors proposed mathematical models of data synchronization for most possible cases for both crisp and fuzzy synchronization. In Chapter 6, various thermal and chemical diffusion problems have been handled with uncertain bounded parameters. The involved parameters are taken as interval/fuzzy. Fuzzy finite element method (FFEM) based on limit form of arithmetic rules and explicit finite difference method (EFDM) are included. In order to handle the incomplete information under fuzzy uncertainty in the mathematical modeling, some generalization has been proposed for arbitrary order differential equations of Riemann–Liouville type with respect to another function in Chapter 7. Chapter 8 addresses nanofluid flow between two inclined planes known as Jeffery–Hamel problem in an uncertain environment. Here nanoparticle volume fraction is taken as an uncertain parameter in terms of fuzzy number. Further, Chapter 9 presents a review of the several techniques for attribute reduction based on fuzzy rough set theory. Degree of dependency and discernibility matrix-based approaches are widely discussed for supervised, semisupervised, and unsupervised information systems. Applications of such techniques in different areas such as image processing, signal processing, and bioinformatics are also presented in Chapter 9.

Chapters 10 and 11 include interval and affine-based research. As regards, Chapter 10 is devoted to theoretical and practical aspects of interval mathematics. The theories of classical intervals and parametric intervals are formally constructed and their mathematical structures are uncovered. Moreover, with a view to treating some problems of the present interval theories, a new alternate theory of intervals, namely, the "theory of universal intervals," is presented and proved to have a nice S-field algebra, which extends the ordinary field of the reals. Further, homotopy perturbation method (HPM) based on affine contractor has been developed in Chapter 11 for evaluating the solution bounds of nonlinear dynamical problems in uncertain environments. Different application problems, viz., Rayleigh equation, Van der Pol Duffing equation, and nonhomogeneous Lane–Emden equation are taken into consideration.

Nano- and microstructural problems are studied in Chapters 12 and 13, respectively. Accordingly, Chapter 12 considers dynamical behavior (free vibration) of Euler–Bernoulli nanobeam under the framework of the strain gradient model. The differential transform method (DTM) is applied to investigate the dynamic behavior of SS and CC boundary conditions. Effects of small-scale parameter and length-scale parameter on frequency parameters are reported. Lie Symmetry Groups approach is implemented in Chapter 13 for analyzing the response of a very nonlinear system represented by a stiff differential equation. The problem under discussion has been the dynamic performance of a microcantilever beam subjected to an electrostatic force, which acts very close to the pull-in potential.

Advanced numerical methods with respect to numerical and application problems are investigated in Chapters 14 and 15 and optimization method in Chapter 16. In this respect, Chapter 14 discusses the mathematical fundamentals and performances of some innovative numerical approaches, namely, the generalized differential quadrature (GDQ) and the generalized integral quadrature (GIQ), while illustrating the procedure to evaluate the weighting coefficients, principal types of discretization, and some applications to simple functions. The accuracy of the GDQ and the GIQ methods are demonstrated through convergence analyses, which can be of great interest for many engineering problems of practical interest. An efficient method is provided in Chapter 15 for solving a nonlinear multidimensional inverse problem. This inverse problem is concerning the diffusion equation with an unknown source control parameter. The proposed method is based on applying the θ-weighted finite difference scheme for time discretization and Chebyshev collocation method for spatial approximation. Chapter 16 deals with the problem of optimal resource allocation when designing control strategies for infectious diseases in developing countries, subject to budgetary constraints. A binary integer programming formulation is developed for this purpose. Different objectives such as minimizing the total risk are considered in the same context, and it is briefly described how to meet the expected objectives by using a goal programming approach.

Artificial intelligence understanding with respect to different application problems is defined and discussed in Chapter 17. The last chapter, viz., Chapter 18, considered the problem to find the zeros of monotone operators. Both direct and iterative methods have been proposed to solve such types of issues. Proximal point algorithm and its different modified form along with their convergence behavior have been studied. In this respect, various numerical and application problems are also addressed.

It is worth mentioning that the present book incorporates a systematic understanding of theory related to the multidiscipline area and the applications. This may prove to be a benchmark for graduate and postgraduate students, teachers, engineers, and researchers in the mentioned subject areas. The book provides comprehensive results up to date and self-contained review of the topic along with application-oriented treatment of the use of newly developed methods of different modeling, simulation, and computation in various domains of engineering and sciences. It is believed that the readers will find this book as an excellent source to have a variety of application problems concerning mathematical models in one place, which may be fruitful and challenging for the future direction of research.

Rourkela, March 2020 *Snehashish Chakraverty*

Acknowledgments

First of all, the editor would like to express his sincere gratitude and thanks to all the authors and coauthors for their time and efforts for the preparation of the chapters. The editor do greatly appreciates their efforts in completing the chapters on time as well as for other related documents. The help and support by all of his PhD students are commendable. The editor would very much like to acknowledge the encouragement, patience, and support provided by his family members, in particular to his parents Late Sri B.K. Chakraborty and Srimati Parul Chakraborty, his wife Shewli, and daughters Shreyati and Susprihaa. The editor of this book is not responsible for the statements made or opinion expressed by the authors in this book. The editor does not hold any responsibility for any omissions or typographical errors.

Finally, this editor is most appreciative to the "Wiley" team for their help and support during the preparation of this book.

1

Connectionist Learning Models for Application Problems Involving Differential and Integral Equations

Susmita Mall, Sumit Kumar Jeswal, and Snehashish Chakraverty

Department of Mathematics, National Institute of Technology Rourkela, Rourkela, Odisha 769008, India

1.1 Introduction

1.1.1 Artificial Neural Network

Artificial intelligence (AI) also coined as machine intelligence is the concept and development of intelligent machines that work and react like humans. In the twentieth century, AI has a great impact starting from human life to various industries. Researchers from various parts of the world are developing new methodologies based on AI so as to make human life more comfortable. AI has different research areas such as artificial neural network (ANN), genetic algorithm (GA), support vector machine (SVM), fuzzy concept, etc.

In 1943, McCulloh and Pitts [1] designed a computational model for neural network based on mathematical concepts. The concept of perceptron has been proposed by Rosenblatt [2]. In the year 1975, Werbos [3] introduced the well-known concept of backpropagation algorithm. After the creation of backpropagation algorithm, different researchers explored the various aspects of neural network.

The concept of ANN inspires from the biological neural systems. Neural network consists of a number of identical units known as artificial neurons. Mainly, neural network comprises three layers, viz., input, hidden, and output. Each neuron in the nth layer has been interconnected with the neurons of the $(n+1)$th layer by some signal. Each connection has been assigned with weight. The output has been calculated after multiplying each input with its corresponding weight. Further, the output passes through an activation function to get the desired output.

ANN has various applications in the field of engineering and science such as image processing, weather forecasting, function approximation, voice recognition, medical diagnosis, robotics, signal processing, etc.

1.1.2 Types of Neural Networks

There are various types of neural network models, but commonly used models have been discussed further.

1. *Feedforward neural network*: In feedforward network, the connection between the neurons does not form a cycle. In this case, the information passes through in one direction.
   i. *Single-layer perceptron*: It is a simple form of feedforward network without having hidden layers, viz., function link neural network (FLNN).
   ii. *Multilayer perceptron*: Multilayer perceptron is feedforward network, which comprises of three layers, viz., input, hidden, and output.
   iii. *Convolutional neural network (CNN)*: CNN is a class of deep neural networks that is mostly used in analyzing visual imagery.

Mathematical Methods in Interdisciplinary Sciences, First Edition. Edited by Snehashish Chakraverty.

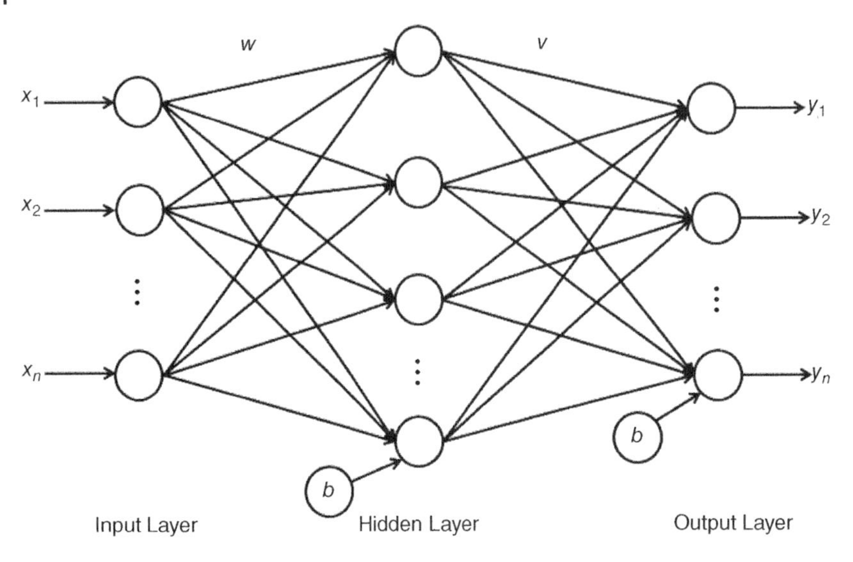

Figure 1.1 General model of ANN.

2. *Feedback or recurrent neural network*: In feedback neural network, the different connections between the neurons form a directed cycle. In this case, the information flow is bidirectional.

The general model of an ANN has been depicted in Figure 1.1.

1.1.3 Learning in Neural Network

There are mainly three types of learning, viz., supervised or learning with a teacher, unsupervised or learning without a teacher, and reinforcement learning. In the supervised case, the desired output of the ANN model is known, but on the other hand, the desired output is unknown in the case of unsupervised learning. In reinforcement learning, the ANN model learns its behavior based on the feedback from the environment. Various learning rules have been developed to train or update the weights. Some of them are mentioned below.

(a) Hebbian learning rule
(b) Perceptron learning rule
(c) Delta learning rule or backpropagation learning rule
(d) Out star learning rule
(e) Correlation learning rule

1.1.4 Activation Function

An output node receives a number of input signals, depending on the intensity of input signals, we summed these input signals and pass them through some function known as activation function. Depending on the input signals, it generates an output and accordingly further actions have been performed. There are different types of activation functions from which some of them are listed below.

1. Identity function
2. Bipolar step function
3. Binary step function
4. Sigmoidal function

- Unipolar sigmoidal function
- Bipolar sigmoidal function

5. Tangent hyperbolic function
6. Rectified linear unit (ReLU)

Although we have mentioned a number of activation functions, but for the sake of completeness, we have just discussed about sigmoidal function below.

1.1.4.1 Sigmoidal Function

Sigmoidal function is a type of mathematical function that takes the form of an S-shaped curve or a sigmoid curve. There are two types of sigmoidal functions depending on their range.

- *Unipolar sigmoidal function*: The unipolar sigmoidal function can be defined as

  $$\phi(x) = \frac{1}{1 + e^{-x}}$$

  In this case, the range of the output lies within the interval [0, 1].
- *Bipolar sigmoidal function*: The bipolar sigmoidal function can be formulated as

  $$\phi(x) = \frac{1 - e^{-x}}{1 + e^{-x}}$$

  The output range for bipolar case lies between [−1, 1]. Figure 1.2 represents the bipolar sigmoidal function.

1.1.5 Advantages of Neural Network

(a) The ANN model works with incomplete or partial information.
(b) Information processing has been done in a parallel manner.
(c) An ANN architecture learns and need not to be programmed.
(d) It can handle large amount of data.
(e) It is less sensitive to noise.

1.1.6 Functional Link Artificial Neural Network (FLANN)

Functional Link Artificial Neural Network (FLANN) is a class of higher order single-layer ANN models. FLANN model is a flat network without an existence of a hidden layer, which makes the training algorithm of the network less complicated. The single-layer FLANN model is developed by Pao [4, 5] for function approximation and pattern recognition with less computational cost and faster convergence rate. In FLANN, the hidden layer is replaced by a

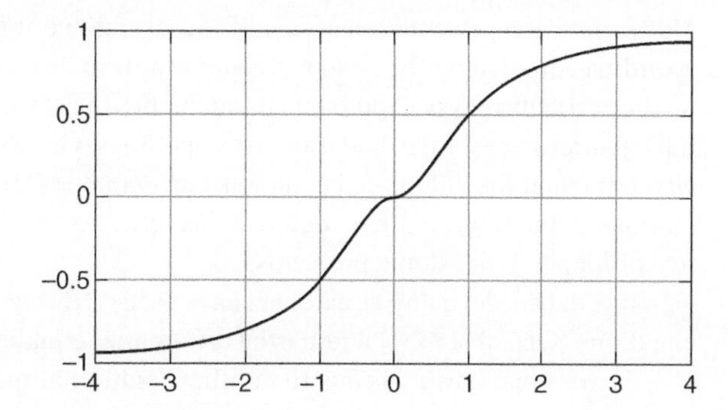

Figure 1.2 Sigmoidal function.

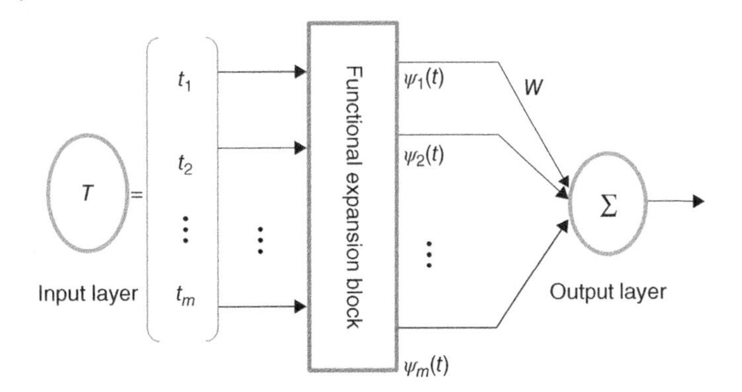

Figure 1.3 Structure of FLANN model.

functional expansion block for artificial enhancement of the input patterns using linearly independent functions such as orthogonal polynomials, trigonometric polynomials, exponential polynomials, etc. Being a single-layer model, its number of parameters is less than multilayer ANN. Thus, FLNN model has less computational complexity and higher convergence speed. In recent years, many researchers have commonly used various types of orthogonal polynomials, viz., Chebyshev, Legendre, Hermite, etc., as a basis of FLANN.

As such, Chebyshev neural network model is widely used for handling various complicated problems, viz., function approximation [6], channel equalization [7], digital communication [8], solving nonlinear differential equations (DEs) [9, 10], system identification [11, 12], etc. Subsequently, Legendre neural network has been used for solving DEs [13], system identification [14], and for prediction of machinery noise [15]. Also, Hermite neural network is used for solving Van der Pol Duffing oscillator equations [16, 17].

Figure 1.3 shows the structure of FLANN model, which consists of an input layer, a set of basis functions, a functional expansion block, and an output layer. The input layer contains $T = [t_1, t_2, t_3, \ldots, t_m]$ input vectors, a set of m basis functions $\{\psi_1, \psi_2, \ldots, \psi_m\} \in B_M$, and weight vectors $W = [w_1, w_2, \ldots, w_m]$ from input to output layer associated with functional expansion block. The output of FLANN with the following input–output relationship may be written as

$$\breve{Y} = \lambda(Z_j) \quad \text{and} \quad Z_j = \sum w_{ji}\psi_j(T)$$

here, $T \in R^m$ and $\breve{Y} = [\breve{y}_1, \breve{y}_2, \ldots, \breve{y}_n] \in R^n$ denote the input and output vectors of FLANN, respectively, $\psi_j(T)$ is the expanded input data with respect to basis functions, and w_{ji} stands for the weight associated with the jth output. The nonlinear differentiable functions, viz., sigmoid, tan hyperbolic, etc., are considered as an activation function $\lambda(\cdot)$ of single-layer FLANN model.

1.1.7 Differential Equations (DEs)

Mathematical representation of scientific and engineering problem(s) is called a mathematical model. The corresponding equation for the system is generally given by differential equation (ordinary and partial), and it depends on the particular physical problem. As such, there exist various methods, namely, exact solution when the differential equations are simple, and numerical methods, viz., Runge–Kutta, predictor–corrector, finite difference, finite element when the differential equations are complex. Although these numerical methods provide good approximations to the solution, they require a discretization of the domain (via meshing), which may be challenging in two or higher dimensional problems.

In recent decade, many researchers have used ANN techniques for solving various types of ordinary differential equations (ODEs) [18–21] and partial differential equations (PDEs) [22–25]. ANN-based approximate solutions of DEs are more advantageous than other traditional numerical methods. ANN method is general and can be applied to solve linear and nonlinear singular initial value problems. Computational complexity does not increase considerably with the increase in number of sampling points and dimension of the problem. Also, we may use ANN model as a black box to get numerical results at any arbitrary point in the domain after training of the model.

1.1.8 Integral Equation

Integral equations are defined as the equations where an unknown function appears under an integral sign. There are mainly two types of integral equations, viz., Fredholm and Volterra. ANN is a novel technique for handling integral equations. As such, some articles related to integral equations have been discussed. Golbabai and Seifollahi [26] proposed a novel methodology based on radial basis function network for solving linear second-order integral equations of Fredholm and Volterra types. An radial basis function (RBF) network has been presented by Golbabai et al. [27] to find the solution of a system of nonlinear integral equations. Jafarian and Nia [28] introduced ANN architecture for solving second kind linear Fredholm integral systems. A novel ANN-based model has been developed by Jafarian et al. [29] for solving a second kind linear Volterra integral equation system. S. Effati and Buzhabadi [30] proposed a feedforward neural network-based methodology for solving Fredholm integral equations of second kind. Bernstein polynomial expansion method has been used by Jafarian et al. [31] for solving linear second kind Fredholm and Volterra integral equation systems. Kitamura and Qing [32] found the numerical solution of Fredholm integral equations of first kind using ANN. Asady et al. [33] constructed a feedforward ANN architecture to solve two-dimensional Fredholm integral equations. Radial basis functions have been chosen to find an iterative solution for linear integral equations of second kind by Golbabai and Seifollahi [34]. Cichocki and Unbehauen [35] gave a neural network-based method for solving a linear system of equations. A multilayer ANN architecture has been proposed by Jeswal and Chakraverty [36] for solving transcendental equation.

1.1.8.1 Fredholm Integral Equation of First Kind

The Fredholm equation of first kind can be defined as

$$f(x) = \int_a^b K(x,t)\phi(t)dt$$

where K is called kernel and ϕ and f are the unknown and known functions, respectively. The limits of integration are constants.

1.1.8.2 Fredholm Integral Equation of Second Kind

The Fredholm equation of second kind is

$$\phi(x) = f(x) + \lambda \int_a^b K(x,t)\phi(t)dt$$

here, λ is an unknown parameter.

1.1.8.3 Volterra Integral Equation of First Kind

The Volterra integral equation of first kind is defined as

$$f(x) = \int_a^x K(x,t)\phi(t)dt$$

1.1.8.4 Volterra Integral Equation of Second Kind

The Volterra Integral equation of second kind may be written as

$$\phi(x) = f(x) + \lambda \int_a^x K(x,t)\phi(t)dt$$

1.1.8.5 Linear Fredholm Integral Equation System of Second Kind

A system of linear Fredholm integral equations of second kind may be formulated as [28]

$$\sum_{i=1}^{n} L_{1i}(u)G_i(u) = g_1(u) + \int_a^b \left(\sum_{i=1}^{n} K_{1i}(v,u) \cdot G_i(v) \right) dv$$
$$\sum_{i=1}^{n} L_{2i}(u)G_i(u) = g_2(u) + \int_a^b \left(\sum_{i=1}^{n} K_{2i}(v,u) \cdot G_i(v) \right) dv$$
$$\vdots$$
$$\sum_{i=1}^{n} L_{ni}(u)G_i(u) = g_n(u) + \int_a^b \left(\sum_{i=1}^{n} K_{ni}(v,u) \cdot G_i(v) \right) dv$$

here, $u, v \in [a, b]$ and $L(u), K(v, u)$ are real functions. $G(u) = [G_1(u), G_2(u), \ldots, G_n(u)]$ is the solution set to be found.

1.2 Methodology for Differential Equations

1.2.1 FLANN-Based General Formulation of Differential Equations

In this section, general formulation of differential equations using FLANN has been described. In particular, the formulations of ODEs are discussed here.

In general, differential equations (ordinary as well as partial) can be written as [18]

$$F(t, y(t), \nabla y(t), \nabla^2 y(t), \ldots, \nabla^n y(t)) = 0, \quad t \in \bar{D} \subseteq R^n \tag{1.1}$$

where F defines the structure of DEs, ∇ stands for differential operator, and $y(t)$ is the solution of DEs. It may be noted that for ODEs $t \in \bar{D} \subset R$ and for PDEs $t = (t_1, t_2, \ldots, t_n) \in \bar{D} \subset R^n$. Let $y_{\text{Lg}}(t, p)$ denotes the trial solution of FLANN (Laguerre neural network, LgNN) with adjustable parameters p and then the above general differential equation changes to the form

$$F(t, y_{\text{Lg}}(t,p), \nabla y_{\text{Lg}}(t,p), \nabla^2 y_{\text{Lg}}(t,p), \ldots) = 0 \tag{1.2}$$

The trial solution $y_{\text{Lg}}(t, p)$ satisfies the initial and boundary conditions and may be formulated as

$$y_{\text{Lg}}(t,p) = K(t) + H(t, N(t,p)) \tag{1.3}$$

where $K(t)$ satisfies the boundary/initial conditions and contains no weight parameters. $N(t, p)$ denotes the single output of FLANN with parameters p and input t. $H(t, N(t, p))$ makes no contribution to the initial or boundary conditions, but it is used in FLANN where weights are adjusted to minimize the error function.

The general form of corresponding error function for the ODE can be expressed as

$$E(t,p) = \sum_{i=1}^{h} \frac{1}{2} \left\{ \frac{d^n y_{\text{Lg}}(t_i,p)}{dt^n} - f\left(t_i, y_{\text{Lg}}(t_i), \frac{dy_{\text{Lg}}(t_i,p)}{dt}, \ldots, \frac{d^{n-1} y_{\text{Lg}}(t_i,p)}{dt^{n-1}} \right) \right\}^2 \tag{1.4}$$

We take some particular cases in Sections 1.2.1.1 and 1.2.1.2.

1.2.1.1 Second-Order Initial Value Problem

Second-order initial value problem may be written as

$$\frac{d^2 y}{dt^2} = f\left(t, y, \frac{dy}{dt} \right) \quad t \in [a, b] \tag{1.5}$$

with initial conditions $y(a) = C, y'(a) = C'$.

The LgNN trial solution is formulated as

$$y_{\mathrm{Lg}}(t, p) = C + C'(t - a) + (t - a)^2 N(t, p) \tag{1.6}$$

The error function is computed as follows

$$E(t, p) = \sum_{i=1}^{h} \frac{1}{2}\left(\frac{d^2 y_{\mathrm{Lg}}(t_i, p)}{dt^2} - f\left[t_i, y_{\mathrm{Lg}}(t_i, p), \frac{dy_{\mathrm{Lg}}(t_i, p)}{dt}\right]\right)^2 \tag{1.7}$$

1.2.1.2 Second-Order Boundary Value Problem

Next, let us consider a second-order boundary value problem as

$$\frac{d^2 y}{dt^2} = f\left(t, y, \frac{dy}{dt}\right) \quad t \in [a, b] \tag{1.8}$$

with boundary conditions $y(a) = A, y(b) = B$.

The LgNN trial solution for the above boundary value problem is

$$y_{\mathrm{Lg}}(t, p) = \frac{bA - aB}{b - a} + \frac{B - A}{b - a} t + (t - a)(t - b)N(t, p) \tag{1.9}$$

In this case, the error function is the same as Eq. (1.7).

1.2.2 Proposed Laguerre Neural Network (LgNN) for Differential Equations

Here, we have included architecture of LgNN, general formulation for DEs, and its learning algorithm with gradient computation of the ANN parameters with respect to its inputs.

1.2.2.1 Architecture of Single-Layer LgNN Model

Structure of single-layer LgNN model has been displayed in Figure 1.4. LgNN is one type of FLANN model based on Laguerre polynomials. LgNN consists of single-input node, single-output node, and functional expansion block of Laguerre orthogonal polynomials. The hidden layer is removed by transforming the input data to a higher dimensional space using Laguerre polynomials. Laguerre polynomials are the set of orthogonal polynomials obtained by solving Laguerre differential equation

$$xy'' + (1 - x)y' + ny = 0 \tag{1.10}$$

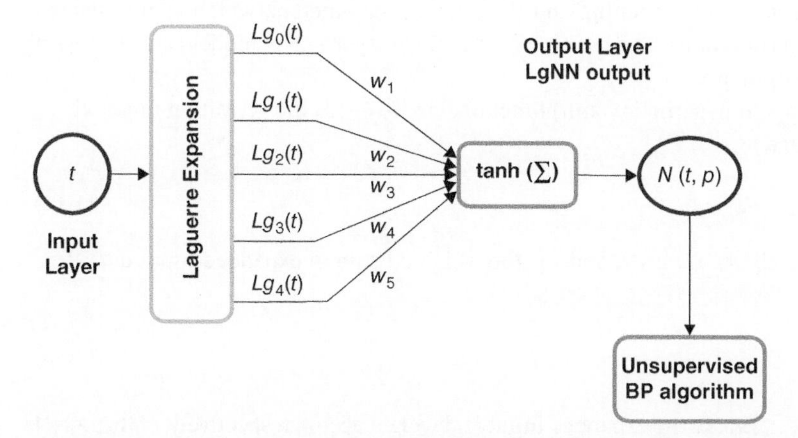

Figure 1.4 Structure of single-layer Laguerre neural network.

These are the first few Laguerre polynomials

$$\left.\begin{aligned} Lg_{0(x)} &= 1 \\ Lg_{1(x)} &= 1 - x \\ Lg_{2(x)} &= \frac{1}{2}(x^2 - 4x + 2) \\ Lg_{3(x)} &= \frac{1}{6}(-x^3 + 9x^2 - 18x + 6) \\ Lg_{4(x)} &= \frac{1}{24}(x^4 - 16x^3 + 72x^2 - 96x + 24) \end{aligned}\right\} \tag{1.11}$$

$$\vdots$$

The higher order Laguerre polynomials may be generated by recursive formula

$$(n + 1)\mathrm{Lg}_{n+1}(t) = (2n + 1 - t)\mathrm{Lg}_n(t) - n\mathrm{Lg}_{n-1}(t) \tag{1.12}$$

Here, we have considered that one input node has h number of data, i.e. $t = (t_1, t_2, \ldots, t_h)^T$. Then, the functional expansion block of input data is determined by using Laguerre polynomials as

$$[\mathrm{Lg}_0(t_1), \mathrm{Lg}_1(t_1), \mathrm{Lg}_2(t_1), \ldots; \mathrm{Lg}_0(t_2), \mathrm{Lg}_1(t_2), \mathrm{Lg}_2(t_2), \ldots; \ldots; \mathrm{Lg}_0(t_h), \mathrm{Lg}_1(t_h), \mathrm{Lg}_2(t_h); \ldots] \tag{1.13}$$

In view of the above discussion, main advantages of the single-layer LgNN model for solving differential equations may be given as follows:

1. The total number of network parameters is less than that of traditional multilayer ANN.
2. It is capable of fast learning and is computationally efficient.
3. It is simple to implement.
4. There are no hidden layers in between the input and output layers.
5. The backpropagation algorithm is unsupervised.
6. No other optimization technique is to be used.

1.2.2.2 Training Algorithm of Laguerre Neural Network (LgNN)

Backpropagation (BP) algorithm with unsupervised training has been applied here for handling differential equations. The idea of BP algorithm is to minimize the error function until the FLANN learned the training data. The gradient descent algorithm has been used for training and the network parameters (weights) are updated by taking negative gradient at each step (iteration). The training of the single-layer LgNN model is to start with random weight vectors from input to output layer.

We have considered the nonlinear tangent hyperbolic (tanh) function, viz., $\frac{e^t - e^{-t}}{e^t + e^{-t}}$, as the activation function.

The output of LgNN may be formulated as

$$N(t, p) = \tanh(s) = \frac{e^s - e^{-s}}{e^s + e^{-s}} \tag{1.14}$$

where t is the input data, p stands for weights of LgNN, and s is the weighted sum of expanded input data. It is formulated as

$$s = \sum_{k=1}^{m} w_k \mathrm{Lg}_{k-1}(t) \tag{1.15}$$

where $\mathrm{Lg}_{k-1}(t)$ and w_k with $k = \{1, 2, 3, \ldots, m\}$ are the expanded input and weight vector, respectively, of the LgNN model.

The weights are modified by taking negative gradient at each step of iteration

$$w_k^{T+1} = w_k^T + \Delta w_k^T = w_k^T + \left(-\rho \frac{\partial E(t,p)^T}{\partial w_k^T} \right) \tag{1.16}$$

here, ρ is the learning parameter, T is the step of iteration, and $E(t, p)$ denotes the error function of LgNN model.

1.2.2.3 Gradient Computation of LgNN

Derivatives of the network output with respect to the corresponding inputs are required for error computation. As such, derivative of $N(t, p)$ with respect to t is

$$\frac{dN(t,p)}{dt} = \sum_{k=1}^{m} \left[1 - \left(\frac{e^{(w_k \mathrm{Lg}_{k-1}(t))} - e^{-(w_k \mathrm{Lg}_{k-1}(t))}}{e^{(w_k \mathrm{Lg}_{k-1}(t))} + e^{-(w_k \mathrm{Lg}_{k-1}(t))}} \right)^2 \right] (w_k \mathrm{Lg}'_{k-1}(t)) \tag{1.17}$$

Similarly, the second derivative of $N(t, p)$ can be computed as

$$\frac{d^2N}{dt^2} = \sum_{k=1}^{m} \left[\begin{array}{l} \left\{ -2 \left(\frac{e^{(w_k \mathrm{Lg}_{k-1}(t))} - e^{-(w_k \mathrm{Lg}_{k-1}(t))}}{e^{(w_k \mathrm{Lg}_{k-1}(t))} + e^{-(w_k \mathrm{Lg}_{k-1}(t))}} \right) \left(\frac{(e^{(w_k \mathrm{Lg}_{k-1}(t))} + e^{-(w_k \mathrm{Lg}_{k-1}(t))})^2 - (e^{(w_k \mathrm{Lg}_{k-1}(t))} - e^{-(w_k \mathrm{Lg}_{k-1}(t))})^2}{(e^{(w_k \mathrm{Lg}_{k-1}(t))} + e^{-(w_k \mathrm{Lg}_{k-1}(t))})^2} \right) \right\} \\ (w_k \mathrm{Lg}'_{k-1}(t))^2 + (w_k \mathrm{Lg}''_{k-1}(t)) \left\{ 1 - \left(\frac{e^{(w_k \mathrm{Lg}_{k-1}(t))} - e^{-(w_k \mathrm{Lg}_{k-1}(t))}}{e^{(w_k \mathrm{Lg}_{k-1}(t))} + e^{-(w_k \mathrm{Lg}_{k-1}(t))}} \right)^2 \right\} \end{array} \right] \tag{1.18}$$

After simplifying Eq. (1.18), we get

$$\frac{d^2N}{dt^2} = \sum_{k=1}^{m} \left[\begin{array}{l} \left(\left\{ 2 \left(\frac{e^{(w_k \mathrm{Lg}_{k-1}(t))} - e^{-(w_k \mathrm{Lg}_{k-1}(t))}}{e^{(w_k \mathrm{Lg}_{k-1}(t))} + e^{-(w_k \mathrm{Lg}_{k-1}(t))}} \right)^3 - 2 \left(\frac{e^{(w_k \mathrm{Lg}_{k-1}(t))} - e^{-(w_k \mathrm{Lg}_{k-1}(t))}}{e^{(w_k \mathrm{Lg}_{k-1}(t))} + e^{-(w_k \mathrm{Lg}_{k-1}(t))}} \right) \right\} (w_k \mathrm{Lg}'_{k-1}(t))^2 \right) \\ + \left(\left\{ 1 - \left(\frac{e^{(w_k \mathrm{Lg}_{k-1}(t))} - e^{-(w_k \mathrm{Lg}_{k-1}(t))}}{e^{(w_k \mathrm{Lg}_{k-1}(t))} + e^{-(w_k \mathrm{Lg}_{k-1}(t))}} \right)^2 \right\} (w_k \mathrm{Lg}''_{k-1}(t)) \right) \end{array} \right] \tag{1.19}$$

where w_k denotes the weights of LgNN and $\mathrm{Lg}'_{k-1}(t)$, $\mathrm{Lg}''_{k-1}(t)$ are the first and second derivatives of Laguerre polynomials, respectively.

Differentiate Eq. (1.6) with respect to t

$$\frac{d^2 y_{\mathrm{Lg}}(t,p)}{dt^2} = 2N(t,p) + 4(t-a)\frac{dN(t,p)}{dt} + (t-a)^2 \frac{d^2N(t,p)}{dt^2} \tag{1.20}$$

Finally, the converged LgNN results may be used in Eqs. (1.6) and (1.9) to obtain the approximate results.

1.3 Methodology for Solving a System of Fredholm Integral Equations of Second Kind

While solving a system of Fredholm integral equations, it leads to a system of linear equations. Therefore, we have proposed an ANN-based method for solving the system of linear equations.

A linear system of equations is written as

$$L_{n \times n} y = m \ \text{ or } \ \sum_{j=1}^{n} l_{ij} y_j = m_i, \quad i = 1, 2, \dots, n \tag{1.21}$$

where $L = [l_{ij}]$ is a $n \times n$ matrix, $y = [y_1, y_2, \dots, y_n]^T$ is a $n \times 1$ unknown vector, and $m = [m_1, m_2, \dots, m_n]^T$ is a $n \times 1$ known vector.

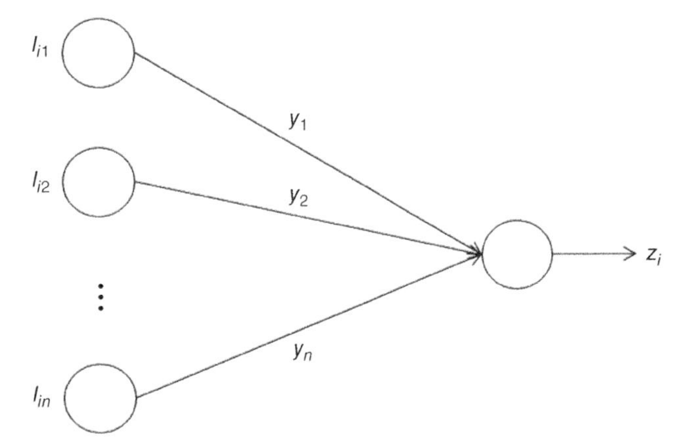

Figure 1.5 ANN model for solving linear systems.

From Eq. (1.21), the vector $L_i^T = (l_{i1}, l_{i2}, \dots, l_{in})$ and m_i have been chosen as the input and target output of the ANN model, respectively. Further, y is the weights of the discussed ANN model, which can be shown in Figure 1.5.

Taking the input, target output, and weights, training of the ANN model has been done until weight vector y matches the target output m.

The ANN output has been calculated as

$$z_i = f\left(\sum_{j=1}^{n} l_{ij} y_j\right) = L_i^T y, \quad i = 1, 2, \dots, n \tag{1.22}$$

where f is the activation function, which is chosen to be the identity function in this case.

The error term can be computed as

$$E = \frac{1}{2}\sum_{i=1}^{n} e_i^2 \tag{1.23}$$

where $e_i = m_i - z_i$.

The weights have been adjusted using the steepest descent algorithm as

$$\Delta y_j = -\eta \frac{\partial E}{\partial y_j} = -\eta \frac{\partial E}{\partial e_i}\frac{\partial e_i}{\partial y_j} = \eta \sum_{i=1}^{n} E_i l_{ij} \tag{1.24}$$

and the updated weight matrix is given by

$$y^{(q+1)} = y^{(q)} + \Delta y_j, \quad j = 1, 2, \dots, n \tag{1.25}$$

where η is the learning parameter and $y^{(q)}$ is the q-th updated weight vector.

An algorithm for solving linear systems has been included in Section 1.3.1.

1.3.1 Algorithm

Step 1: Choose the precision tolerance "tol," learning rate η, random weights, $q = 0$ and $E = 0$;
Step 2: Calculate $z_i = L_i^T y$, $e_i = m_i - z_i$;
Step 3: The error term is given by

$$E = E + \frac{1}{2}\sum_{i=1}^{n} e_i^2$$

and weight updating formula is

$$y^{(q+1)} = y^{(q)} + \Delta y_j$$

where

$$\Delta y_j = \eta \sum_{i=1}^{n} E_i l_{ij}, \quad j = 1, 2, \ldots, n$$

If $E <$ tol, then go to step 2, else go to step 4.
Step 4: Print y.

1.4 Numerical Examples and Discussion

1.4.1 Differential Equations and Applications

In this section, a boundary value problem, a nonlinear Lane-Emden equation, and an application problem of Duffing oscillator equation have been investigated to show the efficiency and powerfulness of the proposed method. In all cases, the feedforward neural network and unsupervised BP algorithm have been used for training. The initial weights from the input layer to the output layer are taken randomly.

Example 1.1 Consider the following linear boundary value problem [37]

$$\frac{d^2y}{dt^2} - t\frac{dy}{dt} = 3y + 4.2t$$

with boundary conditions $y(0) = 0, y(1) = 1.9$.

As mentioned in Eq. (1.9), the LgNN trial solution is expressed as

$$y_{\text{Lg}}(t, p) = 1.9t + t(t-1)N(t, p)$$

The single-layer LgNN model has been trained for 10 equidistant points in [0, 1] and 5 weights with respect to the first 5 Laguerre polynomials. Table 1.1 shows the comparison between the analytical results and LgNN approximate results. The above comparison is displayed in Figure 1.6. Also, Figure 1.7 depicts the graph of error between

Table 1.1 Comparison between analytical and LgNN results (Example 1.1).

Input data	Analytical [37]	LgNN
0	0	0
0.1	0.0910	0.0925
0.2	0.1880	0.1908
0.3	0.2970	0.2941
0.4	0.4240	0.4237
0.5	0.5750	0.5746
0.6	0.7560	0.7549
0.7	0.9730	0.9768
0.8	1.2320	1.2290
0.9	1.5390	1.5430
1.0	1.9000	1.9056

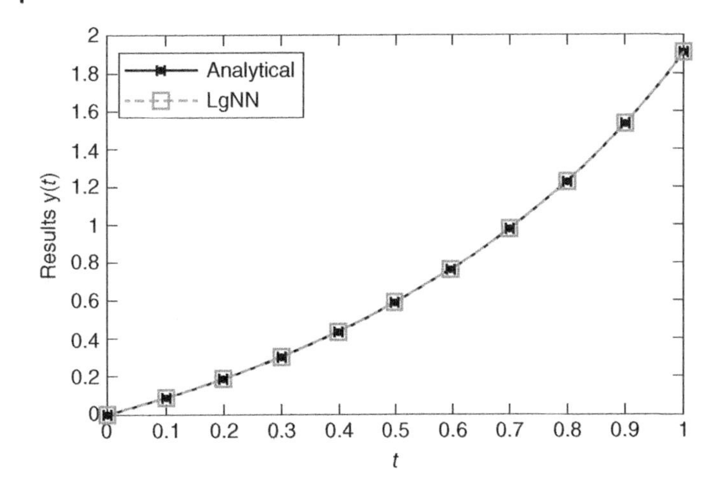

Figure 1.6 Plot of analytical and LgNN results (Example 1.1).

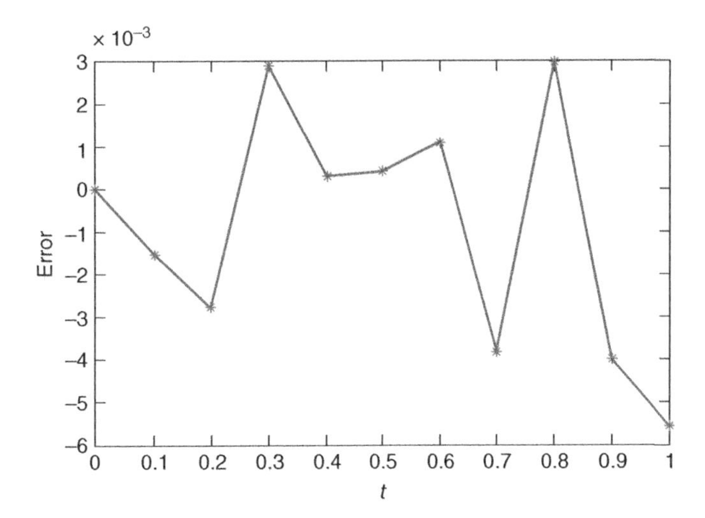

Figure 1.7 Error plot between analytical and LgNN results (Example 1.1).

analytic and LgNN results. From Figures 1.6 and 1.7, one may observe that the approximate results by LgNN method are very close to the analytical results.

Example 1.2 Many problems in quantum mechanics and astrophysics may be modeled by singular nonlinear second-order ODEs as proposed by Lane [38] and Emden [39]. The Lane-Emden equation has been described by various phenomena, viz., thermionic currents, thermal explosions, thermal behavior of a spherical cloud of gas acting under the mutual attraction of its molecules and subject to the classical laws of thermodynamics, etc. [40, 41]. The Lane-Emden equation is challenging to solve not only analytically but numerically because of the presence of singular point at the origin, i.e. $t = 0$. In this investigation, we have used single-layer LgNN method for solving the said equation.

In this example, a nonlinear Lane-Emden equation has been taken [41]

$$\frac{d^2y}{dt^2} + \frac{10}{t}\frac{dy}{dt} + y^{10}(t) = t^{100} + 190t^8, \quad t \in [0, 1]$$

subject to $y(0) = y'(0) = 0$.

Accordingly, the related LgNN trial solution may be written as

$$y_{\mathrm{Lg}}(t, p) = t^2 N(t, p)$$

Ten points in the domain [0, 1] and five weights with respect to five Laguerre orthogonal polynomials have been considered for this problem. As the previous case, the LgNN and analytical results are shown in Table 1.2. The comparison between our proposed LgNN results and analytical results is depicted in Figure 1.8. The error plot is cited in Figure 1.9.

Example 1.3 In this example, we have taken an application problem of Duffing oscillator equation. Nowadays, weak signal detection is a challenging task in the case of early machinery failure signal detection and signal detection. The Van der Pol Duffing and Duffing oscillator equations are widely used to detect weak signals [42]. The

Table 1.2 Comparison between analytical and LgNN results (Example 1.2).

Input data	Analytical [41]	LgNN
0	0	0
0.1	0.000 000 000 1	0.000 000 009 9
0.2	0.000 000 102 4	0.000 000 098 9
0.3	0.000 005 904 9	0.000 005 906 1
0.4	0.000 104 857 6	0.000 105 255 6
0.5	0.000 976 562 5	0.000 977 465 6
0.6	0.006 046 617 6	0.006 089 657 9
0.7	0.028 247 524 9	0.028 628 513 9
0.8	0.107 374 182 4	0.112 070 691 8
0.9	0.348 678 440 1	0.349 183 701 0
1.0	0.999 999 999 9	0.999 898 789 9

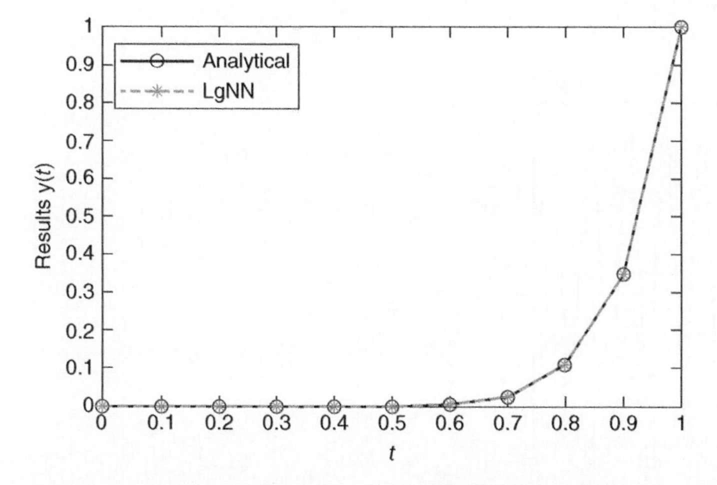

Figure 1.8 Plot of analytical and LgNN results (Example 1.2).

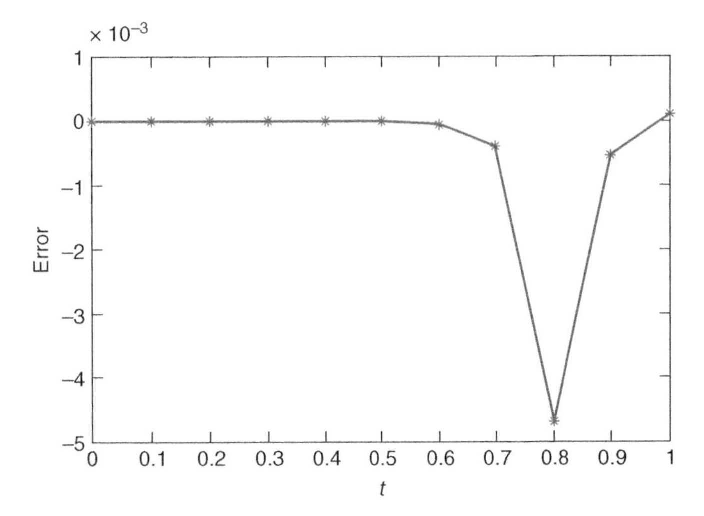

Figure 1.9 Error plot between analytical and LgNN results (Example 1.2).

idea of signal detection is based on the characteristic of a chaotic oscillator; that is, a small periodic signal may be detected by Duffing oscillator via a transition from chaotic to periodic motion [43].

Here, we have considered a Duffing oscillator equation used for weak signal detection [43]

$$\frac{dy^2}{d^2t} + \lambda\frac{dy}{dt} - y + y^3 = \zeta \cos t + \delta(t) \quad \text{subject to } y(0) = 1.0, \;\; y'(0) = 1.0$$

where $\lambda = 0.5$, $\zeta = 0.825$ (amplitude of external exciting periodic force) and $\delta(t) = 0.001 \cos t$ (frequency of external weak signal).

For the above problem, we may write the LgNN trial solution as

$$y_{\text{Lg}}(t, p) = 1.0 + 1.0t + t^2 N(t, p)$$

Again, the network is trained for time $t = 100$ seconds and seven weight vectors from input to output layer. The problem has no analytical solution. Time series plots between time t and displacement $y(t)$ by using Runge–Kutta method (RKM) and LgNN have been shown in Figures 1.10 and 1.11, respectively. Finally, phase plane plots by RKM and LgNN are displayed in Figures 1.12 and 1.13.

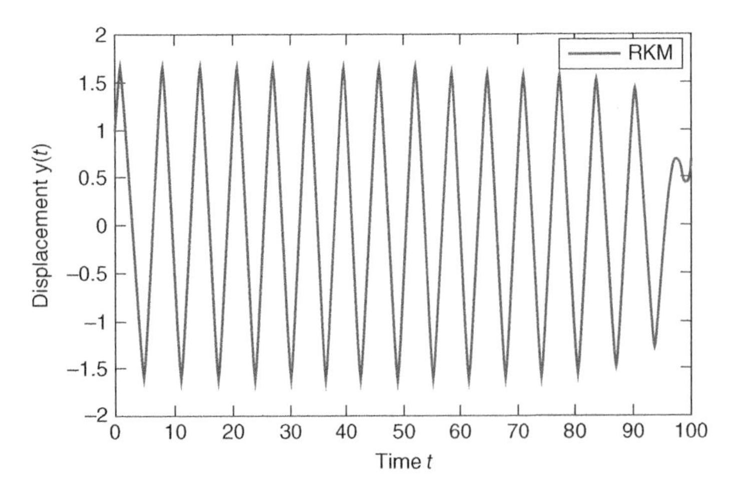

Figure 1.10 Time series plot by RKM (Example 1.3).

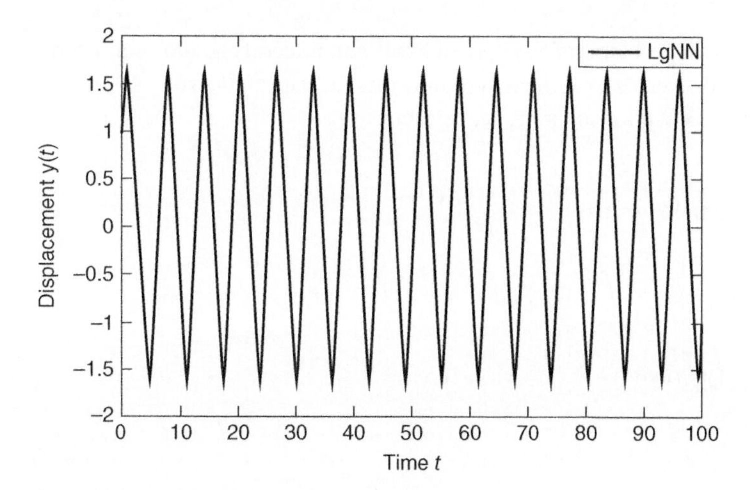

Figure 1.11 Time series plot by LgNN (Example 1.3).

Figure 1.12 Phase plane plot for RKM.

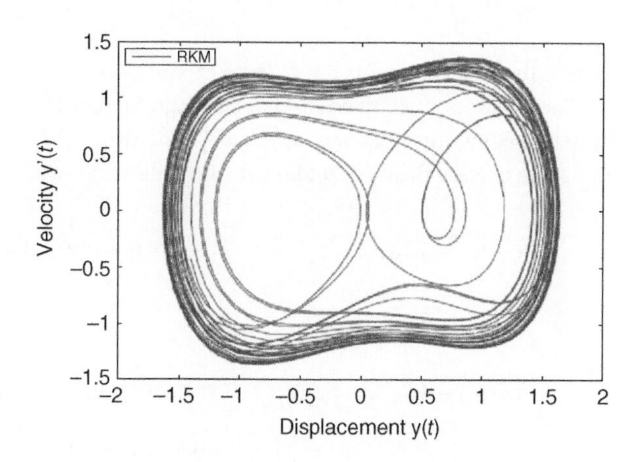

Figure 1.13 Phase plane plot for LgNN.

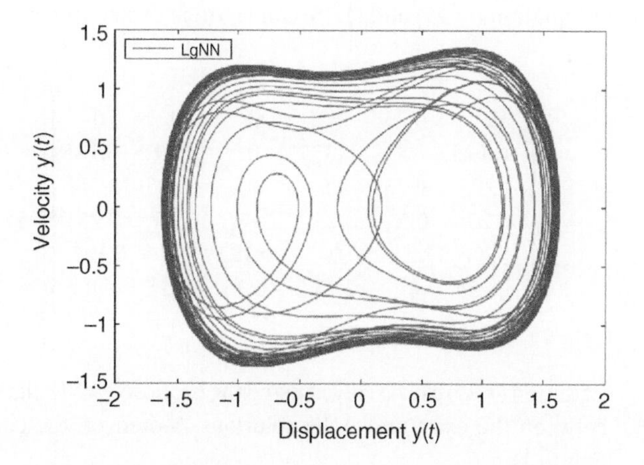

1.4.2 Integral Equations

In this section, we have investigated three example problems of a system of Fredholm integral equations of second kind. The system of integral equations may be converted to a linear system using the trapezoidal rule. As such, a feedforward neural network model has been proposed for solving linear system.

Example 1.4 Let us consider an integral equations system

$$\begin{cases} 2G_1(u) + 3G_2(u) = g_1(u) + \int_0^1 (u+v)G_1(v)dv + \int_0^1 uvG_2(v)dv \\ 3G_1(u) - 4G_2(u) = g_2(u) + \int_0^1 (2uv-1)G_1(v)dv + \int_0^1 (u-v)G_2(s)dv \end{cases} \tag{1.26}$$

with

$$g_1(u) = 2e^u - \frac{3}{u-2} - u(e-1) - u(\ln(4)-1) - 1$$
$$g_2(u) = e - 2u + \ln(4) + 3e^u + \frac{4}{u-2} - u\ \ln(2) - 2$$

where the exact solutions are $G_1(u) = e^u$ and $G_2(u) = \frac{1}{2-u}$.

The integration involved in Eq. (1.26) can be handled using a trapezoidal rule. By dividing the integral region into two equal intervals of width, $h = \frac{1}{2}$. The three different node points are $u_i = \frac{(i-1)}{2}$ and $g_i = g(u_i)$, where $i = 1$, 2, 3. Thus, using the trapezoidal rule, the following equations have been obtained

$$2G_1(u_i) + 3G_2(u_i) = g_1(u_i) + \frac{1}{4}\left(u_iG_1(0) + (2u_i+1)G_1\left(\frac{1}{2}\right) + (u_i+1)G_1(1)\right) + \frac{1}{4}\left(u_iG_2\left(\frac{1}{2}\right) + u_iG_2(1)\right) \tag{1.27}$$

$$\begin{aligned} 3G_1(u_i) - 4G_2(u_i) = g_2(u_i) &+ \frac{1}{4}\left(-G_1(0) + (2u_i-2)G_1\left(\frac{1}{2}\right) + (2u_i-1)G_1(1)\right) \\ &+ \frac{1}{4}\left(u_iG_2(0) + (2u_i-1)G_2\left(\frac{1}{2}\right) + (u_i-1)G_2(1)\right) \end{aligned} \tag{1.28}$$

Equations (1.27) and (1.28) can be transformed into a linear system as

$$\begin{bmatrix} 2 & -0.25 & -0.25 & 3 & 0 & 0 \\ -0.125 & 1.5 & -0.375 & 0 & 2.875 & -0.125 \\ -0.25 & -0.75 & 1.5 & 0 & -0.25 & 2.75 \\ 3.25 & 0.5 & 0.25 & -4 & 0.25 & 0.25 \\ 0.25 & 3.25 & 0 & -0.125 & -4 & 0.125 \\ 0.25 & 0 & 2.75 & -0.25 & -0.25 & -4 \end{bmatrix} \begin{bmatrix} G_1(0) \\ G_1\left(\frac{1}{2}\right) \\ G_1(1) \\ G_2(0) \\ G_2\left(\frac{1}{2}\right) \\ G_2(1) \end{bmatrix} = \begin{bmatrix} 2.5 \\ 3.2452 \\ 5.3320 \\ 3.1046 \\ 3.0375 \\ 3.5663 \end{bmatrix} \tag{1.29}$$

Equation (1.29) can be solved using the proposed ANN algorithm (Section 1.3.1). Table 1.3 contains the comparison between the exact and ANN solutions. Moreover, the convergence plot for the solutions has been depicted in Figure 1.14.

Table 1.3 Comparison between exact and ANN solutions.

u_j	Exact	ANN	u_j	Exact	ANN
	$G_1(u)$			$G_2(u)$	
0	1	1.0071	0	0.5	0.5374
0.5	1.6487	1.6973	0.5	0.6667	0.6979
1	2.7183	2.8084	1	1	1.0249

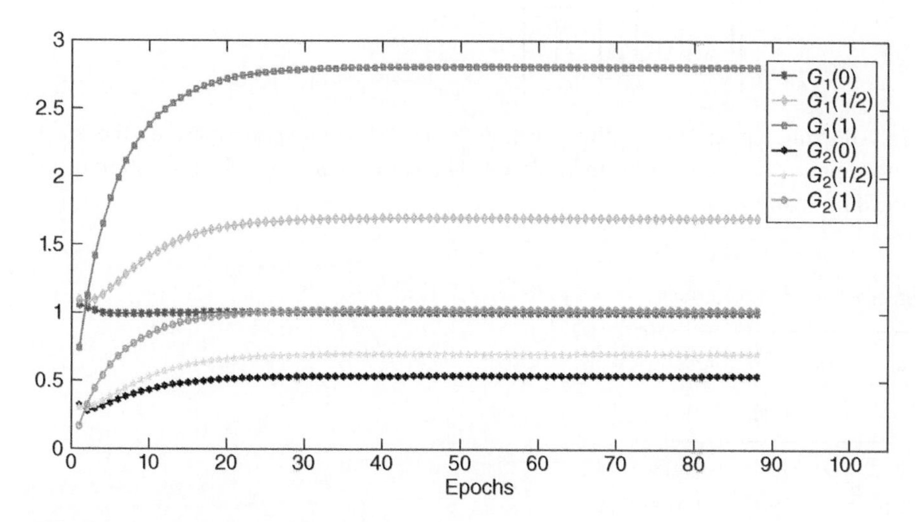

Figure 1.14 Convergence plot of Eq. (1.29).

Example 1.5 In the following example, a system of integral equations has been examined

$$\begin{cases} G_1(u) = g_1(u) + \int_0^1 (u-v)^3 G_1(v)dv + \int_0^1 (u-v)^2 G_2(v)dv \\ G_2(u) = g_2(u) + \int_0^1 (u-v)^4 G_1(v)dv + \int_0^1 (u-v)^3 G_2(v)dv \end{cases} \tag{1.30}$$

whose exact solutions are $G_1(u) = u^2$ and $G_2(u) = -u + u^2 + u^3$ and with

$$g_1(u) = \frac{1}{20} - \frac{11}{30}u + \frac{5}{3}u^2 - \frac{1}{3}u^3$$

$$g_2(u) = -\frac{1}{30} - \frac{41}{60}u + \frac{3}{20}u^2 + \frac{23}{12}u^3 - \frac{1}{3}u^4$$

Similarly (as in the case of Example 1.4), the following equations have been derived using the trapezoidal rule

$$G_1(u_i) = g_1(u_i) + \frac{1}{4}\left(u_i^3 G_1(0) + 2\left(u_i - \frac{1}{2}\right)^3 G_1\left(\frac{1}{2}\right)\right) + \frac{1}{4}\left(u_i^2 G_2(0) + 2\left(u_i - \frac{1}{2}\right)^2 G_2\left(\frac{1}{2}\right)\right) \tag{1.31}$$

$$G_2(u_i) = g_2(u_i) + \frac{1}{4}\left(u_i^4 G_1(0) + 2\left(u_i - \frac{1}{2}\right)^4 G_1\left(\frac{1}{2}\right)\right) + \frac{1}{4}\left(u_i^3 G_2(0) + 2\left(u_i - \frac{1}{2}\right)^3 G_2\left(\frac{1}{2}\right)\right) \tag{1.32}$$

Equations (1.31) and (1.32) can be written in the form of a linear system as

$$\begin{bmatrix} 1 & \frac{1}{16} & 0 & 0 & -\frac{1}{8} & 0 \\ -\frac{1}{32} & 1 & 0 & -\frac{1}{16} & 0 & 0 \\ -\frac{1}{4} & -\frac{1}{16} & 1 & -\frac{1}{4} & -\frac{1}{8} & 0 \\ 0 & -\frac{1}{32} & 0 & 1 & \frac{1}{16} & 0 \\ -\frac{1}{64} & 0 & 0 & -\frac{1}{32} & 1 & 0 \\ -\frac{1}{4} & -\frac{1}{32} & 0 & -\frac{1}{4} & -\frac{1}{16} & 1 \end{bmatrix} \begin{bmatrix} G_1(0) \\ G_1\left(\frac{1}{2}\right) \\ G_1(1) \\ G_2(0) \\ G_2\left(\frac{1}{2}\right) \\ G_2(1) \end{bmatrix} = \begin{bmatrix} \frac{1}{20} \\ \frac{29}{120} \\ \frac{61}{60} \\ -\frac{1}{30} \\ -\frac{19}{160} \\ \frac{61}{60} \end{bmatrix} \tag{1.33}$$

The system (1.33) has been solved using the ANN algorithm (Section 1.3.1). The comparison between the exact and ANN solutions at three node points has been listed in Table 1.4. The convergence plot for the solutions of system (1.33) has been illustrated in Figure 1.15.

Table 1.4 Comparison between exact and ANN solutions.

u_j	Exact	ANN	u_j	Exact	ANN
	$G_1(u)$			$G_2(u)$	
0	0.0000	0.0200	0	0.0000	−0.0184
0.5	0.2500	0.2411	0.5	−0.1250	−0.1191
1	1.0000	1.0173	1	1.0000	1.0172

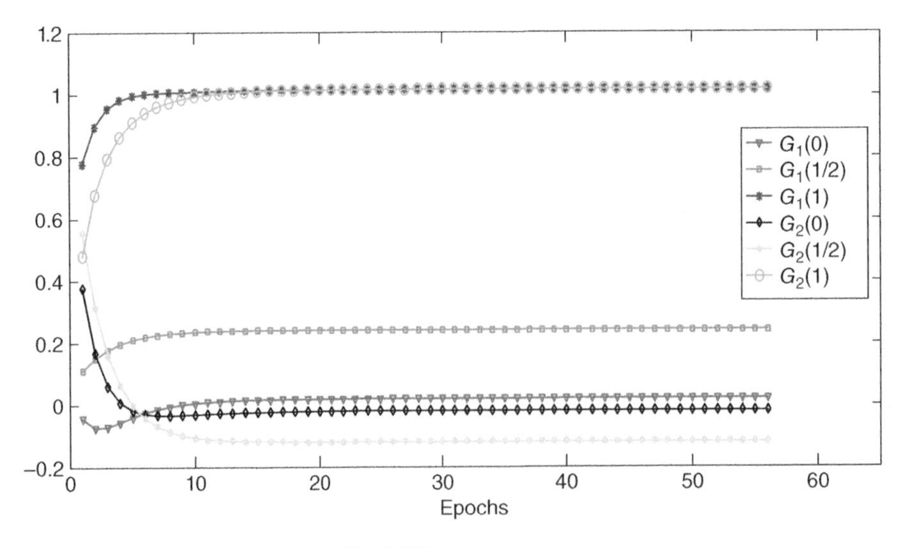

Figure 1.15 Convergence plot of Eq. (1.33).

Example 1.6 Consider the following system of integral equations

$$\begin{cases} G_1(u) = g_1(u) + \int_0^1 \frac{u+v}{3}(G_1(v) + G_2(v))dv \\ G_2(u) = g_2(u) + \int_0^1 uv(G_1(v) + G_2(v))dv \end{cases} \tag{1.34}$$

with

$$g_1(u) = \frac{u}{18} + \frac{17}{36}$$
$$g_2(u) = u^2 - \frac{19}{12}u + 1$$

where the exact solutions are $G_1(u) = u + 1$ and $G_2(u) = u^2 + 1$.

Applying the trapezoidal rule in the region of integration for Eq. (1.34), we have

$$\begin{aligned} G_1(u_i) = g_1(u_i) &+ \frac{1}{4}\left(\frac{u_i}{3}G_1(0) + \frac{2u_i+1}{3}G_1\left(\frac{1}{2}\right) + \frac{u_i+1}{3}G_2(1)\right) \\ &+ \frac{1}{4}\left(\frac{u_i}{3}G_2(0) + \frac{2u_i+1}{3}G_2\left(\frac{1}{2}\right) + \frac{u_i+1}{3}G_2(1)\right) \end{aligned} \tag{1.35}$$

$$G_2(u_i) = g_2(u_i) + \frac{1}{4}\left(u_iG_1\left(\frac{1}{2}\right) + u_iG_1(1)\right) + \frac{1}{4}\left(u_iG_2\left(\frac{1}{2}\right) + u_iG_2(1)\right) \tag{1.36}$$

Equations (1.35) and (1.36) can be written in the form of a linear system as

$$\begin{bmatrix} 0 & -\frac{1}{12} & -\frac{1}{12} & 0 & -\frac{1}{12} & -\frac{1}{12} \\ -\frac{1}{24} & \frac{11}{12} & -\frac{1}{8} & -\frac{1}{24} & -\frac{1}{6} & -\frac{1}{8} \\ -\frac{1}{12} & -\frac{1}{4} & \frac{5}{6} & -\frac{1}{12} & -\frac{1}{4} & -\frac{1}{6} \\ 0 & 0 & 0 & 1 & 0 & 0 \\ 0 & -\frac{1}{8} & -\frac{1}{8} & 0 & \frac{7}{8} & -\frac{1}{8} \\ 0 & -\frac{1}{4} & -\frac{1}{4} & 0 & -\frac{1}{4} & \frac{3}{4} \end{bmatrix} \begin{bmatrix} G_1(0) \\ G_1\left(\frac{1}{2}\right) \\ G_1(1) \\ G_2(0) \\ G_2\left(\frac{1}{2}\right) \\ G_2(1) \end{bmatrix} = \begin{bmatrix} \frac{17}{36} \\ \frac{1}{2} \\ \frac{19}{36} \\ 1 \\ \frac{11}{24} \\ \frac{5}{12} \end{bmatrix} \tag{1.37}$$

ANN procedure has been used to find the solutions of system (1.37). The comparison between the exact and ANN solutions at three node points has been listed in Table 1.5. The convergence plot for the solutions of system (1.37) has been shown in Figure 1.16.

Table 1.5 Comparison between exact and ANN solutions.

u_j	Exact	ANN	u_j	Exact	ANN
	$G_1(u)$			$G_2(u)$	
0	1.0000	1.0704	0	1.0000	1.0000
0.5	1.5000	1.4784	0.5	1.2500	1.3556
1	2.0000	2.1328	1	2.0000	2.2111

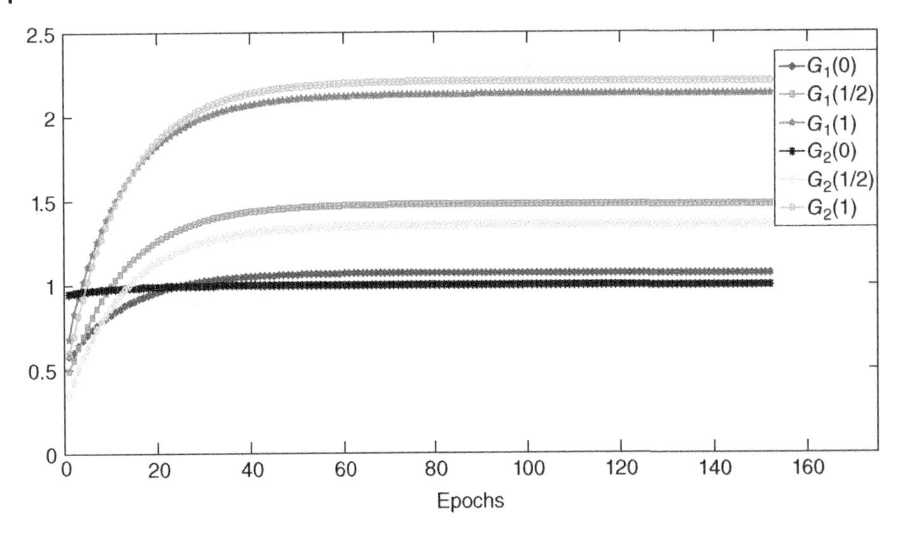

Figure 1.16 Convergence plot of Eq. (1.37).

1.5 Conclusion

As we know, most of the fundamental laws of nature may be formulated as differential equations. Numerous practical problems of mathematics, physics, engineering, economics, etc., can be modeled by ODEs and PDEs. In this chapter, the single-layer FLANN, namely, LgNN model has been developed for solving ODEs. The accuracy of the proposed method has been examined by solving a boundary value problem, a second-order singular Lane-Emden equation, and an application problem of Duffing oscillator equation. In LgNN, the hidden layer is replaced by a functional expansion block of the input data by using Laguerre orthogonal polynomials. The LgNN computed results are shown in terms of tables and graphs.

Systems of Fredholm and Volterra equations are developing great interest for research in the field of science and engineering. In this chapter, an ANN-based method for handling system of Fredholm equations of second kind has been investigated. The system of Fredholm equations has been transformed to a linear system using trapezoidal rule. As such, a single-layer ANN architecture has been developed for solving the linear system. Comparison table between the exact and ANN solutions has been included. Further, convergence plots for solutions of different linear systems have been illustrated to show the efficacy of the proposed ANN algorithm.

References

1 McCulloh, W.S. and Pitts, W. (1943). A logical calculus of the ideas immanent in neural nets. *Bulletin of Mathematical Biophysics* 5 (4): 133–137.

2 Rosenblatt, F. (1958). The perceptron: a probalistic model for information storage and organization in the brain. *Psychological Review* 65 (6): 386–408.

3 Werbos, P.J. (1975). *Beyond Regression: New Tools for Prediction and Analysis in the Behavioral Sciences*. Harvard University.

4 Pao, Y.H. (1989). *Adaptive Pattern Recognition and Neural Networks*. Addison-Wesley Longman Publishing Co., Inc.

5 Pao, Y.H. and Klaseen, M. (1990). The functional link net in structural pattern recognition. *IEEE TENCON'90: 1990 IEEE Region 10 Conference on Computer and Communication Systems. Conference Proceedings*, Hong Kong (24–27 September 1990). IEEE.

6 Lee, T.T. and Jeng, J.T. (1998). The Chebyshev-polynomials-based unified model neural networks for function approximation. *IEEE Transactions on Systems, Man, and Cybernetics Part B: Cybernetics* 28: 925–935.

7 Weng, W.D., Yang, C.S., and Lin, R.C. (2007). A channel equalizer using reduced decision feedback Chebyshev functional link artificial neural networks. *Information Sciences* 177: 2642–2654.

8 Patra, J.C., Juhola, M., and Meher, P.K. (2008). Intelligent sensors using computationally efficient Chebyshev neural networks. *IET Science, Measurement and Technology* 2: 68–75.

9 Mall, S. and Chakraverty, S. (2015). Numerical solution of nonlinear singular initial value problems of Emden–Fowler type using Chebyshev neural network method. *Neurocomputing* 149: 975–982.

10 Mall, S. and Chakraverty, S. (2014). Chebyshev neural network based model for solving Lane–Emden type equations. *Applied Mathematics and Computation* 247: 100–114.

11 Purwar, S., Kar, I.N., and Jha, A.N. (2007). Online system identification of complex systems using Chebyshev neural network. *Applied Soft Computing* 7: 364–372.

12 Patra, J.C. and Kot, A.C. (2002). Nonlinear dynamic system identification using Chebyshev functional link artificial neural network. *IEEE Transactions on Systems, Man, and Cybernetics Part B: Cybernetics* 32: 505–511.

13 Mall, S. and Chakraverty, S. (2016). Application of Legendre neural network for solving ordinary differential equations. *Applied Soft Computing* 43: 347–356.

14 Patra, J.C. and Bornand, C. (2010). Nonlinear dynamic system identification using Legendre neural network. *The 2010 International Joint Conference on Neural Networks (IJCNN)*, Barcelona, Spain (18–23 July 2010). IEEE.

15 Nanda, S.K. and Tripathy, D.P. (2011). Application of functional link artificial neural network for prediction of machinery noise in opencast mines. *Advances in Fuzzy Systems* 2011: 1–11.

16 Mall, S. and Chakraverty, S. (2016). Hermite functional link neural network for solving the Van der Pol-Duffing oscillator equation. *Neural Computation* 28 (8): 1574–1598.

17 Chakraverty, S. and Mall, S. (2017). *Artificial Neural Networks for Engineers and Scientists: Solving Ordinary Differential Equations*. CRC Press/Taylor & Francis Group.

18 Lagaris, I.E., Likas, A., and Fotiadis, D.I. (1998). Artificial neural networks for solving ordinary and partial differential equations. *IEEE Transactions on Neural Networks* 9: 987–1000.

19 Malek, A. and Beidokhti, R.S. (2006). Numerical solution for high order differential equations, using a hybrid neural network – optimization method. *Applied Mathematics and Computation* 183: 260–271.

20 Yazid, H.S., Pakdaman, M., and Modaghegh, H. (2011). Unsupervised kernel least mean square algorithm for solving ordinary differential equations. *Neurocomputing* 74: 2062–2071.

21 Selvaraju, N. and Abdul Samath, J. (2010). Solution of matrix Riccati differential equation for nonlinear singular system using neural networks. *International Journal of Computer Applications* 29: 48–54.

22 Shirvany, Y., Hayati, M., and Moradian, R. (2009). Multilayer perceptron neural networks with novel unsupervised training method for numerical solution of the partial differential equations. *Applied Soft Computing* 9: 20–29.

23 Aarts, L.P. and Van der Veer, P. (2001). Neural network method for solving partial differential equations. *Neural Processing Letters* 14: 261–271.

24 Hoda, S.A. and Nagla, H.A. (2011). Neural network methods for mixed boundary value problems. *International Journal of Nonlinear Science* 11: 312–316.

25 McFall, K. and Mahan, J.R. (2009). Artificial neural network for solution of boundary value problems with exact satisfaction of arbitrary boundary conditions. *IEEE Transactions on Neural Networks* 20: 1221–1233.

26 Golbabai, A. and Seifollahi, S. (2006). Numerical solution of the second kind integral equations using radial basis function networks. *Applied Mathematics and Computation* 174 (2): 877–883.

27 Golbabai, A., Mammadov, M., and Seifollahi, S. (2009). Solving a system of nonlinear integral equations by an RBF network. *Computers & Mathematics with Applications* 57 (10): 1651–1658.

28 Jafarian, A. and Nia, S.M. (2013). Utilizing feed-back neural network approach for solving linear Fredholm integral equations system. *Applied Mathematical Modelling* 37 (7): 5027–5038.

29 Jafarian, A., Measoomy, S., and Abbasbandy, S. (2015). Artificial neural networks based modeling for solving Volterra integral equations system. *Applied Soft Computing* 27: 391–398.

30 Effati, S. and Buzhabadi, R. (2012). A neural network approach for solving Fredholm integral equations of the second kind. *Neural Computing and Applications* 21 (5): 843–852.

31 Jafarian, A., Nia, S.A.M., Golmankhaneh, A.K., and Baleanu, D. (2013). Numerical solution of linear integral equations system using the Bernstein collocation method. *Advances in Difference Equations* 2013 (1): 123.

32 Kitamura, S. and Qing, P. (1989). Neural network application to solve Fredholm integral equations of the first kind. In: *Joint International Conference on Neural Networks*, 589. IEEE.

33 Asady, B., Hakimzadegan, F., and Nazarlue, R. (2014). Utilizing artificial neural network approach for solving two-dimensional integral equations. *Mathematical Sciences* 8 (1): 117.

34 Golbabai, A. and Seifollahi, S. (2006). An iterative solution for the second kind integral equations using radial basis functions. *Applied Mathematics and Computation* 181 (2): 903–907.

35 Cichocki, A. and Unbehauen, R. (1992). Neural networks for solving systems of linear equations and related problems. *IEEE Transactions on Circuits and Systems I: Fundamental Theory and Applications* 39 (2): 124–138.

36 Jeswal, S.K. and Chakraverty, S. (2018). Solving transcendental equation using artificial neural network. *Applied Soft Computing* 73: 562–571.

37 Ibraheem, K.I. and Khalaf, M.B. (2011). Shooting neural networks algorithm for solving boundary value problems in ODEs. *Applications and Applied Mathematics* 6: 187–200.

38 Lane, J.H. (1870). On the theoretical temperature of the sun under the hypothesis of a gaseous mass maintaining its volume by its internal heat and depending on the laws of gases known to terrestrial experiment. *The American Journal of Science and Arts*, 2nd series 4: 57–74.

39 Emden, R. (1907). *Gaskugeln: Anwendungen der mechanischen Warmen-theorie auf Kosmologie and meteorologische Probleme*. Leipzig: Teubner.

40 Chowdhury, M.S.H. and Hashim, I. (2007). Solutions of a class of singular second order initial value problems by homotopy-perturbation method. *Physics Letters A* 365: 439–447.

41 Vanani, S.K. and Aminataei, A. (2010). On the numerical solution of differential equations of Lane-Emden type. *Computers & Mathematics with Applications* 59: 2815–2820.

42 Zhihong, Z. and Shaopu, Y. (2015). Application of van der Pol–Duffing oscillator in weak signal detection. *Computers and Electrical Engineering* 41: 1–8.

43 Jalilvand, A. and Fotoohabadi, H. (2011). The application of Duffing oscillator in weak signal detection. *ECTI Transactions on Electrical Engineering, Electronics, and Communications* 9: 1–6.

2

Deep Learning in Population Genetics: Prediction and Explanation of Selection of a Population

Romila Ghosh[1] *and Satyakama Paul*[2]

[1] *Department of Statistics, Amity University, Kolkata, India*
[2] *Institute of Intelligent Systems, University of Johannesburg, Johannesburg, South Africa*

2.1 Introduction

A major aim in population genetic studies is to draw insights into the evolutionary background of a population. Until decades after its emergence in twentieth century, theories on genetics have been way ahead of the data required to prove the statements. Improvement of the data generating capacity of genomes in the end of twentieth century have enabled analysis of genomic data over related disciplines. Currently, with the advent of whole genome sequencing data, demographic inference can be carried out with more efficient models [1].

In this paper, we focus on using deep learning (DL) to infer on selection on a whole-genome variation data of *Drosophila melanogaster* species. Since the availability of genome sequencing data, several studies on demography and selection have been carried on the populations of *Drosophila*. However, determination or joint inference of both demography and selection has been argued by several researchers to be difficult because of the high influence of selection over shaping the demography of a population [2]. In many of the previous research studies on selection, it is clearly indicated that for *Drosophila* genome, selection I confounded demography [3] and vice versa. Attempts to infer upon selection over *D.melanogaster* populations have been made using maximum-likelihood procedure where selection has been found to be difficult to infer upon, given the presence of demographic parameters [4].

2.2 Literature Review

Machine learning (ML) models are applied in various disciplines and applications, from text processing to quantitative modeling. In population genetics, we are mainly interested in supervised learning algorithms and classification models. If we look at ML approaches developed to infer upon only selection, then notable research studies include methods based on support vector machines (SVMs), boosting, etc. [5].

Pavlidis et al. [6] provide an approach of parameter estimation previously for deploying a classification algorithm to classify between neutral or positive selection, wherein adaptations of the w-statistic for the parameter estimation part and Sweep Finder algorithm for classification are used. Another improvement on the above approach is proposed with SweeD algorithm, which is a modification over Sweep Finder algorithm allowing direct calculations of neutral site frequency (SF) over a demographic model. The SweeD algorithm shows the ability to handle larger samples, thereby resulting in determination of positive selection sweeps with more precision [7]. Another approach to separating selective sweeps using a demographic model is done by developing site frequency

Mathematical Methods in Interdisciplinary Sciences, First Edition. Edited by Snehashish Chakraverty.

(SF) select that improves test scores, where SVMs are trained on normalized scaled site frequency spectrum (SFS) vectors developed from simulated populations [8]. Boosting is a statistical method for improving model predictions of simple learning algorithms. Functional gradient descent (FGD) algorithm [9] is used upon simulated populations showing neutral and selective sweep and to predict upon positive selection [10]. Another method using L2[1] boosting algorithm is proposed [9] to infer upon positive selection over genome-wide single-nucleotide variant (SNV) data and performed using hierarchical classification, where different boosting functions were sequentially applied on the input data [11].

Few works have addressed selection along with population size changes. A maximum likelihood-based approach to identify events inducing low genetic variability and differentiating between demographic bottlenecks and positive selection is described by Galtier et al. [12] over African population data of *D.melanogaster*. The results inferred that the occurrence of positive selection is evident over any demographic bottleneck. A research by Gossmann et al. [13] to estimate the population size of *D.melanogaster* found that population size is correlated with positive selection. Another study [14] on *D.melanogaster* genomic data using the background selection model (BGS) assumes a uniform distribution of deleterious mutations among chromosomes. Upon review of popular models on selection, it is concluded that BGS framework can be used as a null model to study forms of natural selection [15]. Another research [16] on studying the patterns and probabilities of selection footprints under rapid adaptation contrasts model predictions against observed adaptation patterns in *D.melanogaster* population. It concludes that soft sweeps are frequent in rapid adaptation, but complex adaptive patterns also play a significant role in rapid evolution. A study to detect incomplete selective sweeps over African populations of *D.melanogaster* was made by identifying sweeping halotypes (SHs) that carry beneficial alleles that have rapidly increased in frequency [17]. Harris et al. [18] proposed a method to detect hard and soft selective sweeps by identifying multilocus genotypes using the H12 and H1/H2 statistics proposed by Garud et al. [19].

DL is one of the modern approaches being adapted for population genetic inferences. Originally inspired by the connection between neurons in the brain [20], neural network consists of layers of interconnected nodes (neurons), wherein each node is a perceptron that calculates a simple output function from weighted incoming information. In the presence of a well-defined number of neurons and an efficient connectivity pattern, neural networks are very efficient in learning the features of input and provide class discriminations as output or capture the structure of the data, given that enough training data or instances are provided [21].

In genomics, DL models are often used for genomic data classification with the aim to obtain feature classification among populations. A DL method named D-GEX is proposed by Chen et al. [22] to infer upon the expression of target genes from a subset of landmark genes. D-GEX outperforms the linear regression-based approach by a relative improvement of approximately 6.6%. An approach with DL-based algorithmic framework named PEDLA is developed to predict enhancers and is found to achieve 95% accuracy [23]. Another similar framework is devised to identify enhancer and promoter regions in the human genome. A deep three-layer feedforward neural network used to train the data and the model is denoted as DECRES (DEep learning for identifying Cis-Regulatory Elements) [24]. An approach provided by Song and Sheehan [25] uses a combination of approximate Bayesian computation [26] and a DL architecture with 3 hidden layers and 60 neurons to infer upon selection of *D.melanogaster* population where a subset of carefully chosen summary statistics as per their importance is used as the input. Another study by Jones et al. [27] shows a brief overview of the advances of DL models in the field of computational biology where the scope and potential of DL as a tool in genomics, biological image processing, and medical diagnosis is described. Lastly, an approach to classify selective sweeps exploits a supervise ML algorithms termed diploS/HIC. diploS/HIC uses coalescent simulations to generate training data followed by a feature vector calculation and further uses a convolutional neural network (CNN) architecture consisting of three convolution layers and finally a softmax activation layer to get the classification results [28].

---

1 L1 and L2 are common regularization methods used in ML to reduce a model's overfitting.

2.3 Dataset Description

2.3.1 Selection and Its Importance

Selection is the procedure by which certain traits in a population gain reproductive advantage over other traits in the population provided that the traits are heritable. Natural selection is an explanation of how diversity of life increases. Life forms vary and have different reproduction rates, causing populations to shift toward the most successful variations. Mutation is a phenomenon that causes genetic variation in an individual. However, the genetic variation due to mutation is random and mostly change the genetic structure because of which mutation is not visible. Although selection acts on that variation in a very nonrandom way, genetic variants that aid survival and reproduction are much more likely to become dominant than variants that do not.

In an evolutionary scenario, selection is the process by which populations evolve by inheriting adaptive changes. Natural selection increases the mean fitness of a population by allowing more fit individuals to evolve in the later generations, thus producing changes in the frequencies of the genetic alleles associated with variation in fitness. Thus, selection plays an important role in shaping demography by means of adaptation and helps in the study of evolutionary distance between different species' genomes.

In this paper, we use the simulation procedure followed by Song and Sheehan [25] and Peter et al. [29] and use the simulation data provided (as part of the software evoNet). We briefly describe the simulation procedure here for completeness. Measuring the impact of selection, each 100 kb region was divided into three smaller regions: (i) close to the selected site (40–60 kb), (ii) mid-range from the selected site (20–40 and 60–80 kb), and (iii) far from the selected site (0–20 and 80–100 kb). The following statistics are calculated within each of these regions:

(1) *Selection*: If a genomic region is selected, it represents higher frequency than is expected by chance. In other words, the mating is not random, and the Darwinian principle of the survival of the fittest is favoring some region. This region might code for genes that help in fighting diseases or in general help in survival.
(2) *Site frequency spectrum (SFS)*: The SFS is the distribution of allele frequencies of a given set of loci or single-nucleotide polymorphisms (SNPs) in a population. Allele frequency is the relative frequency of an allele at a particular locus in a population. A SNP is a substitution of a single nucleotide that occurs at a specific position in the genome. In the dataset, (SFS_i) is the number of segregating sites where the number of minor allele occurs i times $\forall$ ($i = 1, 2, \ldots, [n/2]$). Enough samples were simulated to get 60 haplotypes. Normalization was done by dividing with the sum of the entries, i.e. 60.
(3) *Identity by state (IBS) tract length distribution*: IBS refers to two identical alleles or two identical segments or similar nucleotide sequences in the genome. IBS tract is a set of continuous genomic regions (denoted by, say L) between a pair of samples delimited by bases where they differ. Similar to length distribution between segregating sites, we count the tract lengths that fall in each bin of a set of 24 equally spaced bins() starting at 0–1500. The tract length distribution is given by:

$$d_m^{\mathrm{IBS}} = \frac{|\, l \in L : l \in \text{bin } m \,|}{|\, L \,|} \tag{2.1}$$

(4) *Linkage disequilibrium (LD) distribution*: LD measures correlation between segregating sites. It is the deviation from the law of Mendel, which states that the identity of two alleles at the loci of the same chromatid is independent. Site alleles should be independent if there is high recombination between the sites. If there is no recombination, alleles should be tightly linked. Here, LD is computed between one site in the selected region and one site in each of the three regions (including the selected region). LD distribution is computed over 10 equally spaced regions, the first one ending at −0.05 and the last bin starting at 0.2.
(5) *Length distribution between segregating sites (BIN)*: The BIN system clusters gene sequences using refined single linkage algorithm (RESL) to produce operational taxonomic units (OTUs) that closely correspond to species [30]. In the dataset [26] [29], 16 equally spaced bins are defined starting from 0 to 300 to compute

the distribution of the number of bases between successive segregating sites. Let B_k be the number of bases between k and the $(k + 1)$ segregating sites. Then, the distribution of these lengths is given by

$$d_j^{\text{BET}} = \frac{|\, k \in \{1, 2, \dots, S-1\} : B_k \in \text{bin}\, j \,|}{S} \tag{2.2}$$

2.4 Objective

A dataset for computing the summary statistics was simulated that is relevant to selection in *Drosophila*. Refer to Elyashiv et al. [31] for a detailed analysis of selection on *Drosophila*. The program developed by Ewing and Hermisson [32] is used for simulating the datasets. Each dataset corresponded to 100 haplotypes. For a particular selection scenario, a 100 kb region was simulated with the selected site (if present) occurring randomly in the middle 20 kb of the region. Baseline effective population size is $N_{\text{ref}} = 100\,000$, a per-base, per-generation mutation rate is $\mu = 8.4 \times 10^{-9}$, and a per-base, per-generation recombination rate r equal to μ. This process was repeated 10 000 times to generate 10 000 rows or instances.

The variables in the dataset are formed mainly by four statistics, SFS, IBS, LD, and BIN. For each statistic name, Close, Mid, and Far represent the genomic subregion where the statistic is calculated. The numbers after each of the colons refer to the position of the statistic within its distribution or order. For example – for the folded SFS statistics, the value after the colon represents a number of minor alleles, for the LD statistic. In the dataset, each region of a gene refers to – 87 SFS statistics, 48 BIN statistics, 75 IBS statistics, and 30 LD statistics. In effect, there are 240 independent variables or predictors in the dataset, and each predictor corresponds to a specific genomic region. It might also be noted that there are no missing values in the dataset. The dependent variable or response is selection – which comprises two classes, neutral region and hard sweep. The description of the selection classes are as follows: neutral region is the class corresponds to no selection in the genome, and hard sweep is the class corresponds to positive selection on a de novo mutation. Figure 2.1 shows the proportional frequency of the two classes of selection, and Figure 2.2 shows the spatial distribution of the two classes across 240 predictors. *t*-Distributed stochastic neighborhood embedding has been used to compress the 240 predictors into a two-dimensional hyperplane.

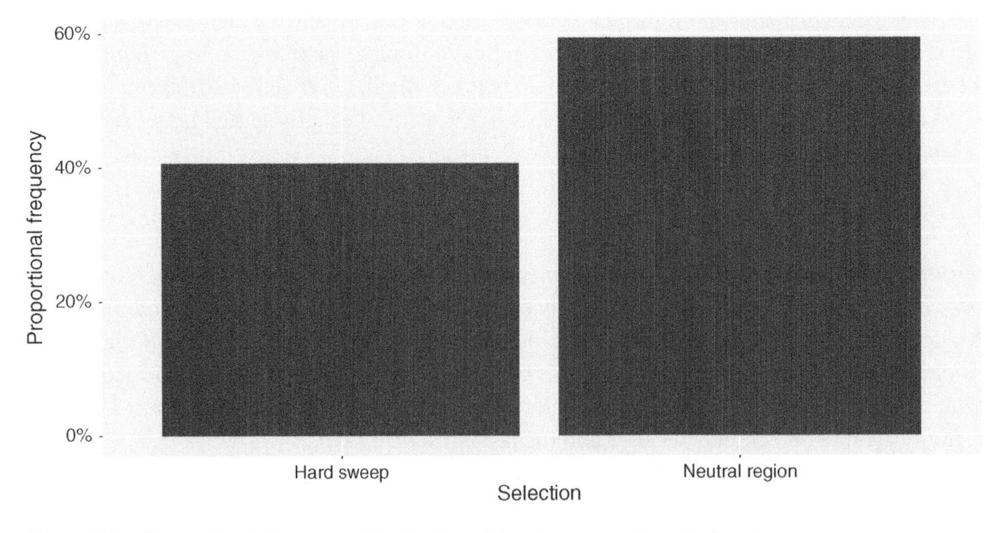

Figure 2.1 Proportional frequency distribution of hard sweep and neutral region.

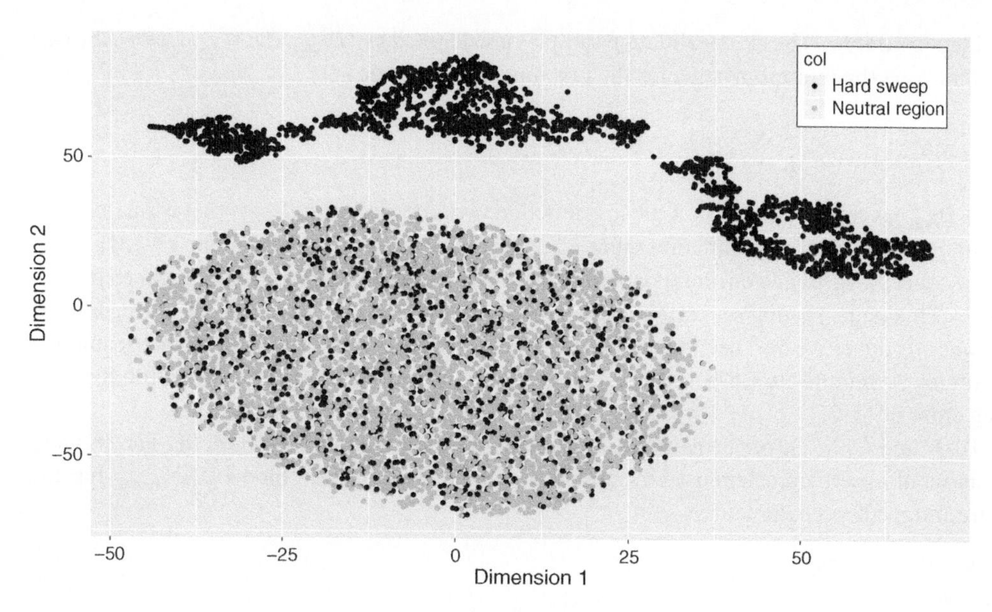

Figure 2.2 Spatial distribution of the two classes of selection.

The objective of this work is to create a classification algorithm (among the family of advanced ML and DL algorithms) that can best predict the two classes of selection. The event here is hard sweep and as shown in Figure 2.1, the event rate is 40.65%.

2.5 Relevant Theory, Results, and Discussions

As discussed in Section 2.4, our objective is to first decide upon a classification algorithm that is best suited to our problem. Because there are multiple ML algorithms and DL architectures to choose from and each of their implementation is time consuming, we use the automatic machine learning (automl) framework to choose the best classification algorithm for the task.

2.5.1 automl

As the name suggests, automl is a framework that automates various components of a ML pipeline such as feature engineering, reduction of high dimensional data, hyperparameter tuning of various kinds of models, etc. However, in this work, we use one of its specific features – selection of the best algorithm for a classification task. In particular, we use the open source H2O automl framework in *R*.[2]

In H2O, the choice set (from which the best algorithm is chosen) consists of random forest (RF) and extremely randomized trees (XRT), generalized linear models (GLMs), gradient boosting machine (GBM), extreme gradient boosted machines (XGBoost), deep neural networks (DNNs), and a stacked ensemble (SE) of all the previous models [33]. It might also be noted that each of these models is built using a random set of relevant hyperparameters.

Let us assume that for the above set of given algorithms A, their corresponding hyperparameters are denoted by $A^{(\cdot)}$. Also, let us assume that the training set is denoted by $D_{\text{train}} = \{(\vec{x}_1, y_1), \ldots, (\vec{x}_n, y_n)\}$ and the training set is split

2 Other open source automl frameworks are provided by Scikit-Learn, TPOT, etc.

into K-fold cross-validation samples. The cross-validation samples are denoted by $\{D^1_{\text{valid}}, \dots, D^K_{\text{valid}}\}$. Now, the best algorithm is chosen that gives the minimum value for the loss function or metric

$$g^*, \vec{A}^*, \vec{\lambda}^* \in \text{argmin}_{g \in G, \vec{A} \in A^+ \vec{\lambda} \in A} \frac{1}{K} \sum_{i=1}^{K} L(P_{g,\vec{A},\vec{\lambda}}(D^{(i)}_{\text{train}}), D^{(i)}_{\text{valid}})$$

where $L(P_{g,\vec{A},\vec{\lambda}}(D_{\text{train}}), D_{\text{valid}})$ is the performance of the loss function over the algorithm $P_{g,\vec{A},\vec{\lambda}}$ that is run over the training data D_{train} and evaluated over the validation set D_{valid} using a random set of hyperparameters [34].

Following the above concept, we divide our total dataset of 10 000 rows randomly into training and testing set in the ratio of 0.8 : 0.2. Thus, our training set comprises 8037 rows and the test set has 1963 rows. In addition, a fivefold cross-validation is done. We also use area under the curve (AUC) as a loss function or metric and the automl pipeline is run for 90 minutes to extract the best classification algorithm. Table 2.1 shows the five best models in decreasing order of AUC.

As can be noted, DNN and GBM outperforms the rest of the algorithms such as RF, GLM, SE, etc. In addition, from the random set of hyperparameters used by automl, it suggests that the best model $\text{DNN}_{\text{model}_1}$ has the following architecture and the hyperparameters:

- number of neurons in the input layer: 240
- number of hidden layers: 3
- neurons in each hidden layer: 50
- activation function in all three hidden layers: rectifier with dropout
- number of neurons in the output layer: 2
- activation function used in the output layer: softmax
- percentage dropout used in all the input and hidden layers: 10
- L_1 and L_2 regularization: 0
- momentum: 0

2.5.2 Hypertuning the Best Model

In Section 2.5.1, automl suggests that the best classification algorithm for this test to be DNN, among the set of six models. In this subsection, we further intend to hypertune the DNN's parameters to see if AUC can be improved above 0.820 390 2. In this regard, our choice set of the hyperparameters consists of the following combinations:

- number of hidden layers: 3
  (1) various combinations in the first, second, and third hidden layers
    - 250, 250, 250
    - 500, 500, 500

Table 2.1 automl results.

Algorithm	AUC
$\text{DNN}_{\text{model}_1}$	0.820 390 2
$\text{GBM}_{\text{model}_2}$	0.819 398 4
$\text{GBM}_{\text{model}_3}$	0.814 290 7
$\text{GBM}_{\text{model}_4}$	0.812 530 5
$\text{GBM}_{\text{model}_5}$	0.812 455 3

- 750, 750, 750

- number of hidden layers: 4
  (1) various combinations in the first, second, third, and fourth hidden layers
    - 30, 30, 30, 30
    - 60, 60, 60, 60
    - 90, 90, 90, 90
    - 120, 120, 120, 120
- number of hidden layers: 5
  (1) 240, 120, 60, 30, 15
- various options of activation function in each neuron each layer:
  - Rectifier
  - Rectifier with dropout
  - tanh
  - tanh with dropout
  - Maxout
  - Maxout with dropout
- various options of input dropout %: 0, 0.05, 0.1, 0.2, 0.3, 0.4, 0.5
- various options of L_1 and L_2 values: 0, 1.0e−06, 2.0e−06, …, 1.0e−04
- various options of learning rates: 0, 01, 0.005, 0.001
- various options of rate of annealing: 1e−8, 1e−7, 1e−6
- various options of epochs: 50, 100, 200, 500

Based on the above wide combinations of hyperparameter values and an early stopping rate of 1e−2 for improvement in the value of AUC (at which the algorithm stops), the developed DNN shows an improvement in the value of AUC – 0.869 831 2 (0.820 390 2 as shown by automl). Also, the architecture of the best DNN model is as follows:

- number of neurons each in the four hidden layers: 30
- activation function in the four hidden layers: Maxout
- activation function in the output layer: Softmax
- dropout % in the input layer: 10
- dropout % in the three hidden layers: 0
- L_1 in the four hidden layers and output: 1e−6
- L_2 in the four hidden layers and output: 0.000 034

Finally, the accuracy metrics on the test set is as shown in Table 2.2. The vertical values are actual and the across ones are predicted. Precision, recall, and overall accuracy are 0.996, 0.38, and 0.85, respectively.

As can be seen from Figure 2.2, as there is a high level of overlap between hard sweep and neutral region classes in the circular blog, we expect the miss-classification to happen in that region. Lastly, Figure 2.3 shows the first 10 most important predictors.

Table 2.2 Accuracy on test data.

	Hard sweep	Neutral region	Error rate in %
Hard sweep	500	296	0.372
Neutral region	0	1167	0.0
Total	500	1463	0.151

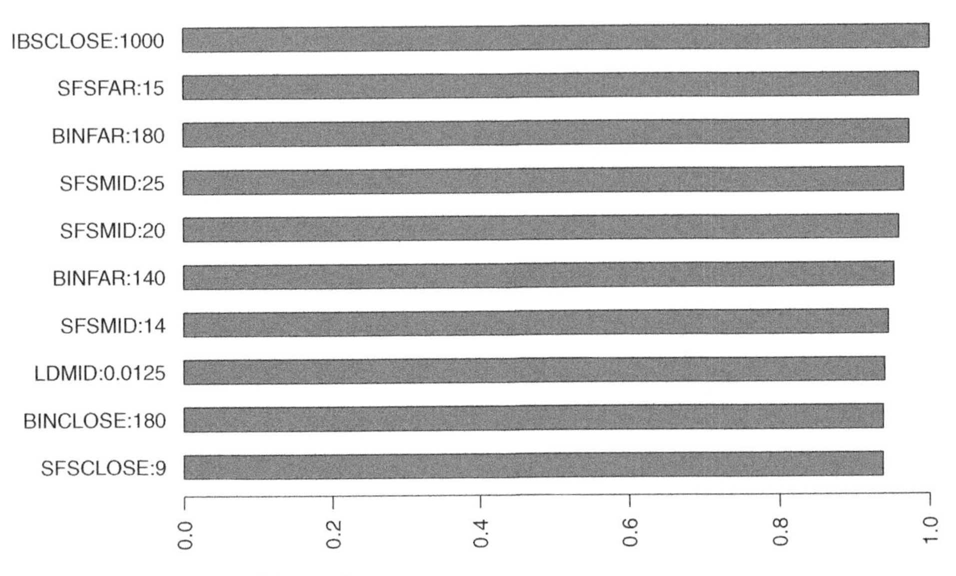

Figure 2.3 Importance of the predictors.

2.6 Conclusion

This research work is a detailed explanation of utilization of the concepts of automl and hyperparameter tuning of a DNN to find the best DNN architecture to classify and predict the selection classes of *D.melanogaster* species. To the best of authors' knowledge , such a detailed explanation of the implementation of automl and parameter hypertuning has not been carried out in the past. In future, we would like to incorporate the concepts of explainable AI to further understand the importance of various predictors in a model agnostic framework.

References

1 Pool, J.E., Hellmann, I., Jensen, J.D., and Neilsen, R. (2010). Population genetic inference from genomic sequence variation. *Genome Research* 20 (3): 291–300.
2 Li, J., Li, H., Jakobsson, M. et al. (2012). Joint analysis of demography and selection in population genetics: where do we stand and where could we go? *Molecular Ecology* 21 (1): 28–44.
3 Sella, G., Petrov, D.A., Prezworski, M., and Andolfatto, P. (2009). Pervasive natural selection in the *Drosophila* genome? *PLoS Genetics* 5 (6): 1–13.
4 González, J., Macpherson, J.M., Messer, P.W., and Petrov, D.A. (2009). Inferring the strength of selection in *Drosophila* under complex demographic models. *Molecular Biology and Evolution* 26 (3): 513–526.
5 Schrider, D.R. and Kern, A.D. (2018). Supervised machine learning for population genetics: a new paradigm. *Trends in Genetics* 34 (4): 301–312.
6 Pavlidis, P., Jensen, J.D., and Stephan, W. (2010). Searching for footprints of positive selection in whole-genome SNP data from nonequilibrium populations. *Genetics* 185 (3): 907–922.
7 Pavlidis, P., živković, D., Stamatakis, A., and Alachiotis, N. (2013). SweeD: likelihood-based detection of selective sweeps in thousands of genomes. *Molecular Biology and Evolution* 30 (9): 2224–2234.

8 Ronen, R., Udpa, N., Halperin, E., and Bafna, V. (2013). Learning natural selection from the site frequency spectrum. *Genetics* 195 (1): 181–193.

9 Bühlmann, P. and Hothorn, T. (2007). Boosting algorithms: regularization, prediction and model fitting. *Statistical Science* 22 (5): 477–505.

10 Lin, K., Li, H., Schlötterer, C., and Futschik, A. (2011). Distinguishing positive selection from neutral evolution: boosting the performance of summary statistics. *Genetics* 187 (1): 229–244.

11 Pybus, M., Luisi, P., Dall'Olio, G.M. et al. (2015). Hierarchical boosting: a machine-learning framework to detect and classify hard selective sweeps in human populations. *Bioinformatics* 31 (24): 3946–3952.

12 Galtier, N., Depaulis, F., and Barton, N.H. (2000). Detecting bottlenecks and selective sweeps from DNA sequence polymorphism. *Genetics* 155 (2) 981–987.

13 Gossmann, T.I., Woolfit, M., and Eyre-Walker, A. (2011). Quantifying the variation in the effective population size within a genome. *Genetics* 189 (4): 1389–1402.

14 Charlesworth, B. (2009). Background selection and patterns of genetic diversity in *Drosophila melanogaster*. *Genetics Research* 68 (2): 131–149.

15 Comeron, J.M. (2017). Background selection as null hypothesis in population genomics: insights and challenges from *Drosophila* studies. *Philosophical Transactions of the Royal Society B: Biological Sciences* 372 (1736) 1–13. (available at https://royalsocietypublishing.org/doi/pdf/10.1098/rstb.2016.0471)

16 Hermisson, J. and Pennings, P.S. (2017). Soft sweeps and beyond: understanding the patterns and probabilities of selection footprints under rapid adaptation. *Methods in Ecology and Evolution* 8 (6): 700–716.

17 Vy, H.M.T., Won, Y.-J., and Kim, Y. (2017). Multiple modes of positive selection shaping the patterns of incomplete selective sweeps over African populations of *Drosophila melanogaster*. *Molecular Biology and Evolution* 34 (11): 2792–2807.

18 Harris, A.M., Garud, N.R., and DeGiorgio, M. (2018). Detection and classification of hard and soft sweeps from unphased genotypes by multilocus genotype identity. *Genetics* 210 (4): 1429–1452.

19 Garud, N.R., Messer, P.W., Buzbas, E.O., and Petrov, D.A. (2015). Recent selective sweeps in North American *Drosophila melanogaster* show signatures of soft sweeps. *PLoS Genetics* 11 (2): 1–32. (available at https://journals.plos.org/plosgenetics/article/file?id=10.1371/journal.pgen.1005004&type=printable)

20 Hopfield, J.J. (1982). Neural networks and physical systems with emergent collective computational abilities. *Proceedings of the National Academy of Sciences of the United States of America* 79 (8): 2554–2558.

21 Hornik, K. (1991). Approximation capabilities of multilayer feedforward networks. *Neural Networks* 4 (2): 251–257.

22 Chen, Y., Li, Y., Narayan, R. et al. (2016). Gene expression inference with deep learning. *Bioinformatics* 32 (12): 1832–1839.

23 Liu, F., Li, H., Ren, C. et al. (2016). PEDLA: predicting enhancers with a deep learning-based algorithmic framework. *Scientific Reports* 6: 28517.

24 Li, Y., Shi, W., and Wasserman, W.W. (2018). Genome-wide prediction of *cis*-regulatory regions using supervised deep learning methods. *BMC Bioinformatics* 19: 202.

25 Song, Y.S. and Sheehan, S. (2016). Deep learning for population genetic inference. *PLoS Computational Biology* 12 (3): 1–28..

26 Beaumont, M.A., Zhang, W., and Balding, D.J. (2002). Approximate Bayesian computation in population genetics. *Genetics* 2025–2035. (available at https://journals.plos.org/ploscompbiol/article/file?id=10.1371/journal.pcbi.1004845&type=printable)

27 Jones, W., Alasoo, K., Fishman, D., and Parts, L. (2017). Computational biology: deep learning. *Emerging Topics in Life Sciences* 1 (3): 257–274.

28 Kern, A.D. and Schrider, D.R. (2018). diploS/HIC: an updated approach to classifying selective sweeps. *G3: Genes Genome Genetics* 8 (6): 1959–1970.

29 Peter, B.M., Huerta-Sanchez, E., and Nielsen, R. (2012). Distinguishing between selective sweeps from standing variation and from a de novo mutation. *PLoS Genetics* 8 (10): 1–14. (available at https://journals.plos.org/plosgenetics/article/file?id=10.1371/journal.pgen.1003011&type=printable)

30 Ratnasingham, S. and Hebert, P.D.N. (2013). A DNA-based registry for all animal species: the barcode index number (BIN) system. *PLoS One* 8 (8).

31 Elyashiv, E., Sattath, S., Hu, T.T. et al. (2016). A genomic map of the effects of linked selection in *Drosophila*. *PLoS Genetics* 12 (8): 1–24. (available at https://journals.plos.org/plosgenetics/article/file?id=10.1371/journal.pgen.1006130&type=printable)

32 Ewing, G. and Hermisson, J. (2010). MSMS: a coalescent simulation program including recombination, demographic structure and selection at a single locus. *Bioinformatics* 26 (16): 2064–2065.

33 AutoML: Automatic Machine Learning. http://docs.h2o.ai/h2o/latest-stable/h2o-docs/automl.html (accessed 01 July 2019).

34 Zoller, M.A. and Huber, M.F. (2019) Benchmark and Survey of Automated Machine Learning Frameworks. https://arxiv.org/pdf/1904.12054.pdf (accessed 01 July 2019).

3

A Survey of Classification Techniques in Speech Emotion Recognition

Tanmoy Roy[1], *Tshilidzi Marwala*[1], *and Snehashish Chakraverty*[2]

[1] *Electrical and Electronic Engineering Science, University of Johannesburg, Johannesburg 2006, South Africa*
[2] *Department of Mathematics, National Institute of Technology, Rourkela, Odisha 769008, India*

3.1 Introduction

Speech recognition models are presently in a very advanced state, and examples such as Siri, Cortana, and Alexa are the technical marvels that not can only hear us comfortably, but those bots can reply to us with equal comfort as well. However, *speech recognition systems* cannot detect the underlying emotion in speech signals. Speech emotion recognition (SER) is the field of study that explores the avenues of detecting emotions concealed in speech signals. The initial study on emotion started as a study of psychology and acoustics of emotions. The first detailed study on emotions was reported way back in 1872 by Charles Darwin [1]. Fairbanks and Pronovost [2] were among the first who studied pitch of voice during simulated emotion. Since the late 1950s, there has been a significant increase in interest by researchers regarding the psychological and acoustic aspects of emotion [3–6]. However, in the year 1995, Picard [7] introduced the term "affective computing," and after that, the study of emotional states has become an integral part of artificial intelligence (AI) research.

SER is a *machine learning* (ML) problem where the speech utterances are classified depending on their underlying emotions. This chapter gives an overview of the prominent classification techniques used in SER. Researchers used different types of classifiers for SER, but in most of the situations, a proper justification is not provided for choosing a particular classification model [8]. Two plausible explanations are that classifiers that are successful in *automatic speech recognition* (ASR) are assumed to be working well in SER (like hidden Markov model [HMM]), and secondly, those classifiers that perform well in most classification problems are chosen [9]. There are two broad categories of classifiers (Figure 3.1) used in SER: the *linear classifiers* and the *non-linear classifiers*. Although many classification models have been used for SER, but few among them become effective and popular among the researchers. Four classification models, namely HMM, *Gaussian mixture model* (GMM), *support vector machine* (SVM), and *deep learning* (DL), have been identified as prominent for this work and they are discussed in Sections 3.4.1–3.4.4.

3.2 Emotional Speech Databases

Researchers are trying to solve SER as an ML problem, and ML approaches are data-driven. Moreover, SER research field is not mature enough to identify the underlying emotion of a random spoken conversion. Thus, the SER research depends heavily on emotional speech databases [10, 11]. The speech databases created by researchers and organizations to support the SER research. The database naturalness, quality of recordings,

Mathematical Methods in Interdisciplinary Sciences, First Edition. Edited by Snehashish Chakraverty.

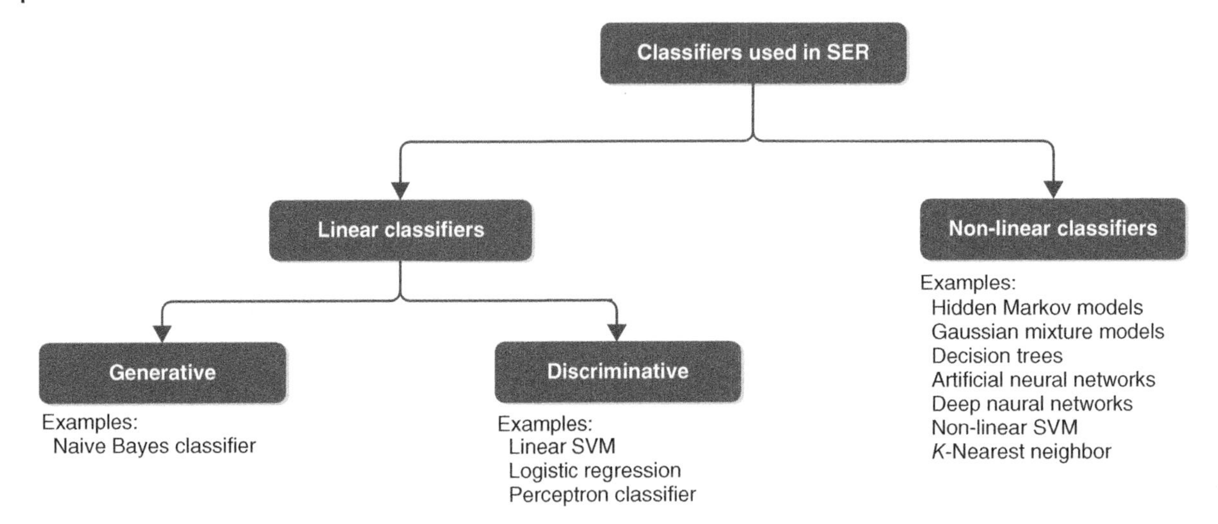

Figure 3.1 Categories of classifiers used in SER along with some examples.

number and type of emotions considered, and speech collection strategy are critical inputs for the classification stage because those features of the database will decide the classification methodology [12–15]. The design of the speech database can have different factors [13, 16]. First of all, the existing databases can be categorized into three categories: (i) simulated by actors, (ii) seminatural, and (iii) natural. The simulated databases, created by enacting emotions by actors, are usually well annotated, adequately labeled, and are of better quality because the recordings are performed in a controlled near noise-free environment. The number of recordings is also usually high for simulated databases. However, acted emotions are not natural enough, and sometimes, an expression of the same emotion varies a lot depending on the actor, which makes the feature selection process very difficult. Seminatural databases are also the collection of the enactions by professional or nonprofessional actors, but here, the actors are trying to keep it as natural as possible. Natural emotional databases are difficult to label because manually labeling a big set of speech recordings is a daunting task, and there is no method available yet to label the emotions automatically. As a result, the number of emotions covered in a natural dataset is low, and the number of data points is also low. Natural recordings usually depict continuous emotions, which can create hurdles during the classification phase because of the presence of overlapping emotions.

3.3 SER Features

Feature selection and feature extraction are vital steps toward building a successful classification model. In SER, various types of features are extracted from the speech signals. Speech signals carry an enormous amount of information apart from the intended message. Researchers agreed that speech signals also carry vital information regarding the emotional state of the speaker [17]. However, researchers are still undecided over the right set of features of the speech signals, which can represent the underlying emotional state. This section contains the details of feature sets that are heavily used so far in SER research and performed well in the classification stage. There are three prominent categories in speech features used in SER: (i) the prosodic features, (ii) the spectral or vocal tract features, and (iii) the excitation source features.

Prosody features are the characteristics of the speech sound generated by the human speech production system, for example, pitch or fundamental frequency (F_0) and energy. Researchers used different derivatives of pitch and energy as various prosody features [18–20]. These are also called continuous features and can be grouped into the following categories [8, 16, 21, 22]: (i) pitch-related features, (ii) formant features, (iii) energy-related features,

(iv) timing features, and (v) articulation features. Several studies tried to establish the relationship between prosody speech features and the underlying patterns of different emotions [21–28].

The features used to represent glottal activity, mainly the vibration of glottal folds, are known as the source or excitation source features. These are also called voice quality features because glottal folds determine the characteristics of the voice. Some researchers believed that the emotional content of an utterance is strongly related to voice quality [21, 29, 30]. Voice quality measures for a speech signal includes harshness, breathiness, and tenseness. The relation of voice quality features with different emotions is not a well-explored area, and researchers have produced contradictory conclusions. For example, Scherer [29] associated anger with tense voice, whereas Murray and Arnott [22] associated anger with a breathy voice.

Spectral features are the characteristics of various sound components generated from different cavities of the vocal tract system. They are also called segmental or system features. Spectral features extracted in the form of

1. ordinary linear predictor coefficients (LPC) [31],
2. one-sided autocorrelation linear predictor coefficients (OSALPC) [32],
3. short-time coherence method (SMC) [33], and
4. least-squares modified Yule–Walker equations (LSMYWE) [34].

There is a particular type of spectral feature called the *cepstral* features, which are extensively used by SER researchers. Cepstral features can be derived from the corresponding linear features such as linear predictor cepstral coefficients (LPCC) is derived from *linear predictor* (LP). Mel-frequency cepstral coefficients are one such cepstral feature that along with its various derivatives is widely used in SER research [35–39].

3.4 Classification Techniques

SER deals with speech signals. The analog (continuous) speech signal is sampled at a specified time interval to get the discrete time speech signal. A discrete time signal can be represented as follows:

$$\begin{aligned} C = \{c_l\}_{l \in \mathbb{L}}, \quad &\text{where } \{c_l\} = \{c_1, c_2, \dots, c_l\} \\ &\{c_l\} \in \mathbb{R} \end{aligned} \tag{3.1}$$

where L is the total number of sample points in the speech signal. First, only the speech utterance section is extracted from the speech sound by using a speech endpoint detection algorithm. In this case, an algorithm proposed by Roy et al. [40] is used.

This speech signal contains various information that can be retrieved for further processing. Emotional states guide human thoughts, and those thoughts are expressed in different forms [41], such as speech. The primary objective of an SER is to find the patterns in speech signals, which can describe the underlying emotions. The pattern recognition task is carried out by different ML algorithms. Features are extracted from the speech signal C in two forms

1. local features by splitting the signals into smaller frames and computing statistics of each frame and
2. global features by calculating statistics on the whole utterance.

Let, there be N number of sample points after the feature extraction process. If local features are computed from C by assuming 10 splits, then there will be 10 data points from C. Now, suppose there is a total of 100 recorded utterances, and each utterance is split into 10 frames, then will be total $(100 \times 10) = 1000$ data points available. When global features are computed, then each utterance will produce one data point. The selection of local or global feature depends on the feature extraction strategy. Now, suppose, N is the number of data points such that $n = 1, 2, 3, \dots, N$, where n is the index. If D number of features are extracted from C, then each data point is a D

Table 3.1 List of literatures on SER grouped by classification models used.

No.	Classifiers	References
1.	Hidden Markov model	[42–51]
2.	Gaussian mixture model	[12,15,39,52–59]
3.	K-Nearest neighbor	[57,60–64]
4.	Support vector machine	[48,49,55,65–73]
5.	Artificial neural network	[43,55,57,74–78]
6.	Bayes classifier	[43,53,60,79]
7.	Linear discriminant analysis	[16,64,80–82]
8.	Deep neural network	[35–37,83–94]

dimensional feature vector. Each utterance in the speech database is labeled properly, so that it can be used for supervised classification. Therefore, the dataset is denoted as

$$X = \{x_n, y_n\}_{n=1}^{N} \tag{3.2}$$

where y_n is the label corresponding to a data point x_n and $X \in \mathbb{R}^{N \times D}$. Once the data is available, the next step is to find a *predictive function* called *predictor*. More specifically, the task of finding a function f is called learning so that $f: X \rightarrow Y$. Different classification models take different approaches to learning. For SER, the prediction task is usually considered as a multiclass classification problem.

Table 3.1 shows a list of classifiers commonly used in SER along with literature references. Although in Table 3.1, there are eight classifiers listed, but not all of them become prominent for SER tasks. In Sections 3.4.1–3.4.4, four most prominent classifiers (HMM, GMM, SVM, and *deep neural network* (DNN)) for SER are discussed to depict the SER-specific implementation technique.

3.4.1 Hidden Markov Model

HMMs are suitable for the sequence classification problems that consist of a process that unfolds in time. That is why, HMM is very successful in ASR systems, where the sequence of the spoken utterance is a time-dependent process. The HMM parameters are tuned in the model training phase to best explain the training data for the known category. The model classifies an unseen pattern based on the highest posterior probability.

HMM comprises two processes. The first processes consist of a first-order Markov chain whose states capture the temporal structure of the data, but these states are not observable that is hidden. The *transition model*, which is a stochastic model, drives the state transition process. Each hidden state has an observation associated with it. The *observation model*, which is again a stochastic model, decides that in a given hidden state, the probability of the occurrence of different observations [95–97].

Figure 3.2 shows a generic HMM. Assuming the length of the observation sequence to be T so that $O = O_0, O_1, \dots, O_{T-1}$ is an observation sequence, and N is the number of hidden states. The observation sequences are derived from C in Eq. (3.1) by computing features of the frames. The state *transition* probability matrix is denoted by A, whereas the *observation* probability matrix is denoted by B. Also, let π_0 be the initial state probability for the hidden Markov chain.

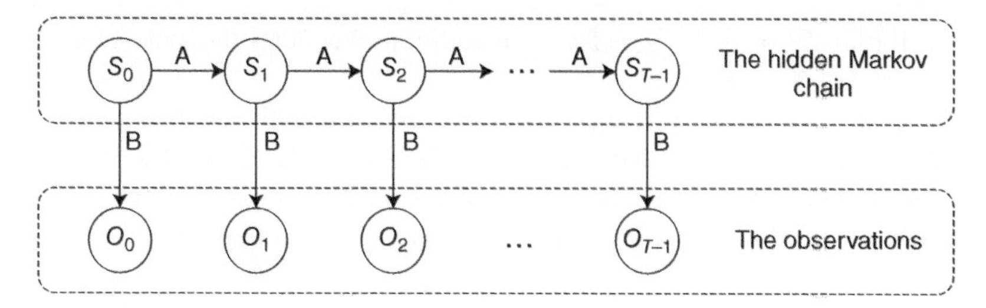

Figure 3.2 Schematic diagram of an HMM, where S_i and O_i are the states and observations, respectively, $i = 0, 1, \ldots, T-1$. A is the *transition model* and B is the *observation model*.

In the training phase, the model parameters are determined. Here, the model is denoted by λ, which contains three parameters A, B, and π; thus, $\lambda = (A, B, \pi)$. The parameters are usually determined using the expectation maximization (EM) algorithm [98] so that the probability of the observation sequence Q is maximum. After the model λ is determined, the probability of an unseen sequence O_u that is $p(O_u|\lambda)$ can be found to get the sequence classification results.

SER researchers used HMM for a long time and used it with various types of feature sets. For example, some researchers [42–45] used prosody features, and some [44–46] used spectral features. Researchers using the HMM achieved the average SER classification accuracy between 75.5% and 78.5% [45, 47–51], which is comparable with other classification techniques, but further improvement possibilities are low. Moreover, that is why HMM has been replaced by other classification techniques in later studies such as SVM, GMM, or DNN.

3.4.1.1 Difficulties in Using HMM for SER

- HMM may follow two types of topology: fully connected or left-to-right. Most ASR systems use left-to-right topology [47], but this topology will not work for SER because a particular token can occur at any stage of the utterance. Therefore, in the case of SER, fully connected topology is more suitable [45]. However, the problem domain of SER is different from ASR, and the sequence in the utterance is the most crucial attribute toward successful ASR, but the SER sequence is not that essential. The whole utterance represents the emotional state and not the sequence of words or silence.
- The optimal number of states required for SER is hard to decide because there is no fixed rule of splitting the speech signal into smaller frames.
- The observation type could be discrete or continuous [95], but for SER, it is hard to decide whether to consider it discrete or continuous.
- In ASR, every spoken word is broken into smaller phonemes that are very neatly handled by the HMM. However, for SER, at least a word should start making some sense, and even a set of words should be reasonable, which is a very different scenario than ASR. Therefore, applying HMM for SER poses another challenge.

3.4.2 Gaussian Mixture Model

An unknown distribution $p(x)$ can be described by a convex combination of K base distributions like Gaussians, Bernoulli's, or Gammas, using mixture models. GMM is the special case of mixture models, where the base distribution is assumed to be Gaussian. GMM is a probabilistic density estimation process where a finite number of K Gaussian distributions of the form $\mathcal{N}(x|\mu_k, \Sigma_k)$ is combined, where x is a D-dimensional vector, i.e. $x \in \mathbb{R}^D$, μ_k is the corresponding *mean* vector and Σ_k is the covariance matrix, such that [99, 100]

$$p(x|\theta) = \sum_{k=1}^{K} \pi_k \mathcal{N}(x|\mu_k, \Sigma_k) \tag{3.3}$$

where π_k are the mixture of weights, such that $0 \leq \pi_k \leq 1$, $\sum_{k=1}^{K} \pi_k = 1$. In addition, θ denotes the collection of parameters of the model $\theta := \{\mu_k, \Sigma_k, \pi_k : k = 1, 2, \ldots, K\}$.

Now, consider the dataset $X = x_1, \ldots, x_N$, extracted similarly as in Eq. (3.2), and it is assumed that x_n, $n = 1, \ldots, N$ are drawn *independent and identically distributed* (i.i.d.) from an unknown distribution $p(x)$. The objective here is to find a good approximation of $p(x)$ by means of a GMM with K mixture components and for that the maximum likelihood estimate (MLE) of the parameters θ need to be obtained [100, 101]. The i.i.d. assumption allows the $p(X \mid \theta)$ to be written [99, 100] as follows:

$$p(X \mid \theta) = \prod_{n=1}^{N} p(x_n \mid \theta) \tag{3.4}$$

where the individual likelihood term $p(x_n \mid \theta)$ is a Gaussian mixture density as in Eq. (3.3). Then, it is required to get the log-likelihood [99, 100]

$$\log p(X \mid \theta) = \sum_{n=1}^{N} \log p(x_n \mid \theta) = \sum_{n=1}^{N} \log \sum_{k=1}^{K} \pi_k \mathcal{N}(x_n \mid \mu_k, \Sigma_k) \tag{3.5}$$

Therefore, the MLE of the model parameters θ that maximize the log-likelihood defined in Eq. (3.5) need to be obtained. The maximum likelihood of the parameters μ_k, Σ_k, π_k is estimated using the EM algorithm, which is a general iterative scheme for learning parameters in mixture models.

GMM is one of the most popular classification techniques among SER researchers, and many research works are based on GMM [12, 15, 39, 52–59]. Although average accuracy achieved not up to the mark, around an average of 74.83–81.94%, but least training time of GMM among the prominent classifiers made it an attractive choice as SER classifier.

GMMs are efficient in modeling multimodal distributions [101] with much have been data points compared to HMMs. Therefore, when global features are extracted from speech for SER, fewer data points are available, but GMMs work better in those scenarios [8]. Moreover, the average training time is minimal for GMM [8].

3.4.2.1 Difficulties in Using GMM for SER

Following difficulties in using GMM have been identified:

- GMMs cannot model temporal structure since x_n are drawn i.i.d. GMM was integrated with the vector autoregressive process to capture the temporal structure in SER [102].
- Deciding the optimal number of Gaussian components is a difficult problem [103].

3.4.3 Support Vector Machine

SVM is fundamentally a two-class or binary classifier. The SVM provides state-of-the-art results in many applications [104]. Possible values for the label or output are usually assumed to be $\{+1, -1\}$ so that the predictor becomes $f: \mathbb{R}^D \rightarrow \{+1, -1\}$, where f is the predictor and D is the dimension of the feature vector. Therefore, given the training dataset of N data points $\{(x_1, y_1), \ldots, (x_N, y_N)\}$, where $x_n \in \mathbb{R}^D$ and $y_n \in \{+1, -1\}$ and $n = 1, 2, \ldots, N$, the problem is to find the f with least classification error. Consider a linear model of the form

$$f(x, w) = w^T x + b \tag{3.6}$$

to solve this binary classification problem, where $w \in \mathbb{R}^D$ is the weight vector and $b \in \mathbb{R}$ is the bias. Also, assume that the dataset is linearly separable in the feature space and the objective here is to find the separating hyperplane that maximizes the margin between the positive and negative examples, which means $w^T x_n + b \geq 0$ when $y_n = +1$ and $w^T x_n + b < 0$ when $y_n = -1$. Now, the requirement that the positive and the negative examples nearest to the hyperplane to be at least one unit away from the hyperplane yields the condition $y_n(w^T x_n + b) \geq 1$ [100]. This condition is known as the canonical representation of the decision hyperplane. Here, the optimization problem is to maximize the distance to the margin, defined in terms of w as $||w||^{-1}$, which is equivalent to minimizing $||w||^2$, that is

$$\begin{aligned} &\underset{w,b}{\arg\min} \quad \frac{1}{2}||w||^2 \\ &\text{subject to} \quad y_n(w^T x_n + b) \geq 1, \quad \forall\ n = 1, \ldots, N \end{aligned} \tag{3.7}$$

Equation (3.7) is known as the *hard* margin, which is an example of quadratic programming. The margin is called *hard* because the formulation does not allow any violation of margin condition.

The assumption of a linearly separable dataset needs to be relaxed for better generalization of the data because, in practice, the class conditional distributions may overlap. This is achieved by the introduction of a *slack variable* $\xi_n \geq 0$ where $n = 1, \ldots, N$, with each training data point [105, 106], which updates the optimization problem as follows

$$\begin{aligned} &\underset{w,b,\xi}{\arg\min} \quad \frac{1}{2}||w||^2 + C \sum_{n==1}^{N} \xi_n \\ &\text{subject to} \quad y_n(w^T x_n + b) \geq 1 - \xi_n \\ &\qquad\qquad\quad \xi_n \geq 0 \\ &\qquad\qquad\quad \forall\ n = 1, \ldots, N \end{aligned} \tag{3.8}$$

where $C > 0$ manages the size of the margin and the total amount of slack that we have. This model allows some data points to be on the wrong side of the hyperplane to reduce the impact of overfitting.

Various methods have been proposed to combine multiple two-class SVMs to build a multiclass classifier. One of the commonly used approaches is the *one-versus-the-rest* approach [107], where K separate SVMs are constructed where K is the number of classes and $K > 2$. The kth model $y_k(X)$ is trained using the data from class C_k as the positive examples and the data from the remaining $K - 1$ classes as negative examples. There is another approach called *one-versus-one*, where all possible pairs of classes are trained in $K(K - 1)/2$ different two-class SVM classifiers. Platt [108] proposed the *directed acyclic graph support vector machine* (DAGSVM), directed acyclic graph SVM.

SVM is extensively used in SER [48, 49, 55, 65–71]. Performance of SVM for SER task in most of the research studies carried out yielded nearly close results, and accuracy is varying around 80% mark. However, Hassan and Damper [69] achieved 92.3% and 94.6% classification accuracy using linear and hierarchical kernels, respectively. They have used a linear kernel instead of a nonlinear *radial basis function* (RBF) kernel because of very high dimensional features space [72]. Hu et al. [73] explored GMM supervector-based SVM with different kernels like *linear*, *RBF*, *polynomial*, and *GMM KL divergence* and found that GMM KL performed the best in classifying emotions.

3.4.3.1 Difficulties with SVM

There is no systematic way to choose the kernel functions, and hence, the separability of transformed features is not guaranteed. Moreover, in SER, complete separation in training data is not recommended to avoid overfitting.

3.4.4 Deep Learning

Deep feedforward networks or multilayer perceptrons (MLPs) are the pure forms of DL models. The objective of an MLP is to approximate some function f^* such that a classifier, $y = f^*(x)$, maps an input x to a category y. An MLP defines a mapping $y = f(x; \theta)$, where θ is a set of parameters and learns the value of the parameters θ that result in the best function approximation. Deep networks are represented as a composition of different functions, and a directed acyclic graph describes how those functions are composed together. For example, there might be three functions $f(1), f(2)$, and $f(3)$ connected in a chain to form $f(x) = f^{(3)}(f^{(2)}(f^{(1)}(x)))$, where $f^{(1)}$ is the first layer of the network, $f^{(2)}$ is the second layer, and so on. The length of the chain gives the *depth* of the model, and this *depth* is behind the name *deep learning*. The last layer of the network is the *output layer*.

The training phase of neural network $f(x)$ is altered to approximate $f^*(x)$. Each training example x has a corresponding label $y \approx f^*(x)$, and training decides the values for θ such that the output layer can produce values close to y say $\hat{y}$. However, the behavior of the hidden layers are not directly specified by training data, and that is why those layers are called *hidden*. The hidden layers bring the nonlinearity into the system by transforming the input $\phi(x)$, where ϕ is a nonlinear transform. The whole transformation process is done in the hidden layers, which provide a new representation of x. Therefore, now, it is required to learn ϕ and the model is now becomes $y = F(x; \theta, w) = \phi(x; \theta)^T w$, where θ is used to learn ϕ and parameters w maps $\phi(x)$ to the desired output. The ϕ is the so-called *activation function* of the hidden layers of the feedforward network [109].

Most modern neural networks are trained using the MLE, which means the cost function $J(\theta)$ is the negative log-likelihood or the cross-entropy function between the training data and the model distribution. Therefore, the cost function becomes [109]

$$J(\theta) = -\mathbb{E}_{x,y \sim \hat{p}_{\text{data}}} \log p_{\text{model}}(y \mid x) \tag{3.9}$$

where p_{model} is the model distribution that varies depending on the selected model, and $\hat{p}_{\text{data}}$ is the target distribution from data. The output distribution determines the choice of the output unit. For example, Gaussian output distribution requires a linear output unit, Bernoulli output distribution requires a *Sigmoid* function, *Softmax* Units for Multinoulli Output Distributions, and so on. However, the choice of a hidden unit is still an active area of research, but rectified linear units (ReLU) are the most versatile ones that work well in most of the scenarios. Logistic sigmoid and hyperbolic tangent are other two options out of many other functions researchers are using.

Therefore, in the forward propagation, $\hat{y}$ is produced and the cost function $J(\theta)$ is computed. Now, the information generated in the form of $J(\theta)$ is appropriately processed so that w parameters can be appropriately chosen. This task is accomplished in two phases, first computing the gradients using the famous back-propagation algorithm, and in the second phase, the w values are updated based on the gradients computed by the backprop algorithm. The w values are updated through methods such as stochastic gradient descent (SGD). The backprop algorithm applies the chain rule recursively to compute the derivatives of the cost function $J(\theta)$.

Different variants of DL exist now, but convolutional neural networks (CNNs) [110, 111] and recurrent neural networks (RNNs) [112] are the most successful ones. Convolutional networks are neural networks that use convolution in place of general matrix multiplication in at least one of their layers, whereas when feedforward neural networks are extended to include feedback connections, they are called RNNs. RNNs are specialized in processing sequential data.

SER researchers have used CNNs [35, 37, 83–87], RNNs [36, 84, 88], or a combination of the two extensively for SER. Shallow one-layer or two-layer CNN structures may not be able to learn effectively the affective features that are discriminative enough to distinguish the subjective emotions [85]. Therefore, researchers are recommending a deep structure. Researchers [36, 37, 89, 90] have studied the effectiveness of attention mechanism.

Researchers are applying end-to-end DL systems in SER [84, 91–94], and most of them use arousal-valence model of emotions. Although using end-to-end DL, the average classification accuracy for arousal is 78.16%, which is decent, for valence, and it is pretty low 43.06%. Among other DNN techniques, very recently, a maximum accuracy of 87.32% is achieved by using a fine-tuned Alex-Net on Emo-DB [85]. Han et al. [113] used extreme learning machine (ELM) for classification, where a DNN takes as input the popular acoustic features within a speech segment and produces segment-level emotion state probability distributions, from which utterance-level features are constructed.

3.4.4.1 Drawbacks of Using Deep Learning for SER

Although researchers are using the DL architectures extensively for solving the SER problem, DL methods pose the following difficulties:

- Implementation of tensor operations is a complicated task, which results in limited resources [109]. There are limited sets of libraries, such as TensorFlow (https://github.com/tensorflow/tensorflow), Theano (https://github.com/Theano/Theano), PyTorch (https://github.com/pytorch/pytorch), MXNet (https://github.com/apache/incubator-mxnet), and CNTK (https://github.com/Microsoft/CNTK), which provide the service.
- Back-propagation often involves the summation of many tensors together, which makes the memory management task difficult and often requires substantial computational resources.
- The introduction of DL methods also increased the feature set dimension for SER manifolds, for example, Wöllmer et al. [88] extracted a total 4843 features.
- One crucial question raised by Lorenzo-Trueba et al. [114] is that how emotional information should be represented as labels for supervised DNN training, e.g. should emotional class and emotional strength be factorized into separate inputs or not?

3.5 Difficulties in SER Studies

SER mystery is not yet solved, and it has been proved to be difficult. Here are the prominent difficulties faced by the researchers.

- The topic called *emotion* is inherently uncertain. Because the very experience of emotion is very subjective, its expression varies largely from person to person. Moreover, there is little consensus over the definition of emotion. These are the major hurdles to proceed with the research [115]. For example, several studies [51–53, 116–119] reported that there is confusion between anger and happiness emotional expression.
- SER is challenging because of the affective gap between the subjective emotions and low-level features [85]. Also, the feature analysis in SER is less studied [113], and researchers are still actively looking for the best feature set.
- Speaker and language dependency of classification results are a concern for building more generic SER systems [120]. The same model gives very different classification results with different datasets. Studies reported [53, 81, 117, 121] the speaker dependency phenomenon and tried to address that issue.
- Standard speech databases are not available for SER research so that new models can be effectively benchmarked. Moreover, the absence of good-quality natural speech emotional databases is hindering the real-life implementation of SER systems.
- Cross-corpora recognition results are low [38, 65, 122]. This indicates that existing models are not generalizing enough for real-life implementation.
- Classification between high-arousal and low-arousal emotions is achieved more accurately, but for other cases, it is low, which needs to be addressed. Moreover, the accuracy of n-way classification with all the emotions in the database is still very low.

3.6 Conclusion

This chapter reviewed different phases of SER. The primary focus is on four prominent classification techniques used in SER to date. HMM was the first technique that has seen some success and then GMM and SVM propelled that progress forward. Also, presently DL techniques, mainly CNN–*long short-term memory* (LSTM) combination, is providing state of the art classification performance. However, things have not changed much in case of selecting a feature set for SER because low-level descriptors (LLD) are still one of the prominent choices, although some researchers in recent times are trying DL techniques for feature learning. The nature of SER databases are changing, and features such as facial expressions and body movements are being included along with the spoken utterances. However, the availability of quality databases is still a challenge.

This work is a survey of current research work in the SER system. Specifically, the classification models used in SER are discussed in more detail, while the features and databases are briefly mentioned. A long list of articles have been referred to during this study, and eight classification models have been identified, which are being used by researchers. However, four classification models have been identified as more effective for SER than the rest. Those four classification models are HMM, GMM, SVM, and DL, which are discussed in the context of SER with a relevant mathematical background. Their drawbacks are also highlighted.

Research results show that *DL models* are performing significantly better than HMM, GMM, and SVM. The classification accuracy for SER has improved with the introduction of DL techniques because of their higher capacity to learn the underlying pattern in the emotional speech signal. The DL architectures are evolving each day with more and more improvement in classification accuracy of SER speech signals. It is observed that the SER field is still facing many challenges, which are barring research outputs from being implemented as an industry-grade product. It is also noticed during this study that research work related to feature set enhancement is much less compared to the works done on enhancing classification techniques. However, the available classification techniques in ML field are in a very advanced state, and with the right feature set, they can yield very high classification accuracy rates. DL as a sub-field of ML, even achieved state-of-the-art classification accuracy in many fields like computer vision, text mining, automatic speech recognition, to name a few. Therefore, the classification technique should not be a hindrance for SER anymore; only the appropriate feature set needs to be fed into the classification system.

References

1 Darwin, C. (1948). *The Expression of Emotion in Man and Animals*. Watts and Co.

2 Fairbanks, G. and Pronovost, W. (1938). Vocal pitch during simulated emotion. *Science* 88 (2286): 382–383. https://doi.org/10.1126/science.88.2286.382.

3 Freedman, D.G., Loring, C.B., and Martin, R.M. (1967). Emotional behavior and personality development. In: *Infancy and Early Childhood*. (ed. Y. Brackbill), 429–502. New York: Free Press.

4 Williams, C.E. and Stevens, K.N. (1969). On determining the emotional state of pilots during flight: an exploratory study. *Aerospace Medicine* 40: 1369–1372.

5 Williams, C.E. and Stevens, K.N. (1972). Emotions and speech: some acoustical correlates. *The Journal of the Acoustical Society of America* 52 (4B): 1238–1250. https://doi.org/10.1121/1.1913238.

6 Williamson, J. (1978). Speech analyzer for analyzing pitch or frequency perturbations in individual speech pattern to determine the emotional state of the person. https://patents.google.com/patent/US4093821.

7 Picard, R.W. (1997). *Affective Computing*. Cambridge, MA: MIT Press. ISBN: 0-262-16170-2.

8 El Ayadi, M., Kamel, M.S., and Karray, F. (2011). Survey on speech emotion recognition: features, classification schemes, and databases. *Pattern Recognition* 44 (3): 572–587.

9 Sreenivasa Rao, K. and Koolagudi, S.G. (2013). *Emotion Recognition using Speech Features*. New York: Springer-Verlag.

10 Gnjatovic, M. and Rosner, D. (2010). Inducing genuine emotions in simulated speech-based human-machine interaction: the nimitek corpus. *IEEE Transactions on Affective Computing* 1 (2): 132–144. https://doi.org/10.1109/T-AFFC.2010.14.

11 Ververidis, D. and Kotropoulos, C. (2006). Emotional speech recognition: resources, features, and methods. *Speech Communication* 48 (9): 1162–1181. https://doi.org/10.1016/j.specom.2006.04.003.

12 Breazeal, C. and Aryananda, L. (2002). Recognition of affective communicative intent in robot-directed speech. *Autonomous Robots* 12 (1): 83–104. https://doi.org/10.1023/A:1013215010749.

13 Campbell, N. (2000). Databases of emotional speech.

14 Engberg, I.S., Hansen, A.V., Andersen, O., and Dalsgaard, P. (1997). Design, recording and verification of a Danish emotional speech database. In: *Proceedings of the 5th European Conference on Speech Communication and Technology*.

15 Slaney, M. and McRoberts, G. (2003). Babyears: a recognition system for affective vocalizations. *Speech Communication* 39: 367–384.

16 Lee, C.M. and Narayanan, S. (2005). Toward detecting emotions in spoken dialogs. *IEEE Transactions on Speech and Audio Processing* 13: 293–303.

17 Ververidis, D. and Kotropoulos, C. (2003). A state of the art review on emotional speech databases. In: *Proceedings of the 1st Richmedia Conference*, 109–119. https://www.researchgate.net/publication/50342391_A_State_of_the_Art_Review_on_Emotional_Speech_Databases

18 Cowie, R. and Cornelius, R.R. (2003). Describing the emotional states that are expressed in speech. *Speech Communication* 40 (1): 5–32. https://doi.org/10.1016/S0167-6393(02)00071-7.

19 Bänziger, T. and Scherer, K.R. (2005). The role of intonation in emotional expressions. *Speech Communication* 46 (3): 252–267. https://doi.org/10.1016/j.specom.2005.02.016. Quantitative Prosody Modelling for Natural Speech Description and Generation.

20 Rao, K.S. and Yegnanarayana, B. (2006). Prosody modification using instants of significant excitation. *IEEE Transactions on Audio, Speech, and Language Processing* 14 (3): 972–980. https://doi.org/10.1109/TSA.2005.858051.

21 Cowie, R., Douglas-Cowie, E., Tsapatsoulis, N et al. (2001). Emotion recognition in human-computer interaction. *IEEE Signal Processing Magazine* 18 (1): 32–80. https://doi.org/10.1109/79.911197.

22 Murray, I. and Arnott, J. (1993). Toward the simulation of emotion in synthetic speech: a review of the literature on human vocal emotion. *The Journal of the Acoustical Society of America* 93: 1097–1108. https://doi.org/10.1121/1.405558.

23 Cowie, R. and Douglas-Cowie, E. (1996). *Automatic statistical analysis of the signal and prosodic signs of emotion in speech* 3 (11): 1989–1992. https://doi.org/10.1109/ICSLP.1996.608027.

24 Banse, R. and Scherer, K. (1996). Acoustic profiles in vocal emotion expression. *Journal of Personality and Social Psychology* 70: 614–636. https://doi.org/10.1037/0022-3514.70.3.614.

25 Oster, A. and Risberg, A. (1986). The identification of the mood of a speaker by hearing impaired listeners. *Speech Transmission Laboratory. Quarterly Progress and Status Reports* 27: 79–90.

26 Beeke, S., Wilkinson, R., and Maxim, J (2009). Prosody as a compensatory strategy in the conversations of people with agrammatism. *Clinical Linguistics & Phonetics* 23: 133–155. https://doi.org/10.1080/02699200802602985.

27 Borchert, M. and Dusterhoft, A. (2005). Emotions in speech - experiments with prosody and quality features in speech for use in categorical and dimensional emotion recognition environments. In: *2005 International Conference on Natural Language Processing and Knowledge Engineering*, 147–151. ISBN: 0-7803-9361-9. https://doi.org/10.1109/NLPKE.2005.1598724.

28 Tao, J., Kang, Y., and Li, A. (2006). Prosody conversion from neutral speech to emotional speech. *IEEE Transactions on Audio, Speech, and Language Processing* 14 (4): 1145–1154. https://doi.org/10.1109/TASL.2006.876113.

29 Scherer, K (1986). Vocal affect expression: a review and a model for future research. *Psychological Bulletin* 99: 143–165. https://doi.org/10.1037//0033-2909.99.2.143.

30 Davitz, J.R. and Beldoch, M. (1964). *The Communication of Emotional Meaning, McGraw-Hill Series in Psychology*. Greenwood Press. ISBN: 9780837185279; https://books.google.co.za/books?id=1ggRAQAAIAAJ.

31 Rabiner, L.R. and Schafer, R.W. (1978). *Digital Processing of Speech Signals, Prentice-Hall Signal Processing Series*. Prentice-Hall. ISBN: 9780132136037. https://books.google.co.za/books?id=YVtTAAAAMAAJ.

32 Hernando, J. and Nadeu, C. (1997). Linear prediction of the one-sided autocorrelation sequence for noisy speech recognition. *IEEE Transactions on Speech and Audio Processing* 5 (1): 80–84. https://doi.org/10.1109/89.554273.

33 Le Bouquin, R. (1996). Enhancement of noisy speech signals: application to mobile radio communications. *Speech Communication* 18 (1): 3–19. https://doi.org/https://doi.org/10.1016/0167-6393(95)00021-6.

34 Bou-Ghazale, S.E., and Hansen, J.H.L. (2000). A comparative study of traditional and newly proposed features for recognition of speech under stress. *IEEE Transactions on Speech and Audio Processing* 8 (4): 429–442. https://doi.org/10.1109/89.848224.

35 Chen, M., He, X., Yang, J., and Zhang, H. (2018). 3-D convolutional recurrent neural networks with attention model for speech emotion recognition. *IEEE Signal Processing Letters* 25 (10): 1440–1444. https://doi.org/10.1109/LSP.2018.2860246.

36 Mirsamadi, S., Barsoum, E., and Zhang, C. (2017). Automatic speech emotion recognition using recurrent neural networks with local attention. In: *2017 IEEE International Conference on Acoustics, Speech and Signal Processing (ICASSP)*, 2227–2231. https://doi.org/10.1109/ICASSP.2017.7952552.

37 Neumann, M. and N.T. Vu (2017). Attentive convolutional neural network based speech emotion recognition: a study on the impact of input features, signal length, and acted speech. *CoRR*, abs/1706.00612. http://arxiv.org/abs/1706.00612.

38 Schuller, B., Vlasenko, B., Eyben, F. et al. (2010). Cross-corpus acoustic emotion recognition: variances and strategies. *IEEE Transactions on Affective Computing* 1 (2): 119–131. https://doi.org/10.1109/T-AFFC.2010.8.

39 Mubarak, O.M., Ambikairajah, E., and Epps, J. (2005). Analysis of an MFCC-based audio indexing system for efficient coding of multimedia sources. In: *Proceedings of the 8th International Symposium on Signal Processing and Its Applications, 2005*, vol. 2, 619–622. https://doi.org/10.1109/ISSPA.2005.1581014.

40 Roy, T., Marwala, T., and Chakraverty, S. (2018). Precise detection of speech endpoints dynamically: a wavelet convolution based approach. *Communications in Nonlinear Science and Numerical Simulation*. https://doi.org/https://doi.org/10.1016/j.cnsns.2018.07.008.

41 Clore, G.L. and Huntsinger, J.R. (2007). How emotions inform judgment and regulate thought. *Trends in Cognitive Sciences* 11 (9): 393–399.

42 Bitouk, D., Verma, R., and Nenkova, A. (2010). Class-level spectral features for emotion recognition. *Speech Communication* 52: 613–625.

43 Fernandez, R. and Picard, R.W. (2003). Modeling drivers' speech under stress. *Speech Communication* 40: 145–159.

44 Zhou, G., Hansen, J.H.L., and Kaiser, J.F. (2001). Nonlinear feature based classification of speech under stress. *IEEE Transactions on Speech and Audio Processing* 9: 201–216.

45 Nwe, T.L., Foo, S.W., and De Silva, L.C. (2003). Speech emotion recognition using hidden Markov models. *Speech Communication* 41: 603–623.

46 Kamaruddin, N. and Wahab, A. (2009). Features extraction for speech emotion. *Journal of Computational Methods in Science and Engineering* 9 (1, 2S1): S1–S12. http://dl.acm.org/citation.cfm?id=1608790.1608791.

47 Schuller, B., Rigoll, G., and Lang, M. (2003). Hidden Markov model-based speech emotion recognition. In: *2003 International Conference on Multimedia and Expo. ICME '03. Proceedings (Cat. No.03TH8698)*, vol. 1, 401–404. https://doi.org/10.1109/ICME.2003.1220939.

48 Kwon, O.-W., Chan, K., Hao, J., and Lee, T.-W. (2003). Eighth European Conference on Speech Communication and Technology.

49 Lee, C.M., Yildirim, S., Bulut, M. et al. (2004). Emotion recognition based on phoneme classes. In: *Proceedings of ICSLP*, 889–892.

50 Philippou-Hübner, D., Vlasenko, B., Grosser, T., and Wendemuth, A. (2010). Determining optimal features for emotion recognition from speech by applying an evolutionary algorithm. In: *INTERSPEECH*, 2358–2361.

51 Lin, J., Wei, W., Wu, C., and Wang, H. (2014). Emotion recognition of conversational affective speech using temporal course modeling-based error weighted cross-correlation model. In: *Signal and Information Processing Association Annual Summit and Conference (APSIPA), 2014 Asia-Pacific*, 1–7. https://doi.org/10.1109/APSIPA.2014.7041621.

52 Jeon, J.H., Xia, R., and Liu, Y. (2011). Sentence level emotion recognition based on decisions from subsentence segments. In: *2011 IEEE International Conference on Acoustics, Speech and Signal Processing (ICASSP)*, 4940–4943. https://doi.org/10.1109/ICASSP.2011.5947464.

53 Lugger, M. and Yang, B. (2007). The relevance of voice quality features in speaker independent emotion recognition. In: *2007 IEEE International Conference on Acoustics, Speech and Signal Processing - ICASSP '07*, vol. 4, IV–17–IV–20. https://doi.org/10.1109/ICASSP.2007.367152.

54 Atassi, H. and Esposito, A. (2008). A speaker independent approach to the classification of emotional vocal expressions. In: *2008 20th IEEE International Conference on Tools with Artificial Intelligence*, vol. 2, 147–152. https://doi.org/10.1109/ICTAI.2008.158.

55 Schuller, B., Rigoll, G., and Lang, M. (2004). Speech emotion recognition combining acoustic features and linguistic information in a hybrid support vector machine-belief network architecture. In: *Proceedings of IEEE ICASSP*, vol. 1, I–577–80. https://doi.org/10.1109/ICASSP.2004.1326051.

56 Neiberg, D., Elenius, K., and Laskowski, K. (2006). Emotion recognition in spontaneous speech using GMMs. *Proceedings of INTERSPEECH*.

57 Wang, Y. and Guan, L. (2004). An investigation of speech-based human emotion recognition. In: *IEEE 6th Workshop on Multimedia Signal Processing, 2004.*, 15–18.

58 Zhou, Y., Sun, Y., Zhang, J., and Yan, Y. (2009). Speech emotion recognition using both spectral and prosodic features. In: *2009 International Conference on Information Engineering and Computer Science*, 1–4. https://doi.org/10.1109/ICIECS.2009.5362730.

59 Luengo, I., Navas, E., Hernáez, I., and Sánchez, J. (2005). Automatic emotion recognition using prosodic parameters. *INTERSPEECH*.

60 Dellaert, F., Polzin, T., and Waibel, A. (1996). Recognizing emotion in speech. In: *Proceedings of the 4th International Conference on Spoken Language*, vol. 3, 1970–1973. https://doi.org/10.1109/ICSLP.1996.608022.

61 Pao, T.-L., Chen, Y.-T., Yeh, J.-H., and Liao, W.-Y. (2005). Combining acoustic features for improved emotion recognition in Mandarin speech. In: *Affective Computing and Intelligent Interaction* (ed. J. Tao, T. Tan, and R.W. Picard), 279–285. Berlin, Heidelberg: Springer-Verlag. ISBN: 978-3-540-32273-3.

62 Yu, F., Chang, E., Xu, Y.-Q., and Shum, H. (2001). Emotion detection from speech to enrich multimedia content. In: *IEEE Pacific Rim Conference on Multimedia*.

63 Petrushin, V. (2000). Emotion recognition in speech signal: experimental study, development, and application. In: *ICSLP 2000*, 222–225.

64 Lee, C.M., Narayanan, S., and Pieraccini, R. (2001). Recognition of negative emotions from the speech signal. In: *IEEE Workshop on Automatic Speech Recognition and Understanding, 2001. ASRU '01.*, 240–243. https://doi.org/10.1109/ASRU.2001.1034632.

65 Sun, R. and Moore, E. (2012). A preliminary study on cross-databases emotion recognition using the glottal features in speech. *INTERSPEECH*.

66 Espinosa, H.P., García, C.A.R., and Pineda, L.V. (2010). Features selection for primitives estimation on emotional speech. In: *2010 IEEE International Conference on Acoustics, Speech and Signal Processing*, 5138–5141. https://doi.org/10.1109/ICASSP.2010.5495031.

67 Wu, S., Falk, T.H., and Chan, W.-Y. (2011). Automatic speech emotion recognition using modulation spectral features. *Speech Communication* 53 (5): 768–785. https://doi.org/https://doi.org/10.1016/j.specom.2010.08.013. Perceptual and Statistical Audition.

68 Pierre-Yves, O. (2003). The production and recognition of emotions in speech: features and algorithms. *International Journal of Human-Computer Studies* 59 (1): 157–183. https://doi.org/https://doi.org/10.1016/S1071-5819(02)00141-6. Applications of Affective Computing in Human-Computer Interaction.

69 Hassan, A. and Damper, R.I. (2010). Multi-class and hierarchical svms for emotion recognition. *INTERSPEECH*.

70 Rozgic, V., Ananthakrishnan, S., Saleem, Shirin et al. (2012). Emotion recognition using acoustic and lexical features. *INTERSPEECH*.

71 Yeh, L.-Y. and Chi, T.-S. (2010). Spectro-temporal modulations for robust speech emotion recognition. *INTERSPEECH*.

72 Hsu, C.W., Chang, C.-C., and Lin, C.-J. (2003). A practical guide to support vector classification.

73 Hu, H., Xu, M., and Wu, W. (2007). GMM supervector based svm with spectral features for speech emotion recognition. In: *2007 IEEE International Conference on Acoustics, Speech and Signal Processing - ICASSP '07*, vol. 4, IV–413–IV–416. https://doi.org/10.1109/ICASSP.2007.366937.

74 Zhu, A. and Luo, Q. (2007). Study on speech emotion recognition system in e-learning. In: *Human-Computer Interaction. HCI Intelligent Multimodal Interaction Environments* (ed. J.A. Jacko), 544–552, Berlin, Heidelberg: Springer-Verlag. ISBN: 978-3-540-73110-8.

75 Petrushin, V. (2000). Emotion in speech: recognition and application to call centers. *Proceedings of Artificial Neural Networks in Engineering*.

76 Nakatsu, R., Solomides, A., and Tosa, N. (1999). Emotion recognition and its application to computer agents with spontaneous interactive capabilities. In: *Proceedings IEEE International Conference on Multimedia Computing and Systems*, vol. 2, 804–808. https://doi.org/10.1109/MMCS.1999.778589.

77 Nicholson, J., Takahashi, K., and Nakatsu, R. (1999). Emotion recognition in speech using neural networks. In: *ICONIP'99. ANZIIS'99 ANNES'99 ACNN'99. 6th International Conference on Neural Information Processing. Proceedings (Cat. No.99EX378)*, vol. 2, 495–501. https://doi.org/10.1109/ICONIP.1999.845644.

78 Tato, R., Santos, R., Kompe, R., and Pardo, J.M. (2002). Emotional space improves emotion recognition. *INTERSPEECH*.

79 Wang, Y., Du, S., and Zhan, Y. (2008). Adaptive and optimal classification of speech emotion recognition. In: *2008 4th International Conference on Natural Computation*, vol. 5, 407–411. https://doi.org/10.1109/ICNC.2008.713.

80 Yildirim, S., Bulut, M., Lee, C.M. et al. (2004). An acoustic study of emotions expressed in speech. *INTERSPEECH*.

81 McGilloway, S., Cowie, R., ED, C. et al. (2000). Approaching automatic recognition of emotion from voice: a rough benchmark. *Proceedings of the ISCA ITRW on Speech and Emotion*.

82 Ververidis, D., Kotropoulos, C., and Pitas, I. (2004). Automatic emotional speech classification. In: *2004 IEEE International Conference on Acoustics, Speech, and Signal Processing*, vol. 1, I–593. https://doi.org/10.1109/ICASSP.2004.1326055.

83 Parthasarathy, S. and Tashev, I. (2018). Convolutional neural network techniques for speech emotion recognition. *2018 16th International Workshop on Acoustic Signal Enhancement (IWAENC)*.

84 Trigeorgis, G., Ringeval, F., Brueckner, R. et al. (2016). Adieu features? End-to-end speech emotion recognition using a deep convolutional recurrent network. In: *2016 IEEE International Conference on Acoustics, Speech and Signal Processing (ICASSP)*, 5200–5204. https://doi.org/10.1109/ICASSP.2016.7472669.

85 Zhang, S., Zhang, S., Huang, T., and Gao, W. (2018). Speech emotion recognition using deep convolutional neural network and discriminant temporal pyramid matching. *IEEE Transactions on Multimedia* 20 (6): 1576–1590. https://doi.org/10.1109/TMM.2017.2766843.

86 Mao, Q., Dong, M., Huang, Z., and Zhan, Y. (2014). Learning salient features for speech emotion recognition using convolutional neural networks. *IEEE Transactions on Multimedia* 16 (8): 2203–2213. https://doi.org/10.1109/TMM.2014.2360798.

87 Huang, Z., Dong, M., Mao, Q., and Zhan, Y. (2014). Speech emotion recognition using CNN. In: *Proceedings of the 22nd ACM International Conference on Multimedia, MM '14*, 801–804. New York, NY, USA: ACM. ISBN: 978-1-4503-3063-3. https://doi.org/10.1145/2647868.2654984.

88 Wöllmer, M., Eyben, F., Reiter, S. et al. (2008). Abandoning emotion classes - towards continuous emotion recognition with modelling of long-range dependencies. *INTERSPEECH*.

89 Huang, C. and Narayanan, S.S. (2017). Deep convolutional recurrent neural network with attention mechanism for robust speech emotion recognition. In: *2017 IEEE International Conference on Multimedia and Expo (ICME)*, 583–588. https://doi.org/10.1109/ICME.2017.8019296.

90 Huang, C.-W. and Narayanan, S. (2016). Attention assisted discovery of sub-utterance structure in speech emotion recognition. *INTERSPEECH*.

91 Tzirakis, P., Zhang, J., and Schuller, B.W. (2018). End-to-end speech emotion recognition using deep neural networks. In: *2018 IEEE International Conference on Acoustics, Speech and Signal Processing (ICASSP)*, 5089–5093. https://doi.org/10.1109/ICASSP.2018.8462677.

92 Han, J., Zhang, Z., Ringeval, F., and Schuller, B. (2017). Reconstruction-error-based learning for continuous emotion recognition in speech. In: *2017 IEEE International Conference on Acoustics, Speech and Signal Processing (ICASSP)*, 2367–2371. https://doi.org/10.1109/ICASSP.2017.7952580.

93 Han, J., Zhang, Z., Ringeval, F., and Schuller, B. (2017). Prediction-based learning for continuous emotion recognition in speech. In: *2017 IEEE International Conference on Acoustics, Speech and Signal Processing (ICASSP)*, 5005–5009. https://doi.org/10.1109/ICASSP.2017.7953109.

94 Schmitt, M., Ringeval, F., and Schuller, B.W. (2016). At the border of acoustics and linguistics: bag-of-audio-words for the recognition of emotions in speech. *INTERSPEECH*.

95 Rabiner, L. and Juang, B. (1986). An introduction to hidden Markov models. *IEEE ASSP Magazine* 3 (1): 4–16. https://doi.org/10.1109/MASSP.1986.1165342.

96 Ephraim, Y. and Merhav, N. (2002). Hidden Markov processes. *IEEE Transactions on Information Theory* 48: 1518–1569. https://doi.org/10.1109/TIT.2002.1003838.

97 Stamp, Mark (2017). A revealing introduction to hidden markov models. In: *Introduction to Machine Learning with Applications in Information Security*, 7–35. Chapman and Hall. ISBN: 9781138626782

98 Dempster, A., Laird, N., and Rubin, D.B. (1977). Maximum likelihood from incomplete data via em algorithm. *Journal of the Royal Statistical Society, Series B* 39: 1–38. https://doi.org/10.1111/j.2517-6161.1977.tb01600.x.

99 Bishop, C.M. (2007). *Pattern Recognition and Machine Learning*. New York: Springer-Verlag. ISBN: 978-0-387-31073-2.

100 Deisenroth, M.P., Faisal, A.A., and Ong, C.S. (2019). *Mathematics for Machine Learning*. Cambridge University Press.

101 Bishop, Christopher M (2005). *Neural Networks for Pattern Recognition*, vol. 227. Oxford university press. ISBN: 978-0-19-853864-6

102 El Ayadi, M., Kamel, M.S., and Karray, F. (2007). Speech emotion recognition using gaussian mixture vector autoregressive models. *2007 IEEE International Conference on Acoustics, Speech and Signal Processing - ICASSP '07*, vol. 4, IV–957. https://doi.org/10.1109/ICASSP.2007.367230.

103 Reynolds, D.A. and Rose, R.C. (1995). Robust text-independent speaker identification using Gaussian mixture speaker models. *IEEE Transactions Speech and Audio Processing* 3: 72–83.

104 Steinwart, I. and Christmann, A. (2008). *Support Vector Machines*. Springer.

105 Bennett, K. and Mangasarian, O.L. (2002). Robust linear programming discrimination of two linearly inseparable sets. *Optimization Methods and Software* 1: (01): 23–34. https://doi.org/10.1080/10556789208805504.

106 Cortes, C. and Vapnik, V. (1995). Support-vector networks. *Machine Learning* 20: 273–297. https://doi.org/10.1007/BF00994018.

107 Vapnik, V.N. (1998). *Statistical Learning Theory*. Wiley.

108 Platt, J. (2000). Probabilities for SV machines. In: *Advances in Large Margin Classifiers* (ed. A.J. Smola, P.L. Bartlett, B. Scholkopf, and D. Shuurmans), 61–73. MIT Press.

109 Goodfellow, I., Bengio, Y., and Courville, A. (2016). *Deep Learning*. MIT Press. http://www.deeplearningbook.org.

110 Lecun, Y., Bottou, L., Bengio, Y., and Haffner, P. (1998). Gradient-based learning applied to document recognition. *Proceedings of the IEEE* 86 (11): 2278–2324. https://doi.org/10.1109/5.726791.

111 LeCun, Y., Boser, B., Denker, J.S. et al. (1989). Backpropagation applied to handwritten zip code recognition. *Neural Computation* 1 (4): 541–551. https://doi.org/10.1162/neco.1989.1.4.541.

112 Rumelhart, D., Hinton, G., and Williams, R. (1986). Learning representations by back-propagating errors. *Nature* 323: 533–536.

113 Han, K., Yu, D., and Tashev, I. (2014). Speech emotion recognition using deep neural network and extreme learning machine. *Proceedings of INTERSPEECH*.

114 Lorenzo-Trueba, J., Henter, G.E., Takaki, S. et al. (2018). Investigating different representations for modeling and controlling multiple emotions in DNN-based speech synthesis. *Speech Communication* 99: 135–143. https://doi.org/https://doi.org/10.1016/j.specom.2018.03.002.

115 Schröder, M. and Cowie, R. (2006). Issues in emotion-oriented computing – towards a shared understanding. *Workshop on Emotion and Computing, HUMAINE*.

116 Amir, N., Kerret, O., and Karlinski, D. (2001). Classifying emotions in speech: a comparison of methods. *INTERSPEECH*.

117 Grimm, M., Kroschel, K., Mower, E., and Narayanan, S. (2007). Primitives-based evaluation and estimation of emotions in speech. *Speech Communication* 49 (10): 787–800. https://doi.org/https://doi.org/10.1016/j.specom.2007.01.010. Intrinsic Speech Variations.

118 Kim, J., Lee, S., and Narayanan, S. (2010). An exploratory study of manifolds of emotional speech. In: *2010 IEEE International Conference on Acoustics, Speech and Signal Processing*, 5142–5145. https://doi.org/10.1109/ICASSP.2010.5495032.

119 Ververidis, D. and Kotropoulos, C. (2005). Emotional speech classification using Gaussian mixture models. In: *2005 IEEE International Symposium on Circuits and Systems*, vol. 3, 2871–2874. https://doi.org/10.1109/ISCAS.2005.1465226.

120 Koolagudi, S.G. and Rao, K.S. (2010). Real life emotion classification using VOP and pitch based spectral features. In: *2010 Annual IEEE India Conference (INDICON)*, 1–4. https://doi.org/10.1109/INDCON.2010.5712728.

121 Kim, J., Park, J., and Oh, Y. (2011). On-line speaker adaptation based emotion recognition using incremental emotional information. In: *2011 IEEE International Conference on Acoustics, Speech and Signal Processing (ICASSP)*, 4948–4951. https://doi.org/10.1109/ICASSP.2011.5947466.

122 Shami, M. and Verhelst, W. (2007). An evaluation of the robustness of existing supervised machine learning approaches to the classification of emotions in speech. *Speech Communication* 49 (3): 201–212. https://doi.org/https://doi.org/10.1016/j.specom.2007.01.006.

4

Mathematical Methods in Deep Learning

Srinivasa Manikant Upadhyayula and Kannan Venkataramanan

CRISIL Global Research & Analytics, CRISIL (A S&P Company), CRISIL House, Central Avenue, Hiranandani Business Park, Powai, Mumbai 400 076, India

4.1 Deep Learning Using Neural Networks

Deep learning allows building quantitative models that constitute a series of processing layers to learn representations of data coupled with multiple levels of abstraction [1]. Neural networks were developed to understand the basic functioning of human brain and the entire central nervous system. Later, the model designed to capture the working of the human nervous system is applied to financial service domains such as link analysis of payments, fraud detection in customer transactions, and anomalies in transactions for potential money laundering.

4.2 Introduction to Neural Networks

A neural network works in the same pattern as that of a neuron in the human nervous system. The fundamental epitome of this learning technique is that it consists of a large number of highly organized and connected neurons working in harmony to solve a specific problem including pattern recognition or data classification. Neural networks are not a recent phenomenon but started before the advent of modern computers. It began with the work of McCulloch and Pitts [2] who created a theoretical representation of neural networks using a combination of human nervous system and mathematics (application of calculus and linear algebra). McCulloch–Pitts networks (or referred as MP networks) represent a finite state automaton embodying the logic of propositions, with quantifiers, in the form of computer programs [3].

With the advent of parallel distributed processing in mid-1980s, Rumelhart, McClelland, and coworker [4] applied the concepts of parallel distributed processing to the neural networks. Their work signaled the dawn of applying advanced techniques in neural networks in the domain of medical research. Qian and Sejnowski [5] presented a novel method predicting the secondary structure of globular proteins based on nonlinear neural network models. The average accuracy of the developed model on a testing set of proteins nonhomologous with the corresponding training set was 64.3%. Kneller et al. [6] have applied neural networks to predict the mapping between protein sequence and secondary structure. By adding neural network units that detect periodicities in the input sequence and use of tertiary structural class, the accuracy for predicting the class of all-α proteins is at 79%. Rost and Sander [7] applied evolutionary information contained in multiple sequence alignments as inputs to neural networks and predicted the secondary structure with significant accuracy. The model developed has demonstrated an overall accuracy of 71.6% in a multiple cross-validation test on 126 unique protein chains.

In the recent years, the applications of neural networks have increased exponentially across domains. Courbariaux et al. [8] introduced a method to train binarized neural networks (BNNs) – neural networks with

Mathematical Methods in Interdisciplinary Sciences, First Edition. Edited by Snehashish Chakraverty.

binary weights and activations at run-time. BNNs drastically reduce memory size and accesses and replace most arithmetic operations with bit-wise operations, which is expected to substantially improve power efficiency. Silver et al. [9] have developed a new approach to computer Go in which deep neural networks are trained by a novel combination of supervised learning from human expert games and reinforcement learning from games of self-play. The neural networks play Go at the level of state-of-the-art Monte Carlo tree search programs that simulate thousands of random games of self-play. Their program Alpha Go achieved a 99.8% winning rate against other Go programs. Esteva et al. [10] have applied deep convolutional neural networks (CNNs) to classify skin lesions, trained end-to-end from images directly, using only pixels and disease labels as inputs using a dataset of 129 450 clinical images. The CNN achieves performance on par with all tested experts across both tasks, demonstrating an artificial intelligence capable of classifying skin cancer with a level of competence comparable to dermatologists.

4.2.1 Artificial Neural Network (ANN)

Artificial neural network (ANN) is a computational model that is based on the structure and functioning of human neural networks, and the models are built using an interconnected network of neurons with multiple hidden layer(s) for processing data from input variables.

ANN consists of input variables, hidden layers, and an output layer. The input layer represents the set of neurons that represent the independent features (or input variables). These data points are passed to the hidden layer. In the hidden layer with "n" neurons, each neuron is assigned a weight and the inputs are multiplied with the designated weights, thereby transforming the input parameters and segregating the data to obtain the desired output. The combination of inputs and weights are then passed to the activation function, which determines the processing of different neurons in the hidden layer and passes the results to the output layer. The output layer considers the output results and displays the final result (Figure 4.1).

A single neuron in the hidden layer obtains information from the set of independent variables from the input data (numbered x_1 to x_n). Each input variable will be assigned a weight w_i (where i represents the value from 1 to n).

Mathematically, a single neuron is represented as

$$z = x_1w_1 + x_2w_2 + x_3w_3 + \dots + x_nw_n + \varepsilon_0$$

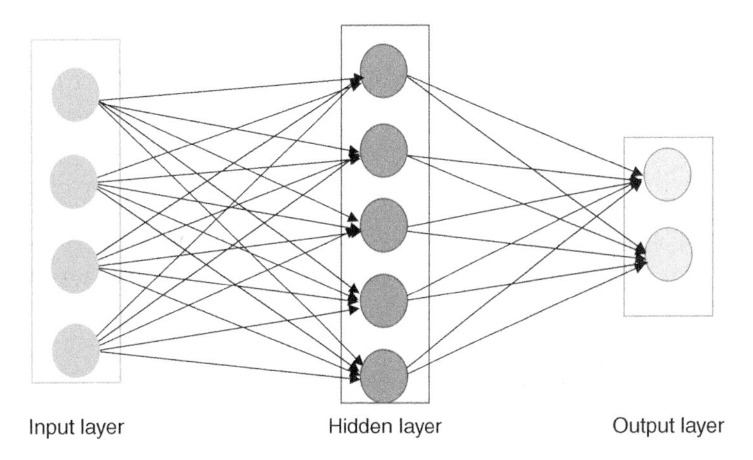

Figure 4.1 Shows the schematic representation of a single hidden layer neural network.

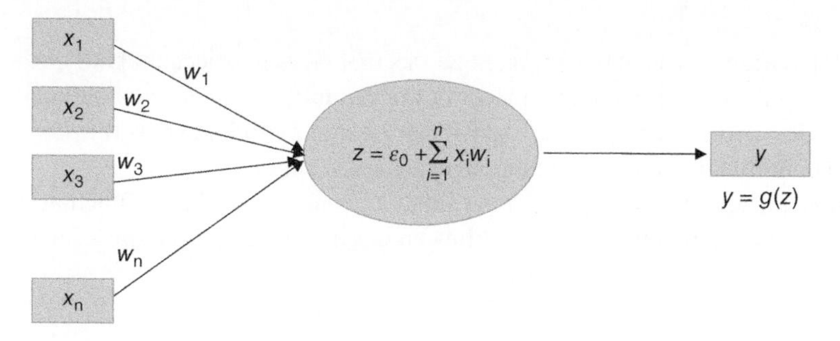

Figure 4.2 Shows the schematic representation of a single neuron of the hidden layer.

In simple terms, it can be written as

$$z = \varepsilon_0 + \sum_{i=1}^{n} x_i w_i$$

where x_i represents the inputs from independent variables,

w_i represents the weights associated with these independent variables, and

ε_0 represents the error or bias associated with the neuron in the hidden layer (Figure 4.2).

For a single hidden layer ANN with "N" neurons, the neurons in the hidden layer are represented as

$$z_1 = \varepsilon_1 + \sum_{i=1}^{n} x_{1i} w_{1i}$$

$$z_2 = \varepsilon_2 + \sum_{i=1}^{n} x_{2i} w_{2i}$$

$$z_3 = \varepsilon_3 + \sum_{i=1}^{n} x_{3i} w_{3i}$$

$$z_N = \varepsilon_N + \sum_{i=1}^{n} x_{Ni} w_{Ni}$$

The entire single hidden layer with "N" neurons with activation function can be represented as

$$Z = g(z_1) + g(z_2) + g(z_3) + \ldots + g(z_N)$$

$$Z = \sum_{i=1}^{N} g(z_i)$$

$$Z = \sum_{j=1}^{N} g\left(\varepsilon_j + \sum_{i=1}^{n} x_i w_{ij} \right)$$

where x_i represents the inputs from independent variables,

w_{ij} represents the weights associated with these independent variables, and

ε_j represents the error or bias associated with each neuron in the hidden layer.

The output y will be represented as an activation function of Z, and it is dependent on the activation function of the neural network.

$$y = g(Z)$$

4.2.1.1 Activation Function

Activation function is defined as the computational logic of the neural networks that takes into account both the input variables and their correspondent weights to determine the impact of the variables on the desired output and segregate the relevant input information necessary to process a particular neuron. Also known as transfer function or cost function, activation function influences the output based on a given set of inputs.

In terms of mathematical representation, an activation function can be either a linear or a nonlinear function. For a linear activation function, the output is linear in nature and lies within a range of $\{-\infty, \infty\}$. In simple terms, it can be represented as

$$y = g(z) = z$$

The identity activation function is the simplest and most commonly linear activation function used in regression problems. The identity function is monotonic in nature, and the derivative of an identity function $g'(z)$ is 1.

However, this kind of activation function does not support when the input variables have complexity in the data or the input data of a variable follow nonlinear patterns. In such cases, a nonlinear activation function is used. It is extremely useful when the data follow a parabolic or exponential curve, and this function makes it easier for the model to adapt to the variety of data points for an input variable.

The most common applications of nonlinear functions are used when the slope follows a differential or derivative function. Also, if the input variable conforms to monotonic functions, then a nonlinear activation function would help in determining the output. Some of the commonly used nonlinear activation functions are – sigmoid function, tanh function, and Rectified Linear Unit (ReLU) function.

4.2.1.2 Logistic Sigmoid Activation Function

A sigmoid function has a characteristic "S"-shaped curve. Also, it is considered to be a special case of logistic function. Mathematically, it is represented as

$$g(z) = \sigma(z) = \frac{1}{1 + e^{-z}}$$

$$g(z) = (1 + e^{-z})^{-1}$$

Graphically, it is presented in Figure 4.3.

For sigmoid activation function, the output is nonlinear in nature and lies within a range of (0,1). Because most of the problems that involve neural networks are classification problems (especially binary classification for areas such as predicting default of a loan or identifying a transaction to be fraud or not), this function is most appropriate

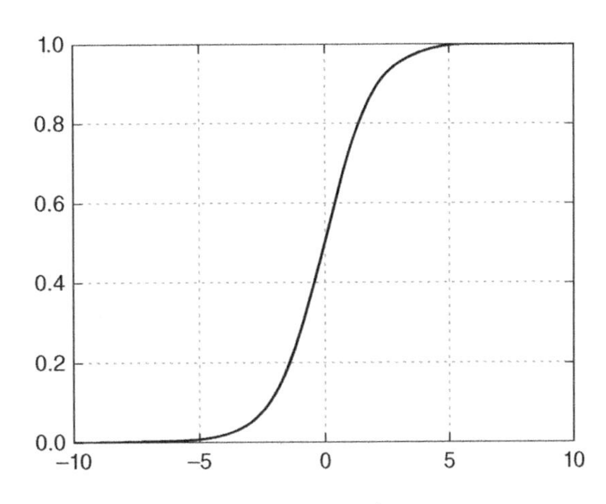

Figure 4.3 Shows the graphical representation of Sigmoid function.

for predicting the probability of the output. Because this function is differentiable and monotonic, the derivative of a sigmoid function $g'(z)$ is

$$y' = g'(z) = (1 + e^{-z})^{-2}(-e^{-z})$$

$$g'^{(z)} = \frac{e^{-z}}{(1 + e^{-z})^2} = \frac{1}{(1 + e^{-z})} \frac{e^{-z}}{(1 + e^{-z})}$$

$$g'(z) = g(z)(1 - g(z))$$

Also, the derivative of this function conforms to a bell-shaped curve or a simple normal distribution, and it is the most appropriate form for calculating the gradients used in the neural networks. The gradients for the layer can be estimated using arithmetic operations such as simple subtraction and multiplication. Also, it is used as an activation function in neural networks to bring nonlinearity into the model. For example, Gershenfeld et al. [11] has used the logistic sigmoid function as an activation function to keep the response of the neural network bounded. The function used is represented as

$$g(z) = \frac{1}{1 + e^{-2\beta z}}$$

4.2.1.3 tanh or Hyperbolic Tangent Activation Function

A sigmoid function-based activation function has its set of limitations during training of the neural networks. For such scenarios where highly negative inputs are fed to the logistic sigmoid function, then the output value is almost zero. This affects the calculations of gradient parameters for the feedforward neural networks when there are large numbers of neurons in the hidden layer and the activation function gets stuck during training.

In such cases, an alternative to the sigmoid activation function is hyperbolic tangent function (tanh). The points in the hyperbolic functions ($\cosh\theta$ and $\sinh\theta$) form a semiequilateral hyperbola. The hyperbolic functions can be defined as the two sides of a right-angled triangle covering the hyperbolic sector. Mathematically, a hyperbolic tangent is represented as

$$\tanh x = \frac{\sinh x}{\cosh x}$$

$$\tanh x = \frac{(e^x - e^{-x})}{(e^x + e^{-x})} = \frac{(e^{2x} - 1)}{(e^{2x} + 1)}$$

Graphically, the function is presented in Figure 4.4.

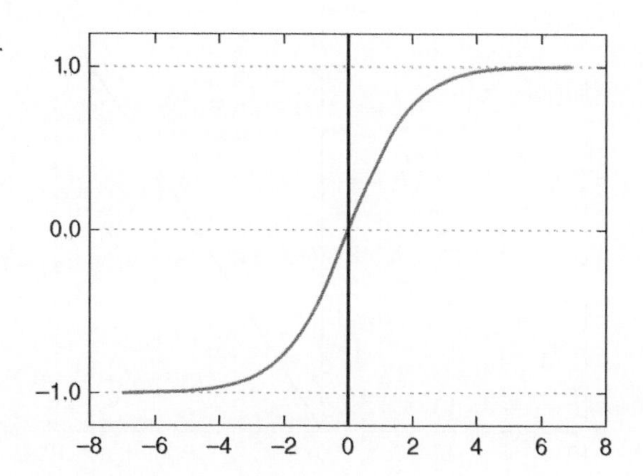

Figure 4.4 Shows the graphical representation of tanh function.

For tanh activation function, the output is similar to sigmoid function ("S" curve) and lies within a range of (−1,1). Like the sigmoid function, this function is differentiable and monotonic, and the derivative of a hyperbolic tangent function $g'(z)$ is

$$y' = g'(z) = \frac{\partial \tanh z}{\partial z}$$

$$g'^{(z)} = 1 - \tanh^2 z$$

$$g'^{(z)} = 1 - g(z)^2$$

Similar to the sigmoid function, the tanh activation function is applied in feedforward neural networks for classification and prediction problems.

4.2.1.4 ReLU (Rectified Linear Unit) Activation Function

The ReLU function is the most used activation function in all forms of neural networks including CNNs [12]. Also known as ramp function, the ReLU function is rectified for all the negative values of the input while it conforms to linear or identity function for all the positive values of the input.

Deep networks with ReLU as an activation function are easy to optimize and train in comparison with the network models with sigmoid-based or tanh-based activation functions because the gradient parameters are able to flow easily when there are multiple hidden layers with a large number of neurons. This has made ReLU a popular activation function and has great applications in speech recognition.

Mathematically, the rectifier in the activation function is defined as

$$f(x) = \max(0, x)$$

where x is the input to the neuron.

In simple terms, this function is represented as

$$f(x) = \begin{cases} 0, & x < 0 \\ x, & x \geq 0 \end{cases}$$

Graphically, the function is presented in Figure 4.5.

This function does not have negative outputs for negative input values. Unlike the above two functions, Sigmoid and tanh, both the ReLU function and its derivative are monotonic in nature and the output lies in the range of

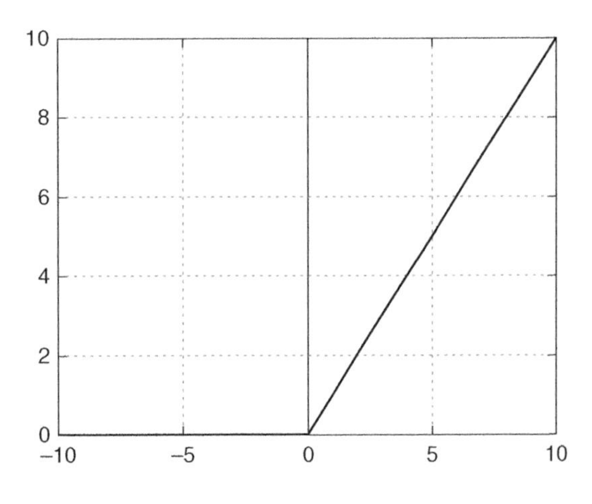

Figure 4.5 Shows the graphical representation of ReLU function.

$[0, \infty)$ for all values of input. The derivative of the ReLU function is

$$y' = g'(z) = \begin{cases} 0, & z < 0 \\ 1, & z \geq 0 \end{cases}$$

However, this function does not include any negative inputs because of which only the positive inputs will be available for training the model. Hence, all negative inputs will not be part of the model, and this drastically affects the predictability and accuracy of the model.

4.3 Other Activation Functions (Variant Forms of ReLU)

4.3.1 Smooth ReLU

In this variant of ReLU, the function is represented as a logistic function

$$f(x) = \log(1 + e^x)$$

The derivative of this variant of ReLU function is

$$y\prime = g\prime(z) = \frac{\partial}{\partial z} \log(1 + e^z)$$

$$g^{\prime(z)} = \frac{e^z}{(1 + e^z)} = \frac{1}{(1 + e^{-z})}$$

4.3.2 Noisy ReLU

In this variant of ReLU, the function is represented as a ramp function with Gaussian noise α

$$f(x) = \max(0, x + \alpha)$$

where $\alpha \sim \eta\ (0, \sigma(x))$

In simple terms, this function is represented as

$$f(x) = \begin{cases} 0, & x < 0 \\ x + \alpha, & x \geq 0 \end{cases}$$

The derivative of this variant of ReLU function is

$$y' = g'(z) = \begin{cases} 0, & z < 0 \\ 1, & z \geq 0 \end{cases}$$

This variant is used in restricted Boltzmann machines for computer vision activities.

4.3.3 Leaky ReLU

In this variant of ReLU, the function is represented as a ramp function with a small, positive gradient for negative inputs

$$f(x) = \begin{cases} 0.01x, & x < 0 \\ x, & x \geq 0 \end{cases}$$

The derivative of this variant of ReLU function is

$$y\prime = g\prime(z) = \begin{cases} 0.01, & z < 0 \\ 1, & z \geq 0 \end{cases}$$

4.3.4 Parametric ReLU

In this variant of ReLU, the function is represented as a ramp function with the coefficient of leakage into a parameter for learning for negative inputs

$$f(x) = \begin{cases} ax, & x < 0 \\ x, & x \geq 0 \end{cases}$$

In cases where $a \leq 1$, then the function is represented as

$$f(x) = \max(ax, x)$$

The derivative of this variant of ReLU function is

$$y\prime = g\prime(z) = \begin{cases} a, & z < 0 \\ 1, & z \geq 0 \end{cases}$$

4.3.5 Training and Optimizing a Neural Network Model

ANN is a supervised learning technique, i.e. the algorithm is trained on labeled data to identify different patterns in the input data that contribute to a given input. For obtaining a specific output, the weights assigned to inputs in each neuron are adjusted accordingly by the model. The higher the weight assigned to a particular input, the more impact the input variable has on the neuron, and this impact will continue to affect the neurons in subsequent layers as well. To show inhibition, negative weights are sometimes assigned to the input variables. This entire process of adjusting the weights of the neuron and obtaining the right set of values to obtain the desired output results is known as training the neural network.

There are multiple algorithms to train a neural network model. The most commonly used algorithm is backpropagation algorithm. This algorithm calculates the error in estimation and correspondingly determines the weights of each layer to obtain the desired output.

4.4 Backpropagation Algorithm

Werbos's [13] backpropagation algorithm provided a breakthrough in the field of neural networks paving way for ANNs. Johansson et al. [14] developed a backpropagation learning for multilayer feedforward neural networks using the conjugate gradient method for improving and optimizing the learning rates. Chen and Jain [15] derived a robust backpropagation learning algorithm that is resistant to the noise effects and is capable of rejecting gross errors during the approximation process.

Yu et al. [16] proposed a general backpropagation algorithm for feedforward neural network learning with time-varying inputs. In this approach, the Lyapunov function is used to analyze the convergence of weights, with the use of the algorithm for minimization of the error function. Khashman [17] proposed a modified backpropagation learning algorithm, with additional emotional weights for the two additional emotional parameters: anxiety and confidence. The proposed neural network was implemented to a facial recognition problem, and the results showed an improved performance with higher recognition rates and faster recognition time in comparison to the results of conventional neural network.

Sapna et al. [18] proposed a novel way of building backpropagation algorithm based on Levenberg–Marquardt algorithm to obtain an intellectual and efficient diabetic prediction method for assisting medical practitioners, special educators, occupational therapists, and psychologists in better assessment of diabetes.

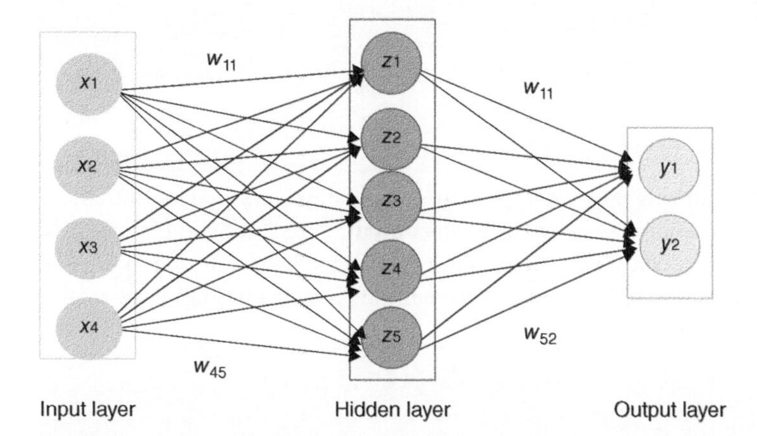

Figure 4.6 Shows the schematic representation of a standard back propagation neural network with the input variables (x_1, x_2, x_3, x_4), hidden layers $(z_1, z_2, z_3, z_4, z_5)$ and output values (y_1, y_2).

Backpropagation algorithm follows a gradient descent approach and implements the chain rule (used in calculus) for calculating the derivative of two or more functions. It is similar to the Gauss–Newton algorithm (Figure 4.6).

For training the neural network, the inputs from input layer (represented by x_i) are multiplied with weights (represented by w_{ij}, where i represents the feature from the input layer and j represents the neuron from the hidden layer). Each neuron is represented by a mathematical function

$$z_1 = \varepsilon_1 + \sum_{i=1}^{n} x_i w_{i1}$$

Similarly, if there are "n" neurons in the hidden layer, then the neurons are represented as

$$z_j = \varepsilon_j + \sum_{i=1}^{n} x_i w_{ij} \quad j \sim (1, n)$$

On applying the activation function on the neurons, the result at each neuron is represented as

$$Z_j = g(z_j) = g\left(\varepsilon_j + \sum_{i=1}^{n} x_i w_{ij}\right)$$

Now, each neuron has its influence on the output (represented by Z_j) and each will be assigned a weightage (represented by W_{jk}, where j represents the neuron from the hidden layer and k represents the values from the output layer). Hence, the first output from "N" neurons represented by Z_1 will be influenced by a combination of results from neuron multiplied by the assigned weightage with the error term E.

$$Z_1 = g(z_1)W_{11} + g(z_2)W_{21} + g(z_3)W_{31} + \ldots + g(z_N)W_{N1} + E_1$$

$$Z_1 = E_1 + \sum_{j=1}^{n} g(z_j)W_{j1}$$

The output y will be represented as an activation function of Z

$$y = g(Z)$$

For "k" values of output obtained, the activation function is represented as

$$y_k = g(Z_k)$$

$$y_k = g\left(E_k + \sum_{j=1}^{n} g(z_j)W_{jk}\right)$$

$$y_k = g\left(E_k + \sum_{j=1}^{n} g\left(\varepsilon_j + \sum_{i=1}^{n} x_i w_{ij}\right) W_{jk}\right)$$

As part of training the neural network and optimizing the model for all the input parameters used for the model, the predicted output obtained from the model is compared with the actual output. Based on the difference, the model is optimized to ensure that the variance between the predicted output and actual output is minimum. For the model, the standard error function is sum of squared difference between the predicted output of the model and the actual output. Mathematically, it is represented as

$$E = \frac{1}{2}\sum_{k=1}^{K} ({y_k}^{\text{predicted}} - {y_k}^{\text{actual}})^2$$

The backpropagation algorithm effectively solved the pattern identification and data classification problems through multiple iterations of training the algorithm and assigning the calculated weights at each node. The constraints for model training and optimization depend on the minimization of error function E and the optimal use of different activation functions.

Training and optimizing are the two important aspects in the process of building a neural network model for solving classification and prediction problems. In order to achieve the best-fit and most optimal model, there are multiple parameters that are to be tuned and optimized. Some of the most important parameters are number of hidden layers, activation function used in the hidden layer, learning rate for gradient descent (for backpropagation algorithm), momentum for gradient descent, number of epochs, and output function. Applying the right combination of optimized parameters to obtain a best-fit model is a tricky process and requires multiple repetition of model training through modification of each parameter for achieving the least error and maximum accuracy.

As part of building an ANN model, the developer (or modeler) has to assess the model's variable selection criteria and transformation process to ensure a strong relationship between the transformed predictor variables and dependent variable. As part of model's variable selection criteria, the developer needs to

1) Review the selected set of input and predictor variables
2) Review all the new features created and its predictive power including categorizing the continuous variables in the input parameters
3) Perform data assessment and profiling for identifying blanks, missing values, and outliers in the data
4) Review the criteria used for determining whether the variable should be included for model building
5) Critical examination of all the methods used for treatment of missing (or special) values to ensure whether such treatments will have an adverse effect on the predictive performance of the ANN model or not.

In order to evaluate the predictive power of all the selected input variables, the following two statistical tests need to be performed:

Statistical test	Test description	Points of consideration
Weight of evidence (WoE)	Measures the strength of a set of categories across different values of the predictor variable to separate "good" and "bad" outcomes	High negative or positive values are an indication of strong variable predictive power
Information value (IV)	Assesses the overall power of a variable in separating "good" and "bad" outcomes by summing the product of WoE and the difference of "good" and "bad" across all categories within the variable	Higher IV levels indicate stronger relationship between the variable and the good/bad odds ratio. Can be used to compare predictive power among competing variables

In order to test and review the model building parameter settings and their impact on the model performance, the following parameters need to be considered:

- *Hidden layers*: Number of hidden layers in the neural network
- *Activation function*: The activation function can be sigmoid, linear, and tanh (hyperbolic tangent)
- *Learning rate*: The learning rate for gradient descent
- *Momentum*: The momentum for gradient descent is a value between 0 and 1 that increases the size of the steps taken toward the minimum by trying to jump from some local minima
- *Learning rate scale*: The learning rate will be multiplied by this scale after every iteration
- *Number of epochs*: Number of iterations for samples
- *Batch size*: The size of the batches taken for training
- *Output function*: The output function can be sigmoid, linear, or ReLU
- *Hidden dropout*: The fraction of hidden layer to be dropped out for model training
- *Visible dropout* : The fraction of drop put in the input layer

These parameter settings are essential to regulate the complexity of the ANN model as they ensure optimization of the computational power, the number of variables involved in model training, and the weights to each variable for obtaining the best-fit model.

4.5 Performance and Accuracy

The following tests can be used to formally assess the performance and accuracy of an ANN model.

Statistical test	Test description	Points of consideration
Traditional performance metrics (from confusion matrix) – accuracy, precision, F-measure, sensitivity, and specificity	The confusion matrix compares the predictions of the model with respect to actual values	The higher the key performance metrics, the better the model performance
Receiver operating characteristic (ROC)	The ROC curve displays the trade-off between sensitivity and specificity	The closer the curve is to the left border and then the top border, the more accurate the model
	The area under the ROC curve is a measure of discriminatory power	The closer the curve is to diagonal, the less accurate is the model
Somers' D	Calculated from the difference between the number of concordant and discordant pairs	Value close to zero indicates random model while value close to 1 indicates higher discriminatory power
Kolmogorov–Smirnov (KS)	The KS test is used to test for differences between distribution functions	Higher values of KS statistic correspond to higher level of discriminatory power
Error attribution analysis	Discrepancies between predicted and actual values for the models show the magnitude of error	Lower is the difference between the predicted and actual values, lower is the error in the model

4.6 Results and Observation

We have developed an ANN model based on synthetic data that closely represents the customer credit transaction data to predict the potential defaulters in the credit card payment.

The current dataset is synthetic data prepared for the purpose of prototype development. Despite a very high performance in the validation dataset, the model may be overfitting while identifying the defaulters. Therefore, owing to human error in the existing system and model overperformance, we have created two new datasets by extremely swapping the potential defaulters (PDs) with genuine customers (GCs) by 35% and 65%. This process ensures randomness to the dataset and reduces the problem of overfitting to satisfactory levels. If the performance of the model deteriorates significantly, then we can assess that the model has captured random noise in the dataset, thereby necessitating model redevelopment. Therefore, this process will indirectly assess the sensitivity of ANN algorithm.

As part of data sampling and feature engineering, we have aggregated the credit card transaction data at customer level to determine customer-level behavior. We obtain customer-level average transaction amount, maximum transaction amount, and number of transactions executed for each of the modes of transfer. We create a new feature based on the average duration and the standard deviation of the durations between any two transactions irrespective of the mode of transfer. We obtained the final dataset with a total of 38 746 values and 27 variables, which can be used for training the model. Although actual data may contain more variables than the synthetic dataset and has more variability, the synthetic dataset is created such that it closely represents the actual data as careful segregation is performed while preparing the dataset.

For variable selection, stepwise logistic regression is used and all 27 variables were considered to build the logistic regression model. Then, bidirectional stepwise logistic regression based on Akaike information criterion (AIC) value was implemented to determine the best combination of independent variables to estimate the dependent variable. Finally, we obtained a model with 13 variables.

We used the variables that were significant for initially training the ANN model. After multiple iterations, including variable addition and deletion, the final model variables were identified. After rigorous tuning, the identified variables based on random trials and prudential tweaking, the final set of parameters was identified for best-fit ANN model. For example, the identified model parameters include

- Number of hidden layers = 2
- Number of neurons in the first layer = 80
- Number of neurons in the second layer = 50
- Learning rate = 0.337
- Momentum = 0.7
- Number of epochs = 170
- Activation function used for hidden layers and output layer = sigmoid function

For comparison, we have tested the performance of this ANN against the standard logistic regression model for predicting the defaulters. We found that both ANN and logistic regression yielded similar Area Under the Curve (AUC) values. Other factors also showed similar results:

- *Logistic regression*: Sensitivity of 93.7%, specificity of 53.76%, and transactional volume reduction of 39.5% with a small misclassification of 0.83% was achieved.
- *ANN*: Sensitivity of 94.8%, specificity of 50.03%, and transactional volume reduction of 37.6% with a small misclassification of 0.65% was achieved.

However, on actual dataset, ANN and other forms of neural network algorithms are likely to predict nonlinearity in the dataset better than logistic regression. The objective of the ANN modeling is to identify potential areas of improvement in identifying potential defaulters in Credit Card domain using neural networks. Although we have used a synthetic dataset, we have tested various assumptions and patterns of the transaction dataset and developed a generalized ANN with backpropagation model. This model can be customized to cater to the requirements of a specific scenario in Credit Risk and Fraud detection domains.

References

1 LeCun, Y., Bengio, Y., and Hinton, G. (2015). Deep learning. *Nature* 521 (7553): 436.

2 McCulloch, W.S. and Pitts, W. (1943). A logical calculus of the ideas immanent in nervous activity. *The Bulletin of Mathematical Biophysics* 5 (4): 115–133.

3 Cowan, J.D. (1990). Neural networks: the early days. In: *Advances in Neural Information Processing Systems*, 828–842.

4 McClelland, J.L., Rumelhart, D.E., and PDP Research Group (1986). Parallel distributed processing. *Explorations in the Microstructure of Cognition* 2: 216–271.

5 Qian, N. and Sejnowski, T.J. (1988). Predicting the secondary structure of globular proteins using neural network models. *Journal of Molecular Biology* 202 (4): 865–884.

6 Kneller, D.G., Cohen, F.E., and Langridge, R. (1990). Improvements in protein secondary structure prediction by an enhanced neural network. *Journal of Molecular Biology* 214 (1): 171–182.

7 Rost, B. and Sander, C. (1994). Combining evolutionary information and neural networks to predict protein secondary structure. *Proteins: Structure, Function, and Bioinformatics* 19 (1): 55–72.

8 Courbariaux, M., Hubara, I., Soudry, D., El-Yaniv, R., & Bengio, Y. (2016). Binarized neural networks: training deep neural networks with weights and activations constrained to+ 1 or −1. arXiv preprint arXiv:1602.02830.

9 Silver, D., Huang, A., Maddison, C.J. et al. (2016). Mastering the game of Go with deep neural networks and tree search. *Nature* 529 (7587): 484.

10 Esteva, A., Kuprel, B., Novoa, R.A. et al. (2017). Dermatologist-level classification of skin cancer with deep neural networks. *Nature* 542 (7639): 115.

11 Gershenfeld, H.K., Neumann, P.E., Li, X. et al. (1999). Mapping quantitative trait loci for seizure response to a GABAA receptor inverse agonist in mice. *Journal of Neuroscience* 19 (10): 3731–3738.

12 Hahnloser, R.H.R., Sarpeshkar, R., Mahowald, M. et al. (2000). Digital selection and analogue amplification coexist in a cortex-inspired silicon circuit. *Nature* 405 (6789): 947.

13 Werbos, P. (1974). Beyond regression: new tools for prediction and analysis in the behavioral sciences. PhD dissertation. Harvard University.

14 Wadell, I., Johansson, H., Sjölander, P. et al. (1991). Fusimotor reflexes influencing secondary muscle spindle afferents from flexor and extensor muscles in the hind limb of the cat. *Journal de Physiologie* 85 (4): 223–234.

15 Chen, D.S. and Jain, R.C. (1994). A robust backpropagation learning algorithm for function approximation. *IEEE Transactions on Neural Networks* 5 (3): 467–479.

16 Yu, X., Efe, M.O., and Kaynak, O. (2002). A general backpropagation algorithm for feedforward neural networks learning. *IEEE Transactions on Neural Networks* 13 (1): 251–254.

17 Khashman, A. (2008). A modified backpropagation learning algorithm with added emotional coefficients. *IEEE Transactions on Neural Networks* 19 (11): 1896–1909.

18 Sapna, S., Tamilarasi, A., and Kumar, M.P. (2012). Backpropagation learning algorithm based on Levenberg Marquardt Algorithm. *Computer Science and Information Technology (CS and IT)* 2: 393–398.

5

Multimodal Data Representation and Processing Based on Algebraic System of Aggregates

Yevgeniya Sulema[1] *and Etienne Kerre*[2]

[1] *Department of Computer Systems Software, Faculty of Applied Mathematics, Igor Sikorsky Kyiv Polytechnic Institute, Kyiv, 03056, Ukraine*
[2] *Department of Applied Mathematics, Computer Science and Statistics, Ghent University, Ghent, B9000, Belgium*

5.1 Introduction

In many cases, we can face with a task of collecting data of different nature obtained (registered, recorded, and measured) in different time slots from sources of different types. For example, long-life medical observation of a patient is such a case: data about a patient's health status can be obtained from a wide range of medical equipment and tools that can be both digital (e.g. a magnetic imaging system) and nondigital (e.g. a mercury thermometer), as well as these datasets can be registered both in paper notes such as the so-called "patient's medical card" and by using electronic medical documenting systems. The next task after data collection is data synchronization. This task is a part of a wider task of data analysis of all collected multimodal datasets in order to get a whole picture on the object (subject, process, and event) status, behavior, further states, etc.

One of the possible ways toward solving the task of multimodal data analysis is presenting data sets describing the same object as a complex data structure called an *aggregate* introduced in the *Algebraic System of Aggregates (ASA)* [1, 2].

Formally, an aggregate A is a tuple of arbitrary tuples, whose elements belong to certain sets [1, 2]:

$$A = [\![M_j | \langle a_i^j \rangle_{i=1}^{n_j}]\!]_{j=1}^{N} = [\![\{A\} | \langle A \rangle]\!] \tag{5.1}$$

where $\{A\}$ is a tuple of sets M_j and $\langle A \rangle$ is a tuple of element tuples $\langle a_i^j \rangle_{i=1}^{n_j}$ corresponding to the tuple of sets ($a_i^j \in M_j$).

If multimodal data are defined (registered, recorded, and measured) in terms of time, an aggregate must have a time value sequence as the first tuple. Such timewise aggregate is called a *multi-image* [1, 2]. The multi-image is a data structure that describes an object in such complete way as it is enabled by the available data. Thus, a *multi-image* is a nonempty aggregate formalized as:

$$I = [\![T, M_1, \dots, M_N | \langle t_1, \dots, t_\tau \rangle, \langle a_1^1, \dots, a_{n_1}^1 \rangle, \dots, \langle a_1^N, \dots, a_{n_N}^N \rangle]\!] \tag{5.2}$$

where T is a set of time values; $\tau \geq n_i,\quad i \in [1, \dots, N]$.

For example, if there are three tuples of the values measured simultaneously, namely, systolic blood pressure (sp), diastolic blood pressure (dp), and heart rate (hr), as well as there is a tuple of time moments when these values are being measured, then we can present all the measurements as one complex data structure – a multi-image I:

$$I = [\![T, M_{\text{sp}}, M_{\text{dp}}, M_{\text{hr}} | \langle 2, 4, 8, 13, 21, 26 \rangle, \langle 185, 166, 175, 166, 171, 152 \rangle,$$
$$\langle 76, 73, 74, 73, 71, 76 \rangle, \langle 74, 81, 76, 93, 97, 97, 96 \rangle]\!]$$

Mathematical Methods in Interdisciplinary Sciences, First Edition. Edited by Snehashish Chakraverty.

However, the composition of multimodal data in one multi-image becomes a nontrivial task if the data values have been measured in different time moments. To process such data sequences, we need to synchronize them at first. This task is especially important in cases when multimodal data are being collected for a relatively long time from sources of different types such as remote sensors, paper archives, cloud storages, etc.

In this chapter, we discuss operations on multimodal data and relations between them, which enable complex representation of multiple multimodal data sequences obtained with respect to time in order to compose multi-images and process them. We pay special attention to multimodal data synchronization by using both crisp and fuzzy approaches.

5.2 Basic Statements of ASA

The ASA is an algebraic system [3–6] that consists of sets (M, F, R), where M is a nonempty set (carrier), whose elements are the elements of the system, F is a set of operations, and R is a set of relations.

As mentioned above, the *carrier* of ASA is an arbitrary set of specific structures called aggregates. Tuple elements in an aggregate A can have both crisp values a_i and fuzzy values $\widetilde{a}_i$.

A tuple element can be empty $\emptyset$; a tuple may be empty as well. For example, in Eq. (5.3), the second tuple is empty.

$$A = [\![M_1, M_2, \dots, M_N | \langle a_1^1, a_2^1, \dots, a_{n_1}^1 \rangle, \langle \emptyset \rangle, \dots \langle a_1^N, a_2^N, \dots, a_{n_N}^N \rangle]\!] \tag{5.3}$$

If an aggregate consists of only empty tuples, it is called an *empty aggregate*: $A = [\![M_j | \langle \emptyset \rangle]\!]_{j=1}^N$.

An aggregate that does not include any component is called a *null-aggregate*: $A_\emptyset = [\![\emptyset \mid \langle \emptyset \rangle]\!]$. The null-aggregate plays a role of a neutral element in ASA.

A tuple element can be undefined, and its notation is _. A tuple may be undefined as well. If an aggregate consists of undefined tuples, it is called an *undefined aggregate*: $A = [\![M_j | \langle \_ \rangle]\!]_{j=1}^N$. The practical meaning of an undefined aggregate is that we can predefine a data structure (aggregate) even if data sequences are currently unavailable, e.g. if remote sensors are off and, thus, data are not received yet; however, we know their type and can predefine the aggregate sets and tuples.

The aggregate *length* $|A|$ is the quantity of tuples in it. The length of the empty aggregate $A = [\![M_j | \langle \emptyset \rangle]\!]_{j=1}^N$ is $|A| = N$. The length of the undefined aggregate $A = [\![M_j | \langle \_ \rangle]\!]_{j=1}^N$ is also $|A| = N$. At the same time, the length of the null-aggregate is $|A_\emptyset| = 0$.

The sequence order of sets and corresponding tuples in an aggregate defines how operations on the aggregate will be fulfilled. In this regard, two aggregates, A_1 and A_2, can be compatible, $A_1 \div A_2$, quasi-compatible, $A_1 \doteq A_2$, incompatible, $A_1 \mathring{=} A_2$, or hiddenly compatible, $A_1 (\div) A_2$.

Aggregates A_1 and A_2 are *compatible* if $\{A_1\} \equiv \{A_2\}$, i.e. both aggregates have the same set of sets and the sequence order of the sets in both aggregates is the same.

Aggregates A_1 and A_2 are *quasi-compatible* if $\{A_1\} \not\equiv \{A_2\}$ and $\{A_1\} \cap \{A_2\} \neq \emptyset$, i.e. the type and sequence order of these aggregates coincide partly.

Aggregates A_1 and A_2 are *incompatible* if $\{A_1\} \cap \{A_2\} = \emptyset$.

Aggregates A_1 and A_2 are *hiddenly compatible* if $\{A_1\} \not\equiv \{A_2\}$, $|A_1| = |A_2| = N$, and $\forall M_j \subset \{A_k\}$, $j = [1, \dots, N]$, $k = [1, 2]$, i.e. both aggregates have the same set of sets, but the order of these sets differs.

For example, let the sets be defined in the following way:

$M_t = [35.0, \dots, 39.9]$ is a set of temperature values (°C);
$M_{\mathrm{hr}} = [50, \dots, 110]$ is a set of heart rate values (bpm);
$M_{\mathrm{sp}} = [80, \dots, 190]$ is a set of systolic pressure values (mmHg);
$M_{\mathrm{dp}} = [55, \dots, 100]$ is a set of diastolic pressure values (mmHg).

Let us consider the following aggregates whose elements belong to these sets:

$A_1 = [\![M_t, M_{\text{hr}} \mid \langle 36.4,36.1,36.3,36.2,36.5,36.3\rangle, \langle 75,76,74,73,75,75\rangle]\!]$;
$A_2 = [\![M_t, M_{\text{hr}} \mid \langle 36.5,36.5,36.8,36.6,36.3,36.4,37.0,36.5\rangle, \langle 74,81,76,93,97,97,96\rangle]\!]$;
$A_3 = [\![M_{\text{sp}}, M_{\text{dp}} \mid \langle 185,166,175,166,171,152\rangle, \langle 76,73,74,73,71,76\rangle]\!]$;
$A_4 = [\![M_t, M_{\text{sp}} \mid \langle 36.5,36.5,36.8,36.6,36.3,36.4\rangle, \langle 177,159,174,155,167,150\rangle]\!]$;
$A_5 = [\![M_{\text{hr}}, M_t \mid \langle 74,81,76,93,97\rangle, \langle 36.5,36.5,36.3,36.4,37.0,36.5\rangle]\!]$.

Then, we can conclude that

- $A_1 \div A_2$ because the first tuple in both A_1 and A_2 belongs to set M_t and the second tuple in both A_1 and A_2 belongs to set M_{hr}.
- $A_1 \mathrel{\mathring{=}} A_3$ because the first tuple of A_1 belongs to set M_t, whereas the first tuple of A_3 belongs to M_{sp}, as well as the second tuple of A_1 belongs to set M_{hr}, whereas the second tuple of A_3 belongs to M_{dp}.
- $A_1 \doteq A_4$ because the first tuple of both A_1 and A_4 belongs to set M_t, but at the same time, the second tuple of A_1 belongs to set M_{hr}, whereas the second tuple of A_4 belongs to M_{sp}.
- $A_1(\div)A_5$ because now they are incompatible, but they will become compatible if we change the order of tuples in one of them (e.g. in A_5): $A_5^{\text{new}} = [\![M_t, M_{hr} | \langle 36.5,36.5,36.3,36.4,37.0,36.5\rangle, \langle 74,81,76,93,97\rangle]\!]$. It can be fulfilled by applying an ordering operation to the elements of this tuple.

5.3 Operations on Aggregates and Multi-images

Operations on aggregates include *arithmetical operations*, *logical operations*, and *ordering operations*. Arithmetical operations include elementwise addition, scalar addition, elementwise subtraction, scalar subtraction, elementwise multiplication, scalar multiplication, elementwise division, and scalar division. The logical operations on aggregates are union; intersection; difference; symmetric difference; and exclusive intersection [1].

The *union* of two aggregates A_1 and A_2 is aggregate B, which includes components of both aggregates and is formed according to the following rule:

1. If aggregates A_1 and A_2 are such as $A_1 \div A_2$ and

$$A_1 = [\![M_1, M_2, \dots, M_N | \langle a_1^1, a_2^1, \dots, a_l^1\rangle, \langle b_1^1, b_2^1, \dots, b_m^1\rangle, \dots, \langle w_1^1, w_2^1, \dots, w_n^1\rangle]\!]$$

$$A_2 = [\![M_1, M_2, \dots, M_N | \langle a_1^2, a_2^2, \dots, a_r^2\rangle, \langle b_1^2, b_2^2, \dots, b_q^2\rangle, \dots, \langle w_1^2, w_2^2, \dots, w_p^2\rangle]\!]$$

then the elements of i-tuple of A_2 are added at the end of i-tuple of A_1:

$$B = A_1 \cup A_2 = [\![M_1, M_2, \dots, M_N | \langle a_1^1, a_2^1, \dots, a_l^1, a_1^2, a_2^2, \dots, a_r^2\rangle, \langle b_1^1, b_2^1, \dots, b_m^1, b_1^2, b_2^2, \dots, b_q^2\rangle, \dots \langle w_1^1, w_2^1, \dots, w_n^1, w_1^2, w_2^2, \dots, w_p^2\rangle]\!] \tag{5.4}$$

2. If aggregates A_1 and A_2 are such as $A_1 \mathrel{\mathring{=}} A_2$ and

$$A_1 = [\![M_1^1, M_2^1, \dots, M_N^1 | \langle a_1, a_2, \dots, a_l\rangle, \langle b_1, b_2, \dots, b_m\rangle, \dots \langle w_1, w_2, \dots, w_n\rangle]\!]$$

$$A_2 = [\![M_1^2, M_2^2, \dots, M_K^2 | \langle c_1, c_2, \dots, c_r\rangle, \langle d_1, d_2, \dots, d_q\rangle, \dots \langle z_1, z_2, \dots, z_p\rangle]\!]$$

then the tuples of A_2 are added at the end of the tuple of tuples of A_1 and the corresponding sets of A_2 are added at the end of the set sequence of A_1:

$$B = A_1 \cup A_2 = [\![M_1^1, M_2^1, \dots, M_N^1, M_1^2, M_2^2, \dots, M_K^2 | \langle a_1, a_2, \dots, a_l\rangle, \langle b_1, b_2, \dots, b_m\rangle, \dots, \langle w_1, w_2, \dots, w_n\rangle, \langle c_1, c_2, \dots, c_r\rangle, \langle d_1, d_2, \dots, d_q\rangle, \dots, \langle z_1, z_2, \dots, z_p\rangle]\!] \tag{5.5}$$

3. If aggregates A_1 and A_2 are such as $A_1 \doteq A_2$ and

$$A_1 = [\![M_1, M_2^1, \dots, M_x, \dots, M_N^1 | \langle a_1^1, a_2^1, \dots, a_l^1 \rangle, \langle b_1, b_2, \dots, b_m \rangle, \dots, \langle f_1^1, f_2^1, \dots, f_t^1 \rangle, \dots, \langle w_1, w_2, \dots, w_n \rangle]\!]$$

$$A_2 = [\![M_1, M_2^2, \dots, M_x, \dots, M_K^2 | \langle a_1^2, a_2^2, \dots, a_r^2 \rangle, \langle d_1, d_2, \dots, d_q \rangle, \dots, \langle f_1^2, f_2^2, \dots, f_v^2 \rangle, \dots, \langle z_1, z_2, \dots, z_p \rangle]\!]$$

then the elements of i-tuple of A_2 are added at the end of i-tuple of A_1 if the elements of these tuples belong to the same set; otherwise, the rule for incompatible aggregates is applied:

$$B = A_1 \cup A_2 = [\![M_1, M_2^1, \dots, M_x, \dots, M_N^1, M_2^2, \dots, M_K^2 | \langle a_1^1, a_2^1, \dots, a_l^1, a_1^2, a_2^2, \dots, a_r^2 \rangle \langle b_1, b_2, \dots, b_m \rangle, \dots, \langle f_1^1, f_2^1, \dots, f_t^1, f_1^2, f_2^2, \dots, f_v^2 \rangle, \dots, \langle w_1, w_2, \dots, w_n \rangle, \langle d_1, d_2, \dots, d_q \rangle, \dots, \langle z_1, z_2, \dots, z_p \rangle]\!] \quad (5.6)$$

The *intersection* of two aggregates A_1 and A_2 is aggregate B, which includes only common components of both aggregates and is formed according to the following rule:

1. If aggregates A_1 and A_2 are such as $A_1 \div A_2$ and

$$A_1 = [\![M_1, M_2, \dots, M_N | \langle a_1^1, a_2^1, \dots, a_l^1 \rangle, \langle b_1^1, b_2^1, \dots, b_m^1 \rangle, \dots, \langle w_1^1, w_2^1, \dots, w_n^1 \rangle]\!]$$

$$A_2 = [\![M_1, M_2, \dots, M_N | \langle a_1^2, a_2^2, \dots, a_r^2 \rangle, \langle b_1^2, b_2^2, \dots, b_q^2 \rangle, \dots, \langle w_1^2, w_2^2, \dots, w_p^2 \rangle]\!]$$

then B includes the elements of both aggregates, which are common for them, in every tuple:

$$B = A_1 \cap A_2 = [\![M_1, M_2, \dots, M_N | \langle a_{l_1}^1, \dots, a_{l_\alpha}^1, a_{r_1}^2, \dots, a_{r_\beta}^2 \rangle, \langle b_{m_1}^1, \dots, b_{m_\gamma}^1, b_{q_1}^2, \dots, b_{q_\delta}^2 \rangle, \dots, \langle w_{n_1}^1, \dots, w_{n_\lambda}^1, w_{p_1}^2, \dots, w_{p_\mu}^2 \rangle]\!] \quad (5.7)$$

where

$$a_{l_i}^1 \in \bar{a}^1, a_{l_i}^1 \in \bar{a}^2, \quad i \in \langle 1, \dots, \alpha \rangle; \qquad a_{r_j}^2 \in \bar{a}^1, a_{r_j}^2 \in \bar{a}^2, \quad j \in \langle 1, \dots, \beta \rangle$$

$$b_{m_k}^1 \in \bar{b}^1, b_{m_k}^1 \in \bar{b}^2, \quad k \in \langle 1, \dots, \gamma \rangle; \qquad b_{q_s}^2 \in \bar{b}^1, b_{q_s}^2 \in \bar{b}^2, \quad s \in \langle 1, \dots, \delta \rangle$$

$$w_{n_u}^1 \in \bar{w}^1, w_{n_u}^1 \in \bar{w}^2, \quad u \in \langle 1, \dots, \lambda \rangle; \qquad w_{p_y}^2 \in \bar{w}^1, w_{p_y}^2 \in \bar{w}^2, \quad y \in \langle 1, \dots, \mu \rangle$$

It means that if we go through elements of $\bar{a}^1 = \langle a_1^1, a_2^1, \dots a_l^1 \rangle$ from its first element a_1^1 to its last element a_l^1 and compare each next element $a_{l_i}^1$ $(1 \le l_i \le l$, where l is the total number of elements in $\bar{a}^1)$ with elements of $\bar{a}^2 = \langle a_1^2, a_2^2, \dots a_r^2 \rangle$, then $a_{l_1}^1$ is the first element of $\bar{a}^1$, which is also present in $\bar{a}^2$, and $a_{l_\alpha}^1$ is the last element of $\bar{a}^1$, which is also present in $\bar{a}^2$.

Because we find *intersection* of $\bar{a}^1$ and $\bar{a}^2$, we also look for $a_{r_1}^2$ and $a_{r_\beta}^2$. To find them, we go through elements of $\bar{a}^2$ from its first element a_1^2 to its last element a_r^2 and compare each next element $a_{r_j}^2$ $(1 \le r_j \le r$, where r is the total number of elements in $\bar{a}^2)$ with elements of $\bar{a}^1$, then $a_{r_1}^2$ is the first element of $\bar{a}^2$, which is also present in $\bar{a}^1$, and $a_{r_\beta}^2$ is the last element of $\bar{a}^2$, which is also present in $\bar{a}^1$. Finally, the result of *intersection* of $\bar{a}^1$ and $\bar{a}^2$ is the tuple with elements $\langle a_{l_1}^1 \dots a_{l_\alpha}^1, a_{r_1}^2 \dots a_{r_\beta}^2 \rangle$.

For example, if $\bar{a}^1 = \langle 9, 5, \mathbf{2}, \mathbf{7}, \mathbf{9}, \mathbf{3}, 6 \rangle$ and $\bar{a}^2 = \langle 1, \mathbf{3}, \mathbf{2}, \mathbf{4}, \mathbf{7}, \mathbf{3}, 8 \rangle$, then $a_{l_1}^1 = \mathbf{2}$ and $a_{l_\alpha}^1 = \mathbf{3}$. At the same time, $a_{r_1}^2 = \mathbf{3}$ as well as $a_{r_\beta}^2 = \mathbf{3}$. Finally, $\bar{a}^1 \cap \bar{a}^2 = \langle \mathbf{2}, \mathbf{7}, \mathbf{3}, \mathbf{3}, \mathbf{2}, \mathbf{7}, \mathbf{3} \rangle$.

2. If aggregates A_1 and A_2 are such as $A_1 \overset{\circ}{=} A_2$, then $B = A_1 \cap A_2 = [\![\emptyset \,|\, \langle \emptyset \rangle]\!] = A_\emptyset$.
3. If aggregates A_1 and A_2 are such as $A_1 \doteq A_2$ and

$$A_1 = [\![M_1, M_2^1, \dots, M_x, \dots, M_N^1 | \langle a_1^1, a_2^1, \dots, a_l^1 \rangle, \langle b_1, b_2, \dots, b_m \rangle, \dots, \langle f_1^1, f_2^1, \dots, f_t^1 \rangle, \dots, \langle w_1, w_2, \dots, w_n \rangle]\!]$$

$$A_2 = [\![M_1, M_2^2, \dots, M_x, \dots, M_K^2 | \langle a_1^2, a_2^2, \dots, a_r^2 \rangle, \langle d_1, d_2, \dots, d_q \rangle, \dots, \langle f_1^2, f_2^2, \dots, f_v^2 \rangle, \dots, \langle z_1, z_2, \dots, z_p \rangle]\!]$$

then B includes elements of both aggregates, which are common for them, only in tuples of common sets; thus, the number of sets shortens. Only considering the common tuples, we proceed as in the first case:

$$B = A_1 \cap A_2 = [\![M_1, \dots, M_x | \langle a^1_{l_1}, \dots, a^1_{l_\alpha}, a^2_{r_1}, \dots, a^2_{r_\beta} \rangle, \dots, \langle f^1_{t_1}, \dots, f^1_{t_\rho}, f^2_{v_1}, \dots, f^2_{v_\omega} \rangle]\!] \tag{5.8}$$

The *difference* of two aggregates A_1 and A_2 is aggregate B, which includes only components present in A_1 and absent in A_2; it is formed according to the following rule:

1. If aggregates A_1 and A_2 are such as $A_1 \div A_2$ and

$$A_1 = [\![M_1, M_2, \dots, M_N | \langle a^1_1, a^1_2, \dots, a^1_l \rangle, \langle b^1_1, b^1_2, \dots, b^1_m \rangle, \dots, \langle w^1_1, w^1_2, \dots, w^1_n \rangle]\!]$$

$$A_2 = [\![M_1, M_2, \dots, M_N | \langle a^2_1, a^2_2, \dots, a^2_r \rangle, \langle b^2_1, b^2_2, \dots, b^2_q \rangle, \dots, \langle w^2_1, w^2_2, \dots, w^2_p \rangle]\!]$$

then B includes the elements of A_1, which are absent in A_2, in every tuple:

$$B = A_1 \backslash A_2 = [\![M_1, M_2, \dots, M_N | \langle a^1_{l_1}, \dots, a^1_{l_\alpha} \rangle, \langle b^1_{m_1}, \dots, b^1_{m_\gamma} \rangle, \dots, \langle w^1_{n_1}, \dots, w^1_{n_\lambda} \rangle]\!] \tag{5.9}$$

where

$$a^1_{l_i} \in \bar{a}^1, a^1_{l_i} \notin \bar{a}^2, \quad i \in \langle 1, \dots, \alpha \rangle$$

$$b^1_{m_k} \in \bar{b}^1, b^1_{m_k} \notin \bar{b}^2, \quad k \in \langle 1, \dots, \gamma \rangle$$

$$w^1_{n_u} \in \bar{w}^1, w^1_{n_u} \notin \bar{w}^2, \quad u \in \langle 1, \dots, \lambda \rangle$$

2. If aggregates A_1 and A_2 are such as $A_1 \mathrel{\mathring{=}} A_2$, then $B = A_1 \backslash A_2 = A_1$.
3. If aggregates A_1 and A_2 are such as $A_1 \doteq A_2$ and

$$A_1 = [\![M_1, M^1_2, \dots, M_x, \dots, M^1_N | \langle a^1_1, a^1_2, \dots, a^1_l \rangle, \langle b_1, b_2, \dots, b_m \rangle, \dots, \langle f^1_1, f^1_2, \dots, f^1_t \rangle, \dots, \langle w_1, w_2, \dots, w_n \rangle]\!]$$

$$A_2 = [\![M_1, M^2_2, \dots, M_x, \dots, M^2_K | \langle a^2_1, a^2_2, \dots, a^2_r \rangle, \langle d_1, d_2, \dots, d_q \rangle, \dots, \langle f^2_1, f^2_2, \dots, f^2_v \rangle, \dots, \langle z_1, z_2, \dots, z_p \rangle]\!]$$

then B includes the elements of A_1, which are absent in A_2, in tuples of common sets and all tuples of sets defined only in A_1:

$$B = A_1 \backslash A_2 = [\![M_1, M^1_2, \dots, M_x, \dots, M^1_N | \langle a^1_{i_1}, \dots, a^1_{i_\alpha} \rangle, \langle b_1, b_2, \dots, b_m \rangle, \dots, \langle f^1_{t_1}, \dots, f^1_{t_\rho} \rangle, \dots, \langle w_1, w_2, \dots, w_n \rangle]\!] \tag{5.10}$$

where

$$a^1_{l_i} \in \bar{a}^1, a^1_{l_i} \notin \bar{a}^2, \quad i \in \langle 1, \dots, \alpha \rangle$$

$$f^1_{t_e} \in \bar{f}^1, f^1_{t_e} \notin \bar{f}^2, \quad e \in \langle 1, \dots, \rho \rangle$$

The *symmetric difference* of two aggregates A_1 and A_2 is aggregate B, which includes both components present in A_1 and absent in A_2 and components present in A_2 and absent in A_1; it is formed according to the following rule:

1. If aggregates A_1 and A_2 are such as $A_1 \div A_2$ and

$$A_1 = [\![M_1, M_2, \dots, M_N | \langle a^1_1, a^1_2, \dots, a^1_l \rangle, \langle b^1_1, b^1_2, \dots, b^1_m \rangle, \dots, \langle w^1_1, w^1_2, \dots, w^1_n \rangle]\!]$$

$$A_2 = [\![M_1, M_2, \dots, M_N | \langle a^2_1, a^2_2, \dots, a^2_r \rangle, \langle b^2_1, b^2_2, \dots, b^2_q \rangle, \dots, \langle w^2_1, w^2_2, \dots, w^2_p \rangle]\!]$$

then B includes the elements of A_1, which are absent in A_2, and elements of A_2, which are absent in A_1, in every tuple:

$$B = A_1 \Delta A_2 = [\![M_1, M_2, \dots, M_N | \langle a^1_{l_1}, \dots, a^1_{l_\alpha}, a^2_{r_1}, \dots, a^2_{r_\beta} \rangle, \langle b^1_{m_1}, \dots, b^1_{m_\gamma}, b^2_{q_1}, \dots, b^2_{q_\delta} \rangle, \dots, \langle w^1_{n_1}, \dots, w^1_{n_\lambda}, w^2_{p_1}, \dots, w^2_{p_\mu} \rangle]\!] \tag{5.11}$$

where

$$a^1_{l_i} \in \bar{a}^1, a^1_{l_i} \notin \bar{a}^2, \quad i \in \langle 1, \dots, \alpha \rangle; \qquad a^2_{r_j} \notin \bar{a}^1, a^2_{r_j} \in \bar{a}^2, \quad j \in \langle 1, \dots, \beta \rangle$$

$$b^1_{m_k} \in \bar{b}^1, b^1_{m_k} \notin \bar{b}^2, \quad k \in \langle 1, \dots, \gamma \rangle; \qquad b^2_{q_s} \notin \bar{b}^1, b^2_{q_s} \in \bar{b}^2, \quad s \in \langle 1, \dots, \delta \rangle$$

$$w^1_{n_u} \in \bar{w}^1, w^1_{n_u} \notin \bar{w}^2, \quad u \in \langle 1, \dots, \lambda \rangle; \qquad w^2_{p_y} \notin \bar{w}^1, w^2_{p_y} \in \bar{w}^2, \quad y \in \langle 1, \dots, \mu \rangle$$

2. If aggregates A_1 and A_2 are such as $A_1 \mathring{=} A_2$ and

$$A_1 = [\![M^1_1, M^1_2, \dots, M^1_N | \langle a_1, a_2, \dots, a_l \rangle, \langle b_1, b_2, \dots, b_m \rangle, \dots, \langle w_1, w_2, \dots, w_n \rangle]\!]$$

$$A_2 = [\![M^2_1, M^2_2, \dots, M^2_K | \langle c_1, c_2, \dots, c_r \rangle, \langle d_1, d_2, \dots, d_q \rangle, \dots, \langle z_1, z_2, \dots, z_p \rangle]\!]$$

then B is equal to the union of A_1 and A_2:

$$B = A_1 \Delta A_2 = [\![M^1_1, M^1_2, \dots, M^1_N, M^2_1, M^2_2, \dots, M^2_K | \langle a_1, a_2, \dots, a_l \rangle, \langle b_1, b_2, \dots, b_m \rangle, \dots, \langle w_1, w_2, \dots, w_n \rangle, \langle c_1, c_2, \dots, c_r \rangle, \langle d_1, d_2, \dots, d_q \rangle, \dots, \langle z_1, z_2, \dots, z_p \rangle]\!] = A_1 \cup A_2 \tag{5.12}$$

3. If aggregates A_1 and A_2 are such as $A_1 \dot{=} A_2$ and

$$A_1 = [\![M_1, M^1_2, \dots, M_x, \dots, M^1_N | \langle a^1_1, a^1_2, \dots, a^1_l \rangle, \langle b_1, b_2, \dots, b_m \rangle, \dots, \langle f^1_1, f^1_2, \dots, f^1_t \rangle, \dots, \langle w_1, w_2, \dots, w_n \rangle]\!]$$

$$A_2 = [\![M_1, M^2_2, \dots, M_x, \dots, M^2_K | \langle a^2_1, a^2_2, \dots, a^2_r \rangle, \langle d_1, d_2, \dots, d_q \rangle, \dots, \langle f^2_1, f^2_2, \dots, f^2_v \rangle, \dots, \langle z_1, z_2, \dots, z_p \rangle]\!]$$

then B includes elements of A_1, which are absent in A_2, and elements of A_2, which are absent in A_1, in tuples of common sets, all tuples of sets defined only in A_1, and all tuples of sets defined only in A_2:

$$B = A_1 \Delta A_2 = [\![M_1, M^1_2, \dots, M_x, \dots, M^1_N, M^2_2, \dots, M^2_K | \langle a^1_{l_1}, \dots, a^1_{l_\alpha}, a^2_{r_1}, \dots, a^2_{r_\beta} \rangle, \langle b_1, b_2, \dots, b_m \rangle, \dots, \langle f^1_{t_1}, \dots, f^1_{t_\rho}, f^2_{v_1}, \dots, f^2_{v_\omega} \rangle, \dots, \langle w_1, w_2, \dots, w_n \rangle, \langle d_1, d_2, \dots, d_q \rangle, \dots, \langle z_1, z_2, \dots, z_p \rangle]\!] \tag{5.13}$$

where

$$a^1_{l_i} \in \bar{a}^1, a^1_{l_i} \notin \bar{a}^2, \quad i \in \langle 1, \dots, \alpha \rangle; \qquad a^2_{r_j} \notin \bar{a}^1, a^2_{r_j} \in \bar{a}^2, \quad j \in \langle 1, \dots, \beta \rangle$$

$$f^1_{t_e} \in \bar{f}^1, f^1_{t_e} \notin \bar{f}^2, \quad e \in \langle 1, \dots, \rho \rangle; \qquad f^2_{v_h} \notin \bar{f}^1, f^2_{v_h} \in \bar{f}^2, \quad h \in \langle 1, \dots, \omega \rangle$$

The *exclusive intersection* of two aggregates A_1 and A_2 is aggregate B, which includes only components of A_1 common for both aggregates and is formed according to the following rule:

1. If aggregates A_1 and A_2 are such as $A_1 \div A_2$ and

$$A_1 = [\![M_1, M_2, \dots, M_N | \langle a^1_1, a^1_2, \dots, a^1_l \rangle, \langle b^1_1, b^1_2, \dots, b^1_m \rangle, \dots, \langle w^1_1, w^1_2, \dots, w^1_n \rangle]\!]$$

$$A_2 = [\![M_1, M_2, \dots, M_N | \langle a^2_1, a^2_2, \dots, a^2_r \rangle, \langle b^2_1, b^2_2, \dots, b^2_q \rangle, \dots, \langle w^2_1, w^2_2, \dots, w^2_p \rangle]\!]$$

then B includes the elements of A_1, which are common for both A_1 and A_2, in every tuple:

$$B = A_1 \neg A_2 = [\![M_1, M_2, \dots, M_N | \langle a^1_{l_1}, \dots, a^1_{l_\alpha} \rangle, \langle b^1_{m_1}, \dots, b^1_{m_\gamma} \rangle, \dots, \langle w^1_{n_1}, \dots, w^1_{n_\lambda} \rangle]\!] \tag{5.14}$$

where

$$a^1_{l_i} \in \bar{a}^1, a^1_{l_i} \in \bar{a}^2, \quad i \in \langle 1, \dots, \alpha \rangle$$

$$b^1_{m_k} \in \bar{b}^1, b^1_{m_k} \in \bar{b}^2, \quad k \in \langle 1, \dots, \gamma \rangle$$

$$w^1_{n_u} \in \bar{w}^1, w^1_{n_u} \in \bar{w}^2, \quad u \in \langle 1, \dots, \lambda \rangle$$

2. If aggregates A_1 and A_2 are such as $A_1 \mathring{=} A_2$, then $B = A_1 \neg A_2 = [\![\emptyset \mid \langle\emptyset\rangle]\!] = A_\emptyset$.
3. If aggregates A_1 and A_2 are such as $A_1 \doteq A_2$ and

$$A_1 = [\![M_1, M_2^1, \ldots, M_x, \ldots, M_N^1 | \langle a_1^1, a_2^1, \ldots, a_l^1\rangle, \langle b_1, b_2, \ldots, b_m\rangle, \ldots, \langle f_1^1, f_2^1, \ldots, f_t^1\rangle, \ldots, \langle w_1, w_2, \ldots, w_n\rangle]\!]$$

$$A_2 = [\![M_1, M_2^2, \ldots, M_x, \ldots, M_K^2 | \langle a_1^2, a_2^2, \ldots, a_r^2\rangle, \langle d_1, d_2, \ldots, d_q\rangle, \ldots, \langle f_1^2, f_2^2, \ldots, f_v^2\rangle, \ldots, \langle z_1, z_2, \ldots, z_p\rangle]\!]$$

then B includes the elements of A_1, which are common for both A_1 and A_2, only in tuples of common sets; thus, the number of sets shortens:

$$B = A_1 \neg A_2 = [\![M_1, \ldots, M_x | \langle a_{l_1}^1, \ldots, a_{l_\alpha}^1\rangle, \ldots, \langle f_{t_1}^1, \ldots, f_{t_\rho}^1\rangle]\!] \tag{5.15}$$

where

$$a_{l_i}^1 \in \bar{a}^1, a_{l_i}^1 \in \bar{a}^2, \quad i \in \langle 1, \ldots, \alpha\rangle$$

$$f_{t_e}^1 \in \bar{f}^1, f_{t_e}^1 \in \bar{f}^2, \quad e \in \langle 1, \ldots, \rho\rangle$$

The difference between *exclusive intersection* and *intersection* is that the result of *intersection* of two aggregates A_1 and A_2 is the aggregate that includes common components of both aggregates. At the same time, the result of *exclusive intersection* is the aggregate that includes components of A_1, which are also present in A_2, but it does not include any components of A_2. For example, if we have two compatible aggregates A_1 and A_2 such as:

$$A_1 = [\![M_t, M_{\text{hr}} | \langle 36.4, 36.1, 36.3, 36.2, 36.5, 36.3\rangle, \langle 75, 76, 74, 73, 75, 75\rangle]\!]$$

$$A_2 = [\![M_t, M_{\text{hr}} | \langle 36.5, 36.5, 36.8, 36.6, 36.3, 36.4, 37.0, 36.5\rangle, \langle 74, 81, 76, 93, 97, 97, 96\rangle]\!]$$

Then, *intersection* gives us $A_1 \cap A_2 = [\![M_t, M_{\text{hr}} \mid \langle 36.4, 36.3, 36.5, 36.3, 36.5, 36.5, 36.3, 36.4, 36.5\rangle, \langle 76, 74, 74, 76\rangle]\!]$. At the same time, *exclusive intersection* is resulted in $A_1 \neg A_2 = [\![M_t, M_{\text{hr}} \mid \langle 36.4, 36.3, 36.5, 36.3\rangle, \langle 76, 74\rangle]\!]$.

If we have two quasi-compatible aggregates A_1 and A_3, where A_3 is defined as:

$$A_3 = [\![M_t, M_{\text{sp}} | \langle 36.5, 36.5, 36.8, 36.6, 36.3, 36.4, 37.0, 36.5\rangle, \langle 177, 159, 174, 155, 167, 150, 177, 135\rangle]\!]$$

Then, we get a result in a similar way but only for the first tuples because they both belong to the same set M_t:

$$A_1 \cap A_3 = [\![M_t | \langle 36.4, 36.3, 36.5, 36.3, 36.5, 36.5, 36.3, 36.4, 36.5\rangle]\!] \text{ and } A_1 \neg A_3 = [\![M_t | \langle 36.4, 36.3, 36.5, 36.3\rangle]\!]$$

The logical operations in ASA are noncommutative because the sequence order is important in tuples; this property distinguishes them from the logical operations on sets.

Ordering operations include sets ordering, ascending sorting, descending sorting, singling, extraction, and insertion [2].

The *sets ordering* operation reorders an aggregate according to a template aggregate. The template aggregate can be arbitrary, undefined, or empty.

Let the aggregate A be defined as:

$$A = [\![M_3, M_1, M_2, \ldots, M_N | \langle a_{i_3}^3\rangle_{i_3=1}^{n_3}, \langle a_{i_1}^1\rangle_{i_1=1}^{n_1}, \langle a_{i_2}^2\rangle_{i_2=1}^{n_2}, \ldots, \langle a_{i_N}^N\rangle_{i_N=1}^{n_N}]\!] \tag{5.16}$$

Besides, let the template aggregate A_{tem} be defined as:

$$A_{\text{tem}} = [\![M_1, M_2, M_3, \ldots, M_N | \langle \text{-} \rangle]\!] \tag{5.17}$$

Then, the result of sets ordering operation on the aggregates A and A_{tem} is aggregate B defined as:

$$B = A \vDash A_{\text{tem}} = [\![\{A_{\text{tem}}\} | \langle A\rangle]\!] = [\![M_1, M_2, M_3, \ldots, M_N | \langle a_{i_1}^1\rangle_{i_1=1}^{n_1}, \langle a_{i_2}^2\rangle_{i_2=1}^{n_2}, \langle a_{i_3}^3\rangle_{i_3=1}^{n_3}, \ldots, \langle a_{i_N}^N\rangle_{i_N=1}^{n_N}]\!] \tag{5.18}$$

The *sets ordering* operation can also be applied to two arbitrary aggregates A_1 and A_2, where A_2 is used as a template. The practical meaning of such operation is that we can reorder one aggregate (A_1) according to another aggregate (A_2). Thus, if $A_1(\div)A_2$, these aggregates become compatible as a result of sets ordering operation and we can further handle them in the same way, e.g. we can compare the first tuple of A_1 with the first tuple of A_2 and they will be comparable – which was not possible before reordering because the "old" first tuple of A_1 and the first tuple of A_2 belonged to different sets.

The result of sets ordering operation depends on the aggregates' compatibility:

- If $A_1 \div A_2$, then sets ordering operation is resulted in aggregate B, which is equal to A_1:

$$B = A_1 \vDash A_2 = [\![\{A_2\}|\langle A_1\rangle]\!] = [\![\{A_1\}|\langle A_1\rangle]\!] = A_1$$

- If $A_1 \doteq A_2$ and there is no hidden compatibility between A_1 and A_2, then the result of sets ordering operation is empty aggregate:
  $B = A_1 \vDash A_2 = [\![\{A_2\} \mid \langle\emptyset\rangle]\!]$.
- If $A_1 \mathring{=} A_2$ and there is no hidden compatibility between A_1 and A_2, then the result of sets ordering operation is aggregate B:

$$B = A_1 \vDash A_2 = [\![M_j^2|\langle a_i^j\rangle_{i=1}^{n_j}]\!]_{j=1}^{N} \tag{5.19}$$

  where $\langle a_i^j\rangle_{i=1}^{n_j} \in M_j^2$, $M_j^2 \subset \{A_2\}$, $\langle a_i^j\rangle_{i=1}^{n_j} \subset \langle A_1\rangle$, $\langle B\rangle \not\equiv \langle A_1\rangle$.
- If $A_1(\div)A_2$, then the result of sets ordering operation is aggregate B such that $B \div A_2$ and $\langle B\rangle \equiv \langle A_1\rangle$:

$$B = A_1 \vDash A_2 = [\![\{A_2\}|\langle a_i^j\rangle_{i=1}^{n_j}]\!]_{j=1}^{N} \tag{5.20}$$

  where $\langle a_i^j\rangle_{i=1}^{n_j} \in \langle A_1\rangle$.

For example, if we have aggregates A_1, A_2, and A_3 such as $A_1 \mathring{=} A_2, A_2(\div) A_3$, which are defined in the following way:

$$A_1 = [\![M_t, M_{\text{hr}}, M_{\text{sp}}|\langle 36.4,36.1,36.3,36.2,36.5,36.3\rangle, \langle 75,76,74,73,75,75\rangle, \langle 163,161,164,165,168\rangle]\!]$$

$$A_2 = [\![M_t, M_{\text{sp}}|\langle 36.5,36.5,36.8,36.6,36.3,36.4,37.0,36.5\rangle, \langle 171,183,175,183,167,167,176\rangle]\!]$$

$$A_3 = [\![M_{\text{sp}}, M_t|\langle 177,159,174,155,167,150,177,135\rangle, \langle 37.5,37.2,36.8,37.0,36.6\rangle]\!]$$

Then, we can obtain the following results of sets ordering operation:

$$A_1 \vDash A_2 = [\![M_t, M_{\text{sp}}\ |\langle 36.4,36.1,36.3,36.2,36.5,36.3\rangle, \langle 163,161,164,165,168\rangle]\!]$$

$$A_3 \vDash A_2 = [\![M_t, M_{\text{sp}}\ |\langle 37.5,37.2,36.8,37.0,36.6\rangle, \langle 177,159,174,155,167,150,177,135\rangle]\!]$$

The sorting operations are *ascending sorting* and *descending sorting*. These operations enable reordering of all tuples according to new – sorted – elements order (ascending or descending) of a certain tuple (called a *primary tuple*) among all tuples of the aggregate.

Let $A_1 = [\![M_j|\langle a_i^j\rangle_{i=1}^{n_j}]\!]_{j=1}^{N}$ and $\exists k$ such as $1 < k < N, k \neq 2$ and $n_1 > n_k > n_N, n_2 = n_k$. Then, the result of ascending sorting operation of A_1 according to the elements of tuple $\bar{a}^k$ is aggregate B such as:

$$\begin{aligned} B = A_1 \uparrow \bar{a}^k = [\![& M_1, M_2, \dots, M_k, \dots, M_N|\langle a_\alpha^1, a_\beta^1, \dots, a_\nu^1, \dots, a_\omega^1, a_{n_k+1}^1, \dots, a_{n_1}^1\rangle, \\ & \langle a_\alpha^2, a_\beta^2, \dots, a_\nu^2, \dots, a_\omega^2\rangle, \dots, \langle a_\alpha^k, a_\beta^k, \dots, a_\nu^k, \dots, a_\omega^k\rangle, \dots, \langle a_\alpha^N, a_\beta^N, \dots, a_\nu^N\rangle]\!] \end{aligned} \tag{5.21}$$

where $a_\alpha^k < a_\beta^k < \dots < a_\nu^k < \dots < a_\omega^k$; $a_m^j \in \langle a_i^j\rangle_{i=1}^{n_j}$, $j = 1 \dots N$, $m \in [\alpha, \beta, \dots, \nu, \dots, \omega]$, $1 \le m \le n$ and $n = n_k$ if $n_j \ge n_k$ or $n = n_j$ if $n_j < n_k$.

The result of descending sorting operation of A_1 according to the elements of tuple $\bar{a}^k$ is aggregate B such as:

$$B = A_1 \downarrow \bar{a}^k = [\![M_1, M_2, \dots, M_k, \dots, M_N | \langle a^1_\omega, \dots, a^1_\nu, \dots, a^1_\beta, a^1_\alpha, a^1_{n_k+1}, \dots, a^1_{n_1} \rangle,$$
$$\langle a^2_\omega, \dots, a^2_\nu, \dots, a^2_\beta, a^2_\alpha \rangle, \dots, \langle a^k_\omega, \dots, a^k_\nu, \dots, a^k_\beta, a^k_\alpha \rangle, \dots, \langle a^N_\nu, \dots, a^N_\beta, a^N_\alpha \rangle]\!] \tag{5.22}$$

If $k = 1$, $k = 2$, or $k = N$, the sorting operation is fulfilled by the same principle.

If $n_1 = n_2 = \dots = n_k = \dots = n_N$, the result of ascending sorting operation is:

$$B = A_1 \uparrow \bar{a}^k = [\![M_j | \langle a^j_\alpha, a^j_\beta, \dots, a^j_\nu, \dots, a^j_\omega \rangle]\!]^N_{j=1} \tag{5.23}$$

and the result of descending sorting operation is:

$$B = A_1 \downarrow \bar{a}^k = [\![M_j | \langle a^j_\omega, \dots, a^j_\nu, \dots, a^j_\beta, a^j_\alpha \rangle]\!]^N_{j=1} \tag{5.24}$$

A variant of the sorting operations is the sorting operations with appending (*ascending sorting with appending* and *descending sorting with appending*), which allow to increase the length of shorter tuples according to the length of the primary tuple by adding a value x ($x \in [\emptyset, \_, q]$, $q \in M_j$, $1 \le j \le N$) either to the end or to the beginning of the shorter tuples.

The result of ascending sorting with appending at the end of the shorter tuples of aggregate A_1 is aggregate B:

$$B = A_1 \uparrow (\bar{a}^k, x) = [\![M_1, M_2, \dots, M_k, \dots, M_N | \langle a^1_\alpha, a^1_\beta, \dots, a^1_\nu, \dots, a^1_\omega, a^1_{n_k+1}, \dots, a^1_{n_1} \rangle,$$
$$\langle a^2_\alpha, a^2_\beta, \dots, a^2_\nu, \dots, a^2_\omega \rangle, \dots, \langle a^k_\alpha, a^k_\beta, \dots, a^k_\nu, \dots, a^k_\omega \rangle, \dots, \langle a^N_\alpha, a^N_\beta, \dots, a^N_\nu, x, \dots, x \rangle]\!] \tag{5.25}$$

where $|\langle a^N_\alpha, a^N_\beta, \dots, a^N_\nu, x, \dots, x \rangle| = n_k$.

The result of ascending sorting with appending at the beginning of the shorter tuples of the aggregate A_1 is aggregate B:

$$B = A_1 \uparrow (x, \bar{a}^k) = [\![M_1, M_2, \dots, M_k, \dots, M_N | \langle a^1_\alpha, a^1_\beta, \dots, a^1_\nu, \dots, a^1_\omega, a^1_{n_k+1}, \dots, a^1_{n_1} \rangle,$$
$$\langle a^2_\alpha, a^2_\beta, \dots, a^2_\nu, \dots, a^2_\omega \rangle, \dots, \langle a^k_\alpha, a^k_\beta, \dots, a^k_\nu, \dots, a^k_\omega \rangle, \dots, \langle x, \dots, x, a^N_\alpha, a^N_\beta, \dots, a^N_\nu \rangle]\!] \tag{5.26}$$

where $|\langle x, \dots, x, a^N_\alpha, a^N_\beta, \dots, a^N_\nu \rangle| = n_k$.

The result of descending sorting with appending at the end of the shorter tuples of aggregate A_1 is aggregate B:

$$B = A_1 \downarrow (\bar{a}^k, x) = [\![M_1, M_2, \dots, M_k, \dots, M_N | \langle a^1_\omega, \dots, a^1_\nu, \dots, a^1_\beta, a^1_\alpha, a^1_{n_k+1}, \dots, a^1_{n_1} \rangle,$$
$$\langle a^2_\omega, \dots, a^2_\nu, \dots, a^2_\beta, a^2_\alpha \rangle, \dots, \langle a^k_\omega, \dots, a^k_\nu, \dots, a^k_\beta, a^k_\alpha \rangle, \dots, \langle a^N_\nu, \dots, a^N_\beta, a^N_\alpha, x, \dots, x \rangle]\!] \tag{5.27}$$

The result of descending sorting with appending at the beginning of the shorter tuples of the aggregate A_1 is aggregate B:

$$B = A_1 \downarrow (x, \bar{a}^k) = [\![M_1, M_2, \dots, M_k, \dots, M_N | \langle a^1_\omega, \dots, a^1_\nu, \dots, a^1_\beta, a^1_\alpha, a^1_{n_k+1}, \dots, a^1_{n_1} \rangle,$$
$$\langle a^2_\omega, \dots, a^2_\nu, \dots, a^2_\beta, a^2_\alpha \rangle, \dots, \langle a^k_\omega, \dots, a^k_\nu, \dots, a^k_\beta, a^k_\alpha \rangle, \dots, \langle x, \dots, x, a^N_\nu, \dots, a^N_\beta, a^N_\alpha \rangle]\!] \tag{5.28}$$

If aggregate A_1 is defined as:

$$A_1 = [\![M_1, \dots, M_k, \dots, M_N | \langle a^1_1, \dots, a^1_m, \dots, a^1_{m+p}, a^1_{m+p+1}, \dots, a^1_{n_1} \rangle, \dots,$$
$$\langle a^k_1, \dots, a^k_m, \dots, a^k_{m+p}, a^k_{m+p+1}, \dots, a^k_{n_k} \rangle, \dots, \langle a^N_1, \dots, a^N_m, \dots, a^N_{m+p}, a^N_{m+p+1}, \dots, a^N_{n_N} \rangle]\!] \tag{5.29}$$

where $1 \le k \le N$, and let $\exists m_l, \forall l$ such as $a^k_m = a^k_{m+1} = \dots = a^k_{m+p}$, $1 \le m \le (n_k - p)$, $1 \le p \le n_k$, then the result of *singling* operation on aggregate A_1 by the tuple $\bar{a}^k$ is aggregate B such as:

$$B = A_1 \parallel \bar{a}^k = [\![M_1, \dots, M_k, \dots, M_N | \langle a^1_1, \dots, a^1_m, a^1_{m+p+1}, \dots, a^1_{n_1} \rangle, \dots,$$
$$\langle a^k_1, \dots, a^k_m, a^k_{m+p+1}, \dots, a^k_{n_k} \rangle, \dots, \langle a^N_1, \dots, a^N_m, a^N_{m+p+1}, \dots, a^N_{n_N} \rangle]\!] \tag{5.30}$$

If there is aggregate A_1 defined as:

$$A_1 = [\![M_1, \ldots, M_k, \ldots, M_N | \langle a^1_{i_1} \rangle^{n_1}_{i_1=1}, \ldots, \langle a^k_1, \ldots, a^k_{m-1}, a^k_m, a^k_{m+1}, \ldots, a^k_{n_k} \rangle, \ldots, \langle a^N_{i_N} \rangle^{n_N}_{i_N=1}]\!] \tag{5.31}$$

then the result of *extraction* operation of the element a^k_m from aggregate A_1 is aggregate B such as:

$$B = A_1 \ltimes a^k_m = [\![M_1, \ldots, M_k, \ldots, M_N | \langle a^1_{i_1} \rangle^{n_1}_{i_1=1}, \ldots, \langle a^k_1, \ldots, a^k_{m-1}, a^k_{m+1}, \ldots, a^k_{n_k} \rangle, \ldots, \langle a^N_{i_N} \rangle^{n_N}_{i_N=1}]\!] \tag{5.32}$$

The operation of *conditional extraction* is defined on the assumption of a given condition such as $a^k_m = d,\ a^k_m < d$ or $a^k_m > d,\ \forall d \in M_k$. For example: $B = A_1 \ltimes a^k_m|_{a^k_m = d}$.

If there is aggregate A_1 defined as:

$$A_1 = [\![M_1, \ldots, M_k, \ldots, M_N | \langle a^1_{i_1} \rangle^{n_1}_{i_1=1}, \ldots, \langle a^k_1, \ldots, a^k_{m-1}, a^k_m, a^k_{m+1}, \ldots, a^k_{n_k} \rangle, \ldots, \langle a^N_{i_N} \rangle^{n_N}_{i_N=1}]\!] \tag{5.33}$$

where $1 \le m \le N$ and let $\exists d$ such as $d \in M_k,\ 1 \le k \le N$, then the result of *insertion* operation of d to aggregate A_1 can be obtained by using two equivalent ways defined in Eqs. (5.34) and (5.35).

$$\begin{aligned} B &= A_1 \rtimes (d \prec a^k_m) = A_1 \rtimes (d \succ a^k_{m-1}) \\ &= [\![M_1, \ldots, M_k, \ldots, M_N | \langle a^1_{i_1} \rangle^{n_1}_{i_1=1}, \ldots, \langle a^k_1, \ldots, a^k_{m-1}, d, a^k_m, a^k_{m+1}, \ldots, a^k_{n_k} \rangle, \ldots, \langle a^N_{i_N} \rangle^{n_N}_{i_N=1}]\!] \end{aligned} \tag{5.34}$$

$$\begin{aligned} B &= A_1 \rtimes (d \succ a^k_m) = A_1 \rtimes (d \prec a^k_{m+1}) \\ &= [\![M_1, \ldots, M_k, \ldots, M_N | \langle a^1_{i_1} \rangle^{n_1}_{i_1=1}, \ldots, \langle a^k_1, \ldots, a^k_{m-1}, a^k_m, d, a^k_{m+1}, \ldots, a^k_{n_k} \rangle, \ldots, \langle a^N_{i_N} \rangle^{n_N}_{i_N=1}]\!] \end{aligned} \tag{5.35}$$

The operation of *conditional insertion* is defined on the assumption of a given condition such as $d = a^k_m,\ d < a^k_m$ or $d > a^k_m,\ \forall d \in M_k$. For example, $B = A_1 \rtimes (d \prec a^k_m)|_{d = a^k_m,\ m = m_1 \ldots m_2}$.

The ordering operations in ASA allow us to reorder both elements in tuples and tuples in aggregates (multi-images).

5.4 Relations and Digital Intervals

Relations in ASA include relations between tuple elements, relations between tuples, and relations between aggregates. Relations between tuple elements are *is greater*, *is less*, *is equal*, *proceeds*, and *succeeds*. The first three relations are based on element value and the last two relations concern elements position in a tuple. Naturally, elements must belong to the same tuple. Relations between tuples enable the following types of tuple comparison: arithmetical comparison, frequency comparison, and interval comparison. Arithmetical comparison is element-wise and based on the following relations: *is strictly greater*, *is majority-vote greater*, *is strictly less*, *is majority-vote less*, *is strictly equal*, and *is majority-vote equal*.

Frequency comparison is based on the following relations: *is thicker*, *is rarer*, and *is equally frequent*. Interval comparison is based on the following temporal relations: *coincides with*, *is before*, *is after*, *meets*, *is met by*, *overlaps*, *is overlapped by*, *during*, *contains*, *starts*, *is started by*, *finishes*, and *is finished by*.

Relations between tuples of aggregates are identical to relations between single tuples defined above: relations of arithmetical comparison, relations of frequency comparison, and relations of interval comparison. However, the possibility of their application depends on the aggregates' compatibility: relations between tuples can be applied only to compatible and quasi-compatible aggregates. Hiddenly compatible aggregates must be first transformed to compatible ones [2] and then a relation between tuples can be applied to them.

Because a multi-image, by definition, is an aggregate, all types of relations defined in ASA can be applied to multi-images as well.

A time tuple in a multi-image, which is defined by (5.2), can be considered as an interval. However, in contrast with a classical interval, the time tuple consists of a finite number of discrete values. To show the difference between the time tuple and a classical interval, let us first discuss a classical approach of interval-based temporal logic.

One of the pioneering works related to interval processing is [7], where Allen presented interval algebra and interval-based temporal logic, including relations between intervals. If X and Y are intervals such as $X = [x^-, x^+]$ and $Y = [y^-, y^+]$, the relations between them are defined as shown in Table 5.1.

In [8], Allen and Hayes axiomatized a theory of time in terms of intervals and the single relation *meet*. They extended Allen's interval-based theory by formally defining the beginnings and endings of intervals, which have properties normally associated with points. The authors distinguished between these point-like objects and the concept of a moment as hypothesized in discrete time models.

In these and other related works [9–13], the relations of Allen's interval-based theory are to be applied to an interval $[x^-, x^+]$, where $x^- \leq x^+$. However, in ASA, we operate with discrete values and we cannot use Allen's interval-based theory approaches directly. For example, when we measure data for further composing of a multi-image, we operate with discrete time values, which being considered all together can be defined as a discrete time interval as it consists of specific discrete time points when the data values have been obtained. Thus, we need to make a transition from an *interval* [7–13] to a *discrete interval*.

Let us define a *discrete interval* as a tuple $\bar{t}$, whose elements are unique discrete values $\bar{t} = \langle t_i \rangle_{i=1}^{n}$ such as that either $t_i < t_{i+1}$ or $t_i > t_{i+1}$ is true for all pairs (t_i, t_{i+1}), $\forall i \in [1 \ldots n-1]$, $t_i \in \mathbb{R}$. Thus, a discrete interval is a strictly increasing or decreasing finite sequence in $\mathbb{R}$.

Then, discrete intervals can be a subject of relations similar to those introduced by Allan in [7]. These temporal relations between two discrete intervals are coincides with, is before, is after, meets, is met by, overlaps, is overlapped by, during, contains, starts, is started by, finishes, and is finished by. Let us define these relations. Note that for consistency with works on Allen's interval algebra, we refer to in this research; hereinafter, we use notion for relations between digital intervals similar to that used in Allen's interval algebra (e.g. e_d).

Table 5.1 Allen's interval algebra relations.

Relation	Notation	Definition
X before Y	$b(X, Y)$	$x^+ < y^-$
X after Y	$bi(X, Y)$	$x^- > y^+$
X equal Y	$e(X, Y)$	$x^- = y^-$ and $x^+ = y^+$
X meets Y	$m(X, Y)$	$x^+ = y^-$
X is met by Y	$mi(X, Y)$	$x^- = y^+$
X overlaps Y	$o(X, Y)$	$x^- < y^-$ and $x^+ > y^-$ and $x^+ < y^+$
X is overlapped by Y	$oi(X, Y)$	$x^- > y^-$ and $x^- < y^+$ and $x^+ > y^+$
X during Y	$d(X, Y)$	$x^- > y^-$ and $x^+ < y^-$
X contains Y	$di(X, Y)$	$x^- < y^-$ and $x^- > y^+$
X starts Y	$s(X, Y)$	$x^- = y^-$ and $x^+ < y^+$
X is started by Y	$si(X, Y)$	$x^- = y^-$ and $x^+ > y^+$
X finishes Y	$f(X, Y)$	$x^- > y^-$ and $x^+ = y^+$
X is finished by Y	$fi(X, Y)$	$x^- < y^-$ and $x^+ = y^+$

For two discrete intervals $\bar{t}^1$ and $\bar{t}^2$, we define:

- The relation $\bar{t}^1$ *coincides with* $\bar{t}^2$ as:

$$e_d(\bar{t}^1, \bar{t}^2) \text{ if } t_1^1 = t_1^2, t_{n_1}^1 = t_{n_2}^2 \text{ and } n_1 = n_2 \tag{5.36}$$

- The relation $\bar{t}^1$ *is before* $\bar{t}^2$ as:

$$b_d(\bar{t}^1, \bar{t}^2) \quad \text{if } t_{n_1}^1 < t_1^2 \tag{5.37}$$

- The relation $\bar{t}^1$ *is after* $\bar{t}^2$ as:

$$bi_d(\bar{t}^1, \bar{t}^2) \quad \text{if } t_1^1 > t_{n_2}^2 \tag{5.38}$$

- The relation $\bar{t}^1$ *meets* $\bar{t}^2$ as:

$$m_d(\bar{t}^1, \bar{t}^2) \quad \text{if } t_{n_1}^1 = t_1^2 \tag{5.39}$$

- The relation $\bar{t}^1$ *is met by* $\bar{t}^2$ as:

$$mi(\bar{t}^1, \bar{t}^2) \quad \text{if } t_1^1 = t_{n_2}^2 \tag{5.40}$$

- The relation $\bar{t}^1$ *overlaps* $\bar{t}^2$ as:

$$o_d(\bar{t}^1, \bar{t}^2) \quad \text{if } t_1^1 < t_1^2 \text{ and } t_{n_1}^1 < t_{n_2}^2 \text{ and } t_1^2 < t_{n_1}^1 \tag{5.41}$$

- The relation $\bar{t}^1$ *is overlapped by* $\bar{t}^2$ as:

$$oi_d(\bar{t}^1, \bar{t}^2) \quad \text{if } t_1^2 < t_1^1 \text{ and } t_{n_2}^2 < t_{n_1}^1 \text{ and } t_1^1 < t_{n_2}^2 \tag{5.42}$$

- The relation $\bar{t}^1$ *during* $\bar{t}^2$ as:

$$d_d(\bar{t}^1, \bar{t}^2) \quad \text{if } t_1^1 > t_1^2 \text{ and } t_{n_1}^1 < t_{n_2}^2 \tag{5.43}$$

- The relation $\bar{t}^1$ *contains* $\bar{t}^2$ as:

$$di_d(\bar{t}^1, \bar{t}^2) \quad \text{if } t_1^1 < t_1^2 \text{ and } t_{n_1}^1 > t_{n_2}^2 \tag{5.44}$$

- The relation $\bar{t}^1$ *starts* $\bar{t}^2$ as:

$$s_d(\bar{t}^1, \bar{t}^2) \quad \text{if } t_1^1 = t_1^2 \text{ and } t_{n_1}^1 < t_{n_2}^2 \tag{5.45}$$

- The relation $\bar{t}^1$ *is started by* $\bar{t}^2$ as:

$$si_d(\bar{t}^1, \bar{t}^2) \quad \text{if } t_1^1 = t_1^2 \text{ and } t_{n_1}^1 > t_{n_2}^2 \tag{5.46}$$

- The relation $\bar{t}^1$ *finishes* $\bar{t}^2$ as:

$$f_d(\bar{t}^1, \bar{t}^2) \quad \text{if } t_1^1 > t_1^2 \text{ and } t_{n_1}^1 = t_{n_2}^2 \tag{5.47}$$

- The relation $\bar{t}^1$ *is finished by* $\bar{t}^2$ as:

$$fi_d(\bar{t}^1, \bar{t}^2) \quad \text{if } t_1^1 < t_1^2 \text{ and } t_{n_1}^1 = t_{n_2}^2 \tag{5.48}$$

Note that since $\bar{t}^1$ and $\bar{t}^2$ are *discrete* intervals, there is no requirement that t_i^1 and t_j^2 $(1 < i < n_1, 1 < j < n_2)$ must coincide in such relations as $e_d(\bar{t}^1, \bar{t}^2)$, $o_d(\bar{t}^1, \bar{t}^2)$, $oi_d(\bar{t}^1, \bar{t}^2)$, $d_d(\bar{t}^1, \bar{t}^2)$, $di_d(\bar{t}^1, \bar{t}^2)$, $s_d(\bar{t}^1, \bar{t}^2)$, $si_d(\bar{t}^1, \bar{t}^2)$, $f_d(\bar{t}^1, \bar{t}^2)$, and $fi_d(\bar{t}^1, \bar{t}^2)$. For example, if we have two tuples $\bar{t}^1 = \langle t_1^1, t_2^1, t_3^1, t_4^1 \rangle$ and $\bar{t}^2 = \langle t_1^2, t_2^2, t_3^2, t_4^2, t_5^2 \rangle$ as depicted in Figure 5.1.

Then, $\bar{t}^1$ *coincides with* $\bar{t}^2$ because $t_1^1 = t_1^2$ and $t_4^1 = t_5^2$.

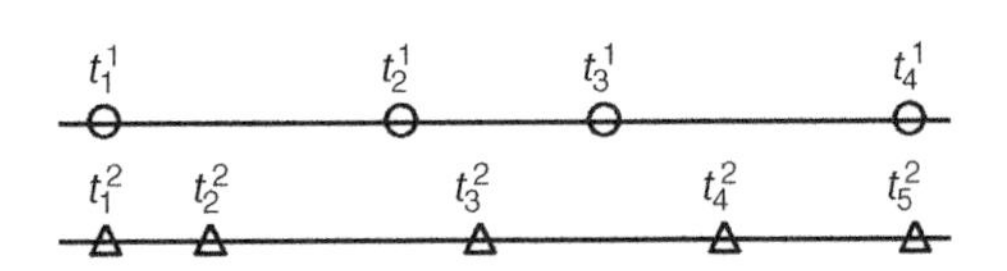

Figure 5.1 An example of two coinciding discrete intervals.

In this work, we also introduce the relation *between*, which can be applied to any number of discrete intervals:

- $w_d^{(\alpha,\beta)}(\bar{t}^1)$ means that a discrete interval $\bar{t}^1$ is *between* the discrete values α and β;
- $w_d^{(\alpha,\beta)}(\bar{t}^1, \bar{t}^2)$ means that both discrete intervals $\bar{t}^1$ and $\bar{t}^2$ are *between* the discrete values α and β;
- $w_d^{(\alpha,\beta)}(\bar{t}^1, \dots, \bar{t}^n)$ means that all discrete intervals $\bar{t}^j (j \in [1 \dots n])$ are *between* the discrete values α and β.

Thus, the relation *between* for two tuples $\bar{t}^1$ and $\bar{t}^2$ can be defined as follows:

$$w_d^{(\alpha,\beta)}(\bar{t}^1, \bar{t}^2) \quad \text{if } \alpha \le t_1^1, \alpha \le t_1^2, t_{n_1}^1 \le \beta, t_{n_2}^2 \le \beta \tag{5.49}$$

where α and β are given values, $\alpha, \beta \in T$; T is the time value set.

The relation *between* for two discrete intervals can be considered as a generalized form of the relation *coincides with*, because the relation *between* (5.49) is equivalent to the relation *coincides with* (5.36) if both $t_1^1 = t_1^2$ and $t_{n_1}^1 = t_{n_2}^2$.

5.5 Data Synchronization

Because a multi-image is a digital representation of a real-world object (process, phenomenon, event, subject), data tuples of different modalities, which are obtained from different sources (sensors, cloud storages, local storages, computing resources, etc.), need to be synchronized between each other to enable constructing the proper multi-image as well as to compare and analyze several multi-images in the same processing procedure. Mathematical models of data synchronization will enable further development of algorithms as well as software for multimodal data processing.

Let us consider two multi-images I_1 and I_2, which present a state of the same object of study and are defined as:

$$I_1 = [\![T, M_1 | \langle t_i^1 \rangle, \langle a_i^1 \rangle]\!]_{i=1}^{n_1}$$
$$I_2 = [\![T, M_2 | \langle t_i^2 \rangle, \langle a_i^2 \rangle]\!]_{i=1}^{n_2} \tag{5.50}$$

We assume that one characteristic of this object has been measured in time moments $\langle t_i^1 \rangle_{i=1}^{n_1}$, and as a result, we obtained the data tuple $\langle a_i^1 \rangle_{i=1}^{n_1}$; another characteristic has been measured in time moments $\langle t_i^2 \rangle_{i=1}^{n_2}$, and as a result, we obtained the data tuple $\langle a_i^2 \rangle_{i=1}^{n_2}$. Since I_1 and I_2 describe different sides of behavior of the same object, our task is to compose a joint multi-image I that consists of the joint time tuple $\bar{t} = \langle t_j \rangle_{j=1}^{n}$ and the synchronized data tuples $\langle d_j^1 \rangle_{j=1}^{n}$ and $\langle d_j^2 \rangle_{j=1}^{n}$. The tuple $\langle d_j^1 \rangle_{j=1}^{n}$ includes values of $\langle a_i^1 \rangle_{i=1}^{n_1}$ and empty elements and is formed according to the rule:

$$d_j^1 = \begin{cases} a_i^1 & \text{if } t_j = t_i^1 \\ \emptyset & \text{otherwise} \end{cases} \tag{5.51}$$

Similarly, the tuple $\langle d_j^2 \rangle_{j=1}^{n}$ includes values of $\langle a_i^2 \rangle_{i=1}^{n_2}$ and empty elements and is formed according to the rule:

$$d_j^2 = \begin{cases} a_i^2 & \text{if } t_j = t_i^2 \\ \emptyset & \text{otherwise} \end{cases} \tag{5.52}$$

In general, the joint multi-image I is formed as a result of three operations, viz., *union* defined by (5.6), *sorting* defined by (5.21), and *singling* defined by (5.30):

$$I = ((I_1 \cup I_2) \uparrow \bar{t}) \parallel \bar{t} = [\![T, M_1, M_2 | \langle t_j \rangle, \langle d_j^1 \rangle, \langle d_j^2 \rangle]\!]_{j=1}^{n} \tag{5.53}$$

The *union* enables consolidation of two multi-images I_1 and I_2 in one multi-image $I_{I_1 \cup I_2}$, but if time tuples in I_1 and I_2, i.e. $\bar{t}^1 = \langle t_i^1 \rangle_{i=1}^{n_1}$ and $\bar{t}^2 = \langle t_i^2 \rangle_{i=1}^{n_2}$, include the same values, then $I_{I_1 \cup I_2}$ will include duplicated elements and, thus, the synchronization of data tuples $\langle a_i^1 \rangle_{i=1}^{n_1}$ and $\langle a_i^2 \rangle_{i=1}^{n_2}$ will be incorrect. To avoid it, we must sort $I_{I_1 \cup I_2}$ by time tuple values by using *sorting* operation and, next, remove duplicated elements in the joint time tuple $\bar{t} = \langle t_j \rangle_{j=1}^{n}$ by using *singling* operation.

Let us consider the following practical task as an example. We assume that two parameters, viz., a temperature and an erythrocyte sedimentation rate, have being measured for the same patient during four-week health status monitoring. Because these measurements are of different nature and fulfilled by different hospital units, we obtain two multi-images, even if measurements of both types were obtained at the same days of a month:

$$I_1 = [\![T, M_t | \langle 2, 9, 16, 23 \rangle, \langle 37.7, 37.5, 37.3, 36.6 \rangle]\!]$$
$$I_2 = [\![T, M_{\text{esr}} | \langle 2, 9, 16, 23 \rangle, \langle 17, 15, 15, 13 \rangle]\!]$$

where

$M_t = [36.0, \ldots, 39.9]$ is a set of temperature values (°C);
$M_{\text{esr}} = [2, \ldots, 20]$ is a set of erythrocyte sedimentation rate values (mm/h);
$T = [1, \ldots, 31]$ is a set of time values (day of a month).

If we try to apply only *union* operation to I_1 and I_2, we obtain wrong synchronization because data of different types do not correspond each other: $I_{I_1 \cup I_2} = I_1 \cup I_2 = [\![T, M_t, M_{\text{esr}} | \langle 2, 9, 16, 23, 2, 9, 16, 23 \rangle, \langle 37.7, 37.5, 37.3, 36.6 \rangle, \langle 17, 15, 15, 13 \rangle]\!]$.

To correct it, we need to sort $I_{I_1 \cup I_2}$ in ascending order of time values and then remove duplicated time values:

$$I = ((I_1 \cup I_2) \uparrow \bar{t}) \parallel \bar{t} = [\![T, M_t, M_{\text{esr}} | \langle 2, 9, 16, 23 \rangle, \langle 37.7, 37.5, 37.3, 36.6 \rangle, \langle 17, 15, 15, 13 \rangle]\!].$$

Now, all data correspond each other, e.g. we can see that on the 16th day of monitoring, the patient had the body temperature 37.3 °C and the erythrocyte sedimentation rate in his blood test was 15 mm/h.

In this example, we considered the particular case when time values $\langle t_i^1 \rangle_{i=1}^{n_1}$ fully coincide with $\langle t_i^2 \rangle_{i=1}^{n_2}$; however, in a general case, time tuples can be connected with any temporal relation defined by (5.36)–(5.49). Thus, to find time values t_j satisfying the condition in (5.51) and/or (5.52), we need to analyze interval relations between time value tuples $\bar{t}^1$ and $\bar{t}^2$ and compose the joint time value tuple $\bar{t}$, which includes all time values from both $\bar{t}^1$ and $\bar{t}^2$ but do not duplicate them if some of time values in $\bar{t}^1$ and $\bar{t}^2$ coincide. To do this, we need to compare value intervals of $\bar{t}^1$ and $\bar{t}^2$.

The simplest case is when the measurements of both $\bar{a}^1$ and $\bar{a}^2$ data tuples have been fulfilled simultaneously (hereinafter, we suppose that the data values are obtained from sensors as a result of measuring certain parameters of a physical process; however, the data can also be obtained as a result of modeling, processing, prediction, simulation, etc.). It means that $\bar{t}^1$ and $\bar{t}^2$ are connected by the relation *coincides with*. In general, the relation *coincides with* between two tuples $\bar{t}^1$ and $\bar{t}^2$ is defined in Eq. (5.36). This relation enables many different subcases depending on a number of values in each data tuple. Let us consider all possible subcases (Figures 5.2–5.4).

The first group of subcases for $e_d(\bar{t}^1, \bar{t}^2)$ is when n_1 is an even value ($n_1 \bmod 2 = 0$) and $\bar{t}^1 \sim \bar{t}^2$, i.e. $\bar{t}^1$ *is equally frequent* to $\bar{t}^2$; it means that $|\bar{t}^1| = |\bar{t}^2|$ (i.e. $n_1 = n_2$).

If $t_i^1 = t_i^2, \forall i \in [1 \ldots n_1]$, then $n = n_1$ and the tuple values of the multi-image I as introduced in (5.2) are defined in (5.54) and depicted in Figure 5.2a.

$$\begin{aligned} t_i &= t_i^1 \\ d_i^1 &= a_i^1 \\ d_i^2 &= a_i^2 \end{aligned} \tag{5.54}$$

The case when $t_i^1 = t_i^2$ and $t_{i+1}^1 > t_{i+1}^2$ can be illustrated by Figure 5.2b. Such mutual alignment of the time moments t_i^1, t_i^2, t_{i+1}^1, and t_{i+1}^2, when data values a_i^1, a_i^2, a_{i+1}^1, and a_{i+1}^2 have been measured, requires synchronization

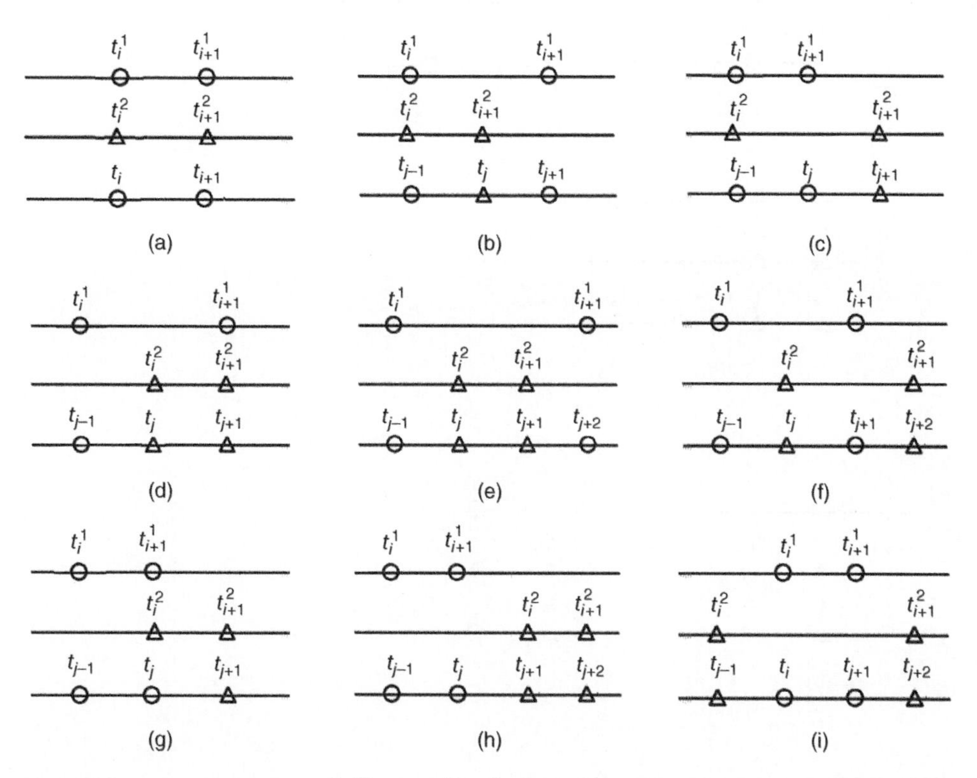

Figure 5.2 The subcases for $e_d(\bar{t}^1, \bar{t}^2)$ when $\bar{t}^1 \sim \bar{t}^2$. t_i^1 and t_{i+1}^1 are elements of the time tuple $\bar{t}^1$; t_i^2 and t_{i+1}^2 are elements of the time tuple $\bar{t}^2$; t_{j-1}, t_j, t_{j+1}, and t_{j+2} are elements of the synchronized time tuple $\bar{t}$. For two measurement processes: (a) measurements of both parameters are simultaneous at the current and the next moments of time; (b) the current measurements of both parameters are simultaneous, but the next measurement of the 2nd parameter happens earlier than the next measurement of the 1st parameter; (c) the current measurements of both parameters are simultaneous, but the next measurement of the 1st parameter happens earlier than the next measurement of the 2nd parameter; (d) the current measurement of the 1st parameter happens earlier than the current measurement of the 2nd parameter, but the next measurements of both parameters are simultaneous; (e) the current measurement of the 1st parameter happens earlier than the current measurement of the 2nd parameter, but the next measurement of the 1st parameter happens later than the next measurement of the 2nd parameter; (f) the current and the next measurements of the 1st parameter happen earlier than the current and the next measurements of the 2nd parameter, respectively; (g) the next measurement of the 1st parameter and the current measurement of the 2nd parameter are simultaneous; (h) both measurements of the 1st parameter happen earlier than both measurements of the 2nd parameter; (i) the current measurement of the 1st parameter happens later than the current measurement of the 2nd parameter, but the next measurement of the 1st parameter happens earlier than the next measurement of the 2nd parameter; (j) the current measurement of the 1st parameter happens later than the current measurement of the 2nd parameter, but the next measurements of both parameters are simultaneous; (k) the current and the next measurements of the 1st parameter happen later than the current and the next measurements of the 2nd parameter, respectively; (l) the current measurement of the 1st parameter and the next measurement of the 2nd parameter are simultaneous; (m) both measurements of the 1st parameter happen later than both measurements of the 2nd parameter.

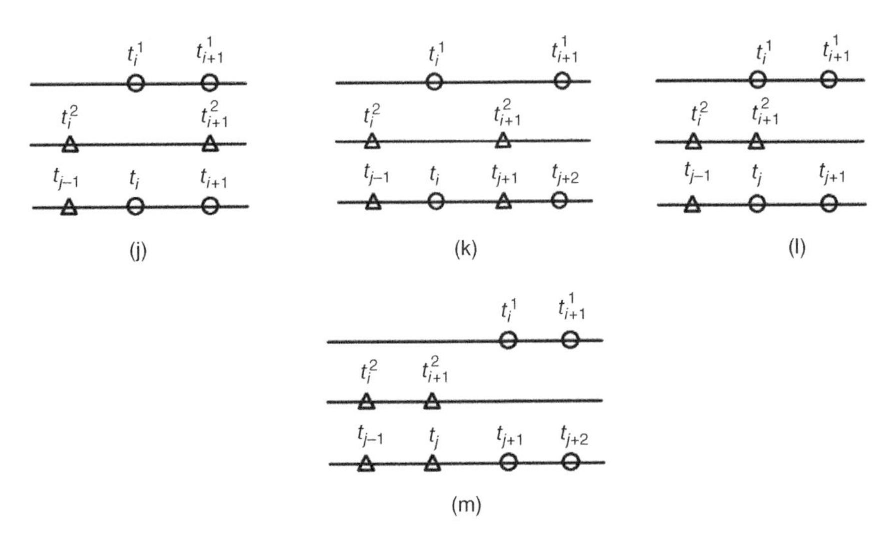

Figure 5.2 *(Continued)*

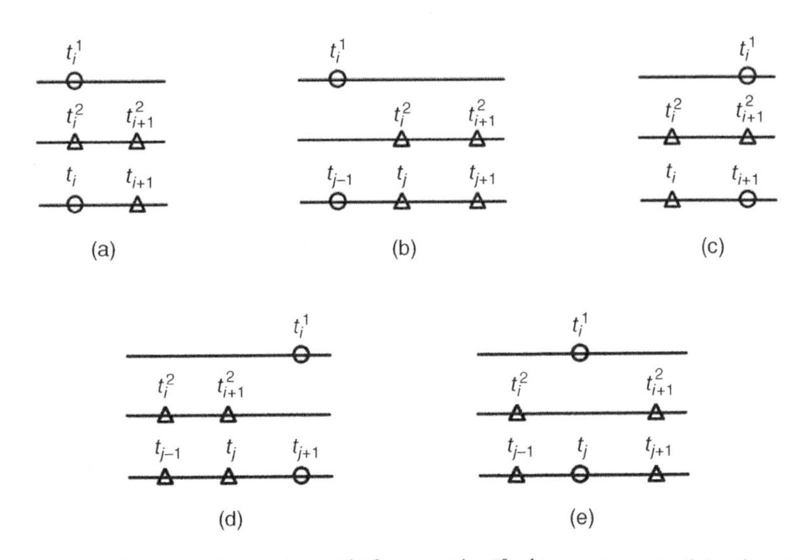

Figure 5.3 The subcases for $e_d(\bar{t}^1, \bar{t}^2)$ when $\bar{t}^1 \lhd \bar{t}^2$. t_i^1 is an element of the time tuple $\bar{t}^1$; t_i^2 and t_{i+1}^2 are elements of the time tuple $\bar{t}^2$; t_{j-1}, t_j, and t_{j+1} are elements of the synchronized time tuple $\bar{t}$. For two measurement processes: (a) the current measurements of both parameters are simultaneous; (b) the current measurement of the 1st parameter happens earlier than the current measurement of the 2nd parameter; (c) the current measurement of the 1st parameter and the next measurement of the 2nd parameter are simultaneous; (d) the current measurement of the 1st parameter happens later than the next measurement of the 2nd parameter; (e) the current measurement of the 1st parameter happens later than the current measurement and earlier than the next measurement of the 2nd parameter.

of tuple elements which can be stated as:

$$
\begin{array}{lll}
t_{j-1} = t_i^1 & t_j = t_{i+1}^2 & t_{j+1} = t_{i+1}^1 \\
d_{j-1}^1 = a_i^1 & d_j^1 = \varnothing & d_{j+1}^1 = a_{i+1}^1 \\
d_{j-1}^2 = a_i^2 & d_j^2 = a_{i+1}^2 & d_{j+1}^2 = \varnothing
\end{array}
\tag{5.55}
$$

It means that in time moment $t_i^1 = t_i^2$, we have both data values a_i^1 and a_i^2 measured (consider the values at the left column above as well as the graphical elements at the left side of Figure 5.2b); in time moment t_{i+1}^2, we have

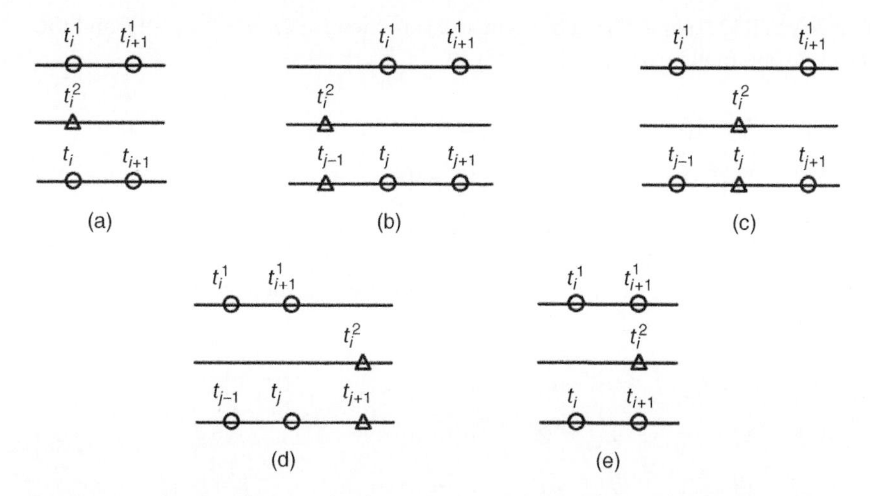

Figure 5.4 The subcases for $e_d(\bar{t}^1, \bar{t}^2)$ when $\bar{t}^1 \rhd \bar{t}^2$. t_i^1 and t_{i+1}^1 are elements of the time tuple $\bar{t}^1$; t_i^2 is an element of the time tuple $\bar{t}^2$; t_{j-1}, t_j, and t_{j+1} are elements of the synchronized time tuple $\bar{t}$. For two measurement processes: (a) the current measurements of both parameters are simultaneous; (b) the current measurement of the 1st parameter happens later than the current measurement of the 2nd parameter; (c) the current measurement of the 2nd parameter happens later than the current measurement and earlier than the next measurement of the 1st parameter; (d) the current measurement of the 2nd parameter happens later than the next measurement of the 1st parameter; (e) the current measurement of the 2nd parameter and the next measurement of the 1st parameter are simultaneous.

only a_{i+1}^2 measured (consider the central column as well as the central part of Figure 5.2b); in time moment t_{i+1}^1, we have only a_{i+1}^1 measured (consider the right column as well as the right side of Figure 5.2b).

The rule for j index calculation is $j = \frac{3i}{2}$, where $i \in [2\ldots(n_1 - 2)]$ and i is an even value ($i \bmod 2 = 0$). The total number of elements in each of the joint tuples, viz., $\langle t_j \rangle_{j=1}^n$, $\langle d_j^1 \rangle_{j=1}^n$, and $\langle d_j^2 \rangle_{j=1}^n$, is $n = \frac{3n_1}{2} - 1$. If we consider the first and the last values of each joint tuple when $\bar{t}^1$ and $\bar{t}^2$ are connected with relation *coincides with*, we obtain the following:

$$
\begin{array}{lllll}
t_1 = t_1^1 & t_{j-1} = t_i^1 & t_j = t_{i+1}^2 & t_{j+1} = t_{i+1}^1 & t_n = t_{n_1}^1 \\
d_1^1 = a_1^1 & d_{j-1}^1 = a_i^1 & d_j^1 = \varnothing & d_{j+1}^1 = a_{i+1}^1 & d_n^1 = a_{n_1}^1 \\
d_1^2 = a_1^2 & d_{j-1}^2 = a_i^2 & d_j^2 = a_{i+1}^2 & d_{j+1}^2 = \varnothing & d_n^2 = a_{n_2}^2
\end{array}
\tag{5.56}
$$

Similarly, if $t_i^1 = t_i^2$ and $t_{i+1}^1 < t_{i+1}^2$, $\forall i \in [2 \ldots (n_1 - 2)]$, $i \bmod 2 = 0$, $j = \frac{3i}{2}$, and $n = \frac{3n_1}{2} - 1$, then the tuple value synchronization can be illustrated by Figure 5.2c. This case requires data synchronization defined in (5.57).

$$
\begin{array}{lllll}
t_1 = t_1^1 & t_{j-1} = t_i^1 & t_j = t_{i+1}^1 & t_{j+1} = t_{i+1}^2 & t_n = t_{n_1}^1 \\
d_1^1 = a_1^1 & d_{j-1}^1 = a_i^1 & d_j^1 = a_{i+1}^1 & d_{j+1}^1 = \varnothing & d_n^1 = a_{n_1}^1 \\
d_1^2 = a_1^2 & d_{j-1}^2 = a_i^2 & d_j^2 = \varnothing & d_{j+1}^2 = a_{i+1}^2 & d_n^2 = a_{n_2}^2
\end{array}
\tag{5.57}
$$

If $t_i^1 < t_i^2$ and $t_{i+1}^1 = t_{i+1}^2$ (Figure 5.2d), $\forall i \in [2\ldots(n_1 - 2)]$ such as $i \bmod 2 = 0$, then $j = \frac{3i}{2}$, $n = \frac{3n_1}{2} - 1$, and the tuple values of the multi-image I are as follows:

$$
\begin{array}{lllll}
t_1 = t_1^1 & t_{j-1} = t_i^1 & t_j = t_i^2 & t_{j+1} = t_{i+1}^1 & t_n = t_{n_1}^1 \\
d_1^1 = a_1^1 & d_{j-1}^1 = a_i^1 & d_j^1 = \varnothing & d_{j+1}^1 = a_{i+1}^1 & d_n^1 = a_{n_1}^1 \\
d_1^2 = a_1^2 & d_{j-1}^2 = \varnothing & d_j^2 = a_i^2 & d_{j+1}^2 = a_{i+1}^2 & d_n^2 = a_{n_2}^2
\end{array}
\tag{5.58}
$$

If $t_i^1 < t_i^2$ and $t_{i+1}^1 > t_{i+1}^2$ (Figure 5.2e), $\forall i \in [2\ldots(n_1-2)]$ such as $i \bmod 2 = 0$, then $j = 2i$, $n = 2(n_1-1)$, and the tuple values of the multi-image I can be obtained by using (5.59).

$$\begin{array}{llllll} t_1 = t_1^1 & t_{j-2} = t_i^1 & t_{j-1} = t_i^2 & t_j = t_{i+1}^2 & t_{j+1} = t_{i+1}^1 & t_n = t_{n_1}^1 \\ d_1^1 = a_1^1 & d_{j-2}^1 = a_i^1 & d_{j-1}^1 = \varnothing & d_j^1 = \varnothing & d_{j+1}^1 = a_{i+1}^1 & d_n^1 = a_{n_1}^1 \\ d_1^2 = a_1^2 & d_{j-2}^2 = \varnothing & d_{j-1}^2 = a_i^2 & d_j^2 = a_{i+1}^2 & d_{j+1}^2 = \varnothing & d_n^2 = a_{n_2}^2 \end{array} \tag{5.59}$$

If $t_i^1 < t_i^2$, $t_{i+1}^1 < t_{i+1}^2$, and $t_{i+1}^1 > t_i^2$ (Figure 5.2f), $\forall i \in [2\ldots(n_1-2)]$ such as $i \bmod 2 = 0$, then $j = 2i$, $n = 2(n_1-1)$, and the tuple values of the multi-image I can be defined as:

$$\begin{array}{llllll} t_1 = t_1^1 & t_{j-2} = t_i^1 & t_{j-1} = t_i^2 & t_j = t_{i+1}^1 & t_{j+1} = t_{i+1}^2 & t_n = t_{n_1}^1 \\ d_1^1 = a_1^1 & d_{j-2}^1 = a_i^1 & d_{j-1}^1 = \varnothing & d_j^1 = a_{i+1}^1 & d_{j+1}^1 = \varnothing & d_n^1 = a_{n_1}^1 \\ d_1^2 = a_1^2 & d_{j-2}^2 = \varnothing & d_{j-1}^2 = a_i^2 & d_j^2 = \varnothing & d_{j+1}^2 = a_{i+1}^2 & d_n^2 = a_{n_2}^2 \end{array} \tag{5.60}$$

If $t_i^1 < t_i^2$, $t_{i+1}^1 < t_{i+1}^2$, and $t_{i+1}^1 = t_i^2$ (Figure 5.2g), $\forall i \in [2\ldots(n_1-2)]$ such as $i \bmod 2 = 0$, then $j = \frac{3i}{2}$, $n = \frac{3n_1}{2} - 1$, and the tuple values of the multi-image I are as follows:

$$\begin{array}{lllll} t_1 = t_1^1 & t_{j-1} = t_i^1 & t_j = t_i^2 & t_{j+1} = t_{i+1}^2 & t_n = t_{n_1}^1 \\ d_1^1 = a_1^1 & d_{j-1}^1 = a_i^1 & d_j^1 = a_{i+1}^1 & d_{j+1}^1 = \varnothing & d_n^1 = a_{n_1}^1 \\ d_1^2 = a_1^2 & d_{j-1}^2 = \varnothing & d_j^2 = a_i^2 & d_{j+1}^2 = a_{i+1}^2 & d_n^2 = a_{n_2}^2 \end{array} \tag{5.61}$$

If $t_{i+1}^1 < t_i^2$ (Figure 5.2h), $\forall i \in [2\ldots(n_1-2)]$ such as $i \bmod 2 = 0$, then $j = 2i$, $n = 2(n_1-1)$, and the tuple values of the multi-image I are defined in (5.62).

$$\begin{array}{llllll} t_1 = t_1^1 & t_{j-2} = t_i^1 & t_{j-1} = t_{i+1}^1 & t_j = t_i^2 & t_{j+1} = t_{i+1}^2 & t_n = t_{n_1}^1 \\ d_1^1 = a_1^1 & d_{j-2}^1 = a_i^1 & d_{j-1}^1 = a_{i+1}^1 & d_j^1 = \varnothing & d_{j+1}^1 = \varnothing & d_n^1 = a_{n_1}^1 \\ d_1^2 = a_1^2 & d_{j-2}^2 = \varnothing & d_{j-1}^2 = \varnothing & d_j^2 = a_i^2 & d_{j+1}^2 = a_{i+1}^2 & d_n^2 = a_{n_2}^2 \end{array} \tag{5.62}$$

If $t_i^1 > t_i^2$ and $t_{i+1}^1 < t_{i+1}^2$ (Figure 5.2i), $\forall i \in [2\ldots(n_1-2)]$ such as $i \bmod 2 = 0$, then $j = 2i$, $n = 2(n_1-1)$, and the tuple values of the multi-image I are as follows:

$$\begin{array}{llllll} t_1 = t_1^1 & t_{j-2} = t_i^2 & t_{j-1} = t_i^1 & t_j = t_{i+1}^1 & t_{j+1} = t_{i+1}^2 & t_n = t_{n_1}^1 \\ d_1^1 = a_1^1 & d_{j-2}^1 = \varnothing & d_{j-1}^1 = a_i^1 & d_j^1 = a_{i+1}^1 & d_{j+1}^1 = \varnothing & d_n^1 = a_{n_1}^1 \\ d_1^2 = a_1^2 & d_{j-2}^2 = a_i^2 & d_{j-1}^2 = \varnothing & d_j^2 = \varnothing & d_{j+1}^2 = a_{i+1}^2 & d_n^2 = a_{n_2}^2 \end{array} \tag{5.63}$$

If $t_i^1 > t_i^2$ and $t_{i+1}^1 = t_{i+1}^2$ (Figure 5.2j), $\forall i \in [2\ldots(n_1-2)]$ such as $i \bmod 2 = 0$, then $j = \frac{3i}{2}$, $n = \frac{3n_1}{2} - 1$, and the tuple values of the multi-image I are defined in (5.64).

$$\begin{array}{lllll} t_1 = t_1^1 & t_{j-1} = t_i^2 & t_j = t_i^1 & t_{j+1} = t_{i+1}^1 & t_n = t_{n_1}^1 \\ d_1^1 = a_1^1 & d_{j-1}^1 = \varnothing & d_j^1 = a_i^1 & d_{j+1}^1 = a_{i+1}^1 & d_n^1 = a_{n_1}^1 \\ d_1^2 = a_1^2 & d_{j-1}^2 = a_i^2 & d_j^2 = \varnothing & d_{j+1}^2 = a_{i+1}^2 & d_n^2 = a_{n_2}^2 \end{array} \tag{5.64}$$

If $t_i^1 > t_i^2, t_{i+1}^1 > t_{i+1}^2$, and $t_i^1 < t_{i+1}^2$ (Figure 5.2k), $\forall i \in [2\ldots(n_1-2)]$ such as $i \bmod 2 = 0$, then $j = 2i$, $n = 2(n_1-1)$, and the tuple values of the multi-image I can be defined as follows:

$$\begin{array}{llllll} t_1 = t_1^1 & t_{j-2} = t_i^2 & t_{j-1} = t_i^1 & t_j = t_{i+1}^2 & t_{j+1} = t_{i+1}^1 & t_n = t_{n_1}^1 \\ d_1^1 = a_1^1 & d_{j-2}^1 = \emptyset & d_{j-1}^1 = a_i^1 & d_j^1 = \emptyset & d_{j+1}^1 = a_{i+1}^1 & d_n^1 = a_{n_1}^1 \\ d_1^2 = a_1^2 & d_{j-2}^2 = a_i^2 & d_{j-1}^2 = \emptyset & d_j^2 = a_{i+1}^2 & d_{j+1}^2 = \emptyset & d_n^2 = a_{n_2}^2 \end{array} \tag{5.65}$$

If $t_i^1 = t_{i+1}^2$ (Figure 5.2l), $\forall i \in [2\ldots(n_1-2)]$ such as $i \bmod 2 = 0$, then $j = \frac{3i}{2}$, $n = \frac{3n_1}{2} - 1$, and the tuple values of the multi-image I are as follows:

$$\begin{array}{lllll} t_1 = t_1^1 & t_{j-1} = t_i^2 & t_j = t_i^1 & t_{j+1} = t_{i+1}^1 & t_n = t_{n_1}^1 \\ d_1^1 = a_1^1 & d_{j-1}^1 = \emptyset & d_j^1 = a_i^1 & d_{j+1}^1 = a_{i+1}^1 & d_n^1 = a_{n_1}^1 \\ d_1^2 = a_1^2 & d_{j-1}^2 = a_i^2 & d_j^2 = a_{i+1}^2 & d_{j+1}^2 = \emptyset & d_n^2 = a_{n_2}^2 \end{array} \tag{5.66}$$

If $t_i^1 > t_{i+1}^2$ (Figure 5.2m), $\forall i \in [2\ldots(n_1-2)]$ such as $i \bmod 2 = 0$, then $j = 2i$, $n = 2(n_1-1)$, and the tuple values of the multi-image I are defined in (5.67).

$$\begin{array}{llllll} t_1 = t_1^1 & t_{j-2} = t_i^2 & t_{j-1} = t_{i+1}^2 & t_j = t_i^1 & t_{j+1} = t_{i+1}^1 & t_n = t_{n_1}^1 \\ d_1^1 = a_1^1 & d_{j-2}^1 = \emptyset & d_{j-1}^1 = \emptyset & d_j^1 = a_i^1 & d_{j+1}^1 = a_{i+1}^1 & d_n^1 = a_{n_1}^1 \\ d_1^2 = a_1^2 & d_{j-2}^2 = a_i^2 & d_{j-1}^2 = a_{i+1}^2 & d_j^2 = \emptyset & d_{j+1}^2 = \emptyset & d_n^2 = a_{n_2}^2 \end{array} \tag{5.67}$$

The second group of subcases for $e_d(\bar{t}^1, \bar{t}^2)$ is when n_1 is an odd value ($n_1 \bmod 2 \neq 0$) and $\bar{t}^1 \sim \bar{t}^2$, i.e. $\bar{t}^1$ *is equally frequent* to $\bar{t}^2$; it means that $|\bar{t}^1| = |\bar{t}^2|$ (i.e. $n_1 = n_2$). Mutual alignment of i-elements and $(i+1)$-elements in both data sequences for these subcases is similar to the subcases of the first group and, therefore, it is also illustrated in Figure 5.2. However, since n_1 is an odd value, it complicates data synchronization because we need to analyze mutual alignment of $t_{n_1-1}^1$ and $t_{n_2-1}^2$ as well.

If $t_i^1 = t_i^2$ (Figure 5.2a), $\forall i \in [1\ldots n_1]$, then the tuple values can be defined in the same way as it has been set in (5.54).

If $t_i^1 = t_i^2, t_{i+1}^1 > t_{i+1}^2$ (Figure 5.2b), $\forall i \in [2\ldots(n_1-2)]$ such as $i \bmod 2 = 0$ and $t_{n_1-1}^1 = t_{n_2-1}^2$, then $j = \frac{3i}{2}$, $n = \frac{3(n_1-1)}{2}$ and the tuple values of the multi-image I can be defined as follows:

$$\begin{array}{llllll} t_1 = t_1^1 & t_{j-1} = t_i^1 & t_j = t_{i+1}^2 & t_{j+1} = t_{i+1}^1 & t_{n-1} = t_{n_1-1}^1 & t_n = t_{n_1}^1 \\ d_1^1 = a_1^1 & d_{j-1}^1 = a_i^1 & d_j^1 = \emptyset & d_{j+1}^1 = a_{i+1}^1 & d_{n-1}^1 = a_{n_1-1}^1 & d_n^1 = a_{n_1}^1 \\ d_1^2 = a_1^2 & d_{j-1}^2 = a_i^2 & d_j^2 = a_{i+1}^2 & d_{j+1}^2 = \emptyset & d_{n-1}^2 = a_{n_2-1}^2 & d_n^2 = a_{n_2}^2 \end{array} \tag{5.68}$$

If $t_i^1 = t_i^2$, $t_{i+1}^1 > t_{i+1}^2$ (Figure 5.2b), $\forall i \in [2\ldots(n_1-2)]$ such as $i \bmod 2 = 0$, and $t_{n_1-1}^1 < t_{n_2-1}^2$, then $j = \frac{3i}{2}$, $n = \frac{3(n_1-1)}{2} + 1$, and the tuple values of the multi-image I are defined in (5.69).

$$\begin{array}{lllllll} t_1 = t_1^1 & t_{j-1} = t_i^1 & t_j = t_{i+1}^2 & t_{j+1} = t_{i+1}^1 & t_{n-2} = t_{n_1-1}^1 & t_{n-1} = t_{n_2-1}^2 & t_n = t_{n_1}^1 \\ d_1^1 = a_1^1 & d_{j-1}^1 = a_i^1 & d_j^1 = \emptyset & d_{j+1}^1 = a_{i+1}^1 & d_{n-2}^1 = a_{n_1-1}^1 & d_{n-1}^1 = \emptyset & d_n^1 = a_{n_1}^1 \\ d_1^2 = a_1^2 & d_{j-1}^2 = a_i^2 & d_j^2 = a_{i+1}^2 & d_{j+1}^2 = \emptyset & d_{n-2}^2 = \emptyset & d_{n-1}^2 = a_{n_2-1}^2 & d_n^2 = a_{n_2}^2 \end{array} \tag{5.69}$$

If $t_i^1 = t_i^2$, $t_{i+1}^1 > t_{i+1}^2$ (Figure 5.2b), $\forall i \in [2\ldots(n_1-2)]$ such as $i \bmod 2 = 0$, and $t_{n_1-1}^1 > t_{n_2-1}^2$, then $j = \frac{3i}{2}$, $n = \frac{3(n_1-1)}{2} + 1$, and the tuple values of the multi-image I are as follows:

$$\begin{array}{lllllll}
t_1 = t_1^1 & t_{j-1} = t_i^1 & t_j = t_{i+1}^2 & t_{j+1} = t_{i+1}^1 & t_{n-2} = t_{n_2-1}^2 & t_{n-1} = t_{n_1-1}^1 & t_n = t_{n_1}^1 \\
d_1^1 = a_1^1 & d_{j-1}^1 = a_i^1 & d_j^1 = \emptyset & d_{j+1}^1 = a_{i+1}^1 & d_{n-2}^1 = \emptyset & d_{n-1}^1 = a_{n_1-1}^1 & d_n^1 = a_{n_1}^1 \\
d_1^2 = a_1^2 & d_{j-1}^2 = a_i^2 & d_j^2 = a_{i+1}^2 & d_{j+1}^2 = \emptyset & d_{n-2}^2 = a_{n_2-1}^2 & d_{n-1}^2 = \emptyset & d_n^2 = a_{n_2}^2
\end{array} \quad (5.70)$$

If $t_i^1 = t_i^2$, $t_{i+1}^1 < t_{i+1}^2$ (Figure 5.2c), $\forall i \in [2\ldots(n_1-2)]$ such as $i \bmod 2 = 0$, and $t_{n_1-1}^1 = t_{n_2-1}^2$, then $j = \frac{3i}{2}$, $n = \frac{3(n_1-1)}{2}$, and the tuple values of the multi-image I are defined as:

$$\begin{array}{llllll}
t_1 = t_1^1 & t_{j-1} = t_i^1 & t_j = t_{i+1}^1 & t_{j+1} = t_{i+1}^2 & t_{n-1} = t_{n_1-1}^1 & t_n = t_{n_1}^1 \\
d_1^1 = a_1^1 & d_{j-1}^1 = a_i^1 & d_j^1 = a_{i+1}^1 & d_{j+1}^1 = \emptyset & d_{n-1}^1 = a_{n_1-1}^1 & d_n^1 = a_{n_1}^1 \\
d_1^2 = a_1^2 & d_{j-1}^2 = a_i^2 & d_j^2 = \emptyset & d_{j+1}^2 = a_{i+1}^2 & d_{n-1}^2 = a_{n_2-1}^2 & d_n^2 = a_{n_2}^2
\end{array} \quad (5.71)$$

If $t_i^1 = t_i^2$, $t_{i+1}^1 < t_{i+1}^2$ (Figure 5.2c), $\forall i \in [2\ldots(n_1-2)]$ such as $i \bmod 2 = 0$, and $t_{n_1-1}^1 < t_{n_2-1}^2$, then $j = \frac{3i}{2}$, $n = \frac{3(n_1-1)}{2} + 1$, and the tuple values of the multi-image I are stated in (5.72).

$$\begin{array}{lllllll}
t_1 = t_1^1 & t_{j-1} = t_i^1 & t_j = t_{i+1}^1 & t_{j+1} = t_{i+1}^2 & t_{n-2} = t_{n_1-1}^1 & t_{n-1} = t_{n_2-1}^2 & t_n = t_{n_1}^1 \\
d_1^1 = a_1^1 & d_{j-1}^1 = a_i^1 & d_j^1 = a_{i+1}^1 & d_{j+1}^1 = \emptyset & d_{n-2}^1 = a_{n_1-1}^1 & d_{n-1}^1 = \emptyset & d_n^1 = a_{n_1}^1 \\
d_1^2 = a_1^2 & d_{j-1}^2 = a_i^2 & d_j^2 = \emptyset & d_{j+1}^2 = a_{i+1}^2 & d_{n-2}^2 = \emptyset & d_{n-1}^2 = a_{n_2-1}^2 & d_n^2 = a_{n_2}^2
\end{array} \quad (5.72)$$

If $t_i^1 = t_i^2$, $t_{i+1}^1 < t_{i+1}^2$ (Figure 5.2c), $\forall i \in [2\ldots(n_1-2)]$ such as $i \bmod 2 = 0$, and $t_{n_1-1}^1 > t_{n_2-1}^2$, then $j = \frac{3i}{2}$, $n = \frac{3(n_1-1)}{2} + 1$, and the tuple values of the multi-image I are defined in (5.73).

$$\begin{array}{lllllll}
t_1 = t_1^1 & t_{j-1} = t_i^1 & t_j = t_{i+1}^1 & t_{j+1} = t_{i+1}^2 & t_{n-2} = t_{n_2-1}^2 & t_{n-1} = t_{n_1-1}^1 & t_n = t_{n_1}^1 \\
d_1^1 = a_1^1 & d_{j-1}^1 = a_i^1 & d_j^1 = a_{i+1}^1 & d_{j+1}^1 = \emptyset & d_{n-2}^1 = \emptyset & d_{n-1}^1 = a_{n_1-1}^1 & d_n^1 = a_{n_1}^1 \\
d_1^2 = a_1^2 & d_{j-1}^2 = a_i^2 & d_j^2 = \emptyset & d_{j+1}^2 = a_{i+1}^2 & d_{n-2}^2 = a_{n_2-1}^2 & d_{n-1}^2 = \emptyset & d_n^2 = a_{n_2}^2
\end{array} \quad (5.73)$$

If $t_i^1 < t_i^2$, $t_{i+1}^1 = t_{i+1}^2$ (Figure 5.2d), $\forall i \in [2\ldots(n_1-2)]$ such as $i \bmod 2 = 0$, and $t_{n_1-1}^1 = t_{n_2-1}^2$, then $j = \frac{3i}{2}$, $n = \frac{3(n_1-1)}{2}$, and the tuple values of the multi-image I are as follows:

$$\begin{array}{llllll}
t_1 = t_1^1 & t_{j-1} = t_i^1 & t_j = t_i^2 & t_{j+1} = t_{i+1}^1 & t_{n-1} = t_{n_1-1}^1 & t_n = t_{n_1}^1 \\
d_1^1 = a_1^1 & d_{j-1}^1 = a_i^1 & d_j^1 = \emptyset & d_{j+1}^1 = a_{i+1}^1 & d_{n-1}^1 = a_{n_1-1}^1 & d_n^1 = a_{n_1}^1 \\
d_1^2 = a_1^2 & d_{j-1}^2 = \emptyset & d_j^2 = a_i^2 & d_{j+1}^2 = a_{i+1}^2 & d_{n-1}^2 = a_{n_2-1}^2 & d_n^2 = a_{n_2}^2
\end{array} \quad (5.74)$$

If $t_i^1 < t_i^2$, $t_{i+1}^1 = t_{i+1}^2$ (Figure 5.2d), $\forall i \in [2\ldots(n_1-2)]$ such as $i \bmod 2 = 0$, and $t_{n_1-1}^1 < t_{n_2-1}^2$, then $j = \frac{3i}{2}$, $n = \frac{3(n_1-1)}{2} + 1$, and the tuple values of the multi-image I can be defined as follows:

$$\begin{array}{lllllll}
t_1 = t_1^1 & t_{j-1} = t_i^1 & t_j = t_i^2 & t_{j+1} = t_{i+1}^1 & t_{n-2} = t_{n_1-1}^1 & t_{n-1} = t_{n_2-1}^2 & t_n = t_{n_1}^1 \\
d_1^1 = a_1^1 & d_{j-1}^1 = a_i^1 & d_j^1 = \emptyset & d_{j+1}^1 = a_{i+1}^1 & d_{n-2}^1 = a_{n_1-1}^1 & d_{n-1}^1 = \emptyset & d_n^1 = a_{n_1}^1 \\
d_1^2 = a_1^2 & d_{j-1}^2 = \emptyset & d_j^2 = a_i^2 & d_{j+1}^2 = a_{i+1}^2 & d_{n-2}^2 = \emptyset & d_{n-1}^2 = a_{n_2-1}^2 & d_n^2 = a_{n_2}^2
\end{array} \quad (5.75)$$

If $t_i^1 < t_i^2$, $t_{i+1}^1 = t_{i+1}^2$ (Figure 5.2d), $\forall i \in [2\ldots(n_1-2)]$ such as $i \bmod 2 = 0$, and $t_{n_1-1}^1 > t_{n_2-1}^2$, then $j = \frac{3i}{2}$, $n = \frac{3(n_1-1)}{2} + 1$, and the tuple values of the multi-image I can be stated as follows:

$$\begin{array}{lllllll} t_1 = t_1^1 & t_{j-1} = t_i^1 & t_j = t_i^2 & t_{j+1} = t_{i+1}^1 & t_{n-2} = t_{n_2-1}^2 & t_{n-1} = t_{n_1-1}^1 & t_n = t_{n_1}^1 \\ d_1^1 = a_1^1 & d_{j-1}^1 = a_i^1 & d_j^1 = \emptyset & d_{j+1}^1 = a_{i+1}^1 & d_{n-2}^1 = \emptyset & d_{n-1}^1 = a_{n_1-1}^1 & d_n^1 = a_{n_1}^1 \\ d_1^2 = a_1^2 & d_{j-1}^2 = \emptyset & d_j^2 = a_i^2 & d_{j+1}^2 = a_{i+1}^2 & d_{n-2}^2 = a_{n_2-1}^2 & d_{n-1}^2 = \emptyset & d_n^2 = a_{n_2}^2 \end{array} \tag{5.76}$$

If $t_i^1 < t_i^2, t_{i+1}^1 > t_{i+1}^2$ (Figure 5.2e), $\forall i \in [2\ldots(n_1-2)]$ such as $i \bmod 2 = 0$, and $t_{n_1-1}^1 = t_{n_2-1}^2$, then $j = 2i$, $n = 2n_1 - 3$, and the tuple values of the multi-image I are as follows:

$$\begin{array}{lllllll} t_1 = t_1^1 & t_{j-2} = t_i^1 & t_{j-1} = t_i^2 & t_j = t_{i+1}^2 & t_{j+1} = t_{i+1}^1 & t_{n-1} = t_{n_1-1}^1 & t_n = t_{n_1}^1 \\ d_1^1 = a_1^1 & d_{j-2}^1 = a_i^1 & d_{j-1}^1 = \emptyset & d_j^1 = \emptyset & d_{j+1}^1 = a_{i+1}^1 & d_{n-1}^1 = a_{n_1-1}^1 & d_n^1 = a_{n_1}^1 \\ d_1^2 = a_1^2 & d_{j-2}^2 = \emptyset & d_{j-1}^2 = a_i^2 & d_j^2 = a_{i+1}^2 & d_{j+1}^2 = \emptyset & d_{n-1}^2 = a_{n_2-1}^2 & d_n^2 = a_{n_2}^2 \end{array} \tag{5.77}$$

If $t_i^1 < t_i^2$, $t_{i+1}^1 > t_{i+1}^2$ (Figure 5.2e), $\forall i \in [2\ldots(n_1-2)]$ such as $i \bmod 2 = 0$, and $t_{n_1-1}^1 < t_{n_2-1}^2$, then $j = 2i$, $n = 2(n_1 - 1)$, and the tuple values of the multi-image I are defined in (5.78).

$$\begin{array}{llllllll} t_1 = t_1^1 & t_{j-2} = t_i^1 & t_{j-1} = t_i^2 & t_j = t_{i+1}^2 & t_{j+1} = t_{i+1}^1 & t_{n-2} = t_{n_1-1}^1 & t_{n-1} = t_{n_2-1}^2 & t_n = t_{n_1}^1 \\ d_1^1 = a_1^1 & d_{j-2}^1 = a_i^1 & d_{j-1}^1 = \emptyset & d_j^1 = \emptyset & d_{j+1}^1 = a_{i+1}^1 & d_{n-2}^1 = a_{n_1-1}^1 & d_{n-1}^1 = \emptyset & d_n^1 = a_{n_1}^1 \\ d_1^2 = a_1^2 & d_{j-2}^2 = \emptyset & d_{j-1}^2 = a_i^2 & d_j^2 = a_{i+1}^2 & d_{j+1}^2 = \emptyset & d_{n-2}^2 = \emptyset & d_{n-1}^2 = a_{n_2-1}^2 & d_n^2 = a_{n_2}^2 \end{array} \tag{5.78}$$

If $t_i^1 < t_i^2$, $t_{i+1}^1 > t_{i+1}^2$ (Figure 5.2e), $\forall i \in [2\ldots(n_1-2)]$ such as $i \bmod 2 = 0$, and $t_{n_1-1}^1 > t_{n_2-1}^2$, then $j = 2i$, $n = 2(n_1 - 1)$, and the tuple values of the multi-image I are as follows:

$$\begin{array}{llllllll} t_1 = t_1^1 & t_{j-2} = t_i^1 & t_{j-1} = t_i^2 & t_j = t_{i+1}^2 & t_{j+1} = t_{i+1}^1 & t_{n-2} = t_{n_2-1}^2 & t_{n-1} = t_{n_1-1}^1 & t_n = t_{n_1}^1 \\ d_1^1 = a_1^1 & d_{j-2}^1 = a_i^1 & d_{j-1}^1 = \emptyset & d_j^1 = \emptyset & d_{j+1}^1 = a_{i+1}^1 & d_{n-2}^1 = \emptyset & d_{n-1}^1 = a_{n_1-1}^1 & d_n^1 = a_{n_1}^1 \\ d_1^2 = a_1^2 & d_{j-2}^2 = \emptyset & d_{j-1}^2 = a_i^2 & d_j^2 = a_{i+1}^2 & d_{j+1}^2 = \emptyset & d_{n-2}^2 = a_{n_2-1}^2 & d_{n-1}^2 = \emptyset & d_n^2 = a_{n_2}^2 \end{array} \tag{5.79}$$

If $t_i^1 < t_i^2, t_{i+1}^1 < t_{i+1}^2, t_{i+1}^1 > t_i^2$ (Figure 5.2f), $\forall i \in [2\ldots(n_1-2)]$ such as $i \bmod 2 = 0$, and $t_{n_1-1}^1 = t_{n_2-1}^2$, then $j = 2i$, $n = 2n_1 - 3$, and the tuple values of the multi-image I are defined in (5.80).

$$\begin{array}{lllllll} t_1 = t_1^1 & t_{j-2} = t_i^1 & t_{j-1} = t_i^2 & t_j = t_{i+1}^1 & t_{j+1} = t_{i+1}^2 & t_{n-1} = t_{n_1-1}^1 & t_n = t_{n_1}^1 \\ d_1^1 = a_1^1 & d_{j-2}^1 = a_i^1 & d_{j-1}^1 = \emptyset & d_j^1 = a_{i+1}^1 & d_{j+1}^1 = \emptyset & d_{n-1}^1 = a_{n_1-1}^1 & d_n^1 = a_{n_1}^1 \\ d_1^2 = a_1^2 & d_{j-2}^2 = \emptyset & d_{j-1}^2 = a_i^2 & d_j^2 = \emptyset & d_{j+1}^2 = a_{i+1}^2 & d_{n-1}^2 = a_{n_2-1}^2 & d_n^2 = a_{n_2}^2 \end{array} \tag{5.80}$$

If $t_i^1 < t_i^2, t_{i+1}^1 < t_{i+1}^2, t_{i+1}^1 > t_i^2$ (Figure 5.2f), $\forall i \in [2\ldots(n_1-2)]$ such as $i \bmod 2 = 0$, and $t_{n_1-1}^1 < t_{n_2-1}^2$, then $j = 2i$, $n = 2(n_1 - 1)$, and the tuple values of the multi-image I can be obtained as follows:

$$\begin{array}{llllllll} t_1 = t_1^1 & t_{j-2} = t_i^1 & t_{j-1} = t_i^2 & t_j = t_{i+1}^1 & t_{j+1} = t_{i+1}^2 & t_{n-2} = t_{n_1-1}^1 & t_{n-1} = t_{n_2-1}^2 & t_n = t_{n_1}^1 \\ d_1^1 = a_1^1 & d_{j-2}^1 = a_i^1 & d_{j-1}^1 = \emptyset & d_j^1 = a_{i+1}^1 & d_{j+1}^1 = \emptyset & d_{n-2}^1 = a_{n_1-1}^1 & d_{n-1}^1 = \emptyset & d_n^1 = a_{n_1}^1 \\ d_1^2 = a_1^2 & d_{j-2}^2 = \emptyset & d_{j-1}^2 = a_i^2 & d_j^2 = \emptyset & d_{j+1}^2 = a_{i+1}^2 & d_{n-2}^2 = \emptyset & d_{n-1}^2 = a_{n_2-1}^2 & d_n^2 = a_{n_2}^2 \end{array} \tag{5.81}$$

If $t_i^1 < t_i^2$, $t_{i+1}^1 < t_{i+1}^2$, $t_{i+1}^1 > t_i^2$ (Figure 5.2f), $\forall i \in [2\ldots(n_1-2)]$ such as $i \bmod 2 = 0$, and $t_{n_1-1}^1 > t_{n_2-1}^2$, then $j = 2i$, $n = 2(n_1-1)$, and the tuple values of the multi-image I are defined in (5.82).

$$\begin{matrix} t_1 = t_1^1 & t_{j-2} = t_i^1 & t_{j-1} = t_i^2 & t_j = t_{i+1}^1 & t_{j+1} = t_{i+1}^2 & t_{n-2} = t_{n_2-1}^2 & t_{n-1} = t_{n_1-1}^1 & t_n = t_{n_1}^1 \\ d_1^1 = a_1^1 & d_{j-2}^1 = a_i^1 & d_{j-1}^1 = \varnothing & d_j^1 = a_{i+1}^1 & d_{j+1}^1 = \varnothing & d_{n-2}^1 = \varnothing & d_{n-1}^1 = a_{n_1-1}^1 & d_n^1 = a_{n_1}^1 \\ d_1^2 = a_1^2 & d_{j-2}^2 = \varnothing & d_{j-1}^2 = a_i^2 & d_j^2 = \varnothing & d_{j+1}^2 = a_{i+1}^2 & d_{n-2}^2 = a_{n_2-1}^2 & d_{n-1}^2 = \varnothing & d_n^2 = a_{n_2}^2 \end{matrix} \quad (5.82)$$

If $t_i^1 < t_i^2$, $t_{i+1}^1 < t_{i+1}^2$, $t_{i+1}^1 = t_i^2$ (Figure 5.2g), $\forall i \in [2\ldots(n_1-2)]$ such as $i \bmod 2 = 0$, and $t_{n_1-1}^1 = t_{n_2-1}^2$, then $j = \frac{3i}{2}$, $n = \frac{3(n_1-1)}{2}$, and the tuple values of the multi-image I are as follows:

$$\begin{matrix} t_1 = t_1^1 & t_{j-1} = t_i^1 & t_j = t_i^2 & t_{j+1} = t_{i+1}^2 & t_{n-1} = t_{n_1-1}^1 & t_n = t_{n_1}^1 \\ d_1^1 = a_1^1 & d_{j-1}^1 = a_i^1 & d_j^1 = a_{i+1}^1 & d_{j+1}^1 = \varnothing & d_{n-1}^1 = a_{n_1-1}^1 & d_n^1 = a_{n_1}^1 \\ d_1^2 = a_1^2 & d_{j-1}^2 = \varnothing & d_j^2 = a_i^2 & d_{j+1}^2 = a_{i+1}^2 & d_{n-1}^2 = a_{n_2-1}^2 & d_n^2 = a_{n_2}^2 \end{matrix} \quad (5.83)$$

If $t_i^1 < t_i^2$, $t_{i+1}^1 < t_{i+1}^2$, $t_{i+1}^1 = t_i^2$ (Figure 5.2g), $\forall i \in [2\ldots(n_1-2)]$ such as $i \bmod 2 = 0$, and $t_{n_1-1}^1 < t_{n_2-1}^2$, then $j = \frac{3i}{2}$, $n = \frac{3(n_1-1)}{2} + 1$, and the tuple values of the multi-image I can be obtained by using (5.84).

$$\begin{matrix} t_1 = t_1^1 & t_{j-1} = t_i^1 & t_j = t_i^2 & t_{j+1} = t_{i+1}^2 & t_{n-2} = t_{n_1-1}^1 & t_{n-1} = t_{n_2-1}^2 & t_n = t_{n_1}^1 \\ d_1^1 = a_1^1 & d_{j-1}^1 = a_i^1 & d_j^1 = a_{i+1}^1 & d_{j+1}^1 = \varnothing & d_{n-2}^1 = a_{n_1-1}^1 & d_{n-1}^1 = \varnothing & d_n^1 = a_{n_1}^1 \\ d_1^2 = a_1^2 & d_{j-1}^2 = \varnothing & d_j^2 = a_i^2 & d_{j+1}^2 = a_{i+1}^2 & d_{n-2}^2 = \varnothing & d_{n-1}^2 = a_{n_2-1}^2 & d_n^2 = a_{n_2}^2 \end{matrix} \quad (5.84)$$

If $t_i^1 < t_i^2$, $t_{i+1}^1 < t_{i+1}^2$, $t_{i+1}^1 = t_i^2$ (Figure 5.2g), $\forall i \in [2\ldots(n_1-2)]$ such as $i \bmod 2 = 0$, and $t_{n_1-1}^1 > t_{n_2-1}^2$, then $j = \frac{3i}{2}$, $n = \frac{3(n_1-1)}{2} + 1$, and the tuple values of the multi-image I are as follows:

$$\begin{matrix} t_1 = t_1^1 & t_{j-1} = t_i^1 & t_j = t_i^2 & t_{j+1} = t_{i+1}^2 & t_{n-2} = t_{n_2-1}^2 & t_{n-1} = t_{n_1-1}^1 & t_n = t_{n_1}^1 \\ d_1^1 = a_1^1 & d_{j-1}^1 = a_i^1 & d_j^1 = a_{i+1}^1 & d_{j+1}^1 = \varnothing & d_{n-2}^1 = \varnothing & d_{n-1}^1 = a_{n_1-1}^1 & d_n^1 = a_{n_1}^1 \\ d_1^2 = a_1^2 & d_{j-1}^2 = \varnothing & d_j^2 = a_i^2 & d_{j+1}^2 = a_{i+1}^2 & d_{n-2}^2 = a_{n_2-1}^2 & d_{n-1}^2 = \varnothing & d_n^2 = a_{n_2}^2 \end{matrix} \quad (5.85)$$

If $t_{i+1}^1 < t_i^2$ (Figure 5.2h), $\forall i \in [2\ldots(n_1-2)]$ such as $i \bmod 2 = 0$ and $t_{n_1-1}^1 = t_{n_2-1}^2$, then $j = 2i$, $n = 2n_1 - 3$, and the tuple values of the multi-image I can be found as defined in (5.86).

$$\begin{matrix} t_1 = t_1^1 & t_{j-2} = t_i^1 & t_{j-1} = t_{i+1}^1 & t_j = t_i^2 & t_{j+1} = t_{i+1}^2 & t_{n-1} = t_{n_1-1}^1 & t_n = t_{n_1}^1 \\ d_1^1 = a_1^1 & d_{j-2}^1 = a_i^1 & d_{j-1}^1 = a_{i+1}^1 & d_j^1 = \varnothing & d_{j+1}^1 = \varnothing & d_{n-1}^1 = a_{n_1-1}^1 & d_n^1 = a_{n_1}^1 \\ d_1^2 = a_1^2 & d_{j-2}^2 = \varnothing & d_{j-1}^2 = \varnothing & d_j^2 = a_i^2 & d_{j+1}^2 = a_{i+1}^2 & d_{n-1}^2 = a_{n_2-1}^2 & d_n^2 = a_{n_2}^2 \end{matrix} \quad (5.86)$$

If $t_{i+1}^1 < t_i^2$ (Figure 5.2h), $\forall i \in [2\ldots(n_1-2)]$ such as $i \bmod 2 = 0$, and $t_{n_1-1}^1 < t_{n_2-1}^2$, then $j = 2i$, $n = 2(n_1-1)$, and the tuple values of the multi-image I are stated as follows:

$$\begin{matrix} t_1 = t_1^1 & t_{j-2} = t_i^1 & t_{j-1} = t_{i+1}^1 & t_j = t_i^2 & t_{j+1} = t_{i+1}^2 & t_{n-2} = t_{n_1-1}^1 & t_{n-1} = t_{n_2-1}^2 & t_n = t_{n_1}^1 \\ d_1^1 = a_1^1 & d_{j-2}^1 = a_i^1 & d_{j-1}^1 = a_{i+1}^1 & d_j^1 = \varnothing & d_{j+1}^1 = \varnothing & d_{n-2}^1 = a_{n_1-1}^1 & d_{n-1}^1 = \varnothing & d_n^1 = a_{n_1}^1 \\ d_1^2 = a_1^2 & d_{j-2}^2 = \varnothing & d_{j-1}^2 = \varnothing & d_j^2 = a_i^2 & d_{j+1}^2 = a_{i+1}^2 & d_{n-2}^2 = \varnothing & d_{n-1}^2 = a_{n_2-1}^2 & d_n^2 = a_{n_2}^2 \end{matrix} \quad (5.87)$$

If $t^1_{i+1} < t^2_i$ (Figure 5.2h), $\forall i \in [2\ldots(n_1-2)]$ such as $i \bmod 2 = 0$ and $t^1_{n_1-1} > t^2_{n_2-1}$, then $j = 2i$, $n = 2(n_1-1)$, and the tuple values of the multi-image I can be defined as:

$$\begin{array}{llllllll} t_1 = t^1_1 & t_{j-2} = t^1_i & t_{j-1} = t^1_{i+1} & t_j = t^2_i & t_{j+1} = t^2_{i+1} & t_{n-2} = t^2_{n_2-1} & t_{n-1} = t^1_{n_1-1} & t_n = t^1_{n_1} \\ d^1_1 = a^1_1 & d^1_{j-2} = a^1_i & d^1_{j-1} = a^1_{i+1} & d^1_j = \emptyset & d^1_{j+1} = \emptyset & d^1_{n-2} = \emptyset & d^1_{n-1} = a^1_{n_1-1} & d^1_n = a^1_{n_1} \\ d^2_1 = a^2_1 & d^2_{j-2} = \emptyset & d^2_{j-1} = \emptyset & d^2_j = a^2_i & d^2_{j+1} = a^2_{i+1} & d^2_{n-2} = a^2_{n_2-1} & d^2_{n-1} = \emptyset & d^2_n = a^2_{n_2} \end{array} \tag{5.88}$$

If $t^1_i > t^2_i, t^1_{i+1} < t^2_{i+1}$ (Figure 5.2i), $\forall i \in [2\ldots(n_1-2)]$ such as $i \bmod 2 = 0$, and $t^1_{n_1-1} = t^2_{n_2-1}$, then $j = 2i$, $n = 2n_1 - 3$, and the tuple values of the multi-image I are as follows:

$$\begin{array}{lllllll} t_1 = t^1_1 & t_{j-2} = t^2_i & t_{j-1} = t^1_i & t_j = t^1_{i+1} & t_{j+1} = t^2_{i+1} & t_{n-1} = t^1_{n_1-1} & t_n = t^1_{n_1} \\ d^1_1 = a^1_1 & d^1_{j-2} = \emptyset & d^1_{j-1} = a^1_i & d^1_j = a^1_{i+1} & d^1_{j+1} = \emptyset & d^1_{n-1} = a^1_{n_1-1} & d^1_n = a^1_{n_1} \\ d^2_1 = a^2_1 & d^2_{j-2} = a^2_i & d^2_{j-1} = \emptyset & d^2_j = \emptyset & d^2_{j+1} = a^2_{i+1} & d^2_{n-1} = a^2_{n_2-1} & d^2_n = a^2_{n_2} \end{array} \tag{5.89}$$

If $t^1_i > t^2_i, t^1_{i+1} < t^2_{i+1}$ (Figure 5.2i), $\forall i \in [2\ldots(n_1-2)]$ such as $i \bmod 2 = 0$, and $t^1_{n_1-1} < t^2_{n_2-1}$, then $j = 2i$, $n = 2(n_1-1)$, and the tuple values of the multi-image I are defined as:

$$\begin{array}{llllllll} t_1 = t^1_1 & t_{j-2} = t^2_i & t_{j-1} = t^1_i & t_j = t^1_{i+1} & t_{j+1} = t^2_{i+1} & t_{n-2} = t^1_{n_1-1} & t_{n-1} = t^2_{n_2-1} & t_n = t^1_{n_1} \\ d^1_1 = a^1_1 & d^1_{j-2} = \emptyset & d^1_{j-1} = a^1_i & d^1_j = a^1_{i+1} & d^1_{j+1} = \emptyset & d^1_{n-2} = a^1_{n_1-1} & d^1_{n-1} = \emptyset & d^1_n = a^1_{n_1} \\ d^2_1 = a^2_1 & d^2_{j-2} = a^2_i & d^2_{j-1} = \emptyset & d^2_j = \emptyset & d^2_{j+1} = a^2_{i+1} & d^2_{n-2} = \emptyset & d^2_{n-1} = a^2_{n_2-1} & d^2_n = a^2_{n_2} \end{array} \tag{5.90}$$

If $t^1_i > t^2_i, t^1_{i+1} < t^2_{i+1}$ (Figure 5.2i), $\forall i \in [2\ldots(n_1-2)]$ such as $i \bmod 2 = 0$, and $t^1_{n_1-1} > t^2_{n_2-1}$, then $j = 2i$, $n = 2(n_1-1)$, and the tuple values of the multi-image I are stated in (5.91).

$$\begin{array}{llllllll} t_1 = t^1_1 & t_{j-2} = t^2_i & t_{j-1} = t^1_i & t_j = t^1_{i+1} & t_{j+1} = t^2_{i+1} & t_{n-2} = t^2_{n_2-1} & t_{n-1} = t^1_{n_1-1} & t_n = t^1_{n_1} \\ d^1_1 = a^1_1 & d^1_{j-2} = \emptyset & d^1_{j-1} = a^1_i & d^1_j = a^1_{i+1} & d^1_{j+1} = \emptyset & d^1_{n-2} = \emptyset & d^1_{n-1} = a^1_{n_1-1} & d^1_n = a^1_{n_1} \\ d^2_1 = a^2_1 & d^2_{j-2} = a^2_i & d^2_{j-1} = \emptyset & d^2_j = \emptyset & d^2_{j+1} = a^2_{i+1} & d^2_{n-2} = a^2_{n_2-1} & d^2_{n-1} = \emptyset & d^2_n = a^2_{n_2} \end{array} \tag{5.91}$$

If $t^1_i > t^2_i, t^1_{i+1} = t^2_{i+1}$ (Figure 5.2j), $\forall i \in [2\ldots(n_1-2)]$ such as $i \bmod 2 = 0$, and $t^1_{n_1-1} = t^2_{n_2-1}$, then $j = \frac{3i}{2}$, $n = \frac{3(n_1-1)}{2}$, and the tuple values of the multi-image I are defined in (5.92).

$$\begin{array}{llllll} t_1 = t^1_1 & t_{j-1} = t^2_i & t_j = t^1_i & t_{j+1} = t^1_{i+1} & t_{n-1} = t^1_{n_1-1} & t_n = t^1_{n_1} \\ d^1_1 = a^1_1 & d^1_{j-1} = \emptyset & d^1_j = a^1_i & d^1_{j+1} = a^1_{i+1} & d^1_{n-1} = a^1_{n_1-1} & d^1_n = a^1_{n_1} \\ d^2_1 = a^2_1 & d^2_{j-1} = a^2_i & d^2_j = \emptyset & d^2_{j+1} = a^2_{i+1} & d^2_{n-1} = a^2_{n_2-1} & d^2_n = a^2_{n_2} \end{array} \tag{5.92}$$

If $t^1_i > t^2_i, t^1_{i+1} = t^2_{i+1}$ (Figure 5.2j), $\forall i \in [2\ldots(n_1-2)]$ such as $i \bmod 2 = 0$, and $t^1_{n_1-1} < t^2_{n_2-1}$, then $j = \frac{3i}{2}$, $n = \frac{3(n_1-1)}{2} + 1$, and the tuple values of the multi-image I are as follows:

$$\begin{array}{lllllll} t_1 = t^1_1 & t_{j-1} = t^2_i & t_j = t^1_i & t_{j+1} = t^1_{i+1} & t_{n-2} = t^1_{n_1-1} & t_{n-1} = t^2_{n_2-1} & t_n = t^1_{n_1} \\ d^1_1 = a^1_1 & d^1_{j-1} = \emptyset & d^1_j = a^1_i & d^1_{j+1} = a^1_{i+1} & d^1_{n-2} = a^1_{n_1-1} & d^1_{n-1} = \emptyset & d^1_n = a^1_{n_1} \\ d^2_1 = a^2_1 & d^2_{j-1} = a^2_i & d^2_j = \emptyset & d^2_{j+1} = a^2_{i+1} & d^2_{n-2} = \emptyset & d^2_{n-1} = a^2_{n_2-1} & d^2_n = a^2_{n_2} \end{array} \tag{5.93}$$

If $t_i^1 > t_i^2, t_{i+1}^1 = t_{i+1}^2$ (Figure 5.2j), $\forall i \in [2\ldots(n_1-2)]$ such as $i \bmod 2 = 0$, and $t_{n_1-1}^1 > t_{n_2-1}^2$, then $j = \frac{3i}{2}$, $n = \frac{3(n_1-1)}{2} + 1$, and the tuple values of the multi-image I can be defined as:

$$\begin{array}{lllllll}
t_1 = t_1^1 & t_{j-1} = t_i^2 & t_j = t_i^1 & t_{j+1} = t_{i+1}^1 & t_{n-2} = t_{n_2-1}^2 & t_{n-1} = t_{n_1-1}^1 & t_n = t_{n_1}^1 \\
d_1^1 = a_1^1 & d_{j-1}^1 = \varnothing & d_j^1 = a_i^1 & d_{j+1}^1 = a_{i+1}^1 & d_{n-2}^1 = \varnothing & d_{n-1}^1 = a_{n_1-1}^1 & d_n^1 = a_{n_1}^1 \\
d_1^2 = a_1^2 & d_{j-1}^2 = a_i^2 & d_j^2 = \varnothing & d_{j+1}^2 = a_{i+1}^2 & d_{n-2}^2 = a_{n_2-1}^2 & d_{n-1}^2 = \varnothing & d_n^2 = a_{n_2}^2
\end{array} \tag{5.94}$$

If $t_i^1 > t_i^2, t_{i+1}^1 > t_{i+1}^2, t_i^1 < t_{i+1}^2$ (Figure 5.2k), $\forall i \in [2\ldots(n_1-2)]$ such as $i \bmod 2 = 0$, and $t_{n_1-1}^1 = t_{n_2-1}^2$, then $j = 2i$, $n = 2n_1 - 3$, and the tuple values of the multi-image I are defined in (5.95).

$$\begin{array}{lllllll}
t_1 = t_1^1 & t_{j-2} = t_i^2 & t_{j-1} = t_i^1 & t_j = t_{i+1}^2 & t_{j+1} = t_{i+1}^1 & t_{n-1} = t_{n_1-1}^1 & t_n = t_{n_1}^1 \\
d_1^1 = a_1^1 & d_{j-2}^1 = \varnothing & d_{j-1}^1 = a_i^1 & d_j^1 = \varnothing & d_{j+1}^1 = a_{i+1}^1 & d_{n-1}^1 = a_{n_1-1}^1 & d_n^1 = a_{n_1}^1 \\
d_1^2 = a_1^2 & d_{j-2}^2 = a_i^2 & d_{j-1}^2 = \varnothing & d_j^2 = a_{i+1}^2 & d_{j+1}^2 = \varnothing & d_{n-1}^2 = a_{n_2-1}^2 & d_n^2 = a_{n_2}^2
\end{array} \tag{5.95}$$

If $t_i^1 > t_i^2, t_{i+1}^1 > t_{i+1}^2, t_i^1 < t_{i+1}^2$ (Figure 5.2k), $\forall i \in [2\ldots(n_1-2)]$ such as $i \bmod 2 = 0$, and $t_{n_1-1}^1 < t_{n_2-1}^2$, then $j = 2i$, $n = 2(n_1 - 1)$, and the tuple values of the multi-image I are as follows:

$$\begin{array}{llllllll}
t_1 = t_1^1 & t_{j-2} = t_i^2 & t_{j-1} = t_i^1 & t_j = t_{i+1}^2 & t_{j+1} = t_{i+1}^1 & t_{n-2} = t_{n_1-1}^1 & t_{n-1} = t_{n_2-1}^2 & t_n = t_{n_1}^1 \\
d_1^1 = a_1^1 & d_{j-2}^1 = \varnothing & d_{j-1}^1 = a_i^1 & d_j^1 = \varnothing & d_{j+1}^1 = a_{i+1}^1 & d_{n-2}^1 = a_{n_1-1}^1 & d_{n-1}^1 = \varnothing & d_n^1 = a_{n_1}^1 \\
d_1^2 = a_1^2 & d_{j-2}^2 = a_i^2 & d_{j-1}^2 = \varnothing & d_j^2 = a_{i+1}^2 & d_{j+1}^2 = \varnothing & d_{n-2}^2 = \varnothing & d_{n-1}^2 = a_{n_2-1}^2 & d_n^2 = a_{n_2}^2
\end{array} \tag{5.96}$$

If $t_i^1 > t_i^2, t_{i+1}^1 > t_{i+1}^2, t_i^1 < t_{i+1}^2$ (Figure 5.2k), $\forall i \in [2\ldots(n_1-2)]$ such as $i \bmod 2 = 0$, and $t_{n_1-1}^1 > t_{n_2-1}^2$, then $j = 2i$, $n = 2(n_1 - 1)$, and the tuple values of the multi-image I can be obtained as follows:

$$\begin{array}{llllllll}
t_1 = t_1^1 & t_{j-2} = t_i^2 & t_{j-1} = t_i^1 & t_j = t_{i+1}^2 & t_{j+1} = t_{i+1}^1 & t_{n-2} = t_{n_2-1}^2 & t_{n-1} = t_{n_1-1}^1 & t_n = t_{n_1}^1 \\
d_1^1 = a_1^1 & d_{j-2}^1 = \varnothing & d_{j-1}^1 = a_i^1 & d_j^1 = \varnothing & d_{j+1}^1 = a_{i+1}^1 & d_{n-2}^1 = \varnothing & d_{n-1}^1 = a_{n_1-1}^1 & d_n^1 = a_{n_1}^1 \\
d_1^2 = a_1^2 & d_{j-2}^2 = a_i^2 & d_{j-1}^2 = \varnothing & d_j^2 = a_{i+1}^2 & d_{j+1}^2 = \varnothing & d_{n-2}^2 = a_{n_2-1}^2 & d_{n-1}^2 = \varnothing & d_n^2 = a_{n_2}^2
\end{array} \tag{5.97}$$

If $t_i^1 = t_{i+1}^2$ (Figure 5.2l), $\forall i \in [2\ldots(n_1-2)]$ such as $i \bmod 2 = 0$ and $t_{n_1-1}^1 = t_{n_2-1}^2$, then $j = \frac{3i}{2}$, $n = \frac{3(n_1-1)}{2}$, and the tuple values of the multi-image I are defined in (5.98).

$$\begin{array}{llllll}
t_1 = t_1^1 & t_{j-1} = t_i^2 & t_j = t_i^1 & t_{j+1} = t_{i+1}^1 & t_{n-1} = t_{n_1-1}^1 & t_n = t_{n_1}^1 \\
d_1^1 = a_1^1 & d_{j-1}^1 = \varnothing & d_j^1 = a_i^1 & d_{j+1}^1 = a_{i+1}^1 & d_{n-1}^1 = a_{n_1-1}^1 & d_n^1 = a_{n_1}^1 \\
d_1^2 = a_1^2 & d_{j-1}^2 = a_i^2 & d_j^2 = a_{i+1}^2 & d_{j+1}^2 = \varnothing & d_{n-1}^2 = a_{n_2-1}^2 & d_n^2 = a_{n_2}^2
\end{array} \tag{5.98}$$

If $t_i^1 = t_{i+1}^2$ (Figure 5.2l), $\forall i \in [2\ldots(n_1-2)]$ such as $i \bmod 2 = 0$, and $t_{n_1-1}^1 < t_{n_2-1}^2$, then $j = \frac{3i}{2}$, $n = \frac{3(n_1-1)}{2} + 1$, and the tuple values of the multi-image I are as follows:

$$\begin{array}{lllllll}
t_1 = t_1^1 & t_{j-1} = t_i^2 & t_j = t_i^1 & t_{j+1} = t_{i+1}^1 & t_{n-2} = t_{n_1-1}^1 & t_{n-1} = t_{n_2-1}^2 & t_n = t_{n_1}^1 \\
d_1^1 = a_1^1 & d_{j-1}^1 = \varnothing & d_j^1 = a_i^1 & d_{j+1}^1 = a_{i+1}^1 & d_{n-2}^1 = a_{n_1-1}^1 & d_{n-1}^1 = \varnothing & d_n^1 = a_{n_1}^1 \\
d_1^2 = a_1^2 & d_{j-1}^2 = a_i^2 & d_j^2 = a_{i+1}^2 & d_{j+1}^2 = \varnothing & d_{n-2}^2 = \varnothing & d_{n-1}^2 = a_{n_2-1}^2 & d_n^2 = a_{n_2}^2
\end{array} \tag{5.99}$$

If $t_i^1 = t_{i+1}^2$ (Figure 5.2l), $\forall i \in [2\ldots(n_1-2)]$ such as $i \bmod 2 = 0$, and $t_{n_1-1}^1 > t_{n_2-1}^2$, then $j = \frac{3i}{2}$, $n = \frac{3(n_1-1)}{2} + 1$, and the tuple values of the multi-image I can be obtained by using (5.100).

$$\begin{array}{lllllll} t_1 = t_1^1 & t_{j-1} = t_i^2 & t_j = t_i^1 & t_{j+1} = t_{i+1}^1 & t_{n-2} = t_{n_2-1}^2 & t_{n-1} = t_{n_1-1}^1 & t_n = t_{n_1}^1 \\ d_1^1 = a_1^1 & d_{j-1}^1 = \emptyset & d_j^1 = a_i^1 & d_{j+1}^1 = a_{i+1}^1 & d_{n-2}^1 = \emptyset & d_{n-1}^1 = a_{n_1-1}^1 & d_n^1 = a_{n_1}^1 \\ d_1^2 = a_1^2 & d_{j-1}^2 = a_i^2 & d_j^2 = a_{i+1}^2 & d_{j+1}^2 = \emptyset & d_{n-2}^2 = a_{n_2-1}^2 & d_{n-1}^2 = \emptyset & d_n^2 = a_{n_2}^2 \end{array} \tag{5.100}$$

If $t_i^1 > t_{i+1}^2$ (Figure 5.2m), $\forall i \in [2\ldots(n_1-2)]$ such as $i \bmod 2 = 0$, and $t_{n_1-1}^1 = t_{n_2-1}^2$, then $j = 2i$, $n = 2n_1 - 3$, and the tuple values of the multi-image I are as follows:

$$\begin{array}{lllllll} t_1 = t_1^1 & t_{j-2} = t_i^2 & t_{j-1} = t_{i+1}^2 & t_j = t_i^1 & t_{j+1} = t_{i+1}^1 & t_{n-1} = t_{n_1-1}^1 & t_n = t_{n_1}^1 \\ d_1^1 = a_1^1 & d_{j-2}^1 = \emptyset & d_{j-1}^1 = \emptyset & d_j^1 = a_i^1 & d_{j+1}^1 = a_{i+1}^1 & d_{n-1}^1 = a_{n_1-1}^1 & d_n^1 = a_{n_1}^1 \\ d_1^2 = a_1^2 & d_{j-2}^2 = a_i^2 & d_{j-1}^2 = a_{i+1}^2 & d_j^2 = \emptyset & d_{j+1}^2 = \emptyset & d_{n-1}^2 = a_{n_2-1}^2 & d_n^2 = a_{n_2}^2 \end{array} \tag{5.101}$$

If $t_i^1 > t_{i+1}^2$ (Figure 5.2m), $\forall i \in [2\ldots(n_1-2)]$ such as $i \bmod 2 = 0$, and $t_{n_1-1}^1 < t_{n_2-1}^2$, then $j = 2i$, $n = 2(n_1 - 1)$, and the tuple values of the multi-image I can be found as defined in (5.102).

$$\begin{array}{llllllll} t_1 = t_1^1 & t_{j-2} = t_i^2 & t_{j-1} = t_{i+1}^2 & t_j = t_i^1 & t_{j+1} = t_{i+1}^1 & t_{n-2} = t_{n_1-1}^1 & t_{n-1} = t_{n_2-1}^2 & t_n = t_{n_1}^1 \\ d_1^1 = a_1^1 & d_{j-2}^1 = \emptyset & d_{j-1}^1 = \emptyset & d_j^1 = a_i^1 & d_{j+1}^1 = a_{i+1}^1 & d_{n-2}^1 = a_{n_1-1}^1 & d_{n-1}^1 = \emptyset & d_n^1 = a_{n_1}^1 \\ d_1^2 = a_1^2 & d_{j-2}^2 = a_i^2 & d_{j-1}^2 = a_{i+1}^2 & d_j^2 = \emptyset & d_{j+1}^2 = \emptyset & d_{n-2}^2 = \emptyset & d_{n-1}^2 = a_{n_2-1}^2 & d_n^2 = a_{n_2}^2 \end{array} \tag{5.102}$$

If $t_i^1 > t_{i+1}^2$ (Figure 5.2m), $\forall i \in [2\ldots(n_1-2)]$ such as $i \bmod 2 = 0$ and $t_{n_1-1}^1 > t_{n_2-1}^2$, then $j = 2i$, $n = 2(n_1 - 1)$, and the tuple values of the multi-image I are as follows:

$$\begin{array}{llllllll} t_1 = t_1^1 & t_{j-2} = t_i^2 & t_{j-1} = t_{i+1}^2 & t_j = t_i^1 & t_{j+1} = t_{i+1}^1 & t_{n-2} = t_{n_2-1}^2 & t_{n-1} = t_{n_1-1}^1 & t_n = t_{n_1}^1 \\ d_1^1 = a_1^1 & d_{j-2}^1 = \emptyset & d_{j-1}^1 = \emptyset & d_j^1 = a_i^1 & d_{j+1}^1 = a_{i+1}^1 & d_{n-2}^1 = \emptyset & d_{n-1}^1 = a_{n_1-1}^1 & d_n^1 = a_{n_1}^1 \\ d_1^2 = a_1^2 & d_{j-2}^2 = a_i^2 & d_{j-1}^2 = a_{i+1}^2 & d_j^2 = \emptyset & d_{j+1}^2 = \emptyset & d_{n-2}^2 = a_{n_2-1}^2 & d_{n-1}^2 = \emptyset & d_n^2 = a_{n_2}^2 \end{array} \tag{5.103}$$

The third group of subcases (Figure 5.3) for $e_d(\bar{t}^1, \bar{t}^2)$ is when $\bar{t}^1 \lhd \bar{t}^2$, i.e. $\bar{t}^1$ *is rarer* than $\bar{t}^2$; it means that $|\bar{t}^1| < |\bar{t}^2|$. Because ratio R between $|\bar{t}^1|$ and $|\bar{t}^2|$ can be various, let us assume that for every i-element in $\bar{t}^1$, there are two elements in $\bar{t}^2$, except t_1^1 and $t_{n_1}^1$, which coincide with t_1^2 and $t_{n_2}^2$, respectively (Figure 5.3). It means that $n_2 \bmod 2 = 0$ and $R = \frac{n_1}{2(n_1-1)}$.

If $t_i^1 = t_i^2$ (Figure 5.3a), $\forall i \in [2\ldots(n_2-2)]$ such as $i \bmod 2 = 0$, then $n = n_2$ and the tuple values of the multi-image I can be defined as:

$$\begin{array}{llll} t_1 = t_1^1 & t_i = t_i^1 & t_{i+1} = t_{i+1}^2 & t_n = t_{n_2}^1 \\ d_1^1 = a_1^1 & d_i^1 = a_i^1 & d_{i+1}^1 = \emptyset & d_n^1 = a_{n_1}^1 \\ d_1^2 = a_1^2 & d_i^2 = a_i^2 & d_{i+1}^2 = a_{i+1}^2 & d_n^2 = a_{n_2}^2 \end{array} \tag{5.104}$$

If $t_i^1 < t_i^2$ (Figure 5.3b), $\forall i \in [2\ldots(n_2-2)]$ such as $i \bmod 2 = 0$, then $j = \frac{3i}{2}$, $n = \frac{3n_2}{2} - 1$, and the tuple values of the multi-image I are as follows:

$$\begin{array}{lllll} t_1 = t_1^1 & t_{j-1} = t_i^1 & t_j = t_i^2 & t_{j+1} = t_{i+1}^2 & t_n = t_{n_2}^1 \\ d_1^1 = a_1^1 & d_{j-1}^1 = a_i^1 & d_j^1 = \emptyset & d_{j+1}^1 = \emptyset & d_n^1 = a_{n_1}^1 \\ d_1^2 = a_1^2 & d_{j-1}^2 = \emptyset & d_j^2 = a_i^2 & d_{j+1}^2 = a_{i+1}^2 & d_n^2 = a_{n_2}^2 \end{array} \tag{5.105}$$

If $t_i^1 = t_{i+1}^2$ (Figure 5.3c), $\forall i \in [2 \ldots (n_2 - 2)]$ such as $i \bmod 2 = 0$, then $n = n_2$ and the tuple values of the multi-image I are defined in (5.106).

$$\begin{array}{llll} t_1 = t_1^1 & t_i = t_i^2 & t_{i+1} = t_i^1 & t_n = t_{n_2}^1 \\ d_1^1 = a_1^1 & d_i^1 = \varnothing & d_{i+1}^1 = a_i^1 & d_n^1 = a_{n_1}^1 \\ d_1^2 = a_1^2 & d_i^2 = a_i^2 & d_{i+1}^2 = a_{i+1}^2 & d_n^2 = a_{n_2}^2 \end{array} \tag{5.106}$$

If $t_i^1 > t_{i+1}^2$ (Figure 5.3d), $\forall i \in [2 \ldots (n_2 - 2)]$ such as $i \bmod 2 = 0$, then $j = \frac{3i}{2}$, $n = \frac{3n_2}{2} - 1$, and the tuple values of the multi-image I are as follows:

$$\begin{array}{lllll} t_1 = t_1^1 & t_{j-1} = t_i^2 & t_j = t_{i+1}^2 & t_{j+1} = t_i^1 & t_n = t_{n_2}^1 \\ d_1^1 = a_1^1 & d_{j-1}^1 = \varnothing & d_j^1 = \varnothing & d_{j+1}^1 = a_i^1 & d_n^1 = a_{n_1}^1 \\ d_1^2 = a_1^2 & d_{j-1}^2 = a_i^2 & d_j^2 = a_{i+1}^2 & d_{j+1}^2 = \varnothing & d_n^2 = a_{n_2}^2 \end{array} \tag{5.107}$$

If $t_i^1 > t_i^2$ and $t_i^1 < t_{i+1}^2$ (Figure 5.3e), $\forall i \in [2 \ldots (n_2 - 2)]$ such as $i \bmod 2 = 0$, then $j = \frac{3i}{2}$, $n = \frac{3n_2}{2} - 1$, and the tuple values of the multi-image I are defined as:

$$\begin{array}{lllll} t_1 = t_1^1 & t_{j-1} = t_i^2 & t_j = t_i^1 & t_{j+1} = t_{i+1}^2 & t_n = t_{n_2}^1 \\ d_1^1 = a_1^1 & d_{j-1}^1 = \varnothing & d_j^1 = a_i^1 & d_{j+1}^1 = \varnothing & d_n^1 = a_{n_1}^1 \\ d_1^2 = a_1^2 & d_{j-1}^2 = a_i^2 & d_j^2 = \varnothing & d_{j+1}^2 = a_{i+1}^2 & d_n^2 = a_{n_2}^2 \end{array} \tag{5.108}$$

The forth group of subcases for $e_d(\bar{t}^1, \bar{t}^2)$ is when $\bar{t}^1 \rhd \bar{t}^2$, i.e. $\bar{t}^1$ *is thicker* than $\bar{t}^2$; it means that $|\bar{t}^1| > |\bar{t}^2|$. Similar to the third group ($\bar{t}^1 \lhd \bar{t}^2$), ratio R between $|\bar{t}^1|$ and $|\bar{t}^2|$ can vary. Thus, we assume that for every i-element in $\bar{t}^2$, there are two elements in $\bar{t}^1$, except t_1^2 and $t_{n_2}^2$, which coincide with t_1^1 and $t_{n_1}^1$, respectively (Figure 5.4). It means that $n_1 \bmod 2 = 0$ and $R = \frac{n_2}{2(n_2-1)}$.

If $t_i^1 = t_i^2$ (Figure 5.4a), $\forall i \in [2 \ldots (n_1 - 2)]$ such as $i \bmod 2 = 0$, then $n = n_1$ and the tuple values of the multi-image I can be defined as:

$$\begin{array}{llll} t_1 = t_1^1 & t_i = t_i^1 & t_{i+1} = t_{i+1}^1 & t_n = t_{n_1}^1 \\ d_1^1 = a_1^1 & d_i^1 = a_i^1 & d_{i+1}^1 = a_{i+1}^1 & d_n^1 = a_{n_1}^1 \\ d_1^2 = a_1^2 & d_i^2 = a_i^2 & d_{i+1}^2 = \varnothing & d_n^2 = a_{n_2}^2 \end{array} \tag{5.109}$$

If $t_i^1 > t_i^2$ (Figure 5.4b), $\forall i \in [2 \ldots (n_1 - 2)]$ such as $i \bmod 2 = 0$, then $j = \frac{3i}{2}$, $n = \frac{3n_1}{2} - 1$, and the tuple values of the multi-image I are as follows:

$$\begin{array}{lllll} t_1 = t_1^1 & t_{j-1} = t_i^2 & t_j = t_i^1 & t_{j+1} = t_{i+1}^1 & t_n = t_{n_1}^1 \\ d_1^1 = a_1^1 & d_{j-1}^1 = \varnothing & d_j^1 = a_i^1 & d_{j+1}^1 = a_{i+1}^1 & d_n^1 = a_{n_1}^1 \\ d_1^2 = a_1^2 & d_{j-1}^2 = a_i^2 & d_j^2 = \varnothing & d_{j+1}^2 = \varnothing & d_n^2 = a_{n_2}^2 \end{array} \tag{5.110}$$

If $t_i^1 < t_i^2$ and $t_{i+1}^1 > t_i^2$ (Figure 5.4c), $\forall i \in [2 \ldots (n_1 - 2)]$ such as $i \bmod 2 = 0$, then $j = \frac{3i}{2}$, $n = \frac{3n_1}{2} - 1$, and the tuple values of the multi-image I are defined in (5.111).

$$\begin{array}{lllll} t_1 = t_1^1 & t_{j-1} = t_i^1 & t_j = t_i^2 & t_{j+1} = t_{i+1}^1 & t_n = t_{n_1}^1 \\ d_1^1 = a_1^1 & d_{j-1}^1 = a_i^1 & d_j^1 = \varnothing & d_{j+1}^1 = a_{i+1}^1 & d_n^1 = a_{n_1}^1 \\ d_1^2 = a_1^2 & d_{j-1}^2 = \varnothing & d_j^2 = a_i^2 & d_{j+1}^2 = \varnothing & d_n^2 = a_{n_2}^2 \end{array} \tag{5.111}$$

If $t^1_{i+1} < t^2_i$ (Figure 5.4d), $\forall i \in [2\ldots(n_1-2)]$ such as $i \bmod 2 = 0$, then $j = \frac{3i}{2}$, $n = \frac{3n_1}{2} - 1$, and the tuple values of the multi-image I are as follows:

$$\begin{array}{lllll} t_1 = t^1_1 & t_{j-1} = t^1_i & t_j = t^1_{i+1} & t_{j+1} = t^2_i & t_n = t^1_{n_1} \\ d^1_1 = a^1_1 & d^1_{j-1} = a^1_i & d^1_j = a^1_{i+1} & d^1_{j+1} = \emptyset & d^1_n = a^1_{n_1} \\ d^2_1 = a^2_1 & d^2_{j-1} = \emptyset & d^2_j = \emptyset & d^2_{j+1} = a^2_i & d^2_n = a^2_{n_2} \end{array} \tag{5.112}$$

If $t^1_{i+1} = t^2_i$ (Figure 5.4e), $\forall i \in [2\ldots(n_1-2)]$ such as $i \bmod 2 = 0$, then $n = n_1$ and the tuple values of the multi-image I are defined in (5.113).

$$\begin{array}{llll} t_1 = t^1_1 & t_i = t^1_i & t_{i+1} = t^1_{i+1} & t_n = t^1_{n_1} \\ d^1_1 = a^1_1 & d^1_i = a^1_i & d^1_{i+1} = a^1_{i+1} & d^1_n = a^1_{n_1} \\ d^2_1 = a^2_1 & d^2_i = \emptyset & d^2_{i+1} = a^2_i & d^2_n = a^2_{n_2} \end{array} \tag{5.113}$$

Thus, we presented all possible subcases for the relation *coincides with*.

The case, which is more general than defined above, is when the same number of measurements of both $\bar{a}^1$ and $\bar{a}^2$ have been fulfilled in the same period of time from the moment α to the moment β but not simultaneously. It means that $\bar{t}^1$ and $\bar{t}^2$ are connected by the relation *between* defined in (5.49). The difference between the relations *coincides with* and *between* consists in defining the first and the last values of tuples in a multi-image. The first values of tuples are to be defined based on comparison of t^1_1 and t^2_1:

$$t_1 = \begin{cases} t^1_1 & \text{if } t^1_1 \le t^2_1 \\ t^2_1 & \text{if } t^1_1 > t^2_1 \end{cases} \quad d^1_1 = \begin{cases} a^1_1 & \text{if } t^1_1 \le t^2_1 \\ \emptyset & \text{if } t^1_1 > t^2_1 \end{cases} \quad d^2_1 = \begin{cases} \emptyset & \text{if } t^1_1 < t^2_1 \\ a^2_1 & \text{if } t^1_1 \ge t^2_1 \end{cases} \tag{5.114}$$

The last values of tuples are to be defined based on comparison of $t^1_{n_1}$ and $t^2_{n_2}$:

$$t_n = \begin{cases} t^1_{n_1} & \text{if } t^1_{n_1} \le t^2_{n_2} \\ t^2_{n_2} & \text{if } t^1_{n_1} > t^2_{n_2} \end{cases} \quad d^1_n = \begin{cases} a^1_{n_1} & \text{if } t^1_{n_1} \le t^2_{n_2} \\ \emptyset & \text{if } t^1_{n_1} > t^2_{n_2} \end{cases} \quad d^2_n = \begin{cases} \emptyset & \text{if } t^1_{n_1} < t^2_{n_2} \\ a^2_{n_2} & \text{if } t^1_{n_1} \ge t^2_{n_2} \end{cases} \tag{5.115}$$

The rest of tuple elements are to be defined in the same way as stated in (5.54)–(5.113).

The case, when measuring of $\bar{a}^1$ forestalls measuring of $\bar{a}^2$, means that $\bar{t}^1$ and $\bar{t}^2$ are connected by the relation *is before* defined in (5.37). Thus, if $b_d(\bar{t}^1, \bar{t}^2)$, then $\forall i_1 \in [1\ldots n_1]$, $\forall i_2 \in [1\ldots n_2]$:

$$\begin{array}{ll} t_{i_1} = t^1_{i_1} & t_{i_2+n_1} = t^2_{i_2} \\ d^1_{i_1} = a^1_{i_1} & d^1_{i_2+n_1} = \emptyset \\ d^2_{i_1} = \emptyset & d^2_{i_2+n_1} = a^2_{i_2} \end{array} \tag{5.116}$$

The case, when measuring of $\bar{a}^1$ succeeds measuring of $\bar{a}^2$, means that $\bar{t}^1$ and $\bar{t}^2$ are connected by the relation *is after* defined in (5.38). Thus, if $bi_d(\bar{t}^1, \bar{t}^2)$, then $\forall i_1 \in [1\ldots n_1]$, $\forall i_2 \in [1\ldots n_2]$:

$$\begin{array}{ll} t_{i_2} = t^2_{i_2} & t_{i_1+n_2} = t^1_{i_1} \\ d^1_{i_2} = \emptyset & d^1_{i_1+n_2} = a^1_{i_1} \\ d^2_{i_2} = a^2_{i_2} & d^2_{i_1+n_2} = \emptyset \end{array} \tag{5.117}$$

The case, when measuring of $\bar{a}^1$ forestalls measuring of $\bar{a}^2$, but measuring of the last value of $\bar{a}^1$ coincides with measuring of the first value of $\bar{a}^2$, means that $\bar{t}^1$ and $\bar{t}^2$ are connected by the relation *meets* defined in (5.39). Thus, if $m_d(\bar{t}^1, \bar{t}^2)$, then $\forall i_1 \in [1 \ldots (n_1 - 1)]$, $\forall i_2 \in [2 \ldots n_2]$:

$$\begin{array}{lll} t_{i_1} = t^1_{i_1} & t_{n_1} = t^1_{n_1} & t_{i_2+n_1-1} = t^2_{i_2} \\ d^1_{i_1} = a^1_{i_1} & d^1_{n_1} = a^1_{n_1} & d^1_{i_2+n_1-1} = \varnothing \\ d^2_{i_1} = \varnothing & d^2_{n_1} = a^2_1 & d^2_{i_2+n_1-1} = a^2_{i_2} \end{array} \tag{5.118}$$

The case, when measuring of $\bar{a}^1$ succeeds measuring of $\bar{a}^2$, but measuring of the first value of $\bar{a}^1$ coincides with measuring of the last value of $\bar{a}^2$, means that $\bar{t}^1$ and $\bar{t}^2$ are connected by the relation *is met by* defined in (5.40). Thus, if $mi_d(\bar{t}^1, \bar{t}^2)$, then $\forall i_1 \in [2 \ldots n_1]$, $\forall i_2 \in [1 \ldots (n_2 - 1)]$:

$$\begin{array}{lll} t_{i_2} = t^2_{i_2} & t_{n_2} = t^2_{n_2} & t_{i_1+n_2-1} = t^1_{i_1} \\ d^1_{i_2} = \varnothing & d^1_{n_2} = a^1_1 & d^1_{i_1+n_2-1} = a^1_{i_1} \\ d^2_{i_2} = a^2_{i_2} & d^2_{n_2} = a^2_{n_2} & d^2_{i_1+n_2-1} = \varnothing \end{array} \tag{5.119}$$

The case, when measuring of $\bar{a}^1$ forestalls measuring of $\bar{a}^2$, but measuring of K last values of $\bar{a}^1$ coincides with measuring of K first values of $\bar{a}^2$, means that $\bar{t}^1$ and $\bar{t}^2$ are connected by the relation *overlaps* defined in (5.41). Thus, if $o_d(\bar{t}^1, \bar{t}^2)$, then $\forall i_1 \in [1 \ldots (n_1 - K)]$, $\forall i_2 \in [(K+1) \ldots n_2]$, $\forall k \in [1 \ldots K]$:

$$\begin{array}{lll} t_{i_1} = t^1_{i_1} & t_{n_1+k-K} = t^2_k & t_{n_1+i_2-K} = t^2_{i_2} \\ d^1_{i_1} = a^1_{i_1} & d^1_{n_1+k-K} = a^1_{n_1+k-K} & d^1_{n_1+i_2-K} = \varnothing \\ d^2_{i_1} = \varnothing & d^2_{n_1+k-K} = a^2_k & d^2_{n_1+i_2-K} = a^2_{i_2} \end{array} \tag{5.120}$$

The case, when measuring of $\bar{a}^1$ succeeds measuring of $\bar{a}^2$, but measuring of K first values of $\bar{a}^1$ coincides with measuring of K last values of $\bar{a}^2$, means that $\bar{t}^1$ and $\bar{t}^2$ are connected by the relation *is overlapped by* defined in (5.42). Thus, if $oi_d(\bar{t}^1, \bar{t}^2)$, then $\forall i_1 \in [(K+1) \ldots n_1]$, $\forall i_2 \in [1 \ldots (n_2 - K)]$, and $\forall k \in [1 \ldots K]$:

$$\begin{array}{lll} t_{i_2} = t^2_{i_2} & t_{n_2+k-K} = t^1_k & t_{i_1+n_2-K} = t^1_{i_1} \\ d^1_{i_2} = \varnothing & d^1_{n_2+k-K} = a^1_k & d^1_{i_1+n_2-K} = a^1_{i_1} \\ d^2_{i_2} = a^2_{i_2} & d^2_{n_2+k-K} = a^2_{n_2+k-K} & d^2_{i_1+n_2-K} = \varnothing \end{array} \tag{5.121}$$

The case, when measuring of $\bar{a}^1$ starts later and at the same time it finishes earlier than measuring of $\bar{a}^2$ does, means that $\bar{t}^1$ and $\bar{t}^2$ are connected by the relation *during* defined in (5.43). Thus, if $d_d(\bar{t}^1, \bar{t}^2)$, therefore, $\bar{t}^2 = \bar{t}^{2,\mathrm{I}} \cup \bar{t}^{2,\mathrm{II}} \cup \bar{t}^{2,\mathrm{III}}$ such as $b_d(\bar{t}^{2,\mathrm{I}}, \bar{t}^1)$, $w_d^{(t^1_1, t^1_{n_1})}(\bar{t}^{2,\mathrm{II}}, \bar{t}^1)$, and $bi_d(\bar{t}^{2,\mathrm{III}}, \bar{t}^1)$. Then, $\forall\, k_1 \in [1 \ldots K_1]$, $\forall\, k_2 \in [K_2 \ldots n_2]$ such as $t^2_{K_1} < t^1_1$ and $t^2_{K_2} > t^1_{n_1}$, we can define the first values $t_{k_1} (t_{k_1} \in \bar{t}^{2,\mathrm{I}})$, $d^1_{k_1}$, $d^2_{k_1}$, and the last values $t_{k_2} (t_{k_2} \in \bar{t}^{2,\mathrm{III}})$, $d^1_{k_2}$, $d^2_{k_2}$ of the resulting tuples $\bar{t}$, $\bar{d}^1$, and $\bar{d}^2$ as defined in (5.122). The rest of tuple elements ($t_k \in \bar{t}^{2,\mathrm{II}}$, d^1_k, d^2_k) of the resulting tuples $\bar{t}$, $\bar{d}^1$, and $\bar{d}^2$ can be obtained accordingly to subcases defined in (5.54)–(5.115). Then, the length of the resulting tuple $\bar{t}$ is $n = K_1 + m + n_2 - K_2 + 1$, where m is the length of the tuple obtained as a result of synchronization of tuples $\bar{t}^{2,\mathrm{II}}$ and $\bar{t}^1$.

$$\begin{array}{ll} t_{k_1} = t^2_{k_1} & t_{n-n_2+k_2} = t^2_{k_2} \\ d^1_{k_1} = \varnothing & d^1_{n-n_2+k_2} = \varnothing \\ d^2_{k_1} = a^2_{k_1} & d^2_{n-n_2+k_2} = a^2_{k_2} \end{array} \tag{5.122}$$

The case, when measuring of $\bar{a}^1$ starts earlier and at the same time it finishes later than measuring of $\bar{a}^2$ does, means that $\bar{t}^1$ and $\bar{t}^2$ are connected by the relation *contains* defined in (5.44). Thus, if $di_d(\bar{t}^1, \bar{t}^2)$, therefore, $\bar{t}^1 =$

$\bar{t}^{1,\mathrm{I}} \cup \bar{t}^{1,\mathrm{II}} \cup \bar{t}^{1,\mathrm{III}}$ such as $b_d(\bar{t}^{1,\mathrm{I}}, \bar{t}^2)$, $w_d^{(t_1^2, t_{n_2}^2)}(\bar{t}^{1,\mathrm{II}}, \bar{t}^2)$, and $bi_d(\bar{t}^{1,\mathrm{III}}, \bar{t}^2)$. Then, $\forall\, k_1 \in [1 \ldots K_1]$ and $\forall\, k_2 \in [K_2 \ldots n_1]$ such as $t_{K_1}^1 < t_1^2$ and $t_{K_2}^1 > t_{n_2}^2$, we can define the first values t_{k_1} ($t_{k_1} \in \bar{t}^{1,\mathrm{I}}$), $d_{k_1}^1$, and $d_{k_1}^2$ and the last values t_{k_2} ($t_{k_2} \in \bar{t}^{1,\mathrm{III}}$), $d_{k_2}^1$, and $d_{k_2}^2$ of the resulting tuples $\bar{t}$, $\bar{d}^1$, and $\bar{d}^2$ as defined in (5.123). The rest of tuple elements ($t_k \in \bar{t}^{1,\mathrm{II}}$, d_k^1, d_k^2) of the resulting tuples $\bar{t}$, $\bar{d}^1$, and $\bar{d}^2$ can be obtained accordingly to subcases defined in (5.54)–(5.115). Then, the length of the resulting tuple $\bar{t}$ is $n = K_1 + m + n_1 - K_2 + 1$, where m is the length of the tuple obtained as a result of synchronization of tuples $\bar{t}^{1,\mathrm{II}}$ and $\bar{t}^2$.

$$\begin{aligned} t_{k_1} &= t_{k_1}^1 & t_{n-n_1+k_2} &= t_{k_2}^1 \\ d_{k_1}^1 &= a_{k_1}^1 & d_{n-n_1+k_2}^1 &= a_{k_2}^1 \\ d_{k_1}^2 &= \varnothing & d_{n-n_1+k_2}^2 &= \varnothing \end{aligned} \tag{5.123}$$

The case, when measuring of both $\bar{a}^1$ and $\bar{a}^2$ start simultaneously, but measuring of $\bar{a}^1$ finishes earlier than measuring of $\bar{a}^2$ does, means that $\bar{t}^1$ and $\bar{t}^2$ are connected by the relation *starts* defined in (5.45). Thus, if $s_d(\bar{t}^1, \bar{t}^2)$, therefore, $\bar{t}^2 = \bar{t}^{2,\mathrm{I}} \cup \bar{t}^{2,\mathrm{II}}$ such as $w_d^{(t_1^1, t_{n_1}^1)}(\bar{t}^{2,\mathrm{I}}, \bar{t}^1)$ and $bi_d(\bar{t}^{2,\mathrm{II}}, \bar{t}^1)$. Then, $\forall k_2 \in [K \ldots n_2]$ such as $t_K^2 > t_{n_1}^1$, we can define the first value t_1($t_1 \in \bar{t}^{2,\mathrm{I}}$), d_1^1, and d_1^2 and the last values t_{k_2} ($t_{k_2} \in \bar{t}^{2,\mathrm{II}}$), $d_{k_2}^1$, and $d_{k_2}^2$ of the resulting tuples $\bar{t}$, $\bar{d}^1$, and $\bar{d}^2$ as defined in (5.124). The rest of tuple elements ($t_{k_1} \in \bar{t}^{2,\mathrm{I}}$, $d_{k_1}^1$, $d_{k_1}^2$, and $\forall k_1 \in [2 \ldots m]$, where m is the length of the tuple obtained as a result of synchronization of tuples $\bar{t}^{2,\mathrm{I}}$ and $\bar{t}^1$) of the resulting tuples $\bar{t}$, $\bar{d}^1$, and $\bar{d}^2$ can be obtained accordingly to subcases defined in (5.54)–(5.115). Then, the length of the resulting tuple $\bar{t}$ is $n = m + n_2 - K + 1$.

$$\begin{aligned} t_1 &= t_1^1 & t_{n-n_2+k_2} &= t_{k_2}^2 \\ d_1^1 &= a_1^1 & d_{n-n_2+k_2}^1 &= \varnothing \\ d_1^2 &= a_1^2 & d_{n-n_2+k_2}^2 &= a_{k_2}^2 \end{aligned} \tag{5.124}$$

The case, when measuring of both $\bar{a}^1$ and $\bar{a}^2$ start simultaneously, but measuring of $\bar{a}^1$ finishes later than measuring of $\bar{a}^2$ does, means that $\bar{t}^1$ and $\bar{t}^2$ are connected by the relation *is started by* defined in (5.46). Thus, if $si_d(\bar{t}^1, \bar{t}^2)$, therefore, $\bar{t}^1 = \bar{t}^{1,\mathrm{I}} \cup \bar{t}^{1,\mathrm{II}}$ such as $w_d^{(t_1^2, t_{n_2}^2)}(\bar{t}^{1,\mathrm{I}}, \bar{t}^2)$ and $bi_d(\bar{t}^{1,\mathrm{II}}, \bar{t}^2)$. Then, $\forall k_2 \in [K \ldots n_1]$ such as $t_K^1 > t_{n_2}^2$, we can define the first value t_1($t_1 \in \bar{t}^{1,\mathrm{I}}$), d_1^1, d_1^2, and the last values t_{k_2} ($t_{k_2} \in \bar{t}^{1,\mathrm{II}}$), $d_{k_2}^1$, and $d_{k_2}^2$ of the resulting tuples $\bar{t}$, $\bar{d}^1$, and $\bar{d}^2$ as defined in (5.125). The rest of tuple elements ($t_{k_1} \in \bar{t}^{1,\mathrm{I}}$, $d_{k_1}^1$, $d_{k_1}^2$, and $\forall\, k_1 \in [2 \ldots m]$, where m is the length of the tuple obtained as a result of synchronization of tuples $\bar{t}^{1,\mathrm{I}}$ and $\bar{t}^2$) of the resulting tuples $\bar{t}$, $\bar{d}^1$, and $\bar{d}^2$ can be obtained accordingly to subcases defined in (5.54)–(5.115). Then, the length of the resulting tuple $\bar{t}$ is $n = m + n_1 - K + 1$.

$$\begin{aligned} t_1 &= t_1^1 & t_{n-n_1+k_2} &= t_{k_2}^1 \\ d_1^1 &= a_1^1 & d_{n-n_1+k_2}^1 &= a_{k_2}^1 \\ d_1^2 &= a_1^2 & d_{n-n_1+k_2}^2 &= \varnothing \end{aligned} \tag{5.125}$$

The case, when measuring of both $\bar{a}^1$ and $\bar{a}^2$ finish simultaneously, but measuring of $\bar{a}^1$ starts later than measuring of $\bar{a}^2$ does, means that $\bar{t}^1$ and $\bar{t}^2$ are connected by the relation *finishes* defined in (5.47). Thus, if $f_d(\bar{t}^1, \bar{t}^2)$, therefore, $\bar{t}^2 = \bar{t}^{2,\mathrm{I}} \cup \bar{t}^{2,\mathrm{II}}$ such as $b_d(\bar{t}^{2,\mathrm{I}}, \bar{t}^1)$ and $w_d^{(t_1^1, t_{n_1}^1)}(\bar{t}^{2,\mathrm{II}}, \bar{t}^1)$. Then, $\forall\, k_1 \in [1 \ldots K]$, such as $t_K^2 < t_1^1$, we can define the first values t_{k_1} ($t_{k_1} \in \bar{t}^{2,\mathrm{I}}$), $d_{k_1}^1$, and $d_{k_1}^2$ and the last value t_n ($t_n \in \bar{t}^{2,\mathrm{II}}$), d_n^1, and d_n^2 of the resulting tuples $\bar{t}$, $\bar{d}^1$, and $\bar{d}^2$ as defined in (5.126). The rest of tuple elements ($t_{k_2} \in \bar{t}^{2,\mathrm{II}}$, $d_{k_2}^1$, $d_{k_2}^2$, and $\forall k_2 \in [(K+1) \ldots (n-1)]$, where the length of the resulting tuple $\bar{t}$ is $n = K + m$; m is the length of the tuple obtained as a result of synchronization of tuples $\bar{t}^{2,\mathrm{II}}$ and $\bar{t}^1$) of the resulting tuples $\bar{t}$, $\bar{d}^1$, and $\bar{d}^2$ can be obtained accordingly to subcases defined in

(5.54)–(5.115).

$$\begin{array}{ll} t_{k_1} = t^2_{k_1} & t_n = t^1_{n_1} \\ d^1_{k_1} = \varnothing & d^1_n = a^1_{n_1} \\ d^2_{k_1} = a^2_{k_1} & d^2_n = a^2_{n_2} \end{array} \tag{5.126}$$

The case, when measuring of both $\bar{a}^1$ and $\bar{a}^2$ finish simultaneously, but measuring of $\bar{a}^1$ starts earlier than measuring of $\bar{a}^2$ does, means that $\bar{t}^1$ and $\bar{t}^2$ are connected by the relation *is finished by* defined in (5.48). Thus, if $fi_d(\bar{t}^1, \bar{t}^2)$, therefore, $\bar{t}^1 = \bar{t}^{1,\mathrm{I}} \cup \bar{t}^{1,\mathrm{II}}$ such as $b_d(\bar{t}^{1,\mathrm{I}}, \bar{t}^2)$ and $w_d^{(t^2_1, t^2_{n_2})}(\bar{t}^{1,\mathrm{II}}, \bar{t}^2)$. Then, $\forall k_1 \in [1 \dots K]$, such as $t^1_K < t^2_1$, we can define the first values t_{k_1} $(t_{k_1} \in \bar{t}^{1,\mathrm{I}})$, $d^1_{k_1}$, and $d^2_{k_1}$ and the last value t_n $(t_n \in \bar{t}^{1,\mathrm{II}})$, d^1_n, and d^2_n of the resulting tuples $\bar{t}$, $\bar{d}^1$, and $\bar{d}^2$ as defined in (5.127). The rest of tuple elements ($t_{k_2} \in \bar{t}^{1,\mathrm{II}}$, $d^1_{k_2}$, and $d^2_{k_2}$, $\forall k_2 \in [(K+1) \dots (n-1)]$, where the length of the resulting tuple $\bar{t}$ is $n = K + m$; m is the length of the tuple obtained as a result of synchronization of tuples $\bar{t}^{1,\mathrm{II}}$ and $\bar{t}^2$) of the resulting tuples $\bar{t}$, $\bar{d}^1$, and $\bar{d}^2$ can be obtained accordingly to subcases defined in (5.54)–(5.115).

$$\begin{array}{ll} t_{k_1} = t^1_{k_1} & t_n = t^2_{n_2} \\ d^1_{k_1} = a^1_{k_1} & d^1_n = a^1_{n_1} \\ d^2_{k_1} = \varnothing & d^2_n = a^2_{n_2} \end{array} \tag{5.127}$$

As one can see, in the case of crisp time value processing, there is a great number of cases for data synchronization. Thus, in applications where the crisp time values are not strictly required, it is reasonable to use fuzzy synchronization.

5.6 Fuzzy Synchronization

For fuzzy data synchronization of multi-images, we need fuzzy equivalents of temporal relations defined by (5.36)–(5.49). There is a number of works on this topic, in particular, [11, 14–18]. In this research, we apply the approach formulated in [11, 18]. Let us present it in detail.

The degree $bb^{\ll}_{(\alpha,\beta)}(\bar{t}^1, \bar{t}^2)$ to which the beginning of $\bar{t}^1$ is approximately $\alpha (\alpha \in \mathbb{R})$ time units before the beginning of $\bar{t}^2$, where β $(\beta \geq 0)$ indicates a degree of tolerance, is a fuzzification generalization of strict inequality $t^1_1 < t^2_1$; it is defined as follows:

$$bb^{\ll}_{(\alpha,\beta)}(\bar{t}^1, \bar{t}^2) = \bar{t}^1 \circ_{T_w} (L^{\ll}_{(\alpha,\beta)} \rhd_{I_{T_w}} \bar{t}^2) \tag{5.128}$$

where the operation $\circ_T$ means composition of fuzzy relations as defined in (5.129); the operation $\rhd_I$ means superproduct as defined in (5.130) according to [19]; T_w is Łukasiewicz t-norm defined in (5.131); I_{T_w} is a Łukasiewicz implicator defined in (5.132); $L^{\ll}_{(\alpha,\beta)}$ is the fuzzy relation between time points t^1_1 and t^2_1 defined in (5.133).

The sup-T composition of fuzzy relations is defined as:

$$(R_1 \circ_T R_2)(x, z) = \sup_{y \in Y} T(R_1(x, y), R_2(y, z)) \tag{5.129}$$

where R_1 is a fuzzy relation from a set X to a set Y; R_2 is a fuzzy relation from a set Y to a set Z; $x \in X, y \in Y, z \in Z$.

The I-superproduct of fuzzy relations is defined as:

$$R_1 \rhd_I R_2(x, z) = \inf_{y \in Y} I(R_2(y, z), R_1(x, y)) \tag{5.130}$$

The Łukasiewicz triangular norm and its corresponding implicator are given as:

$$T_w(a, b) = \max(0, a + b - 1) \tag{5.131}$$

where $a, b \in [0, 1]$.

$$I_{T_w}(a, b) = \min(1 - a + b, 1) \tag{5.132}$$

The fuzzy relation $L^{\ll}_{(\alpha,\beta)}$ models a metric constraint between time points and is given as:

$$L^{\ll}_{(\alpha,\beta)}(t_1^1, t_1^2) = \begin{cases} 1 & \text{if } t_1^2 - t_1^1 > \alpha + \beta \\ 0 & \text{if } t_1^2 - t_1^1 \leq \alpha \\ \dfrac{t_1^2 - t_1^1 - \alpha}{\beta} & \text{otherwise} \end{cases} \tag{5.133}$$

where α is the distance (in time units) between t_1^1 and t_1^2 and β is the degree of tolerance.

The degree $bb^{\leqslant}_{(\alpha,\beta)}(\bar{t}^1, \bar{t}^2)$ to which the beginning of $\bar{t}^1$ is approximately α time units before or the same time as the beginning of $\bar{t}^2$ is a generalization of the nonstrict inequality $t_1^1 \leq t_1^2$; it is defined as follows:

$$bb^{\leqslant}_{(\alpha,\beta)}(\bar{t}^1, \bar{t}^2) = 1 - bb^{\ll}_{(\alpha,\beta)}(\bar{t}^2, \bar{t}^1) \tag{5.134}$$

The degree $ee^{\ll}_{(\alpha,\beta)}(\bar{t}^1, \bar{t}^2)$ to which the end of $\bar{t}^1$ is approximately α time units before the end of $\bar{t}^2$ is a generalization of the strict inequality $t_{n_1}^1 < t_{n_2}^2$; it is defined as follows:

$$ee^{\ll}_{(\alpha,\beta)}(\bar{t}^1, \bar{t}^2) = (\bar{t}^1 \lhd_{I_{T_w}} L^{\ll}_{(\alpha,\beta)}) \circ_{T_w} \bar{t}^2 \tag{5.135}$$

where the operation $\lhd_I$ means subproduct as defined in (5.136) according to [19].

$$R_1 \lhd_I R_2(x, z) = \inf_{y \in Y} I(R_1(x, y), R_2(y, z)) \tag{5.136}$$

The degree $ee^{\leqslant}_{(\alpha,\beta)}(\bar{t}^1, \bar{t}^2)$ to which the end of $\bar{t}^1$ is approximately α time units before or the same time as the end of $\bar{t}^2$ is a generalization of the nonstrict inequality $t_{n_1}^1 \leq t_{n_2}^2$; it is defined as follows:

$$ee^{\leqslant}_{(\alpha,\beta)}(\bar{t}^1, \bar{t}^2) = 1 - ee^{\ll}_{(\alpha,\beta)}(\bar{t}^2, \bar{t}^1) \tag{5.137}$$

The degree $be^{\ll}_{(\alpha,\beta)}(\bar{t}^1, \bar{t}^2)$ to which the beginning of $\bar{t}^1$ is approximately α time units before the end of $\bar{t}^2$ is a generalization of the strict inequality $t_1^1 < t_{n_2}^2$; it is defined as follows:

$$be^{\ll}_{(\alpha,\beta)}(\bar{t}^1, \bar{t}^2) = \bar{t}^1 \circ_{T_w} L^{\ll}_{(\alpha,\beta)} \circ_{T_w} \bar{t}^2 \tag{5.138}$$

The degree $be^{\leqslant}_{(\alpha,\beta)}(\bar{t}^1, \bar{t}^2)$ to which the beginning of $\bar{t}^1$ is approximately α time units before or the same time as the end of $\bar{t}^2$ is a generalization of the nonstrict inequality $t_1^1 \leq t_{n_2}^2$; it is defined as follows:

$$be^{\leqslant}_{(\alpha,\beta)}(\bar{t}^1, \bar{t}^2) = 1 - be^{\ll}_{(\alpha,\beta)}(\bar{t}^2, \bar{t}^1) \tag{5.139}$$

The degree $eb^{\ll}_{(\alpha,\beta)}(\bar{t}^1, \bar{t}^2)$ to which the end of $\bar{t}^1$ is approximately α time units before the beginning of $\bar{t}^2$ is a generalization of the strict inequality $t_{n_1}^1 < t_1^2$; it is defined as follows:

$$eb^{\ll}_{(\alpha,\beta)}(\bar{t}^1, \bar{t}^2) = \bar{t}^1 \lhd_{I_{T_w}} L^{\ll}_{(\alpha,\beta)} \rhd_{I_{T_w}} \bar{t}^2 \tag{5.140}$$

The degree $eb^{\leqslant}_{(\alpha,\beta)}(\bar{t}^1, \bar{t}^2)$ to which the end of $\bar{t}^1$ is approximately α time units before or the same time as the beginning of $\bar{t}^2$ is a generalization of the nonstrict inequality $t_{n_1}^1 \leq t_1^2$; it is defined as follows:

$$eb^{\leqslant}_{(\alpha,\beta)}(\bar{t}^1, \bar{t}^2) = 1 - eb^{\ll}_{(\alpha,\beta)}(\bar{t}^2, \bar{t}^1) \tag{5.141}$$

As one can see in Section 5.5, crisp data synchronization in the case when $e_d(\bar{t}^1, \bar{t}^2)$ is quite complicated because of many subcases. We can simplify data synchronization if we consider the degree of closeness between time moments t_i^1 and t_i^2. Such degree can be estimated as $eb^{\ll}_{(\alpha,\beta)}(t_i^1, t_i^2)$. Let us prove that if Δt is a minimal distance between t_i^1 and t_i^2 (we assume Δt as a time measurement error), then $eb^{\ll}_{(\alpha,\beta)}(t_i^1, t_i^2) = 0$ and, thus, t_i^1 and t_i^2 can be considered as one fuzzy moment of time $\tilde{t}_i$.

According to (5.140), $eb^{\ll}_{(\alpha,\beta)}(t_i^1, t_i^2) = t_i^1 \lhd_{I_{T_W}} L^{\ll}_{(\alpha,\beta)}(t_i^1, t_i^2) \rhd_{I_{T_W}} t_i^2$. From expressions (5.130) and (5.136), we obtain the following:

$$eb^{\ll}_{(\alpha,\beta)}(t_i^1, t_i^2) = \inf_{t_i^1 \in \bar{t}^1}(\inf_{t_i^2 \in \bar{t}^2} I_{T_w}(t_i^2, I_{T_w}(t_i^1, L^{\ll}_{(\alpha,\beta)}(t_i^1, t_i^2)))) \tag{5.142}$$

Taking into account (5.132), we have:

$$\begin{aligned} I_{T_w}(t_i^1, L^{\ll}_{(\alpha,\beta)}(t_i^1, t_i^2)) &= \min(1 - 1 + L^{\ll}_{(\alpha,\beta)}(t_i^1, t_i^2), 1) \\ &= L^{\ll}_{(\alpha,\beta)}(t_i^1, t_i^2) \end{aligned} \tag{5.143}$$

Similarly,

$$\begin{aligned} I_{T_w}(t_i^2, L^{\ll}_{(\alpha,\beta)}(t_i^1, t_i^2)) &= \min(1 - 1 + L^{\ll}_{(\alpha,\beta)}(t_i^1, t_i^2), 1) \\ &= L^{\ll}_{(\alpha,\beta)}(t_i^1, t_i^2) \end{aligned} \tag{5.144}$$

Thus, we can rewrite (5.142) as follows:

$$eb^{\ll}_{(\alpha,\beta)}(t_i^1, t_i^2) = \inf_{t_i^1 \in \bar{t}^1}(\inf_{t_i^2 \in \bar{t}^2} L^{\ll}_{(\alpha,\beta)}(t_i^1, t_i^2)) \tag{5.145}$$

Then, we have from (5.133) that:

$$L^{\ll}_{(\alpha,\beta)}(t_i^1, t_i^2) = \begin{cases} 1 \text{ if } t_i^2 - t_i^1 > \alpha + \beta \\ 0 \text{ if } t_i^2 - t_i^1 \le \alpha \\ \dfrac{t_i^2 - t_i^1 - \alpha}{\beta} \text{ otherwise} \end{cases} \tag{5.146}$$

Since $t_i^2 = t_i^1 + \Delta t$, we have $t_i^2 - t_i^1 < \alpha$ in (5.146) because Δt is a time measurement error, i.e. it is the least possible distance between t_i^1 and t_i^2.

Thus, $L^{\ll}_{(\alpha,\beta)}(t_i^1, t_i^2) = 0$ and therefore $eb^{\ll}_{(\alpha,\beta)}(t_i^1, t_i^2) = 0$. From practical point of view, it means that if there is a time measurement error in two time value sequences $\bar{t}^1$ and $\bar{t}^2$ (actually, it can happen in many cases if the measurements are taken simultaneously but by different executors, e.g. atmosphere pressure is being measured in one research unit and the sun's radiation is being measured by another research unit), we may neglect this time measurement error.

Now let the distance between time moments t_i^1 and t_i^2 exceed Δt and be equal to α. Then, in (5.146), we have $t_i^2 - t_i^1 = \alpha$; thus, $L^{\ll}_{(\alpha,\beta)}(t_i^1, t_i^2) = 0$ and therefore $eb^{\ll}_{(\alpha,\beta)}(t_i^1, t_i^2) = 0$. Thus, the conclusion is that we again can consider these two time moments t_i^1 and t_i^2 as one fuzzy moment of time $\tilde{t}_i$.

Let us consider different cases (Figure 5.5) for α between neighboring values in $\bar{t}^1$ and $\bar{t}^2$ when $eb^{\ll}_{(\alpha,\beta)}(t_i^1, t_i^2) = 0$ in order to simplify data value synchronization defined by (5.54)–(5.113). Let α_1 be a distance between t_i^1 and t_i^2, let α_2 be a distance between t_i^2 and t_{i+1}^1, and let α_3 be a distance between t_{i+1}^1 and t_{i+1}^2. Besides, let α be a minimal distance between such t_i^1 and t_i^2 that we can consider them as one vague time moment to be presented as one fuzzy value $\tilde{t}$ in the time tuple of a multi-image.

If both $\alpha_1 < \alpha$ and $\alpha_3 < \alpha$ are true (Figure 5.5a), then regardless of the value of α_2, the data synchronization can be simplified to (5.147).

$$\begin{aligned} &\tilde{t}_i = t_i^1 && \tilde{t}_{i+1} = t_{i+1}^1 \\ &d_i^1 = a_i^1 && d_{i+1}^1 = a_{i+1}^1 \\ &d_i^2 = a_i^2 && d_{i+1}^2 = a_{i+1}^2 \end{aligned} \tag{5.147}$$

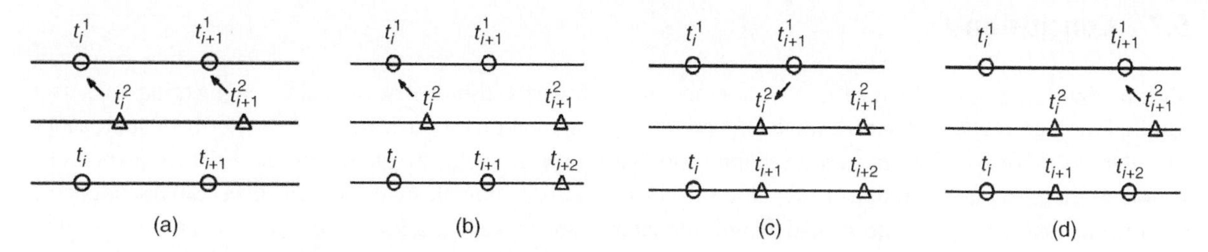

Figure 5.5 Fuzzy synchronization cases. t_i^1 and t_{i+1}^1 are elements of the time tuple $\bar{t}^1$; t_i^2 and t_{i+1}^2 are elements of the time tuple $\bar{t}^2$; t_j, t_{j+1}, and t_{j+2} are elements of the synchronized time tuple $\bar{t}$. For two measurement processes: (a) measurements of both parameters are almost simultaneous at the current and the next moments of time; (b) the current measurements of both parameters are almost simultaneous; (c) the next measurement of the 1st parameter and the current measurement of the 2nd parameter are almost simultaneous; (d) the next measurements of both parameters are almost simultaneous.

If both $\alpha_1 < \alpha$ and $\alpha_3 > \alpha$ are true (Figure 5.5b), then regardless of the value of α_2, the data synchronization can be simplified to (5.148).

$$\begin{array}{lll} \tilde{t}_i = t_i^1 & t_{i+1} = t_{i+1}^1 & t_{i+2} = t_{i+1}^2 \\ d_i^1 = a_i^1 & d_{i+1}^1 = a_{i+1}^1 & d_{i+2}^1 = \varnothing \\ d_i^2 = a_i^2 & d_{i+1}^2 = \varnothing & d_{i+2}^2 = a_{i+1}^2 \end{array} \tag{5.148}$$

If both $\alpha_1 > \alpha$ and $\alpha_2 < \alpha$ are true (Figure 5.5c), then regardless of the value of α_3, the data synchronization can be simplified to (5.149).

$$\begin{array}{lll} t_i = t_i^1 & \tilde{t}_{i+1} = t_i^2 & t_{i+2} = t_{i+1}^2 \\ d_i^1 = a_i^1 & d_{i+1}^1 = a_{i+1}^1 & d_{i+2}^1 = \varnothing \\ d_i^2 = \varnothing & d_{i+1}^2 = a_i^2 & d_{i+2}^2 = a_{i+1}^2 \end{array} \tag{5.149}$$

If either $\alpha_1 > \alpha > \alpha_2 > \alpha_3$ or both $\alpha_1 > \alpha > \alpha_3$ and $\alpha_2 > \alpha$ (Figure 5.5d) is true, the data synchronization can be simplified to (5.150).

$$\begin{array}{lll} t_i = t_i^1 & t_{i+1} = t_i^2 & \tilde{t}_{i+2} = t_{i+1}^1 \\ d_i^1 = a_i^1 & d_{i+1}^1 = \varnothing & d_{i+2}^1 = a_{i+1}^1 \\ d_i^2 = \varnothing & d_{i+1}^2 = a_i^2 & d_{i+2}^2 = a_{i+1}^2 \end{array} \tag{5.150}$$

Thus, fuzzy synchronization allows us to simplify synchronization of data values measured at neighboring moments of time when the time difference is small enough that we can neglect this difference. We considered $eb^{\ll}_{(\alpha,\beta)}(t_i^1, t_i^2)$ as the degree of closeness between time moments t_i^1 and t_i^2 but other relation degrees defined in (5.128), (5.134), (5.135), and (5.137)–(5.141) can be employed depending on a specific condition of the data synchronization task.

When the data sequences have been synchronized, we can compose a multi-image of the object of study and process this multi-image (which is the complex representation of all information we know about this object) employing the logical operations, ordering operations, arithmetical operations, and relations, which are defined in ASA, according to a specific logic and by using specific methods of solving a given task related to the object of study.

5.7 Conclusion

Multimodal data can be obtained in a wide range of applications in industry, medicine, engineering, etc., as a result of measurement, computing, modeling, prediction, and other data processing. The majority of approaches consider multimodal data sequences as separate ones like temperature data sequence, pressure data sequence, etc. However, complex representation of all multimodal data sequences as a sole mathematical object can be conducive to more effective solving of tasks related to multimodal data analysis and processing. In this chapter, we presented the apparatus of ASA that enables such complex representation and processing of multimodal data in the form of aggregates and multi-images.

Multi-images data synchronization is an important task because it ensures correct interpretation of multimodal data and their correlation in terms of time. Multimodal data synchronization can be fulfilled by using relations defined in ASA. After synchronization, we obtain the multi-image, which is a complete timewise data structure, ready for further complex data analysis in such tasks as data modeling, prediction, classification, clustering, and other tasks. Further data processing can be fulfilled by using a domain-specific program language such as ASAMPL [20].

References

1 Dychka, I. and Sulema, Y. (2019). Ordering operations in algebraic system of aggregates for multi-image data processing. *Research Bulletin of the National Technical University of Ukraine "Kyiv Polytechnic Institute"* 1: 15–23.

2 Dychka, I. and Sulema, Y. (2018). Logical operations in algebraic system of aggregates for multimodal data representation and processing. *Research Bulletin of the National Technical University of Ukraine "Kyiv Polytechnic Institute"* 6: 44–52.

3 Maltsev, A.I. (1970). *Algebraic Systems*. Nauka.

4 Fraenkel, A.A., Bar-Hillel, Y., Levy, A. et al. (1973). *Foundations of Set Theory*. Elsevier.

5 Kulik, B.A., Zuenko, A.A., and Fridman, A.Y. (2010). *Algebraic Approach to Intellectual Processing of Data and Knowledge*. Izdatelstvo Politekhnicheskogo Universiteta.

6 Petrovsky, A.B. (2003). *Space of Sets and Multi-sets*. Editorial URSS.

7 Allen, J.F. (1983). Maintaining knowledge about temporal intervals. *Communications of the ACM* 26: 832–843. https://doi.org/10.1145/182.358434.

8 Allen, J.F. and Hayes, P.J. (1989). Moments and points in an interval-based temporal logic. *Computational Intelligence* 5 (3): 225–238. https://doi.org/10.1111/j.1467-8640.1989.tb00329.x.

9 Nebel, B. and Bürckert, H.-J. (1995). Reasoning about temporal relations: a maximal tractable subclass of Allen's interval algebra. *Journal of the ACM* 42 (1): 43–66.

10 Allen, J.F. and Ferguson, G. (1997). Actions and events in interval temporal logic. In: *Spatial and Temporal Reasoning*, 205–245. Springer https://doi.org/10.1007/978-0-585-28322-7_7.

11 Schockaert, S., De Cock, M., and Kerre, E. (2010). *Reasoning About Fuzzy Temporal and Spatial Information from the Web*, Intelligent Information Systems, vol. 3. World Scientific.

12 Bozzelli, L. et al. (2016). Interval vs. point temporal logic model checking: an expressiveness comparison. *Proceedings of the 36th IARCS Annual Conference on Foundations of Software Technology and Theoretical Computer Science (FSTTCS 2016)*, Chennai, India (13–15 December 2016). Chennai: Chennai Mathematical Institute.

13 Grüninger, M. and Li, Z. (2017). The time ontology of Allen's interval algebra. *Proceedings of 24th International Symposium on Temporal Representation and Reasoning (TIME 2017)*, Article No. 16, 16–18 October

2017, Mons, Belgium. LIPICS Schloss Dagstuhl – Leibniz-Zentrum für Informatik GmbH, Dagstuhl Publishing, Saarbrücken/Wadern,Germany.

14 Guesgen, H.W., Hertzberg, J. and Philpott, A. (1994). Towards implementing fuzzy Allen relations. *Proceedings of the ECAI-94 Workshop on Spatial and Temporal Reasoning*. August 8–12 Amsterdam, The Netherlands.

15 Bodenhofer, U. A new approach to fuzzy orderings. *Tatra Mountains Mathematical Publications* 16: 21–29.

16 Badaloni, S. and Giacomin, M. (2000). A fuzzy extension of Allen's interval algebra. *Lecture Notes in Artificial Intelligence* 1792: 155–165.

17 Dubois, D., Hadj Ali, A., and Prade, H. (2003). Fuzziness and uncertainty intemporal reasoning. *Journal of Universal Computer Science* 9: 1168–1194.

18 Schockaert, S., De Cock, M., and Kerre, E. (2008). Fuzzifying Allen's temporal interval relations. *IEEE Transactions on Fuzzy Systems* 16 (2): 517–533.

19 Bandler, W. and Kohout, L.J. (1980). Fuzzy relational products as a tool for analysis and synthesis of the behaviour of complex natural and artificial systems. In: *Fuzzy Sets: Theory and Application to Policy Analysis and Information Systems* (eds. S.K. Wang and P.P. Chang), 341–367. New York: Plenum.

20 Sulema, Y. (2018). ASAMPL: programming language for mulsemedia data processing based on algebraic system of aggregates. In: *Interactive Mobile Communication Technologies and Learning. IMCL 2017*, Advances in Intelligent Systems and Computing, vol. 725, 431–442. Springer https://doi.org/10.1007/978-3-319-75175-7_43.

6

Nonprobabilistic Analysis of Thermal and Chemical Diffusion Problems with Uncertain Bounded Parameters

Sukanta Nayak[1], Tharasi Dilleswar Rao[2], and Snehashish Chakraverty[2]

[1] *Department of Mathematics, Amrita School of Engineering, Amrita Vishwa Vidyapeetham, Coimbatore, Tamil Nadu 641112, India*
[2] *Department of Mathematics, National Institute of Technology Rourkela, Rourkela, Odisha 769008, India*

6.1 Introduction

Diffusion plays a major role in the field of thermal and chemical engineering. It may arise in a wide range of problems, viz., heat transfer, fluid flow, and chemical diffusion. These problems are governed by various linear and nonlinear differential equations. The problem with involved parameters, coefficients, and initial and boundary conditions greatly affects the solution of differential equations. In addition, the parameters used in the modeled physical problem are not crisp (or exact) because of the mechanical defect, experimental and measurement error, etc. In this context, the problem has to be defined with uncertain bounded parameters, which make it challenging to investigate. However, the uncertainties are handled by various authors using probability density functions or statistical methods, but these methods need plenty of data and also may not consider the vague or imprecise parameters. Accordingly, one may use interval and/or fuzzy computation in the analysis of the problems.

As such, this chapter comprises interval and/or fuzzy uncertainties along with well-known numerical methods, viz., finite element and explicit finite difference methods (EFDMs) to investigate heat and gas diffusion problems. The interval and/or fuzzy finite element formulation for tapered fin and finite difference formulation for radon diffusion mechanism in uncertain environment are discussed in details.

6.2 Preliminaries

6.2.1 Interval Arithmetic

As the operating boundary conditions and involved parameters of the system are uncertain in nature, we may take such values as interval/fuzzy numbers. Initially, if the interval values are considered, then the arithmetic rules may be written as [1]

$$[\underline{x}, \bar{x}] = \{x : x \in \mathcal{R}, \underline{x} \le x \le \bar{x}\} \tag{6.1}$$

where $\underline{x}$ and $\bar{x}$ are the left and right values of the interval, respectively. For the same interval, mid value is $m = \frac{\underline{x}+\bar{x}}{2}$ and width $w = \bar{x} - \underline{x}$.

The traditional interval arithmetic for two arbitrary intervals $[\underline{x}, \bar{x}]$ and $[\underline{y}, \bar{y}]$ is defined as follows.

Addition:

$$[\underline{x}, \bar{x}] + [\underline{y}, \bar{y}] = [\underline{x} + \underline{y}, \bar{x} + \bar{y}] \tag{6.2}$$

Mathematical Methods in Interdisciplinary Sciences, First Edition. Edited by Snehashish Chakraverty.

Subtraction:

$$[\underline{x}, \bar{x}] - [\underline{y}, \bar{y}] = [\underline{x} - \bar{y}, \bar{x} - \underline{y}] \tag{6.3}$$

Multiplication:

$$[\underline{x}, \bar{x}] \times [\underline{y}, \bar{y}] = [\min\{\underline{x}\underline{y}, \underline{x}\bar{y}, \bar{x}\underline{y}, \bar{x}\bar{y}\}, \max\{\underline{x}\underline{y}, \underline{x}\bar{y}, \bar{x}\underline{y}, \bar{x}\bar{y}\}] \tag{6.4}$$

Division:

$$[\underline{x}, \bar{x}] \div [\underline{y}, \bar{y}] = [\min\{\underline{x} \div \underline{y}, \underline{x} \div \bar{y}, \bar{x} \div \underline{y}, \bar{x} \div \bar{y}\}, \max\{\underline{x} \div \underline{y}, \underline{x} \div \bar{y}, \bar{x} \div \underline{y}, \bar{x} \div \bar{y}\}] \tag{6.5}$$

Because of the complexity of the traditional interval arithmetic, the same can be modified into the following parametric form [2].

Consider $[\underline{x}, \bar{x}]$ be an arbitrary interval, then $[\underline{x}, \bar{x}]$ can be written as

$$[\underline{x}, \bar{x}] = \left\{\underline{x} + \frac{w}{n} = l \mid n \in [1, \infty)\right\} \tag{6.6}$$

where w is the width of the interval.

If all the values of the interval are in R^+ or R^-, then the arithmetic rules [3] may be written as

$$[\underline{x}, \bar{x}] + [\underline{y}, \bar{y}] = [\min\{\lim_{n\to\infty} l_1 + \lim_{n\to\infty} l_2, \lim_{n\to 1} l_1 + \lim_{n\to 1} l_2\}, \max\{\lim_{n\to\infty} l_1 + \lim_{n\to\infty} l_2, \lim_{n\to 1} l_1 + \lim_{n\to 1} l_2\}] \tag{6.7}$$

$$[\underline{x}, \bar{x}] - [\underline{y}, \bar{y}] = [\min\{\lim_{n\to\infty} l_1 - \lim_{n\to 1} l_2, \lim_{n\to 1} l_1 - \lim_{n\to\infty} l_2\}, \max\{\lim_{n\to\infty} l_1 - \lim_{n\to 1} l_2, \lim_{n\to 1} l_1 - \lim_{n\to\infty} l_2\}] \tag{6.8}$$

$$[\underline{x}, \bar{x}] \times [\underline{y}, \bar{y}] = [\min\{\lim_{n\to\infty} l_1 \times \lim_{n\to\infty} l_2, \lim_{n\to 1} l_1 \times \lim_{n\to 1} l_2\}, \max\{\lim_{n\to\infty} l_1 \times \lim_{n\to\infty} l_2, \lim_{n\to 1} l_1 \times \lim_{n\to 1} l_2\}] \tag{6.9}$$

$$[\underline{x}, \bar{x}] \div [\underline{y}, \bar{y}] = [\min\{\lim_{n\to\infty} l_1 \div \lim_{n\to 1} l_2, \lim_{n\to 1} l_1 \div \lim_{n\to\infty} l_2\}, \max\{\lim_{n\to\infty} l_1 \div \lim_{n\to 1} l_2, \lim_{n\to 1} l_1 \div \lim_{n\to\infty} l_2\}] \tag{6.10}$$

In spite of the complex representation, the beauty of this arithmetic is that for computational purpose, the one parametric representation is quite handy to use. In addition to this, the above form of the interval arithmetic has great utility over the traditional interval arithmetic. Using the above transformation, the interval values are transformed into crisp form and then mathematical limit is operated. The generalized versions of transformed crisp values may easily be handled. This arithmetic may be extended for fuzzy numbers, which is described in the succeeding paragraph.

6.2.2 Fuzzy Number and Fuzzy Arithmetic

A fuzzy number is defined as the normalized convex fuzzy set $\widetilde{A} \subseteq R$, which is piecewise continuous. An arbitrary fuzzy number $\widetilde{A} = [a^L, a^N, a^R]$ is said to be triangular fuzzy number (TFN) when the membership function is given by

$$\mu_{\widetilde{A}}(x) = \begin{cases} 0, & x \le a^L; \\ \dfrac{x - a^L}{a^N - a^L}, & a^L \le x \le a^N; \\ \dfrac{a^R - x}{a^R - a^N}, & a^N \le x \le a^R; \\ 0, & x \ge a^R. \end{cases} \tag{6.11}$$

The fuzzy numbers may be represented as an ordered pair form $[\underline{f}(\alpha), \bar{f}(\alpha)]$, $0 \le \alpha \le 1$, where $\underline{f}(\alpha)$ and $\bar{f}(\alpha)$ are left and right monotonic increasing and decreasing functions over $[0, 1]$, respectively. The TFN $\widetilde{A} = [a^L, a^N, a^R]$ may be transformed into interval form by using $\alpha-$cut as follows (Figure 6.1).

$$\widetilde{A} = [a^L, a^N, a^R] = [a^L + (a^N - a^L)\alpha, a^R - (a^R - a^N)\alpha], \alpha \in [0, 1] \tag{6.12}$$

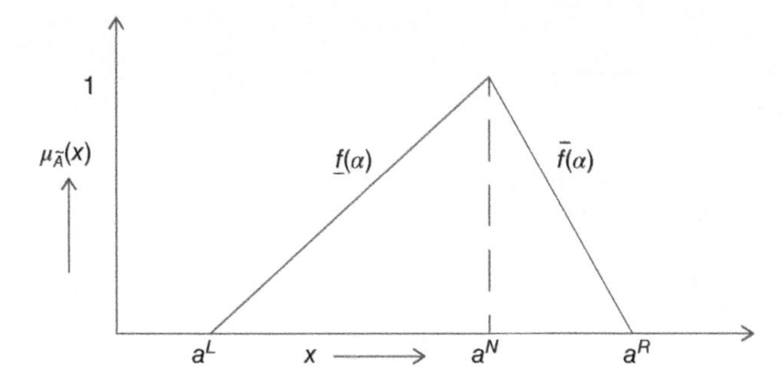

Figure 6.1 Triangular fuzzy number (TFN).

Let us consider two fuzzy numbers $x = [\underline{x}(\alpha), \bar{x}(\alpha)]$ and $y = [\underline{y}(\alpha), \bar{y}(\alpha)]$ and a scalar k then

(a) $x = y$ if and only if $\underline{x}(\alpha) = \underline{y}(\alpha)$ and $\bar{x}(\alpha) = \bar{y}(\alpha)$.
(b) $x + y = [\underline{x}(\alpha) + \underline{y}(\alpha), \bar{x}(\alpha) + \bar{y}(\alpha)]$.
(c) $kx = \begin{cases} [k\underline{x}(\alpha), k\bar{x}(\alpha)], & k \geq 0, \\ [k\bar{x}(\alpha), k\underline{x}(\alpha)], & k \geq 0. \end{cases}$

Combining both the fuzzy number and interval arithmetic, we may represent the fuzzy arithmetic as defined below [4, 5].

$$[\underline{x}(\alpha), \bar{x}(\alpha)] + [\underline{y}(\alpha), \bar{y}(\alpha)] = \\ \times [\min\{\lim_{n\to\infty} m_1 + \lim_{n\to\infty} m_2, \lim_{n\to 1} m_1 + \lim_{n\to 1} m_2\}, \max\{\lim_{n\to\infty} m_1 + \lim_{n\to\infty} m_2, \lim_{n\to 1} m_1 + \lim_{n\to 1} m_2\}] \tag{6.13}$$

$$[\underline{x}(\alpha), \bar{x}(\alpha)] - [\underline{y}(\alpha), \bar{y}(\alpha)] = \\ \times [\min\{\lim_{n\to\infty} m_1 - \lim_{n\to 1} m_2, \lim_{n\to 1} m_1 - \lim_{n\to\infty} m_2\}, \max\{\lim_{n\to\infty} m_1 - \lim_{n\to 1} m_2, \lim_{n\to 1} m_1 - \lim_{n\to\infty} m_2\}] \tag{6.14}$$

$$[\underline{x}(\alpha), \bar{x}(\alpha)] \times [\underline{y}(\alpha), \bar{y}(\alpha)] = \\ \times [\min\{\lim_{n\to\infty} m_1 \times \lim_{n\to\infty} m_2, \lim_{n\to 1} m_1 \times \lim_{n\to 1} m_2\}, \max\{\lim_{n\to\infty} m_1 \times \lim_{n\to\infty} m_2, \lim_{n\to 1} m_1 \times \lim_{n\to 1} m_2\}] \tag{6.15}$$

$$[\underline{x}(\alpha), \bar{x}(\alpha)] \div [\underline{y}(\alpha), \bar{y}(\alpha)] = \\ \times [\min\{\lim_{n\to\infty} m_1 \div \lim_{n\to 1} m_2, \lim_{n\to 1} m_1 \div \lim_{n\to\infty} m_2\}, \max\{\lim_{n\to\infty} m_1 \div \lim_{n\to 1} m_2, \lim_{n\to 1} m_1 \div \lim_{n\to\infty} m_2\}] \tag{6.16}$$

where $[\underline{x}(\alpha), \bar{x}(\alpha)] = \left\{\underline{x}(\alpha) + \frac{\bar{x}(\alpha) - \underline{x}(\alpha)}{n} = m \mid \underline{x}(\alpha) \leq m \leq \bar{x}(\alpha), n \in [1, \infty)\right\}$.

Here, the uncertain parameters are handled by using fuzzy values in place of classical (crisp) values. Therefore, we may combine the proposed fuzzy arithmetic and finite element method to develop fuzzy finite element method (FFEM) and the same can be used to solve various problems of engineering and science.

6.2.3 Parametric Representation of Fuzzy Number

The parametric approach is used to transform the interval-based system of equations into crisp form [6, 7]. As such, the interval $\widetilde{A} = [\underline{x}, \bar{x}]$ may be transformed into crisp form as $\widetilde{A} = \beta\ (\bar{x} - \underline{x}) + \underline{x}$, where $0 \leq \beta \leq 1$, it can also be written as

$$\widetilde{A} = 2\beta\Delta\widetilde{x} + \underline{x} \quad (\Delta\widetilde{x} = \text{Radius of } \widetilde{A})$$

In order to obtain the lower and upper bounds of the solution in parametric form, we may substitute $\beta = 0$ and 1, respectively. Similar concept may be used for the multivariable uncertainty.

Table 6.1 Finite difference schemes for PDEs.

Schemes	Derivative approximation of $c(x, t)$	
Forward	**Derivatives with respect to** x	**Derivatives with respect to** y
	$\frac{\partial c}{\partial x} \approx \frac{c_{i+1,j} - c_{i,j}}{\Delta x}$	$\frac{\partial c}{\partial t} \approx \frac{c_{i,j+1} - c_{i,j}}{\Delta t}$
	$\frac{\partial^2 c}{\partial x^2} \approx \frac{c_{i+2,j} - 2c_{i+1,j} + c_{i,j}}{(\Delta x)^2}$	$\frac{\partial^2 c}{\partial t^2} \approx \frac{c_{i,j+2} - 2c_{i,j+1} + c_{i,j}}{(\Delta t)^2}$
Center	$\frac{\partial c}{\partial x} \approx \frac{c_{i+1,j} - c_{i-1,j}}{2\Delta x}$	$\frac{\partial c}{\partial t} \approx \frac{c_{i,j+1} - c_{i,j-1}}{2\Delta t}$
	$\frac{\partial^2 c}{\partial x^2} \approx \frac{c_{i+1,j} - 2c_{i,j} + c_{i-1,j}}{(\Delta x)^2}$	$\frac{\partial^2 c}{\partial t^2} \approx \frac{c_{i,j+1} - 2c_{i,j} + c_{i,j-1}}{(\Delta t)^2}$
Backward	$\frac{\partial c}{\partial x} \approx \frac{c_{i,j} - c_{i-1,j}}{\Delta x}$	$\frac{\partial c}{\partial t} \approx \frac{c_{i,j} - c_{i,j-1}}{\Delta t}$
	$\frac{\partial^2 c}{\partial x^2} \approx \frac{c_{i,j} - 2c_{i-1,j} + c_{i-2,j}}{(\Delta x)^2}$	$\frac{\partial^2 c}{\partial t^2} \approx \frac{c_{i,j} - 2c_{i,j-1} + c_{i,j-2}}{(\Delta t)^2}$

6.2.4 Finite Difference Schemes for PDEs

The numerical solutions of differential equations based on finite difference provide us with the values at discrete grid points in the domain of $x - t$ plane. The spacing of the grid points in the x and t-directions are assumed to be uniform as Δx and Δt. It is not necessary that Δx and Δy be uniform or equal to each other always. The grid points in the x-direction are represented by index i and similarly j in the t direction. As such, the difference schemes for partial differential equations (PDEs) [8] are presented in Table 6.1.

To evaluate the derivative term $\frac{\partial^2 c}{\partial x \partial t}$, let us write the x-direction as a central difference of t-derivatives, and further, we make use of central difference to find out the y-derivatives. Thus, we obtain

$$\frac{\partial^2 c}{\partial x \partial t} = \frac{\partial}{\partial x}\left(\frac{\partial c}{\partial t}\right) = \frac{\left(\frac{\partial c}{\partial t}\right)_{i+1,j} - \left(\frac{\partial c}{\partial t}\right)_{i-1,j}}{2(\Delta x)} \cong \frac{1}{4\Delta x \Delta t}(c_{i+1,j+1} + c_{i-1,j-1} - c_{i+1,j-1} - c_{i-1,j+1})$$

Similarly, other difference schemes may also be written.

6.3 Finite Element Formulation for Tapered Fin

Let us consider a tapered fin with plane surfaces on the top and bottom. The fin also loses heat to the ambient via the tip. The thickness of the fin varies linearly from d_1 at the base to d_2 at the tip. The width b remains constant throughout the fin and L is the length.

Here, a typical element e_1 is having nodes i and j, respectively, as shown in Figure 6.2. Then, the corresponding area (A_i and A_j) and perimeter (P_i and P_j) for nodes i and j, respectively, are (Figures 6.3 and 6.4)

$$A_i = bd_i; \quad A_j = bd_j; \quad P_i = 2(b + d_i); \quad P_j = 2(b + d_j) \tag{6.17}$$

The shape functions for each element are $\left(1 - \frac{x}{L}\right)$ and $\frac{x}{L}$, and the area of the fin varies linearly. Therefore, the area A of tapered fin may be expressed as follows.

$$A = A_i\left(1 - \frac{x}{L}\right) + A_j\left(\frac{x}{L}\right) = A_i - \frac{A_i - A_j}{L}x \tag{6.18}$$

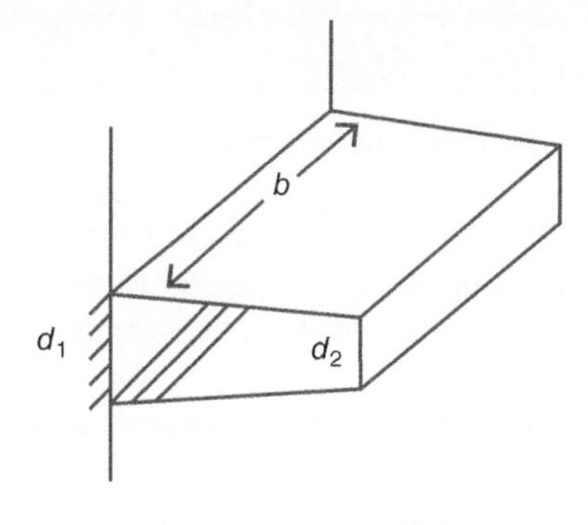

Figure 6.2 Model diagram of tapered fin.

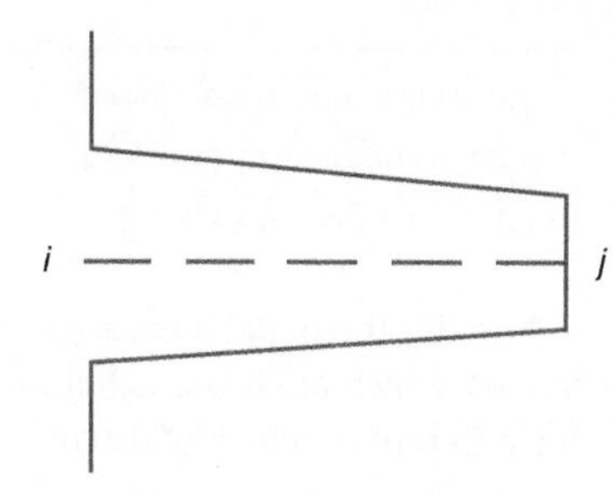

Figure 6.3 Tapered fin having two nodes.

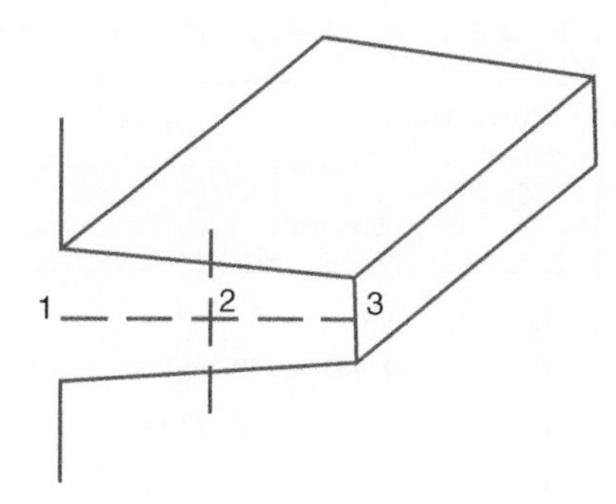

Figure 6.4 Two-element discretization of tapered fin.

Similarly, perimeter P of tapered fin may be written as

$$P = P_i\left(1 - \frac{x}{L}\right) + P_j\left(\frac{x}{L}\right) = P_i - \frac{P_i - P_j}{L}x \tag{6.19}$$

The stiffness matrix for the corresponding tapered fin is [9]

$$[K] = \frac{k}{l}\left(\frac{A_i + A_j}{2}\right)\begin{bmatrix} 1 & -1 \\ -1 & 1 \end{bmatrix} + \frac{hl}{12}\begin{bmatrix} 3P_i + P_j & P_i + P_j \\ P_i + P_j & P_i + 3P_j \end{bmatrix} \tag{6.20}$$

and the force vector is [9]

$$\{f\} = \frac{Gl}{6}\begin{Bmatrix} 2A_i + A_j \\ A_i + 2A_j \end{Bmatrix} - \frac{ql}{6}\begin{Bmatrix} 2P_i + P_j \\ P_i + 2P_j \end{Bmatrix} + \frac{hT_a l}{6}\begin{Bmatrix} 2P_i + P_j \\ P_i + 2P_j \end{Bmatrix} + hT_a A\begin{Bmatrix} 0 \\ 1 \end{Bmatrix} \tag{6.21}$$

where G is the heat source per unit volume, q is the heat flux, h is the heat transfer coefficient, k is the thermal conductivity, T_a is the ambient temperature, and l is the length of each element. Here, in Eq. (6.21), the last term is valid only for the element at the end face with area A. For all other elements, the last term of Eq. (6.21) is zero.

Next considering the above finite element formulation of tapered fin, we have taken a numerical example of the same and obtained the corresponding results under uncertain environment.

Case Study 6.1

Heat Transfer Problem

Here, we have considered a tapered fin where the thickness, d, varies from the base 5 mm to the tip 3 mm (Figure 6.2). The base temperature is maintained at 100 ° C. The total length of the fin is 30 mm and width b is 7 mm. Further corresponding data for this problem are provided in Table 6.2.

Table 6.2 Fuzzy parameters (triangular fuzzy number).

Parameters	Notations	Crisp value	Fuzzy value
Heat transfer coefficient (W/m^2 ° C)	h	120	[114, 120, 126]
Thermal conductivity (W/m^2 ° C)	k	200	[190, 200, 210]
Ambient temperature (°C)	T_a	25	[22, 25, 28]

Initially, the nodal temperatures of tapered fin under conduction–convection system are analyzed for crisp values, which are obvious but investigated for the sake of completeness. The computed nodal temperatures for different numbers of discretization of the same domain are given in Table 6.3.

Table 6.3 Nodal temperatures of tapered fin (crisp value).

Temperatures (°C)	Elements			
	2 Elements	4 Elements	8 Elements	16 Elements
T_1	100	100	100	100
T_2	**91.007 664 998 651 592**	94.937 095 487 730 303	97.328 106 375 902 465	98.626 957 581 524 096
T_3	87.571 508 339 270 409	**91.109 597 346 422 376**	94.955 120 707 462 086	97.327 766 293 459 192
T_4		88.648 840 251 112 034	92.890 199 197 205 462	96.103 218 668 388 038
T_5		87.801 196 102 602 987	**91.146 885 724 102 461**	94.954 329 954 740 174
T_6			89.744 193 857 748 456	93.882 360 998 980 673
T_7			88.708 092 402 078 719	92.888 844 764 162 783
T_8			88.069 110 423 918 502	91.975 167 276 236 917
T_9			87.878 310 328 979 694	**91.143 938 636 050 024**
T_{10}				90.397 712 947 648 927
T_{11}				89.739 474 615 781 631
T_{12}				89.172 696 736 601 765
T_{13}				88.701 410 360 529 565
T_{14}				88.330 287 073 839 798
T_{15}				88.064 738 007 646 042
T_{16}				87.911 033 256 628 144
T_{17}				87.876 446 853 988 256

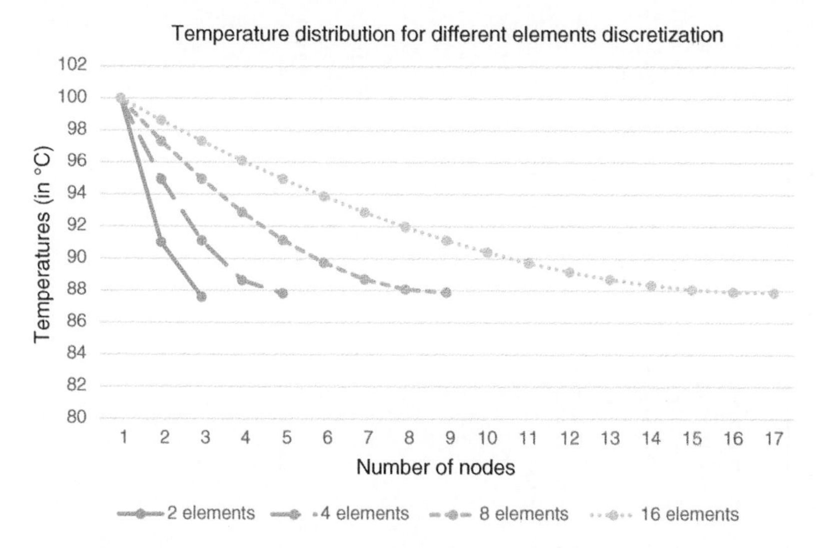

Figure 6.5 Nodal temperatures of tapered fin (crisp values).

Corresponding nodal temperatures for different numbers of discretization have been presented graphically in Figure 6.5.

Now, considering the imprecise values as fuzzy, mentioned above, we get a series of nodal temperatures depending on the value of α. When the value of α becomes 1 (that is crisp), we get the series of nodal temperatures that are presented in Table 6.3. Similarly, when the value of α is zero, then the obtained results are mentioned in Table 6.4. It is worth mentioning that the value of $\alpha = 0$ gives the interval results.

Corresponding nodal temperatures of Table 6.4 for different numbers of discretization with fuzzy parameters are also shown in Figure 6.6.

Next, as described in the introduction, numerical modeling of a chemical diffusion problem, viz., radon diffusion mechanism, in an uncertain scenario is investigated.

6.4 Radon Diffusion and Its Mechanism

Chemically, radon is a radioactive inert gas, which is formed by decay of ^{226}Ra. It has been used as a trace gas in terrestrial, hydrogeological, and atmospheric studies. Radon is usually found in igneous rocks, soils, and groundwater, and inhalation of radon causes lung cancer [10, 11]. Radon problem is a major concern in cold climate country. As such, measuring radon concentration levels in homes, mines, etc., with accuracy is an important task. Many researchers are working on various types of radon transport mechanisms. The transportation phenomena of radon in soil pore are mainly governed by two physical processes, namely, diffusion and advection. As such, different models have been developed based on these transport processes to study the anomalous behavior of soil radon in geothermal fields [12–14]. These models have been employed to estimate process-driven parameters, viz., diffusion coefficient, carrier gas velocity from the measured data of soil radon. Estimation of these parameters may deviate significantly from the true values, if the uncertainty associated with the various input parameters of the models are not taken into consideration. Radon is also monitored as an earthquake precursor from decades along with the various other precursors [15, 16]. However, prediction of the earthquake based on these precursors is still elusive. One of the main reasons for this is the uncertainty involved in the estimated values of the physical factors [17]. Therefore, modeling radon diffusion with uncertainty parameters is a challenging task [18, 19]. In view of

Table 6.4 Nodal temperatures of tapered fin (triangular fuzzy values).

temperatures (°C)	Elements							
	2 Elements		4 Elements		8 Elements		16 Elements	
	Left	Right	Left	Right	Left	Right	Left	Right
T_1	100	100	100	100	100	100	100	100
T_2	**90.627 713 674 689 574**	**91.387 616 322 613 610**	94.725 068 827 360 559	95.149 122 148 100 105	97.216 187 977 239 898	97.440 024 774 565 089	98.569 699 272 960 946	98.684 215 890 087 131
T_3	87.025 484 917 636 106	88.117 531 760 904 640	**90.733 589 214 346 651**	**91.485 605 478 498 215**	94.742 936 571 523 344	95.167 304 843 400 885	97.216 139 545 763 795	97.439 393 041 154 545
T_4			88.161 485 570 865 807	89.136 194 931 358 446	92.589 700 289 305 100	93.190 698 105 106 108	95.940 138 038 821 786	96.266 299 297 954 276
T_5			87.263 982 393 115 398	88.338 409 812 090 745	**90.770 483 576 744 653**	**91.523 287 871 460 539**	94.742 742 990 447 283	95.165 916 919 032 966
T_6					89.304 962 311 577 796	90.183 425 403 919 415	93.625 257 377 692 165	94.139 464 620 269 152
T_7					88.220 023 585 389 711	89.196 161 218 768 083	92.589 266 407 343 786	93.188 423 120 981 795
T_8					87.547 982 901 173 384	88.590 237 946 663 947	91.636 273 578 892 556	92.314 060 973 581 107
T_9					87.341 103 368 466 506	88.415 517 289 493 224	**90.768 912 809 597 111**	**91.518 964 462 502 808**
T_{10}							89.989 828 568 346 979	90.805 597 326 950 732
T_{11}							89.302 111 572 323 909	90.176 837 659 239 126
T_{12}							88.709 359 326 651 438	89.636 034 146 552 092
T_{13}							88.215747936824826	89.187072784234260
T_{14}							87.826 117 730 632 674	88.834 456 417 046 994
T_{15}							87.546 075 912 767 620	88.583 400 102 524 678
T_{16}							87.382 120 381 198 320	88.439 946 132 058 282
T_{17}							87.341 790 041 397 388	88.411 103 666 579 493

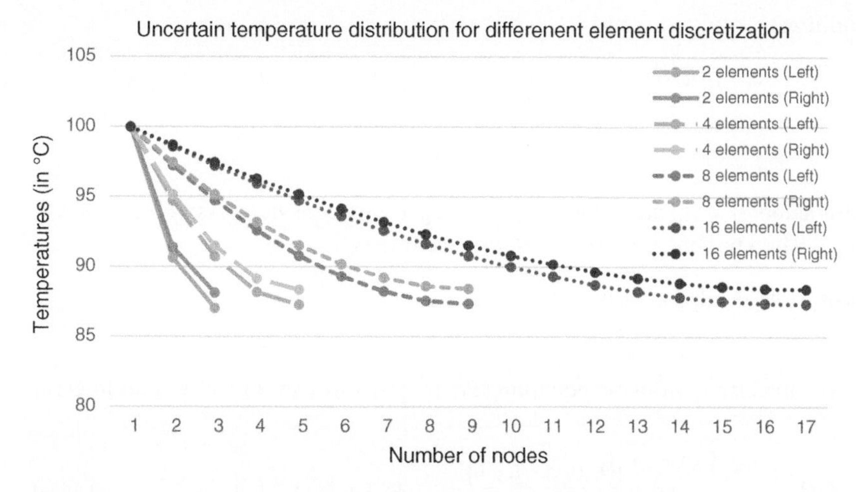

Figure 6.6 Nodal temperatures of tapered fin (left and right values).

this, the targeted chapter investigates the numerical modeling, viz., EFDM to radon transport mechanism in an uncertain scenario.

Radon (Rn) is a radioactive inert gas that can be found in various mediums, viz., soil, water, air, etc., and it is subject to radioactive decay ($Rn^{220}$and Rn^{219}), which can cause radiation exposure problems. The spatial difference of radon concentration that exists between the surface ($x = 0$) and the underground induces a diffusive radon flow toward the surface. As such, the governing differential equations of radon flux density (J) and diffusion equation [13, 14] are given as

$$J = -D\frac{\partial c(x,t)}{\partial x} \tag{6.22}$$

$$\frac{\partial c(x,t)}{\partial t} = D\frac{\partial^2 c(x,t)}{\partial x^2} - \lambda c(x,t) \tag{6.23}$$

where $\lambda = 2.1 \times 10^{-6} \text{s}^{-1}$ (decay constant) and $D =$ diffusion coefficient.

In general, diffusion of radon transforms from higher concentration region to lower concentration region with an initial radon concentration (c_0) and radon uncertain flux density through the soil is zero before the diffusion starts ($t = 0$). Radon flux density increases with time; hence, the initial and boundary conditions for radon diffusion are as follows:

$$c(x,0) = 0, 0 < x < L$$

$$c(0,t) = c_0, t > 0$$

$$J = 0,\ x = L, t > 0$$

6.5 Radon Diffusion Mechanism with TFN Parameters

As discussed, the estimation of process-driven parameters, viz., diffusion coefficient (D) and initial concentration (c_0), from the measured data of soil radon may deviate significantly from the true values. Therefore, by considering D and c_0 as uncertain parameters, we can represent radon diffusion equation as

$$\frac{\partial \tilde{c}(x,t)}{\partial t} = \tilde{D}\frac{\partial^2 \tilde{c}(x,t)}{\partial x^2} - \lambda\tilde{c}(x,t) \tag{6.24}$$

subject to uncertain boundary conditions

$$\widetilde{c}(x,0) = 0, 0 < x < L$$
$$\widetilde{c}(0,t) = \widetilde{c}_0, t > 0$$
$$\widetilde{J} = 0, x = L, t > 0$$

Here, the involved uncertain parameters $\widetilde{D} = (l_1, m_1, n_1)$ and $\widetilde{c}_0 = (l_2, m_2, n_2)$ are considered as TFNs. As such, $\alpha-$ cut is used to transform the fuzzy-based parameters to interval form as follows:

$$\left.\begin{aligned}\widetilde{D} &= [l_1 + \alpha(m_1 - l_1), n_1 - \alpha(n_1 - m_1)] = [\underline{D}(\alpha),\ \bar{D}(\alpha)] \\ \widetilde{c}_0 &= [l_2 + \alpha(m_2 - l_2), n_2 - \alpha(n_2 - m_2)] = [\underline{c}_0(\alpha),\ \bar{c}_0(\alpha)]\end{aligned}\right\} \tag{6.25}$$

Accordingly, by using $\alpha-$ cut in the uncertain diffusion equation (Eq. (6.24)), one can transform into interval form as

$$\left[\frac{\partial \underline{c}(x,t,\alpha)}{\partial t}, \frac{\partial \bar{c}(x,t,\alpha)}{\partial t}\right] = [\underline{D}(\alpha),\ \bar{D}(\alpha)]\left[\frac{\partial^2 \underline{c}(x,t,\alpha)}{\partial x^2}, \frac{\partial^2 \bar{c}(x,t,\alpha)}{\partial x^2}\right] - \lambda[\underline{c}(x,t,\alpha), \bar{c}(x,t,\alpha)] \tag{6.26}$$

subject to interval form initial and boundary conditions

$$\widetilde{c}(x,0,\alpha) = 0, 0 < x < L$$
$$\widetilde{c}(0,t,\alpha) = [\underline{c}_0(\alpha),\ \bar{c}_0(\alpha)], t > 0$$
$$\widetilde{J} = 0, x = L, t > 0$$

Now, the interval form of Eq. (6.26) can be transformed into crisp form by using the parametric concept as

$$\frac{\partial c(x,t,\alpha,\beta)}{\partial t} = D(\alpha,\beta)\frac{\partial^2 c(x,t,\alpha,\beta)}{\partial x^2} - \lambda c(x,t,\alpha,\beta) \tag{6.27}$$

subject to initial and boundary conditions

$$c(x,0,\alpha,\beta) = 0, 0 < x < L$$
$$c(0,t,\alpha,\beta) = c_0(\alpha,\beta), t > 0$$
$$J = 0, x = L, t > 0$$

where

$$\frac{\partial c(x,t,\alpha,\beta)}{\partial t} = \frac{\partial \underline{c}(x,t,\alpha)}{\partial t} + \beta\left(\frac{\partial \bar{c}(x,t,\alpha)}{\partial t} - \frac{\partial \underline{c}(x,t,\alpha)}{\partial t}\right),\ D(\alpha,\beta) = \underline{D}(\alpha) + \beta(\bar{D}(\alpha) - \underline{D}(\alpha))$$
$$\frac{\partial^2 c(x,t,\alpha,\beta)}{\partial x^2} = \frac{\partial^2 \underline{c}(x,t,\alpha)}{\partial x^2} + \beta\left(\frac{\partial^2 \bar{c}(x,t,\alpha)}{\partial x^2} - \frac{\partial^2 c(x,t,\alpha)}{\partial x^2}\right)$$
$$c(x,t,\alpha,\beta) = \underline{c}(x,t,\alpha) + \beta(\bar{c}(x,t,\alpha) - \underline{c}(x,t,\alpha))$$

The next Section 6.5.1 presents numerical modeling of the above-described uncertain radon diffusion problem by using EFDM.

6.5.1 EFDM to Radon Diffusion Mechanism with TFN Parameters

The central difference scheme is used to represent the term $\frac{\partial^2 c(x,t,\alpha,\beta)}{\partial x^2}$ and a forward difference scheme for the term $\frac{\partial c(x,t,\alpha,\beta)}{\partial t}$ in Eq. (6.27) to obtain numerical scheme as

$$c_{i,j+1}(x,t,\alpha,\beta) = a(\alpha,\beta)c_{i-1,j}(x,t,\alpha,\beta) + b(\alpha,\beta)c_{i,j}(x,t,\alpha,\beta) + a(\alpha,\beta)c_{i+1,j}(x,t,\alpha,\beta) \tag{6.28}$$

subject to initial and boundary conditions

$$c_{i,0}(x, 0, \alpha, \beta) = 0, 0 < x < L$$
$$c_{0,j}(0, t, \alpha, \beta) = c_0(\alpha, \beta), t > 0$$
$$c_{N,j}(x, t, \alpha, \beta) = c_{N-1,j}(x, t, \alpha, \beta), x = L, t > 0$$

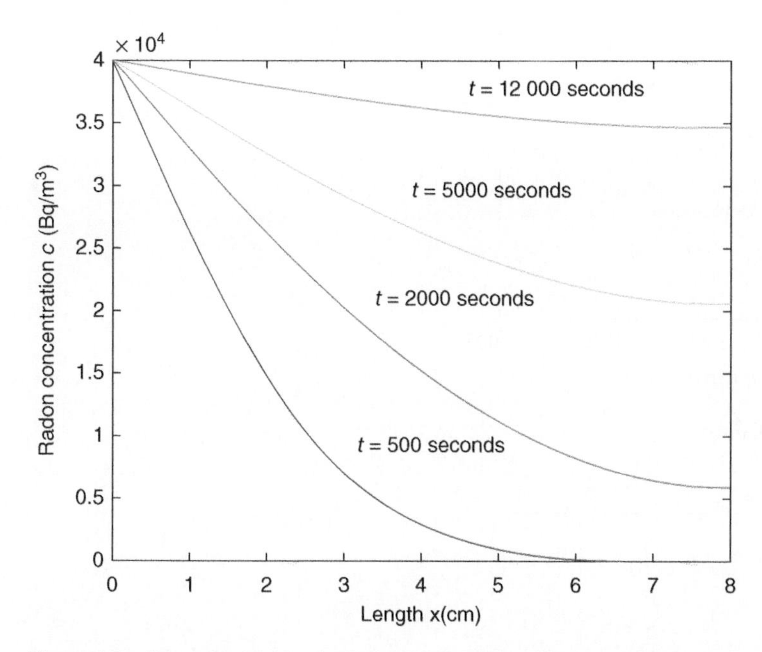

Figure 6.7 Crisp numerical solution of Eq. (6.24) by EFDM at different times (t) and $\alpha = 1$.

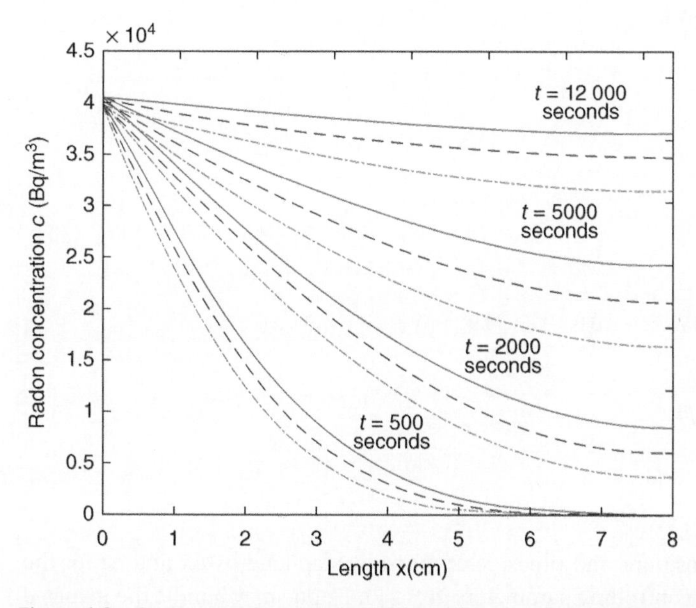

Figure 6.8 Fuzzy numerical solution of Eq. (6.24) by EFDM at different times (t) and α, β.

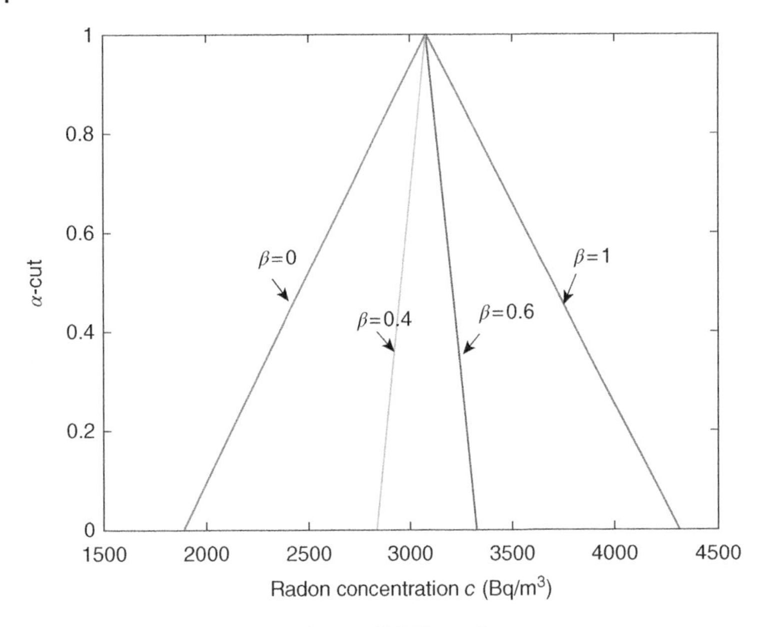

Figure 6.9 Radon concentration at $c(4{,}500s, \alpha, \beta)$.

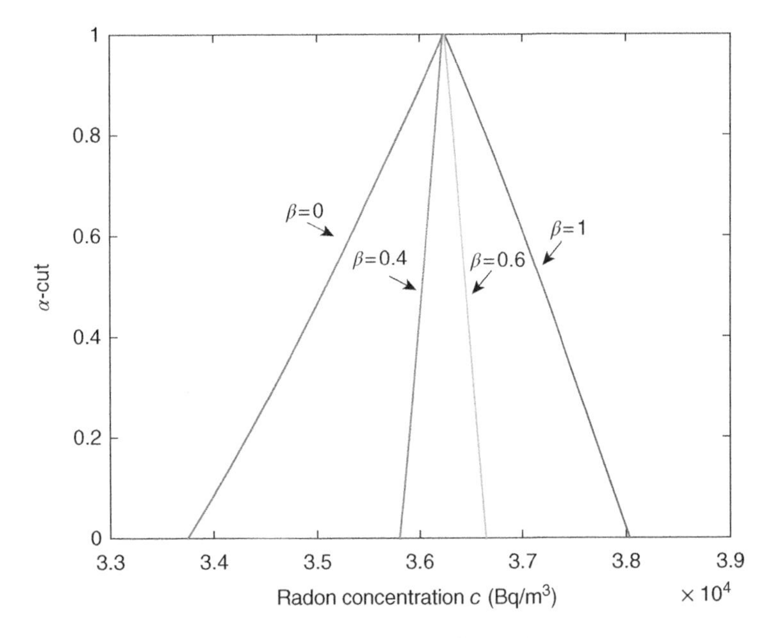

Figure 6.10 Radon concentration at $c(4{,}12\,000s, \alpha, \beta)$.

where

$a(\alpha, \beta) = \frac{D(\alpha,\beta)\Delta t}{\Delta x^2}$ and $b(\alpha, \beta) = (1 - 2a(\alpha, \beta) - \lambda\Delta t)$.

Here, indices i and j represent the discrete positions and times determined by step lengths Δx and Δt for the coordinates x and t, respectively. By varying the controlling parameters $\alpha, \beta \in [0, 1]$, one may handle the involved uncertainty of the considered model (Eq. (6.24)).

Table 6.5 : Radon concentration $c(4, t, \alpha, \beta)$ of Eq. (6.24) by EFDM.

β	α−level	$c((4, t, \alpha, \beta))$			
		t = 500 seconds	t = 2000 seconds	t = 5000 seconds	t = 12 000 seconds
0	0	1 896.91	12 847.48	23 136.86	33 737.18
	0.2	2 125.09	13 374.74	23 816.86	34 312.59
	0.4	2 358.29	13 886.01	24 475.46	34 847.32
	0.6	2 595.55	14 382.79	25 113.69	35 344.87
	0.8	2 836.01	14 866.42	25 732.43	35 808.44
	1	3 078.91	15 338.09	26 332.50	36 240.95
0.4	0	2 836.01	14 866.42	25 732.43	35 808.44
	0.2	2 884.53	14 961.68	25 853.91	35 897.33
	0.4	2 933.03	15 056.46	25 974.65	35 985.01
	0.6	2 981.61	15 150.79	26 094.66	36 071.49
	0.8	3 030.28	15 244.66	26 213.94	36 156.80
	1	3 079.03	15 338.09	26 332.50	36 240.95
0.6	0	3 323.72	15 798.85	26 914.64	36 645.08
	0.2	3 274.67	15 707.52	26 799.61	36 566.41
	0.4	3 225.67	15 615.79	26 683.89	36 486.67
	0.6	3 176.72	15 523.65	26 567.46	36 405.87
	0.8	3 127.84	15 431.08	26 450.34	36 323.97
	1	3 079.03	15 338.09	26 332.50	36 240.95
1	0	4 310.15	17 549.65	29 077.32	38 023.59
	0.2	4 063.25	17 124.34	28 560.30	37 710.58
	0.4	3 816.29	16 691.20	28 027.91	37 377.76
	0.6	3 569.64	16 249.61	27 479.56	37 023.28
	0.8	3 323.72	15 798.85	26 914.64	36 645.08
	1	3 079.03	15 338.09	26 332.50	36 240.95

Case Study 6.2

Radon Diffusion Problem

This section deals with the numerical outcomes of the considered radon diffusion model with TFN parameters, and we applied the above-described EFDM method to the considered model (Eq. (6.24)). In addition, we have considered the crisp/center geometry as in [13]. As such, a length $L = 8$ cm cylindrical soil slab one side was exposed to radon concentration is considered. Accordingly, in these results, the initial concentration, diffusion coefficient, and step lengths of the coordinates are taken as $c_0 = 40\,000$ Bqm^{-3}, $D = 0.005$ cm^2/s, $\Delta x = 0.05$ cm, and $\Delta t = 0.01$ seconds. Further, the TFN parametric values of the involved uncertain parameters as described in the considered model are taken as $\widetilde{c}_0 = (0.004, 0.005, 0.006)$ and $\widetilde{D} = (39\,500, 40\,000, 40\,500)$.

As discussed in the above Section 6.5, α and β are the controlling parameters to handle the involved uncertainty of the considered model Eq. (6.24). As such, one can analyze the behavior or validation of the considered fuzzy radon model by using α and β. Accordingly, some of the outcomes are included and discussed here. Figure 6.7 depicts the center/crisp radon behavior for different times (t) and controlling parameter taken as $\alpha = 1$. In addition, Figure 6.8 depicts the fuzzy numerical solution (lower, center, and upper) radon concentrations at different t by choosing (α, $\beta = 0$) for lower, $\alpha = 1$ for center, and ($\alpha = 0$, $\beta = 1$) to get upper concentration. By this modeling, one can estimate a fuzzy band at every t to investigate the behavior of radon gas.

Further, the involved imprecise parameters of the considered model are taken as TFNs. Therefore, the radon concentration is presented at different times (t) and also for different depths. Accordingly, Figure 6.9 characterizes the radon concentration for all $\alpha = 0.1 : 0.1 : 1$ at a fixed depth ($x = 4$ cm) and time ($t = 500$ seconds). Similarly, the radon concentration for all $\alpha = 0.1 : 0.1 : 1$ at a fixed depth ($x = 4$ cm) and time ($t = 12\,000$ seconds) is also illustrated in Figure 6.10. Accordingly, by using the often values of the controlling parameters α, $\beta \in [0, 1]$ with the depth (x) and time (t), one can investigate the behavior of radon concentration. The radon concentrations of the considered model (Eq. (6.24)) at different t and fixed depth $x = 4$ cm for some α, $\beta \in [0, 1]$ are presented in Table 6.5. From this, one can observe an incremental growth in radon concentration with respect to the increment in t.

6.6 Conclusion

In this chapter, two different alternate approaches have been discussed to handle heat and chemical diffusion problems under an uncertain environment. Especially, the involved parameters, boundary conditions, and initial inputs are carried as uncertain, and accordingly, the uncertain governing differential equations have been investigated. The obtained results give good agreement with special cases, as well as the uncertain spectrum of the obtained solutions is easily illustrated in the given tables and figures. The above study shows that the discussed methods are not limited to the mentioned problems but may be useful to other diffusion and structure problems.

References

1 Moore, R.E., Baker Kearfott, R., and Cloud, M.J. (2009). *Introduction to Interval Analysis*. SIAM.
2 Nayak, S. and Chakraverty, S. (2013). Non-probabilistic approach to investigate uncertain conjugate heat transfer in an imprecisely defined plate. *International Journal of Heat and Mass Transfer* 67: 445–454.
3 Nayak, S. and Chakraverty, S. (2015). Fuzzy finite element analysis of multigroup neutron diffusion equation with imprecise parameters. *International Journal of Nuclear Energy Science and Technology* 9 (1): 1–22.
4 Nayak, S. and Chakraverty, S. (2015). Numerical solution of uncertain neutron diffusion equation for imprecisely defined homogeneous triangular bare reactor. *Sadhana* 40: 2095–2109.
5 Nayak, S. and Chakraverty, S. (2018). Non-probabilistic solution of moving plate problem with uncertain parameters. *Journal of Fuzzy Set Valued Analysis* 2: 49–59.
6 Behera, D. and Chakraverty, S. (2015). New approach to solve fully fuzzy system of linear equations using single and double parametric form of fuzzy numbers. *Sadhana* 40 (1): 35–49.
7 Tapaswini, S. and Chakraverty, S. (2013). Numerical solution of n-th order fuzzy linear differential equations by homotopy perturbation method. *International Journal of Computer Applications* 64 (6): 5–10.
8 Chakraverty, S., Mahato, N., Karunakar, P., and Rao, T.D. (2019). *Advanced Numerical and Semi-Analytical Methods for Differential Equations*. Wiley.
9 Nayak, S., Chakraverty, S., and Datta, D. (2014). Uncertain spectrum of temperatures in a non-homogeneous fin under imprecisely defined conduction-convection system. *Journal of Uncertain Systems* 8 (2): 123–135.
10 Nazaroff, W.W. (1992). Radon transport from soil to air. *Reviews of Geophysics* 30 (2): 137–160.

11 Schery, S.D., Holford, D.J., Wilson, J.L., and Phillips, F.M. (1988). The flow and diffusion of radon isotopes in fractured porous media Part 2, Semi infinite media. *Radiation Protection Dosimetry* 24 (1–4): 191–197.

12 Kozak, J.A., Reeves, H.W., and Lewis, B.A. (2003). Modeling radium and radon transport through soil and vegetation. *Journal of Contaminant Hydrology* 66 (3): 179–200.

13 Savović, S., Djordjevich, A., Peter, W.T., and Nikezić, D. (2011). Explicit finite difference solution of the diffusion equation describing the flow of radon through soil. *Applied Radiation and Isotopes* 69 (1): 237–240.

14 Kohl, T., Medici, F., and Rybach, L. (1994). Numerical simulation of radon transport from subsurface to buildings. *Journal of Applied Geophysics* 31 (1): 145–152.

15 Planinić, J., Radolić, V., and Vuković, B. (2004). Radon as an earthquake precursor. *Nuclear Instruments and Methods in Physics Research Section A: Accelerators, Spectrometers, Detectors and Associated Equipment* 530 (3): 568–574.

16 Fleischer, R.L. (1997). Radon and earthquake prediction. In: *Radon Measurements by Etched Track Detectors: Applications in Radiation Protection, Earth Sciences and the Environment* (eds. A. Durrani Saeed and I. Radomir), 285. World Scientific.

17 Chakraverty, S., Sahoo, B.K., Rao, T.D. et al. (2018). Modelling uncertainties in the diffusion-advection equation for radon transport in soil using interval arithmetic. *Journal of Environmental Radioactivity* 182: 165–171.

18 Rao, T.D. and Chakraverty, S. (2017). Modeling radon diffusion equation in soil pore matrix by using uncertainty based orthogonal polynomials in Galerkin's method. *Coupled Systems Mechanics* 6 (4): 487–499.

19 Rao, T.D. and Chakraverty, S. (2018). Modeling radon diffusion equation by using fuzzy polynomials in Galerkin's method. In: *Recent Advances in Applications of Computational and Fuzzy Mathematics*, 75–93. Singapore: Springer.

7

Arbitrary Order Differential Equations with Fuzzy Parameters

Tofigh Allahviranloo and Soheil Salahshour

Faculty of Engineering and Natural Sciences, Bahcesehir University, Istanbul, Turkey

7.1 Introduction

In the last decades, some generalization of theory of ordinary differential equations has been considered to the arbitrary order differential equations by many researchers, the so-called theory of arbitrary order differential equations (often called as fractional order differential equations [FDEs]) [1–5].

Because of the ability for modeling real phenomena, arbitrary order differential equations have been applied in various fields such as control systems, biosciences, bioengineering [6–8], and references therein.

The extension of fractional calculus is done in several directions, in which two of them are common; the first is proposing several types of fractional integral and fractional derivatives, which is reviewed completely by Teodoro et al. [9], and the second is considering the uncertainty concepts in the structure of FDEs (we emphasize on fuzzy uncertainty) [10–19]. A combination of these two points of view led to propose some new model and obtain new types of solutions and approximations [20, 21].

Based on this approach, we provided some new types of arbitrary order differential equations with respect to another function involving fuzzy uncertainty.

In fact, using the definitions for arbitrary order integral and derivative of Riemann–Liouville type with respect to another function proposed in [1, 22], and using the theory of fuzzy FDEs established in [12, 13, 19, 23], we introduced some new definitions and results for arbitrary order differential equations with respect to another function for the fuzzy-valued functions based on the generalized Hukuhara differentiability on the absolutely continuous space.

To solve such uncertain problems, one of the powerful tools is fuzzy Laplace transform. Therefore, for the first time in the literature, we proposed arbitrary order differential equations with respect to another function using fuzzy parameters (initial values and the unknown solutions). Then, the generalized fuzzy Laplace transform is applied to obtain the Laplace transform of arbitrary order integral and derivative of fuzzy-valued functions to solve linear FDEs. To obtain the large class of solutions for FDEs, the concept of generalized Hukuhara differentiability is applied.

7.2 Preliminaries

In this section, we recall and present some notions used in the article. Let R_F be the set of all fuzzy numbers on the set of all real numbers R. In fact, B is a fuzzy number if it is normal, convex, and semicontinuous.

Mathematical Methods in Interdisciplinary Sciences, First Edition. Edited by Snehashish Chakraverty.

The r-cut representation of B is $B(r) = [B_1(r), B_2(r)]$, where $[B_1(r), B_2(r)]$ is closed and bounded interval, $0 < r \leq 1$. For $r = 0$, $B(0) = \{y \in R | \bar{B}(y) > 0\}$.

Using Zadeh's extension principle, sum and product are defined as follows:

$$A + B = \sup_{t \in R} \min\{A(t), B(s-t)\}, \quad s \in R$$

and

$$(\eta \cdot A)(y) = \begin{cases} A\left(\dfrac{y}{\eta}\right), & \eta > 0 \\ \widetilde{0}, & \eta = 0 \end{cases}$$

where $\widetilde{0} \in R_F$. Moreover, the distance between two fuzzy numbers is define as:

$$D(A, B) = \sup_{r \in [0,1]} d_H(A(r), B(r))$$

where

$$d_H(A(r), B(r)) = \max\{|A_1(r) - B_1(r)|, |A_2(r) - B_2(r)|\}$$

In fact, it is the Hausdorff–Pompeiu distance.

Also, we denote the diameter of A by $d(A)$, and it is computed as $d(A(r)) = A_2(r) - A_1(r)$.

Using diameter, the definition of increasing and decreasing functions is connected to the fuzzy-valued functions [23, 24]. Indeed, the fuzzy-valued functions f is d-monotone if $d(f(x, r))$ is monotone, for every $r \in [0, 1]$.

In the text, we used the generalized Hukuhara difference [25], which is state as follows:

$$A \ominus_g B = C \Leftrightarrow \begin{cases} (1) \quad A = B + C \\ (2) \quad B = A - C \end{cases}$$

Using this definition for difference between two fuzzy-valued functions, the generalized Hukuhara derivative is introduced as follows [25]:

$$X'_g(t) = \lim_{h \to 0} \frac{X(t+h) \ominus_g X(t)}{h}, \quad t, t+h \in (a, b)$$

provided that the generalized Hukuhara difference exists.

Set $C([a, b], R_F)$ as the set of all continuous fuzzy-valued functions, $C_\eta([a, b], R_F)$ is the set of all fuzzy-valued functions, where $(\cdot - a)^\eta x(\cdot) \in C([a, b], R_F)$. Also, let $L([a, b], R_F)$ is the set of all Lebesgue integrable fuzzy-valued functions.

Moreover, the definition of absolutely continuous fuzzy-valued functions is given in [24], which is denoted by $AC([a, b], R_F)$.

7.3 Arbitrary Order Integral and Derivative for Fuzzy-Valued Functions

Here, we recall some basic definition about arbitrary order integral and derivative under fuzzy uncertainty of order β, $0 < \beta \leq 1$, involving Riemann–Liouville type.

Then, we are going to generalize these concepts to the arbitrary order integral and derivative with respect to another function for the fuzzy-valued functions.

Definition 7.1 Let $y \in L([a, b], R_F)$, then the arbitrary order integral of Riemann–Liouville type (of order β, $0 < \beta \leq 1$) is defined as follows [13, 19]:

$$({}_{a^+}I^\beta y)(t) = \frac{1}{\Gamma(\beta)} \int_a^t (t-x)^{\beta-1} y(x) dx, \quad t > a$$

Let $y(t, r) = [y_1(t, r), y_2(t, r)]$, then the r-cut representation of fractional integral based on the lower–upper functions is obtained as follows [12, 19]:

$$({}_{a^+}I^{\beta}y)(t,r) = \frac{1}{\Gamma(\beta)}\left[\int_a^t (t-x)^{\beta-1}y_1(x,r)dx, \int_a^t (t-x)^{\beta-1}y_2(x,r)dx\right]$$

Set $v_\beta : [a, b] \to R_F$, where $v_\beta(t) = ({}_{a^+}I^{\beta}v)(t)$ and $v \in L([a, b], R_F)$, $0 < \beta \leq 1$. Also, consider the function $v_{1-\beta}(t) = ({}_{a^+}I^{1-\beta}v)(t)$, $t \in [a, b]$ (for more detail, see [23, 24]).

Definition 7.2 Arbitrary order derivative for the fuzzy-valued function $f \in AC([a, b], R_F)$ involving Riemann–Liouville type using generalized Hukuhara difference is defined as follows [19]:

$$ {}^{RL}_{a^+}f^{(\beta)}(t) = (f_{1-\beta})'_g(t), \quad 0 < \beta \leq 1, \quad t \in (a, b]$$

where $f_{1-\beta}(t) = \frac{1}{\Gamma(1-\beta)}\int_a^t (t-x)^{-\beta}f(x)dx$.

Now, we introduce some new definitions for the fuzzy arbitrary order integral and derivative of Riemann–Liouville type with respect to another function h for the fuzzy-valued function $f \in AC([a, b], R_F)$. Indeed, for the deterministic case, see [1, 22] and references therein.

Suppose that the real-valued function h, $(h(t) \in R)$, has the following properties:

- $h(t)$ be a strictly increasing function, i.e. $h'(t) > 0$,
- $h'(t)$ is continuous on (a, b).

Definition 7.3 The arbitrary order integral (of order β) for the fuzzy-valued function $f \in L([a, b], R_F)$ with respect to the function h is defined as follows:

$$({}_{a^+}I_h^{\beta}f)(t) = \frac{1}{\Gamma(\beta)}\int_a^t (h(t)-h(x))^{\beta-1}f(x)h'(x)dx, \quad t > a$$

Indeed, the r-cut representation of ${}_{a^+}I_h^{\beta}f$, for $f(t, r) = [f_1(t, r), f_2(t, r)]$ based on the lower–upper functions f_1 and f_2 can be easily obtained as follows:

$$({}_{a^+}I_h^{\beta}f)(t,r) = \frac{1}{\Gamma(\beta)}\left[\int_a^t (h(t)-h(x))^{\beta-1}f_1(x,r)h'(x)dx, \int_a^t (h(t)-h(x))^{\beta-1}f_2(x,r)h'(x)dx\right]$$

Note that for the special case $h(t) = t$, we obtain the same results for fuzzy fractional integral [19, 24].

Definition 7.4 We define the arbitrary order derivative (of order β) for the fuzzy-valued function $f \in AC([a, b], R_F)$ with respect to h as follows:

$$ {}^{RL}_{a^+}f^{(\beta)}(t) = \frac{1}{h'(t)}(f_{1-\beta,h})'_g(t), \quad 0 < \beta \leq 1, \quad t \in (a, b]$$

where $f_{1-\beta,h}(t) = \frac{1}{\Gamma(1-\beta)}\int_a^t (h(t)-h(x))^{-\beta}f(x)h'(x)dx$.

Lemma 7.1 Suppose that for the fuzzy-valued function $f \in AC([a, b], R_F)$, then $f_{1-\beta,h}(t) \in AC([a, b], R_F)$.

Proof: Using the fact that $f \in AC([a, b], R_F)$, we have $({}_{a^+}I_h^{\beta}f_1)(t, r)$ and $({}_{a^+}I_h^{\beta}f_2)(t, r)$ are absolutely continuous on $(a, b]$. Therefore, we deduce that $f_{1-\beta,h}(t) \in AC([a, b], R_F)$. ■

Remark 7.1 If $h(t) = t$, then it is easy to verify that ${}^{RL}_{a^+}f_h^{(\beta)}(t) \to (f)'_g(t)$, $\beta \to 1^-$. Therefore, ${}^{RL}_{a^+}f_h^{(\beta)}(t)$ can be considered as generalization of ${}^{RL}_{a^+}f^{(\beta)}(t)$.

In order to obtain main results, consider the following assumptions:

(H_1) Suppose that the following functions are increasing:

- $d(f(x,r))$
- $\frac{1}{\Gamma(\beta)}\int_a^t (h(t)-h(x))^{\beta-1}d(f(x,r))h'(x)dx$
- $\frac{1}{\Gamma(\beta)}\int_a^t (h(t)-h(x))^{-\beta}d(f(x,r))h'(x)dx$

Now, using the above assumptions, we obtain r-cut representation of arbitrary order derivative of fuzzy-valued function f with respect to h.

Lemma 7.2 Suppose that $f \in AC([a,b], R_F)$ is d-increasing or d-decreasing on $[a, b]$ and $f_{1-\beta,h}(t)$ is d-increasing on $(a, b]$, then

$$ {}^{RL}_{a^+}f_h^{(\beta)}(t,r) = [{}^{RL}_{a^+}f_{1,h}^{(\beta)}(t,r), {}^{RL}_{a^+}f_{2,h}^{(\beta)}(t,r)] $$

and if $f_{1-\beta,h}$ d-decreasing on $(a, b]$, we have:

$$ {}^{RL}_{a^+}f_h^{(\beta)}(t,r) = [{}^{RL}_{a^+}f_{2,h}^{(\beta)}(t,r), {}^{RL}_{a^+}f_{1,h}^{(\beta)}(t,r)] $$

Proof: Using assumption $f \in AC([a,b], R_F)$, we conclude that $({}_{a^+}I_h^{1-\beta}f_1)(t,r)$ and $({}_{a^+}I_h^{\beta}f_2)(t,r)$ are absolutely continuous on $(a, b]$. Hence, $f_{1-\beta,h} \in AC([a,b], R_F)$ and therefore $(f_{1-\beta,h})'_g(t)$ exists on $(a, b]$. Therefore, ${}^{RL}_{a^+}f_h^{(\beta)}(t)$ exists on $(a, b]$. If f is d-increasing on $[a, b]$, using (H_1), we conclude that $f_{1-\beta,h}(t)$ is d-decreasing on $(a, b]$.

Therefore,

$$ \begin{aligned} {}^{RL}_{a^+}f_h^{(\beta)}(t,r) &= \frac{1}{h'(t)}(f_{1-\beta,h})'_g(t,r) = \frac{1}{h'(t)}\left[\frac{d}{dt}({}_{a^+}I_h^{1-\beta}f_1)(t,r), \frac{d}{dt}({}_{a^+}I_h^{1-\beta}f_2)(t,r)\right] \\ &= [({}^{RL}_{a^+}f_{1,h}^{(\beta)})(t,r), ({}^{RL}_{a^+}f_{2,h}^{(\beta)})(t,r)], \quad t > a, \quad 0 < r \le 1 \end{aligned} $$

If $f_{1-\beta,h}(t)$ is d-decreasing on $(a, b]$, we obtain

$$ \begin{aligned} {}^{RL}_{a^+}f_h^{(\beta)}(t,r) &= \frac{1}{h'(t)}\left[\frac{d}{dt}({}_{a^+}I_h^{1-\beta}f_2)(t,r), \frac{d}{dt}({}_{a^+}I_h^{1-\beta}f_1)(t,r)\right] \\ &= [{}^{RL}_{a^+}f_{2,h}^{(\beta)}(t,r), {}^{RL}_{a^+}f_{1,h}^{(\beta)}(t,r)], \quad t > a, \quad 0 < r \le 1 \end{aligned} $$

which completes the proof. ■

7.4 Generalized Fuzzy Laplace Transform with Respect to Another Function

Hence, we are going to extend the concept of fuzzy Laplace transform. On the other hand, we obtain fuzzy Laplace transform with respect to another function h for the fuzzy-valued function f.

Definition 7.5 Suppose that f: $[a,+\infty) \to R_F$ be a fuzzy-valued function and h: $[a,+\infty) \to R_F$ be a real-valued function, where $h'(t) > 0$ on $(a,+\infty)$. The generalized fuzzy Laplace transform of f with respect to h is defined as follows:

$$ L_h\{f(t)\}(s) = \int_a^{+\infty} e^{-s(h(t)-h(a))}f(t)h'(t)dt $$

whenever the integral is valid.

We provide the sufficient conditions for existing the generalized fuzzy Laplace transform for the fuzzy-valued function f.

Definition 7.6 Let $f: [a,+\infty) \to R_F$, then f is of $h(t)$ exponential order if there exists κ, θ, η (they are non-negative) such that:

$$D(f(t),\widetilde{0}) \leq \kappa \exp(\theta h(t)), \quad t \geq \eta$$

Theorem 7.1 Let f is a piecewise continuous fuzzy-valued function of h(t) exponential order, then generalized fuzzy Laplace transform exists for $s > \theta$.

Proof: It is straightforward. ∎

Theorem 7.2 The generalized fuzzy Laplace transforms for the fuzzy-valued functions X_1, X_2 with respect to h is linear, i.e.

$$L_h\{p_1X(t) + p_2Y(t)\} = p_1L_h\{X(t)\} + p_2L_h\{Y(t)\}$$

where p_1 and p_2 are constants.

Proof: It is straightforward. ∎

For clarification, we obtain some generalized fuzzy Laplace transform with respect to h for some fuzzy-valued elementary functions. Indeed, for $\widetilde{c} = c \in R$, the results reduce to the given examples in [22].

Example 7.1 Let $f(t) = \widetilde{c}$, where $\widetilde{c} \in R_F$, then $L_h\{\widetilde{c}\}(s) = \frac{\widetilde{c}}{s}$, $s > 0$.

Example 7.2 Let $g(t) = \widetilde{c}(h(t) - h(a))^\gamma$, then $L_h\{g(t)\}(s) = \frac{\widetilde{c}\Gamma(\gamma)}{s^\gamma}$, $s > 0$.

Example 7.3 Let $Z(t) = \widetilde{c}\ \exp(\theta h(t))$, then $L_h\{Z(t)\}(s) = \frac{\widetilde{c}\ \exp(\theta h(a))}{s-\theta}$, $s > \theta$.

Lemma 7.3 Suppose that $f^{[1]}(t) = \frac{1}{h'(t)}(f)'_g(t)$. Then, if $f \in C([a, b], R_F)$ and of h(t) exponential order, then we have the following:

(1) If f is d-increasing, then $L_h\{f^{[1]}(t)\}(s) = sL_h\{f(t)\}(s) \ominus f(a)$
(2) If f is d-decreasing, then $L_h\{f^{[1]}(t)\}(s) = -f(a) \ominus (-1)sL_h\{f(t)\}(s)$
where $f^{[1]}(t)$ is piecewise continuous over every subset of $[a, b]$.

Proof: Using the definition of generalized Hukuhara differentiability, the proof is straightforward. ∎

Remark 7.2 The results of Lemma 7.3 can be summarized as $L_h\{f^{[1]}(t)\}(s) = sL_n\{f(t)\}(s)\ominus_g f(a)$.

Now, we state the generalized convolution with respect to function h as follows:

$$(f^*{}_h g)(t) = \int_a^t f(s)g(h^{-1}(h(t) + h(a) - h(s)))h'(s)ds$$

Then, it is easy to verify that for fuzzy-valued function f, g (both of them are of $h(t)$ exponential order and piecewise continuous), we have:

$$L_h\{f^*{}_h g\} = L_h\{f\} \cdot L_h\{g\}$$

Using this concept, we obtain a new result for generalized fuzzy Laplace transform of generalized fuzzy arbitrary order integral.

Theorem 7.3 Let f be a piecewise continuous fuzzy-valued function $h(t)$ exponential order. Then,

$$L_h\{({}_{a^+}I_h^{\beta}f)(t)\} = \frac{L_h\{f(t)\}}{s^{\beta}}$$

Proof: Since arbitrary order integral of fuzzy-valued function f can possess the generalized fuzzy Laplace transform with respect to h, then we have:

$$L_h\{({}_{a^+}I_h^{\beta}f)(t,r)\} = [L_h\{({}_{a^+}I_h^{\beta}f_1)(t,r)\}, L_h\{({}_{a^+}I_h^{1-\beta}f_2)(t,r)\}] = \left[\frac{L_h\{f_1(t,r)\}}{s^{\beta}}, \frac{L_h\{f_2(t,r)\}}{s^{\beta}}\right]$$
$$= \frac{L_h\{f(t,r)\}}{s^{\beta}}, \quad 0 < r \le 1$$

which completes the proof. ■

Theorem 7.4 Suppose that $f \in AC([a, b], R_F)$, $h \in C([a, b], R_F)$ such that $h'(t) > 0$, and $({}_{a^+}I_h^{1-\beta}f)(t)$ be of $h(t)$ exponential order. Then, we have the following:

(1) If f is d-increasing, then we have:

$$L_h\{{}_{a^+}^{RL}f_h^{(\beta)}(t,r)\} = s^{\beta}L_h\{f(t)\} \ominus ({}_{a^+}I^{1-\beta}f)(a^+)$$

(2) If f is d-decreasing, then we have:

$$L_h\{{}_{a^+}^{RL}I_h^{(\beta)}f(t,r)\} = -({}_{a^+}I^{1-\beta}f)(a^+) \ominus (-1)s^{\beta}L_h\{f(t)\}$$

Proof:

(1) Let f is d-increasing, then we obtain:

$$\begin{aligned}
L_h\{{}_{a^+}^{RL}f_h^{(\beta)}(t,r)\} &= [L_h\{({}_{a^+}I_h^{1-\beta}f_1)^{[1]}(t,r)\}, L_h\{({}_{a^+}I_h^{1-\beta}f_2)^{[1]}(t,r)\}] \\
&= [sL_h\{({}_{a^+}I_h^{1-\beta}f_1)^{[1]}(t,r)\} - ({}_{a^+}I_h^{1-\beta}f_1)(a^+,r), sL_h\{({}_{a^+}I_h^{1-\beta}f_2)^{[1]}(t,r)\} - ({}_{a^+}I_h^{1-\beta}f_2)(a^+,r)] \\
&= [s^{\beta}L_h\{f_1(t,r)\} - ({}_{a^+}I_h^{1-\beta}f_1)(a^+,r), s^{\beta}L_h\{f_2(t,r)\} - ({}_{a^+}I_h^{1-\beta}f_2)(a^+,r)] \\
&= s^{\beta}L_h\{f(t,r)\} \ominus ({}_{a^+}I^{1-\beta}f)(a^+,r)
\end{aligned}$$

(2) Also, in a completely similar manner, if f is d-decreasing, we have:

$$\begin{aligned}
L_h\{{}_{a^+}^{RL}I_h^{(\beta)}f(t,r)\} &= [L_h\{({}_{a^+}I_h^{1-\beta}f_2)^{[1]}(t,r)\}, L_h\{({}_{a^+}I_h^{1-\beta}f_1)^{[1]}(t,r)\}] \\
&= [sL_h\{({}_{a^+}I_h^{1-\beta}f_2)^{[1]}(t,r)\} - ({}_{a^+}I_h^{1-\beta}f_2)(a^+,r), sL_h\{({}_{a^+}I_h^{1-\beta}f_1)^{[1]}(t,r)\} - ({}_{a^+}I_h^{1-\beta}f_1)(a^+,r)] \\
&= [s^{\beta}L_h\{f_2(t,r)\} - ({}_{a^+}I_h^{1-\beta}f_2)(a^+,r), s^{\beta}L_h\{f_1(t,r)\} - ({}_{a^+}I_h^{1-\beta}f_1)(a^+,r)] \\
&= -({}_{a^+}I^{1-\beta}f)(a^+,r) \ominus (-1)s^{\beta}L_h\{f(t,r)\}
\end{aligned}$$

which completes the proof. ■

Remark 7.3 For unifying the fact of theorem, we have:

$$L_h\{{}_{a^+}^{RL}f_h^{(\beta)}(t,r)\} = s^{\beta}L_h\{f(t)\} \ominus_g ({}_{a^+}I^{1-\beta}f)(a^+).$$

In the theory and application of arbitrary order calculus, the Mittag-Leffler (ML) function has some significant influences.

Let $E_{\alpha}(x) = \sum_{k=0}^{\infty} \frac{x^k}{\Gamma(\alpha k+1)}$ and $E_{\alpha,\beta}(x) = \sum_{k=0}^{\infty} \frac{x^k}{\Gamma(\alpha k+\beta)}$, $\mathrm{Re}(\alpha) > 0, \quad \mathrm{Re}(\beta) > 0.$

Now, we derive the generalized fuzzy Laplace transform with respect to h of composition ML function and fuzzy-valued function. Indeed, the following results are a direct generalization of **Lemma 4.2** in [22].

Lemma 7.4 Let $\widetilde{C} \in R_F$, $\eta \in \mathrm{R}$, $Re(\beta) > 0$, and $\left|\frac{\eta}{s^\beta}\right| < 1$, then

(1) $L_h\{\widetilde{C}E_\beta(\eta(h(t)-h(a))^\beta)\} = \frac{\widetilde{C}s^{\beta-1}}{s^\beta-\eta}$

(2) $L_h\{\widetilde{C}(h(t)-h(a))^{\beta-1}E_{\gamma,\beta}(\eta(h(t)-h(a))^\gamma)\} = \frac{\widetilde{C}s^{\gamma-\beta}}{s^\gamma-\eta}$

Proof: It is straightforward. ■

Now, we can obtain the solution of the following arbitrary order Cauchy problem under generalized fuzzy Riemann–Liouville fractional differentiability with fuzzy parameters:

$$\begin{cases} {}^{RL}_{a^+}f_h^{(\beta)}(t) = \eta f(t) + K(t), & t > a, \quad 0 < \beta \le 1 \\ ({}_{a^+}I_h^{1-\beta}f)(a^+) = \widetilde{C} \in R_F, & \eta \in R, \quad K \in R_F \end{cases}$$

Using generalized fuzzy Laplace transform with respect to h.

Case 7.1 Let $\eta \in R_+ = (0, +\infty)$ and f be a d-increasing function, then by applying generalized fuzzy Laplace transform on both sides of the problem, we have:

$$L_h\{{}^{RL}_{a^+}f_h^{(\beta)}(t,r)\} = L_h\{\eta f(t,r)\} + L_h\{K(t,r)\}$$

Therefore, we have:

$$\begin{cases} L_h\{{}^{RL}_{a^+}f_{1,h}^{(\beta)}(t,r)\} = \eta L_h\{f_1(t,r)\} + L_h\{K_1(t,r)\} \\ L_h\{{}^{RL}_{a^+}f_{2,h}^{(\beta)}(t,r)\} = \eta L_h\{f_2(t,r)\} + L_h\{K_2(t,r)\} \end{cases}$$

Then, we have:

$$\begin{cases} S^\beta L_h\{f_1(t,r)\} - C_1(r) = \eta L_h\{f_1(t,r)\} + L_h\{K_1(t,r)\} \\ S^\beta L_h\{f_2(t,r)\} - C_2(r) = \eta L_h\{f_2(t,r)\} + L_h\{K_2(t,r)\} \end{cases}$$

After simple calculations, we obtain:

$$\begin{cases} L_h\{f_1(t,r)\} = \dfrac{L_h\{K_1(t,r)\}}{S^\beta - \eta} + \dfrac{C_1(r)}{S^\beta - \eta} \\ L_h\{f_2(t,r)\} = \dfrac{L_h\{K_2(t,r)\}}{S^\beta - \eta} + \dfrac{C_2(r)}{S^\beta - \eta} \end{cases}$$

Then, we have:

$$\begin{cases} L_h\{f_1(t,r)\} = L_h\{C_1(h(t)-h(a))^{\beta-1}E_{\beta,\beta}(\eta(h(t)-h(a))^\beta) + (h(t)-h(a))^{\beta-1}E_{\beta,\beta}(\eta(h(t)-h(a))^\beta) *_h K_1(t,r)\} \\ L_h\{f_2(t,r)\} = L_h\{C_2(h(t)-h(a))^{\beta-1}E_{\beta,\beta}(\eta(h(t)-h(a))^\beta) + (h(t)-h(a))^{\beta-1}E_{\beta,\beta}(\eta(h(t)-h(a))^\beta) *_h K_2(t,r)\} \end{cases}$$

Finally, we obtain the lower–upper functions of solution $f(t,r) = [f_1(t,r), f_2(t,r)]$ as follows:

$$\begin{cases} f_1(t,r) = C_1(r)(h(t)-h(a))^{\beta-1}E_{\beta,\beta}(\eta(h(t)-h(a))^\beta) + \int_a^t (h(t)-h(s))^{\beta-1}E_{\beta,\beta}(\eta(h(t)-h(s))^\beta)K_1(s)h'(s)ds \\ f_2(t,r) = C_2(r)(h(t)-h(a))^{\beta-1}E_{\beta,\beta}(\eta(h(t)-h(a))^\beta) + \int_a^t (h(t)-h(s))^{\beta-1}E_{\beta,\beta}(\eta(h(t)-h(s))^\beta)K_2(s)h'(s)ds \end{cases}$$

Case 7.2 Let $\eta \in (-\infty, 0)$. In order to obtain d-decreasing solution, similar to Case 7.1, one can easily obtain:

$$f(t,r) = C(r)(h(t)-h(a))^{\beta-1}E_{\beta,\beta}(\eta(h(t)-h(a))^\beta) \ominus (-1)\int_a^t (h(t)-h(s))^{\beta-1}E_{\beta,\beta}(\eta(h(t)-h(s))^\beta)K(s)h'(s)ds$$

Remark 7.4 Indeed, using generalized Hukuhara difference, we can unify the solution as follows:

$$f(t)\ominus_g C(h(t)-h(a))^{\beta-1}E_{\beta,\beta}(\eta(h(t)-h(a))^{\beta})=\int_a^t (h(t)-h(s))^{\beta-1}E_{\beta,\beta}(\eta(h(t)-h(s))^{\beta})K(s)h'(s)ds$$

As a summing up, in order to handle the incomplete information under fuzzy uncertainty in the mathematical modeling, we proposed some generalization of arbitrary order differential equations of Riemann–Liouville type with respect to another function for the first time in the literature.

For this purpose, using generalized fuzzy Laplace transform, the explicit solution of the fuzzy problem is obtained. In this way, some new and useful results regarding the relation between d-monotone solution and type of differentiability with other function are derived.

Indeed, as we stated in Section 7.1, we combine the common fuzzy fractional differentiability and the theory of derivatives with respect to another function to produce some new type of arbitrary order differentiability [22, 23] and as a result propose a new modeling for real phenomena under fuzzy uncertainty.

Indeed, for future research, we hope to extend some other type of arbitrary integral and differentiability with respect to another function. Notice that, before proceed, we need to investigate the necessary and sufficient conditions for existence and uniqueness of solutions of arbitrary order differential equations with fuzzy parameters, which will be considered in the near future.

References

1 Kilbas, A.A., Srivastava, H.M., and Trujillo, J.J. (2006). *Theory and Applications of Fractional Differential Equations*. Amsterdam: Elsevier Science B.V.

2 Kiryakova, V. (1994). *Generalized Fractional Calculus and Applications*. Harlow: Longman Scientific & Technical, co-published in the United States with Wiley, New York.

3 Lakshmikantham, V. and Vatsala, A.S. (2008). Basic theory of fractional differential equations. *Nonlinear Analysis: Theory, Methods & Applications* 69: 2677–2682.

4 Miller, K.S. and Ross, B. (1993). *An Introduction to the Fractional Calculus and Differential Equations*. New York: Wiley.

5 Podlubny, I. (1999). *Fractional Differential Equation*. San Diego, CA: Academic Press.

6 Baleanu, D., Machado, J.A.T., and Luo, A. (2012). *Fractional Dynamics and Control*. New York: Springer-Verlag.

7 Ionescu, C., Lopes, A., Copot, D. et al. (2017). The role of fractional calculus in modeling biological phenomena: a review. *Communications in Nonlinear Science and Numerical Simulation* 15: 141–159.

8 Krijnen, M.E., van Ostayn, R.A.J., and HosseinNia, H. (2018). The application of fractional order control for an air-based contactless actuation system. *ISA Transactions* 82: 172–183.

9 Sales Teodoro, G., Machado, J.A.T., and Capelas de Oliveira, E. (2019). A review of definitions of fractional derivatives and other operators. *Journal of Computational Physics* 388: 195–208.

10 Allahviranloo, T., Armand, A., and Gouyandeh, Z. (2014). Fuzzy fractional differential equations under generalized fuzzy Caputo derivative. *Journal of Intelligent & Fuzzy Systems* 26: 1481–1490.

11 Allahviranloo, T., Salahshour, S., and Abbasbandy, S. (2012). Explicit solutions of fractional differential equations with uncertainty. *Soft Computing* 16: 297–302.

12 Arshad, S. and Lupulescu, V. (2011). On the fractional differential equations with uncertainty. *Nonlinear Analysis: Theory, Methods & Applications* 74: 3685–3693.

13 Agarwal, R., Arshad, S., O'Regan, D., and Lupulescu, V. (2012). Fuzzy fractional integral equations under compactness type condition. *Fractional Calculus and Applied Analysis* 15: 572–590.

14 Hoa, N.V. (2015). Fuzzy fractional functional differential equations under Caputo gH-differentiability. *Communications in Nonlinear Science and Numerical Simulation* 22: 1134–1157.

15 Hoa, N.V., Vu, H., and Duc, T.M. (2019). Fuzzy fractional differential equations under Caputo–Katugampola fractional derivative approach. *Fuzzy Sets and Systems*: 70–99.

16 Malinowski, M.T. (2015). Random fuzzy fractional integral equations – theoretical foundations. *Fuzzy Sets and Systems* 265: 39–62.

17 Mazandarani, M. and Kamyad, A.V. (2013). Modified fractional Euler method for solving Fuzzy Fractional Initial Value Problem. *Communications in Nonlinear Science and Numerical Simulation* 18: 12–21.

18 Ngo, V.H. (2015). Fuzzy fractional functional integral and differential equations. *Fuzzy Sets and Systems* 280: 58–90.

19 Salahshour, S., Allahviranloo, T., and Abbasbandy, S. (2012). Solving fuzzy fractional differential equations by fuzzy Laplace transforms. *Communications in Nonlinear Science and Numerical Simulation* 17: 1372–1381.

20 Salahshour, S., Ahmadian, A., Abbasbandy, S., and Baleanu, D. (2018). M-fractional derivative under interval uncertainty: theory, properties and applications. *Chaos, Solitons & Fractals* 117: 84–93.

21 Salahshour, S., Ahmadian, A., and Baleanu, D. (2017). Variation of constant formula for the solution of interval differential equations of non-integer order. *The European Physical Journal Special Topics* 226: 3501–3512.

22 Jarad, F. and Abdeljawad, T. (2020). Generalized fractional derivatives and Laplace transform. *Discrete and Continuous Dynamical Systems Series S* 13: 709–722.

23 Ngo, H.V., Lupulescu, V., and O'Regan, D. (2018). A note on initial value problems for fractional fuzzy differential equations. *Fuzzy Sets and Systems* 347: 54–69.

24 Lupulescu, V. (2015). Fractional calculus for interval-valued functions. *Fuzzy Sets and Systems* 265: 63–85.

25 Bede, B. and Stefanini, L. (2013). Generalized differentiability of fuzzy-valued functions. *Fuzzy Sets and Systems* 230: 119–141.

8

Fluid Dynamics Problems in Uncertain Environment

Perumandla Karunakar[1], Uddhaba Biswal[2], and Snehashish Chakraverty[2]

[1]*Department of Mathematics, Amrita School of Engineering, Amrita Vishwa Vidyapeetham, Chennai, Tamil Nadu, 601103, India*
[2]*Department of Mathematics, National Institute of Technology Rourkela, Rourkela, Odisha, 769008, India*

8.1 Introduction

In general, fluid flow between two plates is one of the important problems in the field of fluid dynamics. The study of fluid flow in convergent and divergent channels is very important because of their vast application in engineering and industrial field such as heat exchangers for milk flowing, cold drawing operation in polymer industry, etc. Many authors have shown interest to study two-dimensional incompressible fluid flow between two inclined planes. Jeffery [1] and Hamel [2] were the first persons to discuss about this problem, so it is known as Jeffery–Hamel problem. Jeffery–Hamel problems for heat transfer fluids have been studied by different authors by using different numerical methods. However, the heat transfers fluids, viz., water, oil, and ethylene glycol mixture, are poor heat transfer fluids. With the increasing global competition, industries have a strong need to develop advanced heat transfer fluid with significantly higher thermal conductivity than that of water, oil, etc. It is well known that metals in room temperature have higher thermal conductivity than that of fluids. Moreover, the thermal conductivity of metallic liquids is much greater than that of nonmetallic liquids. Therefore, the fluids that contain suspended metallic particles could be expected to have higher thermal conductivity than that of existing pure fluids. In 1995, Choi [3] have proposed a new class of fluid, termed as nanofluid, i.e. a fluid with suspended nanoparticles in the base fluids and reported that the nanofluids have superior thermal properties as compared to the base fluids. As such, in recent years, Jeffery–Hamel problems for nanofluids have taken attention of many researchers.

In fluid dynamics, most of the problems are related to nonlinear differential equations. Because of this nonlinearity, they do not admit analytical solution. In such situations, researchers relay on numerical methods. Moghimi et al. [4] presented the application of homotopy analysis method (HAM) to Jeffery–Hamel problem for fluid flow between two nonparallel planes. Esmaeilpour and Ganji [5] have used optimal homotopy asymptotic method to analyze the solution of fluid flow between two rigid nonparallel planes. Rostami et al. [6] have used two methods, namely, homotopy perturbation method (HPM) and Akbari–Ganji's method, to solve Jeffery–Hamel problem in the presence of magnetic field and nanoparticle. Differential transformation method (DTM) has been used by Umavathi and Shekar [7] to study Jeffery–Hamel flow of nanofluid with magnetic effect. Hatami and Ganji [8] used weighted residual methods, viz., Galerkin's, least square, and collocation methods to investigate Jeffery–Hamel flow in the presence of magnetic field and nanoparticles. Moradi et al. [9] have used DTM to study the effect of nanoparticle on Jeffery–Hamel flow in the absence of magnetic field.

Mathematical Methods in Interdisciplinary Sciences, First Edition. Edited by Snehashish Chakraverty.

From the literature review, Jeffery–Hamel problem with nanofluid has been investigated for crisp case only. It is worth mentioning here that every particular application of nanofluid depends on the physical properties of nanoparticle as well as base fluid, viz., effective viscosity, thermal conductivity, density, etc., and value of these parameters depend on the value of nanoparticle volume fraction. There exist few models for these parameters of nanofluid derived by different researchers with the assumption that nanoparticles are uniformly distributed over the considered base fluid. However, in practical case, it may not be possible. Therefore, nanoparticle volume fraction may be taken as an uncertain parameter. As per our knowledge, there is no paper dealing with the Jeffery–Hamel flow for an uncertain case. In this regard, we are motivated to investigate Jeffery–Hamel problem in an uncertain environment, which makes it more challenging. Here, nanoparticle volume fraction is taken as an uncertain parameter in term of fuzzy number. A semianalytical method known as HPM has been used to solve the governing fuzzy differential equation related to the Jeffery–Hamel problem for nanofluid.

Shallow water wave (SWW) equations are widely used for many physical phenomenon such as simulation of tsunami-wave propagation, shock waves, tidal flows, and coastal waves. To describe SWW, many models have been introduced, such as Boussinesq equations, Korteweg–de Vries (KdV) equations, and Kadomtsev–Petviashvili (KP) equations, etc. In this regard, another set of coupled differential equations that describe SWW are coupled Whitham–Broer–Kaup (CWBK) equations. These are given by Whitham [10], Broer [11], and Kaup [12]. Further, many works are reported for solving CWBK equations. As such, Xie et al. [13] obtained four pair of solutions of CWBK equations using hyperbolic function method. New generalized transformation has been proposed to find the exact solutions of CWBK equations in shallow water by Yan and Zhang [14]. Yıldırım et al. [15] and Ganji et al. [16] applied HPM to find explicit and numerical travelling wave solutions of CWBK equations. Approximate travelling wave solutions of CWBK equations with the help of reconstruction of variational iteration method (RVIM) have been obtained in [17].

In the above-discussed literature, the constants related to diffusion power are crisp numbers, but they may not be crisp always because these are measured values. To handle these involved uncertainties here, we have considered them as interval numbers, which transform the governing CWBK equations to interval CWBK equations. Solving interval differential equation is a challenging task. As such, in this chapter, homotopy perturbation transform method (HPTM) [18–22] has been applied to handle uncertain differential equations with the help of parametric concept.

8.2 Preliminaries

In this section, we will discuss some basic concept of interval/fuzzy theory and some notations that we have used in further discussion.

8.2.1 Fuzzy Set

A fuzzy set $\widetilde{S}$ is a set consisting of ordered pairs of the elements s of a universal set say S' and their membership value is written as [23–25]

$$\widetilde{S} = \{(s, m(s)) : s \in S', m(s) \in [0, 1]\}$$

where $m(s)$ is a defined membership function for the fuzzy set $\widetilde{S}$.

8.2.2 Fuzzy Number

Fuzzy number is a fuzzy set that is convex, normalized, and defined on real line R. Moreover, its membership function must be piecewise continuous. There are different types of fuzzy numbers based on membership function, viz., triangular, Gaussian, quadratic, exponential fuzzy number, etc. Here, we have used triangular fuzzy number (TFN) and membership function of a TFN $\widetilde{S} = [c, d, e]$ is defined as [23–25]

$$m(s) = \begin{cases} 0, & s \leq c \\ \dfrac{s-c}{d-c}, & c \leq s \leq d \\ \dfrac{e-s}{e-d}, & d \leq s \leq e \\ 0, & s \geq e \end{cases}$$

8.2.3 δ-Cut

δ-Cut of a fuzzy set is defined as the crisp set given by $\widetilde{S}_\delta = \{s \in S' : m(s) \geq \delta\}$.

By using δ-cut, TFN $\widetilde{S} = [c, d, e]$ may be converted into interval form as [24, 25] $\widetilde{S} = [c, d, e] = [(d-c)\delta + c, e - (e-d)\delta], \quad \delta \in [0, 1]$.

8.2.4 Parametric Approach

In general, an interval $I = [\underline{I}, \overline{I}]$ may be transformed into crisp form by the help of parametric concept as [24, 25]

$$I = \gamma(\overline{I} - \underline{I}) + \underline{I}$$

where γ is a parameter that lies in the closed interval [0, 1].

It can also be written as $I = 2\gamma\Delta I + \underline{I}$, where $\Delta I = \frac{\overline{I}-\underline{I}}{2}$ is the radius of I.

8.3 Problem Formulation

In this section, two problems, namely, Jeffery–Hamel problem and CWBK shallow water equations, have been discussed.

As such, Figure 8.1 represents the schematic diagram of the Jeffery–Hamel problem where two plane rigid walls are inclined at an angle of 2ω. Let us consider a system of cylindrical polar coordinates (r, θ, z) in which steady two-dimensional flow of an incompressible conducting viscous fluid from a source or sink at channel forms by the nonparallel walls. The plane walls are considered to be convergent if $\omega < 0$ and divergent if $\omega > 0$. We have assumed purely radial motion that has no change in the flow parameter along the z-axis. Here, the flow depends on r and θ so that $v = (u(r, \theta), 0)$, and moreover, it is assumed that there is no magnetic effect on the z-direction. Now, the continuity equation, Navier–Stokes equation, and Maxwell equations in polar coordinate may be written as

$$\frac{\rho_{\text{nf}}}{r}\frac{\partial}{\partial r}(ru(r,\theta)) = 0 \tag{8.1}$$

$$\begin{aligned} u(r,\theta)\frac{\partial}{\partial r}(ru(r,\theta)) = \frac{-1}{\rho_{\text{nf}}}\frac{\partial P}{\partial r} + \upsilon_{\text{nf}}\Bigg(\frac{\partial^2}{\partial r^2}(u(r,\theta)) + \frac{1}{r}\frac{\partial}{\partial r}(u(r,\theta)) \\ + \frac{1}{r^2}\frac{\partial^2}{\partial \theta^2}(u(r,\theta)) - \frac{u(r,\theta)}{r^2}\Bigg) - \frac{\sigma B_0^2}{\rho_{\text{nf}} r^2}u(r,\theta) \end{aligned} \tag{8.2}$$

$$\frac{1}{r\rho_{\text{nf}}}\frac{\partial P}{\partial \theta} - \frac{2\upsilon_{\text{nf}}}{r^2}\frac{\partial}{\partial \theta}(u(r,\theta)) = 0 \tag{8.3}$$

where $u(r, \theta)$ is the velocity, P is the fluid pressure, B_0 is the electromagnetic induction, σ is the conductivity of the fluid, and ρ_{nf} and υ_{nf} stand for effective density and kinematic viscosity of nanofluid, respectively.

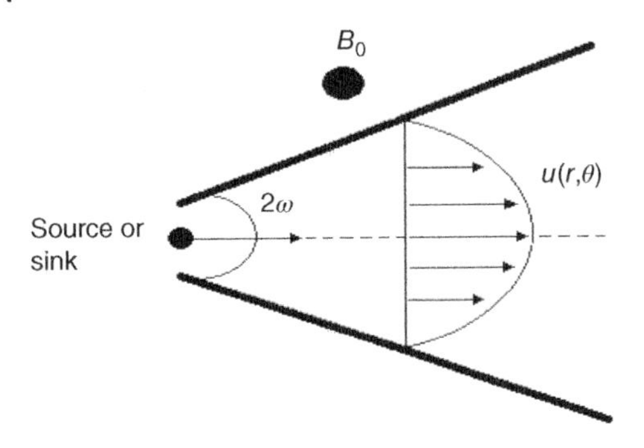

Figure 8.1 Geometry of the problem.

The boundary conditions are

at the center line of the channel: $\dfrac{\partial u}{\partial \theta} = 0$, and
at the boundary of the channel: $u = 0$.

The effective density and kinematic viscosity of nanofluid may be given as [9]

$$\rho_{\rm nf} = \rho_f(1-\phi) + \rho_s\phi, \quad \upsilon_{\rm nf} = \frac{\mu_{\rm nf}}{\rho_{\rm nf}} \tag{8.4}$$

where ϕ stands for nanoparticle volume fraction. $\mu_{\rm nf}$ is the effective dynamic viscosity of nanofluid, and by Brinkman's model [26], it may be given as

$$\mu_{\rm nf} = \frac{\mu_f}{(1-\phi)^{2.5}} \tag{8.5}$$

Considering only radial flow, from continuity equation, we may have

$$ru(r,\theta) = f(\theta) \Rightarrow u(r,\theta) = \frac{f(\theta)}{r} \tag{8.6}$$

Now introduce dimensionless degree as $\eta = \frac{\theta}{\omega}$ and dimensionless form of velocity parameter may be obtained as

$$F(\eta) = \frac{f(\theta)}{f_{\max}} \tag{8.7}$$

By using Eqs. (8.4)–(8.7) and eliminating the pressure term from Eqs. (8.2) and (8.3), nondimensional governing ordinary differential equation for the Jeffery–Hamel problem may be obtained as

$$F'''(\eta) + 2\omega \cdot Re \cdot A_1 \cdot (1-\phi)^{2.5} F(\eta)F'(\eta) + (4-(1-\phi)^{2.5}(Ha)^2)\omega^2 F'(\eta) = 0 \tag{8.8}$$

where Re denotes the Reynolds number, Ha stands for the Hartmann number based on electromagnetic parameter, and A_1 is the ratio of effective density of nanofluid to density of base fluid. These parameters are introduced as

$$Re = \frac{f_{\max}\rho_f\omega}{\mu_f} = \frac{U_{\max} r \rho_f \omega}{\mu_f}, \quad (Ha)^2 = \frac{\sigma B_0^2}{\mu_f}, \quad \text{and } A_1 = \frac{\rho_{\rm nf}}{\rho_f} = (1-\phi) + \frac{\rho_s}{\rho_f}\phi$$

and the reduced boundary conditions are

$$F(0) = 1, \quad F'(0) = 0, \quad F(1) = 0 \tag{8.9}$$

Here, these boundary conditions mean that maximum values of velocity are obtained at centerline $\eta = 0$, and hence, its derivative will be zero at that point. $F(1) = 0$ says that no-slip condition at boundary is considered.

Next, CWBK equations that describe SWW equations which read as [10–17]

$$u_t + uu_x + v_x + \beta u_{xx} = 0 \tag{8.10}$$

$$v_t + (uv)_x + \alpha u_{xxx} - \beta v_{xx} = 0 \tag{8.11}$$

Table 8.1 Some properties of the considered fluid and nanoparticles [27].

Material	Symbol	Density (kg/m^3)	Thermal conductivity (W/m K)
Copper	Cu	8933	401
Sodium alginate	SA	989	0.6376

It may be worth mentioning here that a small change in the value of nanoparticle volume fraction may affect the numerical solution of the considered problem. Similarly, in the case of CWBK equations, the constants α and β representing diffusion power may also not be crisp always. As such, we are motivated to handle such challenging problems in uncertain environment by considering volume fraction ϕ as a fuzzy number and α and β as interval numbers.

Accordingly, fuzzy form of the governing differential equation (8.8) may be written as

$$\widetilde{F}'''(\eta) + 2\omega \cdot Re \cdot \widetilde{A}_1 \cdot (1 - \widetilde{\phi})^{2.5} \widetilde{F}(\eta)\widetilde{F}'(\eta) + (4 - (1 - \widetilde{\phi})^{2.5}(Ha)^2)\omega^2 \widetilde{F}'(\eta) = 0 \tag{8.12}$$

with boundary conditions as

$$F(0) = 1, \quad F'(0) = 0, \quad F(1) = 0 \tag{8.13}$$

here, "~" represents the fuzzy form.

And the interval form of CWBK equations is

$$\frac{\partial}{\partial t}[\underline{u}, \overline{u}] + [\underline{u}, \overline{u}]\frac{\partial}{\partial x}[\underline{u}, \overline{u}] + \frac{\partial}{\partial x}[\underline{v}, \overline{v}] + [\underline{\beta}, \overline{\beta}]\frac{\partial^2}{\partial x^2}[\underline{u}, \overline{u}] = 0 \tag{8.14}$$

$$\frac{\partial}{\partial t}[\underline{v}, \overline{v}] + \frac{\partial}{\partial x}([\underline{u}, \overline{u}][\underline{v}, \overline{v}]) + [\underline{\alpha}, \overline{\alpha}]\frac{\partial^3}{\partial x^3}[\underline{u}, \overline{u}] - [\underline{\beta}, \overline{\beta}]\frac{\partial^2}{\partial x^2}[\underline{v}, \overline{v}] = 0 \tag{8.15}$$

Some physical properties of base fluid and nanoparticle have been presented in Table 8.1 [27].

8.4 Methodology

In this section, two efficient semianalytical methods, viz., HPM and HPTM, which are useful in solving the above-discussed fluid-related problems, have been briefly illustrated.

8.4.1 Homotopy Perturbation Method

To delineate briefly the idea of HPM, let us consider the differential equation [27–30]

$$A(u) - f(x) = 0, \quad x \in \Omega \tag{8.16}$$

with given boundary condition

$$B(u, \partial u/\partial x) = 0, \quad x \in \Gamma \tag{8.17}$$

where A is a differential operator that can be divided into two parts, viz., linear (L) and nonlinear (N), B stands for boundary operator, $f(x)$ is a known analytical function, and Γ is the boundary of the domain Ω.

By splitting A into linear and nonlinear part, Eq. (8.16) may be written as

$$L(u) + N(u) - f(x) = 0, \quad x \in \Omega \tag{8.18}$$

Now, we construct a homotopy $v(r, q): \Omega \times [0, 1] \rightarrow R$ satisfying

$$H(v, q) = (1 - q)[L(v) - L(u_0)] + q[A(v) - f(x)] = 0 \tag{8.19}$$

where q is an embedding parameter lies between 0 and 1 and u_0 is an initial approximation satisfying boundary condition Eq. (8.17).

From Eq. (8.19), one may observe that

when $q = 0$, $L(v) = L(u_0)$ and for $q = 1$, $A(v) - f(x) = 0$ that is, when q converges to 1, we may get approximate solution of Eq. (8.16).

As q is a small parameter, the solution of Eq. (8.19) can be expressed as a power series in q

$$v = v_0 + qv_1 + q^2v_2 + q^3v_3 + \cdots$$

By setting $q = 1$ results the best approximation of Eq. (8.16) that is

$$u = \lim_{q \to 1} v.$$

8.4.2 Homotopy Perturbation Transform Method

Here, Laplace transform method and HPM are combined to have a method called HPTM for solving nonlinear differential equations. It is also called as Laplace homotopy perturbation method (LHPM).

Let us consider a general nonlinear partial differential equation with source term $g(x, t)$ to illustrate the basic idea of HPTM as follows [18–22]

$$Du(x, t) + Ru(x, t) + Nu(x, t) = g(x, t) \tag{8.20}$$

subject to initial conditions

$$u(x, 0) = h(x), \quad \frac{\partial}{\partial t}u(x, 0) = f(x) \tag{8.21}$$

where D is the linear differential operator $D = \frac{\partial^2}{\partial t^2}\left(\text{or}\frac{\partial}{\partial t}\right)$, R is the linear differential operator whose order is less than that of D, and N is the nonlinear differential operator.

The HPTM methodology consists of mainly two steps. The first step is applying Laplace transform on both sides of Eq. (8.20) and the second step is applying HPM where decomposition of nonlinear term is done using He's polynomials.

First by operating Laplace transform on both sides of (8.20), we obtain

$$\mathcal{L}[Du(x, t)] = -\mathcal{L}[Ru(x, t)] - \mathcal{L}[Nu(x, t)] + \mathcal{L}[g(x, t)]$$

Assuming that D is a second-order differential operator and using differentiation property of Laplace transform, we get

$$s^2\mathcal{L}[u(x, t)] - sh(x) - f(x) = -\mathcal{L}[Ru(x, t)] - \mathcal{L}[Nu(x, t)] + \mathcal{L}[g(x, t)]$$

$$\mathcal{L}[u(x, t)] = \frac{h(x)}{s} + \frac{f(x)}{s^2} + \frac{1}{s^2}\mathcal{L}[g(x, t)] - \frac{1}{s^2}\mathcal{L}[Ru(x, t)] - \frac{1}{s^2}\mathcal{L}[Nu(x, t)] \tag{8.22}$$

Applying inverse Laplace transform on both sides of (8.22), we have

$$u(x, t) = G(x, t) - \mathcal{L}^{-1}\left[\frac{1}{s^2}\mathcal{L}[Ru(x, t)] + \frac{1}{s^2}\mathcal{L}[Nu(x, t)]\right] \tag{8.23}$$

where $G(x, t)$ is the term arising from first three terms of right-hand side of (8.22).

Next, to apply HPM, first, we need to assume the solution as a series that contains embedding parameter $p \in [0, 1]$ as

$$u(x,t) = \sum_{n=0}^{\infty} p^n u_n(x,t) \tag{8.24}$$

and the nonlinear term may be decomposed using He's polynomials as

$$Nu(x,t) = \sum_{n=0}^{\infty} p^n H_n(u) \tag{8.25}$$

where $H_n(u)$ represents the He's polynomials [18–22] which are defined as follows:

$$H_n(u_0, u_1, \dots, u_n) = \frac{1}{n!}\left[\frac{\partial^n}{\partial p^n} N\left(\sum_{i=0}^{\infty} p^i u_i\right)\right]_{p=0}, \quad n = 0, 1, 2, 3 \dots \tag{8.26}$$

Substituting Eqs. (8.24) and (8.25) in Eq. (8.23) and combining Laplace transform with HPM, one may obtain the following expression:

$$\sum_{n=0}^{\infty} p^n u_n(x,t) = G(x,t) - p\left(\mathcal{L}^{-1}\left[\frac{1}{s^2}\mathcal{L}\left[R\sum_{n=0}^{\infty} p^n u_n(x,t)\right] + \frac{1}{s^2}\mathcal{L}\left[\sum_{n=0}^{\infty} p^n H_n(u)\right]\right]\right) \tag{8.27}$$

Comparing the coefficients of like powers of "p" on both sides of (8.27), we may obtain the following successive approximations:

$$p^0: \ u_0(x,t) = G(x,t)$$

$$p^1: \ u_1(x,t) = -\mathcal{L}^{-1}\left[\frac{1}{s^2}\mathcal{L}[Ru_0(x,t)] + \frac{1}{s^2}\mathcal{L}[H_0(u)]\right]$$

$$p^2: \ u_2(x,t) = -\mathcal{L}^{-1}\left[\frac{1}{s^2}\mathcal{L}[Ru_1(x,t)] + \frac{1}{s^2}\mathcal{L}[H_1(u)]\right]$$

$$p^3: \ u_3(x,t) = -\mathcal{L}^{-1}\left[\frac{1}{s^2}\mathcal{L}[Ru_2(x,t)] + \frac{1}{s^2}\mathcal{L}[H_2(u)]\right]$$

$$\vdots$$

$$p^n: \ u_n(x,t) = -\mathcal{L}^{-1}\left[\frac{1}{s^2}\mathcal{L}[Ru_{n-1}(x,t)] + \frac{1}{s^2}\mathcal{L}[H_{n-1}(u)]\right]$$

$$\vdots$$

Finally, the solution of the differential equation (8.20) may be obtained as follows:

$$u(x,t) = \lim_{p\to 1} u_n(x,t) = u_0(x,t) + u_1(x,t) + u_2(x,t) + \cdots \tag{8.28}$$

In the next session, we apply HPM and HPTM to governing equations of Jeffery–Hamel problem and CWBK equation in both crisp and uncertain environments.

8.5 Application of HPM and HPTM

8.5.1 Application of HPM to Jeffery–Hamel Problem

To handle fuzziness involved in Eq. (8.12), first we have used δ-cut to convert the fuzzy differential equation given in (8.12) into interval form and then parametric approach may be used to convert the interval form into crisp form.

For simplicity, by putting $\widetilde{a} = 2\omega \cdot Re \cdot \widetilde{A}_1 \cdot (1 - \widetilde{\phi})^{2.5}$ and $\widetilde{b} = (4 - (1 - \widetilde{\phi})^{2.5}(Ha)^2)\omega^2$ in Eq. (8.12), we may have the fuzzy differential equation as

$$\widetilde{F}'''(\eta) + \widetilde{a} \cdot \widetilde{F}(\eta)\widetilde{F}'(\eta) + \widetilde{b} \cdot \widetilde{F}'(\eta) = 0 \tag{8.29}$$

Now, by using δ-cut for fuzzy form, Eq. (8.29) may be converted into interval form as

$$[\underline{F'''}(\delta,\eta), \overline{F'''}(\delta,\eta)] + [\underline{a}(\delta), \overline{a}(\delta)] \cdot [\underline{F}(\delta,\eta), \overline{F}(\delta,\eta)] \cdot [\underline{F}'(\delta,\eta), \overline{F'}(\delta,\eta)] + [\underline{b}(\delta), \overline{b}(\delta)] \cdot [\underline{F}'(\delta,\eta), \overline{F'}(\delta,\eta)] = 0 \tag{8.30}$$

Further, by introducing parametric concept for involved intervals in differential equation (8.30), we may have its crisp form as

$$\begin{aligned}
&\{\gamma(\overline{F'''}(\delta,\eta) - \underline{F'''}(\delta,\eta)) + \underline{F'''}(\delta,\eta)\} \\
&\quad + \{\gamma(\overline{a}(\delta) - \underline{a}(\delta)) + \underline{a}(\delta)\}.\{\gamma(\overline{F}(\delta,\eta) - \underline{F}(\delta,\eta)) + \underline{F}(\delta,\eta)\} \cdot \{\gamma(\overline{F'}(\delta,\eta) - \underline{F}'(\delta,\eta)) + \underline{F}'(\delta,\eta)\} \\
&\quad + \{\gamma(\overline{b}(\delta) - \underline{b}(\delta)) + \underline{b}(\delta)\} \cdot \{\gamma(\overline{F'}(\delta,\eta) - \underline{F}'(\delta,\eta)) + \underline{F}'(\delta,\eta)\} = 0
\end{aligned} \tag{8.31}$$

where δ and γ are parameters that lie between 0 and 1.

Let us denote

$$\begin{aligned}
\gamma(\overline{F}(\delta,\eta) - \underline{F}(\delta,\eta)) + \underline{F}(\delta,\eta) &= F(\gamma,\delta,\eta) \\
\gamma(\overline{F'}(\delta,\eta) - \underline{F}'(\delta,\eta)) + \underline{F}'(\delta,\eta) &= F'(\gamma,\delta,\eta) \\
\gamma(\overline{F'''}(\delta,\eta) - \underline{F'''}(\delta,\eta)) + \underline{F'''}(\delta,\eta) &= F'''(\gamma,\delta,\eta) \\
\gamma(\overline{a}(\delta) - \underline{a}(\delta)) + \underline{a}(\delta) &= a(\gamma,\delta)
\end{aligned}$$

and

$$\gamma(\overline{b}(\delta) - \underline{b}(\delta)) + \underline{b}(\delta) = b(\gamma,\delta)$$

By using these notations, Eq. (8.31) may be written as

$$F'''(\gamma,\delta,\eta) + a(\gamma,\delta)F(\gamma,\delta,\eta)F'(\gamma,\delta,\eta) + b(\gamma,\delta)F'(\gamma,\delta,\eta) = 0 \tag{8.32}$$

with boundary conditions $F(0) = 1, F'(0) = 0, F(1) = 0$.

Now, we apply HPM to solve Eq. (8.32). Homotopy for Eq. (8.32) may be constructed as

$$\begin{aligned}
&H(q,\eta) = (1-q)F'''(\gamma,\delta,\eta) + q(F'''(\gamma,\delta,\eta) + a(\gamma,\delta)F(\gamma,\delta,\eta)F'(\gamma,\delta,\eta) + b(\gamma,\delta)F'(\gamma,\delta,\eta)) = 0 \\
&\Rightarrow H(q,\eta) = F'''(\gamma,\delta,\eta) + q(a(\gamma,\delta)F(\gamma,\delta,\eta)F'(\gamma,\delta,\eta) + b(\gamma,\delta)F'(\gamma,\delta,\eta)) = 0
\end{aligned} \tag{8.33}$$

According to this method, assumed series solution of Eq. (8.33) may be written as

$$F(\gamma,\delta,\eta) = F_0(\gamma,\delta,\eta) + qF_1(\gamma,\delta,\eta) + q^2F_2(\gamma,\delta,\eta) + q^3F_3(\gamma,\delta,\eta) + \cdots \tag{8.34}$$

Afterward, our goal is to find the unknown functions used in the assumed series solution (8.34), viz., $F_0(\gamma,\delta,\eta)$, $F_1(\gamma,\delta,\eta), F_2(\gamma,\delta,\eta), F_3(\gamma,\delta,\eta)\ldots$.

Substituting Eq. (8.34) in Eq. (8.33) and collecting the coefficients of various powers of q, we may get

$$\begin{aligned}
&F_0'''(\gamma,\delta,\eta)q^0 + \\
&[F_1'''(\gamma,\delta,\eta) + b(\gamma,\delta,\eta)F_0'(\gamma,\delta,\eta) + a(\gamma,\delta,\eta)F_0(\gamma,\delta,\eta)F_0'(\gamma,\delta,\eta)]q^1 + \\
&[F_2'''(\gamma,\delta,\eta) + b(\gamma,\delta,\eta)F_1'(\gamma,\delta,\eta) + a(\gamma,\delta,\eta)\{F_1(\gamma,\delta,\eta)F_0'(\gamma,\delta,\eta) + F_1'(\gamma,\delta,\eta)F_0(\gamma,\delta,\eta)\}]q^2 +
\end{aligned}$$

$$[F_3'''(\gamma,\delta,\eta)+b(\gamma,\delta,\eta)F_2'(\gamma,\delta,\eta)+a(\gamma,\delta,\eta)\{F_2(\gamma,\delta,\eta)F_0'(\gamma,\delta,\eta)+F_1(\gamma,\delta,\eta)F_1'(\gamma,\delta,\eta)$$
$$+F_2'(\gamma,\delta,\eta)F_0(\gamma,\delta,\eta)\}]q^3+\cdots=0 \tag{8.35}$$

Further, by equating the coefficient of various powers of q with zero and using proper boundary conditions, we may get the functions $F_0(\gamma,\delta,\eta), F_1(\gamma,\delta,\eta), F_2(\gamma,\delta,\eta), F_3(\gamma,\delta,\eta)\ldots$ explicitly.

Equating the coefficient of q^0 in Eq. (8.35) to zero, we may have

$$F_0'''(\gamma,\delta,\eta)=0 \tag{8.36}$$

with boundary conditions as $F_0(0)=1,\ F_0'(0)=0,\ F_0(1)=0.$

By solving Eq. (8.36) with boundary conditions, we may obtain

$$F_0(\gamma,\delta,\eta)=1-\eta^2 \tag{8.37}$$

Further, by equating the coefficient of q^1 with zero, we may have

$$F_1'''(\gamma,\delta,\eta)+b(\gamma,\delta,\eta)F_0'(\gamma,\delta,\eta)+a(\gamma,\delta,\eta)F_0(\gamma,\delta,\eta)F_0'(\gamma,\delta,\eta)=0$$

$$\Rightarrow F_1'''(\gamma,\delta,\eta)=-[b(\gamma,\delta,\eta)F_0'(\gamma,\delta,\eta)+a(\gamma,\delta,\eta)F_0(\gamma,\delta,\eta)F_0'(\gamma,\delta,\eta)] \tag{8.38}$$

with boundary conditions as $F_1(0)=0,\ F_1'(0)=0,\ F_1(1)=0.$

By solving Eq. (8.38) with boundary conditions, we may obtain

$$F_1(\gamma,\delta,\eta)=-\frac{a(\gamma,\delta,\eta)}{60}\eta^6+\frac{a(\gamma,\delta,\eta)+b(\gamma,\delta,\eta)}{12}\eta^4-\left(\frac{2a(\gamma,\delta,\eta)}{15}+\frac{b(\gamma,\delta,\eta)}{6}\right)\frac{\eta^2}{2} \tag{8.39}$$

Again from coefficient of q^2 with boundary conditions as $F_2(0)=0,\ F_2'(0)=0,\ F_2(1)=0$, we may obtain

$$\begin{aligned}F_2(\gamma,\delta,\eta)&=\frac{a^2(\gamma,\delta,\eta)}{5400}\eta^{10}+\frac{a^2(\gamma,\delta,\eta)+a(\gamma,\delta,\eta)b(\gamma,\delta,\eta)}{560}\eta^8\\&-\left(\frac{a^2(\gamma,\delta,\eta)}{200}+\frac{a(\gamma,\delta,\eta)b(\gamma,\delta,\eta)}{120}+\frac{b^2(\gamma,\delta,\eta)}{360}\right)\eta^6\\&+\left(\frac{a^2(\gamma,\delta,\eta)}{180}+\frac{a(\gamma,\delta,\eta)b(\gamma,\delta,\eta)}{80}+\frac{b^2(\gamma,\delta,\eta)}{144}\right)\eta^4\\&-\left(\frac{163a^2(\gamma,\delta,\eta)}{75\,600}+\frac{a(\gamma,\delta,\eta)b(\gamma,\delta,\eta)}{168}+\frac{b^2(\gamma,\delta,\eta)}{240}\right)\eta^2\end{aligned} \tag{8.40}$$

By proceeding like this for coefficients of various powers of q with appropriate boundary conditions, we may have the functions $F_3(\gamma,\delta,\eta), F_4(\gamma,\delta,\eta), F_5(\gamma,\delta,\eta)\ldots.$

Hence, the approximate solution of Eq. (8.12) may be given as

$$\begin{aligned}F(\gamma,\delta,\eta)&=\lim_{q\to1}\{F_0(\gamma,\delta,\eta)+qF_1(\gamma,\delta,\eta)+q^2F_2(\gamma,\delta,\eta)+q^3F_3(\gamma,\delta,\eta)+\cdots\}\\&=F_0(\gamma,\delta,\eta)+F_1(\gamma,\delta,\eta)+F_2(\gamma,\delta,\eta)+F_3(\gamma,\delta,\eta)+\cdots\end{aligned} \tag{8.41}$$

Here, three-term approximate solution of Eq. (8.12) may be found as

$$F(\gamma,\delta,\eta)=F_0(\gamma,\delta,\eta)+F_1(\gamma,\delta,\eta)+F_2(\gamma,\delta,\eta)$$

where $F_0(\gamma,\delta,\eta)$, $F_1(\gamma,\delta,\eta)$, and $F_2(\gamma,\delta,\eta)$ are given in Eqs. (8.37), (8.39), and (8.40), respectively.

It may be noted that by increasing the number of terms, one may get more appropriate approximate results. Moreover, here, the fuzzy solutions are control by the parameters δ and γ.

8.5.2 Application of HPTM to Coupled Whitham–Broer–Kaup Equations

Now, we use HPTM to find the solution of CWBK equations given in (8.10) and (8.11) [10–17] subject to initial conditions as [15, 16]

$$u(x,0) = \lambda - 2k\sqrt{\alpha+\beta^2}\coth(k(x+x_0)) \tag{8.42}$$

$$v(x,0) = -2k^2(\alpha+\beta\sqrt{\alpha+\beta^2}+\beta^2)\text{cosech}^2(k(x+x_0)) \tag{8.43}$$

Applying the Laplace transform to Eqs. (8.10) and (8.11), we get

$$L\{u_t\} + L\{uu_x + v_x + \beta u_{xx}\} = 0 \tag{8.44}$$

$$L\{v_t\} + L\{(uv)_x + \alpha u_{xxx} - \beta v_{xx}\} = 0 \tag{8.45}$$

Simplifying (8.44) and (8.45), we obtain

$$L\{u(x,t)\} = \frac{1}{s}u(x,0) - \frac{1}{s}L\{uu_x + v_x + \beta u_{xx}\} \tag{8.46}$$

$$L\{v(x,t)\} = \frac{1}{s}v(x,0) - \frac{1}{s}L\{(uv)_x + \alpha u_{xxx} - \beta v_{xx}\} \tag{8.47}$$

Taking inverse Laplace transform on both sides of (8.46) and (8.47), we may have

$$u(x,t) = u(x,0) - L^{-1}\left\{\frac{1}{s}L\{uu_x + v_x + \beta u_{xx}\}\right\} \tag{8.48}$$

$$v(x,t) = v(x,0) - L^{-1}\left\{\frac{1}{s}L\{(uv)_x + \alpha u_{xxx} - \beta v_{xx}\}\right\} \tag{8.49}$$

Now applying HPM to (8.48) and (8.49) gives

$$\sum_{n=0}^{\infty} p^n u_n(x,t) = u(x,0) - p\left(L^{-1}\left\{\frac{1}{s}L\left\{\sum_{n=0}^{\infty} p^n H_{1n}(u) + \sum_{n=0}^{\infty} p^n (v_n(x,t))_x + \beta\sum_{n=0}^{\infty} p^n (u_n(x,t))_{xx}\right\}\right\}\right) \tag{8.50}$$

$$\sum_{n=0}^{\infty} p^n v_n(x,t) = v(x,0) - p\left(L^{-1}\left\{\frac{1}{s}L\left\{\sum_{n=0}^{\infty} p^n H_{2n}(uv) + \alpha\sum_{n=0}^{\infty} p^n (u_n(x,t))_{xxx} - \beta\sum_{n=0}^{\infty} p^n (v_n(x,t))_{xx}\right\}\right\}\right) \tag{8.51}$$

where $H_{1n}(u)$ and $H_{2n}(uv)$ are He's polynomials for the nonlinear terms uu_x and $(uv)_x$, respectively. First, few terms of $H_{1n}(u)$ and $H_{2n}(uv)$ are given by

$$\begin{aligned} H_{10}(u) &= u_0(u_0)_x \\ H_{11}(u) &= u_0(u_1)_x + u_1(u_0)_x \\ H_{12}(u) &= u_0(u_2)_x + u_1(u_1)_x + u_2(u_0)_x \end{aligned} \tag{8.52}$$

$$\begin{aligned} H_{20}(uv) &= u_0(v_0)_x + v_0(u_0)_x \\ H_{21}(uv) &= u_0(v_1)_x + u_1(v_0)_x + v_0(u_1)_x + v_1(u_0)_x \\ H_{22}(uv) &= u_0(v_2)_x + u_1(v_1)_x + u_2(v_0)_x + v_0(u_1)_x + v_1(u_2)_x + v_2(u_0)_x \end{aligned} \tag{8.53}$$

Comparing the coefficients of like powers of p on both sides of Eqs. (8.50) and (8.51), we obtain

$$p^0: \begin{aligned} u_0(x,t) &= u(x,0) \\ v_0(x,t) &= v(x,0) \end{aligned} \tag{8.54}$$

$$p^1: \begin{aligned} u_1(x,t) &= -L^{-1}\left[\frac{1}{s}L\{H_{10}(u) + (v_0)_x + \beta(u_0)_{xx}\}\right] = f_1 t \\ v_1(x,t) &= -L^{-1}\left[\frac{1}{s}L\{H_{20}(uv) + \alpha(u_0)_{xxx} - \beta(v_0)_{xx}\}\right] = g_1 t \end{aligned} \tag{8.55}$$

$$p^2: \begin{array}{l} u_2(x,t) = -L^{-1}\left[\frac{1}{s}L\{H_{11}(u) + (v_1)_x + \beta(u_1)_{xx}\}\right] = f_2\frac{t^2}{2} \\ v_2(x,t) = -L^{-1}\left[\frac{1}{s}L\{H_{21}(uv) + \alpha(u_1)_{xxx} - \beta(v_1)_{xx}\}\right] = g_2\frac{t^2}{2} \end{array} \tag{8.56}$$

$$p^3: \begin{array}{l} u_3(x,t) = -L^{-1}\left[\frac{1}{s}L\{H_{12}(u) + (v_2)_x + \beta(u_2)_{xx}\}\right] = f_3\frac{t^3}{6} \\ v_2(x,t) = -L^{-1}\left[\frac{1}{s}L\{H_{22}(uv) + \alpha(u_2)_{xxx} - \beta(v_2)_{xx}\}\right] = g_3\frac{t^3}{6} \end{array} \tag{8.57}$$

$$\vdots$$

where

$$f_1 = -\{u_0(u_0)_x + (v_0)_x + \beta(u_0)_{xx}\},$$

$$f_2 = -\{u_0(u_1)_x + u_1(u_0)_x + (v_1)_x + \beta(u_1)_{xx}\},$$

$$f_3 = -\{u_0(u_2)_x + u_1(u_1)_x + u_2(u_0)_x + (v_2)_x + \beta(u_2)_{xx}\},$$

$$g_1 = -\{u_0(v_0)_x + v_0(u_0)_x + \alpha(u_0)_{xxx} - \beta(v_0)_{xx}\},$$

$$g_2 = -\{u_0(v_1)_x + u_1(v_0)_x + v_0(u_1)_x + v_1(u_0)_x + \alpha(u_1)_{xxx} - \beta(v_1)_{xx}\},$$

$$g_3 = -\{u_0(v_2)_x + u_1(v_1)_x + u_2(v_0)_x + v_0(u_1)_x + v_1(u_2)_x + v_2(u_0)_x + \alpha(u_2)_{xxx} - \beta(v_2)_{xx}\}.$$

Solutions of CWBK equations (8.10) and (8.11) may be obtained as

$$u(x,t) = u(x,0) + f_1 t + f_2\frac{t^2}{2} + f_3\frac{t^3}{6} + \cdots \tag{8.58}$$

$$v(x,t) = v(x,0) + g_1 t + g_2\frac{t^2}{2} + g_3\frac{t^3}{6} + \cdots \tag{8.59}$$

Exact solutions of CWBK equations (8.10) and (8.11) may be given as [15, 16]

$$u(x,t) = \lambda - 2k\sqrt{\alpha + \beta^2}\ \coth(k(x + x_0) - \lambda t) \tag{8.60}$$

$$v(x,t) = -2k^2(\alpha + \beta\sqrt{\alpha + \beta^2} + \beta^2)\mathrm{cosech}^2(k(x + x_0) - \lambda t) \tag{8.61}$$

Next, we solve interval CWBK equations (8.14) and (8.15) using HPTM with the help of parametric approach defined above. In this regard, first, we transform the interval CWBK equations to crisp form using parametric approach, and then, these transformed equations have been solved using HPTM.

The interval numbers $[\underline{\alpha}, \overline{\alpha}]$ and $[\underline{\beta}, \overline{\beta}]$ may be written in crisp form using parametric approach as

$$[\underline{\alpha}, \overline{\alpha}] = \gamma(\overline{\alpha} - \underline{\alpha}) + \underline{\alpha} \ \text{ and } \ [\underline{\beta}, \overline{\beta}] = \gamma(\overline{\beta} - \underline{\beta}) + \underline{\beta} \tag{8.62}$$

Further, for simplicity denoting

$$[\underline{\alpha}, \overline{\alpha}] = \gamma(\overline{\alpha} - \underline{\alpha}) + \underline{\alpha} = \alpha(\gamma) \ \text{ and } \ [\underline{\beta}, \overline{\beta}] = \gamma(\overline{\beta} - \underline{\beta}) + \underline{\beta} = \beta(\gamma) \tag{8.63}$$

Similarly, interval solutions $[\underline{u}, \overline{u}]$ and $[\underline{v}, \overline{v}]$ of (8.14) and (8.15) may be denoted as

$$[\underline{u}, \overline{u}] = \gamma(\overline{u} - \underline{u}) + \underline{u} = u(x,t,\gamma) \ \text{ and } \ [\underline{v}, \overline{v}] = \gamma(\overline{v} - \underline{v}) + \underline{v} = v(x,t,\gamma) \tag{8.64}$$

Using Eqs. (8.63) and (8.64), the interval CWBK equations (8.14) and (8.15) may be transformed to crisp form as

$$\frac{\partial}{\partial t}u(x,t,\gamma) + u(x,t,\gamma)\frac{\partial}{\partial x}u(x,t,\gamma) + \frac{\partial}{\partial x}v(x,t,\gamma) + \beta(\gamma)\frac{\partial^2}{\partial x^2}u(x,t,\gamma) = 0 \tag{8.65}$$

$$\frac{\partial}{\partial t}v(x,t,\gamma) + \frac{\partial}{\partial x}(u(x,t,\gamma)v(x,t,\gamma)) + \alpha(\gamma)\frac{\partial^3}{\partial x^3}u(x,t,\gamma) - \beta(\gamma)\frac{\partial^2}{\partial x^2}v(x,t,\gamma) = 0 \tag{8.66}$$

subject to initial conditions

$$u(x,0) = \lambda - 2k\sqrt{\alpha(\gamma) + [\beta(\gamma)]^2}\coth(k(x + x_0)) \tag{8.67}$$

$$v(x,0) = -2k^2(\alpha(\gamma) + \beta(\gamma)\sqrt{\alpha(\gamma) + [\beta(\gamma)]^2} + [\beta(\gamma)]^2)\text{cosech}^2(k(x + x_0)) \tag{8.68}$$

Next, we apply HPTM to (8.65) and (8.66) subject to initial conditions (8.67) and (8.68).

Applying Laplace transform to Eqs. (8.65) and (8.66), we obtain

$$L\left\{\frac{\partial}{\partial t}u(x,t,\gamma)\right\} + L\left\{u(x,t,\gamma)\frac{\partial}{\partial x}u(x,t,\gamma) + \frac{\partial}{\partial x}v(x,t,\gamma) + \beta(\gamma)\frac{\partial^2}{\partial x^2}u(x,t,\gamma)\right\} = 0 \tag{8.69}$$

$$L\left\{\frac{\partial}{\partial t}v(x,t,\gamma)\right\} + L\left\{\frac{\partial}{\partial x}(u(x,t,\gamma)v(x,t,\gamma)) + \alpha(\gamma)\frac{\partial^3}{\partial x^3}u(x,t,\gamma) - \beta(\gamma)\frac{\partial^2}{\partial x^2}v(x,t,\gamma)\right\} = 0 \tag{8.70}$$

Simplifying (8.69) and (8.70), we get

$$L\{u(x,t,\gamma)\} = \frac{1}{s}u(x,0,\gamma) - \frac{1}{s}L\left\{u(x,t,\gamma)\frac{\partial}{\partial x}u(x,t,\gamma) + \frac{\partial}{\partial x}v(x,t,\gamma) + \beta(\gamma)\frac{\partial^2}{\partial x^2}u(x,t,\gamma)\right\} \tag{8.71}$$

$$L\{v(x,t,\gamma)\} = \frac{1}{s}v(x,0,\gamma) - \frac{1}{s}L\left\{\frac{\partial}{\partial x}(u(x,t,\gamma)v(x,t,\gamma)) + \alpha(\gamma)\frac{\partial^3}{\partial x^3}u(x,t,\gamma) - \beta(\gamma)\frac{\partial^2}{\partial x^2}v(x,t,\gamma)\right\} \tag{8.72}$$

Taking inverse Laplace transform on both sides of (8.71) and (8.72) will give

$$u(x,t,\gamma) = u(x,0,\gamma) - L^{-1}\left\{\frac{1}{s}L\left\{u(x,t,\gamma)\frac{\partial}{\partial x}u(x,t,\gamma) + \frac{\partial}{\partial x}v(x,t,\gamma) + \beta(\gamma)\frac{\partial^2}{\partial x^2}u(x,t,\gamma)\right\}\right\} \tag{8.73}$$

$$v(x,t,\gamma) = v(x,0,\gamma) - L^{-1}\left\{\frac{1}{s}L\left\{\frac{\partial}{\partial x}(u(x,t,\gamma)v(x,t,\gamma)) + \alpha(\gamma)\frac{\partial^3}{\partial x^3}u(x,t,\gamma) - \beta(\gamma)\frac{\partial^2}{\partial x^2}v(x,t,\gamma)\right\}\right\} \tag{8.74}$$

Next, applying HPM, we get

$$\sum_{n=0}^{\infty}p^n u_n(x,t,\gamma) = u(x,0,\gamma) - L^{-1}\left\{\frac{1}{s}L\left\{\sum_{n=0}^{\infty}p^n H_{3n}\{u(x,t,\gamma)\}\right.\right.$$
$$\left.\left. + \frac{\partial}{\partial x}\sum_{n=0}^{\infty}p^n v_n(x,t,\gamma) + \beta(\gamma)\frac{\partial^2}{\partial x^2}\sum_{n=0}^{\infty}p^n u_n(x,t,\gamma)\right\}\right\} \tag{8.75}$$

$$\sum_{n=0}^{\infty}p^n v_n(x,t,\gamma) = v(x,0,\gamma) - L^{-1}\left\{\frac{1}{s}L\left\{\sum_{n=0}^{\infty}p^n H_{4n}\{u(x,t,\gamma)v(x,t,\gamma)\}\right.\right.$$
$$\left.\left. + \alpha(\gamma)\frac{\partial^3}{\partial x^3}\sum_{n=0}^{\infty}p^n u_n(x,t,\gamma) - \beta(\gamma)\frac{\partial^2}{\partial x^2}\sum_{n=0}^{\infty}p^n v_n(x,t,\gamma)\right\}\right\} \tag{8.76}$$

where H_{3n} and H_{4n} are He's polynomials like H_{1n} and H_{2n}.

Next, proceeding like crisp case, we may obtain the solutions of (8.14) and (8.15) as

$$u(x,t,\gamma) = \lim_{p\to 1} u_n(x,t,\gamma) = u_0(x,t,\gamma) + u_1(x,t,\gamma) + u_2(x,t,\gamma) + \cdots \tag{8.77}$$

$$v(x,t,\gamma) = \lim_{p\to 1} v_n(x,t,\gamma) = v_0(x,t,\gamma) + v_1(x,t,\gamma) + v_2(x,t,\gamma) + \cdots \tag{8.78}$$

8.6 Results and Discussion

Results obtained for Jeffery–Hamel problem and CWBK equations have been discussed in this section. Also, the solutions obtained by the present methods, viz., HPM and HPTM, are validated by comparing with existing/exact solutions in special cases.

Table 8.2 Velocity profile when $Ha = 0$, $Re = 110$, $\omega = 3^0$, $\phi = 0$ by taking different numbers of terms in series solution.

	$F(\gamma, \delta, \eta)$ in different numbers of terms in series solution (8.41)					
η	2 terms	3 terms	4 terms	5 terms	6 terms	7 terms
0	1.000 000 00	1.000 000 00	1.000 000 00	1.000 000 00	1.000 000 00	1.000 000 00
0.1	0.987 396 04	0.986 785 04	0.986 819 50	0.986 809 78	0.986 810 40	0.986 810 53
0.2	0.950 007 65	0.947 670 97	0.947 833 59	0.947 793 01	0.947 795 51	0.947 795 99
0.3	0.889 071 54	0.884 182 46	0.884 631 18	0.884 531 89	0.884 537 57	0.884 538 59
0.4	0.806 536 55	0.798 643 03	0.799 613 72	0.799 412 91	0.799 423 44	0.799 425 16
0.5	0.704 895 19	0.693 876 84	0.695 667 03	0.695 297 16	0.695 315 62	0.695 318 30
0.6	0.586 947 69	0.572 843 40	0.575 774 80	0.575 137 28	0.575 169 95	0.575 174 12
0.7	0.455 498 74	0.438 248 76	0.442 623 66	0.441 592 19	0.441 650 02	0.441 656 68
0.8	0.312 986 75	0.292 176 24	0.298 237 97	0.296 676 10	0.296 773 43	0.296 784 57
0.9	0.161 045 77	0.135 769 96	0.143 660 04	0.141 455 87	0.141 603 05	0.141 623 09
0.0	−0.000 000 0	−0.031 017 12	−0.021 335 46	−0.024 215 76	−0.024 028 34	−0.023 989 46

Table 8.2 presents the solution of Eq. (8.12) for different numbers of terms of the series solution (8.41) when $Ha = 0$, $Re = 110$, $\omega = 3^0$, $\widetilde{\phi} = [0, 0.1, 0.2]$, and $\delta = \gamma = 0$. It is well known that residual error is defined as the error obtained by substitute approximate solution in the governing differential equation. Figure 8.2 depicts the residual error plots for different numbers of terms in the approximate series solution (8.41). In Table 8.3, we have compared the present result with the existing result by DTM when $Ha = 0$, $Re = 50$, $\omega = 5^0$, $\phi = 0$.

Velocity profiles of SA–Cu nanofluid with different values of nanoparticle volume fraction and fixed values $Ha = 500$, $Re = 50$, $\omega = 5^0$ have shown in Figure 8.3. Figure 8.4 present the effect of Hartmann number on nondimensional velocity of SA–Cu nanofluid flow between the inclined planes. Effect of Reynolds number

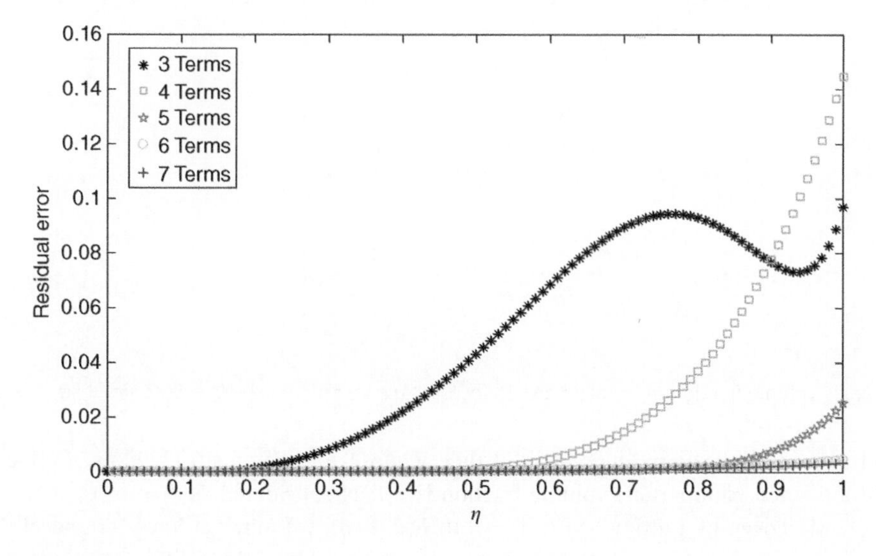

Figure 8.2 Residual error plot for different numbers of terms in series solution (8.41) when $Ha = 2000$, $Re = 50$, $\omega = 3^0$, $\phi = 0.01$.

Table 8.3 Comparison of velocity profile of present results with existing results by DTM when $Ha = 0, Re = 50, \omega = 5^0$, $\phi = 0$.

	Present results for velocity in different numbers of terms in (8.41)					Existing result
η	3 terms	4 terms	5 terms	6 terms	7 terms	DTM [9]
0	1.0	1.0	1.0	1.0	1.0	1.0
0.1	0.982 614 12	0.982 398 66	0.982 416 1	0.982 445 74	0.982 452 04	0.982 431
0.2	0.931 812 53	0.931 082 99	0.931 168 73	0.931 283 62	0.931 306 24	0.931 226
0.3	0.851 479 9	0.850 263 83	0.850 496 56	0.850 737 31	0.850 779 2	0.850 611
0.4	0.747 505 94	0.746 172 58	0.746 627 58	0.747 003 2	0.747 058 09	0.746 791
0.5	0.626 998 88	0.626 077 02	0.626 767 86	0.627 244 49	0.627 299 38	0.626 948
0.6	0.497 356 84	0.497 240 89	0.498 079 88	0.498 587 82	0.498 629 56	0.498 234
0.7	0.365 347 54	0.366 050 3	0.366 867 81	0.367 326 52	0.367 348 59	0.366 966
0.8	0.236 359 72	0.237 463 79	0.238 083 44	0.238 427 36	0.238 432 12	0.238 124
0.9	0.113 978 73	0.114 830 55	0.115 147 47	0.115 334 15	0.115 330 53	0.115 152
0.0	−0.000 000 0	−0.000 000 0	−0.000 000 0	−0.000 000 0	−0.000 000 0	−0.000 000 0

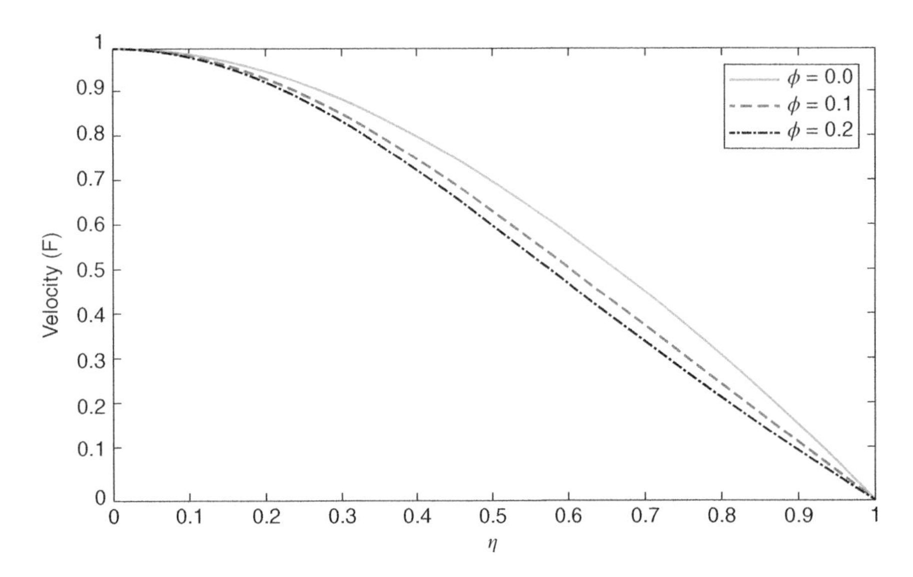

Figure 8.3 Effect of nanoparticle volume fraction on velocity profile for SA–Cu nanofluid when $Ha = 500$, $Re = 50, \omega = 5^0$.

on dimensionless velocity has been displayed in Figure 8.5. Further, Figure 8.6 represents fuzzy plots of velocity profile for different values of η when nanoparticle volume fraction is a TFN considered as $\widetilde{\phi} = [0, 0.1, 0.2]$ and $Ha = 100$, $Re = 50$, $\omega = 3^0$. Moreover, by putting $\delta = 0$, $\gamma = 0$ in $F(\gamma, \delta, \eta)$, we may get lower bound of the nondimensional velocity, whereas by putting $\delta = 0, \gamma = 1$, we may get an upper bound for different values of η. Solution bounds for velocity profile when $Ha = 100$, $Re = 50, \omega = 3^0$, and $\widetilde{\phi} = [0, 0.1, 0.2]$ are given in Table 8.4.

HPTM solutions (8.58) and (8.59) of CWBK equations given in (8.10) and (8.11) are compared with exact solutions (8.60) and (8.61) for $\alpha = 0, \beta = 0.5, k = 0.2, L = 0.005$, and $t = 1$ in Figure 8.7, whereas Figure 8.8 compares the same for $\alpha = 0.5$, $\beta = 0$, $k = 0.2$, $L = 0.005$, and $t = 1$. Solutions (8.77) and (8.78) of the interval CWBK

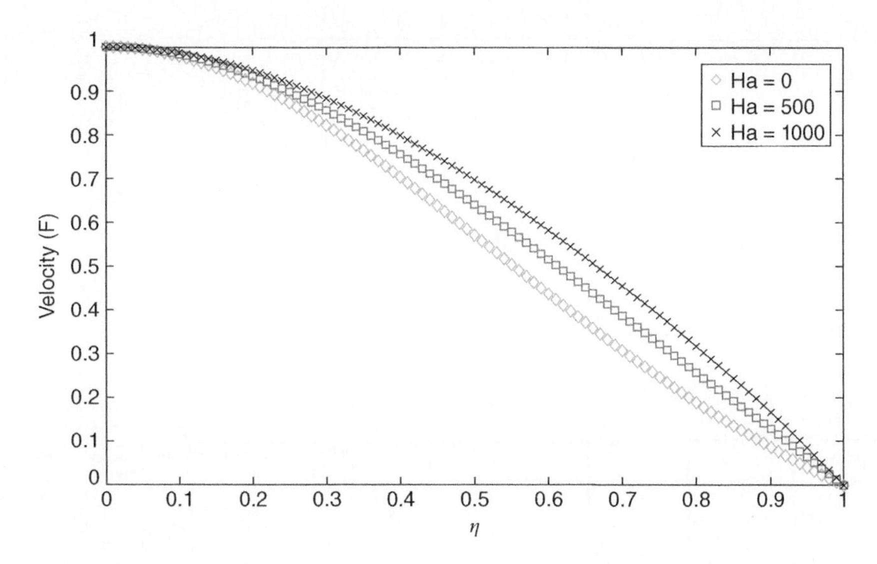

Figure 8.4 Effect of Hartmann number on the velocity profile for SA–Cu nanofluid when $Re = 50$, $\omega = 5^0$, $\phi = 0.1$.

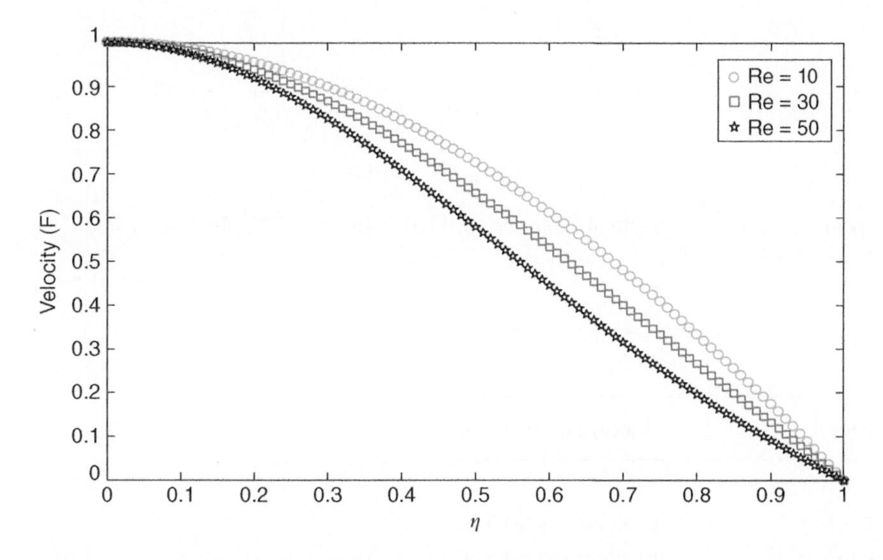

Figure 8.5 Effect of Reynolds number on the velocity profile for SA–Cu nanofluid when $Ha = 50$, $\omega = 5^0$, $\phi = 0.1$.

equations (8.14) and (8.15) are presented in Figure 8.9 for $\alpha = [0.1, 0.9]$, $\beta = 0$, $k = 0.2$, $L = 0.005$, and $t = 0.5$. Finally, Figure 8.10 shows the interval solutions (8.77) and (8.78) for $\alpha = 0$, $\beta = [0.1, 0.9]$, $k = 0.2$, $L = 0.005$, and $t = 0.5$.

From Table 8.2, one may see the convergence of the result with increase in number of terms in the series solution. It may be observed from Figure 8.2 that by increasing number of terms in series solution (8.41), residual error decreases and it tends to zero, which confirms the convergence of the series solution. The present results by HPM are in good agreement with the existing DTM solution [9], which may be seen in Table 8.3. From Figure 8.3, one may conclude that by increase in value of nanoparticle volume fraction, there is a decrease in the velocity profile of SA–Cu nanofluid in between the considered channel. It may also be seen from Figure 8.4 that increase in Hartmann number causes an increase in the velocity profile. Figure 8.5 confirms that the velocity profile for

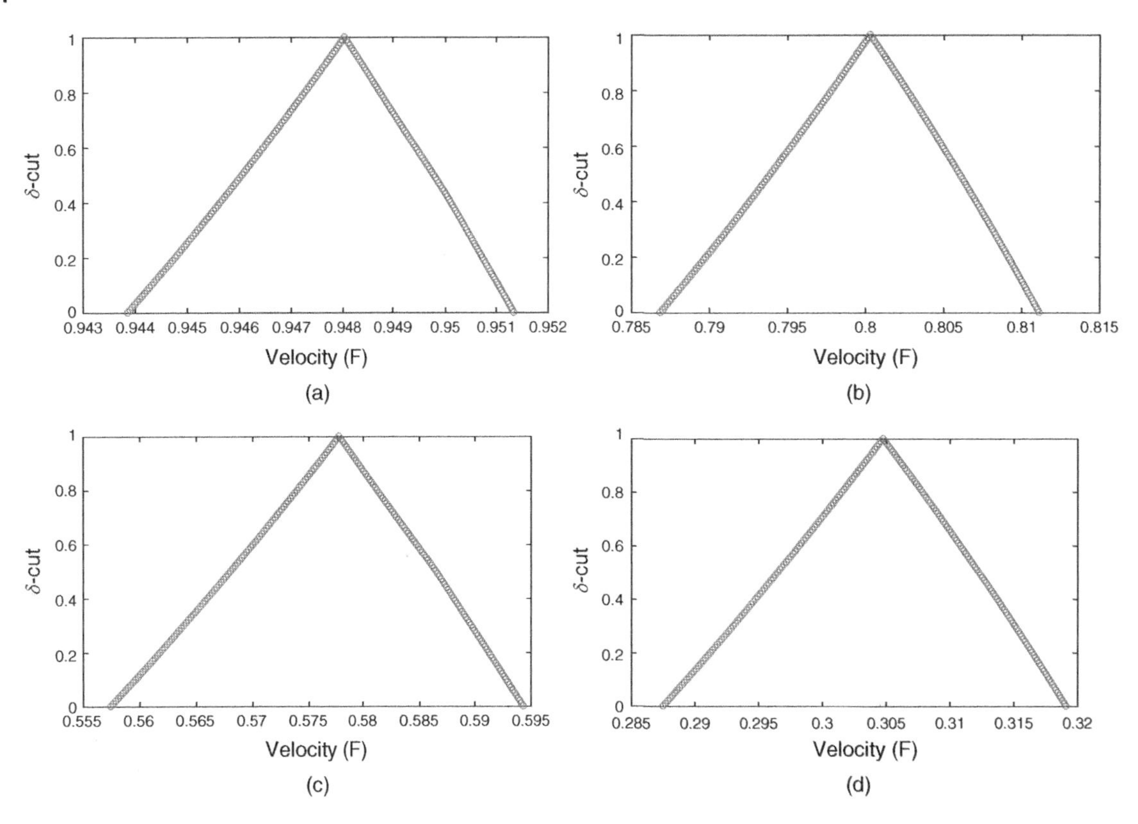

Figure 8.6 Fuzzy plot of the velocity profile for SA–Cu nanofluid when $\tilde{\phi} = [0, 0.1, 0.2]$ and $Ha = 100$, $Re = 50$, $\omega = 3^0$. (a) $\eta = 0.2$, (b) $\eta = 0.4$, (c) $\eta = 0.6$, and (d) $\eta = 0.8$.

Table 8.4 Solution bounds for SA–Cu nanofluid when $Ha = 100, Re = 50$, $\omega = 3^0$, and $\tilde{\phi} = [0, 0.1, 0.2]$.

η	Velocity (lower bound)	Velocity (upper bound)
0.0	1.0	1.0
0.1	0.985 773 307 662 548	0.987 740 276 941 951
0.2	0.943 859 937 040 970	0.951 341 473 381 224
0.3	0.876 452 283 883 681	0.891 904 052 869 956
0.4	0.786 868 265 777 731	0.811 132 250 151 742
0.5	0.679 120 068 532 976	0.711 157 455 666 914
0.6	0.557 432 029 106 776	0.594 323 008 426 450
0.7	0.425 783 780 430 919	0.462 953 665 930 697
0.8	0.287 532 866 251 400	0.319 129 944 261 829
0.9	0.145 141 078 684 186	0.164 480 872 391 153
1.0	0.000 000	0.000 000

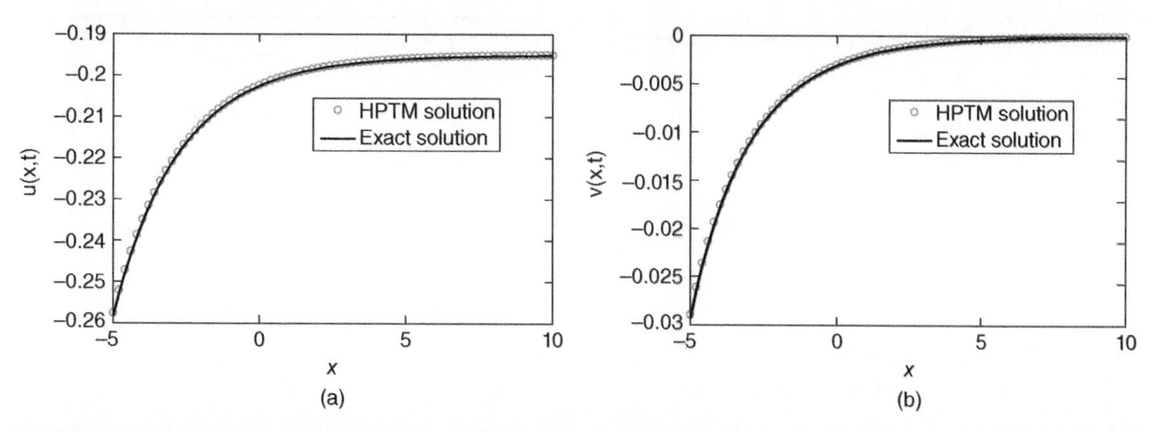

Figure 8.7 Comparison of solution of CWBK equations (8.10) and (8.11) with exact solution for $\alpha = 0$, $\beta = 0.5$, $k = 0.2$, and $L = 0.005$.

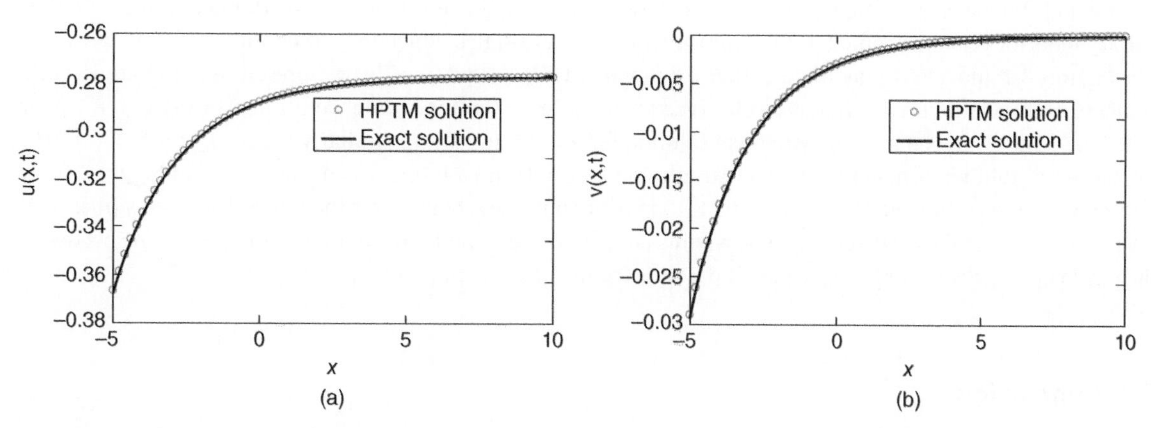

Figure 8.8 Comparison of solution of CWBK equations (8.10) and (8.11) with exact solution for $\alpha = 0.5$, $\beta = 0$, $k = 0.2$, $L = 0.005$, and $t = 1$.

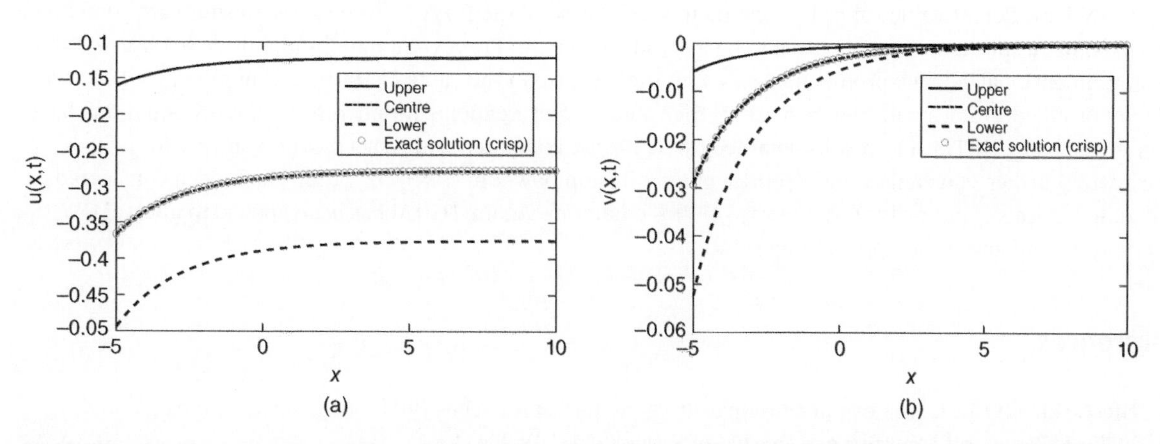

Figure 8.9 Solution of interval CWBK equations (8.14) and (8.15) along with exact solution for $\alpha = [0.1, 0.9]$, $\beta = 0$, $k = 0.2$, and $L = 0.005$.

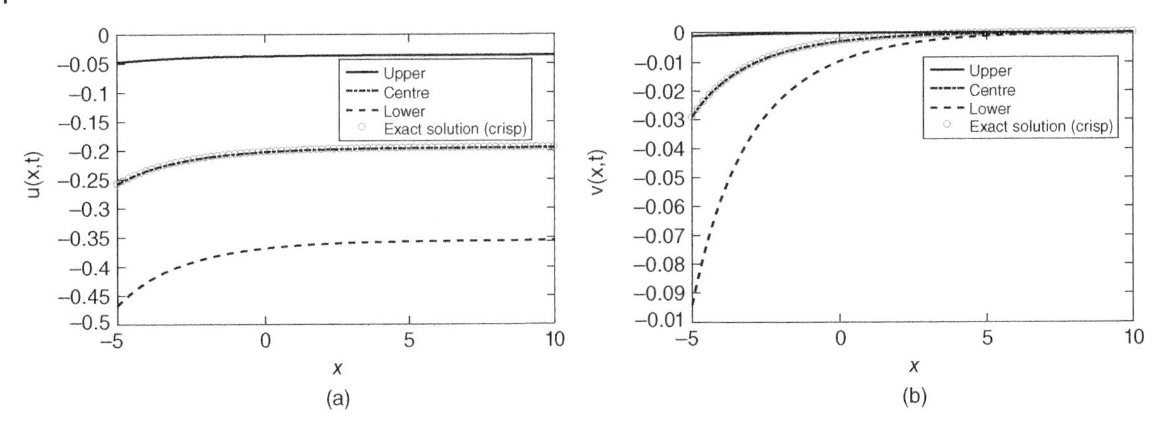

Figure 8.10 Solution of interval CWBK equations (8.14) and (8.15) along with exact solution for $\alpha = 0$, $\beta = [0.1, 0.9]$, $k = 0.2$, and $L = 0.005$.

SA–Cu nanofluid has decreasing nature with increase in the Reynolds number. It is worth mentioning here that the fuzzy plots for velocity profile for different values of η are also TFN, which may be confirmed from Figure 8.6. From Figures 8.7 and 8.8, it may be seen that solutions of CWBK equations by the present method are in good agreement with exact solutions. It may be observed from Figure 8.9 that the center solution of (8.14) and (8.15) for $\alpha = [0.1, 0.9]$ and $\beta = 0$ is matching with crisp solution for $\alpha = 0.5$ and $\beta = 0$. Similarly, for $\alpha = 0$ and $\beta = [0.1, 0.9]$ also, the center solution is matching with crisp solution for $\alpha = 0$ and $\beta = 0.5$, which can be seen in Figure 8.10. A worth mentioning point from Figures 8.9 and 8.10 may be that lower bounds for the interval solutions $[\underline{u}, \overline{u}]$ and $[\underline{v}, \overline{v}]$ are at $\alpha = 0.9$ and $\beta = 0.9$, respectively, whereas upper bounds $\overline{u}$ and $\overline{v}$ are at $\alpha = 0.1$ and $\beta = 0.1$, respectively. This may be due to decreasing nature of u and v with the increase in values of α and β.

8.7 Conclusion

HPM has been applied successfully using the concept of δ - cut and parametric approach to solve Jeffery–Hamel problem in uncertain environment by taking ϕ as a TFN. The obtained fuzzy velocity for this problem is also found to be TFN for different values of η. The results in special case of the fuzzy solution (crisp results) are compared with existing results, and they are found to be in good agreement. For SA–Cu nanofluid, it may be observed that the nondimensional velocity profile increases with the increase in the value of Hartmann number. However, the velocity profile decreases with the increase in the value of ϕ or Reynolds number. Next, CWBK equations have been solved using HPTM. The results obtained by HPTM are compared with exact solution and are found to be in agreement. Further, constants α and β representing diffusion power in CWBK equations have been considered as uncertain in terms of interval to form interval CWBK equations. Again, HPTM has been applied to interval CWBK equations to find lower and upper bound solutions.

References

1 Jeffery, G.B. (1915). L. The two-dimensional steady motion of a viscous fluid. *The London, Edinburgh, and Dublin Philosophical Magazine and Journal of Science* 29 (172): 455–465.

2 Hamel, G. (1917). Spiralförmige Bewegungen zäher Flüssigkeiten. *Jahresbericht der Deutschen Mathematiker-Vereinigung* 25: 34–60.

3 Choi, S.U.S. (1995). Enhancing conductivity of fluids with nanoparticles. *ASME Fluids Engineering Division* 231: 99–105.

4 Moghimi, S.M., Domairry, G., Soleimani, S. et al. (2011). Application of homotopy analysis method to solve MHD Jeffery–Hamel flows in non-parallel walls. *Advances in Engineering Software* 42 (3): 108–113.

5 Esmaeilpour, M. and Ganji, D.D. (2010). Solution of the Jeffery–Hamel flow problem by optimal homotopy asymptotic method. *Computers & Mathematics with Applications* 59 (11): 3405–3411.

6 Rostami, A., Akbari, M., Ganji, D., and Heydari, S. (2014). Investigating Jeffery-Hamel flow with high magnetic field and nanoparticle by HPM and AGM. *Open Engineering* 4 (4): 357–370.

7 Umavathi, J.C. and Shekar, M. (2013). Effect of MHD on Jeffery-Hamel flow in nanofluids by differential transform method. *International Journal of Engineering Research and Applications* 3 (5): 953–962.

8 Hatami, M. and Ganji, D.D. (2014). MHD nanofluid flow analysis in divergent and convergent channels using WRMs and numerical method. *International Journal of Numerical Methods for Heat and Fluid Flow* 24 (5): 1191–1203.

9 Moradi, A., Alsaedi, A., and Hayat, T. (2013). Investigation of nanoparticles effect on the Jeffery–Hamel flow. *Arabian Journal for Science and Engineering* 38 (10): 2845–2853.

10 Whitham, G.B. (1967). Variational methods and applications to water waves. *Proceedings of the Royal Society of London Series A* 299 (1456): 6–25.

11 Broer, L.J.F. (1975). Approximate equations for long water waves. *Applied Scientific Research* 31 (5): 377–395.

12 Kaup, D.J. (1975). A higher-order water-wave equation and the method for solving it. *Progress of Theoretical Physics* 54 (2): 396–408.

13 Xie, F., Yan, Z., and Zhang, H. (2001). Explicit and exact traveling wave solutions of Whitham–Broer–Kaup shallow water equations. *Physics Letters A* 285 (1–2): 76–80.

14 Yan, Z. and Zhang, H. (2001). New explicit solitary wave solutions and periodic wave solutions for Whitham–Broer–Kaup equation in shallow water. *Physics Letters A* 285 (5–6): 355–362.

15 Yıldırım, S.T., Mohyud-Din, A., and Demirli, G. (2010). Traveling wave solutions of Whitham–Broer–Kaup equations by homotopy perturbation method. *Journal of King Saud University, Science* 22 (3): 173–176.

16 Ganji, D.D., Rokni, H.B., Sfahani, M.G., and Ganji, S.S. (2010). Approximate traveling wave solutions for coupled Whitham–Broer–Kaup shallow water. *Advances in Engineering Software* 41 (7–8): 956–961.

17 Imani, A.A., Ganji, D.D., Rokni, H.B. et al. (2012). Approximate traveling wave solution for shallow water wave equation. *Applied Mathematical Modelling* 36 (4): 1550–1557.

18 Madani, M., Fathizadeh, M., Khan, Y., and Yildirim, A. (2011). On the coupling of the homotopy perturbation method and Laplace transformation. *Mathematical and Computer Modelling* 53 (9–10): 1937–1945.

19 Khan, Y. and Wu, Q. (2011). Homotopy perturbation transform method for nonlinear equations using He's polynomials. *Computers & Mathematics with Applications* 61 (8): 1963–1967.

20 Aminikhah, H. (2012). The combined Laplace transform and new homotopy perturbation methods for stiff systems of ODEs. *Applied Mathematical Modelling* 36 (8): 3638–3644.

21 Arshad, S., Sohail, A., and Maqbool, K. (2016). Nonlinear shallow water waves: a fractional order approach. *Alexandria Engineering Journal* 55 (1): 525–532.

22 Chakraverty, S., Mahato, N.R., Karunakar, P., and Rao, T.D. (2019). *Advanced Numerical and Semi-analytical Methods for Differential Equations*. Wiley.

23 Hanss, M. (2005). *Applied Fuzzy Arithmetic: An Introduction with Engineering Applications*. New York: Springer.

24 Chakraverty, S., Tapaswini, S., and Behera, D. (2016). *Fuzzy Arbitrary Order System: Fuzzy Fractional Differential Equations and Applications*. Hoboken, NJ: Wiley.

25 Chakraverty, S. and Perera, S. (2018). *Recent Advances in Applications of Computational and Fuzzy Mathematics*. Singapore: Springer.

26 Brinkman, H.C. (1952). The viscosity of concentrated suspensions and solutions. *The Journal of Chemical Physics* 20 (4): 571.

27 Biswal, U., Chakraverty, S., and Ojha, B.K. (2019). Natural convection of non-Newtonian nanofluid flow between two vertical parallel plates. *International Journal of Numerical Methods for Heat and Fluid Flow* 29: 1984–2008.

28 He, J.H. (1999). Homotopy perturbation technique. *Computer Methods in Applied Mechanics and Engineering* 178 (3–4): 257–262.

29 He, J.H. (2005). Application of homotopy perturbation method to nonlinear wave equations. *Chaos, Solitons & Fractals* 26 (3): 695–700.

30 Karunakar, P. and Chakraverty, S. (2018). Solving shallow water equations with crisp and uncertain initial conditions. *International Journal of Numerical Methods for Heat and Fluid Flow* 28 (12): 2801–2815.

9

Fuzzy Rough Set Theory-Based Feature Selection: A Review

Tanmoy Som[1], Shivam Shreevastava[2], Anoop Kumar Tiwari[3], and Shivani Singh[4]

[1]*Department of Mathematical Sciences, Indian Institute of Technology (Banaras Hindu University), Varanasi, Uttar Pradesh, 221005, India*
[2]*Department of Mathematics, SBAS, Galgotias University, Gautam Buddha Nagar, Uttar Pradesh, 201310, India*
[3]*Department of Computer Science and Engineering, Dr. K. N. Modi University, Tonk, Rajasthan, 304021, India*
[4]*DST-Centre for Interdisciplinary Mathematical Sciences, Institute of Science, Banaras Hindu University, Varanasi, Uttar Pradesh, 221005, India*

9.1 Introduction

Dimensionality reduction [1–3] has been an intriguing field of research in machine learning, signal processing, medical image processing, bioinformatics, etc. The generation of large volume of real-valued datasets from various domains increases the requirement of dimensionality reduction in order to produce the most informative features. Dimensionality reduction likewise improves the performances of fast storage systems as well as prediction algorithms. Feature selection is one of the dimensionality reduction procedures, which preserves the striking qualities of the database systems. It shrinks exceptionally highly correlated features, which may result in lowering the overall accuracy. Attribute selection strategies center around progressively significant and non-redundant features. It has been implemented in numerous areas such as text categorization and web classification [4].

Rough set [5, 6] based methodology is a standout among the most significant procedures of feature selection that gains information from the dataset itself. Rough set concept does not require any outside information for the most informative feature selection. It deals with the vagueness in the database. The element of rough set getting closer to feature selection or attribute reduction is the way toward choosing a subset of features from the original set of features, which form patterns in a given dataset. The subset should be basic and adequate to portray the ideas and preserve a reasonably high accuracy in representing the original attributes. However, this technique can be implemented to discrete information systems only. Hence, discretization strategies are required so as to handle the real-valued information system before feature selection, and this may result in loss of some data information.

So as to handle this issue, fuzzy rough set (FRS) [7, 8]-based technique is used to determine both vagueness and uncertainty available in the dataset. Fuzzy sets [9] and rough sets are combined to create FRS and it provides a key course in reasoning with uncertainty for real-world information systems. FRS idea has been actualized to outperform the inadequacies of the traditional rough set technique in different viewpoints. FRS is utilized to maintain a strategic distance from loss of data because of discretization.

Since its presentation in 1965, fuzzy set theory significantly affected the manner in which we speak to and process with ambiguous data. So far, it has progressed toward becoming part of the bigger worldview of soft computing, an accumulation of procedures that are tolerant to the regular characteristics of imperfect data and knowledge such as uncertainty, imprecision, vagueness, and partial truth, and this holds fast nearer to the human personality than conventional hard computing techniques. Amid the most recent decades, new techniques have developed that generalize the original fuzzy set (which is additionally called type 1 fuzzy set in this specific situation). Type-2 fuzzy sets [10], FRSs [7, 8], interval-valued fuzzy sets [11], and intuitionistic fuzzy sets [12] share all intents and purpose

Mathematical Methods in Interdisciplinary Sciences, First Edition. Edited by Snehashish Chakraverty.

that they would all be able to be formally described by membership functions taking values in a partially ordered set, which is never again equivalent to (however an expansion of) the set of membership degrees used in fuzzy set theory. The presentation of such new, summing-up speculations is regularly joined by extensive discourses on issues, for example, the choice of terminology and the additional estimation of the speculation.

A variety of techniques for attribute reduction in FRS environment are widely discussed by many researchers. Many of them used the concept of degree of dependency-based attribute selection, in which attributes are selected on the basis of dependency of decision attribute over the set of conditional attributes. The concept of discernibility matrix is also developed and used by various researchers for FRS-based attribute reduction. In this chapter, we discuss these two approaches in a wider sense. Apart from these two approaches, some other techniques are also presented for such attribute reduction. We consider all the three types of decision systems, i.e. supervised (all the class labels are available), semisupervised (some of the class labels are available and rest of them are not available), and unsupervised (all the class labels are missing). Attribute reduction of decision systems with missing attribute values are also discussed. Various application areas induced from FRS-based approaches for attribute reduction include especially classification problems. Finally, we present some limitations of FRS-based techniques and suggest some measures to overcome those limitations.

The rest of the chapter is structured as follows. Some preliminaries on rough set, fuzzy set, and fuzzy rough set are discussed in Section 9.2. In Section 9.3, a detailed survey on fuzzy rough-assisted attribute reduction for supervised decision systems is discussed. Various techniques for feature selection of semisupervised and unsupervised decision systems are presented in Section 9.4. In Section 9.5, the approaches for decision systems with missing attribute values are added. We discuss the application of fuzzy rough set theory (FRST) in decision-making problems in Section 9.6. In Section 9.7, we present limitations of FRS-based approaches. In Section 9.8, we conclude our work.

9.2 Preliminaries

In this section, some basic definitions necessary for this chapter are given. We discuss the concepts of rough set theory (RST) and fuzzy set theory. Attribute reduction based on these two concepts is also discussed.

9.2.1 Rough Set Theory

RST can be used to extract knowledge from a domain in a concise way, which, even while reducing the amount of actual knowledge involved, can retain the information content. Discernibility is an important concept of RST, which can be used for feature selection.

Definition 9.1 ***Information System [13]*** A quadruple (U, A, V, f) is said to be an information system, where U (the universe of discourse) is a nonempty set of finite objects, A is a nonempty finite set of attributes, V is the set of attribute values, and f is an information function from $U \rightarrow V$. If $A = C \cup D$ such that $C \cap D = \emptyset$, where C is the set of conditional attributes and D is the set of decision attributes, then (U, A, V, f) is known as a decision system.

9.2.1.1 Rough Set

For any $P \subseteq A$, there is an associated equivalence relation R_P:

$$R_P = \{(x, y) \in U^2 | \forall a \in P, a(x) = a(y)\}$$

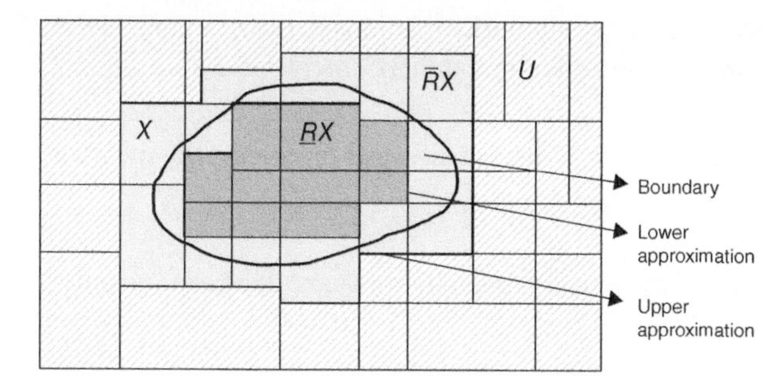

Figure 9.1 Rough set approximations.

If $(x, y) \in R_P$, then x and y are said to be indiscernible by attributes from P. $[x]_P$ denotes the equivalence classes of the P-indiscernibility relation [5]. Let $X \subseteq U$; X can be approximated using the P-lower and P-upper approximations of X, where the lower and upper approximations are defined as below:

$$R_P \downarrow X = \{x \in U \mid [x]_P \subseteq X\}$$

$$R_P \uparrow X = \{x \in U \mid [x]_P \cap X \neq \emptyset\}$$

The tuple $(R_P \downarrow X, R_P \uparrow X)$ is called a rough set. In Figure 9.1, $\underline{R}X$ represents $R_P \downarrow X$ and $\overline{R}X$ represents $R_P \uparrow X$.

9.2.1.2 Rough Set-Based Feature Selection

Let $(U, C \cup D, V, f)$ be a decision system with $Q \in D$ as a decision attribute. Its equivalence classes $[x]_{R_Q}$ are called decision classes. Given $P \subseteq C$, then P-positive region (POS_P) comprises those objects from U for which the values of P allow to predict the decision class clearly [14]:

$$\text{POS}_P(Q) = \bigcup_{x \in X} R_P \downarrow [x]_{R_Q}$$

Indeed, if $x \in \text{POS}_P$, it means that whenever an object has the same values as x for the attributes in P, it will also belong to the same decision class as x. The predictive ability with respect to Q of the attributes in P is then measured by the following value (degree of dependency of Q on P):

$$\gamma_P(Q) = \frac{|\text{POS}_P(Q)|}{|U|}$$

$(U, C \cup D, V, f)$ is called consistent if $\gamma_C(Q) = 1$.

Definition 9.2 ***Reduct [14]*** A subset P of C is called the degree of dependency-based reduct if it satisfies $\gamma_P = \gamma_C$ with $\gamma_{P \setminus \{a\}} \neq \gamma_P, \forall a \in P$.

Initially, the reduct set is an empty set. We add conditional attributes one by one in reduct set and calculate degree of dependencies of decision attribute over the obtained reduct set. The proposed algorithm selects only those conditional attributes, which cause a maximum increase in the degree of dependency of decision attribute.

9.2.2 Fuzzy Set Theory

RST is widely discussed by many researchers for attribute reduction and classification problems. However, the traditional rough set is inefficient to handle attributes with real or continuous values, which are more common than nominal attribute values in the real world. To handle the knowledge uncertainty concepts, i.e. fuzziness and indiscernibility, FRST has become popular as a generalization of crisp rough sets for classification problems.

FRS uses similarity degrees for fuzzy data values of objects to generate similarity relations in place of equivalence relations used by RST. This relaxation in equivalence relations has created researchers' interest in FRS.

Definition 9.3 ***Fuzzy Set [15]*** A fuzzy set in U is a mapping $\mu : U \rightarrow [0, 1]$ and $\mu_A(x)$ represents the membership grade of x in A.

Example 9.1 Let $E = \{1, 2, \ldots, 90\}$ that relates to the age of patients of a hospital in years. Then, fuzzy subset A = "Youth"$\subset E$ can be determined by using a membership function μ_A,

$$\mu_A = \begin{cases} \frac{1}{1+\left(\frac{1}{5}x-5\right)^2}, & 25 < x \\ 1, & \text{otherwise} \end{cases}$$

Properties

For two fuzzy sets A_1 and A_2 in X,

(1) $A_1 \subseteq A_2$ iff $(\forall x \in X)(\mu_{A_1}(x) \leq \mu_{A_2}(x))$.
(2) $\mu_{A_1 \cup A_2}(x) = \sup\{\mu_{A_1}(x), \mu_{A_2}(x)\}$
(3) $\mu_{A_1 \cap A_2}(x) = \inf\{\mu_{A_1}(x), \mu_{A_2}(x)\}$
(4) $\mu_{A_1^c}(x) = 1 - \mu_{A_1}(x)$
(5) If X is finite, the cardinality of the fuzzy set A_1 is calculated as:

$$|A_1| = \sum_{x \in X} \mu_{A_1}(x)$$

9.2.2.1 Fuzzy Tolerance Relation

A fuzzy relation in U is a fuzzy set defined on $U \times U$. For all $y \in X$, the fuzzy set R_y is the R-foreset of y, defined by

$$R_y(x) = R(x, y)$$

for all x in U. If R is a reflexive fuzzy relation, that is,

$$R(x, x) = 1$$

and R is a symmetric fuzzy relation, that is,

$$R(x, y) = R(y, x)$$

For all $x, y \in X$, then R is called a fuzzy tolerance relation [16].

Definition 9.4 ***Fuzzy Triangular Norm [4]*** A triangular norm or t-norm T is an increasing, associative, and commutative mapping from $[0, 1] \times [0, 1] \rightarrow [0, 1]$ satisfying $T(1, x) = x, \forall x \in [0, 1]$. A few widely used t-norms are $T_M(x, y) = \min\{x, y\}$ and $T_L(x, y) = \max\{0, x + y - 1\}$ (Lukasiewicz t-norm), for x, y in $[0, 1]$.

Definition 9.5 ***Fuzzy Implicator [4]*** An implicator is any mapping $I : [0, 1] \times [0, 1] \rightarrow [0, 1]$, satisfying $I(0, 0) = 1$ and $I(1, x) = x, \forall x \in [0, 1]$. Moreover, I needs to be decreasing in its first and increasing in its second component. A few widely used implicators are $I_M(x, y) = \max\{1 - x, y\}$ (Kleene-Dienes implicator) and $I_L(x, y) = \min\{1, 1 - x + y\}$ (Lukasiewicz implicator) $\forall x, y \in [0, 1]$.

9.2.2.2 Fuzzy Rough Set Theory

The FRST is an alternate method for feature selection [7, 8]. It proposes to calculate the similarity between the objects using a fuzzy relation R in U, i.e. $R : U \times U \rightarrow [0, 1]$ is a mapping that assigns to each distinct pair of objects and their corresponding degree of similarity. Given a set $X \subseteq U$ and a fuzzy similarity relation R, the lower and upper approximation of X by R can be calculated in several ways. A general definition is as follows:

$$(R \downarrow X)(x) = \inf_{y \in U} I(R(x, y), X(y))$$

$$(R \uparrow X)(x) = \sup_{y \in U} T(R(x, y), X(y))$$

Here, I is a fuzzy implicator and T is a fuzzy t-norm and $X(y) = 1$, for $y \in X$, otherwise $X(y) = 0$. The pair ($R \downarrow X$, $R \uparrow X$) is known as FRS.

9.2.2.3 Degree of Dependency-Based Fuzzy Rough Attribute Reduction

The rough set-based feature selection works effectively only in datasets with discrete values, whereas most data in day-to-day life contain real values. Hence, the FRS-based feature selection was introduced, which takes into consideration the similarity between the values of objects for a particular attribute. It helps in modeling data with the uncertainty involved in it.

Now, positive region can be defined by Jensen and Shen [4],

$$\mu_{\text{POS}_P(Q)} = \sup_{X \in U/Q} \mu_{R_P \downarrow X}$$

Here, $P \subseteq C, X \subseteq U$, and Q is the set of decision attributes. $\mu_{R_P \downarrow X}$ and $\mu_{\text{POS}_P(Q)}$ are the membership grades of lower approximation and positive region, respectively. Now, degree of dependency can be given by

$$\gamma_P(Q) = \frac{|\mu_{\text{POS}_P(Q)}|}{|U|} = \frac{\sum_{x \in U} \mu_{\text{POS}_P(Q)}(x)}{|U|}$$

The process of calculating reduct set is the same as in the case of rough set approach.

9.2.2.4 Discernibility Matrix-Based Fuzzy Rough Attribute Reduction

Let $\boldsymbol{A} = (U, A, V, f)$ is an information system and DM($\mathbf{A}$) denotes $n \times n$ matrix (m_{ij}), such that $m_{ij} = \{a \in \boldsymbol{A} : a(x_i) \neq a(x_j)\}$ for $i, j = 1, 2, \ldots, n$, then DM($\mathbf{A}$) is known as a discernibility matrix of $\boldsymbol{A}$. A discernibility function $f(\boldsymbol{A})$ is a function of m Boolean variables $\overline{a_1}, \overline{a_2}, \ldots, \overline{a_m}$ corresponding to the attributes $a_1, a_2, \ldots, a_m$, respectively, and is defined as follows [17]:

$$f(\boldsymbol{A})(\overline{a_1}, \overline{a_2}, \ldots, \overline{a_m}) = \wedge\{\vee(m_{ij}) : 1 \leq j < i \leq n\}$$

where $\vee(m_{ij})$ is the disjunction of all variables $\overline{a}$ such that $a \in m_{ij}$.

Let $F(\boldsymbol{A})$ denotes the reduced disjunctive form of $f(\boldsymbol{A})$, which can be obtained from $f(\boldsymbol{A})$ by applying the absorption and multiplication laws as many times as possible, then there exist k and $X_l \subseteq A$ for $l = 1, 2, \ldots, k$ such that $F(\boldsymbol{A}) = (\wedge X_1) \vee \ldots \vee (\wedge X_k)$, where every element in X_l only appears one time. We have RED($\boldsymbol{A}$) $= \{X_1, X_2, \ldots, X_k\}$.

9.3 Fuzzy Rough Set-Based Attribute Reduction

The fundamental aspects of RST for knowledge acquisition involve the searching for some particular subsets of conditional attributes, called reducts. Attribute reduction is an important issue of RST as it is an imperative preprocessing step in data mining, which is aimed at finding a small set of rules from the dataset with determined target. However, it is not well suited for real-valued information systems as it requires discretization, which results in loss of some information. To handle this drawback of RST, Dubois and Prade combined rough set with fuzzy set and introduced FRS [7, 8]. After that, several researchers proposed different techniques for fuzzy rough-assisted attribute reduction and applied it on classification problems. In this section, we discuss such techniques.

9.3.1 Degree of Dependency-Based Approaches

Degree of dependency-based approaches for fuzzy rough feature selection are most used techniques by researchers. In 2004, Jensen and Shen established a novel FRS model by combining the concept of fuzzy and rough set theories and presented attribute reduction technique based on this model to greatly reduce data redundancy. They illustrated this approach with a simple example [18]. Finally, the proposed approach was implemented to the problem of web categorization and the results were very promising as it produced the reduced datasets with minimal information loss. They further proposed a feature selection technique based on FRSs to avoid the information loss because of quantization of the underlying numerical features. This approach maintained dataset semantics. This technique performed better than traditional entropy, random-based, and principal component analysis (PCA) methods. This technique was applied to complex system monitoring and the experimental results justified the approach of the work [19].

In 2005, Bhatt and Gopal discussed about the poorly designed termination criteria of FRS-assisted feature selection algorithm and disclosed that this technique is not convergent on many real-valued datasets [20]. Furthermore, they explained that because of increment in the number of input variables and the size of data patterns, the computational complexity of the algorithm increases exponentially. Based on the natural properties of fuzzy t-norm and t-conorm, they have put forward the concept of FRSs on compact computational domain, which is then utilized to improve the computational efficiency of fuzzy rough set attribute reduction (FRSAR) algorithm. Jensen and Shen [21] proposed a new fuzzy rough attribute reduction method to obtain optimal reduct sets by using the concept of ant colony optimization mechanism. This technique was applied to complex system monitoring. This technique was compared with PCA, entropy-based feature selector, and a transformation-based reduction approach. These comparisons were done with the help of support vector machine classifiers.

In real-world problems, data are usually found in hybrid formats. Therefore, a unified data reduction method for hybrid data is required. In 2006, Hu et al. presented an information measure to compute discernibility power of fuzzy as well as a crisp equivalence relation, which is the key notions of the FRS model as well as classical rough set model [22]. Here, a general definition of significance of numeric, nominal, and fuzzy attributes is established on the basis of information measure. Hu's approach redefined the independence of reduct, hybrid attribute subset, and relative reduct. Moreover, two greedy reduction algorithms based on this information measure were constructed for supervised data and unsupervised dimensionality reduction. Experimental results justified that the reducts produced by the proposed algorithms obtained a better performance when compared with the traditional rough set-based techniques.

In 2007, Hu et al. introduced an efficient hybrid attribute reduction technique using a generalized fuzzy rough model [23]. In this discussion, a theoretic framework of fuzzy rough model is presented based on fuzzy relations. Moreover, several attribute significance measures were derived by using the present fuzzy rough model. Furthermore, a forward greedy algorithm for hybrid attribute reduction was constructed. The experimental results showed that the variable precision fuzzy inclusion-based technique in computing decision of the positive region obtained the optimal performance. Here, it can be observed that the reduct sets are least when compared with the other existing methods, but the accuracies are the highest. Jensen and Shen [24, 25] further discussed the way to handle the noisy data and investigated two extensions of rough sets, namely, tolerance rough sets and fuzzy rough sets. His paper presented the way to retain the semantics of the datasets. This approach was applied for the forensic glass fragment identification. Experimental results proved the supremacy of the proposed technique. He also demonstrated a novel approach for feature selection based on FRSs, which retains dataset semantics. This technique was implemented to two challenging domains, namely, web content classification and complex systems monitoring where a feature reducing step was important and compared empirically with several dimensionality reducers. In the conducted experiments, this technique was shown to equal or improved classification accuracy when compared to the results from unreduced data. Various machine learning algorithms were implemented on both reduced as well as unreduced datasets and justified the supremacy of the approach. In addition, it was compared with the results of other attribute reduction technique and it was observed that this technique was outperforming the others.

In the feature selection technique, it is the key aspect that after omission of some features, the decision systems should maintain the discernibility. Degrees of dependency-based approaches are the popular ways to evaluate attribute subsets with respect to this criterion. In the standard approach, attributes are anticipated to be qualitative rather than quantitative attributes. FRS-based approaches were successfully applied to produce the qualitative features. In 2007, Cornelis et al. introduced a more flexible methodology for attribute reduction based on vaguely quantified rough set (VQRS) [26]. This method successfully implemented for real-valued datasets and handled noise available in the datasets.

In 2009, Jensen et al. introduced the concept of rule induction based on FRS by combining rough and fuzzy sets. In his work, if-then rule was applied on reduced datasets as produced by the proposed feature selection algorithm. The algorithm was evaluated against leading classifiers and found to be effective. The problems of imprecision and uncertainty are very common in the real-world data. Therefore, it is always interesting to propose the methods to deal with these problems. Jensen et al. [27] presented three approaches to cope with these problems. They introduced three new approaches for fuzzy rough feature selection based on fuzzy T-transitive similarity relations that improved the problems encountered with fuzzy rough feature selection. The first technique used the similarity relations to construct approximations of decision concepts, which was based on fuzzy lower approximations. Furthermore, it was evaluated through a new measure of dependency between decision and conditional features. The second technique utilized the information in the fuzzy boundary region to lead the feature search process. The third approach extended the methodology of the discernibility matrix to deal with the fuzzy case. Finally, these techniques were successfully applied for the real-world datasets.

The dependency function usually struggles with the robustness with respect to noisy information. In 2010, Hu et al. presented a novel concept of feature selection by combining fuzzy and rough sets to combat this problem [28]. Their article presented a novel FRS model called soft fuzzy rough sets. They proposed a new dependency function, and it was shown that this function is robust to noisy information. This model successfully reduced the influence of noise. Finally, this model was applied on real-valued datasets, which proved the effectiveness of the new model.

Fuzzy rough feature selection techniques are frequently used for supervised learning as it uses the information of decision class labels. However, FRS-based approaches were rarely discussed for unsupervised learning. In 2010, Mac Parthalain and Jensen presented a novel fuzzy rough feature selection algorithm that attempted to deal with unsupervised datasets [29]. This technique was applied for the datasets, where the class labels were not available. This technique maintained the semantics of the information systems. Moreover, there were no requirements of thresholding or domain information.

FRS is a well-known generalization of rough set, which is used to deal with the real-valued features. Attribute reduction for decision systems with real-valued conditional attributes and symbolic decision attributes was successfully performed by different research articles. However, FRS-based feature selection for the decision systems with real-valued decision attributes was rarely discussed.

Mac Parthalain and Jensen [30] proposed a novel instance and feature selection model using FRS fuzzy rough sets. An unsupervised feature selection approach was presented by them that preserved semantics of dataset while reducing the reduct size [31]. Shang and Barnes [32] presented image classifiers using support vector machine for classifying mars dataset by first obtaining a reduced dataset by fuzzy rough feature selection technique. Thus, the model can handle a dataset having many classes, large scale and diverse properties. Derrac et al. [33] accessed the quality of instances by FRS approach and then used wrapper method for instance pruning in 2013 and coined the method as Fuzzy rough prototype selection. Diao et al. [34] extended the idea of feature selection to support classifier ensemble reduction, by treating classifiers as features and transforming ensemble predictions into training samples.

Cornelis et al. [35] developed a feature selection framework with multiadjoint FRSs, in which a family of adjoint fuzzy sets is used to compute lower approximations. A fast forward greedy algorithm for granular structure selection in dynamic updating environment was proposed by Ju et al. [36].

In 2015, Inuiguchi et al. [37] developed the relationship between belief and plausibility functions in Dempster–Shafer theory of evidence and lower and upper approximations in FRST and gave its practical application in various machine learning domains like feature selection. For reducing overhead and complexity, neighborhood approximation and feature grouping were proposed for fuzzy rough feature selection by Jensen and Mac Parthalain [38]. An intension recognition framework based on fuzzy rough feature selection and multitree genetic programming was introduced for brain signal dataset classification by Lee et al. [39]. Onan [40] performed classification of breast cancer data by combining instance selection, feature selection, and classification. A weak gamma evaluator was utilized to remove useless instances. Consistency-based feature selection was used in conjunction with reranking used for feature selection. Fuzzy rough nearest neighbor approach was used for classification. Two novel feature selection methods based on different interpretations of flock of starling algorithm were proposed by Mac Parthalain and Jensen [41]. Qian et al. [42] developed a fuzzy rough feature selection accelerator for enhancing sample and dimensionality reduction. A weighted KNN regression method followed by fuzzy rough distributed prototype selection method for big data was proposed by Vluymans et al. [43]. Zeng et al. [44] presented FRS approach for incremental feature selection in hybrid information system by combining new hybrid distance based on value difference metric and Gaussian kernel.

In 2016, Arunkumar and Ramakrishnan proposed a novel fuzzy rough feature selection using correlation coefficient and applied to cancer microarray data [45]. Then, global heuristic harmony search is used to select feature subset. An invasive weed optimization-based fuzzy rough feature selection was proposed for mammographic risk analysis for early diagnosis of breast cancer by Guo et al. [46]. Jhawar [47] adopted fuzzy rough-based instance selection of gait decision system for analysis of erroneous gait patterns. Fuzzy rough and VQRS-based feature selection was coupled with clonal selection algorithm and artificial immune recognition system for mining agriculture data by Lasisi et al. [48]. Tsang et al. [49] presented a new KNN rule based on FRS for instance and feature selection. Value of lower approximation reflected distances of sample from classification boundary and the value was used to calculate probability of subsampled data. From this weighted subsampled data, probability was estimated using parzen window technique. Thus, semantics of probability was also preserved. Zhang et al. [50] proposed fuzzy rough-based feature selection using information entropy for reducing heterogeneous data.

In 2017, a comparative study of filter, wrapper, and FRS-based feature selection was presented by Kumar et al. [51] based on execution time, number of features in reduced dataset, and classifier accuracy. For genes microarray data, results illustrated that fuzzy rough feature selection is computationally faster and produces data of less dimensionality than correlation-based filter. Badria et al. [52] extracted features from peganumharmala, a wild plant followed by fuzzy rough feature selection and J48 classification to enhance exploration of pharmacological activities from medicinal plants. FRSs and genetic algorithm were used to remove redundant genes from multiclass cancer microarray gene expression datasets by Chinnaswamy and Srinivasan [53]. Fuzzy rough feature selection was performed on dataset preprocessed by kernel filter followed by fuzzy rough ordered weighted average classification by Meenachi et al. [54]. Qu et al. [55] proposed fuzzy rough feature selection for multiclass datasets by generating association rule implied in the labels and considering each set of sublabels as unique class. Su et al. [56] employed ordered weighted averaging aggregation of fuzzy similarity relations to reflect interaction between attributes for better performance of fuzzy rough feature selection. Wang et al. [57] proposed a fitting model for feature selection to prevent misclassification of samples by defining fuzzy decision of sample using fuzzy neighborhood and parameterized fuzzy relation to characterize information granules. Dai et al. [58] proposed two fuzzy rough attribute reduction based on reduced maximal discernibility pair selection and weighted reduced maximal discernibility pair selection. Such algorithm increased the efficiency by ignoring object pairs that are already discerned by the selected attribute subset. Javidi and Mansoury [59] proposed a combination of filter and wrapper gene selection by employing ant colony with information gain as evaluation measure and FRSs.

In 2018, Cheruku et al. [60] applied fuzzy rough feature selection followed by bat optimization for each class to generate fuzzy rules followed by ada-boosting to increase the accuracy of generated rules. Rhee et al. [61] proposed a model for diagnosis of chronic kidney disease by removing irrelevant attributes by fuzzy rough feature selection followed by type II fuzzy classification by implementing the Sugeno index for determining in fuzzy clustering approach. Yager's relative conditional entropy was employed to find significance of attribute, and fuzzy quick reduct algorithm was used for fuzzy set-valued information system by Ahmed et al. [62]. Lee et al. [63] proposed feature selection by hybridizing fuzzy rough dependency degree as evaluation measure selection with binary shuffled frog leaping algorithm as search strategy. The feature selection for lung cancer microarray genes expression data was proposed by Arunkumar and Ramakrishnan [64] in two stages. During the first stage, information gain based entropy was used for dimensionality reduction, while customized similarity based fuzzy rough feature selection was used during second stage. Hamidzadeh et al. [65] proposed FRS-based web robot detection to better characterize and cluster web visitors of three web sites. Han et al. [66] proposed fuzzy rough feature selection using Laplace weighted summation operator to achieve better classification performance by boosting quality of selected attributes. Hu et al. [67] proposed multimodal attribute reduction by employing multikernel FRSs. Kernels are combined to compute fuzzy similarity for multimodal attributes. Then, the selected features are fed to negative binomial distribution to detect genes that are statistically differentially abundant in two or more phenotypes. Li et al. [68] employed two kernel functions, one to reveal similarity between samples and another to access the degree of overlap between labels. Then, these kernel functions were combined to construct robust multilabel kernelized FRS model for feature selection. A novel multilabel fuzzy rough-based feature selection was proposed by Lin et al. [69] by defining score vector to evaluate probability of different class' sample. Then, the distance between the samples was constructed from local sampling, which was robust against noise, for the proposed model. Liu and Pan [70] proposed a hybrid classifier that can eliminate inconsistent instance without knowing cluster number and start point of cluster by fuzzy rough instance selection. Mahanipour et al. [71] applied fuzzy rough feature selection before applying genetic algorithm for construction of features and concluded that their approach generated more discriminative features than previous approaches. Ratnaparkhi et al. [72] employed fuzzy rough feature selection for discrimination of electrocardiogram into six classes. Then, optimal feature subset was fuzzified using Gaussian membership function and finally rule set is generated to obtain appropriate decision class using FRS. Sheeja and Kuriakose [73] proposed a novel feature selection model using FRS based on divergence measure. Tiwari et al. [74] applied fuzzy rough feature selection on balanced training and test datasets followed by ranking and using fuzzy rough attribute evaluator for enhancing prediction of piezophilic protein. A survey on feature selection technique based on rough sets, FRSs, and ant colony optimization on intrusion detection system was presented by Varma et al. [75]. Zhang et al. [76] proposed kernel-based fuzzy rough feature selection for intrusion detection system. Zhang et al. [77] proposed accelerator algorithm using FRS-based information entropy as an evaluation measure for selecting features. Zhang et al. [78] selected representative instances using coverage ability of fuzzy granules. Then, heuristic algorithm for feature selection is given using implication preserving reduction to maintain discriminative information of selected instances. Zhenlei et al. [79] took a random sample of instances from the whole dataset and applied fuzzy rough feature selection in this random sample, thus reducing time of algorithm on the condition of limited classification accuracy loss. Zuo et al. [80] extracted text features and performed fuzzy rough feature selection for increasing the efficiency of detecting child grooming.

In 2019, Arunkumar and Ramakrishnan [81] balanced the cancer dataset using synthetic minority oversampling technique and reduced its dimensionality by correlation method. Then, fuzzy triangular norm with Lukasiewicz fuzzy implicator was used to compute approximations on fuzzy rough quick reduct algorithm. Maini et al. [82] simultaneously applied particle swarm optimization and intelligent dynamic swarm to appropriately select feature subset by distributed sampling-based initialization to pick better seed population. A novel method for fuzzy rough attribute reduction using distance measure with fixed and variable parameter was introduced by Wang et al.

[83]. Zabihimayvan and Doran [84] employed FRS-based feature selection to features extracted from website samples for detection of phishing attack. Zhao et al. [85] proposed a new approach of fuzzy rough feature selection that combines original selection method with membership function determination of fuzzy *c*-means and fuzzy equivalence, thus taking full advantage of information about datasets as well as the difference between datasets.

9.3.2 Discernibility Matrix-Based Approaches

Apart from the degree of dependency-based feature selection technique, discernibility matrix-based approaches also play a very important role in this area. In 2011, He et al. [86] presented an FRS-based attribute reduction approach to deal with the information systems with fuzzy decision values. In this study, a discernibility matrix-based attribute reduction technique was developed. Moreover, two heuristic algorithms were developed and a comparative study was provided with the existing algorithms. Finally, experimental results implied that proposed algorithm of attribute reduction was feasible and valid with respect to general FRS-assisted attribute reduction methods.

In 2013, Dai and Tian proposed FRS model for set-valued information system and applied the model for feature selection based on discernibility matrix approach [87]. Furthermore, based on the concept of fuzzy gain ratio, Dai and Xu [88] introduced feature selection model for enhancing the accuracy of tumor classification. In 2014, Chen and Yang proposed fuzzy rough feature selection by defining discernibility relations for symbolic and real-valued attribute to calculate dependency and thus measure inconsistency between heterogeneous features [89]. In 2016, Wang proved that reduct defined in FRS approach includes the reduct defined in rough set approach and presented a discernibility matrix-based approach to calculate reduct set for set-valued data [90]. Wei et al. proposed fuzzy rough feature selection for set-valued data using the concept of discernibility matrix-based technique [91].

Two incremental fuzzy rough feature selection techniques for large datasets were proposed by Yang et al. [92] in which one technique selected features as sample subset arrive and output final subset when no sample is left while other technique performed feature selection only when no sample is left but updates relative discernibility relation. In 2018, Dai et al. proposed heuristic algorithm for attribute reduction [93]. A dominance relation to measure dominance between attributes based on which definition of fuzzy approximation is extended and uncertainty measurement issue is investigated. Further, attribute importance was measured using new fuzzy conditional entropy. Moreover, Juneja et al. [94] proposed three-stage dimensionality reduction. Nonlinear features are extracted from functional magnetic resonance imaging (fMRI) data from schizophrenia diagnosis from spatially constrained fuzzy clustering of three-dimensional spatial maps. Finally, decision models are learned by support vector machines on features extracted from fuzzy rough feature selection.

In 2019, Xu et al. modified conditional entropy by introducing fuzzy neighborhood granule and rough decisions, and rough uncertainty model was proposed [95]. Also, variable precision model was proposed for selection of parameters. Finally, rough uncertainty metric model was employed for feature selection of gene dataset for tumor diagnosis.

A brief outline of the works on feature selection techniques based on degree of dependency and discernibility matrix approaches by different authors are given in Tables 9.1 and 9.2, respectively.

9.4 Approaches for Semisupervised and Unsupervised Decision Systems

In a technical report on semisupervised learning, Zhu [97] described many semisupervised methods such as graph-based methods, expectation-maximization (EM) with generative mixture models, self-training, cotraining, transductive support vector machines, Gaussian processes, information regularization, and entropy minimization methods in order to deal with unlabeled data. Jensen et al. [96] proposed a novel approach for semisupervised

Table 9.1 Fuzzy rough set-assisted feature selection techniques based on the degree of dependency approach.

S. No.	Authors	Proposed work
1	Jensen and Shen [18]	Established a novel fuzzy rough set model for attribute reduction and implemented to the problem of web categorization.
2	Shen and Jensen [19]	Proposed a fuzzy rough set-based feature selection technique to avoid the information loss and applied it to the complex system monitoring. This technique performed better than traditional entropy, random-based, and PCA methods.
3	Bhatt and Gopal [20]	Presented the concept of fuzzy rough sets on compact computational domain, which is then utilized to improve the computational efficiency of fuzzy rough set attribute reduction algorithm.
4	Jensen and Shen [21]	Proposed a new fuzzy rough attribute reduction method by using the concept of ant colony optimization mechanism and compared with PCA, entropy-based feature selector, and a transformation-based reduction approach with the help of support vector machine classifiers.
5	Hu et al. [22]	Presented an information measure to compute discernibility power of fuzzy as well as a crisp equivalence relation and redefined the independence of reduct, hybrid attribute subset, and relative reduct.
6	Cornelis et al. [26]	Introduced a more flexible methodology for attribute reduction based on vaguely quantified rough set and successfully implemented on real-valued datasets to handle noise.
7	Hu et al. [28]	Introduced soft fuzzy rough sets and a new dependency function, which is robust to noisy information. This model successfully reduced the influence of noise.
8	Mac Parthalain and Jensen [29]	Presented a novel fuzzy rough feature selection algorithm that attempted to deal with unsupervised datasets. There were no requirements of thresholding or domain information.
9	Shang and Barnes [32]	Presented image classifiers using support vector machine for classifying mars dataset by first obtaining a reduced dataset by fuzzy rough feature selection technique.
10	Derrac et al. [33]	Accessed the quality of instances by fuzzy rough set approach and then used wrapper method, for instance, pruning and coined the method as fuzzy rough prototype selection.
11	Cornelis et al. [35]	Developed feature selection framework with multiadjoint fuzzy rough sets, in which a family of adjoint fuzzy sets is used to compute lower approximations.
12	Ju et al. [36]	Proposed a fast forward greedy algorithm for granular structure selection in dynamic updating environment.
13	Inuiguchi et al. [37]	Developed relationship between belief and plausibility functions in Dempster–Shafer theory of evidence and lower and upper approximations in fuzzy rough set theory and gave its practical application in various machine learning domains like feature selection.
14	Jensen et al. [96]	Proposed novel feature selection approach for semisupervised data and claimed that their model gives stable and valid reduct even with dataset containing up to 90% missing class labels.
15	Lee et al. [39]	Presented an intension recognition framework based on fuzzy rough feature selection. Multitree genetic programming was introduced for brain signal dataset classification.
16	Vluymans et al. [43]	Proposed a weighted KNN regression method followed by fuzzy rough distributed prototype selection method for big data.

(continued)

Table 9.1 (Continued)

S. No.	Authors	Proposed work
17	Zeng et al. [44]	Presented fuzzy rough set approach for incremental feature selection in hybrid information system by combining new hybrid distance based on value difference metric and Gaussian kernel.
18	Arunkumar and Ramakrishnan [45]	Proposed a novel fuzzy rough feature selection using correlation coefficient and applied to cancer microarray data.
19	Lasisi et al. [48]	Fuzzy rough and vaguely quantified rough set-based feature selection was coupled with clonal selection algorithm and artificial immune recognition system for mining agriculture data.
20	Tsang et al. [49]	Presented a new KNN rule based on fuzzy rough set for instance and feature selection. From this weighted subsampled data, probability was estimated using Parzen window technique.
21	Zhang et al. [50]	Proposed fuzzy rough-based feature selection using information entropy for reducing heterogeneous data.
22	Kumar et al. [51]	Discussed a comparative study of filter, wrapper, and fuzzy rough set-based feature selection based on execution time, number of features in reduced dataset, and classifier accuracy.
23	Meenachi et al. [54]	Presented a fuzzy rough feature selection technique on dataset preprocessed by kernel filter followed by fuzzy rough-ordered weighted average classification.
24	Qu et al. [55]	Proposed fuzzy rough feature selection for multiclass datasets by generating association rule implied in the labels and considering each set of sublabels as a unique class.
25	Su et al. [56]	Employed ordered weighted averaging aggregation of fuzzy similarity relations to reflect interaction between attributes for better performance of fuzzy rough feature selection.
26	Wang et al. [57]	Proposed a fitting model for feature selection to prevent misclassification of samples by defining fuzzy decision of sample using fuzzy neighborhood and parameterized fuzzy relation to characterize information granules.
27	Anaraki et al. [63]	Proposed feature selection by hybridizing fuzzy rough dependency degree as evaluation measure selection with binary shuffled frog leaping algorithm as search strategy.
28	Dai et al. [93]	Presented two fuzzy rough attribute reduction techniques based on reduced maximal discernibility pair selection and weighted reduced maximal discernibility pair selection.
29	Tiwari et al. [74]	Applied fuzzy rough feature selection on balanced training and test datasets followed by ranking using fuzzy rough attribute evaluator for enhancing prediction of piezophilic protein.

fuzzy rough feature selection where the object labels in a dataset may only be present partially. The methodology in addition has the interesting property that any reducts (generated subsets) are also valid once the entire dataset is labeled. Sheikhpour et al. [98] conferred a survey paper on semisupervised feature selection methods in which two taxonomies of semisupervised strategies are given based on two different viewpoints, which in fact characterize the hierarchical structure of semisupervised feature selection methods. The first viewpoint is based on the basic classification of feature selection methods, while the second one is based on the classification of semisupervised learning methods.

Table 9.2 Fuzzy rough set-assisted feature selection techniques based on the discernibility matrix approach.

S. No.	Authors	Proposed work
1	He et al. [86]	Presented a fuzzy rough set-based attribute reduction approach to deal with the information systems with fuzzy decision values. In this study, a discernibility matrix-based attribute reduction technique was developed.
2	Dai and Tian [87]	Proposed fuzzy rough set model for set-valued information system and applied the model for feature selection based on discernibility matrix approach.
3	Dai and Xu [88]	Introduced feature selection model for enhancing the accuracy of tumor classification based on the concept of fuzzy gain ratio.
4	Chen and Yang [89]	Proposed fuzzy rough feature selection by defining discernibility relations for symbolic and real-valued attribute to calculate dependency and thus measure the inconsistency between heterogeneous features.
5	Wang [90]	Proved that reduct defined in fuzzy rough set approach includes the reduct defined in rough set approach and presented a discernibility matrix-based technique to calculate reduct for set-valued data.
6	Wei et al. [91]	Proposed fuzzy rough feature selection for set-valued data using the concept of discernibility matrix-based technique.
7	Dai et al. [93]	Presented a heuristic algorithm for attribute reduction via a dominance relation to measure dominance between attributes based on which definition of fuzzy approximation is extended and uncertainty measurement issue investigated.
8	Yang et al. [92]	Discussed two incremental fuzzy rough feature selection techniques for large datasets in which one selected features as sample subset arrive and output final subset when no sample is left while other performs feature selection only when no sample is left but updates relative discernibility relation.
9	Xu et al. [95]	Modified conditional entropy by introducing fuzzy neighborhood granule and rough decisions, and proposed rough uncertainty model. Variable precision model was also proposed for selection of parameters.

Pal et al. [99] presented an article to demonstrate the neuro-fuzzy approaches for both feature extraction and selection under unsupervised learning. To compute membership values in the transformed space, a concept of flexible membership function with the help of weighted distance is introduced and also two new layered networks are proposed in the article. Mitra et al. [100] delineated an unsupervised feature selection algorithm appropriate for datasets, large in dimension as well as size. A new feature similarity measure, called maximum information compression index, is presented in the paper. Dy and Brodley [101] used FSSEM (feature subset selection using expectation-maximization clustering) in order to tackle with two issues involved in feature subset selection algorithm for unlabeled data: the requirement for finding the number of clusters in conjunction with feature selection, and the requirement for normalizing the bias of feature selection criteria relating to dimension. Two different performance measures for computing reducts, scatter separability and maximum likelihood, are also introduced. Mac Parthalain and Jensen [29, 31] proposed some new FRS-based methods to deal with unsupervised feature selection. These methods tackle real-valued data with no involvement of threshold or domain information. Zimek et al. [102] discussed some important characteristics on the "curse of dimensionality" and survey algorithms for unsupervised outlier detection from two categories, namely, considering or not considering subspaces (feature subsets) in the definition of outliers. Ganivada et al. [103] proposed a granular neural network for characterizing salient features of data using the theory of fuzzy set and FRS.

9.5 Decision Systems with Missing Values

Sometimes while collecting data from various sources, some attribute values cannot be determined or found missing in a decision system. In order to find the reduct of such decision systems, either we have to assign some value (according to some rule) to the missing place or ignore those missing attribute values completely. Jensen and Shen [104] proposed a methodology based on interval-valued fuzzy sets using an extension of the fuzzy rough feature selection to deal with missing values in the real-world data. Doquire and Verleysen [105] gave a concept of nearest-neighbor-based mutual information estimator to tackle missing data and feature selection on such types of datasets. The method did not require any imputation algorithm to extract important features. Three missing value imputation methods based on FRSs and their recent extensions, namely, implicator/t-norm-based FRSs, VQRSs, and also ordered weighted average-based rough sets, are presented by Amiri and Jensen [106].

9.6 Applications in Classification, Rule Extraction, and Other Application Areas

Feature selection or attribute reduction has various applications in the fields of data mining, signal processing, image recognition, bioinformatics, text categorization, system monitoring, clustering data, and rule induction (see Figure 9.2). It is the most important preprocessing step in the process of knowledge discovery for databases.

In 2006, Tsai et al. established a Modified Minimization Entropy Principle Algorithm (MMEPA) to construct membership functions of fuzzy sets of linguistic variables [107]. Furthermore, they fuzzified different real-valued datasets by using this concept to construct the rule extraction variable precision rough set model (VP-model), and the extracted rules can obtain a better classification accuracy rate than that of some existing techniques.

In 2007, Kumar et al. presented a novel approach to improve classification efficiency of the standard K-nearest neighbor algorithm by using fuzzy rough uncertainty [108]. The unfussiness and nonparametric uniqueness of the conventional K-nearest neighbor algorithm endure intactness in the proposed algorithm, where the optimal value of K is not needed. Moreover, the produced class confidence value do not inevitably sum up to one, which is interpreted in the form of fuzzy rough ownership values. Hence, the presented algorithm can differentiate between equal evidence and ignorance. Therefore, the semantics of the class confidence values becomes more affluent. It was demonstrated that the proposed classifier oversimplifies the conventional and fuzzy KNN algorithms. The effectiveness of the presented approach is discussed on real-valued datasets.

In 2008, Bhatt and Gopal introduced a novel fuzzy rough classification trees (FRCTs) to solve classification problems from feature-based learning examples [109]. Experimental results proved that the proposed algorithm is performing better than other existing rough decomposition tree induction algorithm and fuzzy decision tree

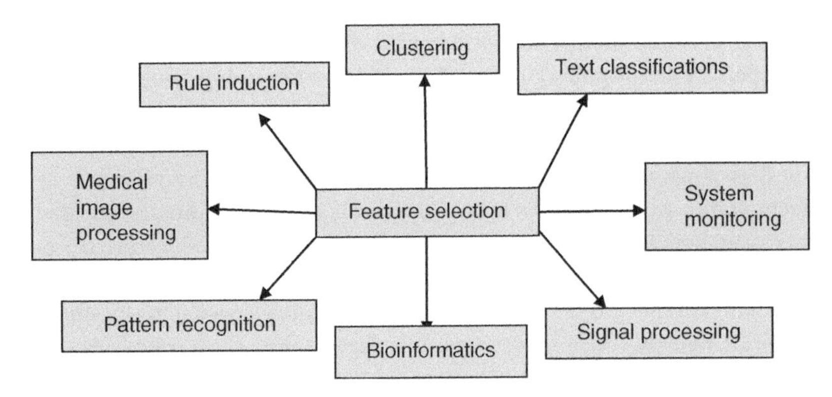

Figure 9.2 Various application areas of feature selection.

generation techniques. In 2010, Chen et al. proposed the concept of FRS-based support vector machines [110]. In his article, fuzzy transitive kernel-based FRSs was introduced. Then, membership values of every training sample were computed using this kernel-based FRS. Finally, hard margin support vector machine was reformulated into FRS by using new constraints based on the above membership values.

In 2010, Zhao et al. proposed the concept of rule-based fuzzy rough classification algorithm [111]. In this work, they presented the way to build a rule-based classifier by using one generalized FRS model by proposing an appropriate learning rule. Firstly, they generalized the FRS model to a robust model with respect to perturbation and misclassification by integrating one controlled threshold into knowledge representation of FRS concept. Secondly, they established a rule for learning and by using the strict mathematical reasoning. Hashemi and Latibari [112] discussed various shortcomings of rough set-based classification algorithm. Furthermore, they introduced FRS classifiers for database mining to avoid these shortcomings. Finally, they performed the statistical analysis to show the power of the proposed algorithm.

In 2011, Ganivada et al. described a novel classification approach for fuzzy classification of patterns [113]. This algorithm was called as a fuzzy rough granular neural network (FRGNN), which was based on the multilayer perceptron using backpropagation. They introduced a new development approach of knowledge extraction from data based on FRS theoretic methods. FRS models are very sensitive to noisy samples, more specifically mislabeled instances. Hu et al. [114] represented a robust FRS-based classification algorithm to solve this problem. They validated this algorithm by performing various experimental results. Jensen and Cornelis [115] proposed a nearest neighbor (NN) algorithm based on FRST to classify test objects and predict their decision value. Moreover, it was presented that the method was robust against noise, and the robustness can be effectively enhanced by invoking the approximations of the VQRS model.

In 2014, Ramentol et al. established a classifier for imbalanced data that uses ordered weighted average aggregation and FRST [116]. In his classification algorithm, different strategies were considered to build a weight vector to take into account data imbalance. The performance of this algorithm was validated by using a wider experimental study.

In 2015, An et al. proposed data distribution aware FRS model [117]. This model considered distribution information and incorporating it in calculating upper and lower fuzzy approximations. The proposed model considered not only the probability density of classes but also similarity between samples. In this article, a robust classifier was constructed with nearest-neighbor classification and prototype covering. In 2016, Vluymans et al. discussed the classification issues assisted with class-imbalanced multi-instance data [118]. Moreover, they handled these issues by proposing FRS-based classification algorithm, which was based on information extracted at instance level and at bag level. Finally, they conducted different experiments to validate their proposed algorithm.

9.7 Limitations of Fuzzy Rough Set Theory

Fuzzy set theory is an integral tool to manage uncertainty; however, it also has certain constraints. These limitations can be highlighted as follows:

- Fuzzy set theory is not fit for taking care of many decision-making problems, for instance, in the casting a ballot issue, where a board of 10 specialists are voting in favor of one item, assume 5 specialists gave the ends "agree," three of them "disagree," and the remaining of the two specialists "abstain." This situation can be viably taken care of by including a nonmembership degree for "disagree" and hesitancy degree for "abstain."
- In the numerous real-world applications, for example, medical diagnosis, sensor data, and so forth, vaguely specified information systems are extremely normal. Fuzzy set theory is executed to deal with such vagueness by generalizing the idea of membership in a set. In a fuzzy set theory, the membership degree of an element in a universe dependably takes a solitary incentive somewhere in the range of 0 and 1; however, those single

values may not totally characterize about the knowledge or the lack of knowledge as the uncertainty is not discovered distinctly in judgment yet additionally in the identification. Accordingly, a few extensions of fuzzy sets are required to deal with the latter uncertainty.

- The FRS approximations characterized above are exceedingly sensitive to noisy values because of the utilization of sup and inf (the generalized existential and universal quantifier, individually). It antagonistically influences the accuracy if there is an occurrence of classification problems.

In spite of the above drawbacks, FRST acted as a powerful technique for attribute reduction and classification problems because of its ability in handling uncertainty and vagueness available in information system. However, the decision-making problems can be suitably handled by using soft set theory or its generalization, which are discussed in Molodtsov [119], Maji et al. [120], or based on its further analysis as in Tripathi and Som [121].

9.8 Conclusion

In this chapter, we have presented a wide survey on techniques of FRS-assisted attributed reduction in order to find better classification accuracy. In this survey, not only supervised decision system is considered but semisupervised and unsupervised decision systems are also discussed. FRS-assisted attribute reduction using degree of dependency and discernibility matrix approaches are mainly focused. We have presented some applications of such techniques in various fields for better understanding. In future, we wish to compare different approaches for classification accuracy of various datasets. Some extensions of fuzzy set theory like intuitionistic fuzzy set and neutrosophic fuzzy set have also been applied for attribute reduction. We intend to implement these extensions for attribute reduction and classification problems.

References

1 Dash, M. and Liu, H. (1997). Feature selection for classification. *Intelligent Data Analysis* 1 (1–4): 131–156.

2 Duda, R. O., Hart, P. E., & Stork, D. G. (1973). Nonparametric techniques. *Pattern classification and scene analysis*, 98–105.

3 Langley, P. (1994). Selection of relevant features in machine learning. In: *Proceedings of the AAAI Fall Symposium on Relevance*, 1–5. New Orleans, LA: AAAI Press.

4 Jensen, R. and Shen, Q. (2008). *Computational Intelligence and Feature Selection: Rough and Fuzzy Approaches*, vol. 8. Wiley.

5 Pawlak, Z. (2012). *Rough Sets: Theoretical Aspects of Reasoning About Data*, vol. 9. Springer Science & Business Media.

6 Pawlak, Z. and Skowron, A. (2007). Rudiments of rough sets. *Information Sciences* 177 (1): 3–27.

7 Dubois, D. and Prade, H. (1990). Rough fuzzy sets and fuzzy rough sets. *International Journal of General Systems* 17 (2–3): 191–209.

8 Dubois, D. and Prade, H. (1992). Putting rough sets and fuzzy sets together. In: *Intelligent Decision Support*, 203–232. Dordrecht: Springer.

9 Zadeh, L.A. (1965). Fuzzy sets. *Information and Control* 8 (3): 338–353.

10 Mizumoto, M. and Tanaka, K. (1976). Some properties of fuzzy sets of type 2. *Information and Control* 31 (4): 312–340.

11 Zadeh, L.A. (1975). The concept of a linguistic variable and its application to approximate reasoning—I. *Information Sciences* 8 (3): 199–249.

12 Atanassov, K.T. (1999). Intuitionistic fuzzy sets. In: *Intuitionistic Fuzzy Sets. Studies in Fuzziness and Soft Computing*, vol. 35, 1–137. Heidelberg: Physica.

13 Huang, S.Y. (ed.) (1992). *Intelligent Decision Support: Handbook of Applications and Advances of the Rough Sets Theory*, vol. 11. Springer Science & Business Media.

14 Tsang, E.C., Chen, D., Yeung, D.S. et al. (2008). Attributes reduction using fuzzy rough sets. *IEEE Transactions on Fuzzy Systems* 16 (5): 1130–1141.

15 Zadeh, L.A., Klir, G.J., and Yuan, B. (1996). *Fuzzy Sets, Fuzzy Logic, and Fuzzy Systems: Selected Papers*, vol. 6. World Scientific.

16 Das, M., Chakraborty, M.K., and Ghoshal, T.K. (1998). Fuzzy tolerance relation, fuzzy tolerance space and basis. *Fuzzy Sets and Systems* 97 (3): 361–369.

17 Skowron, A. and Rauszer, C. (1992). The discernibility matrices and functions in information systems. In: *Intelligent Decision Support*, 331–362. Dordrecht: Springer.

18 Jensen, R. and Shen, Q. (2004). Fuzzy-rough attribute reduction with application to web categorization. *Fuzzy Sets and Systems* 141 (3): 469–485.

19 Shen, Q. and Jensen, R. (2004). Selecting informative features with fuzzy-rough sets and its application for complex systems monitoring. *Pattern Recognition* 37 (7): 1351–1363.

20 Bhatt, R.B. and Gopal, M. (2005). On fuzzy-rough sets approach to feature selection. *Pattern Recognition Letters* 26 (7): 965–975.

21 Jensen, R. and Shen, Q. (2005). Fuzzy-rough data reduction with ant colony optimization. *Fuzzy Sets and Systems* 149 (1): 5–20.

22 Hu, Q., Yu, D., and Xie, Z. (2006). Information-preserving hybrid data reduction based on fuzzy-rough techniques. *Pattern Recognition Letters* 27 (5): 414–423.

23 Hu, Q., Xie, Z., and Yu, D. (2007). Hybrid attribute reduction based on a novel fuzzy-rough model and information granulation. *Pattern Recognition* 40 (12): 3509–3521.

24 Jensen, R. and Shen, Q. (2007). Tolerance-based and fuzzy-rough feature selection. In: *2007 IEEE International Fuzzy Systems Conference*, 1–6. IEEE.

25 Jensen, R. and Shen, Q. (2007). Fuzzy-rough sets assisted attribute selection. *IEEE Transactions on Fuzzy Systems* 15 (1): 73–89.

26 Cornelis, C., De Cock, M., and Radzikowska, A.M. (2007). Vaguely quantified rough sets. In: *International Workshop on Rough Sets, Fuzzy Sets, Data Mining, and Granular-Soft Computing*, 87–94. Berlin, Heidelberg: Springer-Verlag.

27 Jensen, R., Cornelis, C., and Shen, Q. (2009). Hybrid fuzzy-rough rule induction and feature selection. In: *2009 IEEE International Conference on Fuzzy Systems*, 1151–1156. IEEE.

28 Hu, Q., An, S., and Yu, D. (2010). Soft fuzzy rough sets for robust feature evaluation and selection. *Information Sciences* 180 (22): 4384–4400.

29 Mac Parthaláin, N. and Jensen, R. (2010). Measures for unsupervised fuzzy-rough feature selection. *International Journal of Hybrid Intelligent Systems* 7 (4): 249–259.

30 Mac Parthaláin, N. and Jensen, R. (2013). Simultaneous feature and instance selection using fuzzy-rough bireducts. In: *2013 IEEE International Conference on Fuzzy Systems (FUZZ-IEEE)*, 1–8. IEEE.

31 Mac ParthaláIn, N. and Jensen, R. (2013). Unsupervised fuzzy-rough set-based dimensionality reduction. *Information Sciences* 229: 106–121.

32 Shang, C. and Barnes, D. (2013). Fuzzy-rough feature selection aided support vector machines for mars image classification. *Computer Vision and Image Understanding* 117 (3): 202–213.

33 Derrac, J., Verbiest, N., García, S. et al. (2013). On the use of evolutionary feature selection for improving fuzzy rough set based prototype selection. *Soft Computing* 17 (2): 223–238.

34 Diao, R., Chao, F., Peng, T. et al. (2013). Feature selection inspired classifier ensemble reduction. *IEEE Transactions on Cybernetics* 44 (8): 1259–1268.

35 Cornelis, C., Medina, J., and Verbiest, N. (2014). Multi-adjoint fuzzy rough sets: definition, properties and attribute selection. *International Journal of Approximate Reasoning* 55 (1): 412–426.

36 Ju, H., Yang, X., Song, X., and Qi, Y. (2014). Dynamic updating multigranulation fuzzy rough set: approximations and reducts. *International Journal of Machine Learning and Cybernetics* 5 (6): 981–990.

37 Inuiguchi, M., Wu, W.Z., Cornelis, C., and Verbiest, N. (2015). Fuzzy-rough hybridization. In: *Springer Handbook of Computational Intelligence*, 425–451. Berlin, Heidelberg: Springer-Verlag.

38 Jensen, R. and Mac Parthaláin, N. (2015). Towards scalable fuzzy-rough feature selection. *Information Sciences* 323: 1–15.

39 Lee, J.H., Anaraki, J.R., Ahn, C.W., and An, J. (2015). Efficient classification system based on fuzzy-rough feature selection and multitree genetic programming for intension pattern recognition using brain signal. *Expert Systems with Applications* 42 (3): 1644–1651.

40 Onan, A. (2015). A fuzzy-rough nearest neighbor classifier combined with consistency-based subset evaluation and instance selection for automated diagnosis of breast cancer. *Expert Systems with Applications* 42 (20): 6844–6852.

41 Mac Parthaláin, N. and Jensen, R. (2015). Fuzzy-rough feature selection using flock of starlings optimisation. In: *2015 IEEE International Conference on Fuzzy Systems (FUZZ-IEEE)*, 1–8. IEEE.

42 Qian, Y., Wang, Q., Cheng, H. et al. (2015). Fuzzy-rough feature selection accelerator. *Fuzzy Sets and Systems* 258: 61–78.

43 Vluymans, S., Asfoor, H., Saeys, Y. et al. (2015). Distributed fuzzy rough prototype selection for big data regression. In: *2015 Annual Conference of the North American Fuzzy Information Processing Society (NAFIPS) Held Jointly with 2015 5th World Conference on Soft Computing (WConSC)*, 1–6. IEEE.

44 Zeng, A., Li, T., Liu, D. et al. (2015). A fuzzy rough set approach for incremental feature selection on hybrid information systems. *Fuzzy Sets and Systems* 258: 39–60.

45 Arunkumar, C. and Ramakrishnan, S. (2016). A hybrid approach to feature selection using correlation coefficient and fuzzy rough quick reduct algorithm applied to cancer microarray data. In: *2016 10th International Conference on Intelligent Systems and Control (ISCO)*, 1–6. IEEE.

46 Guo, Q., Qu, Y., Deng, A., and Yang, L. (2016). A new fuzzy-rough feature selection algorithm for mammographic risk analysis. In: *2016 12th International Conference on Natural Computation, Fuzzy Systems and Knowledge Discovery (ICNC-FSKD)*, 934–939. IEEE.

47 Jhawar, A., Chan, C.S., Monekosso, D., and Remagnino, P. (2016). Fuzzy-rough based decision system for gait adopting instance selection. In: *2016 IEEE International Conference on Fuzzy Systems (FUZZ-IEEE)*, 1127–1133. IEEE.

48 Lasisi, A., Ghazali, R., Deris, M.M. et al. (2016). Extracting information in agricultural data using fuzzy-rough sets hybridization and clonal selection theory inspired algorithms. *International Journal of Pattern Recognition and Artificial Intelligence* 30 (09): 1660008.

49 Tsang, E.C., Hu, Q., and Chen, D. (2016). Feature and instance reduction for PNN classifiers based on fuzzy rough sets. *International Journal of Machine Learning and Cybernetics* 7 (1): 1–11.

50 Zhang, X., Mei, C., Chen, D., and Li, J. (2016). Feature selection in mixed data: a method using a novel fuzzy rough set-based information entropy. *Pattern Recognition* 56: 1–15.

51 Kumar, C.A., Sooraj, M.P., and Ramakrishnan, S. (2017). A comparative performance evaluation of supervised feature selection algorithms on microarray datasets. *Procedia Computer Science* 115: 209–217.

52 Badria, F.A., Habib, M.M.A., Shoaip, N., and Elmogy, M. (2017). A framework for harmla alkaloid extraction process development using fuzzy-rough sets feature selection and J48 classification. *International Journal of Advanced Computer Research* 7 (33): 213–222.

53 Chinnaswamy, A. and Srinivasan, R. (2017). Hybrid information gain based fuzzy roughset feature selection in cancer microarray data. In: *2017 Innovations in Power and Advanced Computing Technologies (i-PACT)*, 1–6. IEEE.

54 Meenachi, L., Raghul, J.J., Raj, C.M., and Kathiravan, B. (2017). Diagnosis of medical dataset using fuzzy-rough ordered weighted average classification. In: *2017 International Conference on Innovations in Information, Embedded and Communication Systems (ICIIECS)*, 1–5. IEEE.

55 Qu, Y., Rong, Y., Deng, A., and Yang, L. (2017). Associated multi-label fuzzy-rough feature selection. In: *2017 Joint 17th World Congress of International Fuzzy Systems Association and 9th International Conference on Soft Computing and Intelligent Systems (IFSA-SCIS)*, 1–6. IEEE.

56 Su, P., Shen, Q., Chen, T., and Shang, C. (2017). Ordered weighted aggregation of fuzzy similarity relations and its application to detecting water treatment plant malfunction. *Engineering Applications of Artificial Intelligence* 66: 17–29.

57 Wang, C., Qi, Y., Shao, M. et al. (2017). A fitting model for feature selection with fuzzy rough sets. *IEEE Transactions on Fuzzy Systems* 25 (4): 741–753.

58 Dai, J., Hu, H., Wu, W.Z. et al. (2017). Maximal-discernibility-pair-based approach to attribute reduction in fuzzy rough sets. *IEEE Transactions on Fuzzy Systems* 26 (4): 2174–2187.

59 Javidi, M.M. and Mansoury, S. (2017). Diagnosis of the disease using an ant colony gene selection method based on information gain ratio using fuzzy rough sets. *Journal of Particle Science & Technology* 3 (4): 175–186.

60 Cheruku, R., Edla, D.R., Kuppili, V., and Dharavath, R. (2018). RST-BatMiner: a fuzzy rule miner integrating rough set feature selection and bat optimization for detection of diabetes disease. *Applied Soft Computing* 67: 764–780.

61 Rhee, C.M., Ahmadi, S.F., Kovesdy, C.P., and Kalantar-Zadeh, K. (2018). Low-protein diet for conservative management of chronic kidney disease: a systematic review and meta-analysis of controlled trials. *Journal of Cachexia, Sarcopenia and Muscle* 9 (2): 235–245.

62 Ahmed, S., Mafarja, M., Faris, H., and Aljarah, I. (2018). Feature selection using salp swarm algorithm with chaos. In: *Proceedings of the 2nd International Conference on Intelligent Systems, Metaheuristics & Swarm Intelligence*, 65–69. ACM.

63 Anaraki, J.R., Samet, S., Eftekhari, M., and Ahn, C.W. (2018). A fuzzy-rough based binary shuffled frog leaping algorithm for feature selection. arXiv preprint arXiv:1808.00068.

64 Arunkumar, C. and Ramakrishnan, S. (2018). Attribute selection using fuzzy roughset based customized similarity measure for lung cancer microarray gene expression data. *Future Computing and Informatics Journal* 3 (1): 131–142.

65 Hamidzadeh, J., Zabihimayvan, M., and Sadeghi, R. (2018). Detection of Web site visitors based on fuzzy rough sets. *Soft Computing* 22 (7): 2175–2188.

66 Han, X., Qu, Y., and Deng, A. (2018). A Laplace distribution-based fuzzy-rough feature selection algorithm. In: *2018 10th International Conference on Advanced Computational Intelligence (ICACI)*, 776–781. IEEE.

67 Hu, Q., Zhang, L., Zhou, Y., and Pedrycz, W. (2018). Large-scale multimodality attribute reduction with multi-kernel fuzzy rough sets. *IEEE Transactions on Fuzzy Systems* 26 (1): 226–238.

68 Li, Y., Lin, Y., Liu, J. et al. (2018). Feature selection for multi-label learning based on kernelized fuzzy rough sets. *Neurocomputing* 318: 271–286.

69 Lin, Y., Li, Y., Wang, C., and Chen, J. (2018). Attribute reduction for multi-label learning with fuzzy rough set. *Knowledge-Based Systems* 152: 51–61.

70 Liu, Z. and Pan, S. (2018). Fuzzy-rough instance selection combined with effective classifiers in credit scoring. *Neural Processing Letters* 47 (1): 193–202.

71 Mahanipour, A., Nezamabadi-pour, H., and Nikpour, B. (2018). Using fuzzy-rough set feature selection for feature construction based on genetic programming. In: *2018 3rd Conference on Swarm Intelligence and Evolutionary Computation (CSIEC)*, 1–6. IEEE.

72 Ratnaparkhi, A., Bormane, D., and Ghongade, R. (2018). Performance analysis of fuzzy rough assisted classification and segmentation of paper ECG using mutual information and dependency metric. In: *International Conference on ISMAC in Computational Vision and Bio-Engineering*, 789–801. Cham: Springer.

73 Sheeja, T.K. and Kuriakose, A.S. (2018). A novel feature selection method using fuzzy rough sets. *Computers in Industry* 97: 111–116.

74 Tiwari, A.K., Shreevastava, S., Subbiah, K., and Som, T. (2018). Enhanced prediction for piezophilic protein by incorporating reduced set of amino acids using fuzzy-rough feature selection technique followed by SMOTE. In: *International Conference on Mathematics and Computing*, 185–196. Singapore: Springer.

75 Varma, P.R.K., Kumari, V.V., and Kumar, S.S. (2018). A survey of feature selection techniques in intrusion detection system: a soft computing perspective. In: *Progress in Computing, Analytics and Networking*, 785–793. Singapore: Springer.

76 Zhang, Q., Qu, Y., and Deng, A. (2018). Network intrusion detection using kernel-based fuzzy-rough feature selection. In: *2018 IEEE International Conference on Fuzzy Systems (FUZZ-IEEE)*, 1–6. IEEE.

77 Zhang, X., Liu, X., and Yang, Y. (2018). A fast feature selection algorithm by accelerating computation of fuzzy rough set-based information entropy. *Entropy* 20 (10): 788.

78 Zhang, X., Mei, C., Chen, D., and Yang, Y. (2018). A fuzzy rough set-based feature selection method using representative instances. *Knowledge-Based Systems* 151: 216–229.

79 Zhenlei, W., Suyun, Z., Yangming, L. et al. (2018). Fuzzy rough based feature selection by using random sampling. In: *Pacific Rim International Conference on Artificial Intelligence*, 91–99. Cham: Springer.

80 Zuo, Z., Li, J., Anderson, P. et al. (2018). Grooming detection using fuzzy-rough feature selection and text classification. In: *2018 IEEE International Conference on Fuzzy Systems (FUZZ-IEEE)*, 1–8. IEEE.

81 Arunkumar, C. and Ramakrishnan, S. (2019). Prediction of cancer using customised fuzzy rough machine learning approaches. *Healthcare Technology Letters* 6 (1): 13–18.

82 Maini, T., Kumar, A., Misra, R.K., and Singh, D. (2019). Fuzzy rough set-based feature selection with improved seed population in PSO and IDS. In: *Computational Intelligence: Theories, Applications and Future Directions-Volume II*, 137–149. Singapore: Springer.

83 Wang, C., Huang, Y., Shao, M., and Fan, X. (2019). Fuzzy rough set-based attribute reduction using distance measures. *Knowledge-Based Systems* 164: 205–212.

84 Zabihimayvan, M. and Doran, D. (2019). Fuzzy rough set feature selection to enhance phishing attack detection. arXiv preprint arXiv:1903.05675.

85 Zhao, R., Gu, L., and Zhu, X. (2019). Combining fuzzy C-means clustering with fuzzy rough feature selection. *Applied Sciences* 9 (4): 679.

86 He, Q., Wu, C., Chen, D., and Zhao, S. (2011). Fuzzy rough set based attribute reduction for information systems with fuzzy decisions. *Knowledge-Based Systems* 24 (5): 689–696.

87 Dai, J. and Tian, H. (2013). Fuzzy rough set model for set-valued data. *Fuzzy Sets and Systems* 229: 54–68.

88 Dai, J. and Xu, Q. (2013). Attribute selection based on information gain ratio in fuzzy rough set theory with application to tumor classification. *Applied Soft Computing* 13 (1): 211–221.

89 Chen, D. and Yang, Y. (2014). Attribute reduction for heterogeneous data based on the combination of classical and fuzzy rough set models. *IEEE Transactions on Fuzzy Systems* 22 (5): 1325–1334.

90 Wang, C.Y. (2016). A note on a fuzzy rough set model for set-valued data. *Fuzzy Sets and Systems* 294: 44–47.

91 Wei, W., Cui, J., Liang, J., and Wang, J. (2016). Fuzzy rough approximations for set-valued data. *Information Sciences* 360: 181–201.

92 Yang, Y., Chen, D., Wang, H., and Wang, X. (2018). Incremental perspective for feature selection based on fuzzy rough sets. *IEEE Transactions on Fuzzy Systems* 26 (3): 1257–1273.

93 Dai, J., Yan, Y., Li, Z., and Liao, B. (2018). Dominance-based fuzzy rough set approach for incomplete interval-valued data. *Journal of Intelligent & Fuzzy Systems* 34 (1): 423–436.

94 Juneja, A., Rana, B., and Agrawal, R.K. (2018). A novel fuzzy rough selection of non-linearly extracted features for schizophrenia diagnosis using fMRI. *Computer Methods and Programs in Biomedicine* 155: 139–152.

95 Xu, J., Wang, Y., Xu, K., and Zhang, T. (2019). Feature genes selection using fuzzy rough uncertainty metric for tumor diagnosis. *Computational and Mathematical Methods in Medicine* 2019: 6705648.

96 Jensen, R., Vluymans, S., Mac Parthaláin, N. et al. (2015). Semi-supervised fuzzy-rough feature selection. In: *Rough Sets, Fuzzy Sets, Data Mining, and Granular Computing*, 185–195. Cham: Springer.

97 Zhu, X.J. (2005). *Semi-Supervised Learning Literature Survey*. University of Wisconsin-Madison Department of Computer Sciences.

98 Sheikhpour, R., Sarram, M.A., Gharaghani, S., and Chahooki, M.A.Z. (2017). A Survey on semi-supervised feature selection methods. *Pattern Recognition* 64: 141–158.

99 Pal, S.K., De, R.K., and Basak, J. (2000). Unsupervised feature evaluation: a neuro-fuzzy approach. *IEEE Transactions on Neural Networks* 11 (2): 366–376.

100 Mitra, P., Murthy, C.A., and Pal, S.K. (2002). Unsupervised feature selection using feature similarity. *IEEE Transactions on Pattern Analysis and Machine Intelligence* 24 (3): 301–312.

101 Dy, J.G. and Brodley, C.E. (2004). Feature selection for unsupervised learning. *Journal of Machine Learning Research* 5: 845–889.

102 Zimek, A., Schubert, E., and Kriegel, H.P. (2012). A survey on unsupervised outlier detection in high-dimensional numerical data. *Statistical Analysis and Data Mining: The ASA Data Science Journal* 5 (5): 363–387.

103 Ganivada, A., Ray, S.S., and Pal, S.K. (2013). Fuzzy rough sets, and a granular neural network for unsupervised feature selection. *Neural Networks* 48: 91–108.

104 Jensen, R. and Shen, Q. (2009). Interval-valued fuzzy-rough feature selection in datasets with missing values. In: *2009 IEEE International Conference on Fuzzy Systems*, 610–615. IEEE.

105 Doquire, G. and Verleysen, M. (2011). Mutual information for feature selection with missing data. In *Proceedings of the European Symposium on Artificial Neural Networks* (ESANN 2011).

106 Amiri, M. and Jensen, R. (2016). Missing data imputation using fuzzy-rough methods. *Neurocomputing* 205: 152–164.

107 Tsai, Y.C., Cheng, C.H., and Chang, J.R. (2006). Entropy-based fuzzy rough classification approach for extracting classification rules. *Expert Systems with Applications* 31 (2): 436–443.

108 Kumar, M., Grzelakowski, M., Zilles, J. et al. (2007). Highly permeable polymeric membranes based on the incorporation of the functional water channel protein Aquaporin Z. *Proceedings of the National Academy of Sciences of the United States of America* 104 (52): 20719–20724.

109 Bhatt, R.B. and Gopal, M. (2008). FRCT: fuzzy-rough classification trees. *Pattern Analysis and Applications* 11 (1): 73–88.

110 Chen, D., He, Q., and Wang, X. (2010). FRSVMs: fuzzy rough set based support vector machines. *Fuzzy Sets and Systems* 161 (4): 596–607.

111 Zhao, S.Y., Tsang, E.C., Chen, D.G., and Wang, X.Z. (2010). Building a rule-based classifier by using fuzzy rough set technique. *IEEE Transactions on Knowledge and Data Engineering* 22 (5): 624–638.

112 Hashemi, S.K.H. and Latibari, A.J. (2010). Evaluation and identification of walnut heartwood extractives for protection of poplar wood. *BioResources* 6 (1): 59–69.

113 Ganivada, A., Dutta, S., and Pal, S.K. (2011). Fuzzy rough granular neural networks, fuzzy granules, and classification. *Theoretical Computer Science* 412 (42): 5834–5853.

114 Hu, Q., Zhang, L., An, S. et al. (2011). On robust fuzzy rough set models. *IEEE Transactions on Fuzzy Systems* 20 (4): 636–651.

115 Jensen, R. and Cornelis, C. (2011). Fuzzy-rough nearest neighbour classification and prediction. *Theoretical Computer Science* 412 (42): 5871–5884.

116 Ramentol, E., Vluymans, S., Verbiest, N. et al. (2014). IFROWANN: imbalanced fuzzy-rough ordered weighted average nearest neighbor classification. *IEEE Transactions on Fuzzy Systems* 23 (5): 1622–1637.

117 An, S., Hu, Q., Pedrycz, W. et al. (2015). Data-distribution-aware fuzzy rough set model and its application to robust classification. *IEEE Transactions on Cybernetics* 46 (12): 3073–3085.

118 Vluymans, S., Tarragó, D.S., Saeys, Y. et al. (2016). Fuzzy rough classifiers for class imbalanced multi-instance data. *Pattern Recognition* 53: 36–45.

119 Molodtsov, D. (1999). Soft set theory-first results. *Computers & Mathematics with Applications* 37: 19–31.

120 Maji, P.K., Biswas, R., and Roy, R. (2003). Soft set theory. *Computers & Mathematics with Applications* 45: 555–562.

121 Tripathi, V. and Som, T. (2015). The countability aspects of fuzzy soft topological spaces. *Annals of Fuzzy Mathematics and Informatics* 9 (5): 835–842.

10

Universal Intervals: Towards a Dependency-Aware Interval Algebra

Hend Dawood[1] *and Yasser Dawood*[2]

[1]*Department of Mathematics, Faculty of Science, Cairo University, Giza 12613, Egypt*
[2]*Department of Astronomy, Faculty of Science, Cairo University, Giza 12613, Egypt*

> *The most reliable way of carrying out a proof, obviously, is to follow pure logic, a way that, disregarding the particular characteristics of objects, depends solely on those laws upon which all knowledge rests.*
>
> –Gottlob Frege (1848–1925)

10.1 Introduction

Scientific knowledge is not perfect exactitude: it is learning with uncertainty, not eliminating it. We commit errors. But, indeed, we can grasp, measure, and correct our errors and develop ways to deal with uncertainty, which add to our knowledge and make it valuable. Knowledge, then, is not absolute certainty. Knowledge, per contra, is "the tools we develop to purposely get better and better outcomes through our learning about the world" [1]. Many approaches were developed with a view to cope with uncertainty and get reliable knowledge about the world. As examples of these approaches, we can mention *probabilitization*, *fuzzification*, and *intervalization*. For the great degree of reliability it provides, and because of its great importance in many practical applications, intervalization is usually a part of all other methods that deal with uncertainty (see [1] and [2]). Whenever *uncertainty* exists, there is always a need for *interval computations* and intervals arise naturally in a wide variety of disciplines that deal with uncertain data. It is a natural and elegant idea that of expressing uncertain real-valued quantities as real closed intervals. In this very simple and old idea, the field of interval mathematics has its roots from the Greek mathematician Archimedes of Syracuse (287–212 BCE), who used guaranteed lower and upper bounds to compute his constant π [3], to the American mathematician Ramon Edgar Moore (1929–2015), who was the first to define interval analysis in its modern sense and recognize its power as a viable computational tool for intervalizing uncertainty [4]. In between, historically speaking, several distinguished constructions of interval arithmetic by John Charles Burkill, Rosalind Cecily Young, Paul S. Dwyer, Teruo Sunaga, and others (see, e.g. [5–8]) have emphasized the very idea of reasoning about uncertain values through calculating with intervals. By integrating the complementary powers of rigorous mathematics and scientific computing, interval arithmetic is able to offer highly reliable accounts of uncertainty. It should therefore come as no surprise that the interval theory has been fruitfully applied in diverse areas that deal intensely with uncertain quantitative data (see, e.g. [2, 9–12]).

In view of this computational power against error, machine realizations of interval arithmetic are of great importance. It should therefore come as no surprise that there are many software implementations of interval arithmetic. As instances, we may mention INTLAB, Sollya, InCLosure, and others (see, e.g. [13–16]). Fortunately, computers

Mathematical Methods in Interdisciplinary Sciences, First Edition. Edited by Snehashish Chakraverty.

are getting faster, and most existing *parallel processors* provide a tremendous computing power. Therefore, with little extra hardware, it is very possible to make interval computations as fast as floating point computations (for further reading about machine arithmetizations and hardware circuitries for interval arithmetic, see, e.g. [17–22]).

The intended status of the present work is thus to be both an in-depth *introduction* and a *treatise* on the philosophical, mathematical, and practical aspects of interval mathematics. In the body of the work, there is room for novelties which may not be devoid of interest to researchers and specialists. The theories of classical intervals and parametric intervals are formally constructed, and their mathematical structures are uncovered. The notion of interval dependency is formalized by putting on a systematic basis its meaning and thus gaining the advantage of indicating formally the criteria by which it is to be characterized and, accordingly, deducing its fundamental properties in a merely logical manner. Moreover, with a view to treating some problems of the present interval theories, a new alternate theory of intervals, namely, the "*theory of universal intervals*," is presented and proved to have a nice S-field algebra, which extends the ordinary field of the reals.

A number of problems of interval mathematics have motivated the treatise presented in the body of this work. In the first place, in the interval literature, some important parts of interval mathematics are handled in a vaguely naïve manner without formalizing their underlying foundations. These parts are present as basic and indispensable ingredients of both fundamental research and practical applications. This work is rather different in approach and perspective. Our approach is *formal*, by the pursuit of formulating the mathematical concepts in a strictly accurate manner. Our perspective is to put on a *systematic* basis the fundamental notions of interval mathematics by taking the passage from the ambiguous and naïve treatments to the formal technicalities of mathematical logic.

In the second place, a main drawback of interval mathematics is the persisting problem known as the "*interval dependency problem*." This, naturally, confronts us with the question: *Formally, what is interval dependency*? Is it a metaconcept or an object ingredient of interval and fuzzy computations? In other words, what is the fundamental defining properties that characterize the notion of interval dependency as a formal mathematical object? Since the early works on interval mathematics by Burkill [5] and Young [6] in the dawning of the twentieth century, this question has never been touched upon and remained a question still today unanswered. Although the notion of interval dependency is widely used in the interval and fuzzy literature, it is only illustrated by example, without explicit formalization, and no attempt has been made to put on a systematic basis its meaning, that is, to indicate formally the criteria by which it is to be characterized. Here, we attempt to answer this long-standing question.

In the third place, the theory of *parametric interval arithmetic* has an underlying idea that seems elegant and simple, but it is *too simple* to fully account for the notion of interval dependency or to achieve a richer algebraic structure for interval arithmetic. It is therefore imperative both to supply the defect in the parametric approach and to present an alternative theory with a mathematical construction that avoids the defect.

Our approach is *formal*, our perspective is *systematic*, but our concern is *practical*. We describe, in detail, a number of examples that clarify the need for interval computations to cope with uncertainty in real-world systems. The InCLosure commands[1] to compute the results of the examples are described and an InCLosure input file and its corresponding output containing, respectively, the code and results of the examples are also available as a supplementary material to this chapter (see Supplementary Materials). Also, for the text to be self-contained, and with an eye towards making the adopted formal approach clearly comprehensible, our *formalized apparatus* is fixed early in Section 10.3, before moving on to main business of the text. In Section 10.4, we give a systematic characterization of the classical interval theory as a many-sorted algebra over the real field. In Section 10.5, we attempt to answer the long-standing question concerning the defining properties that characterize the notion of interval dependency as a formal mathematical object. We present a complete *systematic formalization* of the notion of interval dependency, by means of the notions of *Skolemization* and *logical quantification dependence*. A novelty of this formalization is the expression of interval dependency as a *logical predicate* (or *relation*) and thus gaining

1 InCLosure (Interval enCLosure) is a language and environment for reliable scientific computing, which is coded entirely in Lisp. Latest version of InCLosure is available for free download via https://doi.org/10.5281/zenodo.2702404.

the advantage of deducing its fundamental properties in a merely logical manner. This result sheds new light on many problems of interval mathematics. Moreover, a breakthrough behind our formalization of interval dependency is that it paves the way and provides the systematic apparatus for developing alternate *dependency-aware* interval theories and computational methods with mathematical constructions that better account for dependencies between the quantifiable uncertainties of the real world. The objective of Section 10.7 is to introduce the theory of parametric intervals and to mathematically examine to what extent it accomplishes its objectives. Based on our formalization of the notion of interval dependency, we attempt, in Sections 10.8 and 10.9, to present an alternate theory of intervals, namely, the "*theory of universal intervals*," with a mathematical construction that tries to avoid some of the defects in the current theories of interval arithmetic, to provide a richer interval algebra, and to better account for the notion of interval dependency. Finally, in Section 10.10, we discuss some further consequences and future prospects concerning the results presented in this chapter.

10.2 The Need for Interval Computations

When modeling and predicting real-life or physical phenomena, we face the problem of *uncertainty* in quantifiable properties. Tackling such uncertainties introduces two different problems: *getting guaranteed bounds* of the exact value of a quantifiable property and *computing an approximation* of the value. The two problems are very different. In many practical applications, the numerical approximations provided by machine real arithmetic are not beneficial. In robotics and control applications, for example, it is important to have guaranteed inclusions of the exact values in order to guarantee stability under uncertainty [1]. The two problems are not equivalent. The implication "*getting guaranteed bounds* $\Rightarrow$ *computing approximations*" is true, but the converse implication is not. Interval arithmetic provides us a reliable way to get guaranteed bounds when modeling physical systems under uncertainty. An interval number (a real closed interval) is then a *reliable enclosure* of an uncertain real-valued quantity, and an interval function is accordingly a reliable enclosure of an uncertain real-valued function.

In order to further clarify the need for the guaranteed enclosures of interval arithmetic, we provide here a brief discussion of the limitations and loss of precision of machine real arithmetic. Machine real numbers have finite decimal places of precision. The *finite* precision provided by modern computers is enough in many real-life applications, and there is scarcely a physical quantity which can be measured beyond the maximum representable value of this precision. So when is the finite precision not enough? The problem arises when doing arithmetic. The operation of "*subtraction*," for example, results, in many situations, in an inevitable loss of precision. Consider, for instance, the expression $x - y$, with $x = 0.963$ and $y = 0.962$. The exact result of $x - y$ is 0.001, but when evaluating this expression on a machine with two significant digits, the values of x and y are rounded *downward* to have the same machine value 0.96, and the machine result of $x - y$ becomes 0, which is a complete loss of precision. Instead, by enclosing the exact values of x and y in interval numbers, with outward rounding, a *guaranteed enclosure* of the exact result of $x - y$ can be obtained easily by computing with machine interval numbers [18].

This problem of machine subtraction has dangerous consequences in numerical computations. To illustrate, consider the problem of calculating the derivative of a real-valued function f at a given point. The method of *finite differences* is a numerical method which can be performed by a computer to approximate the derivative. For a differentiable function f, the first derivative $f^{(1)}$ can be approximated by

$$f^{(1)} \cong \frac{f(x + dx) - f(x)}{dx}.$$

for a *small* nonzero value of dx. As dx approaches zero, the derivative is better approximated. But, as dx gets smaller, the rounding error increases because of the finite precision of machine real arithmetic, and we accordingly get the problematic situation of $f(x + dx) - f(x) = 0$. That is, for small enough values of dx, the derivative will be always computed as *zero*, regardless of the rule of the function f. Using interval enclosures of the function f instead, we can find a way out of this problem, by virtue the *infinite precision* of machine interval arithmetic. For further details on interval enclosures of derivatives, see, e.g. [1] and [18].

Another problem of finite precision arises when *truncating* an infinite operation by a computable finite operation. For example, the exponential function e^x may be written as a Taylor series

$$e^x = 1 + \frac{x^2}{2!} + \frac{x^3}{3!} + \cdots = \sum_{n=0}^{\infty} \frac{x^n}{n!}.$$

In order to compute this infinite series on a machine, we have to *truncate* it to the partial sum

$$S_k = \sum_{n=0}^{k} \frac{x^n}{n!}.$$

for some finite k, and the truncation error then is $|e^x - S_k|$. Using interval bounds for this error term, machine interval arithmetic can provide a *guaranteed enclosure* of the exact value of the exponential function e^x.

The preceding examples shed light on the fact that taking the passage from real arithmetic to interval arithmetic opens the way to the rich technicalities and the infinite precision of interval computations.

10.3 On Some Algebraic and Logical Fundamentals

In order to be able to give a complete systematic formalization of the notion of interval dependency and related notions, we do need readily available a formal apparatus. So, by the pursuit of a strictly accurate formulation of the mathematical concepts, some particular formal approach will be adopted to generate all the results of this work from clear and distinct elementary ideas. Therefore, before moving on to the main business of this chapter, we begin in this section by specifying some notational conventions and formalizing some purely logical and algebraic ingredients we shall need throughout this text (for further details about the notions prescribed here, the reader may consult, e.g. [18, 23–28]).

Most of our notions are characterized in terms of *ordinals* and *ordinal tuples*. So, we first define what an *ordinal* is. An ordinal is the well-ordered set of all ordinals preceding it. That is, for each ordinal n, there exists an ordinal $S(n)$ called the successor of n such that

$$(\forall n)\,(\forall k)\,(k = S(n) \Leftrightarrow (\forall m)\,(m \in k \Leftrightarrow m \in n \vee m = n)).$$

In other words, we have $S(n) = n \cup \{n\}$. Accordingly, the first *infinite* (*transfinite*) *ordinal* is the set $\omega = \{0, 1, 2, \ldots\}$. All ordinals preceding ω (all elements of ω) are *finite ordinals*. The idea of *transfinite counting* (*counting beyond the finite*) is due to Cantor (see [29]).

With the aid of ordinals, the notions of *countably finite*, *countably infinite*, and *uncountably infinite* sets can be characterized as follows. A set S is countably finite if there is a bijective mapping from S onto some finite ordinal $n \in \omega$. A set S is countably infinite (or denumerable) if there is a bijective mapping from S onto the infinite ordinal ω. For example, the set $\{a_0, a_1, a_2\}$ is countably finite because it can be bijectively mapped onto the finite ordinal $3 = \{0, 1, 2\}$, while the set $\{a_0, a_1, a_2, \ldots\}$ is denumerable because it can be bijectively mapped onto the infinite ordinal $\omega = \{0, 1, 2, \ldots\}$. An uncountably infinite set is an infinite set which is not countably infinite. For example, the set $\mathbb{R}$ of real numbers is uncountably infinite.

The notion of an n *-tuple* is characterized in the following definition.

Definition 10.1 ***(Ordinal Tuple)*** For an ordinal $n = S(k)$, an n-tuple (ordinal tuple) is any mapping τ whose domain is n. A finite n-tuple is an n-tuple for some finite ordinal n. That is

$$\begin{aligned}\tau_{S(k)} &= \langle \tau(0), \tau(1), \ldots, \tau(k) \rangle \\ &= \langle (0, \tau(0)), (1, \tau(1)), \ldots, (k, \tau(k)) \rangle.\end{aligned}$$

If $n = 0 = \emptyset$, then, for any set S, there is exactly one mapping (*the empty mapping*) $\tau_\emptyset = \emptyset$ from $\emptyset$ into S.

An important definition we shall need is that of the Cartesian power of a set.

Definition 10.2 ***(Cartesian Power)*** Let $\emptyset$ denote the empty set. For a set S and an ordinal n, the n-th Cartesian power of S is the set S^n of all mappings from n into S, that is

$$S^n = \begin{cases} \{\emptyset\} & n = 0, \\ \text{the set of all } n\text{-tuples of elements of } S & n = 1 \vee 1 \in n. \end{cases}$$

If S is the empty set $\emptyset$, then[2]

$$\emptyset^n = \begin{cases} \{\emptyset\} & n = 0, \\ \emptyset & n = 1 \vee 1 \in n; \end{cases} \text{ and } \emptyset^{\emptyset^n} = \begin{cases} \emptyset & n = 0, \\ \{\emptyset\} & n = 1 \vee 1 \in n. \end{cases}$$

In accordance with the preceding definitions, a set-theoretical relation is a particular type of set. Let S^2 be the binary Cartesian power of a set S. A binary relation on S is a subset of S^2. That is, a set $\mathfrak{R}$ is a binary relation on a set S iff $(\forall \mathbf{r} \in \mathfrak{R})\,((\exists x, y \in S)\,(\mathbf{r} = (x, y))$.

We will continue to follow the formal tradition of Suppes [31] and Tarski [32] in defining, within a set-theoretical framework, the notion of a *finitary relation* and some related concepts. Let $\mathcal{U}^n$ be the n-th Cartesian power of a set $\mathcal{U}$. A set $\mathfrak{R} \subseteq \mathcal{U}^n$ is an n-ary relation on $\mathcal{U}$ iff $\mathfrak{R}$ is a binary relation from $\mathcal{U}^{n-1}$ to $\mathcal{U}$. That is, for $\mathbf{v} = (x_1, \dots, x_{n-1}) \in \mathcal{U}^{n-1}$ and $y \in \mathcal{U}$, an n-ary relation $\mathfrak{R}$ is defined to be $\mathfrak{R} \subseteq \mathcal{U}^n = \{(\mathbf{v}, y) | \mathbf{v} \in \mathcal{U}^{n-1} \wedge y \in \mathcal{U}\}$. In this sense, an n-ary relation is a binary relation (or simply a relation); then its *domain, range, field,* and *converse* are defined to be, respectively dom $(\mathfrak{R}) = \{\mathbf{v} \in \mathcal{U}^{n-1} | \,(\exists y \in \mathcal{U})\,(\mathbf{v}\mathfrak{R}y)\}$, ran $(\mathfrak{R}) = \{y \in \mathcal{U} | \,(\exists \mathbf{v} \in \mathcal{U}^{n-1})\,(\mathbf{v}\mathfrak{R}y)\}$, fld $(\mathfrak{R}) =$ dom $(\mathfrak{R}) \cup$ ran $(\mathfrak{R})$, and $\widehat{\mathfrak{R}} = \{(y, \mathbf{v}) \in \mathcal{U}^n | \mathbf{v}\mathfrak{R}y\}$. It is thus obvious that $y\widehat{\mathfrak{R}}\mathbf{v} \Leftrightarrow \mathbf{v}\mathfrak{R}y$ and $\widehat{\widehat{\mathfrak{R}}} = \mathfrak{R}$.

Two important notions, for the purpose at hand, are the *image* and *preimage* of a set, with respect to an n-ary relation. These are defined as follows [9] and [27].

Definition 10.3 ***(Image and Preimage of a Relation)*** Let $\mathfrak{R}$ be an n-ary relation on a set $\mathcal{U}$, and for $(\mathbf{v}, y) \in \mathfrak{R}$, let $\mathbf{v} = (x_1, \dots, x_{n-1})$, with each x_k is restricted to vary on a set $X_k \subset \mathcal{U}$, that is, $\mathbf{v}$ is restricted to vary on a set $\mathbf{V} \subset \mathcal{U}^{n-1}$. Then, the image of $\mathbf{V}$ (or the image of the sets X_k) with respect to $\mathfrak{R}$, denoted $\mathrm{I}_{\mathfrak{R}}$, is defined to be

$$\begin{aligned} Y = \mathrm{I}_{\mathfrak{R}}(\mathbf{V}) &= \mathrm{I}_{\mathfrak{R}}(X_1, \dots, X_{n-1}) \\ &= \{y \in \mathcal{U} | \,(\exists \mathbf{v} \in \mathbf{V})\,(\mathbf{v}\mathfrak{R}y)\} \\ &= \{y \in \mathcal{U} | \,(\exists_{k=1}^{n-1} x_k \in X_k)\,((x_1, \dots, x_{n-1})\,\mathfrak{R}y)\}, \end{aligned}$$

where the set $\mathbf{V}$, called the preimage of Y, is defined to be the image of Y with respect to the converse relation $\widehat{\mathfrak{R}}$, that is

$$\mathbf{V} = \mathrm{I}_{\widehat{\mathfrak{R}}}(Y) = \{\mathbf{v} \in \mathcal{U}^{n-1} | \,(\exists y \in Y)\,(y\widehat{\mathfrak{R}}\mathbf{v})\}.$$

In accordance with this definition and the fact that $y\widehat{\mathfrak{R}}\mathbf{v} \Leftrightarrow \mathbf{v}\mathfrak{R}y$, we obviously have $Y = \mathrm{I}_{\mathfrak{R}}(\mathbf{V}) \Leftrightarrow \mathbf{V} = \mathrm{I}_{\widehat{\mathfrak{R}}}(Y)$.

Within this set-theoretical framework, a completely general definition of the notion of an *n-ary function* can be formulated. A set f is an n-ary function on a set $\mathcal{U}$ iff f is an $(n+1)$-ary relation on $\mathcal{U}$, and $(\forall \mathbf{v} \in \mathcal{U}^n)\,(\forall y, z \in \mathcal{U})\,(\mathbf{v}fy \wedge \mathbf{v}fz \Rightarrow y = z)$. Thus, an n-ary function is a *many-one* $(n+1)$-ary relation; that is, a relation, with respect to which, any element in its domain is related exactly to one element in its range. Getting down from relations to the particular case of functions, we have at hand the standard notation: $y = f(\mathbf{v})$ in place of $\mathbf{v}fy$. From the fact

2 Amer in [30] used the n-th Cartesian power of $\emptyset$ to define empty structures and axiomatized their first-order theory.

that an n-ary function is a special kind of relation, then all the preceding definitions and results, concerning the *domain*, *range*, *field*, and *converse* of a relation, apply to functions as well.

With some criteria satisfied, a function is called *invertible*. A function f has an inverse, denoted f^{-1}, iff its converse relation $\widehat{f}$ is a function, in which case $f^{-1} = \widehat{f}$. In other words, f is invertible if, and only if, it is an *injection* from its domain to its range, and obviously, the inverse f^{-1} is *unique*, from the fact that the converse relation is always *definable* and *unique*.

By means of the above concepts, we next define the notions of *partial* and *total operations*.

Definition 10.4 ***(Partial and Total Operations)*** Let $\mathcal{S}^n$ be the n-th Cartesian power of a set $\mathcal{S}$. An n-ary (total) operation on $\mathcal{S}$ is a total function $t_n : \mathcal{S}^n \mapsto \mathcal{S}$. An n-ary partial operation in $\mathcal{S}$ is a partial function $p_n : \mathcal{U} \mapsto \mathcal{S}$, where $\mathcal{U} \subset \mathcal{S}^n$. The ordinal n is called the arity of t_n or p_n.

A binary operation is an n-ary operation for $n = 2$. Addition and multiplication on the set $\mathbb{R}$ of real numbers are best-known examples of binary *total* operations, while division is a *partial* operation in $\mathbb{R}$.

A formalized theory is characterized by two things, an object language in which the theory is formalized (the symbolism of the theory) and a set of axioms. The following *metalinguistic* definition characterizes the notion of a formalized theory.

Definition 10.5 ***(Formalized Theories)*** Let $\mathfrak{L}$ be an object formal language. A formalized theory in $\mathfrak{L}$ (or an $\mathfrak{L}$-theory) is a set of $\mathfrak{L}$-sentences which is closed under its associated deductive apparatus. Let $\Lambda_{\mathfrak{T}}$ denote a finite set of $\mathfrak{L}$-sentences, and let φ denote an $\mathfrak{L}$-sentence. The formalized $\mathfrak{L}$-theory $\mathfrak{T}$ of the set $\Lambda_{\mathfrak{T}}$ is the deductive closure of $\Lambda_{\mathfrak{T}}$ under logical consequence, that is

$$\mathfrak{T} = \{\varphi \in \mathfrak{L} | \varphi \text{ is a consequence of } \Lambda_{\mathfrak{T}}\}.$$

The set $\Lambda_{\mathfrak{T}}$ is called the set of axioms (or postulates) of $\mathfrak{T}$.

A model (or an interpretation) of a theory $\mathfrak{T}$ is some particular (algebraic or relational) structure that satisfies every formula of $\mathfrak{T}$. The notion of a structure of a formalized language $\mathfrak{L}$ (or an $\mathfrak{L}$-structure) is characterized in the following definition (see [18] and [27]).

Definition 10.6 ***(Structures)*** Let $\mathfrak{L}$ be a formalized language (possibly with no individual constants). By an $\mathfrak{L}$-structure, we understand a system $\mathfrak{M} = \langle \mathcal{A}; \boldsymbol{F}^{\mathfrak{M}}; \boldsymbol{R}^{\mathfrak{M}} \rangle$, where

- $\mathcal{A}$ is a (possibly empty) set called the individuals universe of $\mathfrak{M}$. The elements of $\mathcal{A}$ are called the individual elements of $\mathfrak{M}$;
- $\boldsymbol{F}^{\mathfrak{M}}$ is a (possibly empty) set of finitary total operations on $\mathcal{A}$. The elements of $\boldsymbol{F}^{\mathfrak{M}}$ are called the $\mathfrak{M}$-operations;
- $\boldsymbol{R}^{\mathfrak{M}}$ is a (possibly empty) set of finitary relations on $\mathcal{A}$. The elements of $\boldsymbol{R}^{\mathfrak{M}}$ are called the $\mathfrak{M}$-relations.

An $\mathfrak{L}$-structure with an empty individual universe is called an *empty* $\mathfrak{L}$ *-structure*.[3] By a *many-sorted* $\mathfrak{L}$ *-structure*, we understand an $\mathfrak{L}$-structure with more than one universe set. By a *relational* $\mathfrak{L}$*-structure*, we understand an $\mathfrak{L}$-structure with a nonempty set of relations and an empty set of functions. By an *algebraic* $\mathfrak{L}$*-structure* (or an $\mathfrak{L}$ *-algebra*), we understand an $\mathfrak{L}$-structure with a nonempty set of functions. An $\mathfrak{L}$-algebra endowed with a compatible ordering relation $\mathfrak{R}$ is called an $\mathfrak{R}$-ordered $\mathfrak{L}$-algebra.

The following definition and its consequence are indispensable.

3 First-order logics with empty structures were first considered by Mostowski [33] and then studied by many logicians (see, e.g. [34, 35], and [30]). Such logics are now referred to as *free logics*.

Definition 10.7 ***(Isomorphism on Structures)*** Let $\mathfrak{A} = \langle \mathcal{A}; \boldsymbol{F}^{\mathfrak{A}}; \boldsymbol{R}^{\mathfrak{A}} \rangle$ and $\mathfrak{B} = \langle \mathcal{B}; \boldsymbol{F}^{\mathfrak{B}}; \boldsymbol{R}^{\mathfrak{B}} \rangle$ be two $\mathfrak{L}$-structures such that $\boldsymbol{F}^{\mathfrak{A}}$ (respectively $\boldsymbol{R}^{\mathfrak{A}}$) is of the same cardinality as $\boldsymbol{F}^{\mathfrak{B}}$ (respectively $\boldsymbol{R}^{\mathfrak{B}}$). An isomorphism from $\mathfrak{A}$ to $\mathfrak{B}$ is a mapping $\iota : \mathcal{A} \hookrightarrow \mathcal{B}$ such that

- ι is a bijection from $\mathcal{A}$ onto $\mathcal{B}$;
- ι is function-preserving, that is

$$(\forall u_1, \dots, u_n \in \mathcal{A})\, (\forall \mathbf{f}^{\mathfrak{A}} \in \boldsymbol{F}^{\mathfrak{A}})\, (\exists \mathbf{f}^{\mathfrak{B}} \in \boldsymbol{F}^{\mathfrak{B}})\, (\iota\, (\mathbf{f}^{\mathfrak{A}}(u_1, \dots, u_n)) \\ = \mathbf{f}^{\mathfrak{B}}(\iota\, (u_1), \dots, \iota\, (u_n)));$$

- ι is relation-preserving, that is

$$(\forall u_1, \dots, u_n \in \mathcal{A})\, (\forall \mathbf{r}^{\mathfrak{A}} \in \boldsymbol{R}^{\mathfrak{A}})\, (\exists \mathbf{r}^{\mathfrak{B}} \in \boldsymbol{R}^{\mathfrak{B}})\, (\mathbf{r}^{\mathfrak{A}}\, (u_1, \dots, u_n) \\ \Leftrightarrow \mathbf{r}^{\mathfrak{B}}\, (\iota\, (u_1), \dots, \iota\, (u_n))).$$

Then, $\mathfrak{A}$ and $\mathfrak{B}$ are isomorphic, in symbols $\mathfrak{A} \simeq \mathfrak{B}$.

Let the mapping $\iota : \mathcal{A} \hookrightarrow \mathcal{B}$ be an isomorphism. For each $u \in \mathcal{A}$, the image $\iota\, (u) \in \mathcal{B}$ is called the isomorphic copy of u (under the isomorphism ι).

Theorem 10.1 ***(Isomorphism is Equivalence)*** *Let $\mathfrak{S}$ be the class of all $\mathfrak{L}$-structures. The isomorphism relation $\simeq$ is an equivalence relation on $\mathfrak{S}$.*

Proof: The proof is easy by showing that the relation $\simeq$ is reflexive, symmetric, and transitive in $\mathfrak{S}$.

That is, isomorphism is a relation between models (interpretations) of a formalized theory. To say that two models of a theory are isomorphic is to say that they have the same *structural form*. In this sense, the isomorphism relation is called "*structural identity*" by some logicians (see, e.g. [36]).

The notion of *inverse elements* is precisely formulated in the following definition.

Definition 10.8 ***(Inverse Elements)*** Let $\langle \mathcal{A}; \bullet; \mathbf{e}_\bullet \rangle$ be an algebra with $\bullet$ is a binary operation on $\mathcal{A}$ and $\mathbf{e}_\bullet$ is the identity element for $\bullet$. We say that every element of $\mathcal{A}$ has an inverse element with respect to the operation $\bullet$ iff

$$(\forall x \in \mathcal{A})\, (\exists y \in \mathcal{A})\, (x \bullet y = \mathbf{e}_\bullet).$$

We conclude this section by characterizing some algebraic structures that will be deployed as intended interpretations of the theories considered in this text.

Definition 10.9 ***(Ringoid)*** A ringoid (or a ring-like structure) is a structure $\mathfrak{R} = \langle \mathcal{R}; +_{\mathcal{R}}, \times_{\mathcal{R}} \rangle$ with $+_{\mathcal{R}}$ and $\times_{\mathcal{R}}$ are total binary operations on the universe set $\mathcal{R}$. The operations $+_{\mathcal{R}}$ and $\times_{\mathcal{R}}$ are called, respectively, the addition and multiplication operations of the ringoid $\mathfrak{R}$.

Definition 10.10 ***(S-Ringoid)*** An S-ringoid (or a subdistributive ringoid) is a ringoid that satisfies at least one of the following subdistributive criteria.

i) $(\forall x, y, z \in \mathcal{R})\, (x \times_{\mathcal{R}} (y +_{\mathcal{R}} z) \subseteq x \times_{\mathcal{R}} y +_{\mathcal{R}} x \times_{\mathcal{R}} z)$,
ii) $(\forall x, y, z \in \mathcal{R})\, ((y +_{\mathcal{R}} z) \times_{\mathcal{R}} x \subseteq y \times_{\mathcal{R}} x +_{\mathcal{R}} z \times_{\mathcal{R}} x)$.

Criteria (i) and (ii) in the preceding definition are called, respectively, left and right subdistributivity (or S-distributivity).

Definition 10.11 ***(Semiring)*** Let $\mathfrak{R} = \langle \mathcal{R}; +_{\mathcal{R}}, \times_{\mathcal{R}} \rangle$ be a ringoid. $\mathfrak{R}$ is said to be a semiring iff

i) $\langle \mathcal{R}; +_{\mathcal{R}} \rangle$ is a commutative monoid with identity element $0_{\mathcal{R}}$,
ii) $\langle \mathcal{R}; \times_{\mathcal{R}} \rangle$ is a monoid with identity element $1_{\mathcal{R}}$,
iii) Multiplication, $\times_{\mathcal{R}}$, left and right distributes over addition, $+_{\mathcal{R}}$,
iv) $0_{\mathcal{R}}$ is an absorbing element for $\times_{\mathcal{R}}$.

A commutative semiring is one whose multiplication is commutative.

Definition 10.12 ***(S-Semiring)*** An S-semiring (or a subdistributive semiring) is an S-ringoid that satisfies criteria (i), (ii), and (iv) in Definition 10.11. A commutative S-semiring is one whose multiplication is commutative.

Noteworthy, the notion of S-semiring is a generalization of the notion of a *near*-semiring; a near-semiring is a ringoid that satisfies the criteria of a semiring except that it is *either* left or right distributive (for detailed discussions of near-semirings and related concepts, the interested reader may consult, e.g. [37–39]).

Finally, let us close this section by making two further definitions we shall need.

Definition 10.13 ***(Number System)*** A number system is a ringoid $\mathfrak{N} = \langle \mathcal{N}; +_{\mathcal{N}}, \times_{\mathcal{N}} \rangle$ with $+_{\mathcal{N}}$ and $\times_{\mathcal{N}}$ are each both commutative and associative, and $\times_{\mathcal{N}}$ distributes over $+_{\mathcal{N}}$.

Definition 10.14 ***(S-Number System)*** An S-number system (or a subdistributive number system) is an S-ringoid $\mathfrak{N} = \langle \mathcal{N}; +_{\mathcal{N}}, \times_{\mathcal{N}} \rangle$ with $+_{\mathcal{N}}$ and $\times_{\mathcal{N}}$ are each both commutative and associative.

10.4 Classical Intervals and the Dependency Problem

With the formalized apparatus of Section 10.3 at our disposal, the main business of this section is to give a systematic characterization of the theory $\mathsf{Th}_{\mathcal{I}}$ of real interval arithmetic. Our adopted strategy to obtain a concrete system of classical interval numbers is to start with the field of real numbers and to "*intervalize*" it, by defining new interval relations and operations. In other words, the theory $\mathsf{Th}_{\mathcal{I}}$ of real intervals will be constructed as a *definitional extension* of the theory of real numbers.

There are many theories of interval arithmetic (see, e.g. [17, 20, 40–45]). We are here interested in characterizing classical interval arithmetic as introduced in, e.g. [1, 4, 12, 18, 46], and [27]. Hereafter and throughout this work, the machinery used, and assumed priori, is the *standard* (*classical*) predicate calculus and axiomatic set theory. Moreover, in all the proofs, elementary facts about operations and relations on the real numbers are usually used without explicit reference.

A theory $\mathsf{Th}_{\mathcal{I}}$ of a *real interval algebra* (a *classical interval algebra* or an *interval algebra over the real field*) is characterized in the following definition (see [1] and [27]).

Definition 10.15 ***(Theory of Real Interval Algebra)*** Take $\sigma = \{+, \times; -, ^{-1}; 0, 1\}$ as a set of nonlogical constants and let $\mathsf{R} = \langle \mathbb{R}; \sigma^{\mathsf{R}} \rangle$ be the totally $\leq$-ordered field of real numbers. The theory $\mathsf{Th}_{\mathcal{I}}$ of an interval algebra over the field R is the theory of a two-sorted structure $\mathfrak{I}_{\mathbb{R}} = \langle \mathcal{I}_{\mathbb{R}}; \mathbb{R}; \sigma^{\mathfrak{I}_{\mathbb{R}}} \rangle$ prescribed by the following set of axioms.

(I1) $(\forall X \in \mathcal{I}_{\mathbb{R}})\,(X = \{x \in \mathbb{R} | (\exists \underline{x} \in \mathbb{R})\,(\exists \overline{x} \in \mathbb{R})\,(\underline{x} \leq_{\mathbb{R}} x \leq_{\mathbb{R}} \overline{x})\})$,
(I2) $(\forall X, Y \in \mathcal{I}_{\mathbb{R}})\,(\circ \in \{+, \times\} \Rightarrow X \circ_{\mathcal{I}_{\mathbb{R}}} Y = \{z \in \mathbb{R} | (\exists x \in X)\,(\exists y \in Y)\,(z = x \circ_{\mathbb{R}} y)\})$,
(I3) $(\forall X \in \mathcal{I}_{\mathbb{R}})\,(\lozenge \in \{-\} \vee (\lozenge \in \{^{-1}\} \wedge 0_{\mathcal{I}_{\mathbb{R}}} \notin X) \Rightarrow \lozenge_{\mathcal{I}_{\mathbb{R}}} X = \{z \in \mathbb{R} | (\exists x \in X)\,(z = \lozenge_{\mathbb{R}} x)\})$.

The sentence (I1) of Definition 10.15 characterizes what an interval number (or a closed $\mathbb{R}$-interval) is. The sentences (I2) and (I3) prescribe, respectively, the binary and unary operations for $\mathbb{R}$-intervals. Hereafter, the uppercase Roman letters X, Y, and Z (with or without subscripts), or equivalently $[\underline{x}, \overline{x}]$, $[\underline{y}, \overline{y}]$, and $[\underline{z}, \overline{z}]$, shall be employed as variable symbols to denote real interval numbers. A *point* (*singleton*) interval number $\{x\}$ shall be denoted by $[x]$. The letters A, B, and C, or equivalently $[\underline{a}, \overline{a}]$, $[\underline{b}, \overline{b}]$, and $[\underline{c}, \overline{c}]$, shall be used to denote constants of $\mathcal{I}_{\mathbb{R}}$. Also, we shall single out the symbols $1_{\mathcal{I}}$ and $0_{\mathcal{I}}$ to denote, respectively, the singleton $\mathbb{R}$-intervals $\{1_{\mathbb{R}}\}$ and $\{0_{\mathbb{R}}\}$. For the purpose at hand, it is convenient to define two proper subsets of $\mathcal{I}_{\mathbb{R}}$: the sets of *symmetric interval numbers* and *point interval numbers*. Respectively, these are defined and denoted by

$$\mathcal{I}_S = \{X \in \mathcal{I}_{\mathbb{R}} | (\exists x \in \mathbb{R})(0 \leq x \wedge X = [-x, x])\}$$
$$\mathcal{I}_{[x]} = \{X \in \mathcal{I}_{\mathbb{R}} | (\exists x \in \mathbb{R})(X = [x, x])\}.$$

From the fact that real intervals are totally $\leq_{\mathbb{R}}$-ordered subsets of $\mathbb{R}$, equality of $\mathbb{R}$-intervals follows immediately from the axiom of extensionality[4] of set theory. That is,

$$[\underline{x}, \overline{x}] =_{\mathcal{I}} [\underline{y}, \overline{y}] \Leftrightarrow \underline{x} =_{\mathbb{R}} \underline{y} \wedge \overline{x} =_{\mathbb{R}} \overline{y}.$$

Since $\mathbb{R}$-intervals are *ordered sets* of real numbers, it follows that the next theorem is derivable from Definition 10.15 (see [1] and [47]).

Theorem 10.2 ***(Interval Operations)*** *For any two interval numbers* $[\underline{x}, \overline{x}]$ *and* $[\underline{y}, \overline{y}]$, *the binary and unary interval operations are formulated in terms of the intervals' endpoints as follows.*

i) $[\underline{x}, \overline{x}] +_{\mathcal{I}} [\underline{y}, \overline{y}] = [\underline{x} +_{\mathbb{R}} \underline{y}, \overline{x} +_{\mathbb{R}} \overline{y}]$,
ii) $[\underline{x}, \overline{x}] \times_{\mathcal{I}} [\underline{y}, \overline{y}] = [\min\{\underline{x}\times_{\mathbb{R}}\underline{y}, \underline{x}\times_{\mathbb{R}}\overline{y}, \overline{x}\times_{\mathbb{R}}\underline{y}, \overline{x}\times_{\mathbb{R}}\overline{y}\}, \max\{\underline{x}\times_{\mathbb{R}}\underline{y}, \underline{x}\times_{\mathbb{R}}\overline{y}, \overline{x}\times_{\mathbb{R}}\underline{y}, \overline{x}\times_{\mathbb{R}}\overline{y}\}]$,
iii) $-_{\mathcal{I}}[\underline{x}, \overline{x}] = [-_{\mathbb{R}}\overline{x}, -_{\mathbb{R}}\underline{x}]$,
iv) $0_{\mathcal{I}} \notin [\underline{x}, \overline{x}] \Rightarrow [\underline{x}, \overline{x}]^{-1_{\mathcal{I}}} = [\overline{x}^{-1_{\mathbb{R}}}, \underline{x}^{-1_{\mathbb{R}}}]$,

where min *and* max *are, respectively, the* $\leq_{\mathbb{R}}$*-minimal and* $\leq_{\mathbb{R}}$*-maximal.*

Wherever there is no confusion, we shall drop the subscripts $\mathcal{I}$ and $\mathbb{R}$. By Definition 10.4, it is obvious that all the interval operations, except interval reciprocal, are *total* operations. The additional operations of interval subtraction and division can be defined, respectively, as $X - Y = X + (-Y)$ and $X \div Y = X \times (Y^{-1})$.

Classical interval arithmetic has a number of peculiar algebraic properties: The point intervals $0_{\mathcal{I}}$ and $1_{\mathcal{I}}$ are identity elements for addition and multiplication, respectively; interval addition and multiplication are both commutative and associative; interval addition is cancellative; interval multiplication is cancellative only for zeroless intervals; an interval number is invertible for addition (respectively, multiplication) if and only if it is a point interval (respectively, a nonzero point interval); and interval multiplication left and right *subdistributes* over interval addition (see Definition 10.10). Accordingly, by Definitions 10.12 and 10.14, we have the following theorem and its corollary (see [1] and [18]).

Theorem 10.3 ***(S-Semiring of Classical Intervals)*** *The structure* $\langle \mathcal{I}_{\mathbb{R}}; +_{\mathcal{I}}, \times_{\mathcal{I}}; 0_{\mathcal{I}}, 1_{\mathcal{I}} \rangle$ *of classical interval numbers is a commutative S-semiring.*

Corollary 10.1 ***(Number System of Classical Intervals)*** *The theory of classical intervals defines an S-number system on the set* $\mathcal{I}_{\mathbb{R}}$.

Throughout this text, we shall employ the following theorem and its corollary (see [1] and [9]).

---

4 The axiom of extensionality asserts that two sets are equal if, and only if, they have precisely the same elements, that is, for any two sets $\mathcal{S}$ and $\mathcal{T}$, $\mathcal{S} = \mathcal{T} \Leftrightarrow (\forall z)(z \in \mathcal{S} \Leftrightarrow z \in \mathcal{T})$.

Theorem 10.4 ***(Inclusion Monotonicity in Classical Intervals)*** *Let X_1, X_2, Y_1, and Y_2 be interval numbers such that $X_1 \subseteq Y_1$ and $X_2 \subseteq Y_2$. Then, for any binary operation $\circ \in \{+, \times\}$ and any definable unary operation $\Diamond \in \{-, ^{-1}\}$, we have*

i) $X_1 \circ_{\mathcal{I}} X_2 \subseteq Y_1 \circ_{\mathcal{I}} Y_2$,
ii) $\Diamond_{\mathcal{I}} X_1 \subseteq \Diamond_{\mathcal{I}} Y_1$.

In consequence of this theorem, from the fact that $[x, x] \subseteq X \Leftrightarrow x \in X$, we have the following important special case.

Corollary 10.2 ***(Membership Monotonicity for Classical Intervals)*** *Let X and Y be real interval numbers with $x \in X$ and $y \in Y$. Then, for any binary operation $\circ \in \{+, \times\}$ and any definable unary operation $\Diamond \in \{-, ^{-1}\}$, we have*

i) $x \circ_{\mathbb{R}} y \in X \circ_c Y$,
ii) $\Diamond_{\mathbb{R}} x \in \Diamond_c X$.

If we endow the classical interval algebra $\langle \mathcal{I}_{\mathbb{R}}; +_{\mathcal{I}}, \times_{\mathcal{I}}; 0_{\mathcal{I}}, 1_{\mathcal{I}} \rangle$ with the *compatible* partial ordering $\subseteq$, then we have a partially ordered commutative S-semiring. In addition to ordering intervals by the set inclusion relation $\subseteq$, there are many orders presented in the interval literature. Among these is Moore's partial ordering, which is defined by $[\underline{x}, \overline{x}] <_M [\underline{y}, \overline{y}] \Leftrightarrow \overline{x} <_{\mathbb{R}} \underline{y}$. In contrast to the case for $\subseteq$, Moore's partial ordering $<_M$ is not compatible with the algebraic operations on $\mathcal{I}_{\mathbb{R}}$ (see [1] and [27]).

We conclude the present section with a peculiarity of interval arithmetic that seems quite strange at first. Note that the set-theoretic characterization of the interval operations (Definition 10.15) implies that interval arithmetic considers *all instances* of variables as *independent*. Accordingly, for two interval variables X and Y assigned the same interval constant A, both the interval operations $X \circ_{\mathcal{I}} X$ and $X \circ_{\mathcal{I}} Y$ are equal and they are the same as the image of the *multivariate* real function $f_{\text{ind}}(x, y) = x \circ_{\mathbb{R}} y$, with $x \in A$ and $y \in A$. In fact, this is one of the strengths of interval mathematics: because images of real functions are inclusion monotonic (see, e.g. [9, 27], and [48]), it follows that the image of the function f_{ind} is an enclosure of the image of a unary real function $f_{\text{dep}}(x) = x \circ_{\mathbb{R}} x$, with $x \in A$, and therefore $X \circ_{\mathcal{I}} X = X \circ_{\mathcal{I}} Y$ is a *guaranteed enclosure* of the image of f_{dep}. However, in many situations, this enclosure might be too wide to be useful. This phenomenon is known as the *interval dependency problem*. The notion of interval dependency and the problems thereof will be logically characterized in the next section.

10.5 Interval Dependency: A Logical Treatment

The notion of *dependency* comes from the notion of a *function*. There is scarcely a mathematical theory which does not involve the notion of a function. In ancient mathematics, the idea of *functional dependence* was not expressed explicitly and was not an independent object of research, although a wide range of specific functional relations were known and studied systematically. The concept of a function appears in a rudimentary form in the works of scholars in the Middle Ages, but only in the work of mathematicians in the seventeenth century, and primarily in those of Pierre de Fermat, Rene Descartes, Isaac Newton, and Gottfried Leibniz, did it begin to take shape as an independent concept. Later, in the eighteenth century, Euler had a more general approach to the concept of a function as "*dependence of one variable quantity on another*" [49]. By the year 1834, Lobachevskii was writing: "The general concept of a function requires that a function of x is a number which is given for each x and gradually changes with x. The value of a function can be given either by an analytic expression or by a condition which gives a means of testing all numbers and choosing one of them; or finally a *dependence* can exist and *remain unknown*" [50].

In the theory of real closed intervals, the notion of interval dependency naturally comes from the idea of functional dependence of real variables. Despite the fact that dependency is an essential and useful notion of real variables, interval dependency is the *main unsolved problem* of the classical theory of interval arithmetic and its modern generalizations (see [1] and [27]). This, naturally, confronts us with the question: *Formally, what is interval dependency*? Is it a metaconcept or an object ingredient of interval and fuzzy computations? In other words, what are the fundamental defining properties that characterize the notion of interval dependency as a formal mathematical object? Since the early works on interval mathematics by Burkill [5] and Young [6] in the dawning of the twentieth century, this question has never been touched upon and remained a question still today unanswered. Although the notion of interval dependency is widely used in the interval and fuzzy literature, it is only illustrated by example, without explicit formalization, and no attempt has been made to put on a systematic basis its meaning, that is, to formally indicate the criteria by which it is to be characterized. Here, we attempt to answer this long-standing question. This section, therefore, is devoted to presenting a complete *systematic formalization* of the notion of interval dependency by means of the notions of *Skolemization* and *logical quantification dependence*. A novelty of this formalization is the expression of interval dependency as a *logical predicate* (or *relation*) and thus gaining the advantage of deducing its fundamental properties in a merely logical manner. This result sheds new light on many problems of interval mathematics. Moreover, a breakthrough behind our formalization of interval dependency is that it paves the way and provides the systematic apparatus for developing alternate *dependency-aware* interval theories and computational methods with mathematical constructions that better account for dependencies between quantifiable uncertainties of the real world.

So, in Section 10.5.1, we delve deep into some semantical and syntactical fundamentals concerning the logical formulation of the notion of functional dependence and some related notions. In Section 10.5.2, we put on a systematic basis the notion of interval dependency. This systematic basis shall play an essential role in our discussion of the interval theory, in the present section and later on.[5] Based on the logical machinery presented here, we attempt, in Sections 10.8 and 10.9, to present an alternate theory of intervals, namely, the "*theory of universal intervals*," with a mathematical construction that tries to avoid some of the defects in the current theories of interval arithmetic and to better account for the notion of interval dependency.

10.5.1 Quantification Dependence and Skolemization

In order to be able to formalize the notion of interval dependency, we delve deep, in this section, into some semantical and syntactical fundamentals concerning the logical formulation of the notion of functional dependence and some related notions.

When scientists observe the world to formulate the defining properties of some physical phenomenon, these defining properties figure as *attributes* (*variables*) depending on some other attributes. Translating this dependence into a formal mathematical language gives rise to the notion of functional dependence: "*a variable* y *is absolutely determined by some given variables* $x_1, \dots, x_n$," or "*a variable* y *is a function of some given variables* $x_1, \dots, x_n$," symbolically $y = f(x_1, \dots, x_n)$. In some cases, such a translation can deterministically result in a certain rule for the function f, for instance, $y = x_1 + \cdots + x_n$. In other cases, we have an approximate rule for f, or we know that a dependence exist, but the rule cannot be determined, in which case, we write the general usual notation $y = f(x_1, \dots, x_n)$, without specifying explicitly a rule for the function f. Therefore, in mathematics, *a dependence is formally a function* (for further exhaustive details about the notion of dependence, from the logical and epistemological viewpoints, see, e.g. [24, 51–53], and [27]).

5 The notions, notations, and abbreviations of this section are indispensable for our mathematical discussion throughout the succeeding sections and hereafter are assumed priori, without further mention.

As is well known, the most elementary part of all mathematical sciences is formal logic. Therefore, getting down to the most elementary fundamentals, it can be clarified that in all mathematical theories, any type of dependence can be reduced to the following simple logical definition (see, e.g. [9] and [27]).

Definition 10.16 ***(Quantification Dependence)*** Let $\mathcal{Q}$ be a quantification matrix[6] and let $\varphi(x_1, \dots, x_m; y_1, \dots, y_n)$ be a quantifier-free formula. For any universal quantification $(\forall x_i)$ and any existential quantification $(\exists y_j)$ in $\mathcal{Q}$, the variable y_j is dependent on the variable x_i in the prenex sentence[7] $\mathcal{Q}\varphi$ iff $(\exists y_j)$ is in the scope of $(\forall x_i)$ in $\mathcal{Q}$. Otherwise, x_i and y_j are independent.

That is, the order of quantifiers in a quantification matrix determines the mutual dependence between the variables in a sentence.

Let us illustrate this by the following two examples.

Example 10.1 Consider the prenex sentence

$$(\exists x)(\forall y)(\exists z)(y = x \circ y \wedge x = z \circ y).$$

which asserts that there exists an identity element x, for the operation $\circ$, with respect to which every element possesses an inverse z.

According to the order in which quantifiers are written, the variable z depends only on y, while there is no dependency between x and y or between x and z.

Example 10.2 In the prenex sentence

$$(\forall x)(\exists y)(\forall z)(\exists u)\varphi(x, y, z, u).$$

the variable y depends on x and the variable u depends on both x and z.

By means of a *Skolem equivalent form* or a *Skolemization*,[8] a quantification dependence is translated into a functional dependence. The notion of a Skolem equivalent form is characterized in the following definition (see, e.g. [27, 55], and [53]).

Definition 10.17 ***(Skolem Equivalent Form)*** Let σ be a sentence that takes the prenex form

$$(\forall_{i=1}^{m} x_i)(\exists_{j=1}^{n} y_j)\varphi(x_1, \dots, x_m; y_1, \dots, y_n).$$

where φ is a quantifier-free formula.

The Skolem equivalent form of σ is defined to be

$$(\exists_{j=1}^{n} f_j)(\forall_{i=1}^{m} x_i)\varphi(x_1, \dots, x_m; f_1, \dots, f_n).$$

where $f_j(x_1, \dots, x_m) = y_j$ are the dependency functions of y_j upon $x_1, \dots, x_m$, for $i \in \{1, \dots, m\}$ and $j \in \{1, \dots, n\}$.

It comes therefore as no surprise that in all mathematics, any instance of a dependence is, in fact, a functional dependence.

In order to clarify the matters, let us consider the following example.

Example 10.3 ***(Skolemization of a Sentence)*** Let a sentence σ take the prenex form

$$(\forall x)(\exists y)(\forall z)(\exists u)\varphi(x, y, z, u).$$

---

6 A *quantification matrix* $\mathcal{Q}$ is a sequence $(Q_1 x_1)\dots(Q_n x_n)$, where $x_1, \dots, x_n$ are variable symbols and each Q_i is $\forall$ or $\exists$.
7 A *prenex sentence* is a sentence of the form $\mathcal{Q}\varphi$, where $\mathcal{Q}$ is a quantification matrix and φ is a quantifier-free formula.
8 *Skolemization* is named after the Norwegian logician Thoralf Skolem (1887–1963), who first presented the notion in [54].

The Skolem equivalent form of σ is

$$(\exists f)(\exists g)(\forall x)(\forall z)\varphi(x, f(x), z, g(x, z)).$$

10.5.2 A Formalization of the Notion of Interval Dependency

With the logical and set-theoretical fundamentals of Sections 10.3 and 10.5.1 at our disposal, this section is devoted to presenting a formalized treatment of the notion of interval dependency, that is, putting on a *systematic basis* its meaning and thus gaining the advantage of indicating formally the criteria by which it is to be characterized and, accordingly, deducing its fundamental properties in a merely logical manner.

Before we proceed, it is convenient here to introduce some notational conventions. By a *finitary* real-valued function in real arguments (in short, a *real function* or $\mathbb{R}$-function), we understand a function $f_{\mathbb{R}}: \mathcal{D}_{\mathbb{R}} \subseteq \mathbb{R}^n \mapsto \mathbb{R}$, and by an *interval function* (or $\mathcal{I}$-function), we understand a function $f_{\mathcal{I}}: \mathcal{D}_{\mathcal{I}} \subseteq \mathcal{I}^n \mapsto \mathcal{I}$. The $\mathbb{R}$-subscripted letters $f_{\mathbb{R}}$, $g_{\mathbb{R}}$, $h_{\mathbb{R}}$ shall be employed to denote real-valued functions, while the $\mathcal{I}$-subscripted letters $f_{\mathcal{I}}$, $g_{\mathcal{I}}$, $h_{\mathcal{I}}$ shall be employed to denote interval-valued functions. If the type of function is clear from its arguments, and if confusion is not likely to ensue, we shall usually drop the subscripts "$\mathbb{R}$" and "$\mathcal{I}$." Thus, we may, for instance, write $f(x_1, \ldots, x_n)$ and $f(X_1, \ldots, X_n)$ for, respectively, a real-valued function and an interval-valued function, both of which are defined by the same rule.

An important notion we shall need is that of the image set of real closed intervals, under an n-ary real-valued function. This notion is a special case of that of the corresponding $(n+1)$-ary relation on $\mathbb{R}$. More precisely, we have the following definition.

Definition 10.18 ***(Image of Real Closed Intervals)*** Let f be an n-ary function on $\mathbb{R}$, and for $(\mathbf{v}, y) \in f$, let $\mathbf{v} = (x_1, \ldots, x_n)$, with each x_k is restricted to vary on a real closed interval $X_k \subset \mathbb{R}$, that is, $\mathbf{v}$ is restricted to vary on a set $\mathbf{V} \subset \mathbb{R}^n$. Then, the image of the closed intervals X_k with respect to f, denoted I_f, is defined to be

$$\begin{aligned} Y = \mathrm{I}_f(\mathbf{V}) &= \mathrm{I}_f(X_1, \ldots, X_n) \\ &= \{y \in \mathbb{R} \mid (\exists \mathbf{v} \in \mathbf{V})(\mathbf{v} f y)\} \\ &= \{y \in \mathbb{R} \mid (\exists_{k=1}^n x_k \in X_k)(y = f(x_1, \ldots, x_n))\} \subseteq \mathbb{R}. \end{aligned}$$

where the set $\mathbf{V}$, called the preimage[9] of Y, is defined to be the image of Y with respect to the converse relation $\widehat{f}$, that is

$$\mathbf{V} = \mathrm{I}_{\widehat{f}}(Y) = \{\mathbf{v} \in \mathbb{R}^n \mid (\exists y \in Y)(y \widehat{f} \mathbf{v})\}.$$

The fact formulated in the following theorem is well-known (see, e.g. [9] and [27]).

Theorem 10.5 ***(Extreme Value Theorem)*** *Let X_k be real closed intervals and let $f(x_1, \ldots, x_n)$ be an n-ary real-valued function with $x_k \in X_k$. If f is continuous in X_k, in symbols* $\mathrm{Cont}(f, X_k)$, *then f must attain its minimum and maximum value, that is*

$$\begin{aligned} (\forall f)(\mathrm{Cont}(f, X_k) \Rightarrow (\exists_{k=1}^n a_k \in X_k)(\exists_{k=1}^n b_k \in X_k)(\forall_{k=1}^n x_k \in X_k) \\ (f(a_1, \ldots, a_n) \le f(x_1, \ldots, x_n) \le f(b_1, \ldots, b_n))). \end{aligned}$$

where $\min_{x_k \in X_k} f = f(a_1, \ldots, a_n)$ and $\max_{x_k \in X_k} f = f(b_1, \ldots, b_n)$ are, respectively, the minimum and maximum of f.

An immediate consequence of Definition 10.18 and Theorem 10.5 is the following important property [27].

---

9 From the fact that the converse relation $\widehat{f}$ is always definable, the preimage of a function f is always definable, regardless of the definability of the inverse function f^{-1}.

Theorem 10.6 ***(Main Theorem of Image Evaluation)*** *Let an n-ary real-valued function f be continuous in the real closed intervals X_k. The (accurate) image $\mathrm{I}_f(X_1, \dots, X_n)$, of X_k, is in turn a real closed interval such that*

$$\mathrm{I}_f(X_1, \dots, X_n) = [\min_{x_k \in X_k} f(x_1, \dots, x_n), \max_{x_k \in X_k} f(x_1, \dots, x_n)].$$

A cornerstone result from the above theorem, that should be stressed at once, is that the *best* way to evaluate the *accurate* image of a continuous real-valued function is to apply minimization and maximization directly to determine the exact lower and upper endpoints of the image. For rational[10] real-valued functions, this optimization problem is, in general, computationally solvable, by applying *Tarski's algorithm*, which is also known as *Tarski's real quantifier elimination* (see, e.g. [56] and [57]). For algebraic[11] real-valued functions, the problem is computable, by applying the cylindrical algebraic decomposition algorithm (CAD algorithm, or *Collins' algorithm*),[12] which is a more effective version of Tarski's algorithm (see, e.g. [59] and [60]).

Before turning to the notion of interval dependency, we first prove the following indispensable result.

Theorem 10.7 ***(Image Inclusions in Prenex Sentences)*** *Let σ_1 and σ_2 be the two prenex sentences such that*

$$\sigma_1 \Leftrightarrow (\forall_{i=1}^{m} x_i \in X_i)\,(\exists_{j=1}^{n} y_j \in Y_j)\,(\exists z)\,(z = f(x_1, \dots, x_m; y_1, \dots, y_n))$$

$$\sigma_2 \Leftrightarrow (\forall_{i=1}^{m} x_i \in X_i)\,(\forall_{j=1}^{n} y_j \in Y_j)\,(\exists z)\,(z = f(x_1, \dots, x_m; y_1, \dots, y_n)).$$

where X_i and Y_j are real closed intervals, and f is a continuous real-valued function with $x_i \in X_i$ and $y_j \in Y_j$.

If $\mathrm{I}_f^{\sigma_1}$ and $\mathrm{I}_f^{\sigma_2}$ are the images of f, respectively, in σ_1 and σ_2, then $\mathrm{I}_f^{\sigma_1} \subseteq \mathrm{I}_f^{\sigma_2}$.

Proof: According to Definition 10.16, in the sentence σ_1, all y_j are dependent on all x_i, and in the sentence σ_2, all x_i and y_j are pairwise independent.

By Definition 10.17, there are some functions $g_j(x_1, \dots, x_m)$ such that σ_1 has the Skolem equivalent form

$$(\exists_{j=1}^{n} g_j)\,(\forall_{i=1}^{m} x_i \in X_i)\,(\exists z)\,(z = f(x_1, \dots, x_m; g_1, \dots, g_n)).$$

Finally, employing Theorem 10.6, we therefore have $\mathrm{I}_f^{\sigma_1} \subseteq \mathrm{I}_f^{\sigma_2}$.

From the fact that existential quantification over a nonempty set $\mathcal{S}$ defines a set $\mathcal{T}$ such that $\mathcal{T} \subseteq \mathcal{S}$, the previous theorem entails, as a special case, the following important result of real analysis.

Corollary 10.3 ***(Inclusion Monotonicity of Real Images)*** *Let $\mathbf{V}_X = (X_1, \dots, X_i, \dots, X_n)$ and $\mathbf{V}_Y = (Y_1, \dots, Y_i, \dots, Y_n)$ be two preimages of a continuous real-valued function f. Then, the image I_f is inclusion monotonic. That is*

$$(\forall_{i=1}^{n} X_i, Y_i)\,(X_i \subseteq Y_i \Rightarrow \mathrm{I}_f(\mathbf{V}_X) \subseteq \mathrm{I}_f(\mathbf{V}_Y)).$$

The following example makes the statement of Theorem 10.7 clear.

Example 10.4 ***(Image Inclusion in Two Prenex Sentences)*** Let σ_1 and σ_2 be the two prenex sentences such that

$$\sigma_1 \Leftrightarrow (\forall x \in [1,2])\,(\exists y \in [1,2])\,(\exists z \in \mathbb{R})\,(z = f(x,y) = y - x)$$

$$\sigma_2 \Leftrightarrow (\forall x \in [1,2])\,(\forall y \in [1,2])\,(\exists z \in \mathbb{R})\,(z = f(x,y) = y - x).$$

---

10 A *rational* real-valued function is a function obtained by means of a finite number of the basic real algebraic operations $\circ_{\mathbb{R}} \in \{+, \times\}$ and $\lozenge_{\mathbb{R}} \in \{-, ^{-1}\}$.

11 An *algebraic* function is a function that satisfies a polynomial equation whose coefficients are polynomials with rational coefficients.

12 The *CAD algorithm* is efficient enough for being one of the most important optimization algorithms of computational real algebraic geometry (see, e.g. [58]).

In the sentence σ_1, the variable y depends on x, and therefore there is some function $g(x)$ such that σ_1 has the Skolem equivalent form

$$(\exists g)(\forall x \in [1,2])(\exists z \in \mathbb{R})(z = f(x, g(x)) = g(x) - x).$$

Let g be the identity function. Consequently, the image of f in σ_1 is $\mathrm{I}_f^{\sigma_1} = \{0\}$.

Obviously, the image of f in σ_2 is $\mathrm{I}_f^{\sigma_2} = [-1, 1]$, and therefore $\mathrm{I}_f^{\sigma_1} \subseteq \mathrm{I}_f^{\sigma_2}$.

Next, we define the notion of an *exact* (or *generalized*) interval operation.

Definition 10.19 ***(Exact Interval Operation)*** Let $\circ_{\mathbb{R}} \in \{+, \times\}$ be a binary real operation, and let $\mathrm{I}_f = \mathrm{I}_f^{\sigma_{\text{Dep}}} \vee \mathrm{I}_f = \mathrm{I}_f^{\sigma_{\text{Ind}}}$, where $\mathrm{I}_f^{\sigma_{\text{Dep}}}$ and $\mathrm{I}_f^{\sigma_{\text{Ind}}}$ are the images of a function f for two real closed intervals X and Y in, respectively, two prenex sentences σ_{Dep} and σ_{Ind} such that

$$\sigma_{\text{Dep}} \Leftrightarrow (\forall x \in X)(\exists y \in Y)(\exists z \in \mathbb{R})(z = f(x, y) = x \circ_{\mathbb{R}} y)$$
$$\sigma_{\text{Ind}} \Leftrightarrow (\forall x \in X)(\forall y \in Y)(\exists z \in \mathbb{R})(z = f(x, y) = x \circ_{\mathbb{R}} y).$$

Then, an exact interval operation $\circ_{\mathcal{J}} \in \{+, \times\}$ is defined by

$$X \circ_{\mathcal{J}} Y = \mathrm{I}_f(X, Y).$$

We then have the following obvious result for the classical interval operations.

Theorem 10.8 ***(Inexactness of Classical Interval Operations)*** *The value of a classical interval operation $X \circ_{\mathcal{I}} Y$ is exact only when the real variables $x \in X$ and $y \in Y$ are independent, that is*

$$X \circ_{\mathcal{I}} Y = \mathrm{I}_f^{\sigma_{\text{Ind}}}(X, Y).$$

Proof: The theorem is immediate from Definition 10.15 of the classical interval algebra.

With the help of the preceding notions and the deductions from them, we are now ready to pass to our formal characterization of the notion of interval dependency.

Definition 10.20 ***(Interval Dependency Relation)*** Let $S_1, \dots, S_m$ be some arbitrary real closed intervals. For two interval variables X and Y, we say that Y is dependent on X, in symbols $Y \mathfrak{D} X$, iff there is some given real-valued function f such that Y is the image of $(X; S_1, \dots, S_m)$ with respect to f. That is

$$Y \mathfrak{D} X \Leftrightarrow Y = \mathrm{I}_f(X; S_1, \dots, S_m).$$

where f is called the dependency function of Y on X. Otherwise, Y is not dependent on X, in symbols $Y \mathfrak{I} X$, that is

$$Y \mathfrak{I} X \Leftrightarrow \neg\, Y \mathfrak{D} X \Leftrightarrow \neg\, Y = \mathrm{I}_f(X; S_1, \dots, S_m).$$

From now on, and throughout the text, the following notational convention shall be adopted. We write $Y \mathfrak{D}_f X$ (with the subscript f) to mean that Y is dependent on X by some given dependency function f, and we write $\mathfrak{I}(X, Y)$ to mean that X and Y are mutually independent. In general, the notation $\mathfrak{I}(X_1, \dots, X_n)$ shall be employed to mean "all $X_1, \dots, X_n$ are pairwise mutually independent." Hereafter, for simplicity of the language, we shall always make use of the following abbreviation.

$$\mathfrak{I}_{k=1}^{n}(X_k) \Leftrightarrow \mathfrak{I}(X_1, \dots, X_n).$$

So, to say that an interval variable Y is dependent on an interval variable X, we must be *given* some real-valued function f such that Y is the image of X under f. This characterization of interval dependency is completely compatible with the concept of functional dependence of real variables: for two real variables x and y, the variable y is *functionally* dependent on x if there is some *given* function f such that $y = f(x)$, and to keep the dependency information, between x and y, in an algebraic expression $x \circ_{\mathbb{R}} y$, it suffices to write $x \circ_{\mathbb{R}} f(x)$. If x and y are mutually dependent by an *idempotence* $y = f(x)$ and $x = g(y)$, then, to keep the dependency information, it suffices to write either $x \circ_{\mathbb{R}} f(x)$ or $g(y) \circ_{\mathbb{R}} y$. In case there is neither such a given function f nor such a given function g, then it is obvious that the real variables x and y are *not* functionally dependent. Definition 10.20 extends this concept to the set of real closed intervals.

The preceding definition, along with two deductions that we shall presently make (Theorem 10.9 and Corollary 10.4), touches the notion of interval dependency in a way which copes with all possible cases. This shall be shown, in detail, in Sections 10.7 and 10.8. For now, to illustrate, let us give the following example.

Example 10.5 ***(Dependency Relation for Two Variables)*** Let X and Y be two interval variables that both are assigned the same individual constant $[0, 1]$. Then, we may have one of the following cases.

i) Y is not dependent on X (there is no given dependency function).
ii) Y is dependent on X, by the identity function $y = f(x) = x$.
iii) Y is dependent on X, by the square function $y = f(x) = x^2$.

This example shows that if two interval *variables* X and Y both are assigned the same *individual constant* (both have the same *value*), it does not necessarily follow that X and Y are *identical*, unless they are dependent by the identity Function.[13] This shall be made precise in Definition 10.27.

As a consequence of our characterization of interval dependency, we have the next immediate theorem that establishes that the interval dependency relation is a *quasi-ordering* relation.

Theorem 10.9 ***(Quasi-Orderness of the Dependency Relation)*** *The interval dependency relation is a quasi-ordering relation on the set of real closed intervals. That is, for any three interval variables X, Y, and Z, the following statements are true:*

i) $\mathfrak{D}$ *is reflexive, in symbols* $(X\mathfrak{D}X)$,
ii) $\mathfrak{D}$ *is transitive, in symbols* $(X\mathfrak{D}Y \wedge Y\mathfrak{D}Z \Rightarrow X\mathfrak{D}Z)$.

In accordance with this theorem and Definition 10.19, we also have the following corollary.

Corollary 10.4 ***(Dependency Relation Properties)*** *For any interval operation $\circ_{\mathcal{J}}$, and for two interval variables X and Y, the following two assertions are true:*

i) $(X\circ_{\mathcal{J}}Y)\mathfrak{D}X$,
ii) $(X\circ_{\mathcal{J}}Y)\mathfrak{D}Y$.

13 The notions of *identity* and *equality* are commonly confused and treated as synonyms. However, they are two *distinct* logical concepts. Despite the fact that equality implies identity in the theory of real numbers, this is not always the case. Two line halves are *equal* but not *identical* (one and the same). Every line *equals* infinitely many other lines, but no line *is* (identical to) any other line (see [61] and [32]). Identity, which is the most fundamental ingredient of any mathematical theory, is characterized by *Leibniz's principle of the identity of indiscernibles*, which states that two entities x and y are identical iff any property of x is also a property of y and vice versa.

The interval dependency problem can now be formulated in the following theorem.

Theorem 10.10 ***(Dependency Problem)*** *Let X_k be real closed intervals and let $f(x_1, \dots, x_n)$ be a continuous real-valued function with $x_k \in X_k$. Evaluating the accurate image of f for the interval numbers X_k, using classical interval arithmetic, is not always possible if there exist X_i and X_j such that $X_j \mathfrak{D} X_i$ for $i \neq j$. That is,*

i) $(\exists f)(\mathrm{I}_f(X_1, \dots, X_n) \neq f(X_1, \dots, X_n))$.
In general,
ii) $(\forall f)(\mathrm{I}_f(X_1, \dots, X_n) \subseteq f(X_1, \dots, X_n))$.

Proof: For (i), it suffices to give a counterexample.

For two interval variables X_1 and X_2 that both are assigned the same individual constant $[-a, a]$, let f be a function defined by the rule $f(x_1, x_2) = x_1 x_2$ with $x_1 \in X_1$ and $x_2 \in X_2$. If $X_2 \mathfrak{D}_g X_1$, with g is the identity function $x_2 = g(x_1) = x_1$, then f has the equivalent rule $f(x) = x^2$, with $x \in [-a, a]$.

According to Theorem 10.6, the (accurate) image of $[-a, a]$ under the real-valued function f is

$$\mathrm{I}_f([-a, a]) = [\min_{x \in [-a,a]} x^2, \max_{x \in [-a,a]} x^2] = [0, a^2].$$

If we evaluate the image of $[-a, a]$ using classical interval arithmetic, by Theorem 10.2, we obtain the interval-valued function,

$$\begin{aligned} f([-a, a]) &= [-a, a] \times [-a, a] \\ &= [-a^2, a^2]. \end{aligned}$$

which is not the actual image of $[-a, a]$ under f, that is, there is some function f, for which

$$\mathrm{I}_f(X_1, \dots, X_n) \neq f(X_1, \dots, X_n).$$

and therefore evaluating the accurate image of real-valued functions is not always possible, using classical interval arithmetic.

Towards proving (ii), let

$$\mathrm{I}_f(X_1, \dots, X_n) = \mathrm{I}_f^{\sigma_1}(X_1, \dots, X_n) \vee \mathrm{I}_f(X_1, \dots, X_n) = \mathrm{I}_f^{\sigma_2}(X_1, \dots, X_n).$$

where $\mathrm{I}_f^{\sigma_1}$ and $\mathrm{I}_f^{\sigma_2}$ are the images of f, respectively, in two prenex sentences σ_1 and σ_2 such that in σ_1, there exist X_i and X_j such that $X_j \mathfrak{D} X_i$ for $i \neq j$, and in σ_2, all X_k are pairwise independent, that is $\mathfrak{I}_{k=1}^{n}(X_k)$. Employing Theorem 10.7, we accordingly have

$$\mathrm{I}_f^{\sigma_1}(X_1, \dots, X_n) \subseteq \mathrm{I}_f^{\sigma_2}(X_1, \dots, X_n).$$

According to Definition 10.15, of the classical interval algebra, all interval variables are assumed to be independent. We consequently have

$$f(X_1, \dots, X_n) = \mathrm{I}_f^{\sigma_2}(X_1, \dots, X_n).$$

Thus,

$$(\forall f)(\mathrm{I}_f(X_1, \dots, X_n) \subseteq f(X_1, \dots, X_n)).$$

and therefore (ii) is verified.

Obviously, the result $[-a^2, a^2]$, obtained using classical interval arithmetic, has an overestimation of

$$|w([-a^2, a^2]) - w([0, a^2])| = a^2.$$

where w is the width of the interval. This overestimated result is due to the fact that the classical interval theory assumes independence of all interval variables, even when dependencies exist.

A numerical example is shown below.

Example 10.6 ***(Overestimation due to Dependency)*** Consider the real-valued function

$$f(x) = x(x-1).$$

with $x \in [0, 1]$.

The actual image of $[0, 1]$ under f is $[-1/4, 0]$. Evaluating the image using classical interval arithmetic, we get

$$f([0,1]) = [0,1] \times ([0,1] - 1) = [-1, 0].$$

which has an overestimation of

$$|w([-1,0]) - w([-1/4,0])| = 3/4.$$

The problem of computing the image $I_f(X_1, \dots, X_n)$, using interval arithmetic, is the main problem of interval computations. This problem is, in general, NP-hard[14] (see, e.g. [62, 63], and [48]). That is, for the classical interval theory, there is no efficient algorithm to make the identity

$$(\forall f)(I_f(X_1, \dots, X_n) = f(X_1, \dots, X_n)).$$

always hold unless NP = P, which is widely believed to be false. However, a considerable scientific effort is put into finding a way out from the interval dependency problem. There are many special methods and algorithms, based on the classical interval theory, that successfully compute useful narrow bounds to the desirable accurate image. In the succeeding section, we shall discuss briefly how to compute useful guaranteed enclosures under interval dependency.

Beyond the techniques based on the classical interval theory, various proposals for possible alternate theories of interval arithmetic were introduced to reduce the dependency effect or to enrich the algebraic structure of interval numbers. Among these alternate theories of intervals, we can mention as examples: *Hansen's generalized intervals*, *Kulisch's complete intervals*, *directed intervals*, *modal intervals*, *parametric intervals*, and others (see, e.g. [20, 40–43], and [9]). In Section 10.7, we will in-depth investigate the theory of parametric intervals, and in Section 10.8, we will present a new alternate theory, namely, the "*theory of universal intervals.*"

10.6 Interval Enclosures Under Functional Dependence

In this section, we give a bit of perspective on how to compute useful interval enclosures under functional dependence. The numerical examples of this section are computed using version 2.0 of InCLosure. The InCLosure commands to compute the results of the examples are described and an InCLosure input file and its corresponding output containing, respectively, the code and results of the examples are also available as a supplementary material to this text (see Supplementary Materials).

As mentioned in the preceding section, there are many ways out of the dependency problem to get narrower enclosures. With a knowledge of regions of monotonicity, most elementary interval functions can be defined to

14 In principle, this result is not necessarily applicable to other theories of interval arithmetic (present or future) because each theory has its peculiar set of algorithms, where each algorithm is a sequence of elementary relations and functions of the foundational level of the theory.

be the exact images of the corresponding real functions. As instances, for an interval number $X = [\underline{x}, \overline{x}]$ and a non-negative integer n, we can define

$$e^X = [e^{\underline{x}}, e^{\overline{x}}]\ ,\ \ln(X) = [\ln(\underline{x}), \ln(\overline{x})] \text{ if } \underline{x} > 0;$$

$$\sqrt{X} = [\sqrt{\underline{x}}, \sqrt{\overline{x}}] \text{ if } \underline{x} \geq 0,\ \sin(X) = [\min_{x \in X}(\sin(x)), \max_{x \in X}(\sin(x))]\ ;$$

$$X^n = \begin{cases} [\underline{x}^n, \overline{x}^n] & \text{iff } \underline{x} > 0 \text{ or } n \text{ is odd,} \\ [\overline{x}^n, \underline{x}^n] & \text{iff } \overline{x} < 0 \text{ and } n \text{ is even,} \\ [0, |X|^n] & \text{iff } 0 \in X \text{ and } n \text{ is even;} \end{cases}$$

where $|X| = \max\{|\underline{x}|, |\overline{x}|\}$ is the absolute value of the interval number X.

Applying the *naive* (*pure*) interval operations on these exact evaluations, we can get better (narrower) enclosures of the images of the algebraic combinations of the corresponding real functions. Moreover, many techniques are used to improve the results obtained from the naive method by reducing the width of the resulting interval. Among these techniques are *centered forms, generalized centered forms, circular complex centered forms, Hansen's method, remainder forms*, the *subdivision method*, and many others (see, e.g. [1, 20, 48, 64–66], and [67]). For instance, we can get *arbitrarily* narrower enclosures of images of real functions using the subdivision method, which is due to Moore (see, e.g. [64] and [65]). This method can be described as follows. Let $X = [\underline{x}, \overline{x}]$ be an interval number and let $w(X) = \overline{x} - \underline{x}$ be the width of X. First, we subdivide the interval X into n subintervals X_i such that

$$X_i = [\underline{x} + (i-1)w(X)/n, \underline{x} + (i)w(X)/n].$$

where $w(X_i) = w(X)/n$. Hence, $X = \cup_{i=1}^{n} X_i$. Then, we evaluate the interval-valued function f for each subinterval $X_i, f(X_i)$. Accordingly, we have [1]

$$I_f(X) \subseteq \cup_{i=1}^{n} f(X_i) \subseteq f(X).$$

That is, the subdivision method produces better enclosures than the naive method. Moreover, the larger the number of subintervals n, the narrower the enclosure of the image $I_f(X)$. Next, we shall describe how to make use of the interval capability of InCLosure to get *arbitrarily* better enclosures of the images of real functions.

Example 10.7 ***(Interval Subdivision in InCLosure)*** Consider the interval function

$$f(X) = X^{10} - X^9 - X^2 - 1.$$

InCLosure provides arbitrarily narrower interval results which are limited only by the computational power of the host machine. We can compute the value of the above interval function at the interval $[-2, 3], f([-2, 3])$, using the following InCLosure command.

```
EvalInt "X^10-X^9-X^2-1" "X=[-2,3]" 1 30
```

This will result in

$$[-19\,693.0, 59\,560.0].$$

The last parameter, "30," in the previous command, is the precision of the result (default is 20). The parameter "1" indicates the number of subintervals if the subdivision method is applied (the number "1" means no subdivision of the interval). To get narrower intervals, the number of subintervals can be increased arbitrarily. Table 10.1 shows InCLosure results obtained for no subdivision, subdivision with 10, 50, 100, 500, and 1000 subintervals of $[-2, 3]$.

Table 10.1 InCLosure result for different numbers of subdivisions.

Number of subintervals	InCLosure result $f([-2,3])$
1 (no subdivision)	$[-19\,693.0, 59\,560.0]$
10	$[-10\,156.256\,835\,937\,5, 55\,227.052\,734\,375]$
50	$[-14.614\,437\,673\,5, 44\,532.444\,024\,131]$
100	$[-3.981\,273\,626\,845\,214\,843\,75, 42\,119.383\,791\,689\,572\,265\,625]$
500	$[-2.201\,068\,282\,169\,522\,690\,76, 39\,938.737\,630\,838\,335\,237\,301]$
1000	$[-2.106\,221\,747\,150\,794\,399\,208\,505\,859\,375, 39\,649.314\,310\,395\,608\,337\,801\,955\,078\,125]$

The exact image of the corresponding real function $f(x) = x^{10} - x^9 - x^2 - 1$ on $[-2, 3]$ is $[-2.025\,16, 39\,356]$. All the results of Table 10.1 are *guaranteed enclosures* of the exact image, and Example 10.7 clearly shows that classical interval arithmetic with the *arbitrarily* narrower interval results of InCLosure thus markedly surpasses the ordinary numerical methods.

10.7 Parametric Intervals: How Far They Can Go

An important and promising theory of interval arithmetic is the theory of parametric intervals. Although it is scarcely mentioned in the interval literature, if at all, parametric interval arithmetic has widespread applications in many scientific fields such as artificial intelligence, fuzzy systems, and granular computing, which require more accuracy and compatibility with the *semantic* of real arithmetic and fuzzy set theory (see, e.g. [56, 68], and [69]).

The term "*parametric interval arithmetic*" is reasonably recent, but the idea perhaps has an earlier root in Cleary's "*logical arithmetic*" which is a logical technique, for real arithmetic in Prolog, that uses constraints over real intervals (see [70] and [71]). Over the past few decades, parametric interval arithmetic has been invented and "reinvented" several times, under different names: "logical intervals," "constrained intervals," "instantiation intervals," "RDM intervals," and "Overestimation-free intervals" (see, e.g. [9, 17, 43, 70, 72, 73], and [74]). The theory of parametric intervals is presented in the literature as an approach to solving the long-standing dependency problem in the classical interval theory, along with the emphasis that parametric interval arithmetic, unlike Moore's classical interval arithmetic, has additive and multiplicative inverse elements and satisfies the distributive law.

Our main purpose in this section is to introduce the theory of parametric intervals and to mathematically examine to what extent it can accomplish these very desirable algebraic properties. We begin, in Section 10.7.1, by formalizing the fundamental notions of the theory of parametric intervals. With the aid of our systematic formalization of the notion of interval dependency as a binary predicate $\mathfrak{D}$ and the related deductions presented in Section 10.5, we turn, in Section 10.7.2, to investigate if the theory of parametric intervals accomplishes its objectives. The investigation of parametric intervals conducted here is *metatheoretical* in nature and based indispensably on the metamathematical apparatus fixed in Section 10.3. That is, most of the results deduced in this section are *metatheorems about* the theory Th_p of parametric intervals.

10.7.1 Parametric Interval Operations: From Endpoints to Convex Subsets

In this section and its successor, we formalize the basic notions and investigate the fundamental properties of the theory Th_p of parametric intervals. The language of the theory Th_p of parametric intervals is an extension of the language of the theory Th_I of classical intervals, by defining new symbols or by introducing new abbreviations to allow for the possibility of expressing new concepts of the theory Th_p.

With the elegant idea that a real closed interval is a *convex* subset[15] of the reals, and motivated by the fact that the *best* way to evaluate the *accurate* image of a continuous real-valued function is to apply minimization and maximization directly to determine the exact lower and upper endpoints of the image; parametric interval arithmetic can be constructed as a simplified type of a min-max optimization problem, with constraints varying in the unit interval.[16]

A definition of a parametric interval can then be formulated as follows.

Definition 10.21 ***(Parametric Interval)*** Let $\underline{x}, \overline{x} \in \mathbb{R}$ such that $\underline{x} \leq \overline{x}$. A parametric interval is defined to be

$$[\underline{x}, \overline{x}] = \{x \in \mathbb{R} | (\exists \lambda_x \in [0,1]) \, (x = (\overline{x} - \underline{x})\lambda_x + \underline{x})\}.$$

where $\min\limits_{\lambda_x}(x) = \underline{x}$ and $\max\limits_{\lambda_x}(x) = \overline{x}$ are, respectively, the lower and upper bounds (endpoints) of $[\underline{x}, \overline{x}]$.

Obviously, Definition 10.21 is equivalent to the definition of a classical interval number, and a parametric interval is a closed and bounded nonempty real interval. However, to simplify the exposition, we shall denote[17] the set of parametric intervals by $\mathcal{P}$, and the uppercase Roman letters X, Y, and Z (with or without subscripts), or equivalently $[\underline{x}, \overline{x}]$, $[\underline{y}, \overline{y}]$, and $[\underline{z}, \overline{z}]$, shall be still employed as variable symbols to denote elements of $\mathcal{P}$. The sets of *point* and *symmetric* parametric intervals shall be denoted by $\mathcal{P}_{[x]}$ and $\mathcal{P}_S$, respectively.

In Definition 10.21, a parametric interval is defined as the image of a continuous real-valued function x of one variable $\lambda_x \in [0,1]$ and two constants $\underline{x}$ and $\overline{x}$. The endpoints, $\underline{x}$ and $\overline{x}$, are, respectively, the minimum and maximum of x with the constraint

$$\begin{aligned} 0 \leq \lambda_x \leq 1 &\Rightarrow 0 \leq (\overline{x} - \underline{x})\lambda_x \leq (\overline{x} - \underline{x}) \\ &\Rightarrow \underline{x} \leq (\overline{x} - \underline{x})\lambda_x + \underline{x} \leq \overline{x} \\ &\Rightarrow \underline{x} \leq x \leq \overline{x}. \end{aligned}$$

Since the endpoints $\underline{x}$ and $\overline{x}$ are known *inputs*, they are parameters and hence the name "*parametric interval arithmetic*," whereas λ_x is a variable that is constrained between 0 and 1. The binary parametric interval operations can be guaranteed to be *continuous* by introducing two constrained variables $\lambda_x, \lambda_y \in [0,1]$. From the fact that $x \in [\underline{x}, \overline{x}]$ and $y \in [\underline{y}, \overline{y}]$ are continuous real-valued functions of λ_x and λ_y respectively, the result of a parametric interval operation shall be the image of the continuous function

$$x \circ_{\mathbb{R}} y = ((\overline{x} - \underline{x})\lambda_x + \underline{x}) \circ_{\mathbb{R}} ((\overline{y} - \underline{y})\lambda_y + \underline{y}).$$

with $\lambda_x, \lambda_y \in [0,1]$ and $\circ_{\mathbb{R}} \in \{+, -, \times, \div\}$ such that $y \neq 0$ if $\circ_{\mathbb{R}}$ is "$\div$."

According to the functional dependence of real variables, Lodwick [43] defines two types of parametric interval operations, namely, "*dependent operations*" and "*independent operations*." The dependent and independent parametric interval operations can be characterized in the following two definitions [72].

---

15 Let $\mathcal{V}$ be a vector space over an ordered field $\langle \mathbb{F}; +_{\mathbb{F}}, \times_{\mathbb{F}}; \leq_{\mathbb{F}} \rangle$. A set C in $\mathcal{V}$ is said to be convex iff

$$(\forall x, y \in C) \, (\forall \lambda \in [0_{\mathbb{F}}, 1_{\mathbb{F}}]) \, (((1 - \lambda)x + \lambda y) \in C) \, .$$

16 Note that various parametrizations can be employed to characterize a parametric interval. As an instance, Elishakoff and Miglis used a *trigonometric* parametrization in [73]. Here, we use the simple *linear* parametrization adopted in [17, 72], and [43].

17 Although we have *many* theories of intervals being discussed throughout the text, we always deal with the *same* set of real closed intervals. However, for legibility and brevity, we shall employ *different notations* for the set of interval numbers, for example, $\mathcal{I}$, $\mathcal{P}$, $\mathcal{U}$, and so forth, according to what theory of intervals is being discussed. Therefore, we can write, for instance, the brief expressions "*addition on* $\mathcal{I}$" and "*addition on* $\mathcal{P}$" instead, respectively, of the expressions "*classical interval addition*" and "*parametric interval addition.*"

Definition 10.22 ***(Parametric Dependent Operations)*** For any parametric interval $[\underline{x}, \overline{x}]$, there exists a parametric interval $[\underline{z}, \overline{z}]$ such that

$$\begin{aligned}[\underline{z}, \overline{z}] &= [\underline{x}, \overline{x}] \circ_{\text{dep}} [\underline{x}, \overline{x}] \\ &= \{z \in \mathbb{R} | (\exists x \in [\underline{x}, \overline{x}])\, (z = x \circ_{\mathbb{R}} x)\} \\ &= \{z \in \mathbb{R} | (\exists \lambda_x \in [0, 1])\, (z = ((\overline{x} - \underline{x})\lambda_x + \underline{x}) \circ_{\mathbb{R}} ((\overline{x} - \underline{x})\lambda_x + \underline{x}))\}.\end{aligned}$$

where

$$\begin{aligned}\underline{z} &= \min_{\lambda_z} (z) = \min_{\lambda_x} (x \circ_{\mathbb{R}} x) \\ \overline{z} &= \max_{\lambda_z} (z) = \max_{\lambda_x} (x \circ_{\mathbb{R}} x).\end{aligned}$$

and $\circ \in \{+, -, \times, \div\}$such that $0 \notin [\underline{x}, \overline{x}]$ if $\circ$is "÷."

Definition 10.23 ***(Parametric Independent Operations)*** For any two parametric intervals $[\underline{x}, \overline{x}]$ and $[\underline{y}, \overline{y}]$, there exists a parametric interval $[\underline{z}, \overline{z}]$ such that

$$\begin{aligned}[\underline{z}, \overline{z}] &= [\underline{x}, \overline{x}] \circ_{\text{ind}} [\underline{y}, \overline{y}] \\ &= \{z \in \mathbb{R} | (\exists x \in [\underline{x}, \overline{x}])\, (\exists y \in [\underline{y}, \overline{y}])\, (z = x \circ_{\mathbb{R}} y)\} \\ &= \{z \in \mathbb{R} | (\exists \lambda_x \in [0, 1])\, (\exists \lambda_y \in [0, 1]) \\ &\qquad (z = ((\overline{x} - \underline{x})\lambda_x + \underline{x}) \circ_{\mathbb{R}} ((\overline{y} - \underline{y})\lambda_y + \underline{y}))\}.\end{aligned}$$

where

$$\begin{aligned}\underline{z} &= \min_{\lambda_z}(z) = \min_{\lambda_x, \lambda_y}(x \circ_{\mathbb{R}} y) \\ \overline{z} &= \max_{\lambda_z}(z) = \max_{\lambda_x, \lambda_y}(x \circ_{\mathbb{R}} y).\end{aligned}$$

and $\circ \in \{+, -, \times, \div\}$such that $0 \notin [\underline{y}, \overline{y}]$ if $\circ$ is "÷."

It is obvious, from Definitions 10.22 and 10.23, that parametric interval arithmetic is a *mathematical programming problem*, and therefore, parametric interval operations can be easily performed by any constraint solver such as GeCode[18], HalPPC[19], and MINION[20]. The minimization and maximization are well-defined, attained, and the resultant $[\underline{z}, \overline{z}]$ is in turn a parametric interval.

10.7.2 On the Structure of Parametric Intervals: Are They Properly Founded?

With the aid of our systematic formalization of the notion of interval dependency as a binary predicate $\mathfrak{D}$ and the related deductions presented in Section 10.5, and by means of the notions prescribed in Sections 10.3 and 10.7.1, we now turn to investigate if the theory of parametric intervals accomplishes its objectives. The investigation conducted in this section is *metatheoretical* in nature and based indispensably on the formalized apparatus fixed in Section 10.3. That is, most of the deductions of this section are *metatheorems about* the theory Th_p of parametric intervals.

In the first place, we must generally ask: *what exactly is the algebraic structure of parametric interval arithmetic?* The interval literature does not provide an answer for this question. However, the parametric approach to interval analysis is usually introduced with the emphasis that parametric interval arithmetic, unlike Moore's classical

18 http://www.gecode.org/.
19 http://sourceforge.net/projects/halppc.
20 http://minion.sourceforge.net/.

interval arithmetic, has additive and multiplicative inverse elements, satisfies the distributive law, and explicitly keeps track of functional dependencies (see, e.g. [43, 73], and [74]). For example, on page 1 in [43], Lodwick says:

> Unlike (classical) interval arithmetic, constrained interval arithmetic has an additive inverse, a multiplicative inverse and satisfies the distributive law. This means that the algebraic structure of constrained interval arithmetic is different than that of (classical) interval arithmetic.

and then presents proofs for the following three statements:

i) Additive inverse. $(\forall X \in \mathcal{P})\,(X -_{\text{dep}} X = [0,0])$.
ii) Multiplicative inverse. $(\forall X \in \mathcal{P})\,(0 \notin X \Rightarrow X \div_{\text{dep}} X = [1,1])$.
iii) Distributive law. $(\forall X, Y, Z \in \mathcal{P})\,(Z \times (X + Y) = Z \times X + Z \times Y)$.

The first two statements are derivable by simple substitution in Definition 10.22 for parametric dependent operations, hence the subscript "dep." For the third statement, the matter is much more complicated, and therefore, we dropped the subscripts for the operation symbols of addition and multiplication.

Getting down to particulars with the above three statements, we must turn to ask, then, the corresponding three questions:

(1) Is the statement "$A -_{\text{dep}} A = [0,0]$" equivalent to "$(-A)$ is the inverse element of A with respect to the operation $+$ on the set $\mathcal{P}$, according to the dependent operation $X +_{\text{dep}} X$"?
(2) Is the statement "$A \div_{\text{dep}} A = [1,1]$" equivalent to "$(A^{-1})$ is the inverse element of A with respect to the operation $\times$ on the set $\mathcal{P}$, according to the dependent operation $X \times_{\text{dep}} X$"?
(3) Does the distributive law hold, according to the dependent and independent operations?

In the sequel, we prove that the answers of the above three questions are all *negative*. On the face of it, the theory of parametric intervals seems to fit squarely into its objectives, but, however the elegance of its underlying idea, we shall argue both that the fit is problematic, and that its mathematical formulation constitutes a serious algebraic defect.

A careful formal investigation of the parametric interval operations will show that the problems of theory of parametric intervals are "foundational" in nature. As discussed in Section 10.4, classical interval arithmetic considers all *numerically equal* intervals as independent, which is not necessarily true as shown in Section 10.5. On the other hand, Definition 10.22 considers only dependency by *identity* and it implies that two numerically equal intervals are always dependent, which is obviously not true. (Take, for example, the interval extension of $f\,(x, y) = x - y$ with each of the variables x and y varies *independently* on the same interval $[1, 2]$.) A consequence of this issue is that Definition 10.23 will not consider numerically equal intervals even if they are independent. In other words, the domain of the dependent operations is the set of pairs (X, X), and the domain of the independent operations is the set of pairs (X, Y) where X and Y cannot be assigned the same individual constant.

Recalling Definition 10.20 of the interval dependency relation, it is also convenient to single out two sets of elements of $\mathcal{P}^2$, according to interval dependencies.

Definition 10.24 ***(Dependent Parametric Set)*** $\mathcal{K}_{\text{dep}} = \{(X, Y) \in \mathcal{P}^2 \mid Y \mathfrak{D} X\}$.

Definition 10.25 ***(Independent Parametric Set)*** $\mathcal{K}_{\text{ind}} = \{(X, Y) \in \mathcal{P}^2 \mid \mathfrak{I}\,(X, Y)\}$.

With Definitions 10.4, 10.6, 10.8, and 10.13 of, respectively, *partial and total operations*, an *$\mathfrak{L}$-structure*, *inverse elements*, and a *number system* at our disposal, we are now ready to prove our statements about the theory of parametric intervals. We begin by investigating what type of algebraic operations the parametric operations are.

Theorem 10.11 ***(Partiality of Parametric Dependent Operations)*** *Parametric dependent addition and multiplication are partial operations in the set $\mathcal{P}$.*

Proof: For $\circ_{\text{dep}} \in \{+, \times\}$, from Definition 10.22, we have $\circ_{\text{dep}}: \text{Id}_{\mathcal{P}} \to \mathcal{P}$, where

$$\text{Id}_{\mathcal{P}} = \{(X, X) | \ X \in \mathcal{P}\}.$$

Obviously, the set $\text{Id}_{\mathcal{P}}$ is the *identity* relation on $\mathcal{P}$, which, by Definition 10.24, is a proper subset of $\mathcal{K}_{\text{dep}}$ and hence a proper subset of $\mathcal{P}^2$. Therefore, according to Definition 10.4, the operations $\circ_{\text{dep}} \in \{+, \times\}$ are partial operations in the set $\mathcal{P}$ of parametric intervals.

One immediate result that this theorem implies is that the parametric dependent operations consider only a single special case of interval dependency, namely, the *dependency by identity*, $X \mathfrak{D} X$. Other cases of interval dependency, characterized in Definition 10.20, are not considered by the parametric dependent operations.

Theorem 10.12 ***(Partiality of Parametric Independent Operations)*** *Parametric independent addition and multiplication are partial operations in the set $\mathcal{P}$.*

Proof: For $\circ_{\text{ind}} \in \{+, \times\}$, Definition 10.23 does not consider independent intervals if they are numerically equal. Accordingly $\circ_{\text{ind}} : \mathcal{V} \to \mathcal{P}$, where $\mathcal{V}$, by Definition 10.25, is a proper subset of $\mathcal{K}_{\text{ind}}$ and hence a proper subset of $\mathcal{P}^2$. Therefore, by Definition 10.4, the operations $\circ_{\text{ind}} \in \{+, \times\}$ are partial operations in the set $\mathcal{P}$ of parametric intervals.

In accordance with the preceding two results, we are now led to the following three theorems, which answer our questions concerning inverse elements and distributivity.

Theorem 10.13 ***(Unprovability of Parametric Additive Inverses)*** *Inverse elements for addition cannot be proved to exist in parametric interval arithmetic. In other words, the statement*

$$(\forall X \in \mathcal{P}) \, (X + (-X) = [0, 0]).$$

is undecidable in the theory $Th_{\mathcal{P}}$ of parametric intervals.

Proof: For any parametric interval X, define the negation of X, in the standard way, to be

$$-X = \{z \in \mathbb{R} | (\exists x \in X) \, (z = -x)\}.$$

Obviously, the relation $(-X) \, \mathfrak{D} X$ is true. But, by Theorem 10.11, the pair $((-X), X) \notin \text{Id}_{\mathcal{P}}$ unless $X = [0, 0]$, and the expression $X + (-X)$ thus is not expressible as a parametric dependent operation. On the other hand, by Theorem 10.12, $((-X), X) \notin \mathcal{K}_{\text{ind}}$ because the predicate $\mathfrak{I}\,((-X), X)$ is not true, and the expression $X + (-X)$ thus is not expressible as a parametric independent operation.

It follows, therefore, that the existence of additive inverses is undecidable in the parametric interval theory.

Theorem 10.14 ***(Unprovability of Parametric Multiplicative Inverses)*** *Inverse elements for multiplication cannot be proved to exist in parametric interval arithmetic. In other words, the statement*

$$(\forall X \in \mathcal{P}) \, (0 \notin X \Rightarrow X \times (X^{-1}) = [1, 1]).$$

is undecidable in the theory $Th_{\mathcal{P}}$ of parametric intervals.

Proof: For any parametric interval X with $0 \notin X$, define the reciprocal of X, in the standard way, to be

$$X^{-1} = \{z \in \mathbb{R} | (\exists x \in X) \, (z = x^{-1})\}.$$

Obviously, the relation $(X^{-1})\,\mathfrak{D}X$ is true. But, by Theorem 10.11, the pair $((X^{-1}), X) \notin \mathrm{Id}_{\mathcal{P}}$ unless $X = [1, 1]$, and the expression $X \times (X^{-1})$ thus is not expressible as a parametric dependent operation. On the other hand, by Theorem 10.12, $((X^{-1}), X) \notin \mathcal{K}_{\text{ind}}$ because the predicate $\mathfrak{I}\,((X^{-1}), X)$ is not true, and the expression $X \times (X^{-1})$ thus is not expressible as a parametric independent operation.

It follows, therefore, that the existence of multiplicative inverses is undecidable in the parametric interval theory.

Theorem 10.15 ***(Unprovability of Parametric Distributivity)*** *The distributive law does not hold in parametric interval arithmetic. In other words, the statement*

$$(\forall X, Y, Z \in \mathcal{P})\,(Z \times (X + Y) = Z \times X + Z \times Y).$$

is not provable in the theory $Th_{\mathcal{P}}$ of parametric intervals.

Proof: Obviously, in the left-hand side

$$Z \times (X + Y).$$

all the variables are mutually independent. Then, applying Definition 10.23 of the parametric independent operations, we obtain the same result as in classical interval arithmetic, that is

$$Z \times_{\text{ind}} (X +_{\text{ind}} Y) = Z \times_{\mathcal{I}} (X +_{\mathcal{I}} Y).$$

Let us now consider the right-hand side

$$(Z \times X) + (Z \times Y).$$

It is clear that the relations $(Z \times X)\mathfrak{D}Z$ and $(Z \times Y)\mathfrak{D}Z$ are true. However, by Theorem 10.11, the pair $((Z \times X), (Z \times Y)) \notin \mathrm{Id}_{\mathcal{P}}$, and the expression $(Z \times X) + (Z \times Y)$ thus is not expressible as a parametric dependent operation. Then, applying Definition 10.23 of the parametric independent operations, we again have the same result as in classical interval arithmetic, that is

$$(Z\times_{\text{ind}}X)+_{\text{ind}}(Z\times_{\text{ind}}Y) = (Z\times_{\mathcal{I}}X)+_{\mathcal{I}}(Z\times_{\mathcal{I}}Y).$$

According to the subdistributive property of the classical interval theory, we also have only the subdistributive law

$$Z\times_{\text{ind}}(X+_{\text{ind}}Y) \subseteq (Z\times_{\text{ind}}X)+_{\text{ind}}(Z\times_{\text{ind}}Y).$$

for parametric interval arithmetic.

It follows, therefore, that distributivity is not provable in the parametric interval theory.

Thus, the preceding three theorems prove that the answers of our questions are all *negative*. The parametric dependent and independent operations do not qualify as *total* operations on $\mathcal{P}$, and in their full extent, do not suffice to cope with interval dependencies except for the special case when the operands are trivially *dependent by identity*, that is, $X\mathfrak{D}X$.

In order to make this clear, we next give an example.

Example 10.8 ***(Inexpressibility of Parametric Intervals)*** Let σ be the prenex sentence such that

$$\sigma \Leftrightarrow (\forall x \in [-1, 1])\,(\exists y \in [0, 1])\,(\exists z \in \mathbb{R})\,(z = y - x).$$

In the sentence σ, the variable y depends on x, and therefore, there is some function $g\,(x)$ such that σ has the Skolem equivalent form

$$(\exists g)\,(\forall x \in [-1, 1])\,(\exists z \in \mathbb{R})\,(z = g\,(x) - x).$$

The dependency function g can be, for instance, the quadratic function, that is $y = g(x) = x^2$. It is clear that the relation $[0,1]\mathfrak{D}[-1,1]$ is true, that is, the interval number $[0,1]$ is dependent on $[-1,1]$. However, the pair $([0,1],[-1,1]) \notin \mathrm{Id}_{\mathcal{P}}$, and the expression $[0,1]-[-1,1]$ thus is not expressible as a parametric dependent operation.

We now pass to our general question concerning the algebraic system of parametric interval arithmetic. The following theorem clarifies an answer.

Theorem 10.16 ***(Undefinability of Parametric Interval Algebra)*** *The theory $Th_{\mathcal{P}}$ of parametric intervals does not define an algebra for addition or multiplication on the set $\mathcal{P}$.*

Proof: By Theorems 10.11 and 10.12, the operations $\circ_{\text{dep}}$ and $\circ_{\text{ind}}$, in $\{+,\times\}$, are partial operations in $\mathcal{P}$, and therefore, according to Definition 10.6, the algebras $\langle\mathcal{P};\circ_{\text{dep}}\rangle$ and $\langle\mathcal{P};\circ_{\text{ind}}\rangle$ are not definable.

That is, the structures $\langle\mathcal{P};\circ_{\text{dep}}\rangle$ and $\langle\mathcal{P};\circ_{\text{ind}}\rangle$ are *undefinable*, for the requirement that an algebraic operation must be *total* on the universe set $\mathcal{P}$.

In consequence of the last theorem, we also have the following important result.

Theorem 10.17 ***(Undefinability of Parametric Number System)*** *The theory $Th_{\mathcal{P}}$ of parametric intervals does not define a number system on the set $\mathcal{P}$.*

Proof: The proof immediately follows from Definitions 10.13 and 10.14, plus Theorem 10.16, by the fact that every number system is an algebra.

Thus, parametric intervals are neither "*numbers*" nor "S*-numbers*," and therefore we cannot talk of "*parametric interval numbers*."

From the above discussion, we can conclude that the underlying idea of parametric interval arithmetic seems elegant and simple, but it is *too simple* to fully account for the notion of interval dependency or to achieve a richer algebraic structure for interval arithmetic. It is therefore imperative both to supply the defect in the parametric approach and to present an alternative theory with a mathematical construction that avoids the defect. The former was attempted in the present section, and the latter is attempted in the next section.

10.8 Universal Intervals: An Interval Algebra with a Dependency Predicate

In the preceding section, we proved that the theory of parametric intervals cannot fully account for the notion of interval dependency, does not define an algebra for interval addition or multiplication, and consequently does not define a number system on the set of parametric intervals. With a view to treating these problems, the present section is devoted to constructing a new arithmetic of interval numbers.

Based on the idea of representing an interval number as a convex set, along with our formalization of the notion of interval dependency (see Section 10.5.2), we attempt, in this section, to present an alternate theory of intervals, namely, the "*theory of universal intervals*," with a mathematical construction that tries to avoid some of the defects in the current theories of interval arithmetic, to provide a richer interval algebra, and to better account for the notion of interval dependency.

We begin, in Sections 10.8.1 and 10.8.2, by defining the key concepts of the universal interval theory, and then we formulate the basic operations and relations for universal interval numbers. In Section 10.9, we carefully construct the algebraic system of universal interval arithmetic, deduce its fundamental properties, and then prove that the universal interval theory constitutes a nice S-field algebra, which extends the ordinary field structure of real numbers.

10.8.1 Universal Intervals, Rational Functions, and Predicates

In this section, we define the key concepts of the theory $\text{Th}_{\mathcal{U}}$ of universal intervals. After formulating the basic operations and relations for universal interval numbers, we introduce the notions of a *universal rational function* and a *universal predicate*. Then, we establish the *fundamental theorem of universal intervals* along with some deductions relating universal interval functions and classical interval functions.

As usual, in all the proofs, elementary facts about operations and relations on the real numbers are usually used without explicit reference. Moreover, the notions, notations, and abbreviations of Section 10.5.2 are indispensable for our mathematical discussion throughout this section and the succeeding sections, and hereafter are assumed priori, without further mention.

Using a simple linear parametrization, we begin by defining a universal interval number as a type of convex set.

Definition 10.26 ***(Universal Interval)*** Let $\underline{x}, \overline{x} \in \mathbb{R}$ such that $\underline{x} \leq \overline{x}$. A universal interval number is a closed and bounded nonempty convex subset of $\mathbb{R}$, that is

$$[\underline{x}, \overline{x}] = \{x \in \mathbb{R} | (\exists \lambda_x \in [0, 1]) (x = (\overline{x} - \underline{x})\lambda_x + \underline{x})\}.$$

where $\min_{\lambda_x}(x) = \underline{x}$ and $\max_{\lambda_x}(x) = \overline{x}$ are, respectively, the lower and upper bounds (endpoints) of $[\underline{x}, \overline{x}]$.

We shall denote the set of universal interval numbers by $\mathcal{U}$. The uppercase Roman letters X, Y, and Z (with or without subscripts), or equivalently $[\underline{x}, \overline{x}]$, $[\underline{y}, \overline{y}]$, and $[\underline{z}, \overline{z}]$, shall be still employed as variable symbols to denote elements of $\mathcal{U}$. The sets of *point* and *symmetric* universal interval numbers shall be denoted, as usual, by symbols $\mathcal{U}_{[x]}$ and $\mathcal{U}_S$, respectively. A *point (singleton)* universal interval $\{x\}$ shall be denoted by $[x]$. In particular, we shall use $1_{\mathcal{U}}$ and $0_{\mathcal{U}}$ to denote, respectively, the singleton universal intervals $\{1_{\mathbb{R}}\}$ and $\{0_{\mathbb{R}}\}$.

By virtue of Definition 10.20, we characterize the equality relation on $\mathcal{U}$, in terms of the dependency relation $\mathfrak{D}$ and the identity function Id, as follows.

Definition 10.27 ***(Equality on Universal Intervals)*** Two universal interval variables X and Y are equal (identical) iff they are dependent by identity, that is

$$X =_{\mathcal{U}} Y \Leftrightarrow X \mathfrak{D}_{\text{Id}} Y.$$

In consequence of this definition and Definition 10.26, we have the following immediate theorem.

Theorem 10.18 ***(Criteria for Universal Equality)*** *Let $[\underline{x}, \overline{x}]$ and $[\underline{y}, \overline{y}]$ be any two universal interval variables. Then*

$$[\underline{x}, \overline{x}] =_{\mathcal{U}} [\underline{y}, \overline{y}] \Leftrightarrow \underline{x} = \underline{y} \wedge \overline{x} = \overline{y} \wedge (\forall x \in [\underline{x}, \overline{x}]) (\exists y \in [\underline{y}, \overline{y}]) (y = Id(x)).$$

Thus, if two universal interval *variables* X and Y both are assigned the same *individual constant (value)*, it does not necessarily follow that X and Y are *equal (identical)*, unless they are dependent by the identity function (recall Example 10.5).

We then characterize the binary and unary algebraic operations for universal interval numbers, respectively, in the following two *set-theoretic* definitions.

Definition 10.28 ***(Binary Universal Operations)*** For any two universal interval numbers X and Y, the binary algebraic operations are defined by

$$X \circ_{\mathcal{U}} Y = \begin{cases} \{z \in \mathbb{R} | (\exists x \in X) (\exists y \in Y) (z = x \circ_{\mathbb{R}} y)\} & \text{if } \ \mathfrak{I}(X, Y) \\ \{z \in \mathbb{R} | (\exists x \in X) (\exists_{l=1}^{m} s_l \in S_l) & \\ \quad (z = x \circ_{\mathbb{R}} f(x; s_1, \ldots, s_m))\} & \text{if } \ Y \mathfrak{D}_f X \end{cases}.$$

where $\circ \in \{+, \times\}$.

Definition 10.29 ***(Unary Universal Operations)*** For any universal interval number X, the unary algebraic operations are defined by

$$\lozenge_{\mathcal{U}} X = \{z \in \mathbb{R} | (\exists x \in X)\,(z = \lozenge_{\mathbb{R}} x)\}.$$

where $\lozenge \in \{-,^{-1}\}$ and $0 \notin X$ if $\lozenge$ is " $^{-1}$. "

Hereafter, if confusion is unlikely, the subscript "$\mathcal{U}$," which stands for "*universal* interval operation," and the subscript "$\mathbb{R}$," in the real relation and operation symbols, may be suppressed.

On comparing Definition 10.28 with the two definitions of dependent and independent parametric interval operations (Definitions 10.22 and 10.23), it might at first seem that the advantage of simplicity lies with the parametric definitions. However, the advantage of Definition 10.28 is that it characterizes, as we shall prove presently, a single "*total binary operation*" on $\mathcal{U}$, for each $\circ \in \{+, \times\}$ and exhibits a uniform approach applicable to *all* cases of interval dependency (specified in Definition 10.20).

With a view to reaching more profound results, we supplement the two preceding definitions by the following characterization of a universal rational function.

Definition 10.30 ***(Universal Rational Functions)*** Let X_k be universal intervals and let $\mathcal{F}$ be a function variable symbol. A universal rational function $\mathcal{F}_{\mathcal{U}}\,(X_1, \dots, X_i, X_j, \dots, X_n)$ is a (multivariate) function obtained by means of a finite number of the universal interval operations such that

$$\mathcal{F}_{\mathcal{U}} = \begin{cases} \{z \in \mathbb{R} |\, (\exists^{n}_{k=1} x_k \in X_k)(z = \\ \quad \mathcal{F}_{\mathbb{R}}(x_1, \dots, x_i, x_j, \dots, x_n))\} & \text{if} \quad \mathfrak{I}^{n}_{k=1}(X_k) \\ \{z \in \mathbb{R} |\, (\exists^{n}_{k=1} x_k \in X_k)\,(\exists^{m}_{l=1} s_l \in S_l)(z = \\ \quad \mathcal{F}_{\mathbb{R}}\,(x_1, \dots, x_i, f_{\mathbb{R}}(x_i; s_1, \dots, s_m), \dots, x_n))\} & \text{if} \quad X_j \mathfrak{D}_f X_i \end{cases}.$$

where $\mathcal{F}_{\mathbb{R}}\,(x_1, \dots, x_i, x_j, \dots, x_n)$ is the corresponding real-valued rational function with $x_k \in X_k$.

As we mentioned before, if the type of function is clear from its arguments, and if confusion is unlikely, we shall usually drop the subscripts "$\mathbb{R}$" and "$\mathcal{U}$" and simply write $\mathcal{F}\,(X_1, \dots, X_n)$ and $\mathcal{F}\,(x_1, \dots, x_n)$ for, respectively, a universal rational function and its corresponding real-valued rational function, which are both defined by the same rule.

By virtue of our definition of a universal interval number as a type of convex set, the evaluation of a universal rational function is a simplified type of *mathematical optimization*, with the constraints are always in the unit interval $[0, 1]$. If there is no dependency between interval numbers, the value of a universal rational function is the same as in classical interval arithmetic. When dependencies exist, we have a different value. Thus, universal interval arithmetic has an algebra different from that of the classical interval theory.

With the help of the notions characterized above, we are now ready to prove the main theorem of this section, and of universal interval arithmetic.

Theorem 10.19 ***(Fundamental Theorem of Universal Intervals)*** *The value of a universal rational function $\mathcal{F}_{\mathcal{U}}(X_1, \dots, X_i, X_j, \dots, X_n)$ is the (accurate) image $\mathrm{I}_{\mathcal{F}}$ of the corresponding real-valued rational function $\mathcal{F}\,(x_1, \dots, x_i, x_j, \dots, x_n)$, with $x_k \in X_k$. That is*

$$\mathcal{F}_{\mathcal{U}} = \begin{cases} \mathrm{I}_{\mathcal{F}}(X_1, \dots, X_i, X_j, \dots, X_n) & \text{if} \quad \mathfrak{I}^{n}_{k=1}(X_k) \\ \mathrm{I}_{\mathcal{F}}(X_1, \dots, X_i, \mathrm{I}_f(X_i; S_1, \dots, S_m), \dots, X_n) & \text{if} \quad X_j \mathfrak{D}_f X_i \end{cases}.$$

Proof: Since a real-valued rational function is continuous, it follows, by Definition 10.30, that a universal rational function is continuous and attains its minimum and maximum values.

Then, by Definitions 10.26 and 10.30, optimizing with respect to all $\lambda \in [0,1]$, we obtain

$$\mathcal{F}_{\mathcal{U}} = \begin{cases} [\min\limits_{x_k \in X_k} \mathcal{F}(x_1, \ldots, x_i, x_j, \ldots, x_n), \\ \quad \max\limits_{x_k \in X_k} \mathcal{F}(x_1, \ldots, x_i, x_j, \ldots, x_n)] & \text{if } \mathfrak{I}_{k=1}^{n}(X_k) \\ [\min\limits_{\substack{x_k \in X_k \\ s_l \in S_l}} \mathcal{F}(x_1, \ldots, x_i, f(x_i; s_1, \ldots, s_m), \ldots, x_n), \\ \quad \max\limits_{\substack{x_k \in X_k \\ s_l \in S_l}} \mathcal{F}(x_1, \ldots, x_i, f(x_i; s_1, \ldots, s_m), \ldots, x_n)] & \text{if } X_j \mathfrak{D}_f X_i \end{cases}.$$

from which we conclude, by Theorem 10.6, that

$$\mathcal{F}_{\mathcal{U}} = \begin{cases} \mathrm{I}_{\mathcal{F}}(X_1, \ldots, X_i, X_j, \ldots, X_n) & \text{if } \mathfrak{I}_{k=1}^{n}(X_k) \\ \mathrm{I}_{\mathcal{F}}(X_1, \ldots, X_i, \mathrm{I}_f(X_i; S_1, \ldots, S_m), \ldots, X_n) & \text{if } X_j \mathfrak{D}_f X_i \end{cases}.$$

and therefore the value of $\mathcal{F}_{\mathcal{U}}$ is the image $\mathrm{I}_{\mathcal{F}}$ of the corresponding real-valued rational function.

Thus, universal interval operations are *exact* (or *generalized*) interval operations (see Definition 10.19), and therefore, we have an exact algebra of universal intervals. That is, it follows, from this theorem, that arithmetical expressions which are *identical* in real arithmetic are identical in universal interval arithmetic because both sides of the identity relation yield the image of the real arithmetical expression. With this result at our disposal, many identities of universal interval arithmetic can be entailed by the corresponding identities of real arithmetic.[21] As examples, we can mention that, for $x \in X$, $y \in Y$, and $z \in Z$, the *commutative* and *distributive* laws for universal interval arithmetic can be immediately established from the corresponding laws of real arithmetic, as follows.

$$x + y = y + x \Rightarrow X + Y = Y + X$$
$$x \times y = y \times x \Rightarrow X \times Y = Y \times X$$
$$z \times (x + y) = z \times x + z \times y \Rightarrow Z \times (X + Y) = Z \times X + Z \times Y.$$

In view of this theorem, we then have the following corollary that gives a new reformulation of the dependency relation.

Corollary 10.5 $Y \mathfrak{D}_f X \Leftrightarrow Y = f_{\mathcal{U}}(X; S_1, \ldots, S_m)$

This corollary assures an important and *peculiar* property of universal interval arithmetic: when a real-valued function is translated into the corresponding universal function, the *semantic* of functional dependence is *completely conserved*, and we have the equivalence

$$y = f(x; s_1, \ldots, s_m) \Leftrightarrow Y = f_{\mathcal{U}}(X; S_1, \ldots, S_m).$$

Accordingly, a universal n-ary predicate is characterized as follows.

Definition 10.31 ***(Universal Predicate)*** Let X_k be universal intervals and let P be an n-ary predicate variable symbol. Then

$$P(X_1, \ldots, X_i, X_j, \ldots, X_n) \wedge X_j \mathfrak{D}_f X_i \Leftrightarrow P(X_1, \ldots, X_i, f(X_i; S_1, \ldots, S_m), \ldots, X_n).$$

21 A similar result for classical interval arithmetic was proved by Moore, for the restricted case when every variable *occurs only once* on each side of the identity relation, or, in other words, when all variables are *functionally independent* (see, e.g. [64, 65], and [12]) Moore's result is entailed by Theorem 10.8.

For instance, let $P(X, T, Y, U)$ be $X + T = Y + U$. Then

$$X + T = Y + U \wedge Y\mathfrak{D}_f X \Leftrightarrow X + T = f(X; S_1, \dots, S_m) + U.$$

Combining Theorem 10.19 with Theorem 10.10, we obtain the following result that establishes the relation between a universal rational function and its corresponding classical interval function.

Theorem 10.20 *Let $\mathcal{F}_{\mathfrak{I}}$ and $\mathcal{F}_{\mathfrak{D}}$ be the values of a universal rational function $\mathcal{F}_{\mathcal{U}}(X_1, \dots, X_n)$, for, respectively, $\mathfrak{I}^n_{k=1}(X_k)$ and $X_j\mathfrak{D}_f X_i$, and let $\mathcal{F}_{\mathcal{I}}(X_1, \dots, X_n)$ be the corresponding classical rational function. Then*

$$\mathcal{F}_{\mathfrak{D}} \subseteq \mathcal{F}_{\mathfrak{I}} = \mathcal{F}_{\mathcal{I}}.$$

and, in general

$$\mathrm{I}_{\mathcal{F}} = \mathcal{F}_{\mathcal{U}} \subseteq \mathcal{F}_{\mathcal{I}}.$$

Proof: According to Theorem 10.7, we have $\mathcal{F}_{\mathfrak{D}} \subseteq \mathcal{F}_{\mathfrak{I}}$. By Theorem 10.19, the value of $\mathcal{F}_{\mathcal{U}}$ is the image $\mathrm{I}_{\mathcal{F}}$ of the corresponding real-valued rational function, which is, by Theorem 10.10, a subset of the classical rational function $\mathcal{F}_{\mathcal{I}}$. The two statements, therefore, are verified.

Thus, by virtue of our formalization of the notion of interval dependency (Definition 10.20), the theory of universal intervals, unlike the classical interval theory and its alternates, copes with *all* possible cases of functional dependence between interval variables and hence the name "*universal* interval arithmetic." This makes our construction differ fundamentally from the interval theories discussed in the previous sections.

10.8.2 The Arithmetic of Universal Intervals

On grounds of the results obtained in Section 10.8.1, we are now ready to formally construct the arithmetic of the theory $\mathrm{Th}_{\mathcal{U}}$ of universal interval numbers. We begin by establishing some theorems concerning the universal interval operations. Then, we provide some numerical examples to illustrate how interval dependencies are *fully* addressed by universal interval arithmetic.

In particular, Theorem 10.19, plus Definitions 10.28 and 10.29, implies the following four easily derivable results.

Theorem 10.21 ***(Universal Interval Addition)*** *For any two universal interval numbers $X = [\underline{x}, \overline{x}]$ and $Y = [\underline{y}, \overline{y}]$, universal interval addition is a total operation, on $\mathcal{U}$, formulated as*

$$X + Y = \begin{cases} [\min\limits_{\lambda_x,\lambda_y}(x+y), \max\limits_{\lambda_x,\lambda_y}(x+y)] \\ \quad = [\underline{x}+\underline{y}, \overline{x}+\overline{y}] & \text{if } \mathfrak{I}(X,Y) \\ [\min\limits_{\lambda_x,\lambda_s}(x+f(x;s_1,\dots,s_m)), \\ \quad \max\limits_{\lambda_x,\lambda_s}(x+f(x;s_1,\dots,s_m))] \\ \quad = X+f(X;S_1,\dots,S_m) & \text{if } Y\mathfrak{D}_f X \end{cases}.$$

Theorem 10.22 ***(Universal Interval Multiplication)*** *For any two universal interval numbers $X = [\underline{x}, \overline{x}]$ and $Y = [\underline{y}, \overline{y}]$, universal interval multiplication is a total operation, on $\mathcal{U}$, formulated as*

$$X \times Y = \begin{cases} [\min\limits_{\lambda_x,\lambda_y}(x\times y), \max\limits_{\lambda_x,\lambda_y}(x\times y)] \\ \quad = [\min\{\underline{x}\underline{y}, \underline{x}\overline{y}, \overline{x}\underline{y}, \overline{x}\overline{y}\}, \max\{\underline{x}\underline{y}, \underline{x}\overline{y}, \overline{x}\underline{y}, \overline{x}\overline{y}\}] & \text{if } \mathfrak{I}(X,Y) \\ [\min\limits_{\lambda_x,\lambda_s}(x\times f(x;s_1,\dots,s_m)), \\ \quad \max\limits_{\lambda_x,\lambda_s}(x\times f(x;s_1,\dots,s_m))] \\ \quad = X\times f(X;S_1,\dots,S_m) & \text{if } Y\mathfrak{D}_f X \end{cases}.$$

Theorem 10.23 ***(Universal Interval Negation)*** *For any universal interval number* $[\underline{x}, \overline{x}]$*, universal interval negation is a total operation, on* $\mathcal{U}$*, formulated as*

$$-[\underline{x}, \overline{x}] = [\min_{\lambda_x}(-x), \max_{\lambda_x}(-x)] = [-\overline{x}, -\underline{x}].$$

Theorem 10.24 ***(Universal Interval Reciprocal)*** *For any universal interval number* $[\underline{x}, \overline{x}] \not\supseteq \{0\}$*, universal interval reciprocal is a partial operation, in* $\mathcal{U}$*, formulated as*

$$[\underline{x}, \overline{x}]^{-1} = [\min_{\lambda_x}(x^{-1}), \max_{\lambda_x}(x^{-1})] = [\overline{x}^{-1}, \underline{x}^{-1}].$$

These results, along with Theorem 10.20, express an important fact of our development: for the case when $\mathfrak{I}(X, Y)$, the value of a universal interval operation is the same as that of the corresponding classical interval operation, that is, $X \circ_{\mathcal{U}} Y = X \circ_{\mathcal{I}} Y$; and for the case when $Y \mathfrak{D}_f X$, a universal interval operation gives a different value, according to the dependency function f. This is why universal interval arithmetic copes with all possible cases of interval dependency. In order to clarify the matters, let, for example, σ be a real sentence that takes the prenex form

$$(\forall x)\,(\exists y)\,(\exists z)\,(\exists u)\,(u = xy + z).$$

The Skolem equivalent form of σ is

$$(\exists f)\,(\exists g)\,(\forall x)\,(\exists u)\,(u = x \times f\,(x) + g\,(x)).$$

Such a sentence, as we proved in Section 10.7.2, is not expressible by the *partial* parametric operations, and its dependency relations are not considered by the classical interval operations. In contrast, the sentence σ can be evaluated in universal interval arithmetic, with its dependency relations are completely coped with. Therefore, it comes as a matter of fact that the theory of universal intervals is *completely compatible* with the *semantic* of real arithmetic. That is, any sentence of real arithmetic can be translated into a *semantically equivalent* sentence of universal interval arithmetic.

To complete our construction of universal interval arithmetic, we next define the *total* operation of "*subtraction*," and the *partial* operations of "*division*" and "*integer exponentiation*," for universal interval numbers.

Definition 10.32 ***(Universal Interval Subtraction)*** For any two universal interval numbers X and Y, universal interval subtraction is defined by

$$X - Y = X + (-Y).$$

Definition 10.33 ***(Universal Interval Division)*** For any $X \in \mathcal{U}$ and any $Y \not\supseteq \{0\}$, universal interval division is defined by

$$X \div Y = X \times (Y^{-1}).$$

Definition 10.34 ***(Integer Exponentiation in Universal Intervals)*** For any universal interval number X and any integer n, the integer exponents of X are defined, in terms of multiplication and reciprocal in $\mathcal{U}$, by the following recursion scheme.

i) $X^0 = 1_{\mathcal{U}}$,
ii) $0 < n \Rightarrow X^n = X^{n-1} \times X$,
iii) $0 \notin X \wedge 0 \leq n \Rightarrow X^{-n} = (X^{-1})^n$.

In view of this definition, we have, as an immediate consequence, the following theorem that prescribes the properties of integer exponents of universal interval numbers.

Theorem 10.25 *For any two universal interval numbers X and Y, and any two positive integers m and n, the following identities hold:*

i) $X^m \times X^n = X^{m+n}$,
ii) $(X^m)^n = X^{m\times n}$,
iii) $(X \times Y)^n = X^n \times Y^n$.

Accordingly, we also have the following easy-deducible corollary.

Corollary 10.6 *The identities of integer exponents (i), (ii), and (iii), in Theorem 10.25, are valid for all* $X \not\supseteq \{0\}$ *and* $Y \not\supseteq \{0\}$, *and any two integers m and n.*

To clarify how interval dependencies are *fully* addressed by the universal interval operations, in a way which is *completely compatible* with the *semantic* of real arithmetic, consider the real-valued square function

$$f(x) = x^2.$$

with $x \in [\underline{x}, \overline{x}]$ and $\underline{x}\overline{x} < 0$. The value of the corresponding classical interval function, according to Theorem 10.2, is given by

$$\begin{aligned} f_I([\underline{x}, \overline{x}]) &= [\underline{x}, \overline{x}]^2 \\ &= [\underline{x}, \overline{x}] \times_I [\underline{x}, \overline{x}] \\ &= [\min\{\underline{x}^2, \underline{x}\overline{x}, \overline{x}^2\}, \max\{\underline{x}^2, \underline{x}\overline{x}, \overline{x}^2\}] \\ &= [\underline{x}\overline{x}, \max\{\underline{x}^2, \overline{x}^2\}]. \end{aligned}$$

which is *not consistent* with the fact that a square is always non-negative. Strictly speaking, the accurate image of the real-valued function f is a proper subset of the value of the corresponding classical interval function f_I, that is

$$I_f([\underline{x}, \overline{x}]) \subset f_I([\underline{x}, \overline{x}]).$$

Now, the value of the corresponding universal interval function, according to Theorem 10.22 and Definition 10.34, is given by

$$\begin{aligned} f_{\mathcal{U}}([\underline{x}, \overline{x}]) &= [\underline{x}, \overline{x}]^2 \\ &= [\underline{x}, \overline{x}] \times_{\mathcal{U}} [\underline{x}, \overline{x}] \\ &= [\min_{x \in [\underline{x}, \overline{x}]} (x^2), \max_{x \in [\underline{x}, \overline{x}]} (x^2)] \\ &= [\min_{\lambda_x \in [0,1]} (((\overline{x} - \underline{x})\lambda_x + \underline{x})^2), \max_{\lambda_x \in [0,1]} (((\overline{x} - \underline{x})\lambda_x + \underline{x})^2)] \\ &= [0, \max\{\underline{x}^2, \overline{x}^2\}]. \end{aligned}$$

which is always non-negative, and we have the identity

$$I_f([\underline{x}, \overline{x}]) = f_{\mathcal{U}}([\underline{x}, \overline{x}]).$$

Therefore, unlike classical interval exponentiation, universal exponentiation is *completely compatible* with the *semantic* of the real-valued function $f(x) = x^2$, with $x \in [\underline{x}, \overline{x}]$.

Therefore, with this construction of the universal interval theory at our disposal, it is not surprising that we can formulate and evaluate interval arithmetic expressions in a way analogous to that of real arithmetic. As a further illustration, let us consider some examples.

Example 10.9 Let σ_1 and σ_2 be the two prenex sentences such that

$$\sigma_1 \Leftrightarrow (\forall x \in [-2,2])\,(\forall y \in [-2,2])\,(\exists z \in \mathbb{R})\,(z = f(x,y) = x \times y)$$
$$\sigma_2 \Leftrightarrow (\forall x \in [-2,2])\,(\exists y \in [-2,2])\,(\exists z \in \mathbb{R})\,(z = f(x,y) = x \times y).$$

It is apparent that the variables x and y are independent in the sentence σ_1, in which case the image of f is $\mathrm{I}_f^{\sigma_1} = [-4,4]$. The value of the corresponding universal interval function, for X and Y both are assigned the same individual constant $[-2,2]$, with $\mathfrak{I}(X,Y)$, is given, according to Theorem 10.22, by

$$\begin{aligned} f_{\mathcal{U}}(X,Y) &= X\times_{\mathcal{U}}Y \\ &= \left[\min_{\substack{x\in[-2,2]\\ y\in[-2,2]}}(x\times y),\ \max_{\substack{x\in[-2,2]\\ y\in[-2,2]}}(x\times y)\right] \\ &= \left[\min_{\substack{\lambda_x\in[0,1]\\ \lambda_y\in[0,1]}}((4\lambda_x-2)\times(4\lambda_y-2)),\ \max_{\substack{\lambda_x\in[0,1]\\ \lambda_y\in[0,1]}}((4\lambda_x-2)\times(4\lambda_y-2))\right] \\ &= [-4,4]. \end{aligned}$$

which is the accurate image of f.

In the sentence σ_2, the variable y depends on x, and therefore there is some function $g(x)$ such that σ_2 has the Skolem equivalent form

$$(\exists g)\,(\forall x \in [-2,2])\,(\exists z \in \mathbb{R})\,(z = f(x,g(x)) = x \times g(x)).$$

Let us consider the following two cases for the dependency function g.

i) g is given to be the identity function $y = g(x) = x$. Then, $f(x) = x^2$, with an image of $\mathrm{I}_f^{\sigma_2} = [0,4]$. The value of the corresponding universal interval function, for X is assigned the value $[-2,2]$, is given, according to Theorem 10.22 and Definition 10.34, by

$$\begin{aligned} f_{\mathcal{U}}(X) &= X^2 \\ &= X\times_{\mathcal{U}}X \\ &= [\min_{x\in[-2,2]}(x^2),\ \max_{x\in[-2,2]}(x^2)] \\ &= [\min_{\lambda_x\in[0,1]}((4\lambda_x-2)^2),\ \max_{\lambda_x\in[0,1]}((4\lambda_x-2)^2)] \\ &= [0,4]. \end{aligned}$$

which is the accurate image of f.

ii) g is given to be the negation function $y = g(x) = -x$. Then, $f(x) = -x^2$, with an image of $\mathrm{I}_f^{\sigma_2} = [-4,0]$. The value of the corresponding universal interval function, for X is assigned the value $[-2,2]$, is given, according to Theorem 10.22 and Definition 10.34, by

$$\begin{aligned} f_{\mathcal{U}}(X) &= -X^2 \\ &= X\times_{\mathcal{U}}(-X) \\ &= [\min_{x\in[-2,2]}(-x^2),\ \max_{x\in[-2,2]}(-x^2)] \\ &= [\min_{\lambda_x\in[0,1]}(-(4\lambda_x-2)^2),\ \max_{\lambda_x\in[0,1]}(-(4\lambda_x-2)^2)] \\ &= [-4,0]. \end{aligned}$$

which is the accurate image of f.

Example 10.10 Let σ be the prenex sentence such that

$$\sigma \Leftrightarrow (\forall x \in [-1,1])\,(\exists y \in [-1,1])\,(\exists z \in \mathbb{R})\,(z = f(x,y) = x + y).$$

It is apparent that the variable y depends on x, and therefore, there is some function $g(x)$ such that σ has the Skolem equivalent form

$$(\exists g)\,(\forall x \in [-1,1])\,(\exists z \in \mathbb{R})\,(z = f(x, g(x)) = x + g(x)).$$

Let us consider the following two cases for the dependency function g.

i) g is given to be the identity function $y = g(x) = x$. Then, $f(x) = 2x$, with an image of $I_f^{\sigma} = [-2, 2]$. The value of the corresponding universal interval function, for X is assigned the value $[-1, 1]$, is given, according to Theorem 10.21, by

$$\begin{aligned} f_{\mathcal{U}}(X) &= X +_{\mathcal{U}} X \\ &= [\min_{x\in[-1,1]}(2x), \max_{x\in[-1,1]}(2x)] \\ &= [\min_{\lambda_x\in[0,1]}(2(2\lambda_x - 1)), \max_{\lambda_x\in[0,1]}(2(2\lambda_x - 1))] \\ &= [-2, 2]. \end{aligned}$$

which is the accurate image of f.

ii) g is given to be the negation function $y = g(x) = -x$. Then, $f(x) = 0$, with an image of $I_f^{\sigma} = [0, 0]$. The value of the corresponding universal interval function, for X is assigned the value $[-1, 1]$, is given, according to Theorem 10.21, by

$$\begin{aligned} f_{\mathcal{U}}(X) &= X +_{\mathcal{U}} (-X) \\ &= [\min_{x\in[-1,1]}(x - x), \max_{x\in[-1,1]}(x - x)] \\ &= [0, 0] = 0_{\mathcal{U}}. \end{aligned}$$

which is the accurate image of f.

The identity $X +_{\mathcal{U}} (-X) = [0, 0]$, in the above example, expresses the fact that *additive inverses* exist in universal interval arithmetic. This fact, along with the fundamental algebraic properties of universal interval arithmetic, shall be established in Section 10.9.

Aside from the important fact that universal interval arithmetic copes with all possible cases of interval dependency, the cornerstone result from the above construction is that each of the universal operations of addition and multiplication, unlike the case with parametric intervals, is a single "*total operation*" on $\mathcal{U}$, not a "*partial operation*," and therefore, we can fix the structures $\langle \mathcal{U}; +_{\mathcal{U}} \rangle$ and $\langle \mathcal{U}; \times_{\mathcal{U}} \rangle$ and study their properties in the standard way. A careful investigation of the algebraic system of universal interval arithmetic is attempted in Section 10.9.

Finally, we close this section by characterizing point operations for universal intervals. A *point* universal interval operation is an operation whose operands are universal intervals, and whose result is a *point* universal interval (or, up to isomorphism, a real number). This is made precise in the following definition.

Definition 10.35 ***(Point Universal Operations)*** Let $\mathcal{U}^n$ be the n-th Cartesian power of $\mathcal{U}$. An n-ary point universal operation, ω_n, is a function that maps elements of $\mathcal{U}^n$ to the set $\mathcal{U}_{[x]}$ of point universal intervals, that is

$$\omega_n : \mathcal{U}^n \mapsto \mathcal{U}_{[x]}.$$

Point operations, for a universal interval number $[\underline{x}, \overline{x}]$, are also optimization functions with respect to the argument $\lambda_x \in [0, 1]$ and have the same results as in classical interval arithmetic. Next, we define some of the point operations for universal interval numbers.

- *Width*: $w([\underline{x}, \overline{x}]) = \max_{\lambda_x}(x) - \min_{\lambda_x}(x) = \overline{x} - \underline{x}$.
- *Radius*: $r([\underline{x}, \overline{x}]) = \frac{w([\underline{x}, \overline{x}])}{2} = \frac{(\overline{x} - \underline{x})}{2}$.
- *Midpoint*: $m([\underline{x}, \overline{x}]) = \frac{\max_{\lambda_x}(x) + \min_{\lambda_x}(x)}{2} = ((\overline{x} - \underline{x})\lambda_x + \underline{x})_{\lambda_x = 1/2} = \frac{(\overline{x} + \underline{x})}{2}$.
- *Absolute value*: $|[\underline{x}, \overline{x}]| = \max\{|\min_{\lambda_x}(x)|, |\max_{\lambda_x}(x)|\} = \max\{|\underline{x}|, |\overline{x}|\}$.
- *Distance*: $d([\underline{x}, \overline{x}], [\underline{y}, \overline{y}]) = \max\{|\underline{x} - \underline{y}|, |\overline{x} - \overline{y}|\}$.

Obviously, for two point universal intervals $[x, x]$ and $[y, y]$, the universal distance operation reduces to the usual distance operation on the reals. That is

$$d([x, x], [y, y]) = \max\{|x - y|, |x - y|\} = |x - y|.$$

It is also straightforward to prove that the universal distance induces a *metric space* $(\mathcal{U}, d)$ on the set $\mathcal{U}$ of universal intervals. The importance of this result is that starting from the distance function for universal intervals, the induced *metric space* is a generalization of the standard metric space of real numbers. Thus, the notions of a sequence, convergence, continuity, and a limit are readily definable for universal intervals in the standard way. These notions give rise to what we may call a "*measure theory for universal intervals.*"

10.9 The S-Field Algebra of Universal Intervals

We shall now make use of the part of the theory developed in Section 10.8 to in-depth investigate the algebraic system of universal intervals. In our definition of a universal interval number, the properties of real numbers are naturally assumed in advance.

Let us first mention that as variants of the proofs presented in this section, most of the identities of universal interval arithmetic can be established from the corresponding identities of real arithmetic. This fact is immediately entailed by Theorem 10.19.

In addition, by virtue of Theorems 10.21 –10.24, along with Theorem 10.20, if $\mathfrak{I}(X, Y)$, then the result of a universal interval operation is the same as that of the corresponding classical interval operation, that is, $X \circ_{\mathcal{U}} Y = X \circ_{\mathcal{I}} Y$. This implies that the algebra of universal interval arithmetic will differ from that of classical interval arithmetic for only when $Y \mathfrak{D}_f X$. Accordingly, in the proofs of this section, we shall consider only the case when $Y \mathfrak{D}_f X$. Moreover, for brevity, and without loss of generality, we shall assume that

$$Y \mathfrak{D}_f X \Leftrightarrow \mathrm{I}_f(X) = Y \Leftrightarrow f_{\mathcal{U}}(X) = Y.$$

Our first results of this section, concerning the isomorphism properties, follow immediately from this fact, plus the isomorphism theorem for classical intervals.

Theorem 10.26 ***(Universal Isomorphicity to Real Numbers)*** *The structure $\langle \mathcal{U}_{[x]}; +_{\mathcal{U}}, \times_{\mathcal{U}}; <_M \rangle$ is isomorphically equivalent to the ordered field $\langle \mathbb{R}; +_{\mathbb{R}}, \times_{\mathbb{R}}; <_{\mathbb{R}} \rangle$ of real numbers.*

That is, up to isomorphism, the algebra of point universal intervals, endowed with Moore's ordering $<_M$, is structurally identical to the ordered field of real numbers.

Theorem 10.27 ***(Universal Extensibility of Classical Intervals)*** *Let $+_{\mathfrak{J}}$ and $\times_{\mathfrak{J}}$ be, respectively, the universal addition and multiplication restricted to the case $\mathfrak{J}(X, Y)$. Then, the structure $\langle \mathcal{U}; +_{\mathfrak{J}}, \times_{\mathfrak{J}} \rangle$ is isomorphically equivalent to the commutative S-semiring $\langle \mathcal{I}; +_{\mathcal{I}}, \times_{\mathcal{I}} \rangle$ of classical interval numbers.*

That is, universal interval arithmetic extends classical interval arithmetic in the sense that if it is the case when $\mathfrak{J}(X, Y)$, then the algebra of the universal interval theory is equivalent to that of the classical interval theory.

An important theorem we shall now prove is the inclusion monotonicity theorem for universal intervals, which asserts that the inclusion relation is compatible with the algebraic operations on the set $\mathcal{U}$ of universal interval numbers.

Theorem 10.28 ***(Inclusion Monotonicity in Universal Intervals)*** *Let X_1, X_2, Y_1, and Y_2 be universal interval numbers such that $X_1 \subseteq Y_1$ and $X_2 \subseteq Y_2$. Then, for any binary operation $\circ \in \{+, \times\}$ and any definable unary operation $\lozenge \in \{-, ^{-1}\}$, we have*

i) $X_1 \circ_{\mathcal{U}} X_2 \subseteq Y_1 \circ_{\mathcal{U}} Y_2$,
ii) $\lozenge_{\mathcal{U}} X_1 \subseteq \lozenge_{\mathcal{U}} Y_1$.

Proof: By Theorem 10.19, the value of a universal interval function is the accurate image of the corresponding real function. Images of continuous real functions are inclusion monotonic by Corollary 10.3, and accordingly the theorem follows.

Analogously to classical intervals, from the fact that $[x, x] \subseteq X \Leftrightarrow x \in X$, the membership monotonicity statements $x \circ_{\mathbb{R}} y \in X \circ_{\mathcal{U}} Y$ and $\lozenge_{\mathbb{R}} x \in \lozenge_{\mathcal{U}} X$ are derivable.

Now, we turn to investigate the algebra of universal interval arithmetic. With the help of the results obtained in Section 10.8, we next deduce the fundamental algebraic properties of the universal interval operations.

Theorem 10.29 ***(Absorbing Element in Universal Intervals)*** *The universal interval number $0_{\mathcal{U}}$ is an absorbing element for universal multiplication, that is*

$$(\forall X \in \mathcal{U})\,(0_{\mathcal{U}} \times X = X \times 0_{\mathcal{U}} = 0_{\mathcal{U}}).$$

Proof: For any universal interval number X, according to Theorem 10.22 and assuming the properties of real multiplication, we have

$$\begin{aligned} 0_{\mathcal{U}} \times X &= [\min_{\lambda_x}(0 \times x), \max_{\lambda_x}(0 \times x)] \\ &= [\min_{\lambda_x}(x \times 0), \max_{\lambda_x}(x \times 0)] \\ &= X \times 0_{\mathcal{U}} = 0_{\mathcal{U}}. \end{aligned}$$

and therefore, the point universal interval $0_{\mathcal{U}}$ *absorbs* any universal interval number X by universal multiplication.

In a manner analogous to the proof of Theorem 10.29, the next two theorems concerning identity elements are easily provable.

Theorem 10.30 ***(Identity for Universal Addition)*** *The universal interval number $0_{\mathcal{U}}$ is both a left and right identity for universal addition, that is*

$$(\forall X \in \mathcal{U})\,(0_{\mathcal{U}} + X = X + 0_{\mathcal{U}} = X).$$

Theorem 10.31 ***(Identity for Universal Multiplication)*** *The universal interval number $1_{\mathcal{U}}$ is both a left and right identity for universal multiplication, that is*

$$(\forall X \in \mathcal{U})\,(1_{\mathcal{U}} \times X = X \times 1_{\mathcal{U}} = X).$$

Theorem 10.32 ***(Commutativity in Universal Intervals)*** *Both universal interval addition and multiplication are commutative, that is*

i) $(\forall X, Y \in \mathcal{U})\,(X + Y = Y + X)$,
ii) $(\forall X, Y \in \mathcal{U})\,(X \times Y = Y \times X)$.

Proof: For $\mathfrak{I}\,(X, Y)$, the result holds analogously to commutativity in the classical interval theory.

The proof for $Y \mathfrak{D}_f X \Leftrightarrow f\,(X) = Y$ is constructed as follows.

(i) For any two universal interval numbers X and Y, according to Theorem 10.21 and assuming the properties of real addition, we have

$$\begin{aligned} X + Y &= X + f\,(X) \\ &= [\min_{\lambda_x}(x + f\,(x)), \max_{\lambda_x}(x + f\,(x))] \\ &= [\min_{\lambda_x}(f\,(x) + x), \max_{\lambda_x}(f\,(x) + x)] \\ &= f\,(X) + X = Y + X. \end{aligned}$$

(ii) In a manner analogous to (i), according to Theorem 10.22, we have

$$\begin{aligned} X \times Y &= X \times f\,(X) \\ &= [\min_{\lambda_x}(x \times f\,(x)), \max_{\lambda_x}(x \times f\,(x))] \\ &= [\min_{\lambda_x}(f\,(x) \times x), \max_{\lambda_x}(f\,(x) \times x)] \\ &= f\,(X) \times X = Y \times X. \end{aligned}$$

Therefore, both addition and multiplication are commutative in $\mathcal{U}$.

Theorem 10.33 ***(Associativity in Universal Intervals)*** *Both universal addition and multiplication are associative, that is*

i) $(\forall X, Y, Z \in \mathcal{U})\,(X + (Y + Z) = (X + Y) + Z)$,
ii) $(\forall X, Y, Z \in \mathcal{U})\,(X \times (Y \times Z) = (X \times Y) \times Z)$.

Proof: For the case when all variables are pairwise independent, the result holds analogously to associativity in the classical interval theory.

For the case when some variables are functionally dependent, without loss of generality, we consider the dependency instance $Y \mathfrak{D}_f X \Leftrightarrow f\,(X) = Y$, and the proof is constructed as follows.

(i) For any three universal interval numbers X, Y, and Z, let $[\underline{s}, \overline{s}] = X + (Y + Z)$ and $[\underline{t}, \overline{t}] = (X + Y) + Z$. According to Theorems 10.19 and 10.21, we have

$$\begin{aligned}[\underline{s}, \overline{s}] &= X + (Y + Z) \\ &= X + (f(X) + Z) \\ &= [\min_{\lambda_x,\lambda_z} (x + (f(x) + z)), \max_{\lambda_x,\lambda_z} (x + (f(x) + z))].\end{aligned}$$

and

$$\begin{aligned}[\underline{t}, \overline{t}] &= (X + Y) + Z \\ &= (X + f(X)) + Z \\ &= [\min_{\lambda_x,\lambda_z} ((x + f(x)) + z), \max_{\lambda_x,\lambda_z} ((x + f(x)) + z)].\end{aligned}$$

Optimizing with respect to $\lambda_x, \lambda_z \in [0, 1]$, and assuming associativity of real addition, we thus get

$$\underline{s} = \min_{\lambda_x,\lambda_z} (x + (f(x) + z)) = \min_{\lambda_x,\lambda_z} ((x + f(x)) + z) = \underline{t}.$$

and

$$\overline{s} = \max_{\lambda_x,\lambda_z} (x + (f(x) + z)) = \max_{\lambda_x,\lambda_z} ((x + f(x)) + z) = \overline{t}.$$

Hence, according to Theorem 10.18, we have $X + (Y + Z) = (X + Y) + Z$.

(ii) In a manner analogous to (i), let $[\underline{s}, \overline{s}] = X \times (Y \times Z)$ and $[\underline{t}, \overline{t}] = (X \times Y) \times Z$. According to Theorems 10.19 and 10.22, we have

$$\begin{aligned}[\underline{s}, \overline{s}] &= X \times (Y \times Z) \\ &= X \times (f(X) \times Z) \\ &= [\min_{\lambda_x,\lambda_z} (x \times (f(x) \times z)), \max_{\lambda_x,\lambda_z} (x \times (f(x) \times z))].\end{aligned}$$

and

$$\begin{aligned}[\underline{t}, \overline{t}] &= (X \times Y) \times Z \\ &= (X \times f(X)) \times Z \\ &= [\min_{\lambda_x,\lambda_z} ((x \times f(x)) \times z), \max_{\lambda_x,\lambda_z} ((x \times f(x)) \times z)].\end{aligned}$$

Optimizing with respect to $\lambda_x, \lambda_z \in [0, 1]$, and assuming associativity of real multiplication, we thus get

$$\underline{s} = \min_{\lambda_x,\lambda_z} (x \times (f(x) \times z)) = \min_{\lambda_x,\lambda_z} ((x \times f(x)) \times z) = \underline{t}.$$

and

$$\overline{s} = \max_{\lambda_x,\lambda_z} (x \times (f(x) \times z)) = \max_{\lambda_x,\lambda_z} ((x \times f(x)) \times z) = \overline{t}.$$

Hence, according to Theorem 10.18, we have $X \times (Y \times Z) = (X \times Y) \times Z$.

Therefore, both addition and multiplication are associative in $\mathcal{U}$.

Hereafter, in all the succeeding theorems, and if not otherwise stated, it should be understood that any two interval variables X and Y can be dependent or independent. The proofs for the case when X and Y are dependent can be simply obtained, in a manner analogous to the preceding theorems, by employing the equivalence $Y \mathfrak{D}_f X \Leftrightarrow f(X) = Y$. The trivial case of dependence by identity $X \mathfrak{D} X$ is obvious.

Theorem 10.34 ***(Cancellativity of Universal Addition)*** *Universal interval addition is cancellative, that is*

$$(\forall X, Y, Z \in \mathcal{U})\,(X + Z = Y + Z \Rightarrow X = Y).$$

Proof: Let X, Y, and Z be in $\mathcal{U}$. Assume that $X + Z = Y + Z$. Then, by Theorem 10.21, we immediately have

$$[\min_{\lambda_x,\lambda_z}(x+z), \max_{\lambda_x,\lambda_z}(x+z)] = [\min_{\lambda_y,\lambda_z}(y+z), \max_{\lambda_y,\lambda_z}(y+z)].$$

which, according to Theorem 10.18, yields

$$\min_{\lambda_x,\lambda_z}(x+z) = \min_{\lambda_y,\lambda_z}(y+z) \;\wedge\; \max_{\lambda_x,\lambda_z}(x+z) = \max_{\lambda_y,\lambda_z}(y+z).$$

Optimizing with respect to $\lambda_x, \lambda_y, \lambda_z \in [0, 1]$, we thus get

$$\min_{\lambda_x}(x) = \min_{\lambda_y}(y) \;\wedge\; \max_{\lambda_x}(x) = \max_{\lambda_y}(y).$$

that is $X = Y$, and therefore addition is cancellative in $\mathcal{U}$.

In contrast to the case for addition, and analogously to the classical interval theory, the following theorem asserts that multiplication is not always cancellative[22] in $\mathcal{U}$.

Theorem 10.35 ***(Cancellativity of Universal Multiplication)*** *A universal interval number Z is cancellable for multiplication if, and only if, it is a zeroless interval, that is*

$$(\forall X, Y \in \mathcal{U})\,((X \times Z = Y \times Z \Rightarrow X = Y) \Leftrightarrow 0 \notin Z).$$

Proof: Let X, Y, and Z be in $\mathcal{U}$. Assume that $X \times Z = Y \times Z \Rightarrow X = Y$. Then, by Theorems 10.18 and 10.22, we have

$$\min_{\lambda_x,\lambda_z}(x \times z) = \min_{\lambda_y,\lambda_z}(y \times z) \;\wedge\; \max_{\lambda_x,\lambda_z}(x \times z) = \max_{\lambda_y,\lambda_z}(y \times z)$$
$$\Rightarrow \min_{\lambda_x}(x) = \min_{\lambda_y}(y) \;\wedge\; \max_{\lambda_x}(x) = \max_{\lambda_y}(y).$$

which yields $z \neq 0$, that is $0 \notin [\underline{z}, \overline{z}]$.

The converse direction is easy to prove, and therefore multiplication, is not cancellative in $\mathcal{U}$ except for the case when $0 \notin [\underline{z}, \overline{z}]$.

An important property peculiar to the theory of universal intervals is that unlike classical interval arithmetic, universal interval arithmetic has inverse elements for addition and multiplication. This property figures in the following two theorems.

Theorem 10.36 ***(Additive Inverses in Universal Intervals)*** *Additive inverses exist in universal interval arithmetic, that is*

$$(\forall X \in \mathcal{U})\,(X + (-X) = 0_{\mathcal{U}}).$$

---

22 The cancellative laws for addition and multiplication are also derivable from Theorems 10.36 and 10.37, by the fact that *every invertible element is cancellable.* The cancellative law for multiplication is also entailed by Theorem 10.39, from the fact that *an element is not cancellable for multiplication iff it is a zero divisor.*

Proof: Let X be any universal interval number. According to Theorem 10.21, we immediately have

$$X + (-X) = [\min_{\lambda_x}(x + (-x)), \max_{\lambda_x}(x + (-x))] = 0_{\mathcal{U}}.$$

and therefore for each $X \in \mathcal{U}$, there is an inverse element $(-X) \in \mathcal{U}$ under universal addition.

Theorem 10.37 ***(Multiplicative Inverses in Universal Intervals)*** *Every zeroless universal interval number is invertible for multiplication on $\mathcal{U}$, that is*

$$(\forall X \in \mathcal{U})\,(0 \notin X \Rightarrow X \times (X^{-1}) = 1_{\mathcal{U}}).$$

Proof: Let X be any zeroless universal interval number, that is, $0 \notin X$. According to Theorem 10.22, we immediately have

$$X \times (X^{-1}) = [\min_{\lambda_x}(x \times (x^{-1})), \max_{\lambda_x}(x \times (x^{-1}))] = 1_{\mathcal{U}}.$$

and therefore for each zeroless $X \in \mathcal{U}$, there is an inverse element $(X^{-1}) \in \mathcal{U}$ under universal multiplication.

The result formulated in the following theorem establishes the additive and multiplicative properties of point universal intervals.

Theorem 10.38 ***(Properties of Point Universal Intervals)*** *Let X and Y be two universal interval numbers, and let A be any arbitrary constant in $\mathcal{U}_{[x]}$. Then:*

i) The sum $X + Y$ is a point universal interval iff each of X and Y is a point universal interval, or $Y = A + (-X)$, that is

$$(\forall X, Y \in \mathcal{U})(X + Y \in \mathcal{U}_{[x]} \Leftrightarrow (X \in \mathcal{U}_{[x]} \wedge Y \in \mathcal{U}_{[x]}) \vee (Y = A + (-X))).$$

ii) The product $X \times Y$ is a point universal interval iff each of X and Y is a point universal interval, or at least one of X and Y is $0_{\mathcal{U}}$, or $Y = A \times (X^{-1})$ with $0 \notin X$, that is

$$(\forall X, Y \in \mathcal{U})(X \times Y \in \mathcal{U}_{[x]} \Leftrightarrow (X \in \mathcal{U}_{[x]} \wedge Y \in \mathcal{U}_{[x]}) \vee (X = 0_{\mathcal{U}} \vee Y = 0_{\mathcal{U}}) \vee (Y = A \times (X^{-1}) \wedge 0 \notin X)).$$

Proof: For (i) and (ii), let X and Y be any two universal interval numbers.

(i) According to Theorem 10.21, we have

$$X + Y = [\min_{\lambda_x,\lambda_y}(x + y), \max_{\lambda_x,\lambda_y}(x + y)].$$

Assume that $X + Y \in \mathcal{U}_{[x]}$. Then, $\min_{\lambda_x,\lambda_y}(x + y) = \max_{\lambda_x,\lambda_y}(x + y)$, which yields that each of X and Y is a point universal interval, or, by Theorem 10.36, $Y = A + (-X)$.

The converse direction is easy to prove.

(ii) In a manner analogous to (i), according to Theorem 10.22, we have

$$X \times Y = [\min_{\lambda_x,\lambda_y}(x \times y), \max_{\lambda_x,\lambda_y}(x \times y)].$$

Assume $X \times Y \in \mathcal{U}_{[x]}$. Then, $\min_{\lambda_x,\lambda_y}(x \times y) = \max_{\lambda_x,\lambda_y}(x \times y)$, which yields that each of X and Y is a point universal interval, or at least one of X and Y is $0_{\mathcal{U}}$, or, by Theorem 10.37, $Y = A \times (X^{-1})$ with $0 \notin X$.

The converse direction is easy to prove.

In consequence of this theorem and Theorem 10.29, the following important property of the algebra of universal intervals is easily derivable.

Theorem 10.39 ***(Zero Divisors in Universal Intervals)*** *Nonzero zero divisors do not exist in universal interval arithmetic, that is*

$$(\forall X, Y \in \mathcal{U})\,(X \times Y = 0_{\mathcal{U}} \Rightarrow X = 0_{\mathcal{U}} \vee Y = 0_{\mathcal{U}}).$$

Thus, like the algebra of real numbers, the algebra of universal intervals has no nonzero zero divisors, that is for each $X \neq 0_{\mathcal{U}}$, there is no $Y \neq 0_{\mathcal{U}}$ such that the identity $X \times Y = 0_{\mathcal{U}}$ holds.

Now, we turn to the very desirable algebraic property of distributivity. Distributivity of universal interval arithmetic is established in the next theorem.

Theorem 10.40 ***(Distributivity in Universal Intervals)*** *Multiplication distributes over addition in universal interval arithmetic, that is*

$$(\forall X, Y, Z \in \mathcal{U})\,(Z \times (X + Y) = Z \times X + Z \times Y).$$

Proof: For any three universal interval numbers X, Y, and Z, let $[\underline{s}, \overline{s}] = Z \times (X + Y)$ and $[\underline{t}, \overline{t}] = Z \times X + Z \times Y$. According to Theorems 10.19, 10.21, and 10.22, we have

$$\begin{aligned}[\underline{s}, \overline{s}] &= Z \times (X + Y)\\ &= [\min_{\lambda_x, \lambda_y, \lambda_z} (z \times (x + y)), \max_{\lambda_x, \lambda_y, \lambda_z} (z \times (x + y))].\end{aligned}$$

and

$$\begin{aligned}[\underline{t}, \overline{t}] &= Z \times X + Z \times Y\\ &= [\min_{\lambda_x, \lambda_y, \lambda_z} (z \times x + z \times y), \max_{\lambda_x, \lambda_y, \lambda_z} (z \times x + z \times y)].\end{aligned}$$

Optimizing with respect to λ_x, λ_y, $\lambda_z \in [0, 1]$, we thus get

$$\underline{s} = \min_{\lambda_x, \lambda_y, \lambda_z} (z \times (x + y)) = \min_{\lambda_x, \lambda_y, \lambda_z} (z \times x + z \times y) = \underline{t}.$$

and

$$\overline{s} = \max_{\lambda_x, \lambda_y, \lambda_z} (z \times (x + y)) = \max_{\lambda_x, \lambda_y, \lambda_z} (z \times x + z \times y) = \overline{t}.$$

Hence, according to Theorem 10.18, we have $Z \times (X + Y) = Z \times X + Z \times Y$, and therefore multiplication distributes over addition in $\mathcal{U}$.

Thus, in contrast to the classical interval theory and its present alternates, universal interval arithmetic does satisfy the distributive law.

We shall now make use of the preceding results to prove the following theorem about the algebraic system of universal interval arithmetic.

Theorem 10.41 ***(Integral Domain of Universal Intervals)*** *The structure $\mathfrak{U} = \langle \mathcal{U}; +_{\mathcal{U}}, \times_{\mathcal{U}}; 0_{\mathcal{U}}, 1_{\mathcal{U}} \rangle$ is an integral domain*[23] *with every zeroless element has a multiplicative inverse.*

23 An *integral domain* is a commutative unital ring with no zero divisors.

Proof: By Theorems, 10.33, 10.32, 10.30, and 10.36, the additive structure $\langle \mathcal{U}; +_{\mathcal{U}}, 0_{\mathcal{U}} \rangle$ is an Abelian group. By Theorems 10.33 and 10.32, the multiplicative structure $\langle \mathcal{U}; \times_{\mathcal{U}}, 1_{\mathcal{U}} \rangle$ is an Abelian monoid. In consequence of Theorem 10.40, $\times_{\mathcal{U}}$ distributes over $+_{\mathcal{U}}$. By Theorem 10.29, $0_{\mathcal{U}}$ is an absorbing element for $\times_{\mathcal{U}}$. Hence, $\mathfrak{U}$ is a commutative unital ring, which has, by Theorem 10.39, no nonzero zero divisors. By Theorem 10.37, every zeroless interval has a multiplicative inverse. The structure $\mathfrak{U}$, of universal interval arithmetic, is therefore an integral domain with every zeroless element has a multiplicative inverse.

A field structure $\langle \mathcal{F}; +_{\mathcal{F}}, \times_{\mathcal{F}}; 0_{\mathcal{F}}, 1_{\mathcal{F}} \rangle$ is an integral domain in which every element $\alpha \neq 0_{\mathcal{F}}$ has a multiplicative inverse. The difference between the field structure and the structure of universal intervals is that for the universal interval algebra, we have the condition $0 \notin \alpha$ instead of $\alpha \neq 0_{\mathcal{F}}$. Such an algebraic structure is not usual in mathematics, and it emerges from the fact that the elements of the universe set of the universal interval algebra are themselves *sets* (*interval numbers*). Therefore, to make an explicit stipulation about this new type of algebraic property, it is very convenient to define a new type of algebraic structure, the *set-valued field* (or the S-*field*), that extends the ordinary field structure to the case when elements of the universe set are themselves sets.

Definition 10.36 ***(S-Field)*** Let $\mathfrak{O} = \langle \mathcal{O}; +_{\mathcal{O}}, \times_{\mathcal{O}}; 0_{\mathcal{O}}, 1_{\mathcal{O}} \rangle$ be a field. By a set-valued field (or an S-field) over $\mathfrak{O}$ we understand a field structure $\mathfrak{S} = \langle \mathcal{S}; +_{\mathcal{S}}, \times_{\mathcal{S}}; 0_{\mathcal{S}}, 1_{\mathcal{S}} \rangle$ subject to the following conditions.

i) $\mathcal{S} \subseteq \wp(\mathcal{O})$,
ii) For any $x \in \mathcal{O}$ there is an $X \in \mathcal{S}$ such that $X = \{x\}$,
iii) For $x, y \in \mathcal{O}$, and $\circ \in \{+, \times\}$, $\circ_{\mathcal{S}}$ extends $\circ_{\mathcal{O}}$ such that

$$\{x\} \circ_{\mathcal{S}} \{y\} = \{x \circ_{\mathcal{O}} y\},$$

iv) $0_{\mathcal{S}} = \{0_{\mathcal{O}}\}$such that for each $X \in \mathcal{S}$ the element $0_{\mathcal{O}}$ is the zero element for all $x \in X$,
v) $1_{\mathcal{S}} = \{1_{\mathcal{O}}\}$such that for each $X \in \mathcal{S}$ the element $1_{\mathcal{O}}$ is the unital element for all $x \in X$,
vi) The field axiom

$$(\forall X \in \mathcal{S})\,(0_{\mathcal{S}} \neq X \Rightarrow (\exists Y \in \mathcal{S})\,(X \times_{\mathcal{S}} Y = 1_{\mathcal{S}})).$$

is extended to be

$$(\forall X \in \mathcal{S})\,(0_{\mathcal{S}} \nsubseteq X \Rightarrow (\exists Y \in \mathcal{S})\,(X \times_{\mathcal{S}} Y = 1_{\mathcal{S}})).$$

In view of this definition and Theorem 10.41, the following result can then be concluded.

Theorem 10.42 ***(S-Field of Universal Intervals)*** *The structure* $\mathfrak{U} = \langle \mathcal{U}; +_{\mathcal{U}}, \times_{\mathcal{U}}; 0_{\mathcal{U}}, 1_{\mathcal{U}} \rangle$ *is an S-field.*

Finally, an important immediate result that the preceding theorem implies is the following.

Corollary 10.7 ***(Number System of Universal Intervals)*** *The theory of universal intervals defines an S-number system on the set* $\mathcal{U}$.

Universal intervals thus are S-numbers, and therefore we, can talk of "*universal interval numbers*."

In conclusion, unlike classical interval arithmetic and its present alternates, universal intervals have *additive inverses*, *multiplicative inverses*, and satisfy the property of *distributivity*. By virtue of the algebraic properties proved in this section, universal interval arithmetic possesses a rich S-field algebra, which extends the ordinary field structure of real numbers, and therefore, we do not have to sacrifice the useful properties of ordinary arithmetic. In addition, with our formalization of the notion of interval dependency at disposal (see Section 10.5.2), universal interval operations are defined such that they exhibit a uniform approach applicable to *all cases* of interval dependency. Therefore, on comparing the universal interval theory with other theories of intervals, the main advantage that lies with universal interval arithmetic, over all other theories of intervals, is that the theory of universal intervals is *completely compatible* with the *semantic* of real arithmetic. That is, any sentence of real arithmetic can be translated into a *semantically equivalent* sentence of universal interval arithmetic, without loss of dependency information.

10.10 Guaranteed Bounds or Best Approximation or Both?

In some practical situations, *some overestimation* can be beneficial and important. There are two different problems: *getting guaranteed bounds* of an exact value and *computing an approximation* of the value. The two problems are very different. In robotics and control applications, for example, it is important to have guaranteed inclusions of the exact values in order to guarantee stability under uncertainty.

An interval theory is *theoretically exact* if interval dependencies are *fully* addressed by its algebraic system, and the accurate results can be symbolically obtained. If an interval theory is theoretically exact, we cannot always have computational guaranteed bounds without proper outward rounding because of the limitations of floating point arithmetic. As an example, the classical interval theory is *partially exact* if the computed terms are dependency-free, as for instance, the image of the function $f(x, y) = xy$, with x and y being functionally independent. We cannot always have guaranteed bounds for the image of f, without applying outward rounding. Provided that the universal interval theory is theoretically exact and its algebraic operations are theoretically *overestimation-free*, we can get the best guaranteed inclusions by applying a proper outward rounding.

One issue is that to evaluate a universal interval expression, the minimization and maximization should usually be applied to the whole corresponding *real* expression, in order not to lose the dependency information. If the problem is too large and the optimization is computationally too costly, we may have to *deviate* from the theoretical construction of the theory by dividing the optimizational problem and hence losing some dependency information. For example, in the interval expression

$$X \times Y + X \times Z = [\min_{\lambda_x, \lambda_y, \lambda_z} (xy + xz), \max_{\lambda_x, \lambda_y, \lambda_z} (xy + xz)].$$

we have to apply optimization to the real expression $xy + xz$ as a whole to get the accurate result. Optimizing each of the products xy and xz *separately*, and then summing the results for the individual optimizations, we shall, with this departure from the theoretical construction, lose the dependency information for x.

An approach we can use for overcoming this issue is to slightly alter our construction of the theory of universal intervals by representing an interval by a *3-tuple* $X = [\underline{x}, \overline{x}; \alpha_x]$ called a "*triple interval*," where the third "slot" α_x, that we may call the "*dependency characterizer*," is a symbolic variable used to keep the dependency information for X, as long as there is a *live accessible* value that uses it. The idea is not to lose the dependency information, during a series of calculations, after evaluating *separately* an intermediate interval expression, such that after evaluating a function

$$f([\underline{x}, \overline{x}; \alpha_x], [\underline{y}, \overline{y}; \alpha_y]) = [\underline{z}, \overline{z}; f(\alpha_x, \alpha_y)].$$

the third slot $f(\alpha_x, \alpha_y)$ of the resulting interval keeps the dependency information, and a problem which is computationally too costly can be divided into smaller problems without losing the dependency information. Now, we can rewrite the dependency and identity between two universal *triple* intervals X and Y, respectively, as

$$Y \mathfrak{D}_f X \Leftrightarrow X = [\underline{x}, \overline{x}; \alpha_x] \wedge Y = [\underline{y}, \overline{y}; f(\alpha_x)]$$

$$X = Y \Leftrightarrow X \mathfrak{D}_{\mathrm{Id}} Y$$

$$\Leftrightarrow \underline{x} = \underline{y} \wedge \overline{x} = \overline{y} \wedge \alpha_x = \alpha_y.$$

Triple interval representations are used, with the *classical* interval operations, to evaluate the *particular* interval expressions $X - X$ and $X \div X$ as, respectively, $[0, 0]$ and $[1, 1]$. Although it is limited to such a *dependence by identity*, examples of this usage can be found in algorithmic algebra and artificial intelligence and are successfully and *neatly implementable* in symbolic languages (computer algebra systems), automated theorem provers, and numeric languages (see, e.g. [75, 76], and [69]). The implementation idea is based on constructing an "*interval data structure*" with *three slots*, rather than the usual representation of an interval as a "pair." In symbolic languages, the "*dependency slots*" are added to a *sequence* or a *list* type. In numerical languages, a *variable name* is constructed for each newly added dependency slot and a *counter incremented* (there is a convenient mechanism that is built-in to "Lisp" for this purpose, a "*weak hash table*"). The dependency information can also be *inherited* in any *object-oriented* language. As we (humans) ordinarily do not need to see the extra slot, programs ordinarily will not display it in the computer output (see [75]).

The following code fragment shows how to perform the operations

$$[-1, 1; \alpha_x] - [-1, 1; \alpha_x] = [0, 0; \alpha_x - \alpha_x]$$

$$[-1, 1; \alpha_x] - [-1, 1; \alpha_y] = [-2, 2; \alpha_x - \alpha_y].$$

in the computer algebra system "Macsyma" (or "Maxima") (see, e.g. [77] and [75]).

```
block([x:interval(-1,1)], x-x)
block([x:interval(-1,1),y:interval(-1,1)], x-y)
```

The idea can be implemented similarly in other symbolic languages such as "SymbolicC++" and "Maple" (see, e.g. [78] and [79]). In automated theorem provers, the 3-tuple representation of an interval is usually called a "*block interval*" and the dependency slot is called a "*pending identifier*" (see, e.g. [69]).

Using the triple interval representation with classical interval arithmetic provides some nice results, such as $X - X = [0, 0]$, but the outcome is very limited because of the fact that the classical interval operations are not *dependence-aware*, and interval dependencies cannot therefore be fully addressed via the classical interval theory (and in fact cannot be fully addressed via the present alternate interval theories as well, and this is why the dependency problem is still persisting). In contrast, using the triple interval representation with universal interval arithmetic shall yield more profound results, namely:

(1) Interval dependencies are *fully* addressed by the universal interval operations, and the accurate result can be computationally obtained, if the problem under concern is not computationally too costly.
(2) If the problem under concern is computationally too costly (such as an engineering problem with a large number of constrained variables); the computational cost can be reduced to a minimum by using the triple interval representation to keep the dependency information, and dividing the problem into smaller ones.
(3) From (1) and (2), the *accurate* result can be obtained, along with the computational cost reduced to a *minimum*, by using the triple interval representation with universal interval arithmetic.
(4) Universal interval arithmetic has a nice S-*field* algebra, which extends the ordinary field structure of real numbers.
(5) With universal interval arithmetic equipped with proper *outward rounding*, we can have reasonable *guaranteed bounds* of the exact values whenever needed.

Supplementary Materials

To reproduce the results of the calculations in this chapter, latest version of InCLosure is available for free download via https://doi.org/10.5281/zenodo.2702404 or from the first author's website at: http://scholar.cu.edu.eg/

henddawood/software/InCLosure. An InCLosure input file and its corresponding output containing, respectively, the code and results of the examples are also available as a supplementary material to this chapter, via http://doi.org/10.5281/zenodo.3236736.

Acknowledgments

The authors would like to thank Prof. Nefertiti Megahed for her advices that greatly contributed to polishing this text and for all the discussions we have had on the subject. We would like also to acknowledge Prof. Snehashish Chakraverty, the editor of this book, for his patience and constructive comments, and for his outstanding help with this chapter. Finally, it is a pleasure to acknowledge the courtesy and substantial help which we have received from the staff of Wiley & Sons at all stages of writing this text.

References

1 Dawood, H. (2014). Interval mathematics as a potential weapon against uncertainty. In: *Mathematics of Uncertainty Modeling in the Analysis of Engineering and Science Problems*, Chapter 1, S. Chakraverty, 1–38. Hershey, PA: IGI Global. ISBN: 978-1-4666-4991-0. http://dx.doi.org/10.4018/978-1-4666-4991-0.ch001.

2 Kosheleva, O. and Kreinovich, V. (2019). Physics Need for Interval Uncertainty and How it Explains Why Physical Space is (at least) 3-Dimensional. Technical Report UTEP-CS-19-05. University of Texas at El Paso.

3 Heath, T.L. (ed.) (2009). *The Works of Archimedes: Edited in Modern Notation with Introductory Chapters*. Cambridge: Cambridge University Press. https://doi.org/10.1017/CBO9780511695124.

4 Moore, R.E (1959). Automatic Error Analysis in Digital Computation. Technical Report LMSD-48421. Palo Alto, CA: Lockheed Missiles and Space Company, Lockheed Corporation.

5 Burkill, J.C. (1924). Functions of intervals. *Proceedings of the London Mathematical Society* 2 (1): 275–310.

6 Young, R.C. (1931). The algebra of many-valued quantities. *Mathematische Annalen* 104: 260–290.

7 Dwyer, P.S. (1951). *Linear Computations*. New York: Chapman & Hall.

8 Sunaga, T. (1958). Theory of an interval algebra and its application to numerical analysis. *RAAG Memoirs* 2: 29–46.

9 Dawood, H. (2012). Interval mathematics: foundations, algebraic structures, and applications. Master's thesis. Giza: Department of Mathematics, Faculty of Science, Cairo University. http://dx.doi.org/10.13140/RG.2.2.24252.13449.

10 Fu, C., Liu, Y., and Xiao, Z. (2019). Interval differential evolution with dimension-reduction interval analysis method for uncertain optimization problems. *Applied Mathematical Modelling* 69: 441–452.

11 Kreinovich, V. and Shary, S.P. (2016). Interval methods for data fitting under uncertainty: a probabilistic treatment. *Reliable Computing* 23: 105–141.

12 Moore, R.E., Kearfott, R.B., and Cloud, M.J. (2009). *Introduction to Interval Analysis*. SIAM.

13 Rump, S.M. (1999). INTLAB–INTerval LABoratory. In: *Developments in Reliable Computing* (ed. T. Csendes), 77–104. Dordrecht: Kluwer Academic Publishers.

14 Chevillard, S., Joldes, M., and Lauter, C. (2010). Sollya: an environment for the development of numerical codes. In: *Mathematical Software - ICMS 2010, Lecture Notes in Computer Science*, vol. 6327 (ed. K. Fukuda, J. van der Hoeven, M. Joswig, and N. Takayama), 28–31. Heidelberg: Springer-Verlag.

15 Dawood, H. (2018). InCLosure (Interval enCLosure)–A language and environment for reliable scientific computing. Computer Software, Version 2.0. Giza, Egypt: Department of Mathematics, Faculty of Science, Cairo University. https://doi.org/10.5281/zenodo.2757278. InCLosure Support: http://scholar.cu.edu.eg/henddawood/software/InCLosure.

16 Melquiond, G. (2008). Proving bounds on real-valued functions with computations. In: *International Joint Conference on Automated Reasoning IJCAR, Lecture Notes in Artificial Intelligence*, vol. 5195 (ed. A. Armando, P. Baumgartner, and G. Dowek), 2–17. Springer-Verlag. https://doi.org/10.1007/978-3-540-71070-7_2.

17 Dawood, H. (2011). *Theories of Interval Arithmetic: Mathematical Foundations and Applications*. Saarbrücken: LAP Lambert Academic Publishing. ISBN: 978-3-8465-0154-2.

18 Dawood, H. (2019). On some algebraic and order-theoretic aspects of machine interval arithmetic. *Online Mathematics Journal (OMJ)* 1 (2): 1–13. http://doi.org/10.5281/zenodo.2656089.

19 Kolev, L.V. (1993). *Interval Methods for Circuit Analysis*. World Scientific Publishing Company.

20 Kulisch, U.W. (2008). *Computer Arithmetic and Validity: Theory, Implementation, and Applications*. Walter de Gruyter.

21 Muller, J.-M., Brisebarre, N., De Dinechin, F. et al. (2009). *Handbook of Floating-Point Arithmetic*, 1e. Boston, MA: Birkhäuser.

22 IEEE 1788 Committee (2015). IEEE Standard for Interval Arithmetic. *IEEE Std 1788-2015*, pp. 1–97. https://doi.org/10.1109/IEEESTD.2015.7140721. https://ieeexplore.ieee.org/document/7140721.

23 Barnes, D.W. and Mack, J.M. (1975). *An Algebraic Introduction to Mathematical Logic*, 1e. Springer-Verlag.

24 Dawood, H. and Dawood, Y. (2013). Logical Aspects of Interval Dependency. Technical Report CU-Math-2013-03-LAID. Department of Mathematics, Faculty of Science, Cairo University.

25 Dawood, H. and Dawood, Y. (2014). On Some Order-Theoretic Aspects of Interval Algebras. Technical Report CU-Math-2014-06-OTAIA. Department of Mathematics, Faculty of Science, Cairo University.

26 Dawood, H. and Dawood, Y. (2017). Investigations Into a Formalized Theory of Interval Differentiation. Technical Report CU-Math-2017-03-IFTID. Department of Mathematics, Faculty of Science, Cairo University.

27 Dawood, H. and Dawood, Y. (2019). A logical formalization of the notion of interval dependency: towards reliable intervalizations of quantifiable uncertainties. *Online Mathematics Journal* 1 (3). http://doi.org/10.5281/zenodo.3234184.

28 Menini, C. and Van Oystaeyen, F. (2004). *Abstract Algebra: A Comprehensive Treatment*, 1e. CRC Press.

29 (a)Cantor, G. (1897). Beitrage zur begrundung der transfiniten mengenlehre II. *Mathematische Annalen* 49: 207–246;(b) Translated with introduction and commentary by Jourdain, P.E.B. (1955). *Contributions to the Founding of the Theory of Transfinite Numbers*. New York: Dover Publications.

30 Amer, M.A. (1989). First order logic with empty structures. *Studia Logica* 48 (2): 169–177.

31 Suppes, P. (1972). *Axiomatic Set Theory*. New York: Dover Publications.

32 Tarski, A. (1994). *Introduction to Logic and to the Methodology of the Deductive Sciences*, 4e. New York: Oxford University Press. translated from Polish by Olaf Helmer.

33 Mostowski, A. (1951). On the rules of proof in the pure functional calculus of the first order. *The Journal of Symbolic Logic* 16 (2): 107–111.

34 Van Orman Quine, W. (1954). Quantification and the empty domain. *The Journal of Symbolic Logic* 19 (3): 177–179.

35 Hintikka, J. (1959). Existential presuppositions and existential commitments. *The Journal of Philosophy* 56 (3): 125–137.

36 Corcoran, J. (1980). Categoricity. *History and Philosophy of Logic* 1 (1): 187–207. https://doi.org/10.1080/01445348008837010.

37 van Hoorn, W.G. and van Rootselaar, B. (1967). Fundamental notions in the theory of seminearrings. *Compositio Mathematica* 18 (1–2): 65–78.

38 Pilz, G. (1983). *Near-Rings: The Theory and its Applications, North-Holland Mathematics Studies*, vol. 23. North-Holland Publishing Company.

39 Clay, J.R. (1992). *Nearrings: Geneses and Applications*. Oxford University Press.

40 Hansen, E.R. (1975). A generalized interval arithmetic. In: *Interval Mathematics, Lecture Notes in Computer Science*, vol. 29. (ed. Karl L. E. Nickel), 7–18. Springer-Verlag.

41 Gardenyes, E., Mielgo, H., and Trepat, A. (1985). Modal intervals: reason and ground semantics. In: *Interval Mathematics, Lecture Notes in Computer Science*, vol. 212. (ed. Karl L. E. Nickel), 27–35. Springer-Verlag.

42 Markov, S.M. (1995). On directed interval arithmetic and its applications. *Journal of Universal Computer Science* 1 (7): 514–526.

43 Lodwick, W.A. (1999). Constrained Interval Arithmetic. Technical Report 138. Denver: University of Colorado at Denver, Center for Computational Mathematics.

44 Dawood, H. and Dawood, Y. (2016). Interval Algerbras: A Formalized Treatment. Technical Report CU-Math-2016-06-IAFT. Department of Mathematics, Faculty of Science, Cairo University.

45 Dawood, H. and Dawood, Y. (2013). A Dependency-Aware Interval Algebra. Technical Report CU-Math-2013-09-DAIA. Department of Mathematics, Faculty of Science, Cairo University.

46 Shayer, S. (1965). Interval Arithmetic with Some Applications for Digital Computers. Technical Report LMSD-5136512. Palo Alto, CA: Lockheed Missiles and Space Company, Lockheed Corporation.

47 Dawood, H. and Dawood, Y. (2010). On the Metamathematics of the Theory of Interval Numbers. Technical Report CU-Math-2010-06-MTIN. Giza: Department of Mathematics, Faculty of Science, Cairo University.

48 Rokne, J. and Ratschek, H. (1984). *Computer Methods for the Range of Functions*. Ellis Horwood Publications.

49 Euler, L. (2000). *Foundations of Differential Calculus*. Springer-Verlag.

50 Lobachevskii, N.I. (1951). *Complete Works 5*. Moscow-Leningrad.

51 Armstrong, W.W. (1974). Dependency structures of data base relationships. In: *Proceedings of IFIP Congress*, 580–583.

52 Hintikka, J. (1996). *The Principles of Mathematics Revisited*. Cambridge University Press.

53 Vaananen, J. (2007). *Dependence Logic: A New Approach to Independence Friendly Logic*. Cambridge University Press.

54 Skolem, T.A. (1920). Logisch-kombinatorische untersuchungen über die erfüllbarkeit oder beweisbarkeit mathematischer satze nebst einem theoreme über dichte mengen. Skrifter utgit av Videnskabsselskapet i Kristiania, I. Matematisk-Naturvidenskabelig Klasse No. 4, pp. 1–36. Translated from Norwegian as "*Logico-combinatorial investigations on the satisfiability or provability of mathematical propositions: a simplified proof of a theorem by Loewenheim*" by Stefan Bauer-Mengelberg.

55 Feferman, S. (2006). What kind of logic is independence-friendly logic? In *The Philosophy of Jaakko Hintikka, Library of Living Philosophers*, vol. 30 (ed. R.E. Auxier and L.E. Hahn), 453–469. Open Court Publishing Company.

56 Chen, G. and Pham, T.T. (2000). *Introduction to Fuzzy Sets, Fuzzy Logic, and Fuzzy Control Systems*, 1e, CRC Press.

57 Tarski, A. (1951). *A Decision Method for Elementary Algebra and Geometry*, 2e. University of California Press.

58 Basu, S., Pollack, R., and Roy, M.-F. (2003). *Algorithms in Real Algebraic Geometry*, 1e. Springer-Verlag.

59 Collins, G.E. (1975). Quantifier elimination for the elementary theory of real closed fields by cylindrical algebraic decomposition. In: *Automata Theory and Formal Languages 2nd GI Conference Kaiserslautern, Lecture Notes in Computer Science*, vol. 33. (ed. H. Brakhage), 134–183. Springer-Verlag.

60 McCallum, S. (2001). On propagation of equational constraints in cad-based quantifier elimination. In: *Proceedings of the International Symposium on Symbolic and Algebraic Computation*, 223–231.

61 Corcoran, J. and Ramnauth, A. (2013). Equality and identity. *Bulletin of Symbolic Logic* 19 (3): 255–256.

62 Gaganov, A.A. (1985). Computational complexity of the range of the polynomial in several variables. *Cybernetics* 21: 418–421.

63 Kreinovich, V. (2008). Interval computations as an important part of granular computing: an introduction. In: *Handbook of Granular Computing*, Chapter 1, 1e (ed. W. Pedrycz, A. Skowron, and V. Kreinovich). Wiley-Interscience. 3–32.

64 Moore, R.E. (1966). *Interval Analysis*. Prentice Hall.

65 Moore, R.E. (1979). *Methods and Applications of Interval Analysis, SIAM studies in Applied Mathematics*, vol. 2. Philadelphia, PA: SIAM.

66 Alefeld, G. and Herzberger, J. (1983). *Introduction to Interval Computation*, 1e. New York: Academic Press.

67 Alefeld, G. and Mayer, G. (2000). Interval analysis: theory and applications. *Journal of Computational and Applied Mathematics* 121 (1): 421–464.

68 Lodwick, W.A. (2008). Fundamentals of interval analysis and linkages to fuzzy set theory. In: *Handbook of Granular Computing*, Chapter 3, 1e. W. Pedrycz, A. Skowron, and V. Kreinovich). Wiley-Interscience. 55–79.

69 Pichler, F. (2007). *Computer Aided Systems Theory*. Springer-Verlag.

70 Cleary, J.G. (1987). Logical arithmetic. *Future Computing Systems* 2 (2): 125–149.

71 Cleary, J.G. (1993). Proving the Existence of Solutions in Logical Arithmetic. Technical report 93/5. Hamilton, New Zealand: University of Waikato, Department of Computer Science.

72 Dawood, H. and Dawood, Y. (2019). Parametric intervals: more reliable or foundationally problematic? *Online Mathematics Journal* 1 (3). http://doi.org/10.5281/zenodo.3234186.

73 Elishakoff, I. and Miglis, Y. (2012). Overestimation-free computational version of interval analysis. *International Journal for Computational Methods in Engineering Science and Mechanics* 13 (5): 319–328. https://doi.org/10.1080/15502287.2012.683134.

74 Piegat, A. and Landowski, M. (2017). Is an interval the right result of arithmetic operations on intervals? *International Journal of Applied Mathematics and Computer Science* 27 (3): 575–590. https://doi.org/10.1515/amcs-2017-0041.

75 Fateman, R.J. (2009). *Interval Arithmetic, Extended Numbers and Computer Algebra Systems*. University of California at Berkeley.

76 Keene, S. (1988). *Object-Oriented Programming in Common Lisp: A Programmer's Guide to CLOS*. Addison-Wesley Publishing Company.

77 Chu, D. and Barnes, D.J. (2010). *Introduction to Modeling for Biosciences*. Springer-Verlag.

78 Robertson, J.S. (1996). *Engineering Mathematics with Maple*. New York: McGraw-Hill.

79 Tan, K.S., Steeb, W.-H., and Hardy, Y. (2008). *Computer Algebra With Symbolic C++*. World Scientific.

11

Affine-Contractor Approach to Handle Nonlinear Dynamical Problems in Uncertain Environment

Nisha Rani Mahato, Saudamini Rout, and Snehashish Chakraverty

Department of Mathematics, National Institute of Technology Rourkela, Rourkela, Odisha 769008, India

11.1 Introduction

Nonlinear dynamical problems form backbone in several fields of science and engineering, viz., structural mechanics, fluid dynamics, control theory, robotics, seismology, circuit analysis, etc. In this regard, nonlinear ordinary differential equations (NODEs) are taken into consideration in this chapter. The general form of NODE governed from different science and engineering problems is given as

$$L(y) + N(y) = F(t) \tag{11.1}$$

subject to the n initial conditions (ICs), $y(t) = a_1$ and $y^{(i)}(t) = a_{i+1}$ for $i = 1, 2, \ldots, n$. Here, $L(y)$ and $N(y)$ are the respective linear and nonlinear parts of the NODE (11.1), and $F(t)$ is the external force vector applied on the nonlinear problem. In actual practice, the material properties of the modeled NODEs are taken as crisp (exact) values for ease handling of the nonlinear systems. Few literature studies regarding the computation of solutions for nonlinear dynamical problems having crisp parameters using various semianalytical methods are discussed initially. He [1] proposed a nonlinear analytical technique, viz., homotopy perturbation method (HPM) for solving NODEs with crisp parameters. Further, Lyapunov's artificial small parameter method (LASPM) has been discussed by Lyapunov [2] to solve nonlinear differential equations. Further, He [3] presented a semianalytic technique, viz., variational iteration method to handle autonomous ordinary differential systems. HPM and its applications to nonlinear wave equations have also been discussed by He [4]. Akbari et al. [5] presented an algebraic method, viz., Akbari–Ganji method (AGM), for solving NODEs such as Van der Pol, Rayleigh, and Duffing equations. Some notes on using the HPM procedure for solving time-dependent differential equations have been discussed by Babolian et al. [6]. Chowdhury and Hashim [7] gave various solutions of a class of singular second-order initial value problems using HPM. Another aspect for computation of solutions for NODEs has been developed based on artificial neural networks. Chebyshev neural network-based model has been proposed by Mall and Chakraverty [8] for solving Lane–Emden-type equations. Also, Mall and Chakraverty [9] developed a Hermite functional link neural network to solve the Van der Pol–Duffing oscillator equation.

As mentioned above, the parameters and variables associated with the material properties with respect to (11.1) are considered as crisp values. However, in actual practice, we may have incomplete information about the variables and parameters being a result of errors in measurements, observations, or it may be due to maintenance-induced error. As such, we may have only vague, imprecise, and incomplete information about the variables and parameters rather than the crisp values. Because of such vagueness and imprecision, these variables and parameters are uncertain in nature. Traditionally, these uncertain parameters are handled using probabilistic approach. However, probabilistic methods sometimes may not deliver reliable results without

Mathematical Methods in Interdisciplinary Sciences, First Edition. Edited by Snehashish Chakraverty.

sufficient experimental data. Other than probabilistic approach, interval analysis and fuzzy set theory are other aspects of handling uncertainties. In fuzzy set theory, the fuzzy numbers may be easily converted to corresponding interval forms using the α-cut method. As such, interval arithmetic (IA) may be considered as a powerful tool to handle uncertainties with more ease. R.E. Moore [10–12] introduced the concepts of interval analysis in early 1960s. Interval computations, algebra of intervals, interval matrices and functions, interval differential, and integral equations have also been discussed in Moore [12].

Further, the dependency problem in standard interval arithmetic is a major hurdle while handling uncertainties with interval computations. Because of this scenario, the overestimation occurs, which is comparatively larger than the exact range of uncertain parameters. In this regard, affine arithmetic (AA) is a recent approach to handle uncertainties in a more efficient manner. As such, few literature studies are available on interval dependency problem, affine arithmetic, and its applications. In 1993, AA was first introduced by Comba and Stolfi [13] as an improved version of classical interval arithmetic to handle uncertainties. Few years later, Stolfi and De Figueiredo [14] discussed the drawbacks of standard interval arithmetic because of the presence of interval overestimation problem in it. Further, the reason behind the dependency problem in standard interval arithmetic and also the concept and applications of AA are presented by De Figueiredo and Stolfi [15]. Skalna and Hladík [16] proposed a new algorithm for Chebyshev minimum-error multiplication of reduced affine forms. Parametric interval linear systems with nonaffine dependencies have been solved using a direct method by Skalna [17]. Ceberio et al. [18] and Ludäscher et al. [19] handled interval-type and affine arithmetic-type techniques for expert systems having uncertainties. Affine arithmetic, its implementation, and improvements are discussed by Rump and Kashiwagi [20]. Akhmerov [21] illustrated an interval-affine Gaussian algorithm for constrained systems. Utilizing the best multiplication in AA, a dividing method is developed by Miyajima and Kashiwagi [22]. Vaccaro and Canizares [23], Wang et al. [24], and Gu et al. [25] discussed about the affine arithmetic-based algorithm for uncertain power flow and optimal power flow studies. Few authors also developed modified AA and surveyed its accuracy with general AA. Shou et al. [26] and Soares [27] worked on modified AA. Further, there also exist other efficient methods to handle interval dependency problem in classical interval arithmetic. Contractor-based approach is one of the well-known numerical approaches used to overcome the dependency problem. In this regard, few literature studies are present by some authors. In 2009, Chabert and Jaulin [28] introduced contractor programming. Applications of interval analysis with examples in parameter and state estimation, robust control, and robotics have been given by Jaulin et al. [29]. Rego et al. [30] presented set-based state estimation of nonlinear systems using constrained zonotopes and interval arithmetic.

There exist very few literature studies regarding the solution of nonlinear dynamical problems in uncertain environment. Chakraverty and Mahato [31] developed a method, viz., Laplace–Pade parametric HPM, to solve nonlinear oscillators having uncertain parameters. Enclosures of uncertain nonlinear equations using set inversion via interval analysis (SIVIA) Monte Carlo approach is discussed by Mahato et al. [32]. Moreover, Rout and Chakraverty [33] proposed an affine approach to solve nonlinear dynamical problems of structures in uncertain environment. To the best of our knowledge, there may not exist any work to solve uncertain nonlinear dynamics by using affine-contractor approach.

In this regard, this chapter proposes a semianalytic method, viz., HPM, based on affine-contractor for evaluating the solution bounds of nonlinear dynamical problems in uncertain environment. Initially, all the uncertain parameters of the interval nonlinear ordinary differential equation (INODE) in the form of closed intervals are converted into its affine forms having different noise symbols $\varepsilon_i \in [-1, 1]$ for $i = 1, 2, \ldots$. Then, the affine nonlinear ordinary differential equation (ANODE) having uncertain ICs also transformed in the form of affine numbers that has been solved by using affine HPM. Here, the solutions are in the form of noise symbols. Further, to obtain the enclosures of the solution, affine-contractor HPM has been proposed. In this respect, the present section of this chapter contains the introduction and literature survey. In Section 11.2, classical interval arithmetic and its properties are discussed. Then, the dependency problem in the case of standard interval arithmetic is mentioned in Section 11.3. In this regard, affine arithmetic, contractor, SIVIA, etc., are incorporated in Sections 11.4

and 11.5. Section 11.6 introduces the proposed methods affine HPM and affine-contractor HPM for calculating the enclosures of INODE. Some illustrative application problems, viz., Rayleigh equation (for unforced NODE), Van der Pol–Duffing equation (for forced NODE), and nonhomogeneous Lane–Emden equation, are discussed in Section 11.7 based on the proposed algorithm. Finally, the chapter ends with overall conclusions in Section 11.8.

11.2 Classical Interval Arithmetic

In this section, interval systems and their basic notations, set operations in interval systems, standard interval computations, and the algebra of intervals are incorporated.

11.2.1 Intervals

Interval is a floating-point representation of a real quantity "r", which may be denoted as $[r] = [\inf(r), \sup(r)]$ or $[r] = [\underline{r}, \overline{r}]$. The intervals are subsets of the real number set $\mathbb{R}$ and may be defined as

$$[r] = [\underline{r}, \overline{r}] = \{a \in \mathbb{R} \mid \underline{r} \leq a \leq \overline{r}, \quad \text{where} \quad \underline{r}, \overline{r} \in \mathbb{R} \text{ and } \underline{r} \leq \overline{r}\}$$

In this regard, some of the terminologies of interval are illustrated below.

Center: The center of an interval $[r] = [\underline{r}, \overline{r}]$ may be defined as

$$r_c = \frac{1}{2}(\underline{r} + \overline{r})$$

Width: The width of an interval $[r] = [\underline{r}, \overline{r}]$ may be defined as

$$r_w = \overline{r} - \underline{r}$$

Radius: The radius of an interval $[r] = [\underline{r}, \overline{r}]$ may be defined as

$$r_\Delta = \frac{1}{2}(\overline{r} - \underline{r})$$

An interval can be represented in the form of its center and radius as $[r] = [r_c - r_\Delta, r_c + r_\Delta]$. It may be noted that two intervals $[r] = [\underline{r}, \overline{r}]$ and $[s] = [\underline{s}, \overline{s}]$ are said to be equal if and only if $\underline{r} = \underline{s}$ and $\overline{r} = \overline{s}$. Further, the interval $[r] = [\underline{r}, \overline{r}]$ is known as degenerate interval $\{r\}$ in which $\underline{r} = \overline{r}$. The degenerate interval is identical with the real number r.

11.2.2 Set Operations of Interval System

For intervals $[r] = [\underline{r}, \overline{r}]$ and $[s] = [\underline{s}, \overline{s}]$, the set operations of the interval system are illustrated as follows.

Interval intersection: The intersection of two intervals $[r]$ and $[s]$ (having common points) is illustrated as

$$[r] \cap [s] = \{x\colon x \in [r] \text{ and } x \in [s]\} = [\max\{\underline{r}, \underline{s}\}, \min\{\overline{r}, \overline{s}\}] \tag{11.2}$$

Further, if $\overline{s} < \underline{r}$ or $\overline{r} < \underline{s}$ (that is having no common points), then the intersection of the intervals is empty,

$$[r] \cap [s] = \emptyset$$

Interval union: The union of two intervals having common points may be defined as

$$[r] \cup [s] = \{x\colon x \in [r] \text{ or } x \in [s]\} = [\min\{\underline{r}, \underline{s}\}, \max\{\overline{r}, \overline{s}\}] \tag{11.3}$$

Interval hull: The interval hull of two intervals may be represented as

$$[r]\underline{\cup}[s] = [\min\{\underline{r}, \underline{s}\}, \max\{\overline{r}, \overline{s}\}] \tag{11.4}$$

Moreover, it may be noted that

$$[r] \cup [s] \subseteq [r]\underline{\cup}[s] \tag{11.5}$$

11.2.3 Standard Interval Computations

Let $[r] = [\underline{r}, \overline{r}]$ and $[s] = [\underline{s}, \overline{s}]$ be two interval representations. Then, standard interval computation may be defined as

$$[r] * [s] = \{r * s \mid r \in [r], s \in [s]\}$$

where "$* = \{+, -, \times, /\}$". Keeping this in view, all the operations of standard interval arithmetic may be described as follows (Moore et al. [12]).

- *Addition*:

$$[r] + [s] = [\underline{r} + \underline{s}, \overline{r} + \overline{s}] \tag{11.6}$$

- *Subtraction*:

$$[r] - [s] = [\underline{r} - \overline{s}, \overline{r} - \underline{s}] \tag{11.7}$$

  In particular, $[r] - [r] = [\underline{r} - \overline{r}, \overline{r} - \underline{r}] \neq \{0\}$.
- *Scalar multiplication*:

$$\begin{cases} c \cdot [r] = [c \cdot \underline{r}, c \cdot \overline{r}], & c \geq 0 \\ c \cdot [r] = [c \cdot \overline{r}, c \cdot \underline{r}], & c < 0 \end{cases} \tag{11.8}$$

- *Multiplication*:

$$[r] \cdot [s] = [\min\{X([r], [s])\}, \max\{X([r], [s])\}] \tag{11.9}$$

  where, $X([r], [s]) = (\underline{rs}, \underline{r}\overline{s}, \overline{r}\underline{s}, \overline{rs})$.
- *Reciprocal*:

$$\frac{1}{[r]} = \begin{cases} \left[\frac{1}{\overline{r}}, \frac{1}{\underline{r}}\right], & 0 \notin [r] \\ (-\infty, \infty), & 0 \in [r] \end{cases} \tag{11.10}$$

- *Division*:

$$\frac{[r]}{[s]} = [r] \cdot \frac{1}{[s]} = \begin{cases} [\underline{r}, \overline{r}] \cdot \left[\frac{1}{\overline{s}}, \frac{1}{\underline{s}}\right], & 0 \notin [s] \\ (-\infty, \infty), & 0 \in [s] \end{cases} \tag{11.11}$$

  In particular, $\frac{[r]}{[r]} \neq 1$.
- *Power*:
  - If $m > 0$ is an odd integer, then

$$[r]^m = [\underline{r}^m, \overline{r}^m] \tag{11.12a}$$

– If $m > 0$ is an even integer, then

$$[r^m] = \begin{cases} [\underline{r}^m, \overline{r}^m], & [r] > 0 \\ [\overline{r}^m, \underline{r}^m], & [r] < 0 \\ [0, \max\{\underline{r}^m, \overline{r}^m\}], & 0 \in [r] \end{cases} \tag{11.12b}$$

11.2.4 Algebraic Properties of Interval

For intervals $[r]$, $[s]$, and $[t]$, the following algebraic properties of intervals may be considered.

Commutative: Intervals are commutative under both addition and multiplication as

$$[r] + [s] = [s] + [r] \text{ and } [r] \cdot [s] = [s] \cdot [r]$$

respectively.

Associative: Both interval addition and multiplication are associative as

$$[r] + ([s] + [t]) = ([r] + [s]) + [t] \text{ and } [r] \cdot ([s] \cdot [t]) = ([r] \cdot [s]) \cdot [t]$$

respectively.

Identity: The additive identity and multiplicative identity elements are the degenerate intervals 0 and 1, respectively.

$$[r] + 0 = 0 + [r] = [r]$$
$$[r] \cdot 1 = 1 \cdot [r] = [r]$$

Inverse: There do not exist any additive or multiplicative inverses except for degenerate intervals. $-[r]$ is not an additive inverse of the interval $[r]$. Because $[r] + (-[r]) = [\underline{r} - \overline{r}, \overline{r} - \underline{r}] \neq \{0\}$. Similarly, as $\frac{[r]}{[r]} \neq \{1\}$, then, $\frac{1}{[r]}$ is not a multiplicative inverse of $[r]$.

Subdistributive: The subdistributive law for intervals is illustrated as

$$[r] \cdot ([s] + [t]) \subseteq [r] \cdot [s] + [r] \cdot [t] \tag{11.13}$$

It may be noted that the interval system does not obey the distributive law of classical real number system.

Cancelation law: The cancelation law holds in the case of interval addition, but it does not hold for interval multiplication, that is

$$[r] + [t] = [s] + [t] \Rightarrow [r] = [s]$$

11.3 Interval Dependency Problem

The dependency problem in interval analysis is a major hurdle while handling uncertainties with interval computations. Standard interval arithmetic treats all its operands independently over their ranges. However, the case when the operands partially depend on each other, interval computation results to comparatively wide intervals than exact solution. This situation of overestimation of the resulting solutions is generally called as the "interval dependency problem," because of this scenario, for long iterative calculations and complex computations, the overestimation of range increases rapidly with each iteration.

For instance, the subtraction routine in standard interval arithmetic cannot assume that the two given intervals denote the same ideal quantity because there may be the case where two independent quantities have the same range. Thus, in interval arithmetic, subtraction may be responsible for its dependency problem.

Subtraction of an interval $[r] = [\underline{r}, \overline{r}]$ with itself may be given as

$$[x] = [r] - [r] = [\underline{r} - \overline{r}, \overline{r} - \underline{r}] \neq \{0\} \tag{11.14}$$

From the terminologies of the interval mentioned above, the width of the above interval is obtained as $x_w = (\overline{r} - \underline{r}) - (\underline{r} - \overline{r}) = 2(\overline{r} - \underline{r})$, which is twice the width of the operand interval instead of being zero.

11.4 Affine Arithmetic

AA is recently developed to handle uncertainties in an efficient manner because of the presence of interval dependency problem in standard interval arithmetic. AA is a model for self-validated numerical computation that aims to attack the dependency problems in interval computations. AA keeps track of first-order correlations between computed and input quantities. These correlations are automatically exploited in primitive operations with the result that in many cases, AA is able to produce interval estimates that are much better than the ones obtained with standard interval arithmetic. Moreover, AA also implicitly provides a geometric representation for the joint range of related quantities that can be exploited to increase the efficiency of interval methods to be used in the theoretical and applied research.

The affine representation of the ideal quantity "r" may be denoted as $\hat{r}$ and represented as

$$\hat{r} = r_0 + \sum_{i=1}^{k} r_i \varepsilon_i = r_0 + r_1\varepsilon_1 + r_2\varepsilon_2 + \ldots + r_k\varepsilon_k \tag{11.15}$$

which is a linear polynomial of the real variables ε_i for $i = 1, 2, \ldots, k$ that are known as noise symbols. These unknown symbolic real values lie in a particular interval $\mathbb{U} = [-1, 1]$. Further, a particular affine representation must have noise symbols independent to each other, but there may be some of the common noise symbols present in two different affine representations. Every affine form may have different number of noise symbols in its representations. During the time of affine computations, new noise symbols may also be generated.

The first term of the affine representation r_0 is known as the central value of the affine form $\hat{r}$ and all the associated coefficients r_i of each noise symbols ε_i (for $i = 1, \ldots, k$) are finite real numbers for $i = 0, 1, \ldots, k$ and are called as partial deviation of the affine form $\hat{r}$.

11.4.1 Conversion Between Interval and Affine Arithmetic

The interval $[r] = [\underline{r}, \overline{r}]$ may be transformed into its affine representation as

$$\hat{r} = r_0 + r_m\varepsilon_m \tag{11.16}$$

where r_0 and r_m are the center and radius of the given interval $[r]$, respectively. They are represented as

$$r_0 = \frac{1}{2}(\underline{r} + \overline{r}) \text{ and } r_m = \frac{1}{2}(\overline{r} - \underline{r})$$

Further, the uncertainty present in the ideal quantity r of the interval $[r]$ is expressed by a new noise symbol ε_m, which should not be occurred in any other existing affine representations.

Conversely, let $\hat{r}$ be an affine representation given as

$$\hat{r} = r_0 + \sum_{i=1}^{k} r_i \varepsilon_i = r_0 + r_1\varepsilon_1 + \ldots + r_k\varepsilon_k \tag{11.17}$$

The interval bounds of the given affine representation may be obtained as

$$[r] = [r_0 - d_r, r_0 + d_r] \tag{11.18}$$

where d_r is known as the total deviation of the affine representation and may be defined as

$$d_r = \sum_{i=1}^{k} |r_i| = |r_1| + |r_2| + \ldots + |r_k|$$

Moreover, the transformed interval bound $[r]$ from the affine $\hat{r}$ is the smallest interval, which contains all the possible values of the affine in spite of the fact that each one of the noise symbol ε_i ranges independently over the interval $[-1, 1]$. It may be noted that all the correlation information present in the affine representation may be discarded after its conversion into interval bounds.

11.4.2 Affine Operations

Affine representations of two intervals $[r]$ and $[s]$ are given as follows:

$$\hat{r} = r_0 + \sum_{i=1}^{k} r_i\varepsilon_i \text{ and } \hat{s} = s_0 + \sum_{i=1}^{k} s_i\varepsilon_i$$

Thus, the binary operations of the affine forms (De Figueiredo and Stolfi [15]) can be illustrated as follows:

- *Addition*:

$$\hat{r} + \hat{s} = (r_0 + s_0) + \sum_{i=1}^{k}(r_i + s_i)\varepsilon_i \tag{11.19}$$

- *Subtraction*:

$$\hat{r} - \hat{s} = (r_0 - s_0) + \sum_{i=1}^{k}(r_i - s_i)\varepsilon_i \tag{11.20}$$

  Particularly, $\hat{r} - \hat{r} = 0$.
- *Scalar multiplication*:

$$c \cdot \hat{r} = (c \cdot r_0) + \sum_{i=1}^{k}(c \cdot r_i)\varepsilon_i \tag{11.21}$$

  where $c \in \mathbb{R}$.
- *Multiplication*:

$$\hat{r} \cdot \hat{s} = \left(r_0 + \sum_{i=1}^{k} r\varepsilon_i\right) \cdot \left(s_0 + \sum_{i=1}^{k} s_i\varepsilon_i\right) = (r_0 s_0) + \sum_{i=1}^{k}(r_0 s_i + r_i s_0)\varepsilon_i + \sum_{i=1}^{k} r_i\varepsilon_i \cdot \sum_{i=1}^{k} s_i\varepsilon_i \tag{11.22}$$

  Therefore,

$$\hat{r} \cdot \hat{s} = (r_0 s_0) + \sum_{i=1}^{k}(r_0 s_i + r_i s_0)\varepsilon_i + \omega_m\varepsilon_m$$

  where

$$|\omega_m| \geq \left|\sum_{i=1}^{k} r_i\varepsilon_i \cdot \sum_{i=1}^{k} s_i\varepsilon_i\right|, \varepsilon_i \in [-1, 1].$$

  It may be noted that while performing affine multiplication, a new noise symbol (ε_m) is generated and $|\omega_m|$ is an upper bound for the approximated error.

- *Division*:

$$\frac{\hat{r}}{\hat{s}} = \hat{r} \cdot \frac{1}{\hat{s}} = \frac{r_0}{s_0} + \frac{1}{\hat{r}} \sum_{i=1}^{k} \left(r_i - \frac{r_0}{s_0} s_i \right) \varepsilon_i \tag{11.23}$$

provided $\hat{s} \neq \{0\}$ (Skalna [17]). Particularly, $\frac{\hat{r}}{\hat{r}} = 1$.

Next, we consider an example to show that how affine arithmetic is able to overcome the "interval dependency problem."

Example 11.1 Let us consider a nonlinear function such that

$$g(s) = s_1^2 + s_2^2 - s_1 s_2 \quad \forall s \in [1, 2] \times [3, 4] \tag{11.24}$$

The exact solution of the function is given by $g(s) \in [6.75, 13]$. By standard interval arithmetic, we may have,

$$\begin{aligned} g(s)([s]) &= [s_1^2] + [s_2^2] - [s_1][s_2] \\ &= [1, 2]^2 + [3, 4]^2 - [1, 2][3, 4] \\ &= [2, 17] \end{aligned}$$

That is,

$$[g(s)]_{\text{IA}} \in [2, 17] \tag{11.25}$$

Further, the interval bounds of the variables may be converted into its affine forms as follows:

$$s_1 \in [1, 2] \Rightarrow \hat{s}_1 = 1.5 + 0.5\varepsilon_1 \tag{11.26a}$$

and

$$s_2 \in [3, 4] \Rightarrow \hat{s}_2 = 3.5 + 0.5\varepsilon_2 \tag{11.26b}$$

where ε_1 and ε_2 are the noise symbols of the affine representations and $\varepsilon_i \in \mathbb{U} = [-1, 1]$ for $i = 1, 2$. Then, applying different operations of affine arithmetic, the affine form of the given function $g(s)$ may be calculated as

$$\begin{aligned} g(s) &= \hat{s}_1^2 + \hat{s}_2^2 - \hat{s}_1 \hat{s}_2 \\ &= (1.5 + 0.5\varepsilon_1)^2 + (3.5 + 0.5\varepsilon_2)^2 - (1.5 + 0.5\varepsilon_1)(3.5 + 0.5\varepsilon_2) \\ &= 9.25 - 0.25\varepsilon_1 + 2.75\varepsilon_2 + 0.25\varepsilon_3 + 0.25\varepsilon_4 + 0.25\varepsilon_5 \end{aligned}$$

where ε_3, ε_4, and ε_5 are newly generated noise symbols during the affine operation to evaluate the solution and $\varepsilon_i \in \mathbb{U} = [-1, 1]$ for $i = 3, 4, 5$.

Thus, the corresponding interval bounds of the affine form associated with function $g(s)$ is obtained as

$$[g(s)]_{\text{AA}} = [9.25 - 3.75, 9.25 + 3.75] = [5.5, 13] \tag{11.27}$$

where $g_0 = 9.25$ and $d_g = 3.75$. Therefore,

$$g(s) \in [g(s)]_{\text{AA}} \subseteq [g(s)]_{\text{IA}}$$

It may be clearly noted that affine arithmetic results in comparatively tighter bounds than standard interval arithmetic.

11.5 Contractor

A contractor C (Chabert and Jaulin [28] and Jaulin et al. [29]) associated with a set $S \subset \mathbb{R}^n$ over domain $\mathbb{D}$ is an operator

$$C: \mathbb{IR}^n \to \mathbb{IR}^n$$

satisfying the following properties:

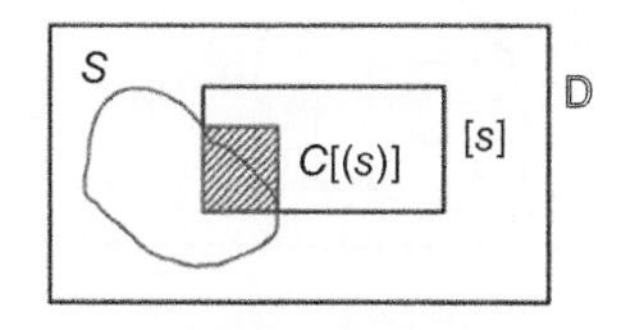

Figure 11.1 Contractor of [s].

- *Contraction*: $C([s]) \subset [s], \quad \forall [s] \in \mathbb{IR}^n$,
- *Completeness*: $C([s]) \cap S = [s] \cap S, \quad \forall [s] \in \mathbb{IR}^n$.

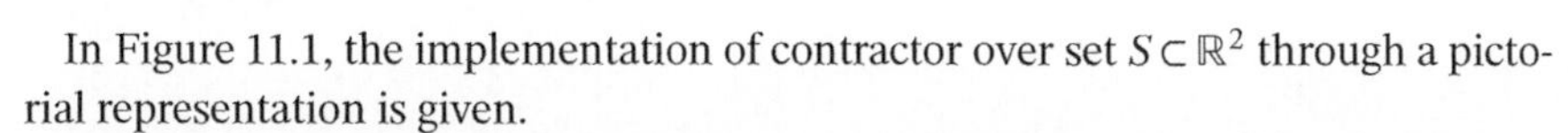

In Figure 11.1, the implementation of contractor over set $S \subset \mathbb{R}^2$ through a pictorial representation is given.

There exist various types of contractors, viz., minimal contractor, forward–backward contractor, and fixed point contractor. These contractors may be illustrated as follows:

Minimal contractor: The minimal contractor may be referred as C_{ml}, which is associated with the set S if there exists a contractor C such that

$$C([s]) \subset C_{\text{ml}}([s]) \quad \text{for} \quad [s] \in \mathbb{IR}^n$$

Therefore,

$$C([s]) \cap C_{\text{ml}}([s]) = [s] \cap [S] \quad \text{for} \quad [s] \in \mathbb{IR}^n$$

Forward–backward contractor: The forward–backward contractor is based on constraint

$$g(s) = 0, \quad \text{where} \quad s \in [s] \text{ and } [s] \in \mathbb{IR}^n$$

Fixed-point contractor: The fixed-point contractor associated with ψ is implemented with respect to the constraint $g(s) = 0$ as $s = \psi(s)$, where $s \in [s] \in \mathbb{IR}^n$. Then,

$$s \in [s] \text{ and } s = \psi(s) \Rightarrow s \in [s] \text{ and } s \in \psi([s])$$

$$\Rightarrow s \in [s] \cap [\psi]([s])$$

In the case of implementation of forward–backward contractor along with fixed-point contractor, helps in computation of forward–backward contractor until the fixed interval vector $C^n([s]) = C^{n+1}([s]) = [s_{\in}]$ with $\in$-width is reached.

11.5.1 SIVIA

The set inversion of a given set $S \subset \mathbb{R}^n$ with respect to a function $g: \mathbb{R}^n \to \mathbb{R}^m$ is expressed as follows:

$$S = g^{-1}(T) = \{s \in \mathbb{R}^n \mid t = g(s) \in T\}, \quad \text{where} \quad T \subset \mathbb{R}^m$$

The set inversion is the case of interval analysis [4] is described as follows:

$$g^{-1}([t]) \cap [s] = \{s \in [s] \mid \exists t \in [t], t = g(s)\} \tag{11.28}$$

where

$$g: \mathbb{R}^n \to \mathbb{R}^m \text{ and } [s] \times [t] \in \mathbb{R}^{n+m}$$

In the case of SIVIA [10], an initial search set $[s_0]$ may be assumed to contain the required set S. Then, by using subpavings with contractors as given in Figure 11.2, the desired enclosure of the solution set S is obtained based on the inclusion properties given below:

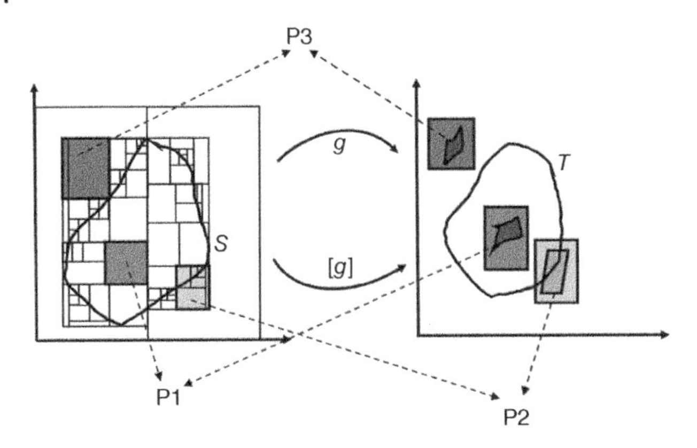

Figure 11.2 Set inversion via interval analysis (SIVIA).

P1: $[g]([s]) \subset T \Rightarrow [s] \in S$, then $[s]$ is a solution;
P2: $[g]([s]) \cap T = \emptyset \Rightarrow [s] \cap S = \emptyset$, then $[s]$ is not a solution;
P3: $[g]([s]) \cap T \neq \emptyset$ and $[g]([s]) \not\subset T$, then $[s]$ is an undetermined solution.

Bisection with respect to the width $w([s]) = \max_i s_{iw}$ for $i = 1, 2, \ldots, n$, where $[s] = \{s_1, s_2, \ldots, s_n\}$ is performed till $\omega(C([s])) > \varepsilon$.

Let us consider an example to show the implementation of contractor-based SIVIA.

Example 11.2 Consider a circle $x^2 + y^2 = r^2$ having width $r \in [3, 4]$ and with the initial search set as $[-5, 5] \times [-5, 5]$.

Then, using PyIbex library (Chabert and Jaulin [28]), the enclosure is depicted in Figure 11.3.

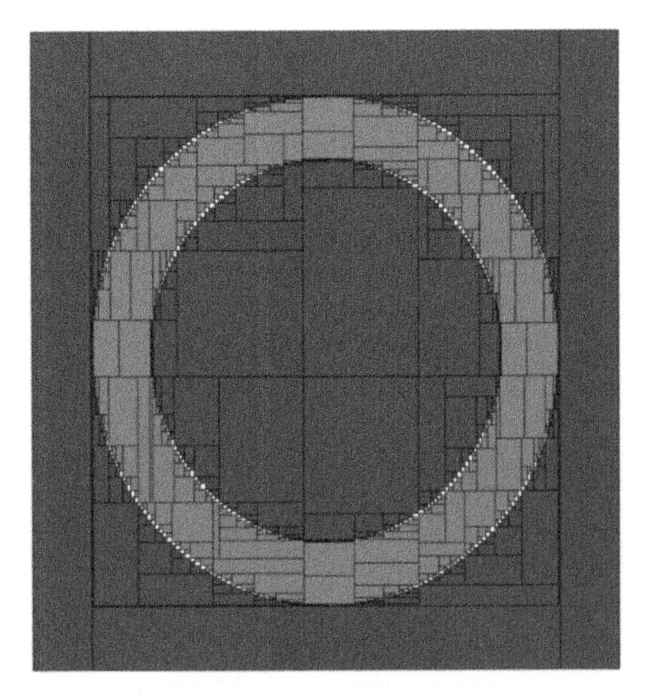

Figure 11.3 SIVIA of circle with width [3, 4].

Table 11.1 Comparison of enclosures obtained using IA, AA, and SIVIA.

Enclosure of $g(s) = y$			
IA	AA	SIVIA	
$[g(s)]_{IA}$	$[g(s)]_{AA}$	Precision, ε	$[g(s)]_C$
[2, 17]	[5.5, 13]	1	[3.1362, 15.1083]
		0.5	[4.9967, 13.275]
		0.1	[6.3884, 10.6755]
		0.05	[6.5825, 10.3576]
		0.01	[6.7128, 10.0783]
		0.005	[6.7308, 10.0438]
		0.001	[6.7461, 10.0038]

In order to have better understanding of the implementation of contractors for computation enclosures of uncertain nonlinear equations, we have again considered Example 11.1. Accordingly, we obtain the constraint satisfaction problem (CSP) associated with (11.24), where variables $\mathcal{V} = \{y, s_1, s_2\}$ subject to constraint $\mathcal{E} = \{e_1\}$ over the domain $\mathbb{D} = [1, 2] \times [3, 4]$ such that

$$e_1 : y = s_1^2 + s_2^2 - s_1 s_2$$

Using SIVIA based on forward–backward contractors, we obtained the enclosure of y. The comparative enclosures with respect to Eqs. (11.25, 11.27) for different precisions $\varepsilon = 1, 0.1, 0.01$ are illustrated in Table 11.1.

It may be seen from Table 11.1 that contractors based on SIVIA, results to tighter enclosure for better precision compared to IA and AA. Based on the efficiency of contractor approach, we have further applied affine as well as contractor approach for solving dynamic problems governed by nonlinear differential equations subject to uncertain parameters. In this regard, classical HPM has been combined with AA and contractors in Section 11.6.

11.6 Proposed Methodology

Let us again consider the general form of NODE given in (11.1),

$$L(y) + N(y) = F(t)$$

subject to the ICs in uncertain environment. Here, the uncertainty with respect to NODE is considered in the form of affine representations (11.17). In particular, the affine forms of ICs may be illustrated as,

$$y(0) = [r_1] = [\underline{r_1}, \overline{r_1}] = \frac{1}{2}(\underline{r_1} + \overline{r_1}) + \frac{1}{2}(\overline{r_1} - \underline{r_1})\varepsilon_1 \tag{11.29a}$$

$$\dot{y}(0) = [r_2] = [\underline{r_2}, \overline{r_2}] = \frac{1}{2}(\underline{r_2} + \overline{r_2}) + \frac{1}{2}(\overline{r_2} - \underline{r_2})\varepsilon_2 \tag{11.29b}$$

$$\vdots$$

$$y^{(n)}(0) = [r_n] = [\underline{r_n}, \overline{r_n}] = \frac{1}{2}(\underline{r_n} + \overline{r_n}) + \frac{1}{2}(\overline{r_n} - \underline{r_n})\varepsilon_n \tag{11.29c}$$

and so on, where $\varepsilon_i \in [-1, 1]$ for $i = 1, 2, \ldots$ are noise symbols of the affine forms. The number of conditions (initial and/or boundary) may depend on the highest order of the NODE. In HPM, a homotopy

$$v(t, p) : \mathbb{R} \times [0, 1] \to \mathbb{R}$$

is constructed satisfying

$$\mathcal{H}(v, p) : (1 - p)[L(v) - L(y_0)] + p[L(v) + N(v) - F(t)] = 0 \tag{11.30}$$

where $p \in [0, 1]$ is known as the embedding parameter and $y_0(t)$ is an initial guess satisfying the ICs of Eq. (11.1), where $t \in \Omega$.

Equivalently, Eq. (11.30) may be illustrated as

$$\mathcal{H}(v, p) : L(v) - L(y_0) + p[L(y_0) + N(v) - F(t)] = 0 \tag{11.31}$$

Therefore, during the deformation of p, we may have

$$\mathcal{H}(v, 0) : L(v) - L(y_0) \quad \text{for} \quad p = 0 \tag{11.32a}$$

$$\mathcal{H}(v, 1) : L(v) + N(v) - F(t) \quad \text{for} \quad p = 1 \tag{11.32b}$$

Equation (11.32a) has $y_0(t)$ as one of its solutions and also for "L" to be linear, $y_0(t)$ is the only solution. Thus,

$$v(t, 0) = y_0(t) \tag{11.33a}$$

and

$$v(t, 1) = y(t) \tag{11.33b}$$

Since p varies from 0 to 1, $v(t, p)$ also varies from $y_0(t)$ to $y(t)$. This process is known as "*deformation*". By assuming the embedding parameter $p \in [0, 1]$ comparatively small and incorporating the classical perturbation technique, we may get the solution of Eqs. (11.33a) and (11.33b) in the form of a power series of p. That is,

$$v(t, p) = \sum_{i=0}^{\infty} (v_i(t) p^i) \tag{11.34}$$

where $v_i(t)$s' are the coefficients yet to be determined. Further, for $p = 1$, we may have

$$v(t) = \sum_{i=0}^{\infty} (v_i(t)) \tag{11.35}$$

which is the approximate solution of (11.1).

To determine the coefficients, the homotopy of (11.1) may be obtained by substituting the power series (11.34) in Eq. (11.30) as

$$(1 - p)\left[L\left(\sum_{i=0}^{\infty} (v_i(t) p^i)\right) - L(y_0(t))\right] + p\left[L\left(\sum_{i=0}^{\infty} (v_i(t) p^i)\right) + N\left(\sum_{i=0}^{\infty} (v_i(t) p^i)\right) - F(t)\right] = 0 \tag{11.36}$$

Now, by equating the coefficients of powers of p from both sides of (11.36), we may have a system of differential equations for $i = 0, 1, \ldots$ as,

$$p^0 : L(v_0) - L(y_0) = 0 \tag{11.37a}$$

subject to ICs $v_0(t_0) = \hat{a}_1, v_0'(t_0) = \hat{a}_2, \ldots, v_0^{(n)}(t_0) = \hat{a}_n$.

$$p^1 : L(v_1) + L(y_0) + N(v_0) - F(t) = 0 \tag{11.37b}$$

subject to ICs $v_1(t_0) = 0, v_1'(t_0) = 0, \dots, v_1^{(n)}(t_0) = 0$

$$\vdots$$

$$p^i : L(v_i) + \text{coeff}(N(v), p^{i-1}) = 0 \tag{11.37c}$$

subject to ICs $v_i(t_0) = 0, v_i'(t_0) = 0, \dots, v_i^{(n)}(t_0) = 0$ for $i = 2, 3, \dots, k$.

The coeff($N(v)$, p^{i-1}) stands for the coefficients of p^i in the nonlinear part $N(v(t))$ of the NODE. Thus, the unknown coefficients may be determined by solving Eqs. (11.37a)–(11.37c).

Accordingly, the above-mentioned affine HPM procedure has been incorporated in the following Algorithm 11.1.

Algorithm 11.1 Affine Homotopy Perturbation Method

Input: Linear part $L(y)$; nonlinear part $N(y)$; force $F(x)$; series truncation limit l;

ICs $\mathbf{y}(x_0) = \{y(x_0) = a_1, y'(x_0) = a_2, \dots, y^{(n)}(x_0) = a_n\}$

Output: Affine solution $y(x)|_{\varepsilon_1,\varepsilon_2,\dots,\varepsilon_m \in [-1,1]} = y(x; \varepsilon_1, \varepsilon_2, \dots, \varepsilon_m)$

Steps:

1. Interval to affine conversion using (11.16)
   Convert uncertain coefficients to affine form
   Convert uncertain ICs to affine form
2. Assume initial solution

$$y_0(x) = \sum_{j=1}^{n} a_j (x - x_0)^{j-1}$$

3. Homotopy $\mathcal{H}(v, p)$

$$(1-p)\left[L\left(\sum_{i=0}^{l}(v_i p^i)\right) - L(y_0)\right] + p\left[L\left(\sum_{i=0}^{l}(v_i p^i)\right) + N\left(\sum_{i=0}^{l}(v_i p^i)\right) - F(x)\right] = 0$$

4. Obtain system of differential equations

$$p^0 : L(v_0) - L(y_0) = 0$$

subject to ICs $v_0(t_0) = a_1, v_0'(t_0) = a_2, \dots, v_0^{(n)}(t_0) = a_n$

$$p^1 : L(v_1) + L(y_0) + N(v_0) - F(x) = 0$$

subject to ICs $v_1(t_0) = 0, v_1'(t_0) = 0, \dots, v_1^{(n)}(t_0) = 0$

$$\vdots$$

$$p^i : L(v_i) + \text{coeff}(N(v), p^{i-1}) = 0$$

subject to ICs $v_i(t_0) = 0, v_i'(t_0) = 0, \dots, v_i^{(n)}(t_0) = 0$ for $i = 2, 3, \dots, k$

5. Solve system of differential equations in step 4 to obtain truncated series solution of v_i for $i = 0, 1, \dots, k$.

$$y(t; \varepsilon_1, \varepsilon_2, \dots, \varepsilon_m) \approx \sum_{i=0}^{k} v_i(t; \varepsilon_1, \varepsilon_2, \dots, \varepsilon_m)$$

6. Obtain affine solution using Laplace transform and Pade approximation
   Laplace transform

$$y_{\mathcal{L}}(s; \varepsilon_1, \varepsilon_2, \ldots, \varepsilon_m) = \mathcal{L}\{y(t; \varepsilon_1, \varepsilon_2, \ldots, \varepsilon_m), t, s\}$$

   Pade approximation

$$y_P(s; \varepsilon_1, \varepsilon_2, \ldots, \varepsilon_m) = \text{pade}\left\{y_{\mathcal{L}}\left(\frac{1}{s}; \varepsilon_1, \varepsilon_2, \ldots, \varepsilon_m\right), [m, q]\right\}$$

   where, $2 \le m, q \le m + q = O(y(t))$
   Inverse Laplace transform

$$y_{\mathcal{L}^{-1}}(t; \varepsilon_1, \varepsilon_2, \ldots, \varepsilon_m) = \mathcal{L}^{-1}\left\{y_P\left(\frac{1}{s}; \varepsilon_1, \varepsilon_2, \ldots, \varepsilon_m\right), s, t\right\}$$

   Affine solution

$$y(x; \varepsilon_1, \varepsilon_2, \ldots, \varepsilon_m) = y_{\mathcal{L}^{-1}}(t; \varepsilon_1, \varepsilon_2, \ldots, \varepsilon_m)$$

In Algorithm 11.1, the Laplace operator "$\mathcal{L}$" is applied to the truncated series solution followed by Pade approximation in order to transform the truncated series to meromorphic function for exhibiting the periodic nature over large time domain. Finally, the inverse Laplace operator "$\mathcal{L}^{-1}$" is applied to retain the affine solution $y(t; \varepsilon_1, \varepsilon_2, \ldots, \varepsilon_m)$.

Next, we have incorporated the implementation of Algorithm 11.1 using a simple nonlinear differential equation in Example 11.3.

Example 11.3 Consider a simple uncertain nonlinear differential equation (crisp form given in He [1]),

$$y'(x) + y^2(x) = 0 \tag{11.38}$$

subject to IC $y(0) = a_1$ where $x \in \Omega$ and $a_1 \in [0.7, 1.1]$.

Using Step 1 of Algorithm 11.1, we initially compute the affine forms of interval uncertainty as $\hat{a}_1 = [0.7, 1.1] = 0.9 + 0.2\varepsilon_1$. Then, step 2 results to initial solution

$$y_0(x) = 0.9 + 0.2\varepsilon_1 \tag{11.39}$$

with respect to (11.38) and $y_0'(x) = 0$. Then, the homotopy $\mathcal{H}(v, p)$ using (11.36) is considered as

$$(1 - p)[L(v) - L(y_0)] + p[L(v) + N(v) - F(x)] = 0$$

where $L(v) = v'$, $N(v) = v^2$, and $F(x) = 0$. Further, using step 3, the homotopy reduces to

$$(1 - p)\left[\sum_{i=0}^{\infty}(v_i'p^i) - y_0'\right] + p\left[\sum_{i=0}^{5}(v_i'p^i) + \left(\sum_{i=0}^{\infty}(v_ip^i)\right)^2\right] = 0$$

where $v = \sum_{i=0}^{\infty}(v_ip^i)$. Then, based on step 4, we obtain the system of differential equations for computing series solution of each v_i where $i = 0, 1, \ldots, k$ as,

$$p^0: v_0' = 0 \quad \text{subject to IC} \quad v_0(0) = 0.9 + 0.2\varepsilon_1 \tag{11.40a}$$

$$p^1: v_1' + v_0' = 0 \quad \text{subject to IC} \quad v_1(0) = 0 \tag{11.40b}$$

$$p^2: v_2' + 2v_0v_1 = 0 \quad \text{subject to IC} \quad v_2(0) = 0 \tag{11.40c}$$

$$\vdots$$

$$p^i: v_i' + \text{coeff}(v_i^2, p^{i-1}) = 0 \quad \text{subject to IC} \quad v_i(0) = 0 \tag{11.40d}$$

for $i = 3, 4, \ldots, k$.

It is worth mentioning that the affine variables $\varepsilon_1 \in [-1, 1]$. Then, by solving the system of differential equations given in Eqs. (11.40a)–(11.40d) to obtain v_i for $i = 0, 1, \ldots, k$. In particular, the truncated series solutions of $O(5)$ in terms of affine variables $\varepsilon_1 \in [-1, 1]$ using Eqs. (11.40a)–(11.40d) are obtained as

$$v_0(x; \varepsilon_1) = 0.9 + 0.2\varepsilon_1 \tag{11.41a}$$

$$v_1(x; \varepsilon_1) = -0.81x - 0.36x\varepsilon_1 - 0.04x\varepsilon_1^2 \tag{11.41b}$$

$$v_2(x; \varepsilon_1) = 0.729x^2 + 0.486x^2\varepsilon_1 + 0.108x^2\varepsilon_1^2 + 0.008x^2\varepsilon_1^3 \tag{11.41c}$$

$$v_3(x; \varepsilon_1) = -0.6561x^3 - 0.5832x^3\varepsilon_1 - 0.1944x^3\varepsilon_1^2 - 0.0288x^3\varepsilon_1^3 - 0.0016x^3\varepsilon_1^4 \tag{11.41d}$$

$$v_4(x; \varepsilon_1) = 0.5905x^4 + 0.6561x^4\varepsilon_1 + 0.2916x^4\varepsilon_1^2 + 0.0648x^4\varepsilon_1^3 + 0.0072x^4\varepsilon_1^4 + 0.0003x^4\varepsilon_1^5 \tag{11.41e}$$

and

$$v_5(x; \varepsilon_1) = -0.5314x^5 - 0.7086x^5\varepsilon_1 - 0.3937x^5\varepsilon_1^2 - 0.1166x^5\varepsilon_1^3 - 0.0194x^5\varepsilon_1^4 - 0.0017x^5\varepsilon_1^5 - 0.0001x^5\varepsilon_1^6 \tag{11.41f}$$

respectively. Accordingly, by considering the truncated series solutions of $O(5)$, we have $v_i(x; \varepsilon_1) = 0$ for $i > 5$. Then, by substituting $v_i(x; \varepsilon_1)$ for $i = 0, 1, \ldots, 5$, we obtain the solution as

$$\begin{aligned}\hat{y}(x) = v(x; \varepsilon_1) &= \sum_{i=0}^{5}(v_i(x; \varepsilon_1))\\ &= 0.9 + 0.2\varepsilon_1 - x(0.81 + 0.36\varepsilon_1 + 0.04\varepsilon_1^2)\\ &\quad + x^2(0.729 + 0.486\varepsilon_1 + 0.108\varepsilon_1^2 + 0.008\varepsilon_1^3) - x^3(0.6561 + 0.5832\varepsilon_1\\ &\quad + 0.1944\varepsilon_1^2 + 0.0288\varepsilon_1^3 + 0.0016\varepsilon_1^4)\\ &\quad + x^4(0.5905 + 0.6561\varepsilon_1 + 0.2916\varepsilon_1^2 + 0.0648\varepsilon_1^3 + 0.0072\varepsilon_1^4 + 0.0003\varepsilon_1^5)\\ &\quad - x^5(0.5314 + 0.7086\varepsilon_1 + 0.3937\varepsilon_1^2 + 0.1166\varepsilon_1^3 + 0.0194\varepsilon_1^4 + 0.0017\varepsilon_1^5 + 0.0001\varepsilon_1^6)\end{aligned} \tag{11.42}$$

with respect to (11.38) in terms of affine variable ε_1 such that $\varepsilon_1 \in [-1, 1]$. Then, applying Laplace transform and Pade approximation, the affine solution reduces to

$$\begin{aligned}\hat{y}(x) = y(x; \varepsilon_1) &= 0.05((9 + 2\varepsilon_1)e^{-0.2x(9+2\varepsilon_1)}(2\cosh(0.1414x(9 + 2\varepsilon_1))\\ &\quad + 1.4142\sinh(0.1414x(9 + 2\varepsilon_1)))\end{aligned} \tag{11.43}$$

One approach to obtain the enclosure of the solution $\hat{y}(x)$ over the domain $x \in \Omega$ with respect to the affine variable may include AA given in Eqs. (11.19–11.23). Such an approach may involve complicated affine arithmetic operations (in particular, division). In this regard, a contractor-based set inversion procedure to compute the solution enclosure of the differential equation (11.1) as given in Section 11.5.1 is considered.

Then, the solution enclosure of Eq. (11.43) is obtained using contractors with respect to SIVIA given in Section 11.5.1. Accordingly, the solution enclosure plot of the NODE (11.38) is depicted in Figure 11.4.

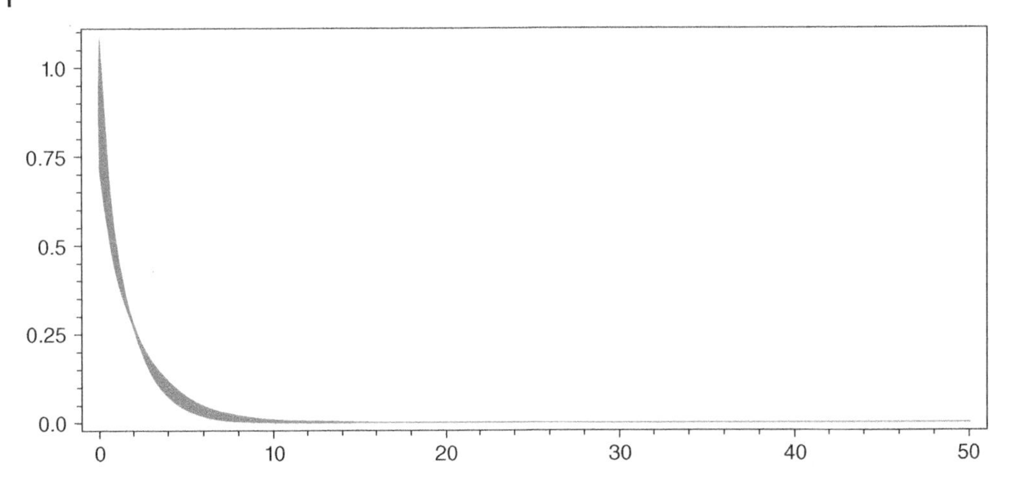

Figure 11.4 Solution enclosure plot of Example 11.3.

11.7 Numerical Examples

In this section, three application problems from different engineering fields are included. The first problem is an unforced nonlinear oscillator, viz., Rayleigh equation. Further, a forced nonlinear oscillator, viz., Van der Pol–Duffing equation, is considered for the second problem. Finally, the behavior of Lane–Emden equation in uncertain environment is studied with respect to affine HPM along with SIVIA-based contractor approach.

11.7.1 Nonlinear Oscillators

11.7.1.1 Unforced Nonlinear Differential Equation

Example 11.4 Let us consider the Rayleigh equation (crisp form in Akbari et al. [5])

$$x''(t) + (0.2 - 0.1x'^2(t))x'(t) + \lambda^2 x(t) = 0 \tag{11.44}$$

subject to uncertain ICs $x(0) = a_1$ and $x'(0) = a_2$, where $t \in \Omega$, $\lambda \in [1.1, 1.3]$, $a_1 \in [0.05, 0.15]$, and $a_2 = 0$ (Chakraverty and Mahato [31]).

Here, we have interval uncertainties in the stiffness parameter $\lambda \in [1.1, 1.3]$ and IC $a_1 \in [0.05, 0.15]$. Using step 1 of Algorithm 11.1, the affine forms of the interval uncertainties λ, a_1 and a_2 have been computed as

$$\hat{\lambda} = [1.1, 1.3] = 1.2 + 0.1\varepsilon_1 \text{ and } \hat{a}_1 = [0.05, 0.15] = 0.1 + 0.05\varepsilon_2.$$

Further, the initial approximation of solution is assumed as

$$x_0(t) = 0.1 + 0.05\varepsilon_2 \tag{11.45}$$

From (11.45), we obtain the first- and second-derivative forms as $x_0'(t) = 0$ and $x_0''(t) = 0$. Then, we obtain the homotopy $\mathcal{H}(v, p)$ as

$$(1 - p)[L(v) - L(y_0)] + p[L(v) + N(v) - F(t)] = 0$$

where $L(v) = v'' + \hat{\lambda}^2 v + 0.2v'$, $N(v) = -0.1v'2v'$, and $F(t) = 0$. Similar to Example 11.3, the series solution is assumed as

$$v = \sum_{i=0}^{k} (v_i p^i)$$

that reduces the homotopy to

$$(1-p)\left[\sum_{i=0}^{k}(v_i''p^i)+\hat{\lambda}^2\sum_{i=0}^{k}(v_i'p^i)+0.2\sum_{i=0}^{k}(v_i'p^i)-(x_0''+\hat{\lambda}^2x_0+0.2x_0')\right]$$
$$+p\left[\sum_{i=0}^{k}(v_i''p^i)+\hat{\lambda}^2\sum_{i=0}^{k}(v_i'p^i)+0.2\sum_{i=0}^{k}(v_i'p^i)\right.$$
$$\left.+0.1\left(\sum_{i=0}^{k}(v_i'p^i)\right)^3\right]=0 \tag{11.46}$$

Then, using step 4, we obtain the system of differential equations, where $\varepsilon_i \in [-1, 1]$ for $i = 1, 2, 3$ as

$$p^0:\ v_0''+\hat{\lambda}^2v_0'+0.2v_0-(x_0''+\hat{\lambda}^2x_0'+0.2x_0)=0$$
$$\text{subject to ICs } v_0(0)=0.1+0.05\varepsilon_2 \text{ and } v_0'(0)=0 \tag{11.47a}$$

$$p^1:\ v_1''+\hat{\lambda}^2v_1'+0.2v_1+(x_0''+\hat{\lambda}^2x_0'+0.2x_0)+0.1{v_0'}^3=0$$
$$\text{subject to IC } v_1(0)=0 \text{ and } v_1'(0)=0 \tag{11.47b}$$

$$p^i:\ v_i''+\hat{\lambda}^2v_i'+0.2v_i+\text{coeff}(N(v),p^{i-1})=0$$
$$\text{subject to IC } v_i(0)=0 \text{ and } v_i'(0)=0 \tag{11.47c}$$

for $i = 2, 3, \ldots, k$. The solution in terms of affine variables $\varepsilon_1, \varepsilon_2 \in [-1, 1]$ is further obtained by solving the system of differential equations given in Eqs. (11.47a)–(11.47c). In particular, we have given the solutions of $O(9)$ with respect to Eqs. (11.47a) and (11.47b) as

$$v_0(t;\varepsilon_1,\varepsilon_2)=0.1+0.05\varepsilon_2$$

and

$$v_1(t;\varepsilon_1,\varepsilon_2,\varepsilon_3)=\frac{143+24\varepsilon_1+\varepsilon_1^2-\sqrt{-143-24\varepsilon_1-\varepsilon_1^2}}{143+24\varepsilon_1+\varepsilon_1^2}(0.05(2+\varepsilon_2)e^{0.1t\,(\sqrt{-143-24\varepsilon_1-\varepsilon_1^2}-1)})$$
$$+\frac{143+24\varepsilon_1+\varepsilon_1^2+\sqrt{-143-24\varepsilon_1-\varepsilon_1^2}}{143+24\varepsilon_1+\varepsilon_1^2}(0.05(2+\varepsilon_2)e^{-0.1t\,(\sqrt{-143-24\varepsilon_1-\varepsilon_1^2}+1)})-0.1-0.5\varepsilon_2,$$

respectively. Accordingly, by substituting $v_i(x;\varepsilon_1,\varepsilon_2)$ for $i = 0, 1$ and $v_i(x;\varepsilon_1,\varepsilon_2) = 0$ for $i \geq 2$, we obtain the affine solution of (11.44) with respect to Laplace transform and Pade approximation as,

$$\hat{x}(t)=v(t;\varepsilon_1,\varepsilon_2)=v_0(t;\varepsilon_1,\varepsilon_2)+v_1(t;\varepsilon_1,\varepsilon_2)$$
$$=0.5(2+\varepsilon_2)e^{-0.1t}\left(\frac{\sinh(0.1t\sqrt{-143-24\varepsilon_1-\varepsilon_1^2})}{\sqrt{-143-24\varepsilon_1-\varepsilon_1^2}}+\cosh(0.1t\sqrt{-143-24\varepsilon_1-\varepsilon_1^2})\right) \tag{11.48}$$

Further, applying SIVIA based on contractors, we have computed the envelope enclosing solution of (11.48) in Figure 11.5.

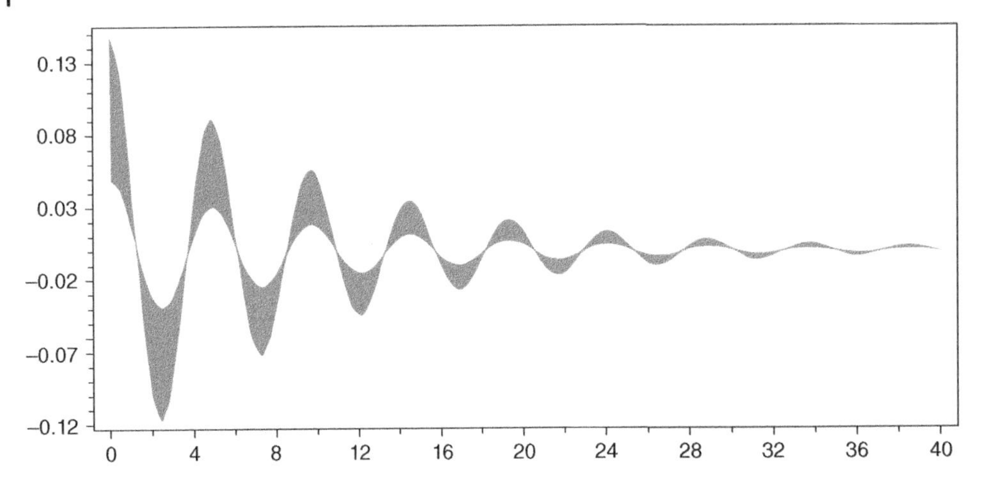

Figure 11.5 Solution enclosure plot of Example 11.4 (Rayleigh equation).

11.7.1.2 Forced Nonlinear Differential Equation

Example 11.5 Let us consider Van der Pol–Duffing equation (crisp form given in Mall and Chakraverty [9]),

$$x''(t) + \gamma(1 - x^2(t))x'(t) + \alpha x(t) + \beta x^3(t) = F\cos(t) \tag{11.49}$$

subject to ICs $x(0) = a_1$ and $x'(0) = a_2$ where $t \in \Omega$, $\gamma = 0.2$, $\alpha = 0.1$, $\beta = 1$, $F = 0.53$, $a_1 \in [0.08, 0.15]$, and $a_2 = -0.2$.

Using step 1 of Algorithm 11.1, we initially compute the corresponding affine form of the interval uncertainty a_1 as

$$\hat{a}_1 = [0.08, 0.15] = 0.115 + 0.035\varepsilon_1$$

Further, with respect to step 2, the initial solution is assumed as

$$x_0(t) = 0.115 + 0.035\varepsilon_1 - 0.2t \tag{11.50}$$

From (11.50), we obtain the first- and second-derivative forms as

$$x_0'(t) = -0.2 \tag{11.51a}$$

and

$$x_0''(t) = 0 \tag{11.51b}$$

Applying (11.36), we obtain the homotopy $\mathcal{H}(v, p)$ as

$$(1 - p)[L(v) - L(y_0)] + p[L(v) + N(v) - F(t)] = 0$$

where $L(v) = v'' + \gamma v' + \alpha v$, $N(v) = -\gamma v^2 v' + \beta v^3$, and $F(t) = F\cos(t)$. Similar to Example 11.2, the series solution is assumed as

$$v = \sum_{i=0}^{k}(v_i p^i)$$

that reduces the homotopy to

$$(1-p)\left[\sum_{i=0}^{k}(v_i''p^i)+\gamma\sum_{i=0}^{k}(v_i'p^i)+\alpha\sum_{i=0}^{k}(v_ip^i)-(x_0''+\gamma x_0'+\alpha x)\right]$$
$$+p\left[\sum_{i=0}^{k}(v_i''p^i)+\gamma\left(1-\left(\sum_{i=0}^{k}(v_ip^i)\right)^2\right)\sum_{i=0}^{k}(v_i'p^i)+\alpha\sum_{i=0}^{k}(v_ip^i)+\beta\left(\sum_{i=0}^{k}(v_ip^i)\right)^3-F\cos(t)\right]=0$$

Then, using step 4, we obtain the system of differential equations, where $\varepsilon_i \in [-1, 1]$ for $i = 1, 2, \ldots, 6$ as

$$p^0:\ v_0''+\gamma v_0'+\alpha v_0-(x_0''+\gamma x_0'+\alpha x_0)=0$$
$$\text{subject to ICs } v_0(0)=0.115+0.035\varepsilon_1 \text{ and } v_0'(0)=-0.2 \tag{11.52a}$$

$$p^1:\ v_1''+\gamma v_1'+\alpha v_1+x_0''+\gamma x_0'+\alpha x_0-\gamma v_0^2v_0'+\beta v_0^3-F\cos(t)=0$$
$$\text{subject to IC } v_1(0)=0 \text{ and } v_1'(0)=0 \tag{11.52b}$$

$$p^i:\ v_i''+\gamma v_i'+\alpha v_i+\text{coeff}(N(v),p^{i-1})=0$$
$$\text{subject to IC } v_i(0)=0 \text{ and } v_i'(0)=0 \tag{11.52c}$$

for $i = 2, 3, \ldots, k$.

The solution in terms of affine variable $\varepsilon_1 \in [-1, 1]$ is further obtained by solving the system of differential equations given in Eqs. (11.52a)–(11.52c). In particular, we have given the solutions with respect to Eqs. (11.52a) and (11.52b) as

$$v_0(t;\varepsilon_1)=0.115+0.035\varepsilon_1-0.2t$$

and

$$v_1(t;\varepsilon_1)=0.2783t^2-0.0136t^3-0.025t^4-0.0026t^2\varepsilon_1-0.0002t^2\varepsilon_1^2-0.000\,02t^2\varepsilon_1^3$$
$$+0.0011t^3\varepsilon_1+0.0001t^3\varepsilon_1^2-0.0004t^4\varepsilon_1$$

respectively. Accordingly, by substituting $v_i(x;\varepsilon_1,\varepsilon_2)$ for $i = 0, 1, \ldots, 6$, we obtain the affine solution of (11.22) as

$$\widehat{x}(t)=v(t;\varepsilon_1,\varepsilon_2,\ldots,\varepsilon_6)=\sum_{i=0}^{6}(v_i(t;\varepsilon_1,\varepsilon_2,\ldots,\varepsilon_6))=0.115+0.035\varepsilon_1-0.2t+0.2783t^2-0.0136t^3-0.025t^4$$
$$-0.0026t^2\varepsilon_1-0.0002t^2\varepsilon_1^2-0.000\,02t^2\varepsilon_1^3+0.0011t^3\varepsilon_1+0.0001t^3\varepsilon_1^2$$
$$-0.0004t^4\varepsilon_1 \tag{11.53}$$

Applying step 6 of 11.1 and contractor programming with respect to SIVIA, the envelope enclosing solution has been illustrated in Figure 11.6.

11.7.2 Other Dynamic Problem

11.7.2.1 Nonhomogeneous Lane–Emden Equation

Example 11.6 Finally, we consider nonhomogeneous Lane–Emden equation (crisp form given in Chowdhury and Hashim [7]; Mall and Chakraverty [8])

$$y''(x)+\frac{2}{x}y'(x)+y^3(x)=6+x^6 \tag{11.54}$$

subject to ICs $y(0) = a_1$ and $y'(0) = a_2$, where $x \in \Omega$, $a_1 \in [0, 0.01]$, and $a_2 = 0$.

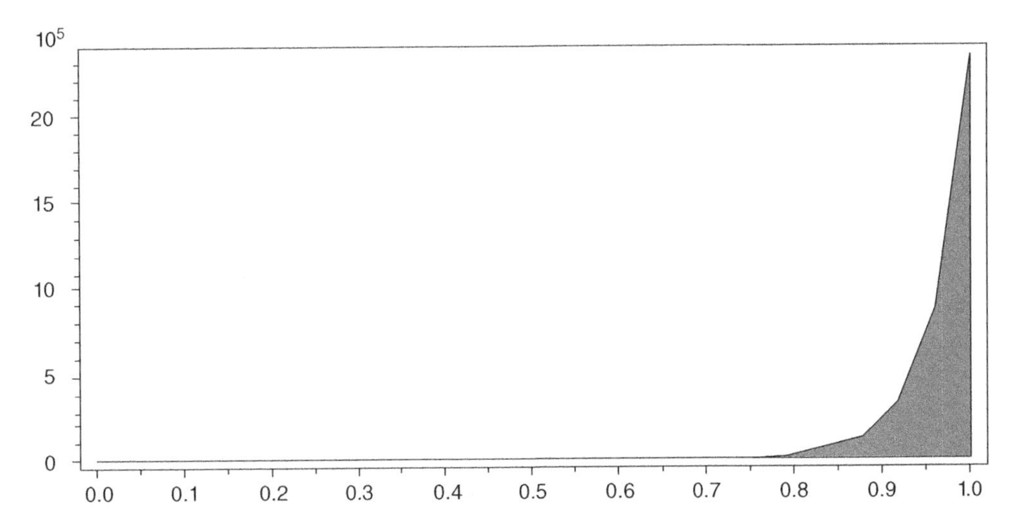

Figure 11.6 Solution enclosure plot of Example 11.5 (Van der Pol–Duffing equation).

Using step 1 of Algorithm 11.1, the affine forms of the interval uncertainty a_1 has been computed as

$$\hat{a}_1 = [0, 0.01] = 0.05 + 0.05\varepsilon_3$$

Further, the initial approximation of solution is assumed as

$$y_0(x) = (0.05 + 0.05\varepsilon_1) \tag{11.55}$$

From (11.55), we obtain the first- and second-derivative forms as

$$y_0'(x) = 0 \tag{11.56a}$$

and

$$y_0''(x) = 0 \tag{11.56b}$$

Applying (11.36), we obtain the homotopy $\mathcal{H}(v, p)$ as

$$(1 - p)[L(v) - L(y_0)] + p[L(v) + N(v) - F(x)] = 0$$

where $L(v) = v'' + \frac{2}{x}v'$, $N(v) = v^3$, and $F(t) = 6 + x^6$. Then, the series solution is assumed as $v = \sum_{i=0}^{k}(v_i p^i)$ that reduces the homotopy to

$$(1 - p)\left[\sum_{i=0}^{k}(v_i'' p^i) + \frac{2}{x}\sum_{i=0}^{k}(v_i' p^i) - \left(y_0'' + \frac{2}{x}y_0'\right)\right] + p\left[\sum_{i=0}^{k}(v_i'' p^i) + \frac{2}{x}\sum_{i=0}^{k}(v_i' p^i) + \left(\sum_{i=0}^{k}(v_i p^i)\right)^3 - 6 - x^6\right] = 0$$

Then, using step 4, we obtain the system of differential equations, where $\varepsilon_1 \in [-1, 1]$ as

$$p^0: v_0'' + \frac{2}{x}v_0' - \left(y_0'' + \frac{2}{x}y_0'\right) = 0$$
$$\text{subject to ICs } v_0(0) = 0.05 + 0.05\varepsilon_1 \text{ and } v_0'(0) = 0 \tag{11.57a}$$

$$p^1: v_1'' + \frac{2}{x}v_1' + \left(y_0'' + \frac{2}{x}y_0'\right) + v_0^3 - 6 - x^6 = 0$$
$$\text{subject to IC } v_1(0) = 0 \text{ and } v_1'(0) = 0 \tag{11.57b}$$

$$p^i: v_i'' + \frac{2}{x}v_i' + \text{coeff}(N(v), p^{i-1}) = 0$$
$$\text{subject to IC } v_i(0) = 0 \text{ and } v_i'(0) = 0 \tag{11.57c}$$

for $i = 2, 3, \ldots, k$.

The solution in terms of affine variable $\varepsilon_1 \in [-1, 1]$ is further obtained by solving the system of differential equations given in Eqs. (11.57a)–(11.57c). In particular, we have given the truncated series solutions of $O(8)$ with respect to Eqs. (11.57a)–(11.57c) as

$$v_0(x; \varepsilon_1) = 0.05 + 0.05\varepsilon_1$$

$$v_1(x; \varepsilon_1) = x^2 - 0.0001x^2\varepsilon_1 - 0.0001x^2\varepsilon_1^2 + 0.013\,89x^8$$

$$v_2(x; \varepsilon_1) = -0.0004x^4 - 0.0075x^4\varepsilon_1 - 0.0004x^4\varepsilon_1^2 + 3.9063x^4\varepsilon_1^4$$

$$v_3(x; \varepsilon_1) = -0.0036x^6 - 0.0036x^6\varepsilon_1$$

$$v_4(x; \varepsilon_1) = -0.0139x^8$$

respectively. Further, $v_i(x; \varepsilon_1) = 0$ for $i \geq 5$.

Accordingly, by substituting $v_i(x; \varepsilon_1)$ for $i = 0, 1, \ldots, 4$, we obtain the affine solution as

$$\widehat{y}(x) = v(x; \varepsilon_1) = \sum_{i=0}^{4}(v_i(x; \varepsilon_1)) \tag{11.58}$$

Then, using Laplace transform and Pade approximation along with SIVIA-based contractor programming, we get the envelope enclosing solution (11.58) of Lane–Emden equation (11.54) in Figure 11.7.

It is worth mentioning that the solution enclosures in Examples 11.3–11.6 contain the corresponding crisp solutions.

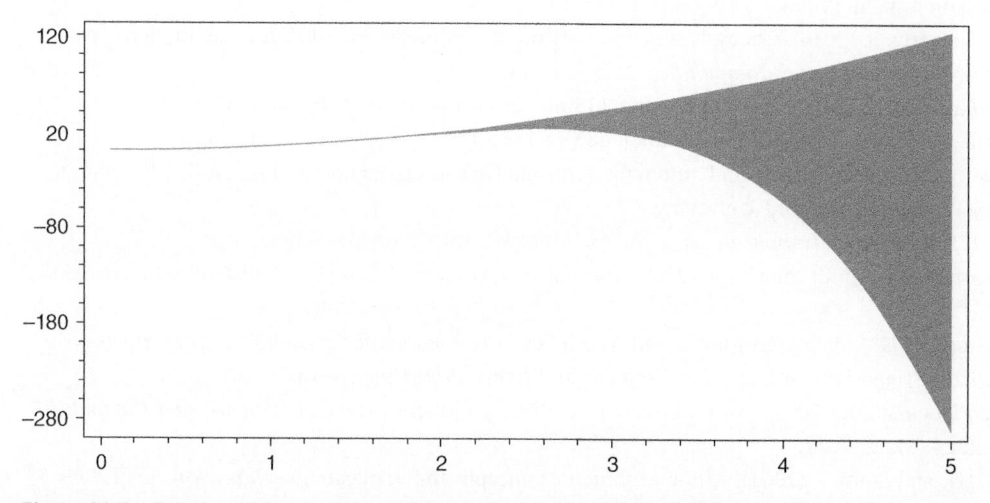

Figure 11.7 Solution enclosure plot of Example 11.6 (Lane–Emden equation).

11.8 Conclusion

In dynamics, several science and engineering problems are generally modeled through NODEs. In this chapter, an affine-contractor-based approach has been proposed to solve nonlinear dynamical problems in uncertain environment. Affine arithmetic is utilized here to handle uncertain or imprecise parameters efficiently because of the presence of dependency problem in the case of classical interval arithmetic. Initially, the solutions of the uncertain NODEs have been determined by using affine HPM technique. Further, the tighter enclosures of the solutions (in the form of noise symbols) have been calculated by using affine-contractor HPM method. Dynamical problems from different fields, viz., nonlinear oscillators such as Rayleigh equation (for unforced NODE) and Van der Pol–Duffing equation (for forced NODE) and other dynamic problems, viz., nonhomogeneous Lane–Emden equation, have been solved to demonstrate the efficiency and accuracy of the proposed method. It may be noticed from the examples that the solution for crisp cases certainly lie within the enclosures of the obtained solution bounds. Therefore, the present procedure indicates its correctness and efficiency.

References

1 He, J.-H. (2003). Homotopy perturbation method: a new nonlinear analytical technique. *Applied Mathematics and Computation* 135 (1): 73–79.

2 Lyapunov, A.M. (1992). The general problem of the stability of motion. *International Journal of Control* 55 (3): 531–534.

3 He, J.-H. (2000). Variational iteration method for autonomous ordinary differential systems. *Applied Mathematics and Computation* 114 (2–3): 115–123.

4 He, J.-H. (2005). Application of homotopy perturbation method to nonlinear wave equations. *Chaos, Solitons & Fractals* 26 (3): 695–700.

5 Akbari, M.R., Ganji, D.D., Majidian, A., and Ahmadi, A.R. (2014). Solving nonlinear differential equations of Vanderpol, Rayleigh and Duffing by AGM. *Frontiers of Mechanical Engineering* 9 (2): 177–190.

6 Babolian, E., Azizi, A., and Saeidian, J. (2009). Some notes on using the homotopy perturbation method for solving time-dependent differential equations. *Mathematical and Computer Modelling* 50 (1–2): 213–224.

7 Chowdhury, M.S.H. and Hashim, I. (2007). Solutions of a class of singular second-order IVPs by homotopy-perturbation method. *Physics Letters A* 365 (5–6): 439–447.

8 Mall, S. and Chakraverty, S. (2014). Chebyshev neural network based model for solving Lane–Emden type equations. *Applied Mathematics and Computation* 247: 100–114.

9 Mall, S. and Chakraverty, S. (2016). Hermite functional link neural network for solving the Van der Pol–Duffing oscillator equation. *Neural Computation* 28 (8): 1574–1598.

10 Moore, R.E. (1962). Interval arithmetic and automatic error analysis in digital computing. Ph.D. Dissertation. Department of Mathematics, Stanford University.

11 Moore, R.E. (1979). *Methods and Applications of Interval Analysis*, vol. 2. SIAM.

12 Moore, R.E., Kearfott, R.B., and Cloud, M.J. (2009). *Introduction to Interval Analysis*. Philadelphia, PA: SIAM Publications.

13 Comba, J.L.D., Stol J. (1993). Affine arithmetic and its applications to computer graphics. In *Proceedings of VI SIBGRAPI (Brazilian Symposium on Computer Graphics and Image Processing)*, 9–18.

14 Stolfi, J. and De Figueiredo, L. (2003). An introduction to affine arithmetic. *Trends in Applied and Computational Mathematics* 4 (3): 297–312.

15 De Figueiredo, L.H. and Stolfi, J. (2004). Affine arithmetic: concepts and applications. *Numerical Algorithms* 37 (1–4): 147–158.

16 Skalna, I. and Hladík, M. (2017). A new algorithm for Chebyshev minimum-error multiplication of reduced affine forms. *Numerical Algorithms* 76 (4): 1131–1152.

17 Skalna, I. (2009). Direct method for solving parametric interval linear systems with nonaffine dependencies. In: *International Conference on Parallel Processing and Applied Mathematics* (eds. R. Wyrzykowski, J. Dongarra, K. Karczewski and J. Wasniewski), 485–494. Springer.

18 Ceberio, M., Kreinovich, V., Chopra, S. et al. (2007). Interval-type and affine arithmetic-type techniques for handling uncertainty in expert systems. *Journal of Computational and Applied Mathematics* 199 (2): 403–410.

19 Ludäscher, B., Nguyen, H.T., Chopra, S. et al. (2007). Interval-type and affine arithmetic-type techniques for handling uncertainty in expert systems. *Journal of Computational and Applied Mathematics* 198 (2): 403–410.

20 Rump, S.M. and Kashiwagi, M. (2015). Implementation and improvements of affine arithmetic. *Nonlinear Theory and Its Applications, IEICE* 6 (3): 341–359.

21 Akhmerov, R.R. (2005). Interval-affine Gaussian algorithm for constrained systems. *Reliable Computing* 11 (5): 323–341.

22 Miyajima, S. and Kashiwagi, M. (2004). A dividing method utilizing the best multiplication in affine arithmetic. *IEICE Electronics Express* 1 (7): 176–181.

23 Vaccaro, A. and Canizares, C.A. (2017). An affine arithmetic-based framework for uncertain power flow and optimal power flow studies. *IEEE Transactions on Power Systems* 32 (1): 274–288.

24 Wang, S., Han, L., and Wu, L. (2015). Uncertainty tracing of distributed generations via complex affine arithmetic based unbalanced three-phase power flow. *IEEE Transactions on Power Systems* 30 (6): 3053–3062.

25 Gu, W., Luo, L., Ding, T. et al. (2014). An affine arithmetic-based algorithm for radial distribution system power flow with uncertainties. *International Journal of Electrical Power & Energy Systems* 58: 242–245.

26 Shou, H., Lin, H., Martin, R., and Wang, G. (2003). Modified affine arithmetic is more accurate than centered interval arithmetic or affine arithmetic. In: *Mathematics of Surfaces* (eds. M.J. Wilson and R.R. Martin), 355–365. Springer.

27 Soares, R.d.P. (2013). Finding all real solutions of nonlinear systems of equations with discontinuities by a modified affine arithmetic. *Computers & Chemical Engineering* 48: 48–57.

28 Chabert, G. and Jaulin, L. (2009). Contractor programming. *Artificial Intelligence* 173: 1079–1100.

29 Jaulin, L., Kieffer, M., Didrit, O., and Walter, E. (2001). *Applied Interval Analysis: With Examples in Parameter and State Estimation, Robust Control and Robotics*, vol. 1. London: Springer-Verlag.

30 Rego, B.S., Raimondo, D.M., and Raffo, G.V. (2018). Set-based state estimation of nonlinear systems using constrained zonotopes and interval arithmetic. In: *2018 European Control Conference (ECC)*, 1584–1589. IEEE.

31 Chakraverty, S. and Mahato, N.R. (2019). Laplace–Pade parametric homotopy perturbation method for uncertain nonlinear oscillators. *Journal of Computational and Nonlinear Dynamics* 14 (9).

32 Mahato, N.R., Jaulin, L., Chakraverty, S., and Dezert, J. (2020). Validated enclosure of uncertain nonlinear equations using SIVIA Monte Carlo. In: *Recent Trends in Wave Mechanics and Vibrations* (eds. S. Chakraverty and P. Biswas), 455–468. Singapore: Springer.

33 Rout, S. and Chakraverty, S. (2020). Affine approach to solve nonlinear eigenvalue problems of structures with uncertain parameters. In: *Recent Trends in Wave Mechanics and Vibrations* (eds. S. Chakraverty and P. Biswas), 407–425. Singapore: Springer.

12

Dynamic Behavior of Nanobeam Using Strain Gradient Model

Subrat Kumar Jena, Rajarama Mohan Jena, and Snehashish Chakraverty

Department of Mathematics, National Institute of Technology Rourkela, Rourkela, Odisha, 769008, India

12.1 Introduction

Dynamic analysis of nanostructures is very crucial for engineering design of several electromechanical devices such as nanoprobes, nanooscillators, nanosensors, etc. This analysis is very fundamental as the experimental study in nanoscale is very tedious and expensive. Also, classical mechanics fails to address the nanoscale effect. In this regard, several nonclassical theories have been introduced by researchers to address the small-scale effect. These theories include strain gradient theory [1], couple stress theory [2], modified couple stress theory [3], micropolar theory, and nonlocal elasticity theory [4]. Investigations related to the dynamical analysis of beams, membranes, nanobeams, nanotubes, nanoribbons, etc., are reported in the literature [5–14].

Akgöz and Civalek [15] analytically studied the static behavior of the Euler–Bernoulli nanobeam under the framework of the modified strain gradient theory and modified couple stress theory. They also investigated the influence of size effect and material parameters on the static response of the beam. Akgöz and Civalek [16] again developed a size-dependent higher order shear deformation beam using modified strain gradient theory, which can address both the microstructural and shear deformation effects. The dynamical behavior is studied analytically using Navier solution. The static and dynamic analyses of size-dependent functionally graded microbeams under the framework of Timoshenko beam theory and strain gradient theory are reported in Ansari et al. [17, 18]. Kahrobaiyan et al. [19] studied the static and dynamic behaviors of size-dependent functionally graded simply supported Euler–Bernoulli microbeams using strain gradient theory. Generalized differential quadrature method has been employed by Khaniki and Hosseini-Hashemi [20] to analyze the buckling behavior of tapered nanobeams using nonlocal strain gradient theory for simply supported boundary condition. Free vibration of a size-dependent functionally graded Timoshenko beam is studied analytically by Li et al. [21] using strain gradient model implementing Navier's solution. Li et al. [22] again investigated the longitudinal vibration of the nonlocal strain gradient rod by using an analytical method as well as the finite element method. Wave propagation for Euler–Bernoulli beams has been presented by Lim et al. [23] using higher order nonlocal theory as well as strain gradient theory. Some other studies of strain gradient model are also reported in the literature [24–27].

As per the title of the chapter, the dynamical behavior of, in particular, free vibration of Euler–Bernoulli strain gradient nanobeam is investigated using the differential transform method (DTM) for SS and CC boundary conditions. Validation and convergence study of the frequency parameters of strain gradient nanobeam are also conducted. Further, the effects of small-scale parameters and length-scale parameters on frequency parameters are reported for the first four modes of frequency parameters through graphical and tabular results.

Mathematical Methods in Interdisciplinary Sciences, First Edition. Edited by Snehashish Chakraverty.

12.2 Mathematical Formulation of the Proposed Model

The equation of motion of Euler–Bernoulli nanobeam can be expressed as [20]

$$\frac{\partial^2 M}{\partial x^2} - P\frac{\partial^2 w}{\partial x^2} - \rho A\frac{\partial^2 w}{\partial t^2} = 0 \tag{12.1}$$

where M is the bending moment that is defined as $M = \int_A z\sigma_{xx}\, dA$, P is the applied compressive force due to mechanical loading, ρ is the mass density, and A is the area of cross section of the beam. For an isotropic beam, the first-order strain gradient model can be written as [20, 23, 24]

$$\left[1 - (e_0a)^2\frac{\partial^2}{\partial x^2}\right]\sigma_{xx} = E\left[1 - l^2\frac{\partial^2}{\partial x^2}\right]\varepsilon_{xx} \tag{12.2}$$

where σ_{xx} is the normal stress, E is Young's modulus, ε_{xx} is the classical strain, e_0a is the nonlocal parameter, and l is the length-scale parameter. Now, multiplying Eq. (12.2) by $z\,dA$ and integrating over the area, we may get

$$M = (e_0a)^2\frac{\partial^2 M}{\partial x^2} - EI\frac{\partial^2 w}{\partial x^2} + l^2 EI\frac{\partial^4 w}{\partial x^4} \tag{12.3}$$

Plugging Eq. (12.1) in Eq. (12.3), the nonlocal bending moment may be written as

$$M = (e_0a)^2\left\{\rho A\frac{\partial^2 w}{\partial t^2} + P\frac{\partial^2 w}{\partial x^2}\right\} - EI\frac{\partial^2 w}{\partial x^2} + l^2 EI\frac{\partial^4 w}{\partial x^4} \tag{12.4}$$

Inserting Eq. (12.4) in the equation of motion of beam, i.e. Eq. (12.1), the governing equation can be obtained as

$$(e_0a)^2\left\{\rho A\frac{\partial^2 w}{\partial t^2} + P\frac{\partial^2 w}{\partial x^2}\right\} - EI\frac{\partial^4 w}{\partial x^4} + l^2 EI\frac{\partial^6 w}{\partial x^6} - P\frac{\partial^2 w}{\partial x^2} - \rho A\frac{\partial^2 w}{\partial t^2} = 0 \tag{12.5}$$

Now assuming the equation of motion as sinusoidal, viz., $w(x, t) = w_0(x)e^{i\omega t}$, Eq. (12.5) can be rewritten as

$$-(e_0a)^2\rho A\omega^2\frac{d^2 w_0}{dx^2} + (e_0a)^2 P\frac{d^4 w_0}{dx^4} - EI\frac{d^4 w_0}{dx^4} + l^2 EI\frac{d^6 w_0}{dx^6} - P\frac{d^2 w_0}{dx^2} + \rho A\omega^2 w_0 = 0 \tag{12.6}$$

Let us consider the following dimensionless parameters as

$X = \frac{x}{L}$ = Dimensionless spatial coordinate

$W = \frac{w_0}{L}$ = Dimensionless transverse displacement

$\beta_1 = \frac{e_0a}{L}$ = Dimensionless nonlocal parameter

$\beta_2 = \frac{l}{L}$ = Dimensionless length – scale parameter

$\lambda^2 = \frac{\rho A\omega^2 L^4}{EI}$ = Frequency parameter

$\overline{P} = \frac{PL^2}{EI}$ = Buckling load parameter.

Using the above dimensionless parameters in Eq. (12.6), the nondimensional form of governing equation can be expressed as

$$-\beta_1^2\lambda^2\frac{d^2 W}{dX^2} + \beta_1^2\overline{P}\frac{d^4 W}{dX^4} - \frac{d^4 W}{dX^4} + \beta_2^2\frac{d^6 W}{dX^6} - \overline{P}\frac{d^2 W}{dX^2} + \lambda^2 W = 0 \tag{12.7}$$

Now by letting $\overline{P} = 0$ in Eq. (12.7), the nondimensional form of transverse vibration equation of nonlocal strain gradient beam can be reduced to

$$\beta_2^2 \frac{d^6 W}{dX^6} - \frac{d^4 W}{dX^4} = \lambda^2 \left(\beta_1^2 \frac{d^2 W}{dX^2} - W \right) \tag{12.8}$$

12.3 Review of the Differential Transform Method (DTM)

DTM is a semianalytical technique that was first introduced by Zhou [28] for solving linear and nonlinear initial value problems arising in electrical circuits. Since then, this technique has been used in many problems arising in different fields of science and engineering. Use of DTM in structural dynamics problem can be found in Chen and Ho [29], Ayaz [30], Ayaz [31], Ozdemir and Kaya [32], Ozdemir and Kaya [33], Balkaya et al. [34], Ozgumus and Kaya [35], Zarepour et al. [36], and Nourifar et al. [37].

Let us consider an analytic function $W(X)$ in a domain D and assume that $X = X_0$ be any point in that domain. Then, the function $W(X)$ can be expressed by a power series having a center located at X_0. The differential transform of the function $W(X)$ can be written as

$$\overline{W}(K) = \frac{1}{K!} \left(\frac{d^K W(X)}{dX^K} \right)_{X=X_0} \tag{12.9}$$

where $W(X)$ is the original function and $\overline{W}(K)$ is the transformed function. The inverse differential transformation is defined as

$$W(X) = \sum_{K=0}^{\infty} (X - X_0)^K \overline{W}(K) \tag{12.10}$$

Combining Eqs. (12.9) and (12.10), we obtain

$$W(X) = \sum_{K=0}^{\infty} \frac{(X - X_0)^K}{K!} \left(\frac{d^K W(X)}{dX^K} \right)_{X=X_0}. \tag{12.11}$$

The function $W(X)$ can be expressed by a finite series, and therefore, Eq. (12.11) can be written as

$$W(X) = \sum_{K=0}^{N} \frac{(X - X_0)^K}{K!} \left(\frac{d^K W(X)}{dX^K} \right)_{X=X_0} \tag{12.12}$$

Implementation of differential transform on some basic functions and the boundary conditions are depicted in Tables 12.1–12.2.

Table 12.1 Implementation of DTM on some basic functions [36].

Original function	Transformed function
$W(X) = W_1(X) \pm W_2(X)$	$\overline{W}(X) = \overline{W}_1(K) \pm \overline{W}_2(K)$
$W(X) = \alpha W_1(X)$	$\overline{W}(X) = \alpha \overline{W}_1(K)$
$W(X) = W_1(X)\, W_2(X)$	$\overline{W}(K) = \sum_{L=0}^{K} \overline{W}_1(L)\, \overline{W}_2(K-L)$
$W(X) = \frac{d^N W_1(X)}{dX^N}$	$\overline{W}(K) = \frac{(K+N)!}{K!} \overline{W}_1(K+N)$
$W(X) = e^{\alpha X}$	$\overline{W}(K) = \frac{\alpha^K}{K!}$

Table 12.2 Implementation of DTM on boundary conditions [36].

$X = 0$		$X = 1$	
Original B.C.	**Transformed B.C.**	**Original B.C.**	**Transformed B.C.**
$W(0) = 0$	$\overline{W}(0) = 0$	$W(1) = 0$	$\sum_{K=0}^{\infty} \overline{W}(K) = 0$
$\frac{dW(0)}{dX} = 0$	$\overline{W}(1) = 0$	$\frac{dW(1)}{dX} = 0$	$\sum_{K=0}^{\infty} K\overline{W}(K) = 0$
$\frac{d^2W(0)}{dX^2} = 0$	$\overline{W}(2) = 0$	$\frac{d^2W(1)}{dX^2} = 0$	$\sum_{K=0}^{\infty} K(K-1)\overline{W}(K) = 0$
$\frac{d^3W(0)}{dX^3} = 0$	$\overline{W}(3) = 0$	$\frac{d^3W(1)}{dX^3} = 0$	$\sum_{K=0}^{\infty} K(K-1)(K-2)\overline{W}(K) = 0$
$\frac{d^4W(0)}{dX^4} = 0$	$\overline{W}(4) = 0$	$\frac{d^4W(1)}{dX^4} = 0$	$\sum_{K=0}^{\infty} K(K-1)(K-2)(K-3)\overline{W}(K) = 0$

12.4 Application of DTM on Dynamic Behavior Analysis

In this section, we will formulate and discuss the dynamic behavior of in particular vibration characteristics of strain gradient nanobeam using DTM to find frequency parameters. Implementing DTM and referring to Table 12.1, the nondimensional form of transverse vibration equation of nonlocal strain gradient beam, i.e. Eq. (12.8), is now reduced into

$$\begin{aligned}&\beta_2^2(K+1)(K+2)(K+3)(K+4)(K+5)(K+6)\,\overline{W}(K+6)-\\&(K+1)(K+2)(K+3)(K+4)\,\overline{W}(K+4) = \beta_1^2\lambda^2(K+1)(K+2)\,\overline{W}(K+2) - \lambda^2\,\overline{W}(K)\end{aligned} \tag{12.13}$$

Now rearranging Eq. (12.13), we obtain the recurrence relation as

$$\overline{W}(K+6) = \frac{(K+1)(K+2)(K+3)(K+4)\,\overline{W}(K+4) + \beta_1^2\lambda^2(K+1)(K+2)\,\overline{W}(K+2) - \lambda^2\,\overline{W}(K)}{\beta_2^2(K+1)(K+2)(K+3)(K+4)(K+5)(K+6)} \tag{12.14}$$

Referring to Table 12.2, the simply supported-simply supported (SS) boundary condition is given as

At $X = 0$:

$$\overline{W}(0) = 0, \overline{W}(1) = C_1, \overline{W}(2) = 0, \overline{W}(3) = C_2, \overline{W}(4) = 0, \overline{W}(5) = C_3 \tag{12.15}$$

At $X = 1$:

$$\sum_{K=0}^{\infty} \overline{W}(K) = 0 \tag{12.16}$$

$$\sum_{K=0}^{\infty} K(K-1)\overline{W}(K) = 0 \tag{12.17}$$

$$\sum_{K=0}^{\infty} K(K-1)(K-2)(K-3)\overline{W}(K) = 0 \tag{12.18}$$

Substituting the values of $\overline{W}(K)$ from Eq. (12.15), into the recurrence relation Eq. (12.14), we obtain

$$\overline{W}(6) = 0$$

$$\overline{W}(7) = \frac{120C_3 + 6\beta_1^2\lambda^2C_2 - \lambda^2C_1}{7!\beta_2^2}$$

$$\overline{W}(8) = 0$$

$$\overline{W}(9) = \frac{840\overline{W}(7) + 20\beta_1^2\lambda^2C_3 - \lambda^2C_2}{60\,480\beta_2^2} = \frac{840\dfrac{120C_3 + 6\beta_1^2\lambda^2C_2 - \lambda^2C_1}{7!\beta_2^2} + 20\beta_1^2\lambda^2C_3 - \lambda^2C_2}{60\,480\beta_2^2}$$

$$\vdots$$

where C_1, C_2, and C_3 are constants. Substituting all the above values of $\overline{W}(K)$ in Eqs. (12.16), (12.17), and (12.18), we will get the system of equations in matrix form as

$$\begin{bmatrix} a_{11}^N(\lambda) & a_{12}^N(\lambda) & a_{13}^N(\lambda) \\ a_{21}^N(\lambda) & a_{22}^N(\lambda) & a_{23}^N(\lambda) \\ a_{31}^N(\lambda) & a_{32}^N(\lambda) & a_{33}^N(\lambda) \end{bmatrix} \begin{bmatrix} C_1 \\ C_2 \\ C_3 \end{bmatrix} = \begin{bmatrix} 0 \\ 0 \\ 0 \end{bmatrix} \tag{12.19}$$

where $a_{ij}^N(\lambda)$, $i, j = 1, 2, 3.$ are the polynomials of λ corresponding to the number of terms as N.

As $C_1 \neq 0$, $C_2 \neq 0$, and $C_3 \neq 0$, Eq. (12.19) implies

$$\begin{vmatrix} a_{11}^N(\lambda) & a_{12}^N(\lambda) & a_{13}^N(\lambda) \\ a_{21}^N(\lambda) & a_{22}^N(\lambda) & a_{23}^N(\lambda) \\ a_{31}^N(\lambda) & a_{32}^N(\lambda) & a_{33}^N(\lambda) \end{vmatrix} = 0 \tag{12.20}$$

Now by solving Eq. (12.20), we get $\lambda = \lambda_i^N$, where $i = 1, 2, 3, \ldots N$ and λ_i^N is the ith mode frequency parameter corresponding to the term N. The value of N for the convergence of the frequency parameter can be obtained from the following relation

$$|\lambda_i^N - \lambda_i^N| \leq \varepsilon \tag{12.21}$$

where λ_i^N is the ith mode frequency corresponding to N, λ_i^{N-1} is the ith mode frequency corresponding to $N-1$, and ε is the degree of precision.

Similarly, by referring to Table 12.2, the Clamped-Clamped (CC) boundary condition can be demonstrated as

At $X = 0$:

$$\overline{W}(0) = 0, \overline{W}(1) = 0, \overline{W}(2) = C_1, \overline{W}(3) = 0, \overline{W}(4) = C_2, \overline{W}(5) = C_3 \tag{12.22}$$

At $X = 1$:

$$\sum_{K=0}^{\infty} \overline{W}(K) = 0 \tag{12.23}$$

$$\sum_{K=0}^{\infty} K\,\overline{W}(K) = 0 \tag{12.24}$$

$$\sum_{K=0}^{\infty} K(K-1)(K-2)\overline{W}(K) = 0 \tag{12.25}$$

Referring to the same procedures as that of SS case, we also obtain the frequency parameter for Clamped-Clamped (CC) boundary condition.

12.5 Numerical Results and Discussion

All the computations for tabular as well as graphical results are carried out using MATLAB tailored code developed by the authors.

12.5.1 Validation and Convergence

Validation and convergence of the present results obtained by DTM are studied in this subsection through graphical as well as tabular results. Setting the length-scale parameter "l" to zero, the strain gradient model is reduced to Eringen's nonlocal model, and the transverse vibration equation of strain gradient nanobeam, i.e. Eq. (12.8), will be converted into vibration equation for nanobeam. In this regard, β_2 has been assigned zero, and the present results have been compared with other well-known results reported in the published literature [38]. Tables 12.3 and 12.4 demonstrate the validation of present results with Wang et al. [38] for SS and CC boundary conditions, respectively. In this comparison, the first three frequency parameters ($\sqrt{\lambda}$) are considered for validation, keeping all other parameters the same as Wang et al. [38]. From these Tables 12.3 and 12.4, we may certainly evident that the present results show admirable agreement with Wang et al. [38]. The convergence of the model is also explored through the graphical results, which are illustrated in Figures 12.1 and 12.2. Both Figures 12.1 and 12.2 represent pointwise convergence of SS and CC boundary conditions, which are plotted by taking $\beta_1 = 1$ and $\beta_2 = 0.5$. From the Figures 12.1 and 12.2, it is revealed that lower mode frequency requires less number of terms than that of

Table 12.3 Comparison of present results with Wang et al. [38] for SS case.

	Present			Ref. [38]		
$\beta_1 = \frac{e_0 a}{L}$	$\sqrt{\lambda_1}$	$\sqrt{\lambda_2}$	$\sqrt{\lambda_3}$	$\sqrt{\lambda_1}$	$\sqrt{\lambda_2}$	$\sqrt{\lambda_3}$
0	3.1416	6.2832	9.4248	3.1416	6.2832	9.4248
0.1	3.0685	5.7817	8.0400	3.0685	5.7817	8.0400
0.3	2.6800	4.3013	5.4422	2.6800	4.3013	5.4422
0.5	2.3022	3.4604	4.2941	2.3022	3.4604	4.2941
0.7	2.0212	2.9585	3.6485	2.0212	2.9585	3.6485

Table 12.4 Comparison of present results with Wang et al. [38] for CC case.

	Present			Ref. [38]		
$\beta_1 = \frac{e_0 a}{L}$	$\sqrt{\lambda_1}$	$\sqrt{\lambda_2}$	$\sqrt{\lambda_3}$	$\sqrt{\lambda_1}$	$\sqrt{\lambda_2}$	$\sqrt{\lambda_3}$
0	4.7300	7.8532	10.9956	4.7300	7.8532	10.9956
0.1	4.5945	7.1402	9.2583	4.5945	7.1402	9.2583
0.3	3.9184	5.1963	6.2317	3.9184	5.1963	6.2317
0.5	3.3153	4.1561	4.9328	3.3153	4.1561	4.9328
0.7	2.8893	3.5462	4.1996	2.8893	3.5462	4.1996

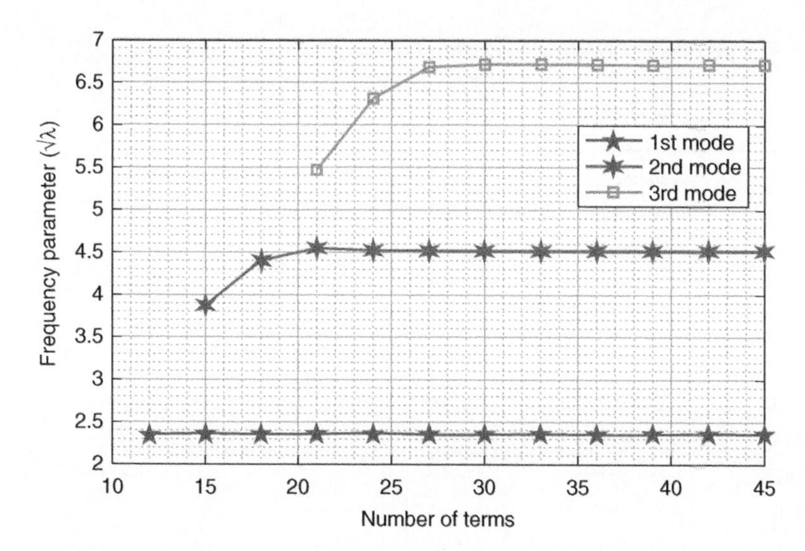

Figure 12.1 Frequency parameter vs. number of terms for SS boundary condition.

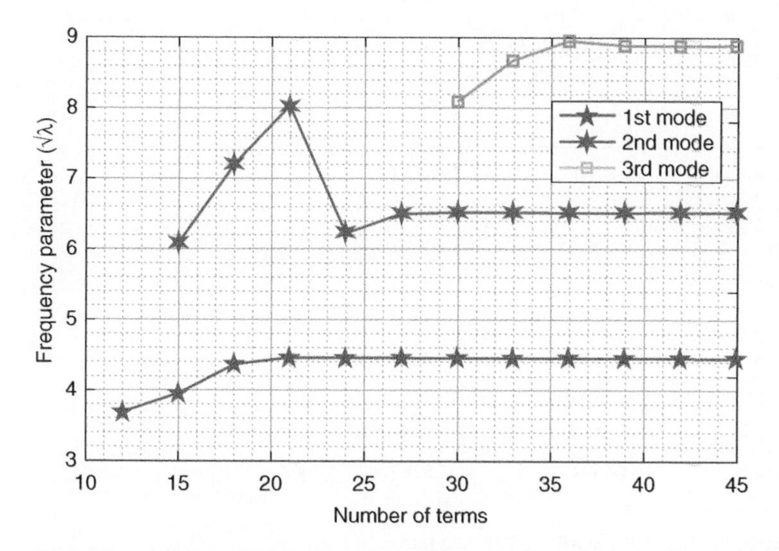

Figure 12.2 Frequency parameter vs. number of terms for CC boundary condition.

higher mode. Approximately 35 number of terms is required for the convergence of third mode frequency of SS boundary condition, whereas 45 number of terms is required for the third mode of CC boundary condition. These figures clearly depict the convergence of the present model by DTM.

12.5.2 Effect of the Small-Scale Parameter

The effect of small-scale parameter on the frequency parameter has been investigated through this subsection. The values of small-scale parameter $\left(\beta_1 = \frac{e_0 a}{L}\right)$ are considered as 0, 0.5, 1, 1.5, 2, 2.5, 3, 3.5, 4, 4.5, and 5. In this regard,

Table 12.5 Effect of small-scale parameter on frequency parameters for SS case.

$\beta_1 = \frac{e_0 a}{L}$	$\sqrt{\lambda_1}$	$\sqrt{\lambda_2}$	$\sqrt{\lambda_3}$	$\sqrt{\lambda_4}$
0	4.2869	11.4086	20.6858	31.6968
0.5	3.1415	6.2831	9.4247	12.5663
1.0	2.3610	4.5230	6.7192	8.9274
1.5	1.9532	3.7058	5.4947	7.2955
2.0	1.6995	3.2132	4.7612	6.3201
2.5	1.5235	2.8756	4.2596	5.6536
3.0	1.3925	2.6258	3.8890	5.1614
3.5	1.2901	2.4315	3.6008	4.7788
4.0	1.2074	2.2747	3.3684	4.4703
4.5	1.1387	2.1448	3.1759	4.2147
5.0	1.0805	2.0349	3.0130	3.9985

Table 12.6 Effect of small-scale parameters on frequency parameters for CC case.

$\beta_1 = \frac{e_0 a}{L}$	$\sqrt{\lambda_1}$	$\sqrt{\lambda_2}$	$\sqrt{\lambda_3}$	$\sqrt{\lambda_4}$
0	8.6506	17.3399	27.8204	39.8553
0.5	6.0226	9.1041	12.4392	15.5264
1.0	4.4687	6.5319	8.9008	11.0353
1.5	3.6854	5.3482	7.2856	9.0190
2.0	3.2032	4.6362	6.3151	7.8134
2.5	2.8698	4.1486	5.6508	6.9896
3.0	2.6222	3.7881	5.1596	6.3812
3.5	2.4290	3.5076	4.7775	5.9081
4.0	2.2729	3.2814	4.4694	5.5267
4.5	2.1435	3.0939	4.2140	5.2108
5.0	2.0338	2.9353	3.9980	4.9434

both the tabular and graphical results are depicted for the first four frequency parameters of SS and CC boundary conditions. Tables 12.5 and 12.6 represent the tabular results for first four frequency parameters for different small-scale parameters. Likewise, Figures 12.3 and 12.4 demonstrate the graphical results for SS and CC boundary conditions. It is witnessed that frequency parameters decrease with the increase of small-scale parameters for both the boundary conditions of all modes. This reduction is very high in case of higher modes than lower modes. Further, to elucidate the nonlocal effect, the response of $\beta_1 = \frac{e_0 a}{L}$ on frequency ratio, which is defined as the ratio of the frequency parameter calculated using nonlocal theory and local theory, is reported as the graphical report in Figures 12.5 and 12.6. These frequency ratios are less than unity and act as an index to predict small-scale effect

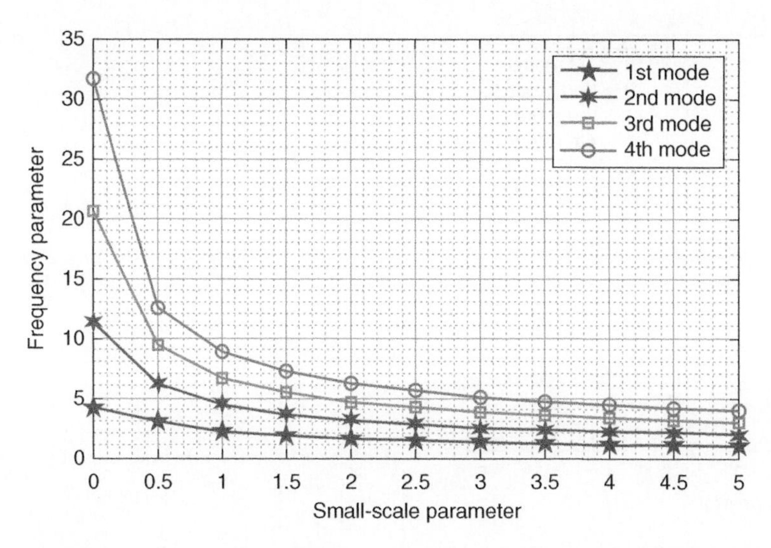

Figure 12.3 Frequency parameter vs. small-scale parameter for SS boundary condition.

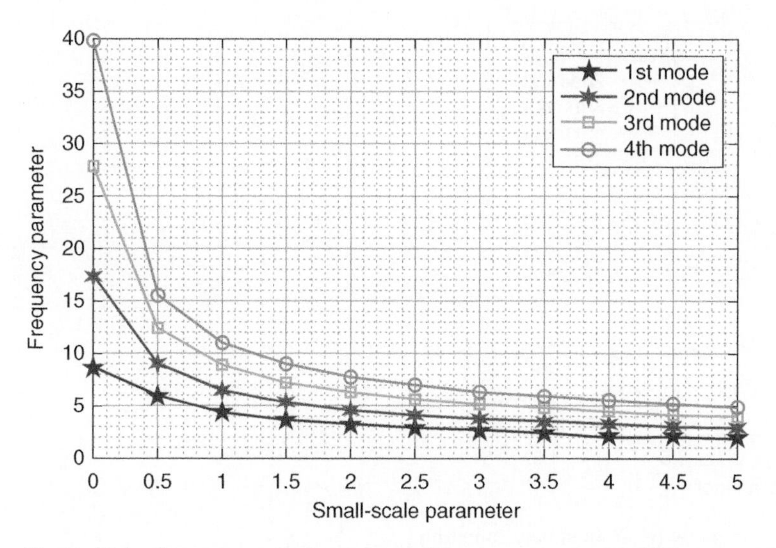

Figure 12.4 Frequency parameter vs. small-scale parameter for CC boundary condition.

on vibration. All the graphical and tabular results are calculated by taking $\beta_2 = 0.5$ with $N = 40$ for SS case and $N = 45$ for CC case.

12.5.3 Effect of Length-Scale Parameter

Length-scale parameter $\left(\beta_2 = \frac{l}{L}\right)$ plays very vital role to study vibration characteristics of strain gradient nanobeam. Tables 12.7 and 12.8 show the variation of first four frequency parameters with length-scale parameter for SS and CC boundary conditions. Similarly, Figures 12.7 and 12.8 represent graphical results for the response of length-scale parameters on frequency parameters. All these graphical and tabular results are computed by

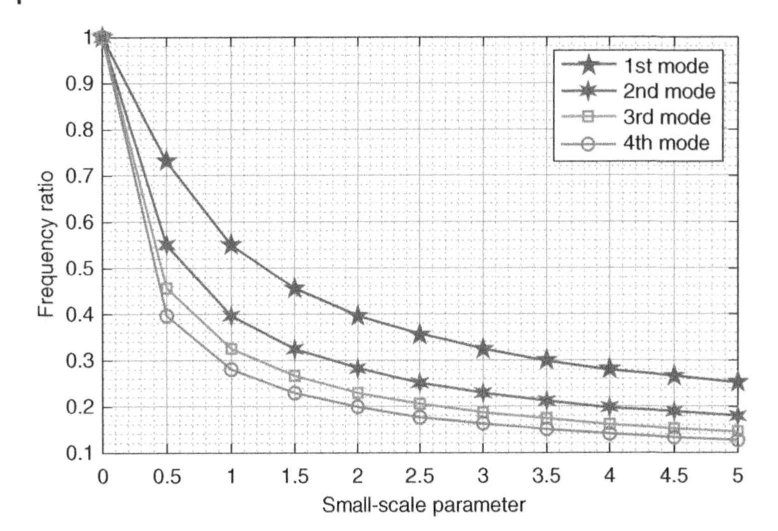

Figure 12.5 Frequency ratio vs. small-scale parameter for SS boundary condition.

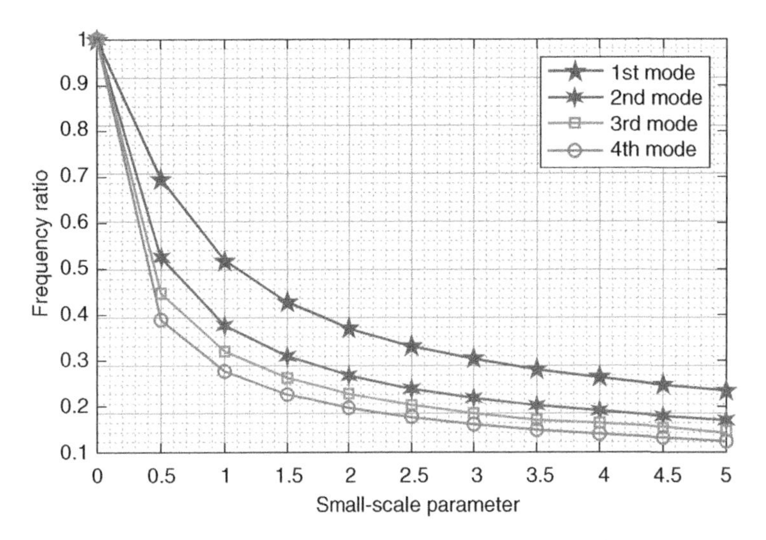

Figure 12.6 Frequency ratio vs. small-scale parameter for CC boundary condition.

considering $\beta_1 = 0.5$. Also, the length-scale parameters are taken as 0, 0.2, 0.4, 0.6, 0.8, 1, 1.2, 1.4, 1.6, 1.8, and 2. From these numerical results, we may clearly note that frequency parameters increase with the increase in length-scale parameter, and this rise is higher in case of higher modes.

12.6 Conclusion

This chapter deals with the study of free vibration of Euler–Bernoulli nanobeam under the framework of the strain gradient model. DTM is applied for the first time to investigate the dynamic behavior of SS and CC boundary conditions. Frequency parameters of strain gradient nanobeam are validated with previously published results of research article showing robust agreement. Also, the convergence of the present results is explored, showing that CC nanobeam requires more points than that of SS boundary condition. Further, the effects of small-scale

Table 12.7 Effect of length-scale parameter on frequency parameters for SS case.

$\beta_2 = \frac{l}{L}$	$\sqrt{\lambda_1}$	$\sqrt{\lambda_2}$	$\sqrt{\lambda_3}$	$\sqrt{\lambda_4}$
0	2.3022	3.4604	4.2940	4.9820
0.2	2.5019	4.3852	6.2725	8.1937
0.4	2.9175	5.6911	8.4804	11.2785
0.6	3.3629	6.8339	10.2901	13.7397
0.8	3.7863	7.8338	11.8424	15.8351
1.0	4.1802	8.7283	13.2196	17.6886
1.2	4.5467	9.5433	14.4690	19.3676
1.4	4.8894	10.2961	15.6203	20.9133
1.6	5.2119	10.9988	16.6932	22.3530
1.8	5.5169	11.6599	17.7017	23.7059
2.0	5.8070	12.2861	18.6562	24.9859

Table 12.8 Effect of length-scale parameter on frequency parameters for CC case.

$\beta_2 = \frac{l}{L}$	$\sqrt{\lambda_1}$	$\sqrt{\lambda_2}$	$\sqrt{\lambda_3}$	$\sqrt{\lambda_4}$
0	2.2905	3.3153	4.2908	4.9328
0.2	4.2032	6.0618	8.1110	10.0197
0.4	5.4552	8.1923	11.1645	13.9184
0.6	6.5506	9.9397	13.6007	16.9875
0.8	7.5091	11.4388	15.6749	19.5915
1.0	8.3665	12.7688	17.5096	21.8916
1.2	9.1477	13.9755	19.1715	23.9736
1.4	9.8693	15.0874	20.7016	25.8896
1.6	10.5428	16.1236	22.1268	27.6738
1.8	11.1766	17.0977	23.4660	29.3501
2.0	11.7768	18.0196	24.7330	30.9359

parameters and length-scale parameters on frequency parameters are reported through graphical and tabular results. The frequency parameters decrease with the increase of small-scale parameters, whereas the trend of frequency parameters is opposite in the case of length-scale parameter.

Acknowledgment

The first author is very much thankful to the Defence Research & Development Organization (DRDO), Ministry of Defence, New Delhi, India (Sanction Code: DG/TM/ERIPR/GIA/17-18/0129/020), and the second author is also thankful to the Department of Science and Technology, Government of India, for providing INSPIRE fellowship (IF170207) to undertake the present research work.

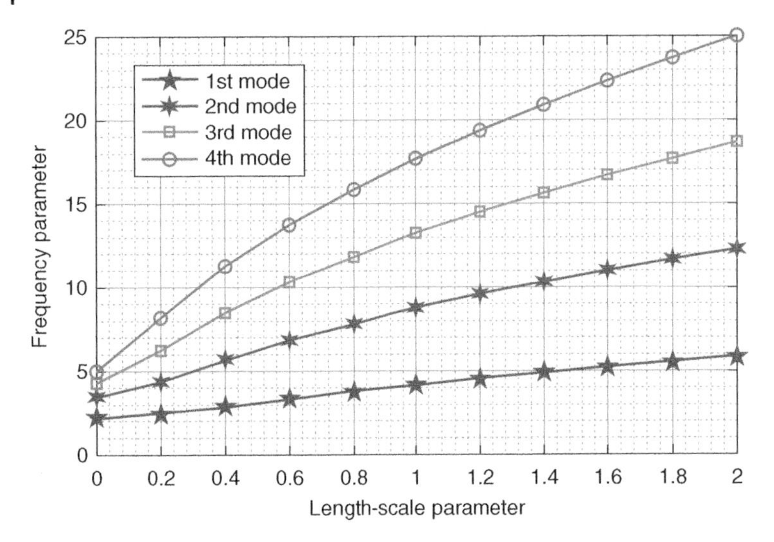

Figure 12.7 Frequency parameter vs. length-scale parameter for SS boundary condition.

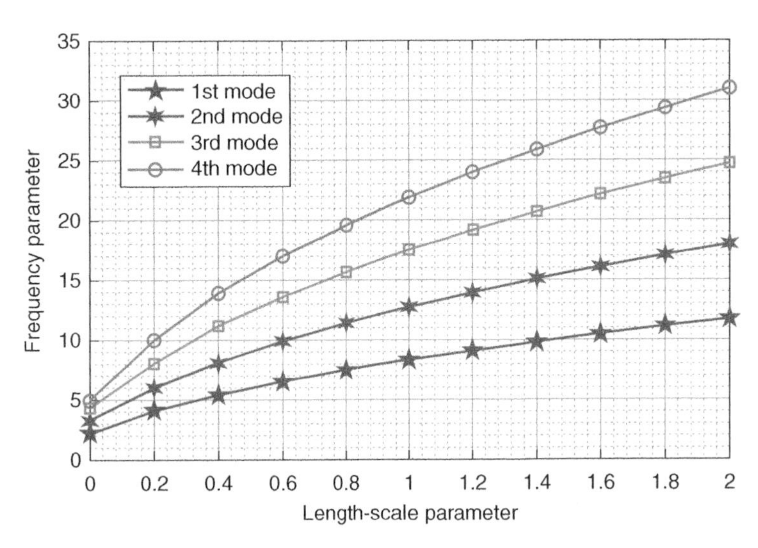

Figure 12.8 Frequency parameter vs. length-scale parameter for CC boundary condition.

References

1 Aifantis, E.C. (1992). On the role of gradients in the localization of deformation and fracture. *International Journal of Engineering Science* 30: 1279–1299.

2 Mindlin, R.D. and Tiersten, H.F. (1962). Effects of couple-stresses in linear elasticity. *Archive for Rational Mechanics and Analysis* 11: 415–448.

3 Yang, F.A.C.M., Chong, A.C.M., DCC, L., and Tong, P. (2002). Couple stress based strain gradient theory for elasticity. *International Journal of Solids and Structures* 39: 2731–2743.

4 Eringen, A.C. (1972). Nonlocal polar elastic continua. *International Journal of Engineering Science* 10: 1–16.

5 Chakraverty, S. and Jena, S.K. (2018). Free vibration of single walled carbon nanotube resting on exponentially varying elastic foundation. *Curved and Layered Structures* 5: 260–272.

6 Jena, S.K. and Chakraverty, S. (2018). Free vibration analysis of Euler–Bernoulli nanobeam using differential transform method. *International Journal of Computational Materials Science and Engineering* 7: 1850020.

7 Jena, S.K. and Chakraverty, S. (2018). Free vibration analysis of variable cross-section single layered graphene nano-ribbons (SLGNRs) using differential quadrature method. *Frontiers in Built Environment* 4: 63.

8 Jena, S.K. and Chakraverty, S. (2018). Free vibration analysis of single walled carbon nanotube with exponentially varying stiffness. *Curved and Layered Structures* 5: 201–212.

9 Jena, R.M. and Chakraverty, S. (2018). Residual power series method for solving time-fractional model of vibration equation of large membranes. *Journal of Applied and Computational Mechanics* 5: 603–615.

10 Jena, S.K. and Chakraverty, S. (2019). Differential quadrature and differential transformation methods in buckling analysis of nanobeams. *Curved and Layered Structures* 6: 68–76.

11 Jena, S.K., Chakraverty, S., Jena, R.M., and Tornabene, F. (2019). A novel fractional nonlocal model and its application in buckling analysis of Euler–Bernoulli nanobeam. *Materials Research Express* 6: 1–17.

12 Jena, S.K., Chakraverty, S., and Tornabene, F. (2019). Vibration characteristics of nanobeam with exponentially varying flexural rigidity resting on linearly varying elastic foundation using differential quadrature method. *Materials Research Express* 6: 1–13.

13 Jena, S.K., Chakraverty, S., and Tornabene, F. (2019). Dynamical behavior of nanobeam embedded in constant, linear, parabolic and sinusoidal types of winkler elastic foundation using first-order nonlocal strain gradient model. *Materials Research Express* 6 (8): 0850f2. 1–23.

14 Jena, R.M., Chakraverty, S., and Jena, S.K. (2019). Dynamic response analysis of fractionally damped beams subjected to external loads using homotopy analysis method. *Journal of Applied and Computational Mechanics* 5: 355–366.

15 Akgöz, B. and Civalek, Ö. (2012). Analysis of micro-sized beams for various boundary conditions based on the strain gradient elasticity theory. *Archive of Applied Mechanics* 82: 423–443.

16 Akgöz, B. and Civalek, Ö. (2013). A size-dependent shear deformation beam model based on the strain gradient elasticity theory. *International Journal of Engineering Science* 70: 1–14.

17 Ansari, R., Gholami, R., and Sahmani, S. (2011). Free vibration analysis of size-dependent functionally graded microbeams based on the strain gradient Timoshenko beam theory. *Composite Structures* 94: 221–228.

18 Ansari, R., Gholami, R., Shojaei, M.F. et al. (2013). Size-dependent bending, buckling and free vibration of functionally graded Timoshenko microbeams based on the most general strain gradient theory. *Composite Structures* 100: 385–397.

19 Kahrobaiyan, M.H., Rahaeifard, M., Tajalli, S.A., and Ahmadian, M.T. (2012). A strain gradient functionally graded Euler–Bernoulli beam formulation. *International Journal of Engineering Science* 52: 65–76.

20 Khaniki, H.B. and Hosseini-Hashemi, S. (2017). Buckling analysis of tapered nanobeams using nonlocal strain gradient theory and a generalized differential quadrature method. *Materials Research Express* 4: 065003.

21 Li, L., Li, X., and Hu, Y. (2016). Free vibration analysis of nonlocal strain gradient beams made of functionally graded material. *International Journal of Engineering Science* 102: 77–92.

22 Li, L., Hu, Y., and Li, X. (2016). Longitudinal vibration of size-dependent rods via nonlocal strain gradient theory. *International Journal of Mechanical Sciences* 115: 135–144.

23 Lim, C.W., Zhang, G., and Reddy, J.N. (2015). A higher-order nonlocal elasticity and strain gradient theory and its applications in wave propagation. *Journal of the Mechanics and Physics of Solids* 78: 298–313.

24 Li, L. and Hu, Y. (2015). Buckling analysis of size-dependent nonlinear beams based on a nonlocal strain gradient theory. *International Journal of Engineering Science* 97: 84–94.

25 Li, X., Li, L., Hu, Y. et al. (2017). Bending, buckling and vibration of axially functionally graded beams based on nonlocal strain gradient theory. *Composite Structures* 165: 250–265.

26 Lu, L., Guo, X., and Zhao, J. (2017). Size-dependent vibration analysis of nanobeams based on the nonlocal strain gradient theory. *International Journal of Engineering Science* 116: 12–24.

27 Şimşek, M. (2016). Nonlinear free vibration of a functionally graded nanobeam using nonlocal strain gradient theory and a novel Hamiltonian approach. *International Journal of Engineering Science* 105: 12–27.

28 Zhou, J.K. (1986). *Differential Transformation and its Application for Electrical Circuits*, 196–102. Huazhong University Press.

29 Chen, C.K. and Ho, S.H. (1999). Solving partial differential equations by two-dimensional differential transform method. *Applied Mathematics and Computation* 106: 171–179.

30 Ayaz, F. (2003). On the two-dimensional differential transform method. *Applied Mathematics and Computation* 143: 361–374.

31 Ayaz, F. (2004). Solutions of the system of differential equations by differential transform method. *Applied Mathematics and Computation* 147: 547–567.

32 Ozdemir, O. and Kaya, M.O. (2006). Flapwise bending vibration analysis of a rotating tapered cantilever Bernoulli–Euler beam by differential transform method. *Journal of Sound and Vibration* 289: 413–420.

33 Ozdemir, O. and Kaya, M.O. (2006). Flapwise bending vibration analysis of double tapered rotating Euler–Bernoulli beam by using the differential transform method. *Meccanica* 41: 661–670.

34 Balkaya, M., Kaya, M.O., and Sağlamer, A. (2009). Analysis of the vibration of an elastic beam supported on elastic soil using the differential transform method. *Archive of Applied Mechanics* 79: 135–146.

35 Ozgumus, O.O. and Kaya, M.O. (2010). Vibration analysis of a rotating tapered Timoshenko beam using DTM. *Meccanica* 45: 33–42.

36 Zarepour, M., Hosseini, S.A., and Ghadiri, M. (2017). Free vibration investigation of nanomass sensor using differential transformation method. *Journal of Applied Physics A* 181: 1–10.

37 Nourifar, M., Keyhani, A., and Aftabi Sani, A. (2018). Free vibration analysis of rotating Euler–Bernoulli beam with exponentially varying cross-section by differential transform method. *International Journal of Structural Stability and Dynamics* 18: 1850024.

38 Wang, C.M., Zhang, Y.Y., and He, X.Q. (2007). Vibration of nonlocal Timoshenko beams. *Nanotechnology* 18: 105401.

13

Structural Static and Vibration Problems

M. Amin Changizi[1] *and Ion Stiharu*[2]

[1]*Mohawk Innovative Technology, Albany, NY 12205, USA*
[2]*Concordia University, Montreal H3G 1M8 QC, Canada*

13.1 Introduction

Although statics and dynamics were defined as disciplines in the classical mechanics about three centuries ago, there are many problems under the above two disciplines that have still not been accurately or fully solved so far. Moreover, the new applications in technologies enforced development of methods that are capable to yield the necessary results for accurate design. Along with time, the accuracy of the solutions has improved, and new phenomena came into the attention of the community. For an example, the simple problem of vibrating cantilever is solved making assumptions that one might not agree with. The stiffness of the cantilever is based on the ratio of the applied force and the deflection of the beam. Such establish value, although comfortable to obtain, may not be agreeable with specific calculations as the tip deflection may not represent the location of interest. The same issue is with the mass of the beam, which, when fixed in one point, may not reflect the same mass in motion as a free beam. On the other hand, the influence of the field force exerted by the electrostatic attraction was of no concern till the microsystems came up.

As past century engineering would accept results within 5% error, such a value is not conceivable now, and the one accepted now will be unacceptable in a few more decades. Both the tradition and fashion are conflicting trends in engineering in the search for accurate solutions. Although the classical approach is part of the traditional training, new methods are introduced. Some of them are accepted by the engineering community as the novel mathematical approaches meet the inertia of the classic approaches that "work" but provide approximate solutions. New methods are proposed, but only the ones that bear the merit of simplicity are successful in being used as standard approaches in engineering. However, the future seems to require some changes in the way the mathematical methods are used. The general use of computers simplifies the introduction of novel mathematical methods. Further, machine learning and the artificial intelligence (AI) will need access to mathematical tools to solve general but also specific problems that require special mathematical approach.

The new mathematical approaches to solve complex statics and dynamics problems have been retrieved from the modern mathematics. Few daring researches took the risk of using such novel methods. Some of them were successful just because their approach would yield more accurate results with a reasonable complexity of the mathematics. Thus, the Lie symmetry groups proved to be of great help to solve mostly nonlinear dynamics problems that were previously solved by various linearization and the application of correction factors. While the search process continues as the intended objective is to find an accurate solution to complex practical problems, the process of "renewal" of the mathematical methods continues.

In this chapter, a comprehensive presentation of the application of Lie symmetry groups is provided. The method is based on defining certain canonical coordinates that satisfy conditions imposed by the application. The Lie point

Mathematical Methods in Interdisciplinary Sciences, First Edition. Edited by Snehashish Chakraverty.

algorithm that is further described helps with the reduction of the order of the ordinary differential equation (ODE). Thus, a second-order differential equation is transformed into a first-order differential equation under the provision that a linear transformation to the specific form of differential equation is found. If such a transformation is not found, the method is not applicable and the equation could not be analytically solved. The new first-order differential equation could eventually be reduced by the Lie symmetry transformation into the solution of the variable following same algorithm. The same condition applies in here as well: a linear transformation such as translation, rotation, or scaling need to be found for the differential equation to be reduced. The solution of the fundamental equation describing one degree of freedom vibrating system is presented. Further, a number of solutions to the nonlinear problem of electrostatic attraction at small-scale devices are presented to witness the capability of the Lie Symmetry Group approach.

13.2 One-parameter Groups

A group represents a point symmetric transformation, which means that each point (x, y) on a curve moves into (x_1, y_1):

$$x_1 = \phi(x, y, \alpha),\ y_1 = \psi(x, y, \alpha) \tag{13.1}$$

where ϕ, ψ are diffeomorphism (C^∞). If any transformation preserves the shape of a curve and it maps the curve on itself, the transformation is called symmetry. If a transformation like (13.1) satisfies the group properties, it is called one-parameter group, while α is called the parameter of the group.

13.3 Infinitesimal Transformation

An infinitesimal transformation for a one-parameter group is defined as:

$$Uf = \xi(x, y)\frac{\partial f}{\partial x} + \eta(x, y)\frac{\partial f}{\partial y} \tag{13.2}$$

where

$$\xi(x, y) = \left.\frac{\partial \phi}{\partial \alpha}\right|_{\alpha=0},\ \eta(x, y) = \left.\frac{\partial \psi}{\partial \alpha}\right|_{\alpha=0},\ f = f(x, y) \tag{13.3}$$

U is the transformation operator. The necessary and sufficient condition that makes a group symmetry group is:

$$Uf = 0 \tag{13.4}$$

This equation will be used to calculate the infinitesimal transformation of an ODE in Section 13.5.

13.4 Canonical Coordinates

Canonical coordinates is a pair functions $r(x, y)$, $s(x, y)$ that satisfies the following conditions [1]:

$$\xi(x, y)r_x + \eta(x, y)r_y = 0$$

$$\xi(x, y)s_x + \eta(x, y)s_y = 1$$

and

$$\begin{bmatrix} r_x & r_y \\ s_x & s_y \end{bmatrix} \neq 0 \tag{13.5}$$

The $s(x, y)$ component of canonical coordinates for a function $f(x, y)$ can be identified from the characteristic equation:

$$\frac{dx}{\xi(x,y)} = \frac{dy}{\eta(x,y)} = ds \tag{13.6}$$

Hence, $s(x, y)$ can be calculated as:

$$s(r,x) = \left(\int \frac{dx}{\xi(x,y(r,x))}\right)\Bigg|_{r=r(x,y)} \tag{13.7}$$

The solution of the below ODE embeds the $r(x, y)$ component:

$$\frac{dy}{dx} = \frac{\eta(x,y)}{\xi(x,y)} \tag{13.8}$$

13.5 Algorithm for Lie Symmetry Point

There are several methods to solve a differential equation by Lie group symmetry. One method is the prolonged vector field [2]. It can be shown that for a second-order differential equation like

$$\frac{d^2y}{dx^2} = \omega\left(x, y, \frac{dy}{dx}\right) \tag{13.9}$$

If an infinitesimal group is applied as an operator on (13.9), both ξ and η in (13.3) must satisfy the below condition:

$$\eta_{xx} + (2\eta_{xy} - \xi_{xx})y' + (\eta_{yy} - \xi_{xy})y'^2 - \xi_{yy}y'^3 + (\eta_y - 2\xi_x - 3\xi_y y')\omega = \xi\omega_x + \eta\omega_y + (\eta_x + (\eta_y - \xi_x)y' - \xi_y y'^2)\omega_{y'} \tag{13.10}$$

By decomposing (13.10) into a system of PDEs, ξ and η can be calculated.

For a first-order differential equation as following:

$$\frac{dy}{dx} = \omega(x,y) \tag{13.11}$$

The following form should be satisfied by ξ and η:

$$\eta_x + (\eta_y - \xi_x)\omega - \xi_y\omega^2 = \xi\omega_x + \eta\omega_y \tag{13.12}$$

13.6 Reduction of the Order of the ODE

A one-parameter group G is assumed to be the symmetry of a differential equation. The relations (13.5), (13.6), and (13.7) can be used to calculate canonical coordinates. Further by defining a new variable $v = \frac{ds}{dr}$ and substituting it in the original ODE, the new ODE will have one order less than the previous ODE.

13.7 Solution of First-Order ODE with Lie Symmetry

If the first-order ODE (13.11) has a Lie symmetry, which can be verified by (13.12), then the ODE (13.11) based on canonical coordinates can be written as:

$$\frac{ds}{dr} = \frac{s_x + \omega(x,y)s_y}{r_x + \omega(x,y)r_y} \tag{13.13}$$

The ODE will be as:

$$\frac{ds}{dr} = \Psi(r) \tag{13.14}$$

The general solution of (13.14) can be expressed as:

$$s - \int \Psi(r)dr = c \tag{13.15}$$

with c as the constant. By substituting (s, r) by (x, y), one can calculate y as a function of x.

13.8 Identification

Lie symmetry method is further used to solve the linear free vibration of a mass-spring-damper system. The equation in (13.16) defines free vibration of a system.

$$\frac{d^2y(t)}{dt^2} + 2\xi\omega_n\frac{dy(t)}{dt} + \omega_n^2 y(t) = 0 \tag{13.16}$$

Equation (13.16) is a homogeneous linear second-order ODE. The solution of this equation is well known, and it can be found in any vibration textbook. Lie point symmetry method will be used below to solve the equation and to demonstrate how the method could be used in solving ODEs. Eq. (13.16) can be rewritten as:

$$\frac{d^2y(t)}{dt^2} = -\left(2\xi\omega_n\frac{dy(t)}{dt} + \omega_n^2 y(t)\right) = 0 \tag{13.17}$$

ω in Eq. (13.9) is as follows for (13.16):

$$\omega\left(x, y, \frac{dy}{dx}\right) = -\left(2\xi\omega_n\frac{dy(t)}{dt} + \omega_n^2 y(t)\right) \tag{13.18}$$

The Lie symmetry point condition in (13.10) should be satisfied. Most of Lie symmetries such as rotation, translation, and scaling could be defined through:

$$\begin{aligned} \xi &= C_1 + C_2x + C_3y \\ \eta &= C_4 + C_5x + C_6y \end{aligned} \tag{13.19}$$

Substitution of (13.19) in (13.10) will yield to:

$$(C_6 - 2C_2 - 3C_3y')(\alpha y' + \beta y) = \beta(C_4 + C_5x + C_6y) + (C_5 + (C_6 - C_2)y' - C_3y'^2)\alpha \tag{13.20}$$

where:

$$\alpha = -2\xi\omega_n, \ \ \beta = -\omega_n^2 \tag{13.21}$$

The variables x, y, y' are independent. By comparing y' coefficient power, it can be shown that:

$$C_2 = C_3 = C_4 = C_5 = 0 \tag{13.22}$$

Hence:

$$\begin{aligned} \xi &= C_1 \\ \eta &= C_6y \end{aligned} \tag{13.23}$$

C_1 and C_2 can be considered as:

$$C_1 = 0, \ \ C_6 = 1 \tag{13.24}$$

Uf in Eq. (13.2) is simplified as:

$$Uf = y \times f_y \tag{13.25}$$

The canonical coordinates can be calculated by (13.6–13.8) as follows:

$$r = t$$
$$s = \ln(y) \tag{13.26}$$

For reducing the order of ODE (13.17):

$$v = \frac{ds}{dr} = \frac{y'}{y} \tag{13.27}$$

$$\frac{d^2s}{dr^2} = \frac{y''}{y} - \frac{y'^2}{y} \tag{13.28}$$

y'' and y' can be calculated from above equations as below:

$$y' = v \times y$$
$$y'' = y \times \left(\frac{dv}{dr} + v^2\right) \tag{13.29}$$

Substitution of (13.29) in (13.16) yields:

$$\frac{dv}{dr} + v^2 + 2\xi\omega_n v + \omega_n^2 = 0 \tag{13.30}$$

This is the reduced form of (13.16) and can be solved by separating variables. In order to illustrate how to use Lie symmetry method to get an answer, this method is used to solve first-order ODE in Eq. (13.30).

Using Lie symmetry as (13.19), which should satisfy (13.12) will result in the following symmetry:

$$\xi = 1$$
$$\eta = 0 \tag{13.31}$$

whereas:

$$\frac{dv}{dr} = -(v^2 + 2\xi\omega_n v + \omega_n^2) = \omega(v, r) \tag{13.32}$$

$$\omega_v = -(2v + 2\xi\omega_n)$$
$$\omega_r = 0 \tag{13.33}$$

Then:

$$Uf = f_r \tag{13.34}$$

In order to avoid any confusion, Eq. (13.32) was rewritten using new variables:

$$\frac{dp}{dq} = -(p^2 + 2\xi\omega_n p + \omega_n^2) = \omega(p, q) \tag{13.35}$$

where:

$$p = v(r)$$
$$q = r \tag{13.36}$$

Canonical coordinates could be calculated from (13.8) and (13.9):

$$\hat{r} = p$$
$$\hat{s} = q \tag{13.37}$$

Equation (13.13) will be calculated as:

$$\frac{d\hat{s}}{d\hat{r}} = \frac{1}{p^2 + 2\xi\omega_n p + \omega_n^2} \tag{13.38}$$

By substituting (13.37) in (13.38):

$$\frac{d\hat{s}}{d\hat{r}} = \frac{1}{\hat{r}^2 + 2\xi\omega_n\hat{r} + \omega_n^2} \tag{13.39}$$

This equation is separable and hence, for $\xi < 1$, the solution of this ODE is:

$$\hat{r} = \omega_n\sqrt{1-\xi^2}\tan(\omega_n(\sqrt{1-\xi^2}\hat{s} - \sqrt{1-\xi^2}C_1) - \omega_n\xi) \tag{13.40}$$

Implementing (13.36 and 13.37) in (13.40) will yield to:

$$v(r) = \omega_n\sqrt{1-\xi^2}\tan(\omega_n(\sqrt{1-\xi^2}r - \sqrt{1-\xi^2}C_1) - \omega_n\xi) \tag{13.41}$$

where C_1 is the constant of integration. Substituting $v(r)$ from (13.26) and r from (13.27) in (13.41) yields:

$$\frac{\frac{dy}{dt}}{y} = \omega_n\sqrt{1-\xi^2}\tan(\omega_n(\sqrt{1-\xi^2}t - \sqrt{1-\xi^2}C_1) - \omega_n\xi) \tag{13.42}$$

The solution of this ODE is:

$$y = C_2e^{-\xi\omega_n t}\cos(\omega_n\sqrt{1-\xi^2}(t + C_1)) \tag{13.43}$$

This form represents the general solution of (13.16). Although Lie symmetry method seems more laborious, then the classical methods used in solving the second-order linear differential equations, it also bears an important advantage vs. the well-known methods. That is, the Lie symmetry group method could yield the solutions of nonlinear differential equations that cannot be solved by other classical approaches.

13.9 Vibration of a Microcantilever Beam Subjected to Uniform Electrostatic Field

For illustration, the problem stated below is solved by the Lie group symmetry method. An electrostatically actuated microcantilever beam, as illustrated in Figure 13.1, can be simplified as a lump mass dynamic model as illustrated in Figure 13.2.The governing nonlinear equation of motion for the dynamics of the system for simplified mass spring damping system, MEMS (micro-electro mechanical system), is:

$$\frac{d^2y(t)}{dt^2} + 2\xi\omega_n\frac{dy(t)}{dt} + \omega_n^2 y(t) = \frac{\varepsilon_0 AV^2}{m(g - y(t))^2} \tag{13.44}$$

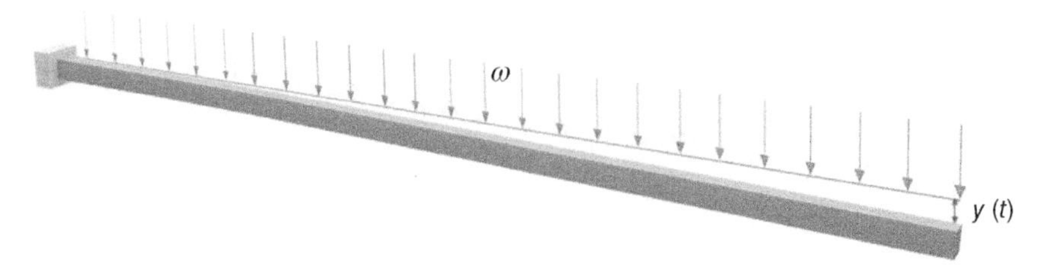

Figure 13.1 Fixed free beam subjected to distributed uniform load (electrostatic attraction).

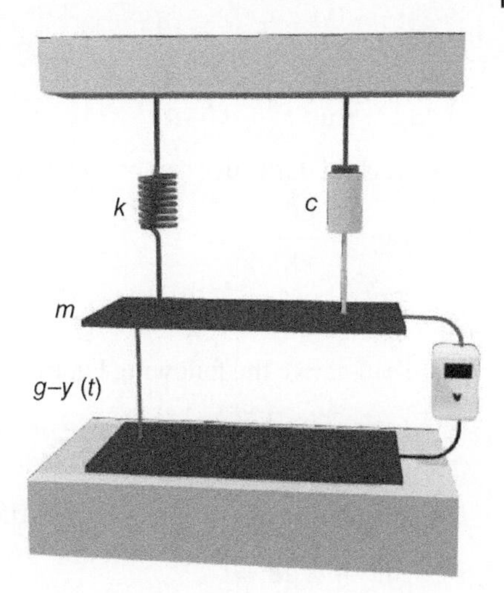

Figure 13.2 The schematic of a mass-spring-damper system of a beam.

where:

$y(t)$ is the deflection of beam
t is time
ξ is the damping factor
ω_n is the natural frequency of beam
ε_0 is the permittivity of air
A is the cross section of the beam
V is the potential difference between the beam and substrate
m is the mass of beam
g is the gap between beam and substrate

The energy balance yields the force (appeared on the right side of the equation), which applies on the two parallel surfaces because of the electrostatic effect. The initial conditions for the ODE are assumed as:

$$y|_{t=0} = 0 \text{ and } \left.\frac{dy}{dt}\right|_{t=0} = v_0 \tag{13.45}$$

Equation (13.44) is a nonlinear equation, and there is no analytical formulation to express so far the result in close form. This equation becomes more complicated when one extends it to extreme microlevel dimensions, which may include very high stiff ODE. In the current work, by using Lie symmetry method, the order of Eq. (13.44) will reduce by one order. Therefore, instead of a second-order ODE, one has to cope with a first-order ODE that by all means is easier to solve. However, the application of Lie symmetry method in this case requires some definitions and terminologies.

13.10 Contact Form for the Equation

Generally, if (x, u) and $(y = u, w = x)$. One can write:

$$w_y = \frac{dw}{dy} \tag{13.46}$$

or

$$dw - w_y dy = dx - w_y du = -w_y\left(du - \frac{1}{w_y}dx\right) = 0 \tag{13.47}$$

therefore:

$$u_x = \frac{1}{w_y} \text{ or } w_y = \frac{1}{u_x} \tag{13.48}$$

The second derivative can be calculated as:

$$w_{yy} = \frac{dw_y}{dy}$$

$$dw_y - w_{yy}dy = -\frac{1}{2}du_x - w_{yy}du \tag{13.49}$$

One can derive the following Eq. from (13.49):

$$dw_y - w_{yy}dy = -\frac{1}{u_x^2}(du_x + w_{yy}u_x^3dx) - w_{yy}(du - u_xdx) \tag{13.50}$$

By considering $w_{xx} = \frac{du_x}{dx}$, one can write:

$$du_x - u_{xx}dx = 0 \tag{13.51}$$

By comparing the first parenthesis of (13.50) with (13.51):

$$u_{xx} = -w_{yy}u_x^3 = -\frac{w_{yy}}{w_y^3} \tag{13.52}$$

13.11 Reducing in the Order of the Nonlinear ODE Representing the Vibration of a Microcantilever Beam Under Electrostatic Field

It is easy to show that Eq. (13.44) has an invariant like (13.19). Simplifying it by the method shown in Eq. (13.7) will yield transformation like (13.31). Applying (13.31) in (13.7) and (13.8) will yield the following:

$$(r, s) = (y, t) \tag{13.53}$$

By defining v as:

$$v = \frac{1}{\frac{dy}{dt}} \tag{13.54}$$

As proved in Eq. (13.9), Eq. (13.54) can be expressed by contact form as:

$$\frac{dv}{dr} = -\frac{\frac{d^2y}{dt^2}}{\left(\frac{dy}{dt}\right)^2} \tag{13.55}$$

or:

$$\frac{d^2y}{d^2t} = -v^{-3}\frac{dv}{dr} \tag{13.56}$$

Substituting (13.55), (13.56), and (13.57) in (13.44) yields:

$$-\frac{dv}{dr} + 2\xi\omega_n v^2 + \left(\omega_n^2 r - \frac{\varepsilon_0 AV^2}{2(r-g)^2 m}\right) v^3 = 0 \tag{13.57}$$

This is a first-order ODE with $v(0) = \frac{1}{v_0}$ as the initial condition. According to the recent investigations, there is no one-parameter group that satisfies its symmetry condition for (13.57). It is the reason for which there is no analytical solution for this ODE. It is shown that a transformation like (13.19) does not have invariant in (13.57).

Hence, (13.57) has no scaling or rotation symmetry. This equation has a singularity (where $r = g$), and it cannot be integrated in close form.

Further, a brief discussion on the variance encountered in the analysis of such type of problems is carried out by Lie group symmetry method. It is expected that not all approaches will produce the same results, fact that will be duly proved below.

13.12 Nonlinear Pull-in Voltage

Pull-in voltage (13.58) is studied based on linearization of Eq. (13.44), whereas nonlinear pull-in voltage was calculated in the following part of this work on the solution of Eq. (13.44).

$$V_{\text{pull-in}} = \sqrt{\frac{8k}{27\varepsilon_0 A}g^3} \tag{13.58}$$

Also, modified pull-in voltage can be calculated as:

$$V_{\text{pull-in}} = \sqrt{\frac{8k_{\text{eff}}}{27\varepsilon_0 A_{\text{eff}}}g^3} \tag{13.59}$$

where:

g is the gap distance between the beam and fixed substrate
ε_0 is the absolute permittivity
A, A_{eff} is the area, effective area
k, k_{eff} is the stiffness, effective stiffness
$A_{\text{eff}}, k_{\text{eff}}$ can be calculated as follows [3]:

$$\begin{aligned} A_{\text{eff}} &= \alpha b_{\text{eff}} l \\ k_{\text{eff}} &= \frac{3}{2}\frac{E^* bh^3}{l^3} \end{aligned} \tag{13.60}$$

where:

$$\begin{aligned} b_{\text{eff}} &= b\left(1 + 0.65\frac{(1-\beta)g}{b}\right) \\ E^* &= \frac{E}{1-\upsilon^2} \\ \alpha &= \frac{4}{\pi}\frac{1-\beta}{\sqrt{1-2\beta}}\text{Arctan}(\sqrt{1-2\beta}) \\ &0.33 \le \beta \le 0.45 \end{aligned} \tag{13.61}$$

where:

b is the width of the beam
E is the Young modulus of elasticity
v is the Poisson ratio of the material
h is the thickness of the beam
l is the length of the beam

The method enables a good grasp of the nonlinear solution of the differential equation. Comparison between linear and nonlinear analysis shows significant difference between the values in the pull-in voltage yield by the solution of the two models. The difference increases by increasing the gap distance. The pull-in voltages of linear and nonlinear analyses are illustrated in Figure 13.3. The numerical values are given in Table 13.1. In this table, for each gap distance, the pull-in voltage was calculated by seven different methods. Second column (linear) shows the pull-in voltage based on Eq. (13.58). Third column (DCNF - Discrete Cantilever Natural Frequency) shows the pull-in voltage value based on Eq. (13.16), where $\omega_n = \sqrt{\frac{k}{m}}, k = \frac{3EI}{l^3}$. Fourth column (CCNF – continue cantilever natural frequency) gives the pull-in voltage derived from Eq. (13.16) by assuming $\omega_n = 3.515\,625\sqrt{\frac{EI}{ml^3}}$. Fifth and seventh columns represent the pull-in voltages derived by Eq. (13.59) for $\beta = 0.33$ and $\beta = 0.45$, respectively.

The sixth and eighth columns provide the pull-in voltages from Eq. (13.16) by modifying the area by Eq. (13.60). As expected and as it can be seen from Figure 13.3, DCNF in all cases yield results that are significantly different from the other models. Nonlinear analysis of CCNF model yields the maximum values for the pull-in voltage and

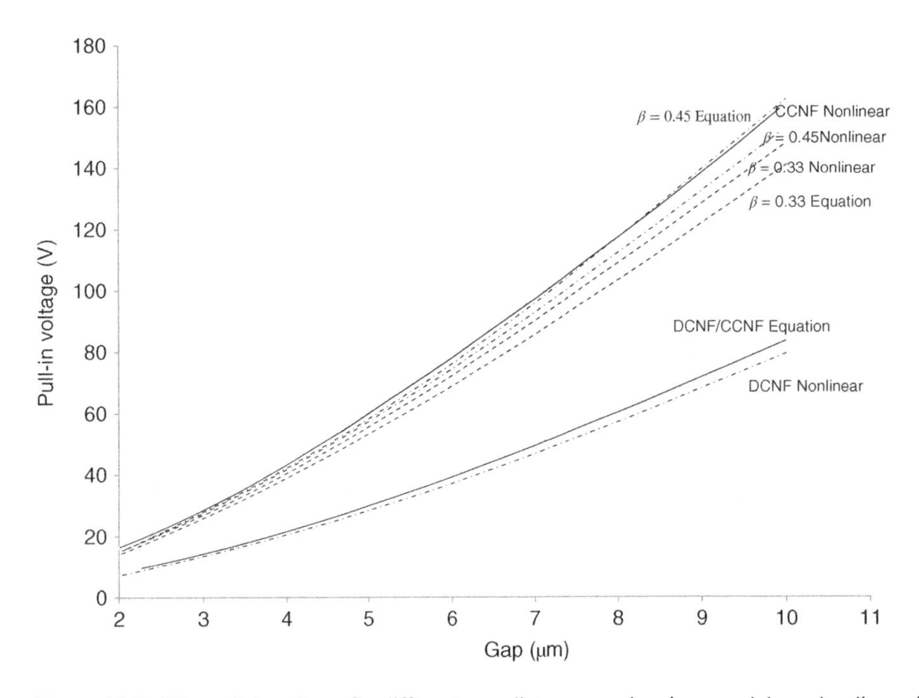

Figure 13.3 The pull-in voltage for different gap distances and various models under discussion.

Table 13.1 Pull-in voltage in linear and nonlinear analysis.

Gap distance (μm)	Linear model (V)	DCNF (V)	CCNF (V)	Modified $\beta = 0.33$ (V)	Nonlinear modified $\beta = 0.33$ (V)	Modified $\beta = 0.45$ (V)	Nonlinear modified $\beta = 0.45$ (V)
2	7.52	7.17	14.57	14.36	13.69	15.38	14.66
4	21.27	20.28	41.22	39.80	37.95	42.78	40.79
6	39.80	37.26	75.73	71.70	68.36	77.32	73.72
8	60.17	57.37	116.59	108.33	103.28	117.16	111.71
10	84.09	80.18	162.94	148.66	141.74	161.24	153.73

the results come almost the same as those modified for $\beta = 0.45$ from the Eq. (13.16). For modified models, in any case, the equation gives a slightly higher value than the nonlinear analysis. Figure 13.3 also reveals that the DCNF model predicts smallest pull-in voltage when compared to the other models, whereas the modified $\beta = 0.45$ yields the largest pull-in voltage. For small gaps, the CCNF model predicts pull-in voltage less than the one obtained for modified $\beta = 0.45$, but by increasing the gap distance, the pull-in voltage by CCNF will be higher than the modified $\beta = 0.45$. This finding is consistent with the dependency of the beam stiffness of various models on the applied potential. The results show that the CCNF model yields the stiffest structure while DCNF produces the least stiff structure regardless of the gap. Modified $\beta = 0.45$ model yields stiffer systems than the modified $\beta = 0.33$ model. The above is confirmed by the resonant frequencies predicted by the models when structures are subjected to low electrostatic forces. Thus, CCNF model exhibits highest natural frequency, while DCNF shows the smallest natural frequency. The behavior of the system very close to the pull-in voltage is another aspect that is studied. For illustration, we shall consider a specific microcantilever and gap distances as indicated below. The results are illustrated in Figures 13.4–13.8, where four sets of results are shown for this gap distances.

Length $l = 200\,\mu\text{m}$
Width $w = 20\,\mu\text{m}$
Thickness $h = 2\,\mu\text{m}$
Density of material $\rho = 2300\ \text{kg/m}^3$
Damping factor $\xi = 0.1$
Young's modulus $E = 169\ \text{GPa}$
Gap $g = 2\,\mu\text{m}$

These four sets are the time response of the system under the application of a step potential close to the pull-in voltage, the phase portrait for the applied potential of 99.999% of the pull-in voltages evaluated from (13.59), the

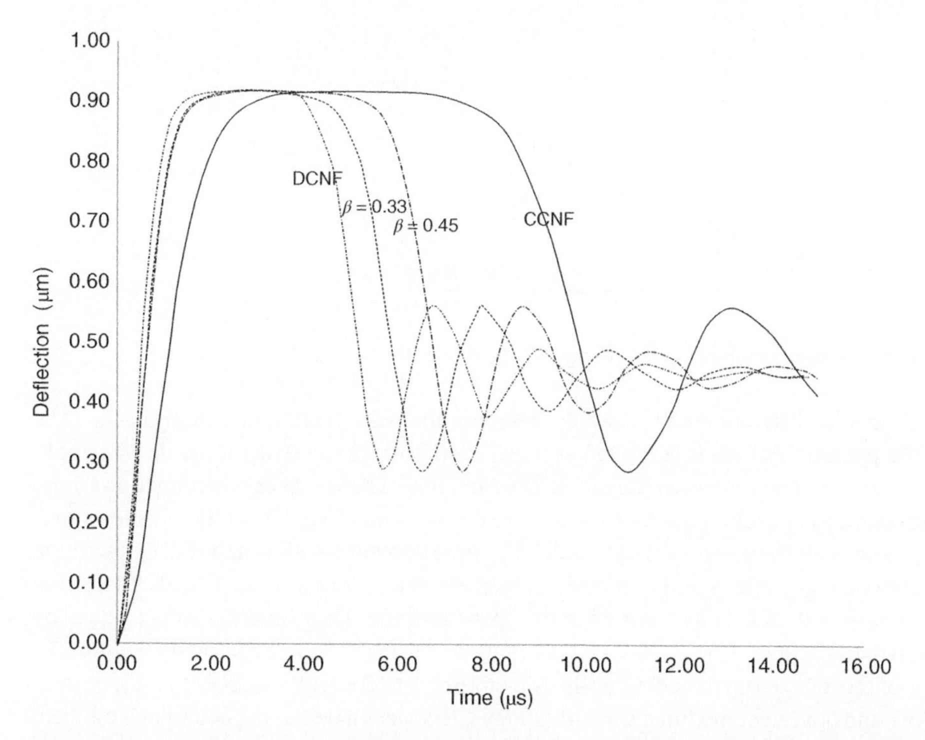

Figure 13.4 Time response of the system near the pull-in voltage for the four selected models.

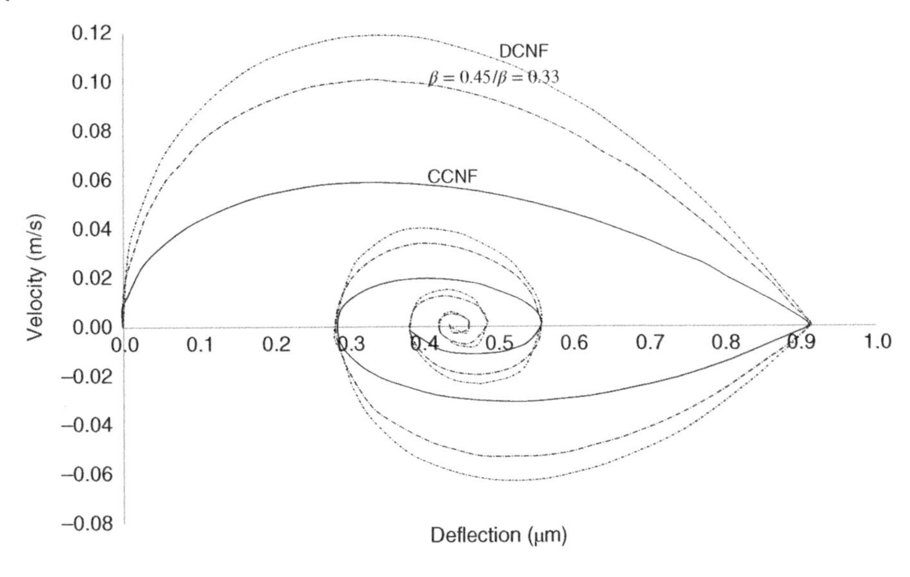

Figure 13.5 Phase portrait of the four models for potentials near the pull-in voltage.

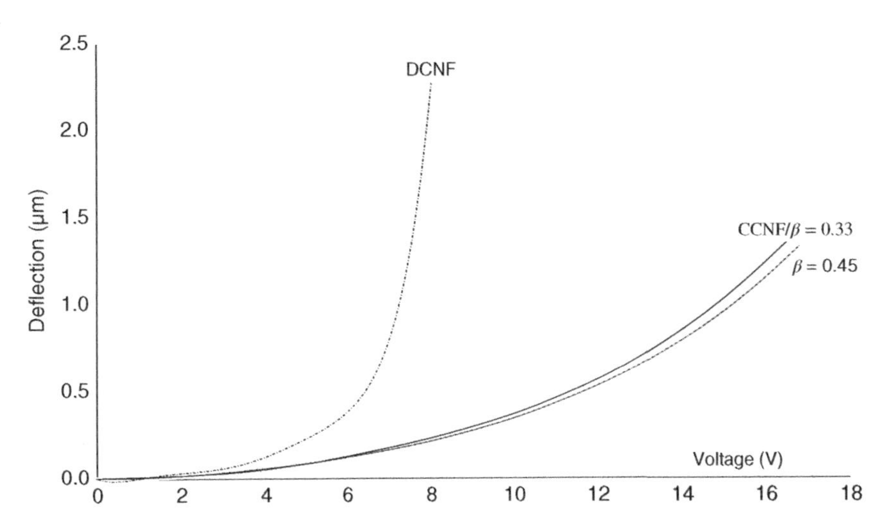

Figure 13.6 Dependency of the deflection with the applied voltage for the four different models.

deflection under various voltages for different models, and the resonant frequency variation yield by each of the four selected models with the potential, which is perceived as a loss of stiffness of the structure under the application of the DC voltage (weakening phenomena observed in experiments). These sets are calculated from the solution of Eq. (13.16). The solutions of each of the four models are derived and plotted. Each graph shows four sets of solutions and each set represents the solution of one model. The time domain solution of the ODE describing the dynamics of a microsystem such as a microcantilever beam subjected to electrostatic forces reveals interesting trend that exhibits different behavior than the one reported in the open literature. These findings were enabled by solving the characteristic differential equation describing this phenomenon using Lie group symmetry. Figure 13.4 illustrates the time response of the four selected models for the gap of 2 μm: DCNF (7.1 V), CCNF (14.5 V), nonlinear modified $\beta = 0.33$ (14.3 V), and nonlinear modified $\beta = 0.45$ (14.6 V). It is seen that after the potential is applied, the system responds with a ramp followed by a flat, which is the same for all models (deflection representing 4/9

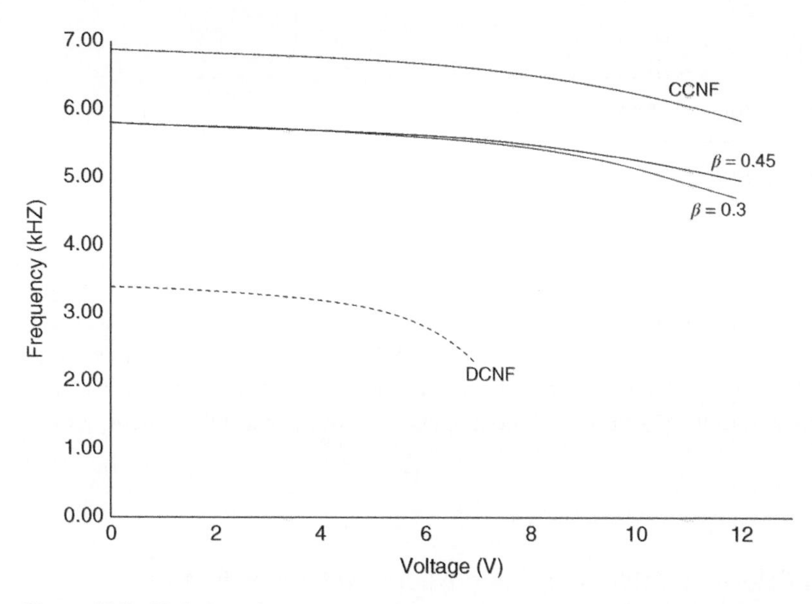

Figure 13.7 Variation of the resonant frequency of the system for the four models with the applied voltage.

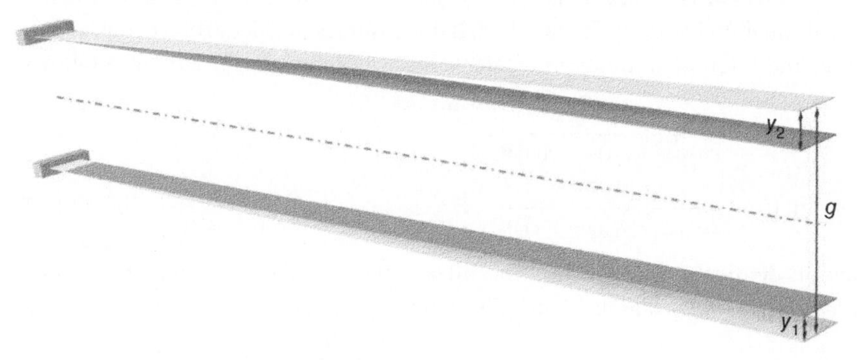

Figure 13.8 Two identical microcantilever beams.

or 45% of the initial gap). It is important to mention here that the quasi static analysis yields a unique limit position that corresponds to a deflection of one-third of the original gap. The duration of the flat, which is a saddle point in the stability of the structure, is dependent on the model. Thus, the duration of the marginal stability is inversely proportional to the instant stiffness; lower stiffness yields longer duration to settlement. Once the system leaves the marginal stability seen in the phase portrait as the saddle point, the equilibrium position is attained regardless of the model in a position representing 2/9 [4] of the initial gap or half of the marginal stability position. Although CCNF model proves to exhibit the lowest resonant frequency at the applied potential, this is due to the fact that the rate of decay of the resonant frequency of the CCNF model is higher than of the other models. It is important to mention that if the critical voltage is assumed in any of the models, no solution is found.

Phase portraits in Figure 13.5 show that the system described by each of the four models is still stable, but it has developed a sharp edge along the x-axis. This edge is the saddle point that is seen in the time response as a flat at the peak deflection point or as previously described as the marginal stability duration. Table 13.2 illustrates the critical distance, which is invariant on the selected model, at which the pull-in occurs as calculated through the presented formulation. All phase diagrams regardless of the model show the saddle point at the critical pull-in distance. Further, the plot indicates a convergence point that is the settling position as illustrated in Figure 13.5.

Table 13.2 Pull-in distance.

	2 μm	4 μm	6 μm	8 μm	10 μm
Pull-in distance (μm)	0.9	1.8	2.7	3.6	4.5
Ratio (%)	45	45	45	45	45

The settlement distance represents 50% of the pull-in distance. The maximum velocity occurs in DCNF, whereas the minimum velocity is encountered in CCNF, while velocities of the modified models are sensitively similar. Figure 13.6 shows the effect of applied voltage on deflection of the beam. As one can see in the figure by reaching to the pull-in voltage deflection increases rapidly. Figure 13.7 shows the decrease of the natural frequency of the beam when increasing the voltage.

13.13 Nonlinear Analysis of Pull-in Voltage of Twin Microcantilever Beams

If the substrate is another identical microcantilever beam like the one above it, the system will be constituted of two beams symmetrically positioned and initially parallel Figure 13.8. If one assumes that deflection of top beam is y_1 and deflection of lower beam is y_2, the differential equations of microcantilever beams can be written as follows:

$$\begin{aligned}\frac{d^2y_1(t)}{dt^2} + 2\xi_1\omega_{1n}\frac{dy_1(t)}{dt} + \omega_{1n}{}^2 y_1(t) &= \frac{\varepsilon_0 A_1 V_1{}^2}{2m_1(g - y_1(t) - y(t)_2)^2} \\ \frac{d^2y_1(t)}{dt^2} + 2\xi_2\omega_{1n}\frac{dy_2(t)}{dt} + \omega_{2n}{}^2 y_2(t) &= \frac{\varepsilon_0 A_2 V_2{}^2}{2m_2(g - y_1(t) - y(t)_2)^2}\end{aligned} \tag{13.62}$$

By assuming that the two beams are identical, one can assume that all parameters of the two beams are the same ($y_2(t) = y_1(t)$, $\xi_2 = \xi_1$, $\omega_{2n} = \omega_{1n}$, $A_2 = A_1$, $V_2 = V_1$, and $m_2 = m_1$). The system of ODEs can thus be reduced to one ODE as follows (Figure 13.9):

$$\frac{d^2y(t)}{dt^2} + 2\xi\omega_n\frac{dy(t)}{dt} + \omega_n^2 y(t) = \frac{\varepsilon_0 AV^2}{2m(g - 2y(t))^2} \tag{13.63}$$

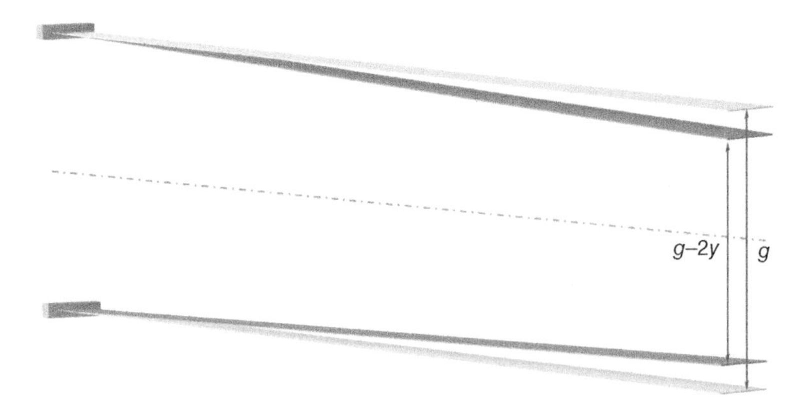

Figure 13.9 Twin microcantilever beams under electrostatic filed.

Initial conditions of beams are the same as in (13.45). Using Lie symmetry method will yield the reduction in the order of differential equation.

$$-\frac{dv}{dr} + 2\xi\omega_n v^2 + \left(\omega_n^2 r - \frac{\varepsilon_0 AV^2}{2m(g-2r)^2}\right) v^3 = 0 \tag{13.64}$$

Eq. (13.64) is a nonlinear first-order ODE and $v(0) = \frac{1}{v_0}$ is the initial condition. The value of pull-in voltage is calculated by assuming that the nonlinear part of Eq. (13.64) to be zero, as follows:

$$V = \sqrt{\frac{2my}{\varepsilon_0 A}}(d-2y)\omega_n \tag{13.65}$$

Neglecting damping part of (13.64) will yield to the following equation:

$$-\frac{dv}{dr} + \left(\omega_n^2 r - \frac{\varepsilon_0 AV^2}{2m(g-2r)^2}\right) v^3 = 0 \tag{13.66}$$

This equation is separable and the solution is:

$$v = \sqrt{\frac{m(g-r)}{m(r-g)(c_1 + \omega^2 r^2) + \varepsilon_0 AV^2}} \tag{13.67}$$

where c_1 is constant. According to Lie symmetry, $v = \frac{1}{\frac{dy(t)}{dt}}$ and $r = y(t)$. Hence,

$$t = \int \sqrt{\frac{m(g-y)}{m(y-g)(c_1 + \omega^2 y^2) + \varepsilon_0 AV^2}} dy + c_2 \tag{13.68}$$

The above integral can be simplified as below:

$$t = \frac{2\sqrt{m}(\text{EllipticF}(j,k)(d-g) + \text{EllipticPi}(j,h,k)(e-d))}{\sqrt{(c-e)(c-g)}} + c_2 \tag{13.69}$$

in which:

$$h = \sqrt{\frac{g-e}{g-d}}, k = h \times \sqrt{\frac{c-d}{c-e}}, j = \frac{1}{h} \times \sqrt{\frac{e-x}{d-x}} \tag{13.70}$$

EllipticF is the incomplete elliptic integral of the first kind and EllipticPi is the incomplete elliptic integral of third kind. Parameters c, d, and e are defined as the values of X, which are the roots of the following equation:

$$m(X-g)(c_1 + \omega^2 X^2) + \varepsilon_0 AV^2 = 0 \tag{13.71}$$

The pull-in voltage is also numerically determined as 129.055 V for a two-beam setup and 182.511 V for a one-beam and rigid substrate. The beam displacements are illustrated in the below graphs. The dimensions of beams are as defined in Section 13.12 except the gap which is $g = 10\,\mu\text{m}$. The pull-in value calculated from (13.65) shows that for the two-beam setup, the error of the exact solution obtained using Lie group symmetry approach with respect to the numerical solution is 0.104%. For the single beam, using almost similar equation, the calculated error increases slightly to 0.449%. The one-beam configuration as the two-beam configuration exhibit same behavior as illustrated in Figures 13.10 and 13.11. When excited with a voltage close to the pull-in voltage, the beams will deflect to a maximum deflection representing 0.45 of the original gap for one beam and half, this is, 0.225 of the original gap for two-beam configuration. After a period of stall, the beams will settle with oscillations vs. a position that corresponds to 0.225 of the original gap for one-beam configuration and 0.1175 for two-beam configuration. The stall is seen as well in the phase diagram for one- and two-beam configurations.

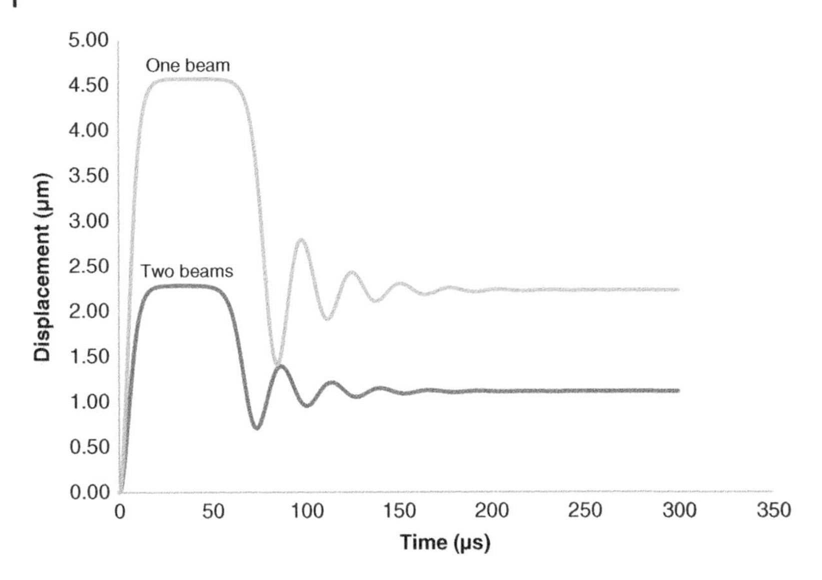

Figure 13.10 The variation of the deflection for one-beam and two-beam scenarios.

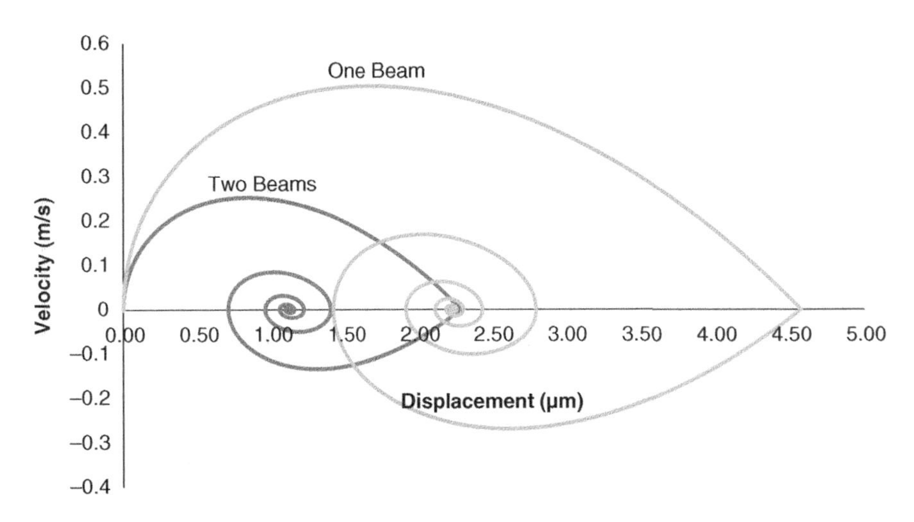

Figure 13.11 Phase diagram of beam.

13.14 Nonlinear Analysis of Pull-in Voltage of Twin Microcantilever Beams of Different Thicknesses

Two nonidentical microcantilever beams are parallel and symmetrically positioned. From the above as shown in Figure 13.8, it can be assumed that the two beams are the same such that $L_1 = L_2$, $b_1 = b_2$, and $E_1 = E_2$ except the thickness of the beams $h_1 \neq h_2$ [5]. Actually, even if certain dimensions are slightly different, the electrostatic attractive force is scaled with regard to the common side of area that is overlapping. After subjecting the beams to electrostatic field, the initial conditions of the beams are still the same as in (13.45). Figure 13.12 represents the two microcantilever beams under the applied force.

Figure 13.12 Nonsymmetric twin set microcantilever beams (exhibit different dimensions) charged at different potentials.

Figure 13.13 The schematic view of a mass-spring-damper system of twin microcantilever beams with different dimensions.

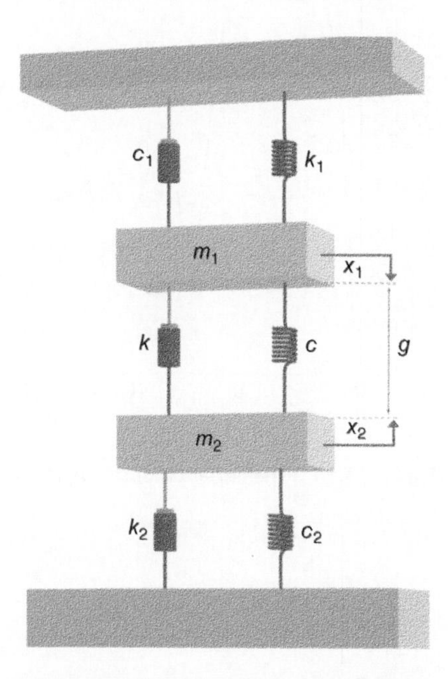

Figure 13.13 represents the schematics of the dynamic model of a nonsymmetric twin set of microcantilever beams under the electrostatic field by using an equivalent mass-spring-damper system:

$$\begin{bmatrix} m_1 & 0 \\ 0 & m_2 \end{bmatrix} \begin{bmatrix} \ddot{x}_1(t) \\ \ddot{x}_2(t) \end{bmatrix} + \begin{bmatrix} c_1 + c & -c \\ -c & c_2 + c \end{bmatrix} \begin{bmatrix} \dot{x}_1(t) \\ \dot{x}_2(t) \end{bmatrix} + \begin{bmatrix} k_1 + k & -k \\ -k & k_2 + k \end{bmatrix} \begin{bmatrix} x_1(t) \\ x_2(t) \end{bmatrix} = \begin{bmatrix} f_1(t) \\ f_2(t) \end{bmatrix} \tag{13.72}$$

The electrostatic attraction between the two parallel beams in the microstructure is notable, and it can be considered as an externally applied force. From the energy balance, the forces f_1, f_2 have been found as:

$$f_1 = f_2 = \frac{\varepsilon_0 A V^2}{2(g - x_1(t) - x_2(t))^2} \tag{13.73}$$

Under specific circumstances, when solvers are not suitable with the type of problem, a numerical approach is considered. The solution of the nonlinear differential equation is stiff (13.72) and is numerically found using

ISODE algorithm [6]. The algorithm makes use of adaptive integration steps such that even when the equations are stiff, that is, close to the pull-in voltage, a correct solution of the problem is still found. The numerical approach was developed for the system of nonlinear ODEs. The results are presented below for the two nonidentical beams.

Two similar beams were studied under the assumption that the below beam thickness is twice of above beam. The dimensions and properties of beams are discussed in Section 13.12, except:

$$g = 10\,\mu\text{m},\ h_1 = \frac{h_2}{2} = 2\ \mu m$$

The numerical solution of the nonlinear system of ODEs in Eq. (13.72) will yield to the following results. Figure 13.14 shows time response of the thinner beam when the voltage is close to the pull-in voltage (145.21 V). Figure 13.15 shows the phase diagram of the same beam. As one can see, the system reaches a saddle point on the deflection close to 4 μm. Figure 13.16 shows the time response of the thicker beam. As it is shown in the curve, the system goes to one stable condition (0.3 μm) and then changes the position and stays on other stable condition (0.5 μm). This phenomenon is illustrated clearly on Figure 13.17.

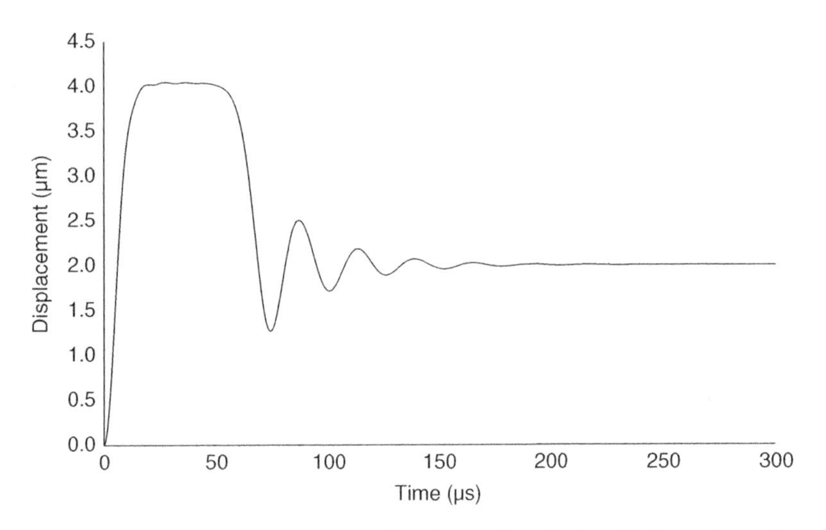

Figure 13.14 The deviation of the deflection vs. time for the thin beam (upper beam).

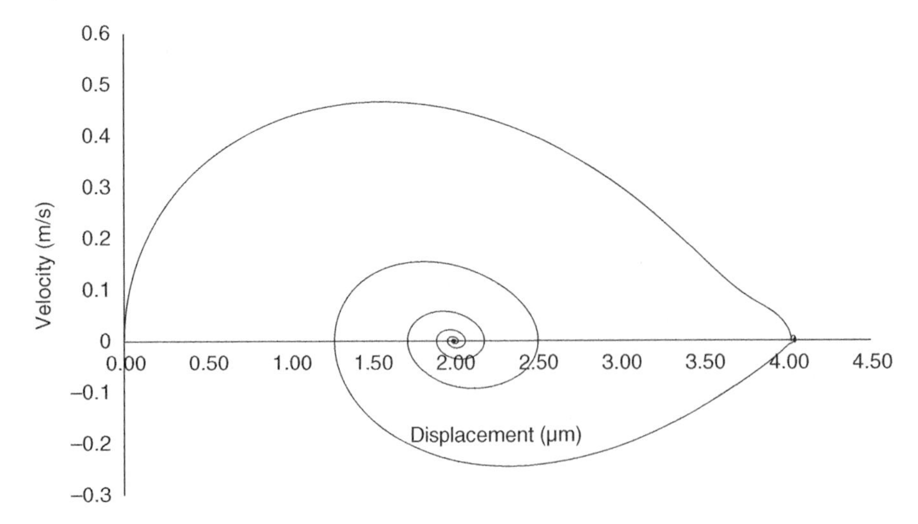

Figure 13.15 Phase diagram of the thin beam (upper beam) – changes in velocity vs. deflection.

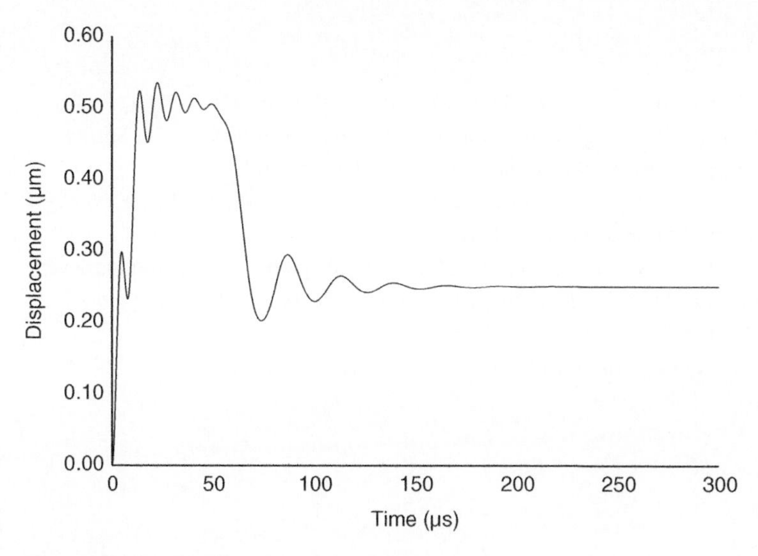

Figure 13.16 The deviation of the deflection vs. time for the thick beam (lower beam).

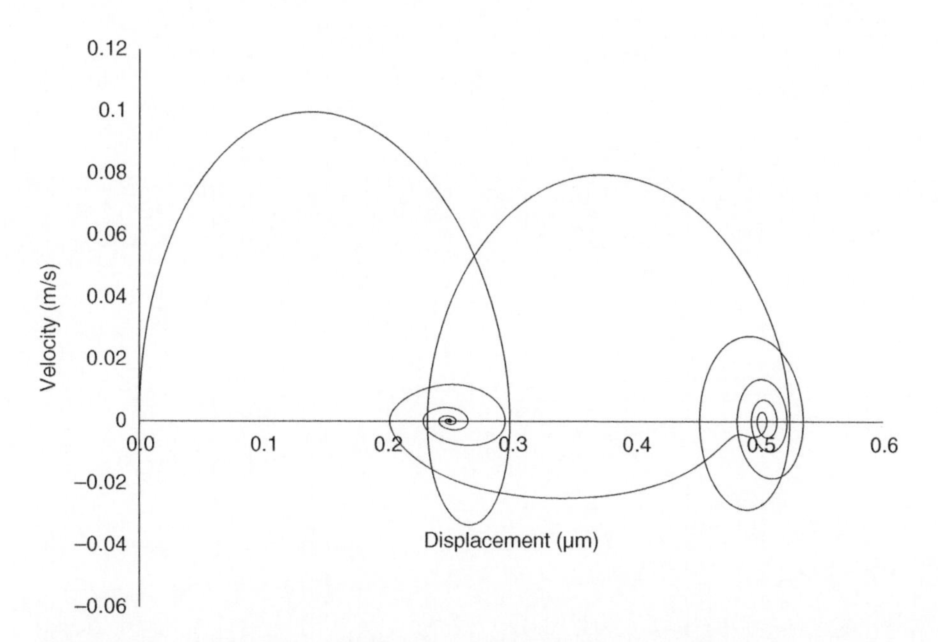

Figure 13.17 Phase diagram of the thick beam (lower beam) – unsteady changes in velocity vs. deflection.

The Lie group symmetry represents a powerful method that enables to reduce the order of the differential equations by one order under the provision that a symmetric transformation is found for the differential equation. The method enables to reveal correlations between parameters that are not possible when using numerical approach. The method is applicable to as many other problems resulting in complex mathematical formulations. If used for solving more practical problems, it may become one of the standard mathematical tools in engineering.

References

1 Arrigo, D.J. (2015). *Symmetry Analysis of Differential Equations: An Introduction*, 1e. Wiley.

2 Bluman, G.W. and Anco, S.C. (2002). *Symmetry and Integration Methods for Differential Equations*, 2e. Springer.

3 Young, W. and Budynas, R. (2011). *Roark's Formulas for Stress and Strain*, 7e. McGraw-Hill.

4 Changizi, M. and Stiharu, I. (2017). Nonlinear analysis of pull-in voltage of twin micro-cantilever beams. *International Journal of Mechanical Engineering and Mechatronics* 4: 11–17.

5 Changizi, M., Sarafrazi, M., and Stiharu, I. (2017). Nonlinear analysis of pull-in voltage of twin micro-cantilever beams with different dimensions. *Proceedings of the 4th International Conference of Control, Dynamic Systems, and Robotics (CDSR'17)*, Toronto (21–23 August 2017). pp. 116-1–116-8.

6 Changizi, M. (2011). Geometry and material nonlinearity effects on static and dynamics performance of MEMS. PhD thesis. Concordia University: Montreal.

14

Generalized Differential and Integral Quadrature: Theory and Applications

Francesco Tornabene and Rossana Dimitri

Department of Innovation Engineering, University of Salento, Lecce, Italy

14.1 Introduction

In the past decades, the simplicity and versatility of the differential quadrature (DQ) method, as proposed by Bellman at the beginning of the 1970s, has gradually increased its application in many branches of engineering and science. Remarkably, the DQ approach represents an efficient numerical tool to solve complex differential equations, and it yields accurate results even with a limited number of collocation points discretizing the domain. This limits many common computational difficulties, especially in a context where nonlinear algorithms and high computational effort are required among complex engineering applications. More in detail, this procedure approximates derivatives of a function at a certain point through a linear sum of all function values in the definition domain, whereby a key aspect is related to the evaluation of weighting coefficients.

In some pioneering works by Bellman and coworkers, two approaches were proposed to compute these weighting coefficients for the first-order derivatives of functions. The first method was based on an algebraic system of equations featuring ill-conditioning drawbacks, whereas the second method was a simple algebraic formulation that assumes the roots of shifted Legendre polynomials as grid points. In the preliminary DQ-based engineering applications, the first Bellman approach has been largely adopted because of its arbitrary selection of grid points [1–3]. Nevertheless, it was found that more than 13 collocation points could not be used because of the ill-conditioning of the algebraic system, which represents a clear numerical limit to its efficiency.

Thus, a generalized differential quadrature (GDQ) method was proposed in literature by Shu [2] to improve the DQ technique in terms of weighting coefficients computation. According to the GDQ approach, the weighting coefficients are related only to the domain discretization and to the order of derivative, independently of the particular problem analyzed, under an arbitrary choice of grid points.

The initial difficulties of the DQ method for the evaluation of the weighting coefficients were overcome by means of the GDQ approach, with the adoption of polynomial approximations and the analysis of the vectorial space of polynomials.

A simple algebraic formulation was presented by Shu [2] to compute the weighting coefficients for the approximation of the first-order derivatives, where any restriction on the grid distribution choice was completely removed. Furthermore, a recursive formula was found to determine the weighting coefficients for higher-order derivatives. When a function is approximated through the Lagrange polynomials, the GDQ technique reverts to the polynomial differential quadrature (PDQ) method. Similarly, Shu [2] presented another simple algebraic formulation, named harmonic differential quadrature (HDQ), where the Fourier polynomials were applied for the evaluation of the weighting coefficients.

In the past 40 years, several researchers have successfully applied the GDQ method in several engineering fields, including the static and dynamic analyses of beams, plates, and shells, or the analysis of waves and viscous fluid

Mathematical Methods in Interdisciplinary Sciences, First Edition. Edited by Snehashish Chakraverty.

flows (henceforth, see [4–14] and references therein). In this chapter, we provide the theoretical fundamentals of both the DQ and GDQ techniques, together with the analysis of the polynomial vector space and the functional approximation. Furthermore, the generalized integral quadrature (GIQ) is derived from the previous ones. We introduce the main properties of the methodologies mentioned above, thus illustrating the procedure for the evaluation of weighting coefficients, principal types of discretization, and some applications to simple functions.

14.2 Differential Quadrature

The DQ method is a numerical technique introduced in 1971, which allows for the solution of complex global and partial differential systems of equations, as typically occurs in many structural, thermal, electrical, and fluid dynamics problems. Since a closed-form solution for differential equations associated with a given problem is not always possible, many approximate methods have been developed in literature with increased accuracy requirements. The main difficulty for an exact closed-form solution of an engineering problem is usually related to the complex boundary conditions associated with the differential system of equations at issue. In many cases, the approximate solution of a problem is represented by its values assumed at some fixed collocation points of a domain. In such a context, the recent DQ method has represented an alternative numerical tool for its accuracy with respect to the more common finite difference method (FDM), finite element method (FEM), or finite volume method (FVM) [14]. The DQ approach belongs to the spectral and pseudospectral methods, and it is interesting for its simplicity and satisfactory results, comparable to the other numerical methods. Originally, this method was used to directly solve linear and nonlinear differential equations with variable coefficients.

The key point of this numerical technique lies in the accurate approximation of a general-order derivative of a smooth function through a linear combination of its values assumed at the selected collocation points discretizing the domain, even for a reduced number of grid points.

14.2.1 Genesis of the Differential Quadrature Method

The first numerical discretization techniques, i.e. the FDM, FEM, or FVM, were developed since the 1950s. Among them, the most widespread approach is based on variational principles (i.e. the FEM), whereby the FDM and the FVM are based on the Taylor's expansion series and the direct application of the law of conservation to physical cells, respectively. Stemming from these approaches, further numerical techniques have been developed in the literature within the so-called class of spectral methods. These discretization schemes are known as methods of weighted residuals, whose key elements represent the basis functions or test functions. Both the spectral methods and FEMs apply the basis functions to approximate numerically the solution. Besides, the basis functions for spectral methods have global properties, whereas the basis functions for FEM-based applications feature local properties, as useful for complex geometries, and they must be specified for each element discretizing the domain. On the other hand, the spectral methods approximate the whole domain, while being able to provide optimal numerical solutions, with a limited number of nodal points.

Bellman and Casti [15] developed the DQ method as an extension of the "integral quadrature" concept. The partial or global derivative of a smooth function is approximated in a neighborhood of a point of its domain, as a weighted linear sum of the function values assumed at the whole points discretizing the domain along the direction of derivation. The key point of the method is the computation of the weighting coefficients for any derivation order, rather than the domain discretization. Bellman and his colleagues suggested two ways of determining these coefficients for the first derivative. The first way is to solve a system of algebraic equations, whereby the second strategy uses a simple algebraic formulation where the mesh points must be selected as roots of the shifted Legendre polynomials.

The first DQ-based applications in literature used these criteria to evaluate the weighting coefficients, although a prevalent use of the first method was observed because of the arbitrary possibility of selecting grid points within the domain. However, for higher-order derivatives in the algebraic system, the matrix associated with the problem becomes ill-conditioned. This highlights why the first DQ applications were carried out limiting the number of grid points up to 13. Several studies were conducted to overcome the difficulties related to the weighting coefficients evaluation. Civan [16] demonstrated that the computational issues of the first approach by Bellman and Casti were due to the Vandermonde matrix. This matrix is associated with the system of algebraic equations, as commonly found in many engineering problems. Some special algorithms, for the solution of the Vandermonde matrix, had been developed by Björck and Pereyra [17]. This algorithm had allowed to compute the weighting coefficients using more than 31 grid points. However, it had not permitted to evaluate the weighting coefficients for order derivatives higher than the first one.

A further improvement in the computation of weighting coefficients was proposed by Quan and Chang [18], who assumed the interpolating Lagrange polynomials as basis polynomials. In this way, an explicit formulation was provided to determine the first- and the second-order derivatives.

A better solution was also proposed by Shu [19] and Shu and Richards [20] based on the polynomial approximation of higher-order derivatives and the analysis of a linear vector space. With this approach, the weighting coefficients for the first-order derivatives can be determined through a simple algebraic formulation without any restriction on the choice of grid points. In addition, the weighting coefficients related to the second- and higher-order derivatives are defined by means of recursive formulae.

The proposed generalization has led to the generalized version of the DQ method, named GDQ method, which is here based on the polynomial approximation, and therefore labeled polynomial differential quadrature method (PDQM). Further approximation formulae were introduced by Shu and Chew [21] and Shu and Xue [22] based on the Fourier expansion series, for the computation of weighting coefficients. In such a case, the approach is called DQ method based on the expansion in Fourier series (*harmonic differential quadrature method* or HDQM).

According to the above-mentioned review, it seems that the mathematical basics of the GDQ method can be found in the analysis of a vector space and on the approximation of linear functions. In this context, several results are available in the scientific and technical literature [8, 14], where the GDQ method has been successfully proposed for different structural analyses because of its accuracy, efficiency, simplicity in use, and low computational cost.

In the past applications, the differential equations did not exceed the second-order derivatives and they did not have more than one boundary condition for each equation of the definition domain. However, for slender beams and thin plates, the differential equations include the fourth-order derivatives, whereby two constraint equations are defined at each side. It is worth observing that the constraint equations become redundant because the problem is defined by a single equation, in view of two boundary conditions. This means that the implementation of those conditions requires a special attention. The problems related to structural mechanics contributed to improve the methodology at issue, as highlighted by the large number of recent works published in the literature [8, 14], including the instability of beams and plates and statics and dynamics of arches and shells with single or double curvature.

14.2.2 Differential Quadrature Law

It is well-known that the term "quadrature" is associated with the approximate evaluation of an integral. Thus, the conventional idea of *integral quadrature* can be simply used to solve differential equations, which makes necessary to introduce the main basic notions, as in the following.

A frequent science and engineering problem involves the evaluation of the integral of a given function $f(x)$ in a closed interval $[a, b]$. If an integral function exists $F(x)$, such that $dF(x)/dx = f(x)$, then the integral value is

$F(b)-F(a)$. Unfortunately, for the solution of problems of practical interest, it is difficult to obtain an explicit expression for $F(x)$, since the values of $f(x)$ can be evaluated at fixed points of the domain.

As can be seen in Figure 14.1, the integral of a function $f(x)$ on $[a, b]$ represents the area under the integral function $f(x)$, and it can be approximated as

$$\int_a^b f(x)dx \cong \sum_{i=1}^{N}\left(\int_a^b l_i(x)dx\right)f(x_i) = \sum_{i=1}^{N} w_i f(x_i) = w_1 f_1 + w_2 f_2 + \cdots + w_N f_N \tag{14.1}$$

where $l_i(x)$ stands for the *Lagrange interpolating polynomials* (whose definition is given in the Section 14.3.1.1) and w_i are the weighting coefficients defined as

$$w_i = \int_a^b l_i(x)dx \tag{14.2}$$

and $f(x_i)$ refers to the values of the function at fixed grid points x_i of the domain. In other words, the domain of the function $f(x)$ is discretized in N nodal points (*discretization grid points*) x_i, $i = 1, 2, \ldots, N$.

Equation (14.1) is named as *integral quadrature*. Usually, the nodal points are assumed to be uniformly distributed. It will be shown that all the conventional quadrature rules can be identified with this important relationship.

Let us consider a one-dimensional problem where a sufficiently smooth function $f(x)$ is defined within a closed interval $[a, b]$ with a discretization of N points of coordinates $a = x_1, x_2, \ldots, x_{N-1}, x_N = b$ (Figure 14.2).

Following the above-mentioned concept of integral quadrature, we can extend the integral quadrature (14.1) concept to compute the derivative of a function $f(x)$. The first-order derivative of a function $f(x)$, with respect to x at an arbitrary nodal point x_i, can be approximated through a weighted linear sum of all the values of the function in the whole domain, as follows:

$$f'(x_i) = f^{(1)}(x_i) = \left.\frac{df(x)}{dx}\right|_{x=x_i} \cong \sum_{j=1}^{N} \varsigma_{ij}^{(1)} f(x_j), \quad i = 1, 2, \ldots, N \tag{14.3}$$

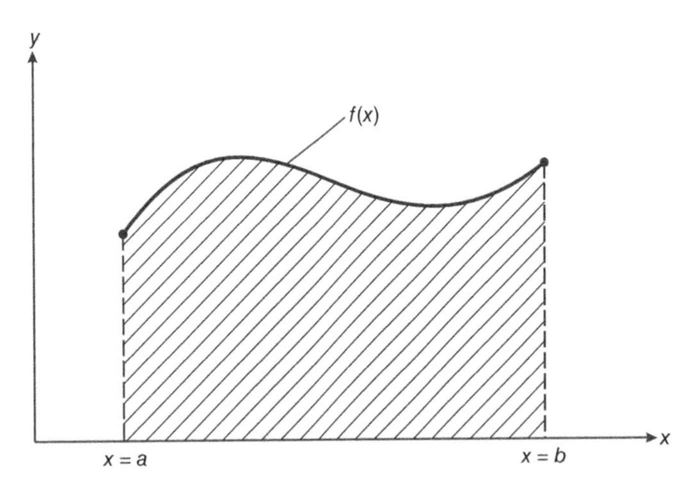

Figure 14.1 Integral of $f(x)$ within a closed interval $[a, b]$.

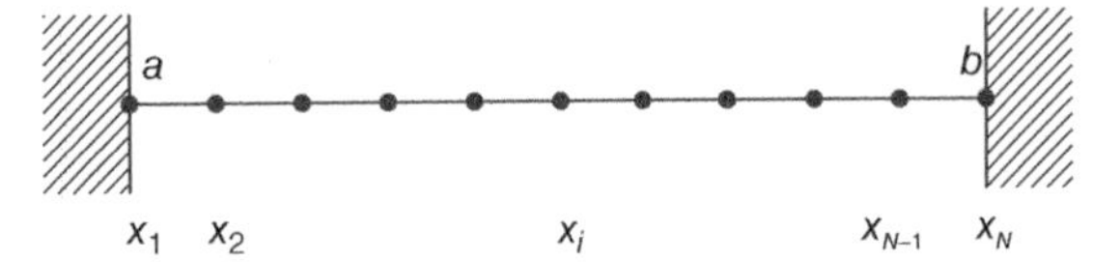

Figure 14.2 One-dimensional representation of the problem.

where the constants $\varsigma_{ij}^{(1)}$ stand for the *weighting coefficients*, $f(x_j)$ refers to values of the function at grid points x_j, and $f'(x_i)$ (or $f^{(1)}(x_i)$) is the first-order derivative of $f(x)$ at x_i.

The coefficients $\varsigma_{ij}^{(1)}$ are characterized by the subscripts i and j. The index i refers to the domain point where the derivative is computed. Equation (14.3) is named *differential quadrature* (DQ) or *differential quadrature law* and defines a *linear operator*.

Starting with this simple application, the key aspects of the present method are related to the *discretization* domain, using N nodal points (mesh or grid definition) and the *evaluation* of weighting coefficients for the summation process. Thus, the critical step of this procedure lies on the definition of weighting coefficients.

14.3 General View on Differential Quadrature

As already stated in the introduction a numerical approach can be distinguished according to the different selection of basis functions and formulation (in a strong or weak form). The present section aims at showing most of the collocation points, basis functions, and weighting coefficients that have been applied in the literature, hitherto, for the DQ-based studies [14]. For the solution of the partial differential equations and their boundary conditions, the unknown function $f(x)$ is approximated by a set of smooth basis functions $\psi_j(x)$ (e.g. the polynomial functions) for $j = 1, 2, \ldots, N$, where N stands for the total number of grid points. From a mathematical standpoint, the approximate solution of a function $f(x)$ can be written as

$$f(x) \cong \sum_{j=1}^{N} \lambda_j \psi_j(x) \tag{14.4}$$

The key point is to solve a partial differential system through the approximation (14.4) at the discrete points x_i for $i = 1, 2, \ldots, N$. Using the so-called differentiation matrix procedure, the derivative of a given function can be easily found in compact form as

$$\mathbf{f}^{(n)} = \mathbf{D}^{(n)}\mathbf{f} \tag{14.5}$$

where

$$\mathbf{f}^{(n)} = \left[\left.\frac{d^n f(x)}{d^n x}\right|_{x_1}, \left.\frac{d^n f(x)}{d^n x}\right|_{x_2}, \ldots, \left.\frac{d^n f(x)}{d^n x}\right|_{x_N} \right]^T \tag{14.6}$$

is the vector of the function derivatives at the whole grid points, whereas

$$\mathbf{f} = [f(x_1), f(x_2), \ldots, f(x_N)]^T \tag{14.7}$$

is the vector of the unknown functions, and $\mathbf{D}^{(n)}$ is the differentiation matrix that must be determined. Thus, the approximation (14.4) can be written in a matrix form as

$$\mathbf{f} = \mathbf{A}\boldsymbol{\lambda} \tag{14.8}$$

where λ is the vector of the unknown coefficients λ_j and $\mathbf{A}$ is the coefficients matrix, being $A_{ij} = \psi_j(x_i)$. Once Eq. (14.4) is defined, its general nth derivative is directly related to the functions $\psi_j(x)$ for linearity reasons because of the independence of the unknown coefficients λ_j on the variable x

$$\frac{d^n f(x)}{dx^n} = \sum_{j=1}^{N} \lambda_j \frac{d^n \psi_j(x)}{dx^n}, \quad \text{for } n = 1, 2, \ldots, N-1 \tag{14.9}$$

rewritten in matrix form as

$$\mathbf{f}^{(n)} = \mathbf{A}^{(n)}\boldsymbol{\lambda}, \quad \text{with } A_{ij}^{(n)} = \left.\frac{d^n \psi_j(x)}{dx^n}\right|_{x_i} \tag{14.10}$$

In order to obtain the differentiation matrix $\mathbf{D}^{(n)}$, the matrix $\mathbf{A}$ should be invertible. This property depends on the basis functions and collocation of grid points. The unknown parameter vector can be determined from Eq. (14.8) by a mathematical invertion of matrix $\mathbf{A}$, namely

$$\boldsymbol{\lambda} = \mathbf{A}^{-1}\mathbf{f} \tag{14.11}$$

Thus, the substitution of Eq. (14.11) into Eq. (14.9) yields to the following relation

$$\mathbf{f}^{(n)} = \mathbf{A}^{(n)}\mathbf{A}^{-1}\mathbf{f} = \mathbf{D}^{(n)}\mathbf{f} \tag{14.12}$$

where the differentiation matrix $\mathbf{D}^{(n)}$ is determined as the matrix product between the inverse of the matrix $\mathbf{A}$ of the basis functions at the grid points $\psi_j(x_i)$ and the derivatives of the same function values $\psi_j^{(n)}(x_i)$, collected in $\mathbf{A}^{(n)}$. In the following, all the mathematical steps are defined for a nth-order derivative,

$$\left.\frac{d^n f(x)}{dx^n}\right|_{x_i} = \sum_{j=1}^{N} D_{ij}^{(n)} f(x_j) \quad \text{with } D_{ij}^{(n)} = (\mathbf{A}^{(n)}\mathbf{A}^{-1})_{ij} \tag{14.13}$$

This approach is similar to the GDQ method, i.e.

$$\left.\frac{d^n f(x)}{dx^n}\right|_{x_i} = \sum_{j=1}^{N} \varsigma_{ij}^{(n)} f(x_j) \tag{14.14}$$

As a result, the coefficients $D_{ij}^{(n)}$ of matrix $\mathbf{D}^{(n)}$ are evaluated as product between an inverse matrix and its derivative, whereas the $\varsigma_{ij}^{(n)}$ coefficients can be computed by means of the recursive formulae by Shu [2]. Although the two methodologies are apparently similar, the last one does not require any matrix inversion to evaluate the weighting coefficients.

In general, a coordinate transformation is required to define the weighting coefficients, for the physical points, within the definition interval. The general approach applied for a coordinate transformation of the domain will be considered in what follows, while focusing to the shifted expression of weighting coefficients, namely

$$\frac{d^n f(x)}{dx^n} = \left(\frac{r_N - r_1}{x_N - x_1}\right)^n \frac{d^n f(x(r))}{dr^n} \quad \begin{array}{l} \text{for} \quad n = 1, 2, \ldots, N-1 \\ \text{with } x(r) = \dfrac{x_N - x_1}{r_N - r_1}(r - r_1) + x_1 \end{array} \tag{14.15}$$

where $x_N - x_1$ stands for the length of the physical domain and $r_N - r_1$ is the domain length that defines the basis polynomial under consideration. Hence, the linear transformation of coordinates within the two intervals reads

$$\varsigma_{ij}^{(n)} = \left(\frac{r_N - r_1}{x_N - x_1}\right)^n \tilde{\varsigma}_{ij}^{(n)} \quad \text{for } i,j = 1, 2, \ldots, N, \;\; n = 1, 2, \ldots, N-1 \tag{14.16}$$

where $\varsigma_{ij}^{(n)}$ and $\tilde{\varsigma}_{ij}^{(n)}$ are the weighting coefficients for the physical and definition domain, respectively. However, despite the validity of Eq. (14.13) for any basis function and discretization grid, the matrix $\mathbf{A}$ can become ill-conditioned for an increasing number of points N because of its similarity to the Vandermonde matrix.

14.3.1 Basis Functions

Several basis functions from the literature are illustrated hereafter to provide a complete useful list, as also summarized in Tables 14.1 and 14.2.

Table 14.1 List of the classic orthogonal polynomials used as basis functions.

Jacobi polynomials

$$\psi_j = J_j^{(\gamma,\delta)}(r) = \frac{(-1)^{j-1}}{2^{j-1}(j-1)!(1-r)^{\gamma}(1+r)^{\delta}} \frac{d^{j-1}}{dr^{j-1}}((1-r)^{j-1+\gamma}(1+r)^{j-1+\delta})$$

$r \in [-1,1], \quad j = 1,2,\ldots,N, \quad \gamma,\delta > -1$

Legendre polynomials

$$\psi_j = L_j(r) = \frac{(-1)^{j-1}}{2^{j-1}(j-1)!} \frac{d^{j-1}}{dr^{j-1}}((1-r^2)^{j-1})$$

$r \in [-1,1] \quad j = 1,2,\ldots,N$

Chebyshev polynomials (first kind)

$$\psi_j = T_j(r) = \cos((j-1)\arccos(r)), \quad r \in [-1,1], \quad j = 1,2,\ldots,N$$

Chebyshev polynomials (second kind)

$$\psi_j = U_j(r) = \frac{\sin(j\ \arccos(r))}{\sin(\arccos(r))}, \quad r \in [-1,1], \quad j = 1,2,\ldots,N$$

Chebyshev polynomials (third kind)

$$\psi_j = V_j(r) = \frac{\cos\left(\frac{(2j-1)\arccos(r)}{2}\right)}{\cos\left(\frac{\arccos(r)}{2}\right)}, \quad r \in [-1,1], \quad j = 1,2,\ldots,N$$

Chebyshev polynomials (fourth kind)

$$\psi_j = W_j(r) = \frac{\sin\left(\frac{(2j-1)\arccos(r)}{2}\right)}{\sin\left(\frac{\arccos(r)}{2}\right)}, \quad r \in [-1,1], \quad j = 1,2,\ldots,N$$

Hermite polynomials

$$\psi_j = H_j(r) = (-1)^{j-1}e^{r^2}\frac{d^{j-1}}{dr^{j-1}}(e^{-r^2}), \quad r \in]-\infty,+\infty[, \quad j = 1,2,\ldots,N$$

Laguerre polynomials

$$\psi_j = G_j(r) = \frac{1}{(j-1)!e^{-r}}\frac{d^{j-1}}{dr^{j-1}}(r^{j-1}e^{-r}), \quad r \in [0,+\infty[, \quad j = 1,2,\ldots,N$$

Lobatto polynomials

$$A_j(r) = \frac{d}{dr}(L_{j+1}(r)) = C_j^{(3/2)}(r) = \frac{j+1}{2}J_j^{(1,1)}(r), \quad r \in [-1,1], \quad j = 1,2,\ldots,N$$

Gegenbauer polynomials

$$\psi_1 = C_1^{(\lambda)}(r) = 1, \quad \psi_j = C_j^{(\lambda)}(r) = \sum_{k=0}^{\phi(j-1)} \frac{(-1)^k(\lambda)_{j-1-k}(2r)^{j-1-2k}}{k!(j-1-2k)!}, \quad \phi(j-1) = \frac{2(j-1)+((-1)^{j-1}-1)}{4}, \quad r \in [-1,1], \quad j = 2,3,\ldots,N$$

Table 14.2 List of the most common basis functions for the functional approximation.

Lagrange polynomials $\psi_j = l_j(r) = \dfrac{\mathbf{L}(r)}{(r_i - r_j)L^{(1)}(r_j)}, \quad r \in]-\infty, +\infty[, \quad j = 1, 2, \ldots, N$ $\mathbf{L}(r) = \prod_{k=1}^{N}(r - r_k), \; \mathbf{L}^{(1)}(r_j) = \prod_{k=1, j\neq k}^{N}(r_j - r_k)$	*Trigonometric Lagrange polynomials* $\psi_j = g_j(r) = \dfrac{\mathcal{G}(r)}{\sin\left(\frac{r - r_j}{2}\right) S^{(1)}(r_j)}, \quad r \in [0, 2\pi], \quad j = 1, 2, \ldots, N$ $\mathcal{G}(r) = \prod_{k=1}^{N} \sin\left(\frac{r - r_k}{2}\right), \; \mathcal{G}^{(1)}(r_j) = \prod_{k=1, j\neq k}^{N} \sin\left(\frac{r_j - r_k}{2}\right)$
Bernstein polynomials $\psi_j = B_j(r) = \dfrac{(N-1)!}{(j-1)!(N-j)!} r^{j-1}(1-r)^{N-j}$ $r \in [0, 1], \quad j = 1, 2, \ldots, N$	*Rational Bernstein functions* $\psi_j = R_j(r) = \dfrac{B_j(r)\omega_j}{\sum_{k=1}^{N} B_k(r)\omega_k}, \quad r \in [0, 1], \quad j = 1, 2, \ldots, N$
Exponential functions $\psi_j = E_j(r) = e^{(j-1)r}, \quad r \in]-\infty, +\infty[, \quad j = 1, 2, \ldots, N$	*Monomial functions* $\psi_j = Z_j(r) = r^{j-1}, \quad r \in]-\infty, +\infty[, \quad j = 1, 2, \ldots, N$
Bessel polynomials $\psi_1 = P_1(r) = 1, \; \psi_j = P_j(r) = \sum_{k=0}^{j-1} \dfrac{(j-1+k)!}{(j-1-k)!k!}\left(\dfrac{r}{2}\right)^k$ $r \in [-\infty, +\infty], \quad j = 2, 3, \ldots, N$	*Cardinal sine function* $\psi_j = S_j = \mathrm{sinc}_j(r) = \dfrac{\sin(\pi(N-1)(r - r_j))}{\pi(N-1)(r - r_j)}$ $r \in [0, 1], \quad j = 1, 2, \ldots, N$
Fourier functions $\psi_1 = F_1(r) = 1, \; \psi_j = F_j(r) = \cos\left(\dfrac{j}{2}r\right)$ for j even $\psi_j = F_j(r) = \sin\left(\dfrac{j-1}{2}r\right)$ for j odd $r \in [0, 2\pi], \quad j = 2, 3, \ldots, N$	*Boubaker polynomials* $\psi_1 = Q_1(r) = 1, \quad r \in [-\infty, +\infty], \quad j = 2, 3, \ldots, N$ $\psi_j = Q_j(r) = \sum_{k=0}^{\phi(j-1)} (-1)^k \binom{j-1-k}{k} \dfrac{j-1-4k}{j-1-k} r^{j-1-2k}$ $\phi(j-1) = \dfrac{2(j-1) + ((-1)^{j-1} - 1)}{4}$

14.3.1.1 Lagrange Polynomials

Lagrange polynomials are applied both in the GDQ approach and in the *Lagrange spectral collocation* approach. The main difference between these two techniques relies on the evaluation of the weighting coefficients. The GDQ method, indeed, uses the above-mentioned recursive formulae found by Shu [2], whereas the spectral methods usually compute the inversion of the coefficient matrix or use a set of exact formulae only for a fixed collocation (the Chebyshev one). The Lagrange polynomials are reported as follows

$$l_j(r) = \frac{\mathbf{L}(r)}{(r - r_j)\mathbf{L}^{(1)}(r_j)}$$

$$\mathrm{L}(r) = \prod_{k=1}^{N}(r - r_k) \quad r \in]-\infty, +\infty[, \quad j = 1, 2, \ldots, N$$

$$\mathbf{L}^{(1)}(r_j) = \prod_{k=1, j\neq k}^{N} (r_j - r_k) \tag{14.17}$$

where r identifies the definition domain of the polynomials and j stands for the jth polynomials. It is worth noticing that each polynomial $l_j(r)$ has degree $N - 1$. The main advantage of these polynomials is that the coefficient matrix **A** corresponds to the identity matrix **I** because the Lagrange basis functions feature the following property

$$l_j(r_i) = \begin{cases} 0 & \text{for } i \neq j \\ 1 & \text{for } i = j \end{cases} \tag{14.18}$$

Thus, matrix **A** is always invertible ($\mathbf{A}^{-1} = \mathbf{I}$), such that Eq. (14.13) becomes

$$\left.\frac{d^n f(x)}{dx^n}\right|_{x_i} = \sum_{j=1}^{N} D_{ij}^{(n)} f(x_j) \quad \text{with } D_{ij}^{(n)} = \varsigma_{ij}^{(n)} = (\mathbf{A}^{(n)}\mathbf{I})_{ij} = A_{ij}^{(n)} \tag{14.19}$$

where

$$A_{ij}^{(n)} = l_j^{(n)}(r_i) \tag{14.20}$$

The computation of the weighting coefficients does not depend on the ill-conditioning of matrix **A** for an increasing number of grid points. For the sake of completeness, we provide the recursive formulae for the evaluation of the weighting coefficients for the first and nth order, whose demonstration can be found in the work by Shu [2]

$$\tilde{\varsigma}_{ij}^{(1)} = \frac{\mathrm{L}^{(1)}(r_i)}{(r_i - r_j)\mathrm{L}^{(1)}(r_j)}$$

$$\tilde{\varsigma}_{ij}^{(n)} = n\left(\tilde{\varsigma}_{ij}^{(1)}\tilde{\varsigma}_{ii}^{(n-1)} - \frac{\tilde{\varsigma}_{ij}^{(n-1)}}{r_i - r_j}\right) \quad \text{for } i \neq j,\ i,j = 1, 2, \ldots, N,\ \ n = 2, 3, \ldots, N-1$$

$$\tilde{\varsigma}_{ii}^{(n)} = -\sum_{j=1, j\neq i}^{N} \tilde{\varsigma}_{ij}^{(n)} \quad \text{for } i = j,\ i,j = 1, 2, \ldots, N,\ \ n = 1, 2, \ldots, N-1 \tag{14.21}$$

14.3.1.2 Trigonometric Lagrange Polynomials

As far as the trigonometric Lagrange polynomials are concerned, we refer to the recursive formulae by Shu [2] as follows

$$
\begin{aligned}
g_j(r) &= \frac{\mathcal{G}(r)}{\sin\left(\frac{r-r_j}{2}\right)\mathcal{G}^{(1)}(r_j)} \\
\mathcal{G}(r) &= \prod_{k=1}^{N} \sin\left(\frac{r-r_k}{2}\right) \quad r \in \left[-\frac{\pi}{4}, \frac{\pi}{4}\right] \text{ or } r \in [0,\pi] \text{ or } r \in [0,2\pi] \quad j = 1,2,\dots,N \\
\mathcal{G}^{(1)}(r_j) &= \prod_{k=1, j\neq k}^{N} \sin\left(\frac{r_j-r_k}{2}\right)
\end{aligned}
\tag{14.22}
$$

where r denotes the definition domain of the polynomial and j refers to the degree of the jth polynomial. Based on a comparative evaluation of Eqs. (14.17) and (14.22), in this case, it is worth noticing the presence of a sine function dependent on the difference between the abscissa r and an arbitrary point of the $r_{k\,grid}$. Besides, the trigonometric Lagrange functions feature the following property

$$
g_j(r_i) = \begin{cases} 0 & \text{for } i \neq j \\ 1 & \text{for } i = j \end{cases}
\tag{14.23}
$$

The weighting coefficients up to the fourth derivation order are here evaluated directly through the recursive formulae by Shu [2]

$$
\begin{aligned}
\tilde{\varsigma}_{ij}^{(1)} &= \frac{\mathcal{G}^{(1)}(r_i)}{2\sin\left(\frac{r_i-r_j}{2}\right)\mathcal{G}^{(1)}(r_j)} \\
\tilde{\varsigma}_{ij}^{(2)} &= \tilde{\varsigma}_{ij}^{(1)}\left(2\tilde{\varsigma}_{ii}^{(1)} - \cot\left(\frac{r_i-r_j}{2}\right)\right) \qquad \text{for } i\neq j,\ i,j=1,2,\dots,N \\
\tilde{\varsigma}_{ij}^{(3)} &= \tilde{\varsigma}_{ij}^{(1)}\left(3\tilde{\varsigma}_{ii}^{(2)} + \frac{1}{2}\right) - \frac{3}{2}\tilde{\varsigma}_{ij}^{(2)}\cot\left(\frac{r_i-r_j}{2}\right) \\
\tilde{\varsigma}_{ij}^{(4)} &= \tilde{\varsigma}_{ij}^{(1)}\left(4\tilde{\varsigma}_{ii}^{(3)} - \tilde{\varsigma}_{ii}^{(1)}\right) + \frac{3}{2}\tilde{\varsigma}_{ij}^{(2)} + \left(\frac{\tilde{\varsigma}_{ij}^{(1)}}{2} - 2\tilde{\varsigma}_{ij}^{(3)}\right)\cot\left(\frac{r_i-r_j}{2}\right) \\
\tilde{\varsigma}_{ii}^{(n)} &= -\sum_{j=1, j\neq i}^{N} \tilde{\varsigma}_{ij}^{(n)} \quad \text{for } i=j,\ i,j=1,2,\dots,N,\ n=1,2,3,4
\end{aligned}
\tag{14.24}
$$

For order derivatives higher than fourth, the matrix product technique must be employed, otherwise the coefficient matrix $\mathbf{A}$ can be inverted. In the last case, the inverse of the coefficient matrix is equal to the identity matrix, $\mathbf{A}^{-1} = \mathbf{I}$, as occurs for the Lagrange polynomials. Thus, the computation of the weighting coefficients is not effected by the ill-conditioning of the coefficient matrix because of the main property (14.23) of the trigonometric Lagrange polynomials.

14.3.1.3 Classic Orthogonal Polynomials

The class of orthogonal polynomials is a family of 2×2 orthogonal polynomials with respect to their inner product. Among them, we cite the so-called classic orthogonal polynomials, e.g. Hermite, Laguerre, and Jacobi (within its particular cases: Gegenbauer [or ultraspherical], Chebyshev, Legendre, and Lobatto) polynomials, because of their large use in literature. These polynomials constitute a basis for the linear vector space V_N, and they can be used for the derivative approximation of a system with partial derivatives.

Jacobi Polynomials

The Jacobi polynomials $J_j^{(\gamma,\delta)}(r)$ satisfy the following differential equation

$$(1-r^2)\frac{d^2 J_j^{(\gamma,\delta)}(r)}{dr^2} + (\delta-\gamma-(\gamma+\delta+2)r)\frac{dJ_j^{(\gamma,\delta)}(r)}{dr} + (j-1)(j+\gamma+\delta)J_j^{(\gamma,\delta)}(r) = 0 \tag{14.25}$$

with $r\in[-1,1]$ and $j=1,2,\ldots,N$. These polynomials include the particular cases of Legendre polynomials $L_j(r)$, Gegenbauer polynomials $C_j^{(\lambda)}(r)$, and Chebyshev polynomials of the first kind $T_j(r)$, second kind $U_j(r)$, third kind $V_j(r)$, and fourth kind $W_j(r)$. The relation between the Jacobi polynomials and its descendants depends on the values assumed by the two parameters δ and γ. The assumption $\gamma=\delta=0$ corresponds to the Legendre polynomials, whereas for $\gamma=\delta=-1/2$, $\gamma=\delta=1/2$, $\gamma=-\delta=-1/2$, and $\delta=-\gamma=-1/2$, the Jacobi polynomials correspond to the Chebyshev of the first, second, third, and fourth kind, respectively. Finally, the Gegenbauer polynomials can be found setting $\lambda=\gamma+1/2=\delta+1/2$. It can be observed that the Legendre and Chebyshev polynomials of the first and second kind are subcases of the Gegenbauer ones. Solving the differential equation (14.25), the Jacobi polynomials are determined as

$$J_j^{(\gamma,\delta)}(r) = \frac{(-1)^{j-1}}{2^{j-1}(j-1)!(1-r)^{\gamma}(1+r)^{\delta}}\frac{d^{j-1}}{dr^{j-1}}((1-r)^{j-1+\gamma}(1+r)^{j-1+\delta}), \quad r\in[-1,1], \quad j=1,2,\ldots,N \tag{14.26}$$

The roots of the Jacobi polynomials in Eq. (14.26) are real and located within the closed interval $r\in[-1,1]$. Otherwise, the Jacobi polynomials are determined through the following recursive formula:

$$\begin{aligned} J_j^{(\gamma,\delta)}(r) &= \frac{(2j+\gamma+\delta-3)((2j+\gamma+\delta-2)(2j+\gamma+\delta-4)r+\gamma^2-\delta^2)J_{j-1}^{(\gamma,\delta)}(r)}{2(j-1)(j+\gamma+\delta-1)(2j+\gamma+\delta-4)} \\ &\quad - \frac{(j+\gamma-2)(j+\delta-2)(2j+\gamma+\delta-2)J_{j-2}^{(\gamma,\delta)}(r)}{(j-1)(j+\gamma+\delta-1)(2j+\gamma+\delta-4)}, \quad \text{for } j=3,4,\ldots,N \end{aligned} \tag{14.27}$$

with $J_1^{(\gamma,\delta)}(r)=1$ and $J_2^{(\gamma,\delta)}(r)=\frac{1}{2}(2(\gamma+1)+(\gamma+\delta+2)(r-1))$. The first three Jacobi polynomials are reported in the following as a function of the two parameters γ, δ, namely

$$\begin{aligned} J_1^{(\gamma,\delta)}(r) &= 1 \\ J_2^{(\gamma,\delta)}(r) &= \frac{1}{2}(2(\gamma+1)+(\gamma+\delta+2)(r-1)) \\ J_3^{(\gamma,\delta)}(r) &= \frac{1}{8}(4(\gamma+1)(\gamma+2)+4(\gamma+\delta+3)(\gamma+2)(r-1)+(\gamma+\delta+3)(\gamma+\delta+4)(r-1)^2) \end{aligned} \tag{14.28}$$

For the evaluation of the weighting coefficients, it is more convenient to write the Jacobi basis functions in the following Lagrange form:

$$f_j(r) = \frac{J_{N+1}^{(\gamma,\delta)}(r)}{(r-r_j)J_{N+1}^{(1)(\gamma,\delta)}(r_j)} \tag{14.29}$$

where $J_{N+1}^{(\gamma,\delta)}(r)$ is the $(N+1)$th Jacobi polynomial defined in Eq. (14.26) and $J_{N+1}^{(1)(\gamma,\delta)}(r_j)$ its first derivative at r_j. The explicit calculations of the weighting coefficients up to the fourth-order derivative are reported below. To this end, the first four derivatives of the Jacobi polynomials are computed in the Lagrangian form (14.29) and read

$$\begin{aligned} \frac{df_j(r)}{dr} &= \frac{J_{N+1}^{(1)(\gamma,\delta)}(r)}{(r-r_j)J_{N+1}^{(1)(\gamma,\delta)}(r_j)} - \frac{J_{N+1}^{(\gamma,\delta)}(r)}{(r-r_j)^2 J_{N+1}^{(1)(\gamma,\delta)}(r_j)} \\ \frac{d^2 f_j(r)}{dr^2} &= \frac{J_{N+1}^{(2)(\gamma,\delta)}(r)}{(r-r_j)J_{N+1}^{(1)(\gamma,\delta)}(r_j)} - \frac{2J_{N+1}^{(1)(\gamma,\delta)}(r)}{(r-r_j)^2 J_{N+1}^{(1)(\gamma,\delta)}(r_j)} - \frac{2J_{N+1}^{(\gamma,\delta)}(r)}{(r-r_j)^3 J_{N+1}^{(1)(\gamma,\delta)}(r_j)} \end{aligned}$$

$$\frac{d^3 f_j(r)}{dr^3} = \frac{J_{N+1}^{(3)(\gamma,\delta)}(r)}{(r-r_j)J_{N+1}^{(1)(\gamma,\delta)}(r_j)} - \frac{3J_{N+1}^{(2)(\gamma,\delta)}(r)}{(r-r_j)^2 J_{N+1}^{(1)(\gamma,\delta)}(r_j)} + \frac{6J_{N+1}^{(1)(\gamma,\delta)}(r)}{(r-r_j)^3 J_{N+1}^{(1)(\gamma,\delta)}(r_j)} - \frac{6J_{N+1}^{(\gamma,\delta)}(r)}{(r-r_j)^4 J_{N+1}^{(1)(\gamma,\delta)}(r_j)}$$

$$\begin{aligned}\frac{d^4 f_j(r)}{dr^4} &= \frac{J_{N+1}^{(4)(\gamma,\delta)}(r)}{(r-r_j)J_{N+1}^{(1)(\gamma,\delta)}(r_j)} - \frac{4J_{N+1}^{(3)(\gamma,\delta)}(r)}{(r-r_j)^2 J_{N+1}^{(1)(\gamma,\delta)}(r_j)} + \frac{12J_{N+1}^{(2)(\gamma,\delta)}(r)}{(r-r_j)^3 J_{N+1}^{(1)(\gamma,\delta)}(r_j)} \\ &\quad - \frac{24J_{N+1}^{(1)(\gamma,\delta)}(r)}{(r-r_j)^4 J_{N+1}^{(1)(\gamma,\delta)}(r_j)} + \frac{24J_{N+1}^{(\gamma,\delta)}(r)}{(r-r_j)^5 J_{N+1}^{(1)(\gamma,\delta)}(r_j)}\end{aligned} \tag{14.30}$$

According to Eq. (14.30), it is clear that higher-order derivatives can be expressed as a function of $J_{N+1}^{(1)(\gamma,\delta)}(r_j)$. Because the Jacobi polynomials are written in a Lagrangian form, after selecting the roots of the Jacobi polynomials as grid points, the following fundamental property is achieved

$$f_j(r_i) = \delta_{ij} \tag{14.31}$$

where δ_{ij} is the Kronecker delta. Thus, the derivatives (14.30) evaluated in r_i assume the following aspect for $r \neq j$

$$\begin{aligned}\frac{df_j(r_i)}{dr} &= \frac{J_{N+1}^{(1)(\gamma,\delta)}(r_i)}{(r_i-r_j)J_{N+1}^{(1)(\gamma,\delta)}(r_j)} \\ \frac{d^2 f_j(r_i)}{dr^2} &= \frac{J_{N+1}^{(2)(\gamma,\delta)}(r_i)}{(r_i-r_j)J_{N+1}^{(1)(\gamma,\delta)}(r_j)} - \frac{2J_{N+1}^{(1)(\gamma,\delta)}(r_i)}{(r_i-r_j)^2 J_{N+1}^{(1)(\gamma,\delta)}(r_j)} \\ \frac{d^3 f_j(r_i)}{dr^3} &= \frac{J_{N+1}^{(3)(\gamma,\delta)}(r_i)}{(r_i-r_j)J_{N+1}^{(1)(\gamma,\delta)}(r_j)} - \frac{3J_{N+1}^{(2)(\gamma,\delta)}(r_i)}{(r_i-r_j)^2 J_{N+1}^{(1)(\gamma,\delta)}(r_j)} + \frac{6J_{N+1}^{(1)(\gamma,\delta)}(r_i)}{(r_i-r_j)^3 J_{N+1}^{(1)(\gamma,\delta)}(r_j)} \\ \frac{d^4 f_j(r_i)}{dr^4} &= \frac{J_{N+1}^{(4)(\gamma,\delta)}(r_i)}{(r_i-r_j)J_{N+1}^{(1)(\gamma,\delta)}(r_j)} - \frac{4J_{N+1}^{(3)(\gamma,\delta)}(r_i)}{(r_i-r_j)^2 J_{N+1}^{(1)(\gamma,\delta)}(r_j)} + \frac{12J_{N+1}^{(2)(\gamma,\delta)}(r_i)}{(r_i-r_j)^3 J_{N+1}^{(1)(\gamma,\delta)}(r_j)} \\ &\quad - \frac{24J_{N+1}^{(1)(\gamma,\delta)}(r_i)}{(r_i-r_j)^4 J_{N+1}^{(1)(\gamma,\delta)}(r_j)}\end{aligned} \tag{14.32}$$

whereby, for $i = j$, they become

$$\begin{aligned}\frac{df_j(r_i)}{dr} &= \frac{J_{N+1}^{(2)(\gamma,\delta)}(r_i)}{2J_{N+1}^{(1)(\gamma,\delta)}(r_j)} \\ \frac{d^2 f_j(r_i)}{dr^2} &= \frac{J_{N+1}^{(3)(\gamma,\delta)}(r_i)}{3J_{N+1}^{(1)(\gamma,\delta)}(r_j)} \\ \frac{d^3 f_j(r_i)}{dr^3} &= \frac{J_{N+1}^{(4)(\gamma,\delta)}(r_i)}{4J_{N+1}^{(1)(\gamma,\delta)}(r_j)} \\ \frac{d^4 f_j(r_i)}{dr^4} &= \frac{J_{N+1}^{(5)(\gamma,\delta)}(r_i)}{5J_{N+1}^{(1)(\gamma,\delta)}(r_j)}\end{aligned} \tag{14.33}$$

Now, let us consider the Chiahara differential problem

$$\begin{aligned}&(2N+\gamma+\delta)(1-r^2)J_{N+1}^{(1)(\gamma,\delta)}(r) - (N(\gamma-\delta) - N(2N+\gamma+\delta)r)J_{N+1}^{(\gamma,\delta)}(r) \\ &\quad - (2(N+\gamma)(N+\delta))J_N^{(\gamma,\delta)}(r) = 0\end{aligned} \tag{14.34}$$

and the Sturm–Liouville ones

$$
\begin{aligned}
&(1-r^2)J_{N+1}^{(2)(\gamma,\delta)}(r)+(\delta-\gamma-(\gamma+\delta+2)r)J_{N+1}^{(1)(\gamma,\delta)}(r)\\
&\quad+N(N+\gamma+\delta+1)J_{N+1}^{(\gamma,\delta)}(r)=0\\
&(1-r^2)J_{N+1}^{(3)(\gamma,\delta)}(r)+(\delta-\gamma-(\gamma+\delta+2)r-2r)J_{N+1}^{(2)(\gamma,\delta)}(r)\\
&\quad+(N(N+\gamma+\delta+1)-(\gamma+\delta+2))J_{N+1}^{(1)(\gamma,\delta)}(r)=0\\
&(1-r^2)J_{N+1}^{(4)(\gamma,\delta)}(r)+(\delta-\gamma-(\gamma+\delta+2)r-4r)J_{N+1}^{(3)(\gamma,\delta)}(r)\\
&\quad+(N(N+\gamma+\delta+1)-2(\gamma+\delta+2)-2)J_{N+1}^{(2)(\gamma,\delta)}(r)=0\\
&(1-r^2)J_{N+1}^{(5)(\gamma,\delta)}(r)+(\delta-\gamma-(\gamma+\delta+2)r-6r)J_{N+1}^{(4)(\gamma,\delta)}(r)\\
&\quad+(N(N+\gamma+\delta+1)-3(\gamma+\delta+2)-6)J_{N+1}^{(3)(\gamma,\delta)}(r)=0
\end{aligned}
\tag{14.35}
$$

Equations (14.34) and (14.35) can be computed at a generic grid point r_i as follows

$$
\begin{aligned}
&(2N+\gamma+\delta)(1-r_i^2)J_{N+1}^{(1)(\gamma,\delta)}(r_i)-(2(N+\gamma)(N+\delta))J_{N}^{(\gamma,\delta)}(r_i)=0\\
&(1-r_i^2)J_{N+1}^{(2)(\gamma,\delta)}(r_i)+(\delta-\gamma-(\gamma+\delta+2)r_i)J_{N+1}^{(1)(\gamma,\delta)}(r_i)=0\\
&(1-r_i^2)J_{N+1}^{(3)(\gamma,\delta)}(r_i)+(\delta-\gamma-(\gamma+\delta+2)r-2r_i)J_{N+1}^{(2)(\gamma,\delta)}(r_i)\\
&\quad+(N(N+\gamma+\delta+1)-(\gamma+\delta+2))J_{N+1}^{(1)(\gamma,\delta)}(r_i)=0\\
&(1-r_i^2)J_{N+1}^{(4)(\gamma,\delta)}(r_i)+(\delta-\gamma-(\gamma+\delta+2)r-4r_i)J_{N+1}^{(3)(\gamma,\delta)}(r_i)\\
&\quad+(N(N+\gamma+\delta+1)-2(\gamma+\delta+2)-2)J_{N+1}^{(2)(\gamma,\delta)}(r_i)=0\\
&(1-r_i^2)J_{N+1}^{(5)(\gamma,\delta)}(r_i)+(\delta-\gamma-(\gamma+\delta+2)r-6r_i)J_{N+1}^{(4)(\gamma,\delta)}(r_i)\\
&\quad+(N(N+\gamma+\delta+1)-3(\gamma+\delta+2)-6)J_{N+1}^{(3)(\gamma,\delta)}(r_i)=0
\end{aligned}
\tag{14.36}
$$

which can be represented in another form using proper definitions easily understandable from the current demonstration. Thus, the Chiahara and Sturm–Liouville differential problems can be rewritten as

$$
\begin{aligned}
&\varpi_1(1-r^2)J_{N+1}^{(1)(\gamma,\delta)}(r)-\varpi_2 J_{N}^{(\gamma,\delta)}(r)=0\\
&(1-r_i^2)J_{N+1}^{(2)(\gamma,\delta)}(r_i)-\varpi_3 J_{N+1}^{(1)(\gamma,\delta)}(r_i)=0\\
&(1-r_i^2)J_{N+1}^{(3)(\gamma,\delta)}(r_i)-(\varpi_3+2r_i)J_{N+1}^{(2)(\gamma,\delta)}(r_i)\\
&\quad+(N(N+\gamma+\delta+1)-(\gamma+\delta+2))J_{N+1}^{(1)(\gamma,\delta)}(r_i)=0\\
&(1-r_i^2)J_{N+1}^{(4)(\gamma,\delta)}(r_i)-(\varpi_3+4r_i)J_{N+1}^{(3)(\gamma,\delta)}(r_i)\\
&\quad+(N(N+\gamma+\delta+1)-2(\gamma+\delta+2)-2)J_{N+1}^{(2)(\gamma,\delta)}(r_i)=0\\
&(1-r_i^2)J_{N+1}^{(5)(\gamma,\delta)}(r_i)+(\varpi_3+6r_i)J_{N+1}^{(4)(\gamma,\delta)}(r_i)\\
&\quad+(N(N+\gamma+\delta+1)-3(\gamma+\delta+2)-6)J_{N+1}^{(3)(\gamma,\delta)}(r_i)=0
\end{aligned}
\tag{14.37}
$$

After some mathematical steps, expressions (14.37) give the following definitions

$$
J_{N+1}^{(1)(\gamma,\delta)}(r_i)=\frac{\varpi_2}{\varpi_1}\frac{J_{N}^{(\gamma,\delta)}(r_i)}{(1-r_i^2)}
$$

$$J_{N+1}^{(2)(\gamma,\delta)}(r_i) = \varpi_3 \frac{J_{N+1}^{(1)(\gamma,\delta)}(r_i)}{(1-r_i^2)} = \frac{\varpi_3\varpi_2}{\varpi_1}\frac{J_N^{(\gamma,\delta)}(r_i)}{(1-r_i^2)^2}$$

$$J_{N+1}^{(3)(\gamma,\delta)}(r_i) = \frac{(\varpi_3+2r_i)J_{N+1}^{(2)(\gamma,\delta)}(r_i) - (N(N+\gamma+\delta+1)-(\gamma+\delta+2))J_{N+1}^{(1)(\gamma,\delta)}(r_i)}{(1-r_i^2)}$$
$$= ((\varpi_3+2r_i)\varpi_3 + (1-r_i^2)(\gamma+\delta+2-N(N+\gamma+\delta+1)))\frac{J_{N+1}^{(1)(\gamma,\delta)}(r_i)}{(1-r_i^2)^2}$$
$$= \varpi_4\frac{J_{N+1}^{(1)(\gamma,\delta)}(r_i)}{(1-r_i^2)^2} = \frac{\varpi_4\varpi_2}{\varpi_1}\frac{J_{N+1}^{(\gamma,\delta)}(r_i)}{(1-r_i^2)^3}$$

$$J_{N+1}^{(4)(\gamma,\delta)}(r_i) = \frac{(\varpi_3+4r_i)J_{N+1}^{(3)(\gamma,\delta)}(r_i) - (N(N+\gamma+\delta+1)-2(\gamma+\delta+2)-2)J_{N+1}^{(2)(\gamma,\delta)}(r_i)}{(1-r_i^2)}$$
$$= ((\varpi_3+4r_i)\varpi_4 + (1-r_i^2)(2(\gamma+\delta+2)+2-N(N+\gamma+\delta+1))\varpi_3)\frac{J_{N+1}^{(1)(\gamma,\delta)}(r_i)}{(1-r_i^2)^3}$$
$$= \varpi_5\frac{J_{N+1}^{(1)(\gamma,\delta)}(r_i)}{(r-r_i^2)^3} = \frac{\varpi_5\varpi_2}{\varpi_1}\frac{J_{N+1}^{(\gamma,\delta)}(r_i)}{(r-r_i^2)^4}$$

$$J_{N+1}^{(5)(\gamma,\delta)}(r_i) = \frac{(\varpi_3+6r_i)J_{N+1}^{(4)(\gamma,\delta)}(r_i) - (N(N+\gamma+\delta+1)-3(\gamma+\delta+2)-6)J_{N+1}^{(3)(\gamma,\delta)}(r_i)}{(1-r_i^2)}$$
$$= ((\varpi_3+6r_i)\varpi_5 + (1-r_i^2)(3(\gamma+\delta+2)+6-N(N+\gamma+\delta+1))\varpi_4)\frac{J_{N+1}^{(1)(\gamma,\delta)}(r_i)}{(r-r_i^2)^4}$$
$$= \varpi_6\frac{J_{N+1}^{(1)(\gamma,\delta)}(r_i)}{(1-r_i^2)^4} = \frac{\varpi_6\varpi_2}{\varpi_1}\frac{J_N^{(\gamma,\delta)}(r_i)}{(1-r_i^2)^5} \quad (14.38)$$

Finally, the weighting coefficients up to the fourth-order derivative can be determined by a combination of Eqs. (14.33) and (14.38). For $i \neq j$ (with $i, j = 1, 2, \ldots, N$ and $\gamma, \delta > -1$), it is

$$\tilde{\varsigma}_{ij}^{(1)} = \frac{J_{N+1}^{(1)(\gamma,\delta)}(r_i)}{(r_i-r_j)J_{N+1}^{(1)(\gamma,\delta)}(r_j)} = \frac{(1-r_j^2)J_N^{(\gamma,\delta)}(r_i)}{(r_i-r_j)(1-r_i^2)J_N^{(\gamma,\delta)}(r_j)}$$

$$\tilde{\varsigma}_{ij}^{(2)} = \frac{J_{N+1}^{(2)(\gamma,\delta)}(r_i)}{(r_i-r_j)J_{N+1}^{(1)(\gamma,\delta)}(r_j)} - \frac{2J_{N+1}^{(1)(\gamma,\delta)}(r_i)}{(r_i-r_j)^2J_{N+1}^{(1)(\gamma,\delta)}(r_j)}$$
$$= \frac{\varpi_3 J_{N+1}^{(1)(\gamma,\delta)}(r_i)}{(r_i-r_j)(1-r_i^2)J_{N+1}^{(1)(\gamma,\delta)}(r_j)} - \frac{2J_{N+1}^{(1)(\gamma,\delta)}(r_i)}{(r_i-r_j)^2J_{N+1}^{(1)(\gamma,\delta)}(r_j)}$$
$$= \frac{\varpi_3(r_i-r_j)-2(1-r_i^2)}{(r_i-r_j)^2(1-r_i^2)}\frac{J_{N+1}^{(1)(\gamma,\delta)}(r_i)}{J_{N+1}^{(1)(\gamma,\delta)}(r_j)} = \frac{(\varpi_3(r_i-r_j)-2(1-r_i^2))(1-r_j^2)J_N^{(\gamma,\delta)}(r_i)}{(r_i-r_j)^2(1-r_i^2)^2J_N^{(\gamma,\delta)}(r_j)}$$

$$\tilde{\varsigma}_{ij}^{(3)} = \frac{J_{N+1}^{(3)(\gamma,\delta)}(r_i)}{(r_i-r_j)J_{N+1}^{(1)(\gamma,\delta)}(r_j)} - \frac{3J_{N+1}^{(2)(\gamma,\delta)}(r_i)}{(r_i-r_j)^2J_{N+1}^{(1)(\gamma,\delta)}(r_j)} + \frac{6J_{N+1}^{(1)(\gamma,\delta)}(r_i)}{(r_i-r_j)^3J_{N+1}^{(1)(\gamma,\delta)}(r_j)}$$
$$= \frac{\varpi_4 J_{N+1}^{(1)(\gamma,\delta)}(r_i)}{(r_i-r_j)(1-r_i^2)^2J_{N+1}^{(1)(\gamma,\delta)}(r_j)} - \frac{3\varpi_3 J_{N+1}^{(1)(\gamma,\delta)}(r_i)}{(r_i-r_j)^2(1-r_i^2)^2J_{N+1}^{(1)(\gamma,\delta)}(r_j)} + \frac{6J_{N+1}^{(1)(\gamma,\delta)}(r_i)}{(r_i-r_j)^3J_{N+1}^{(1)(\gamma,\delta)}(r_j)}$$

$$= \frac{(r_i - r_j)^2 \varpi_4 - 3(r_i - r_j)(1 - r_i^2)\varpi_3 + 6(1 - r_i^2)^2}{(r_i - r_j)^3(1 - r_i^2)^2} \frac{J_{N+1}^{(1)(\gamma,\delta)}(r_i)}{J_{N+1}^{(1)(\gamma,\delta)}(r_j)}$$

$$= \frac{((r_i - r_j)^2 \varpi_4 - 3(r_i - r_j)(1 - r_i^2)\varpi_3 + 6(1 - r_i^2)^2)(1 - r_j^2)J_N^{(\gamma,\delta)}(r_i)}{(r_i - r_j)^3(1 - r_i^2)^3 J_N^{(\gamma,\delta)}(r_j)}$$

$$\tilde{\varsigma}_{ij}^{(4)} = \frac{J_{N+1}^{(4)(\gamma,\delta)}(r_i)}{(r_i - r_j)J_{N+1}^{(1)(\gamma,\delta)}(r_j)} - \frac{4J_{N+1}^{(3)(\gamma,\delta)}(r_i)}{(r_i - r_j)^2 J_{N+1}^{(1)(\gamma,\delta)}(r_j)} + \frac{12J_{N+1}^{(2)(\gamma,\delta)}(r_i)}{(r_i - r_j)^3 J_{N+1}^{(1)(\gamma,\delta)}(r_j)} - \frac{24J_{N+1}^{(1)(\gamma,\delta)}(r_i)}{(r_i - r_j)^4 J_{N+1}^{(1)(\gamma,\delta)}(r_j)}$$

$$= \frac{\varpi_5 J_{N+1}^{(1)(\gamma,\delta)}(r_i)}{(r_i - r_j)(1 - r_i^2)^3 J_{N+1}^{(1)(\gamma,\delta)}(r_j)} - \frac{4\varpi_4 J_{N+1}^{(1)(\gamma,\delta)}(r_i)}{(r_i - r_j)^2(1 - r_i^2)^2 J_{N+1}^{(1)(\gamma,\delta)}(r_j)} + \frac{12\varpi_3 J_{N+1}^{(1)(\gamma,\delta)}(r_i)}{(r_i - r_j)^3(1 - r_i^2)J_{N+1}^{(1)(\gamma,\delta)}(r_j)}$$

$$- \frac{24J_{N+1}^{(1)(\gamma,\delta)}(r_i)}{(r_i - r_j)^4 J_{N+1}^{(1)(\gamma,\delta)}(r_j)}$$

$$= \frac{(r_i - r_j)^3 \varpi_5 - 4(r_i - r_j)^2(1 - r_i^2)\varpi_4 + 12(r_i - r_j)(1 - r_i^2)^2 \varpi_3 - 24(1 - r_i^2)^3}{(r_i - r_j)^4(1 - r_i^2)^3} = \frac{J_{N+1}^{(1)(\gamma,\delta)}(r_i)}{J_{N+1}^{(1)(\gamma,\delta)}(r_j)}$$

$$= \frac{((r_i - r_j)^3 \varpi_5 - 4(r_i - r_j)^2(1 - r_i^2)\varpi_4 + 12(r_i - r_j)(1 - r_i^2)^2 \varpi_3 - 24(1 - r_i^2)^3)(1 - r_j^2)J_N^{(\gamma,\delta)}(r_i)}{(r_i - r_j)^4(1 - r_i^2)^3 J_N^{(\gamma,\delta)}(r_j)} \tag{14.39}$$

and for $i = j$ (with $i, j = 1, 2, \ldots, N$ and $\gamma, \delta > -1$), it reckons

$$\tilde{\varsigma}_{ii}^{(1)} = \frac{J_{N+1}^{(2)(\gamma,\delta)}(r_i)}{2J_{N+1}^{(1)(\gamma,\delta)}(r_j)} = \frac{(\gamma + \delta + 2)r_i + \gamma - \delta}{2(1 - r_i^2)} = \frac{\varpi_3}{2(1 - r_i^2)}$$

$$\tilde{\varsigma}_{ii}^{(2)} = \frac{J_{N+1}^{(3)(\gamma,\delta)}(r_i)}{3J_{N+1}^{(1)(\gamma,\delta)}(r_j)} = \frac{\varpi_3^2 + 2r_i\varpi_3 + (1 - r_i^2)(\gamma + \delta + 2 - N(N + \gamma + \delta + 1))}{3(1 - r_i^2)^2} = \frac{\varpi_4}{3(1 - r_i^2)^2}$$

$$\tilde{\varsigma}_{ii}^{(3)} = \frac{J_{N+1}^{(4)(\gamma,\delta)}(r_i)}{4J_{N+1}^{(1)(\gamma,\delta)}(r_j)}$$

$$= \frac{\varpi_3\varpi_4 + 4r_i\varpi_3 + (1 - r_i^2)(2(\gamma + \delta + 2) + 2 - N(N + \gamma + \delta + 1))\varpi_3}{4(1 - r_i^2)^3} = \frac{\varpi_5}{4(1 - r_i^2)^3}$$

$$\tilde{\varsigma}_{ii}^{(4)} = \frac{J_{N+1}^{(5)(\gamma,\delta)}(r_i)}{5J_{N+1}^{(1)(\gamma,\delta)}(r_j)}$$

$$= \frac{\varpi_3\varpi_5 + 6r_i\varpi_5 + (1 - r_i^2)(3(\gamma + \delta + 2) + 6 - N(N + \gamma + \delta + 1))\varpi_4}{5(1 - r_i^2)^4} = \frac{\varpi_6}{5(1 - r_i^2)^4} \tag{14.40}$$

It should be pointed out that the weighting coefficients provided by Eq. (14.40) are computationally inefficient. Thus, from a numerically standpoint, more accurate results are achieved through the Shu's formula reported as follows [2, 8, 14]

$$\tilde{\varsigma}_{ii}^{(n)} = -\sum_{j=1, j\neq i}^{N} \tilde{\varsigma}_{ij}^{(n)} \quad \text{for } i, j = 1, 2, \ldots, N, \ n = 1, 2, 3, 4 \tag{14.41}$$

Once the coefficients γ, δ are defined and the grid points are selected as roots of the Jacobi polynomial $J_{N+1}^{(\gamma,\delta)}(r)$, we can compute the two expressions (14.39) and (14.40). The weighting coefficients for the first two derivatives coincide with the expressions reported in the works by Quan and Chang [18], here reported for the sake of completeness

$$\tilde{\varsigma}_{ij}^{(1)} = \frac{(1-r_j^2)J_N^{(\gamma,\delta)}(r_i)}{(r_i-r_j)(1-r_i^2)J_N^{(\gamma,\delta)}(r_j)} \quad \text{for } i \neq j \tag{14.42}$$

$$\tilde{\varsigma}_{ij}^{(2)} = -\frac{((r_i-r_j)(\delta-\gamma-(\gamma+\delta+2)r_i)+2(1-r_i^2))(1-r_j^2)J_N^{(\gamma,\delta)}(r_i)}{(r_i-r_j)^2(1-r_i^2)^2J_N^{(\gamma,\delta)}(r_j)} \quad \text{for } i \neq j \tag{14.43}$$

$$\tilde{\varsigma}_{ii}^{(1)} = \frac{(\gamma+\delta+2)r_i+\gamma-\delta}{2(1-r_i^2)}, \quad \text{for } i = j \tag{14.44}$$

$$\tilde{\varsigma}_{ii}^{(2)} = \frac{1}{3(1-r_i^2)^2}((\delta-\gamma-(\gamma+\delta+2)r_i)^2+(1-r_i^2)(\gamma+\delta+2-N(N+\gamma+\delta+1)) - 2r_i(\delta-\gamma-(\gamma+\delta+2)r_i)), \quad \text{for } i = j \tag{14.45}$$

As far as the weighting coefficients of higher-order derivatives are concerned, they can be computed through the matrix multiplication approach, as mentioned above. Moreover, expressions (14.39) and (14.40) are valid within the closed interval $r \in [-1, 1]$. Thus, the coordinate transformations (14.15) and (14.16) must be applied to extend its validity.

Gegenbauer Polynomials

The Gegenbauer polynomials $C_j^{(\lambda)}(r)$, or ultraspherical polynomials, represent a particular case of Jacobi polynomials. They are obtained as solution of the so-called Gegenbauer differential equation, i.e.

$$(1-r^2)\frac{d^2C_j^{(\lambda)}(r)}{dr^2} - (2\lambda+1)r\frac{dC_j^{(\lambda)}(r)}{dr} + (j-1)(j+2\lambda-1)C_j^{(\lambda)}(r) = 0 \tag{14.46}$$

with $r \in [-1, 1]$ and for $j = 1, 2, \ldots, N$. The differential equation (14.46) can be found from (14.25), setting $\lambda = \gamma + 1/2 = \delta + 1/2$. These polynomials can be represented as

$$C_j^{(\lambda)}(r) = \sum_{k=0}^{\phi(j-1)} \frac{(-1)^k(\lambda)_{j-1-k}(2r)^{j-1-2k}}{k!(j-1-2k)!}, \quad r \in [-1, 1], \quad j = 2, 3, \ldots, N \tag{14.47}$$

where $C_1^{(\lambda)}(r) = 1$. The function $\phi(j-1)$ can be defined as

$$\phi(j-1) = \frac{2(j-1)+((-1)^{j-1}-1)}{4} \tag{14.48}$$

where the expression $(\lambda)_m$ in Eq. (14.47) stands for a growing factor of λ with m factors (with m not-negative integers), defined as

$$(\lambda)_m = \lambda(\lambda+1)(\lambda+2)\cdots(\lambda+m-1) = \prod_{k=1}^{m}(\lambda+k-1) \tag{14.49}$$

Otherwise, Gegenbauer polynomials can be determined from the following recursive formula

$$C_j^{(\lambda)}(r) = \frac{2r}{j-1}(j+\lambda-2)C_{j-1}^{(\lambda)}(r) - \frac{1}{j-1}(j+2\lambda-3)C_{j-2}^{(\lambda)}(r) \quad \text{for } j = 3, 4, \ldots, N \tag{14.50}$$

with $C_1^{(\lambda)}(r) = 1$ and $C_2^{(\lambda)}(r) = 2\lambda r$.

Legendre Polynomials

The Legendre polynomials can be determined from Gegenbauer polynomials $C_j^{(\lambda)}(r)$, under the assumption $\lambda = 1/2$. Since the Gegenbauer polynomials represent a subcase of Jacobi polynomials, the Legendre polynomials can also be obtained from the Jacobi polynomials $J_j^{(\gamma,\delta)}(r)$ when $\gamma = \delta = 0$. In the literature, they are generally defined as $J_j^{(0,0)}(r) = L_j(r)$ and can be written in explicit form using the Rodrigues' formulae

$$L_j(r) = \frac{(-1)^{j-1}}{2^{j-1}(j-1)!}\frac{d^{j-1}}{dr^{j-1}}((1-r^2)^{j-1}), \quad r \in [-1,1], \quad j = 1,2,\ldots,N \tag{14.51}$$

Similarly, the Legendre polynomials can be determined according to the following recursive formula

$$L_j(r) = \frac{(2j-3)rL_{j-1}(r) - (j-2)L_{j-2}(r)}{j-1}, \quad j = 3,4,\ldots,N \tag{14.52}$$

where $L_1(r) = 1$ and $L_2(r) = r$. The weighting coefficients can be evaluated using the aforementioned Jacobi DQ or inverting the coefficient matrix according to the classic spectral methods.

Lobatto Polynomials

The Lobatto polynomials $A_j(r)$ can be defined as

$$A_j(r) = \frac{d}{dr}(L_{j+1}(r)) = C_j^{(3/2)}(r) = \frac{j+1}{2}J_j^{(1,1)}(r), \quad r \in [-1,1], \quad j = 1,2,\ldots,N \tag{14.53}$$

whose expression is obtained deriving the Legendre polynomials $L_j(r)$ defined in (14.51). Moreover, these polynomials can be defined from the Gegenbauer polynomials $C_j^{(\lambda)}(r)$ setting $\lambda = 3/2$ or the Jacobi polynomials with $\gamma = \delta = 1$. The weighting coefficients can be evaluated using the aforementioned Jacobi DQ or inverting the coefficient matrix according to the classic spectral methods.

Chebyshev Polynomials of the First Kind

The Chebyshev polynomials of the first kind are included into the set of Jacobi polynomials $J_j^{(\gamma,\delta)}(r)$ because they can be derived setting $\gamma = \delta = -1/2$. In the literature, these polynomials are referred to as $J_j^{(-1/2,-1/2)}(r) = T_j(r)$ and take the following explicit form

$$T_j(r) = \cos((j-1)\arccos(r)), \quad r \in [-1,1], \quad j = 1,2,\ldots,N \tag{14.54}$$

The Chebyshev polynomials of the first kind represent a particular case of Gegenbauer polynomials, and can be written as

$$T_j(r) = \frac{j-1}{2}\lim_{\lambda\to 0}\left(\frac{C_j^{(\lambda)}(r)}{\lambda}\right), \quad j = 2,3,\ldots,N \tag{14.55}$$

The recursive relation reported below can be used for an explicit representation of the Chebyshev polynomials of the first kind, starting from the first two ones $T_1(r) = 1$ and $T_2(r) = r$, namely

$$T_j(r) = 2rT_{j-1}(r) - T_{j-2}(r), \quad j = 3,4,\ldots,N \tag{14.56}$$

The weighting coefficients can be evaluated using the Jacobi DQ or inverting the coefficient matrix according to the classic spectral methods. The latter is known as *Chebyshev spectral method* (or *Chebyshev collocation spectral method*). However, this approach requires a grid point collocation according to the Chebyshev of the first kind grid. The weighting coefficients of the Chebyshev spectral method are defined in the following closed-form

$$\tilde{\varsigma}_{ij}^{(1)} = \frac{(-1)^{i+j}\bar{c}_i}{(r_i - r_j)\bar{c}_j}, \quad \text{for } 1 \le i,\ j \le N,\ i \ne j \tag{14.57}$$

$$\tilde{\varsigma}_{ii}^{(1)} = -\frac{r_i}{2(1-r_i^2)}, \quad \text{for } 2 \leq i \leq N-1 \tag{14.58}$$

$$\tilde{\varsigma}_{11}^{(1)} = -\tilde{\varsigma}_{NN}^{(1)} = \frac{2N^2+1}{6} \tag{14.59}$$

where $\bar{c}_1 = \bar{c}_N = 2$ and $\bar{c}_k = 1$, for $k = 2, 3, \ldots, N-1$. It is worth mentioning that these expressions are valid within the interval $r \in [-1, 1]$; therefore, the coordinate transformation (14.15) and (14.16) must be employed for physical applications. It can be observed that relations (14.57)–(14.59) revert to the ones given by Shu, when we account for the Chebyshev distribution [2].

Chebyshev Polynomials of the Second Kind

Chebyshev polynomials of the second kind represent a particular type of Gegenbauer polynomials $C_j^{(\lambda)}(r)$, for $\lambda = 1$, and a subset of the Jacobi polynomials $J_j^{(\gamma,\delta)}(r)$, when $\gamma = \delta = 1/2$. In the literature, they are indicated as $J_j^{(1/2,1/2)}(r) = C_j^{(1)}(r) = U_j(r)$ and can be written through the following trigonometric form

$$U_j(r) = \frac{\sin(j \arccos(r))}{\sin(\arccos(r))} \quad r \in [-1, 1] \quad j = 1, 2, \ldots, N \tag{14.60}$$

Otherwise, Chebyshev polynomials of the second kind can be applied through the following recursive formula

$$U_j(r) = 2rU_{j-1}(r) - U_{j-2}(r), \quad j = 3, 4, \ldots, N \tag{14.61}$$

with $U_1(r) = 1$ and $U_2(r) = 2r$.

Chebyshev Polynomials of the Third Kind

The Chebyshev polynomials of the third kind is a further subset of Jacobi polynomials $J_j^{(\gamma,\delta)}(r)$, under the assumption $\gamma = -\delta = -1/2$. Chebyshev polynomials of the third kind, usually labeled $V_j(r)$, are described by the following trigonometric relation

$$V_j(r) = \frac{\cos\left(\frac{(2j-1)\arccos(r)}{2}\right)}{\cos\left(\frac{\arccos(r)}{2}\right)}, \quad r \in [-1, 1], \quad j = 1, 2, \ldots, N \tag{14.62}$$

or can be employed using the same recursive formula for polynomials of the first and second kind, while varying the initial conditions. This reads as follows

$$V_j(r) = 2rV_{j-1}(r) - V_{j-2}(r), \quad j = 3, 4, \ldots, N \tag{14.63}$$

with $V_1(r) = 1$ and $V_2(r) = 2r - 1$.

Chebyshev Polynomials of the Fourth Kind

The Chebyshev polynomials of the fourth kind are a particular type of Jacobi polynomials $J_j^{(\gamma,\delta)}(r)$ with $\delta = -\gamma = -1/2$. In the literature, they are labeled $W_j(r)$ and are usually represented through the following trigonometric form

$$W_j(r) = \frac{\sin\left(\frac{(2j-1)\arccos(r)}{2}\right)}{\sin\left(\frac{\arccos(r)}{2}\right)} \quad r \in [-1, 1] \quad j = 1, 2, \ldots, N \tag{14.64}$$

Chebyshev polynomials of the fourth kind can be also determined through the same recursive formula of the first, second, and third kind polynomials, with the only variation in the initial conditions, i.e.

$$W_j(r) = 2rW_{j-1}(r) - W_{j-2}(r) \quad j = 3, 4, \ldots, N \tag{14.65}$$

with $W_1(r) = 1$ and $W_2(r) = 2r + 1$.

The weighting coefficients can be evaluated using the aforementioned Jacobi DQ or by inverting the coefficient matrix according to the classic spectral methods.

14.3.1.4 Monomial Functions

The monomial functions (also named power functions) were the first functions introduced by Bellman in his first works, as the simplest choice that leads to the Vandermonde matrix featuring an ill-conditioning for $N > 13$. Power functions can be defined, in the whole real domain, as

$$Z_j(r) = r^{j-1}, \quad r \in]-\infty, +\infty[, \quad j = 1, 2, \ldots, N \tag{14.66}$$

whose order is directly related to the exponent of variable r. Also in this case, the weighting coefficients can be obtained by inverting the coefficient matrix.

14.3.1.5 Exponential Functions

The exponential functions are another kind of basis represented by

$$E_j(r) = e^{(j-1)r} \quad r \in]-\infty, +\infty[\quad j = 1, 2, \ldots, N \tag{14.67}$$

in the whole real domain, like the monomial functions (14.66), where the weighting coefficients can be determined only by inversion of the coefficient matrix. These exponential functions are well known to become unstable for an increasing number of points within the domain.

14.3.1.6 Bernstein Polynomials

The Bernstein polynomials $B_j^N(r)$, defined in the interval $r \in [0, 1]$, can be written as

$$B_j^N(r) = \frac{(N-1)!}{(j-1)!(N-j)!} r^{j-1}(1-r)^{N-j} \quad r \in [0, 1] \quad j = 1, 2, \ldots, N \tag{14.68}$$

These polynomials are orthogonal and satisfy the Weierstrass theorem, which makes possible their employment in a DQ context. The evaluation of the weighting coefficients is performed, once again, through an inversion of the coefficient matrix.

From expression (14.68), it is possible to define another class of basis functions, known as *rational Bernstein functions* $R_j^N(r)$. These functions get the following form

$$R_j^N(r) = \frac{B_j^N(r)\omega_j}{\sum_{k=1}^{N} B_k^N(r)\omega_k} \quad \text{for } j = 1, 2, \ldots, N \tag{14.69}$$

where ω_j refers to the weight functions, for $j = 1, 2, \ldots, N$. These polynomials are fundamental for the mathematical representation of the *Bézier curves* used to describe free-form surfaces. If the weights ω_j introduced in (14.69) assume the unit value, $\omega_j = 1$ for $j = 1, 2, \ldots, N$, the expression (14.69) takes the same form as reported in Eq. (14.68) because of the partition of unity properties for the set of Bernstein polynomials, i.e.

$$\sum_{k=1}^{N} B_k^N(r) = 1 \tag{14.70}$$

14.3.1.7 Fourier Functions

The Fourier functions $F_j(r)$ represent an alternative definition of the functional approximation because they use harmonic functions instead of the polynomial or monomial form. These bases are expressed as follows

$$F_1(r) = 1, \quad \begin{cases} F_j(r) = \cos\left(\dfrac{j}{2} r\right) & \text{for } j \text{ even} \\ F_j(r) = \sin\left(\dfrac{j-1}{2} r\right) & \text{for } j \text{ odd} \end{cases} \quad r \in [0, 2\pi], \quad j = 2, 3, \dots, N \tag{14.71}$$

within the interval $r \in [0, 2\pi]$, whereby the weighting coefficients are evaluated using the inversion of the coefficient matrix.

14.3.1.8 Bessel Polynomials

Another class of polynomials is represented by the Bessel polynomials, which satisfy the following differential equations

$$r^2 \frac{d^2 P_j(r)}{dr^2} + (2r+2)\frac{dP_j(r)}{dr} - (j-1)jP_j(r) = 0, \quad r \in [-\infty, +\infty], \quad j = 1, 2, \dots, N \tag{14.72}$$

From equation (14.72), the explicit form of the Bessel polynomials can be determined as

$$P_j(r) = \sum_{k=0}^{j-1} \frac{(j-1+k)!}{(j-1-k)!k!}\left(\frac{r}{2}\right)^k, \quad r \in [-\infty, +\infty], \quad j = 2, 3, \dots, N \tag{14.73}$$

with $P_1(r) = 1$. The same polynomials can be easily obtained through the following recursive formula

$$P_j(r) = (2j-3)rP_{j-1}(r) + P_{j-2}(r), \quad \text{for } j = 3, 4, \dots, N \tag{14.74}$$

with $P_1(r) = 1$ and $P_2(r) = 1 + r$.

Also in this case, the weighting coefficients can be determined by inversion of the coefficient matrix, and the exponential functions become unstable increasing the number of grid points within the domain.

14.3.1.9 Boubaker Polynomials

Boubaker polynomials are a sequence of polynomials with integer coefficients, used in several physical applications. These polynomials are defined in the interval $r \in [-\infty, +\infty]$ and read as follows

$$Q_j(r) = \sum_{k=0}^{\phi(j-1)} (-1)^k \binom{j-1-k}{k} \frac{j-1-4k}{j-1-k} r^{j-1-2k} \quad r \in [-\infty, +\infty] \quad j = 2, 3, \dots, N \tag{14.75}$$

with $Q_1(r) = 1$ and $\phi(j-1)$ is a function defined as

$$\phi(j-1) = \frac{2(j-1) + ((-1)^{j-1} - 1)}{4} \tag{14.76}$$

Another possibility to define the Boubaker polynomials is related to the application of the following recursive formula

$$Q_j(r) = rQ_{j-1}(r) - Q_{j-2}(r) \quad \text{for } j = 4, 5, \dots, N \tag{14.77}$$

with $Q_1(r) = 1$, $Q_2(r) = r$, and $Q_3(r) = r^2 + 2$.

Also in this case, the weighting coefficients are determined through a mathematical inversion of the coefficient matrix, whereby the exponential functions become unstable for an increasing number of grid points within the domain.

14.3.2 Grid Distributions

The present section aims at giving a general overview of the most adopted collocations in literature.

14.3.2.1 Coordinate Transformation

The following grids are defined in the interval $\zeta_k \in [0, 1]$, which must be transformed into the real interval $x_k \in [a, b]$. A general coordinate transformation is reported below

$$\begin{aligned} x_k &= \frac{b-a}{d-c}(\zeta_k - c) + a, \quad k = 1, 2, \dots, N \\ x_k &= \frac{b-a}{d-c}(\overline{\zeta}_k - c) + a, \quad k = 1, 2, \dots, N \end{aligned} \tag{14.78}$$

where $\zeta_k, \overline{\zeta}_k \in [c, d]$ are the two discretizations without and with δ-points. In the following, we consider $[c, d] = [0, 1]$, and the relations (14.78) become

$$\begin{aligned} x_k &= (b-a)\zeta_k + a, \quad k = 1, 2, \dots, N \\ x_k &= (b-a)\overline{\zeta}_k + a, \quad k = 1, 2, \dots, N \end{aligned} \tag{14.79}$$

In addition, if a structural component (e.g. a beam) with length ℓ is described in the interval $x_k \in [0, \ell]$, the transformations (14.79) become

$$\begin{aligned} x_k &= \ell\zeta_k, \quad k = 1, 2, \dots, N \\ x_k &= \ell\overline{\zeta}_k, \quad k = 1, 2, \dots, N \end{aligned} \tag{14.80}$$

14.3.2.2 δ-Point Distribution

As previously stated, $\overline{\zeta}_k \in [0, 1]$ refers to the abscissa for the distribution with δ-points. Starting from another distribution without δ-points $\zeta_k \in [0, 1]$, the δ-point distribution is given by the simple relations

$$\overline{\zeta}_1 = 0, \ \overline{\zeta}_2 = \delta, \ \overline{\zeta}_{N-1} = 1 - \delta, \ \overline{\zeta}_N = 1, \ \overline{\zeta}_{k+1} = \zeta_k, \quad \text{for } k = 2, \dots, M-3, \ M = N-2 \tag{14.81}$$

whose expressions are exactly the same as the ones presented before, and can be used for all the distributions reported in what follows.

14.3.2.3 Stretching Formulation

Considering a grid distribution with $\xi_k \in [0, 1]$, and a nonzero constant $\alpha \in \mathbb{R}$, the stretching formulation is defined as

$$\zeta_k = (1-\alpha)(3\xi_k^2 - 2\xi_k^3) + \alpha\xi_k, \quad k = 1, 2, \dots, N \tag{14.82}$$

This technique takes the points in the domain $\xi_k \in [0, 1]$ and stretches them into the domain $\zeta_k \in [0, 1]$. As explained by [2], the real-valued parameter α cannot assume all the values because for some entry, the distribution $\zeta_k \notin [0, 1]$. This means that, once the α parameter is set, the new distribution ζ_k must be checked to belong to the interval $[0, 1]$.

14.3.2.4 Several Types of Discretization

Several grid point distributions are presented hereafter, while providing the proper nomenclature and the definition domain. All these distributions are defined in a dimensionless form, in the domain $\xi_k \in [0, 1]$.

Uniform Distribution

This distribution considers an equally spaced distribution of points in the domain $\xi_k \in [0, 1]$

$$\xi_k = \frac{k-1}{N-1}, \quad k = 1, 2, \dots, N \tag{14.83}$$

This choice is appropriate to define other symmetric distributions presented below, while varying the stretching parameter α.

Harmonic Distribution (Cosine) or "Chebyshev–Gauss–Lobatto Grid Distribution"

The present distribution can be found as the roots of the polynomials $CGL_{N+1}(r) = (1 - r^2)U_{N-1}(r)$. Thus, the points are located as

$$\xi_k = \frac{r_k - r_1}{r_N - r_1}, \quad r_k = \cos\left(\frac{N-k}{N-1}\pi\right), \quad k = 1, 2, \ldots, N, \quad r \in [-1, 1] \tag{14.84}$$

Quadratic Distribution

The present distribution is valid only for an odd number of grid points N, namely

$$\begin{cases} \xi_k = 2\left(\dfrac{k-1}{N-1}\right)^2 & k = 1, 2, \ldots, \dfrac{N+1}{2} \\ \xi_2 = -2\left(\dfrac{k-1}{N-1}\right)^2 + 4\left(\dfrac{k-1}{N-1}\right) - 1 & k = \dfrac{N+1}{2} + 1, \ldots, N-1, N \end{cases} \tag{14.85}$$

Chebyshev Distribution of the First Kind

This distribution is obtained by evaluating the roots of the Chebyshev polynomial of the first kind $T_{N+1}(r)$, i.e.

$$\xi_k = \frac{r_k - r_1}{r_N - r_1}, \quad r_k = \cos\left(\frac{2(N-k)+1}{2N}\pi\right), \quad k = 1, 2, \ldots, N, \quad r \in [-1, 1] \tag{14.86}$$

Chebyshev Distribution of the Second Kind

This distribution is obtained by evaluating the roots of the Chebyshev polynomial of the second kind $U_{N+1}(r)$, i.e.

$$\xi_k = \frac{r_k - r_1}{r_N - r_1}, \quad r_k = \cos\left(\frac{N-k+1}{N+1}\pi\right), \quad k = 1, 2, \ldots, N, \quad r \in [-1, 1] \tag{14.87}$$

Chebyshev Distribution of the Third Kind

This distribution is obtained by evaluating the roots of the Chebyshev polynomial of the third kind $V_{N+1}(r)$, i.e.

$$\xi_k = \frac{r_k - r_1}{r_N - r_1}, \quad r_k = \cos\left(\frac{2(N-k)+1}{2N+1}\pi\right), \quad k = 1, 2, \ldots, N, \quad r \in [-1, 1] \tag{14.88}$$

Chebyshev Distribution of the Fourth Kind

This distribution is obtained by evaluating the roots of the Chebyshev polynomial of the fourth kind $W_{N+1}(r)$, i.e.

$$\xi_k = \frac{r_k - r_1}{r_N - r_1}, \quad r_k = \cos\left(\frac{2(N-k+1)}{2N+1}\pi\right), \quad k = 1, 2, \ldots, N, \quad r \in [-1, 1] \tag{14.89}$$

Legendre Distribution

This distribution is obtained by evaluating the roots of the Legendre polynomial $L_{N+1}(r)$, i.e.

$$\xi_k = \frac{r_k - r_1}{r_N - r_1}, \quad r_k = \text{roots of } L_{N+1}(r) \quad k = 1, 2, \ldots, N \quad r \in [-1, 1] \tag{14.90}$$

Approximate Legendre Distribution

This distribution is obtained by evaluating the roots of the Legendre polynomial $L_{N+1}(r)$, i.e.

$$\xi_k = \frac{r_k - r_1}{r_N - r_1}, \quad r_k = \left(1 - \frac{1}{8N^2} + \frac{1}{8N^3}\right)\cos\left(\frac{4(N-k)+3}{4N+2}\pi\right), \quad k = 1, 2, \ldots, N, \quad r \in [-1, 1] \tag{14.91}$$

Gauss–Legendre or Legendre–Gauss Distribution

This distribution is obtained by evaluating the roots of the polynomial $\mathrm{GL}_{N+1}(r) = \mathrm{LG}_{N+1}(r) = (1-r^2)L_{N-1}(r)$, i.e.

$$\xi_k = \frac{r_k - r_1}{r_N - r_1}, \quad r_k = \text{roots of } \mathrm{GL}_{N+1}(r) = \mathrm{LG}_{N+1}(r), \quad k = 1, 2, \ldots, N, \quad r \in [-1, 1] \tag{14.92}$$

Gauss–Legendre–Radau or Legendre–Gauss–Radau Distribution

This distribution is obtained by evaluating the roots of the polynomial $\mathrm{GLR}_{N+1}(r) = \mathrm{LGR}_{N+1}(r) = L_{N+1}(r) + L_N(r)$, i.e.

$$\xi_k = \frac{r_k - r_1}{r_N - r_1}, \quad r_k = \text{roots of } \mathrm{GLR}_{N+1}(r) = \mathrm{LGR}_{N+1}(r), \quad k = 1, 2, \ldots, N, \quad r \in [-1, 1] \tag{14.93}$$

Chebyshev–Gauss Distribution

This distribution is obtained by evaluating the roots of the polynomial $\mathrm{CG}_{N+1}(r) = (1-r^2)T_{N-1}(r)$, i.e.

$$\xi_k = \frac{r_k - r_1}{r_N - r_1}, \quad r_1 = -1, \ r_N = 1, \ r_k = \cos\left(\frac{2(N-k)-1}{2(N-2)}\pi\right), \quad k = 2, 3, \ldots, N-1, \quad r \in [-1, 1] \tag{14.94}$$

Gauss–Legendre–Lobatto or Legendre–Gauss–Lobatto Distribution

This distribution is obtained by evaluating the roots of the polynomial $\mathrm{GLL}_{N+1}(r) = \mathrm{LGL}_{N+1}(r) = (1-r^2)A_{N-1}(r)$, i.e.

$$\xi_k = \frac{r_k - r_1}{r_N - r_1}, \quad r_k = \text{roots of } \mathrm{GLL}_{N+1}(r) = \mathrm{LGL}_{N+1}(r), \quad k = 1, 2, \ldots, N, \quad r \in [-1, 1] \tag{14.95}$$

Jacobi Distribution

This distribution is obtained by evaluating the roots of the Jacobi polynomial $J_{N+1}^{(\gamma,\delta)}(r)$, i.e.

$$\xi_k = \frac{r_k - r_1}{r_N - r_1}, \quad r_k = \text{roots of } J_{N+1}^{(\gamma,\delta)}(r), \quad k = 1, 2, \ldots, N \quad r \in [-1, 1] \tag{14.96}$$

Chebyshev–Gauss–Radau Distribution

This distribution is obtained by evaluating the roots of the polynomial $\mathrm{CGR}_{N+1}(r) = (1+r)(U_N(r) - U_{N-1}(r))$, i.e.

$$\xi_k = \frac{r_k - r_1}{r_N - r_1}, \quad r_k = \cos\left(\frac{2(N-k)}{2N-1}\pi\right), \quad k = 1, 2, \ldots, N, \quad r \in [-1, 1] \tag{14.97}$$

Gauss–Jacobi or Gauss–Jacobi Distribution

This distribution is obtained by evaluating the roots of the Jacobi–Gauss polynomial $\mathrm{GJ}_{N+1}^{(\gamma,\delta)}(r) = \mathrm{JG}_{N+1}^{(\gamma,\delta)}(r) = (1-r^2)J_{N-1}^{(\gamma,\delta)}(r)$, i.e.

$$\xi_k = \frac{r_k - r_1}{r_N - r_1}, \quad r_k = \text{roots of } \mathrm{GJ}_{N+1}^{(\gamma,\delta)}(r) = \mathrm{JG}_{N+1}^{(\gamma,\delta)}(r), \quad k = 1, 2, \ldots, N, \quad r \in [-1, 1] \tag{14.98}$$

Radau Distribution of the First Kind

This distribution is obtained by evaluating the roots of the polynomial $R_{N+1}^{\mathrm{I}}(r) = (1+r)(L_N(-r) + L_{N-1}(-r))$, i.e.

$$\xi_k = \frac{r_k - r_1}{r_N - r_1}, \quad r_k = \text{roots of } R_N^I, \quad k = 1, 2, \ldots, N, \quad r \in [-1, 1] \tag{14.99}$$

Radau Distribution of the Second Kind

This distribution is obtained by evaluating the roots of the polynomial $R_{N+1}^{\mathrm{II}} = (1-r)(L_N(r) + L_{N-1}(r))$, i.e.

$$\xi_k = \frac{r_k - r_1}{r_N - r_1}, \quad r_k = \text{roots of } R_{N+1}^{\mathrm{II}}, \quad k = 1, 2, \ldots, N, \quad r \in [-1, 1] \tag{14.100}$$

Lobatto Distribution

This distribution is obtained by evaluating the roots of the Lobatto polynomial $A_{N+1}(r)$, i.e.

$$\xi_k = \frac{r_k - r_1}{r_N - r_1}, \quad r_k = \text{roots of } A_{N+1}(r), \quad k = 1, 2, \dots, N, \quad r \in [-1, 1] \tag{14.101}$$

Gauss–Lobatto Distribution

This distribution is obtained by evaluating the roots of the polynomial $\mathrm{GL}_{N+1}(r) = (1 - r^2)A_{N-1}(r)$, i.e.

$$\xi_k = \frac{r_k - r_1}{r_N - r_1}, \quad r_k = \text{roots of } \mathrm{GL}_{N+1}(r), \quad k = 1, 2, \dots, N, \quad r \in [-1, 1] \tag{14.102}$$

Ding Distribution

The distribution according to Ding is located according to the following law

$$\xi_k = \frac{1}{2}\left(1 - \sqrt{2}\cos\left(\frac{\pi}{4} + \frac{\pi}{2}\frac{k-1}{N-1}\right)\right), \quad k = 1, 2, \dots, N \tag{14.103}$$

Elliptic Grid

Let us consider an ellipse centered in the origin of a Cartesian plane, whose semiaxes are a and b in the horizontal and vertical directions (with $a = 1$, $b > 0$). The elliptic grid distribution can be written in terms of the geometric ratio $\kappa = a/b$ ($\kappa \in \mathrm{R}^+$) as follows

$$\xi_k = \frac{r_k - r_1}{r_N - r_1}, \quad r_k = \frac{\kappa \tan\left(-\pi\dfrac{N+1-2k}{2(N-1)}\right)}{\sqrt{1 + \kappa^2 \tan^2\left(-\pi\dfrac{N+1-2k}{2(N-1)}\right)}}, \quad k = 1, 2, 3, \dots, N \tag{14.104}$$

for $r \in [-1, 1]$.

Superelliptic Grid

Let us consider a superellipse centered in the origin of a Cartesian plane, whose semiaxes are a and b in the horizontal and vertical directions (with $a = 1$, $b > 0$). The superelliptic grid distribution can be written in terms of the geometric ratio $\kappa = a/b$ ($\kappa \in \mathrm{R}^+$) and the exponent $n \in \mathrm{N}$ as follows

$$\xi_k = \frac{r_k - r_1}{r_N - r_1}, \quad r_k = \frac{\cos\left(\frac{k-1}{N-1}\pi\right)}{\left(\left|\cos\left(\frac{k-1}{N-1}\pi\right)\right|^n + \kappa^n\left|\sin\left(\frac{k-1}{N-1}\pi\right)\right|\right)^{\frac{1}{n}}}, \quad k = 1, 2, 3, \dots, N \tag{14.105}$$

for $r \in [-1, 1]$.

Cycloidal Grid Distribution

The discrete points can be placed by using a cycloidal grid distribution, namely

$$\xi_k = \frac{r_k - r_1}{r_N - r_1}, \quad r_k = -\frac{1}{\pi}\left(\pi\frac{N+1-2k}{N-1} + \sin\left(\pi\frac{N+1-2k}{N-1}\right)\right), \quad k = 1, 2, 3, \dots, N \tag{14.106}$$

for $r \in [-1, 1]$.

All the expressions (14.83)–(14.106) can be used in the study of structural components using the GDQ method. Unlike the spectral methods, Shu's recursive formulae are valid for any discretization, and they can be used to evaluate the weighting coefficients for the derivative approximation.

The expressions shown above based on the DQ can be applied both in a global and local form. The nth derivative of a function $f(x)$ evaluated at a discrete point x_i can be approximated as the weighted linear sum of the values assumed by the function in the neighborhood of point x_i, i.e.

$$f^{(n)}(x_i) = \left.\frac{d^n f(x)}{dx^n}\right|_{x=x_i} = \sum_{j=1}^{N_i} \varsigma_{ij}^{(n)} f(x_j) \quad i = 1, 2, \dots, N_i \tag{14.107}$$

14.3.3 Numerical Applications: Differential Quadrature

The basic notions of the DQ method are here applied to describe up to the fourth-order derivatives of some functions. These derivatives are here performed using some of the basis functions and grid distributions previously described. When the recursive formulae of the weighting coefficients are not available, the inversion of the coefficient matrix is required. As expectable, the accuracy of the DQ law is verified to increase with the number of grid points N, for all the discretizations shown above.

Here, we analyze the following three functions, defined within the same interval $[\pi/4, \pi]$, namely

$$f(x) = x^5, \quad f(x) = \sqrt{x}, \quad f(x) = \cos(x) \tag{14.108}$$

The error norm is here evaluated as accuracy index, by means of the following relation

$$e_{f^{(n)}} = \left\| \frac{\mathbf{f}^{(n)}_{\langle \mathrm{DQ} \rangle} - \mathbf{f}^{(n)}_{\langle \mathrm{exact} \rangle}}{|\mathbf{f}^{(n)}_{\langle \mathrm{exact} \rangle}|} \right\|_{\infty} = \max_{i=1,2,\dots,N} \left| \frac{\mathbf{f}^{(n)}_{\langle \mathrm{DQ} \rangle} - \mathbf{f}^{(n)}_{\langle \mathrm{exact} \rangle}}{|\mathbf{f}^{(n)}_{\langle \mathrm{exact} \rangle}|} \right|, \quad \text{for } n = 1, 2, 3, 4 \tag{14.109}$$

where $\mathbf{f}^{(n)}_{\langle \mathrm{DQ} \rangle}$ is the vector of the nth derivatives of the function evaluated numerically at all grid points, whereas $\mathbf{f}^{(n)}_{\langle \mathrm{exact} \rangle}$ is the corresponding vector of the exact solutions.

In the following, the derivatives of the functions in (14.108) are computed considering the Lagrange polynomials (PDQ), the Chebyshev polynomials of the first kind, the Hermite polynomials, and the monomial basis functions.

Figures 14.3–14.6 depict the accuracy for the first four derivatives of the function $f(x) = x^5$, in a bilogarithmic scale, where the maximum accuracy is clearly reached for a limited number of grid points. This is mainly related to the polynomial approximation of the function $f(x) = x^5$. Any improvement in accuracy can be noticed for an increasing degree of the approximating polynomial. At the same time, the accuracy seems to decrease for an increased order derivative. The solution obtained with the Lagrange polynomials is more stable than the ones obtained through other approximating polynomials. Hermite polynomials and power functions are characterized by an oscillating behavior for a number of grid points N higher than 51. It can be also observed that the solution found through the Chebyshev polynomials of the first kind diverges completely from the exact solution because of the matrix ill-conditioning that leads to numerical errors during the matrix inversion. The PDQ approach does not suffer this drawback because the weighting coefficients are evaluated in an exact form.

Similar considerations can be repeated for the functions $f(x) = \sqrt{x}$ and $f(x) = \cos(x)$, whose accuracy estimation is reported, for different bases, in Figures 14.7–14.10 and Figures 14.11–14.14, respectively.

In these last two cases, the maximum accuracy is obtained for a higher number of grid points because the present functions are not polynomials. Finally, it can be observed that for a limited number of grid points, the solutions maintain almost the same, independently of the selected basis. This aspect is fundamental for DQ applications related to the domain decomposition technique. The applications reported herein suggest to adopt a mesh with more elements characterized by approximating low-order polynomials.

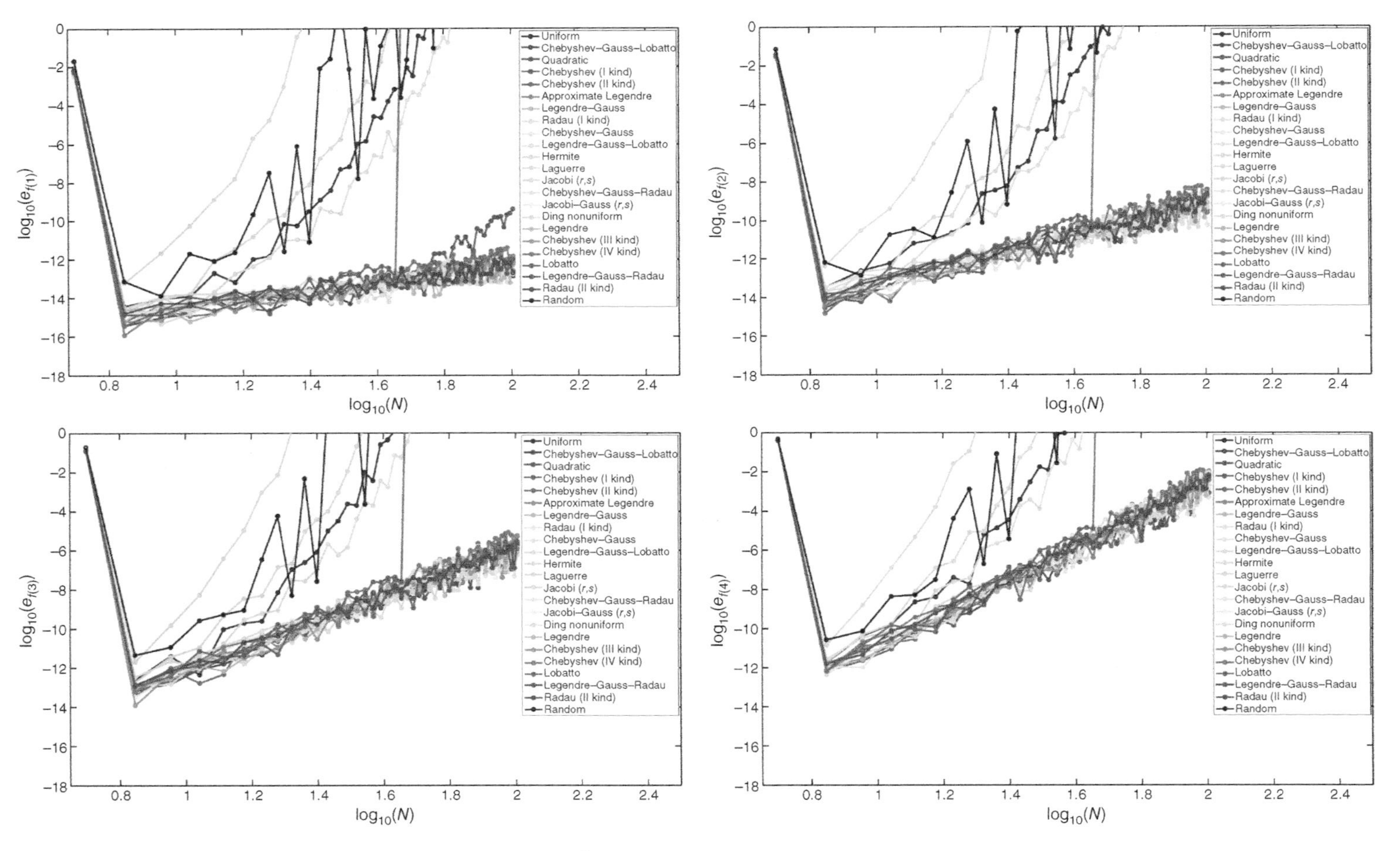

Figure 14.3 Accuracy of the first four derivatives of the function x^5 by the use of the Lagrange polynomials as basis functions.

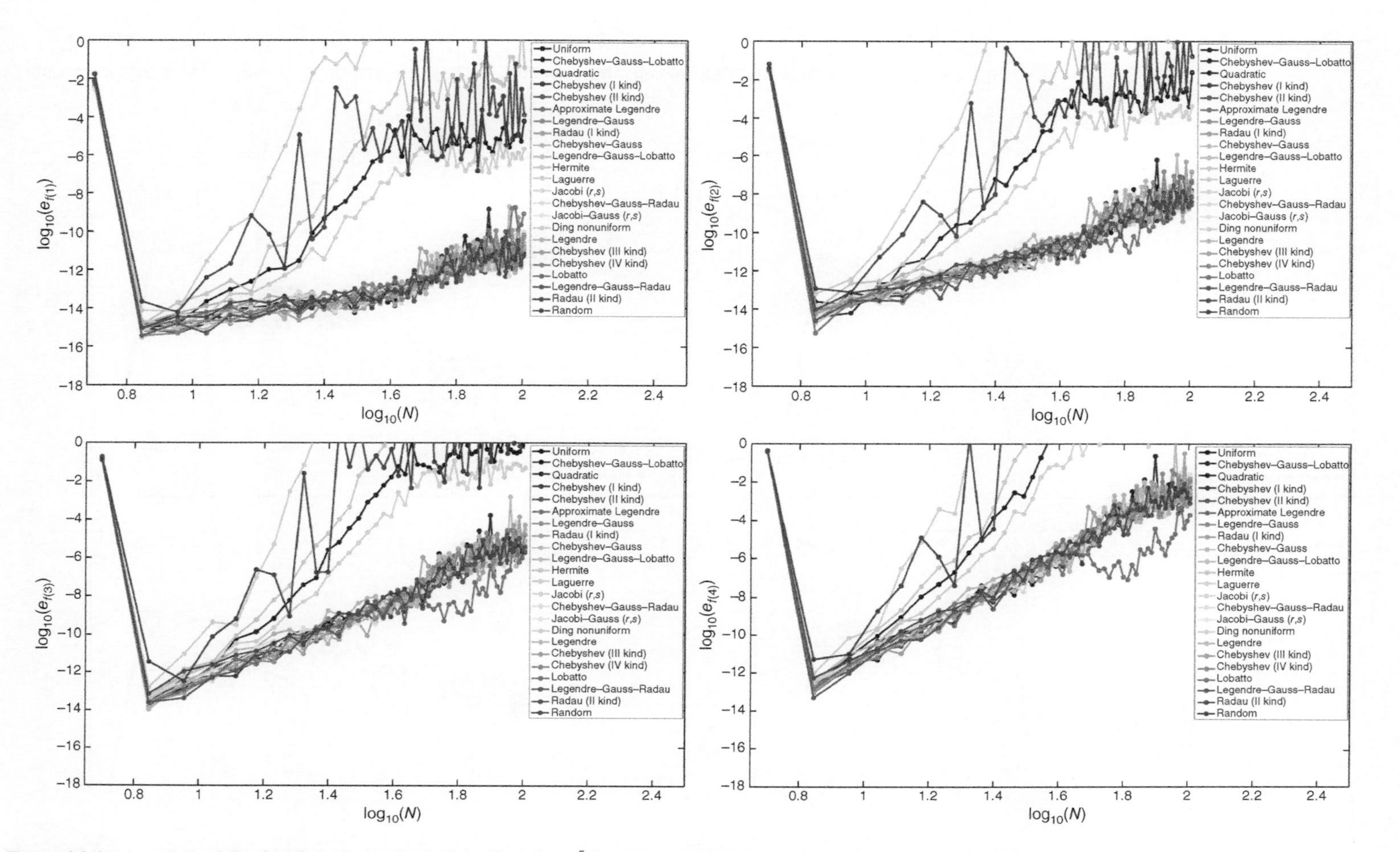

Figure 14.4 Accuracy of the first four derivatives of the function x^5 by the use of the power functions as basis functions.

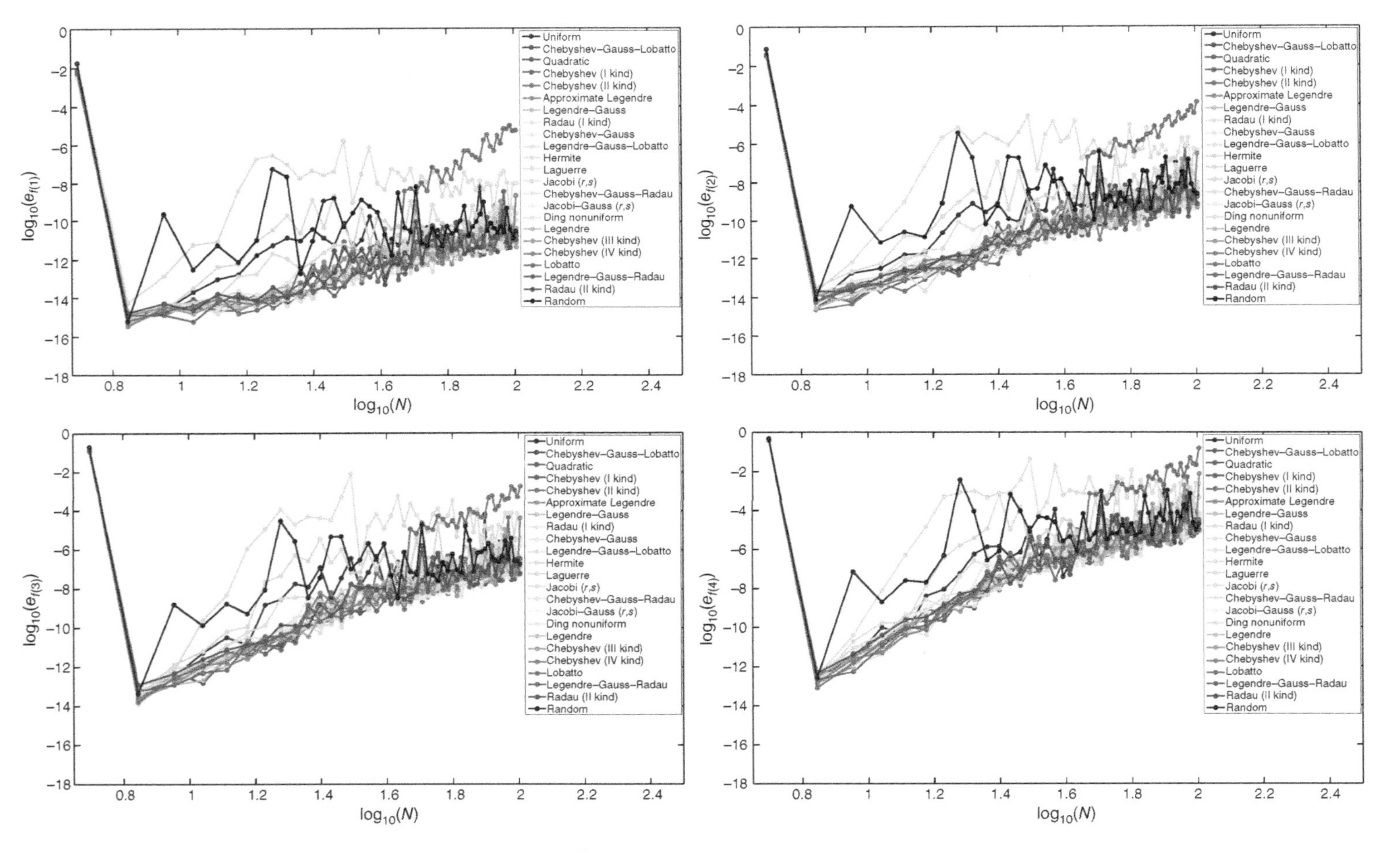

Figure 14.5 Accuracy of the first four derivatives of the function x^5 by the use of the Hermite polynomials as basis functions.

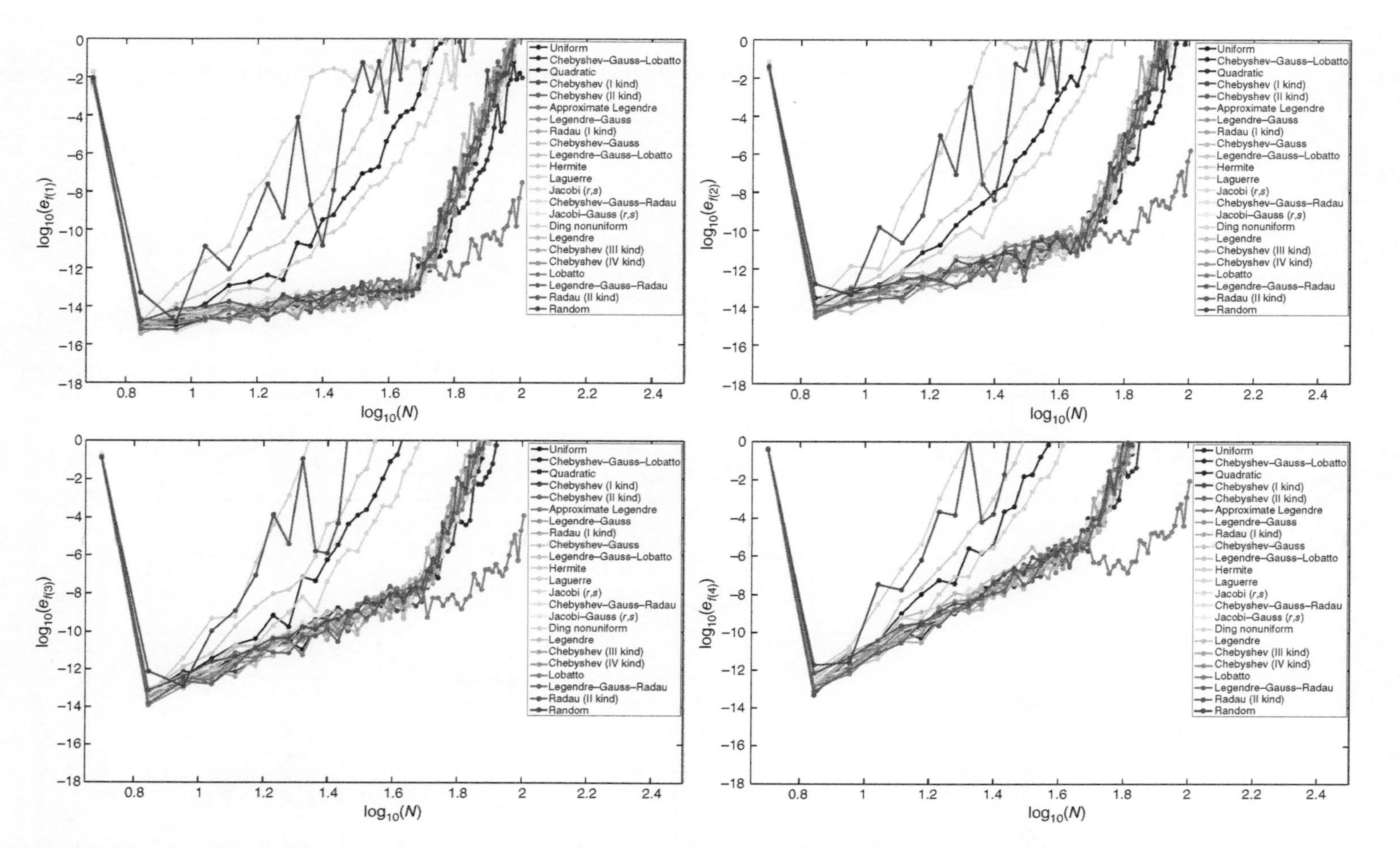

Figure 14.6 Accuracy of the first four derivatives of the function x^5 by the use of the Chebyshev of the first kind polynomials as basis functions.

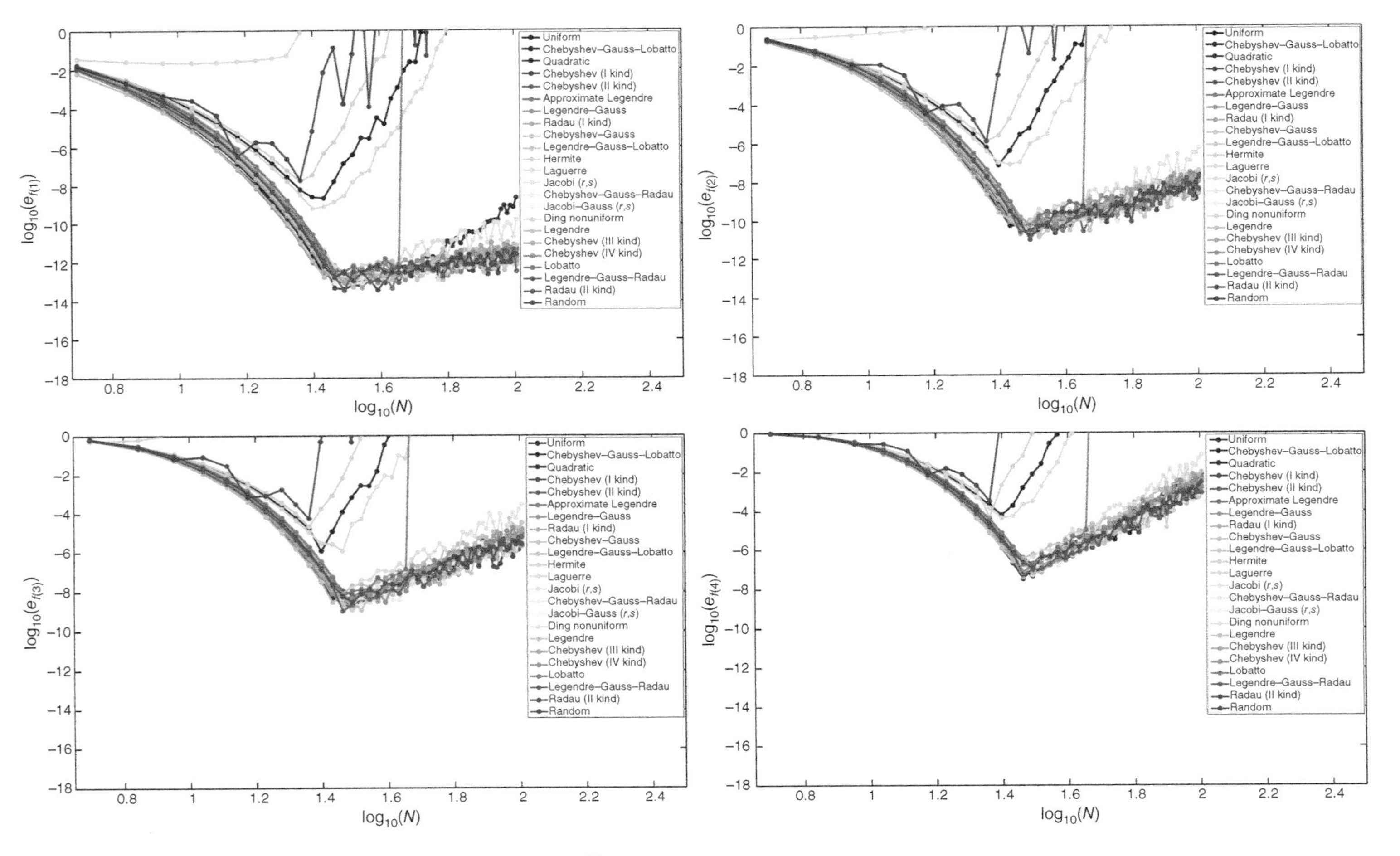

Figure 14.7 Accuracy of the first four derivatives of the function $\sqrt{x}$ by the use of the Lagrange polynomials as basis functions.

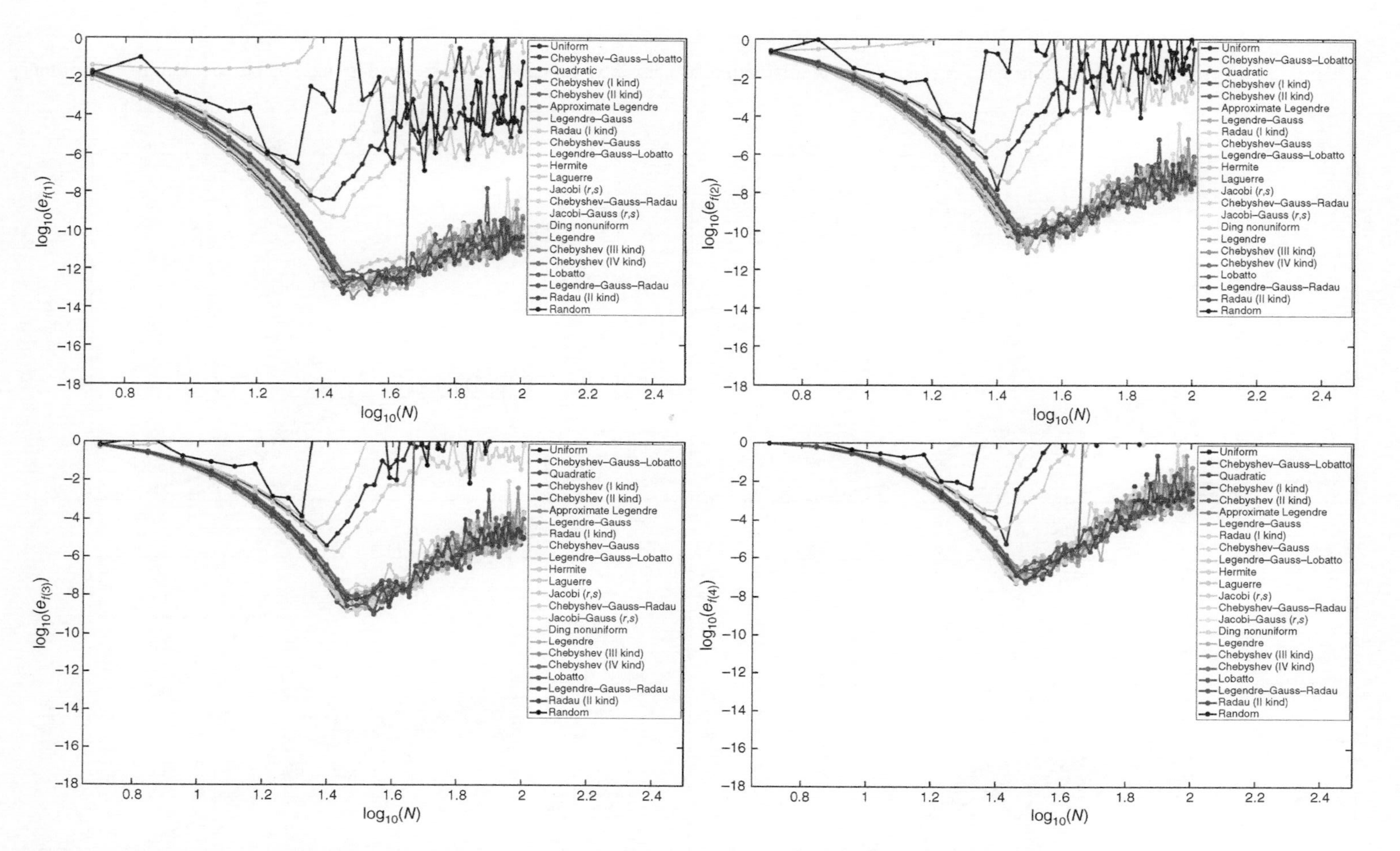

Figure 14.8 Accuracy of the first four derivatives of the function $\sqrt{x}$ by the use of the power functions as basis functions.

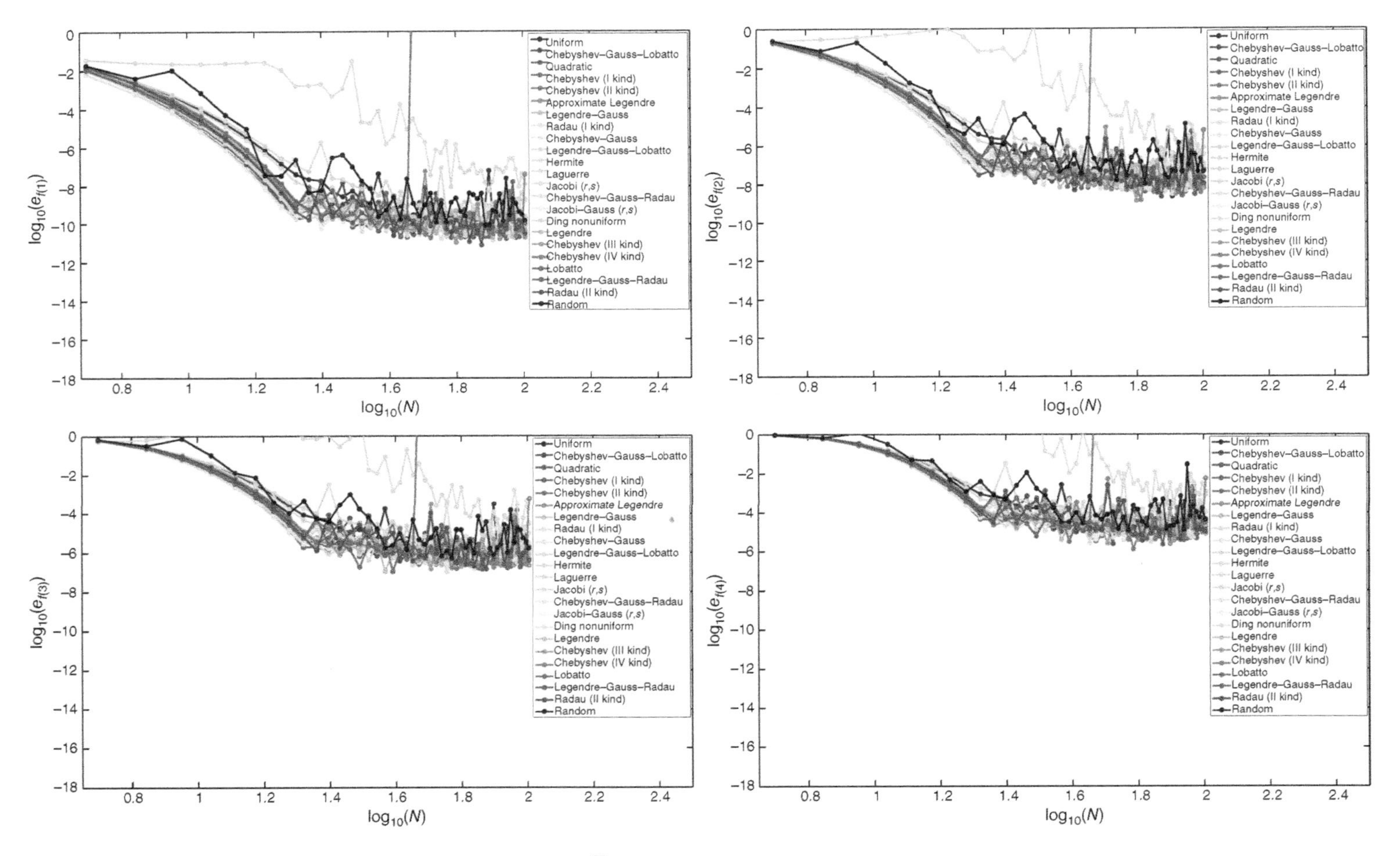

Figure 14.9 Accuracy of the first four derivatives of the function $\sqrt{x}$ by the use of the Hermite polynomials as basis functions.

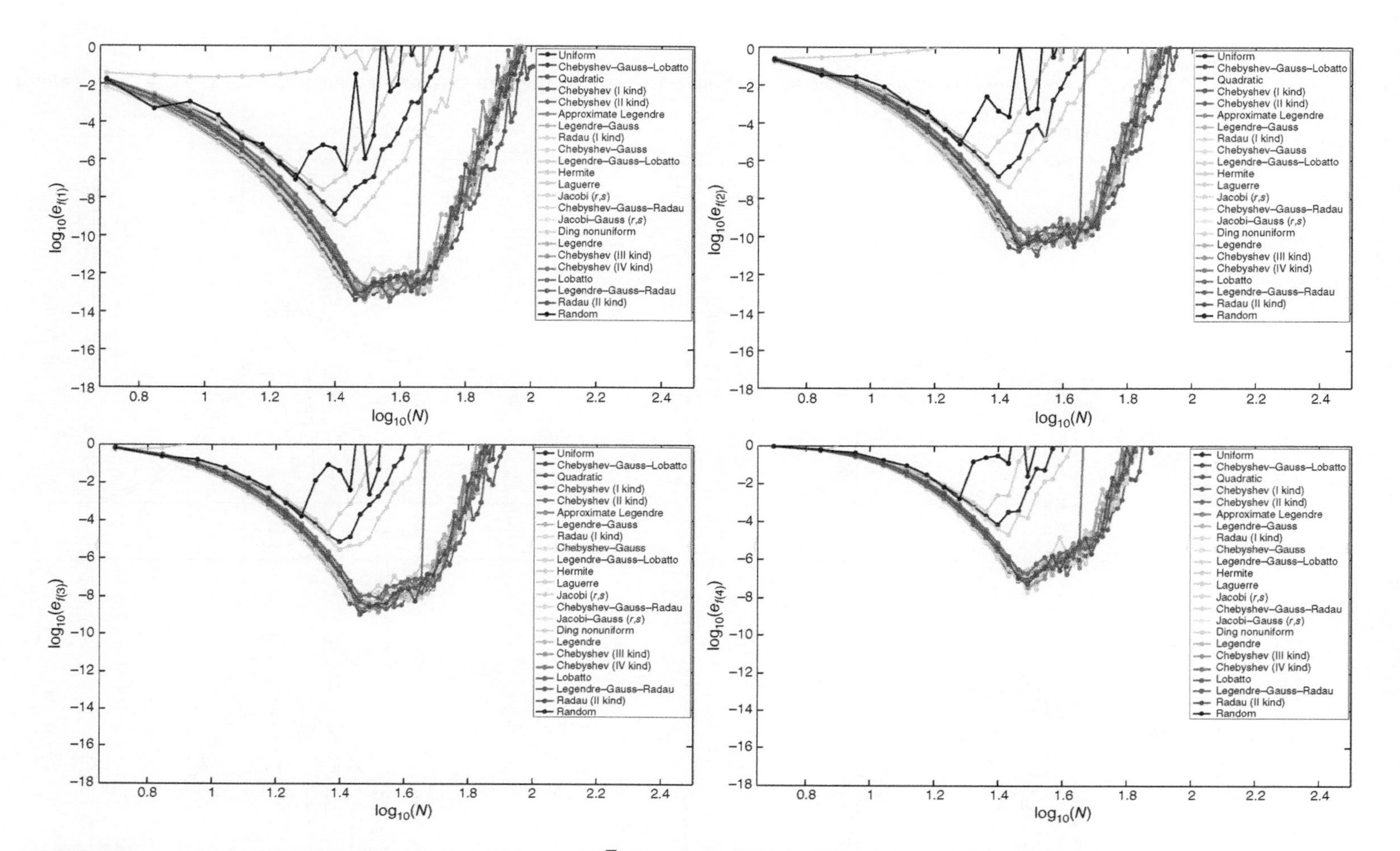

Figure 14.10 Accuracy of the first four derivatives of the function $\sqrt{x}$ using the Chebyshev of the first kind polynomials as basis functions.

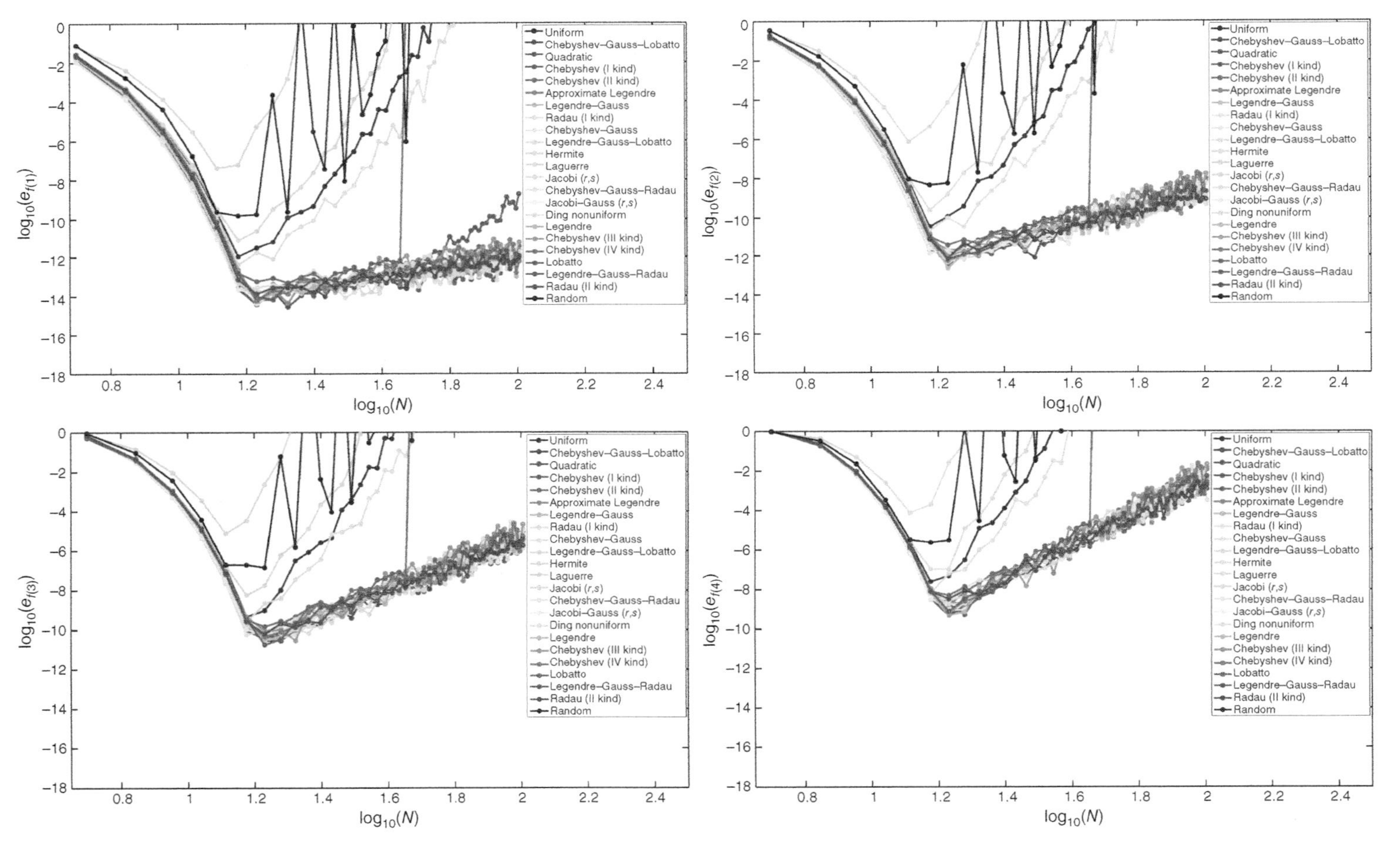

Figure 14.11 Accuracy of the first four derivatives of the function cos(x) using the Lagrange polynomials as basis functions.

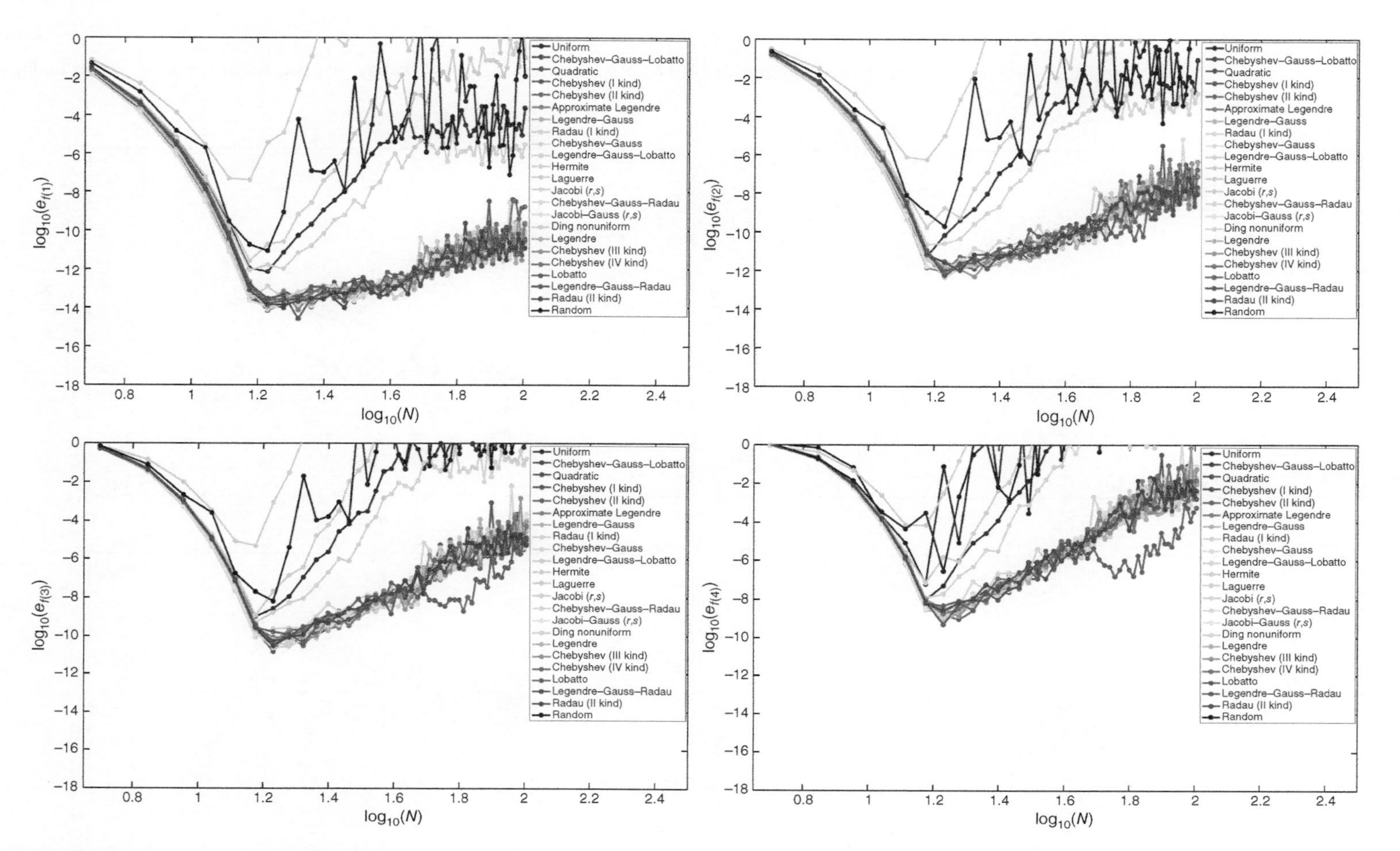

Figure 14.12 Accuracy of the first four derivatives of the function cos(x) by the use of the power functions as basis functions.

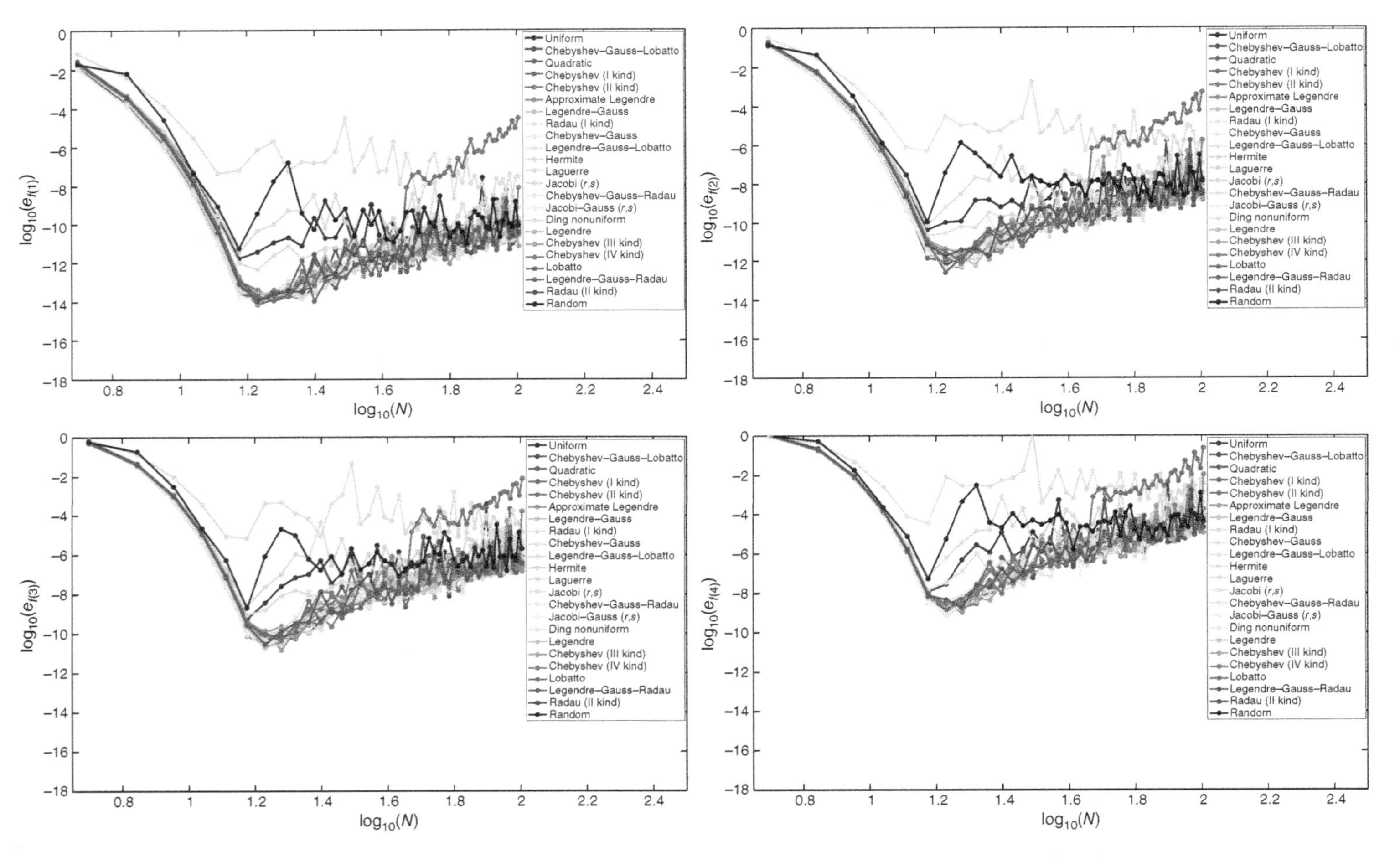

Figure 14.13 Accuracy of the first four derivatives of the function cos(x) by the use of the Hermite polynomials as basis functions.

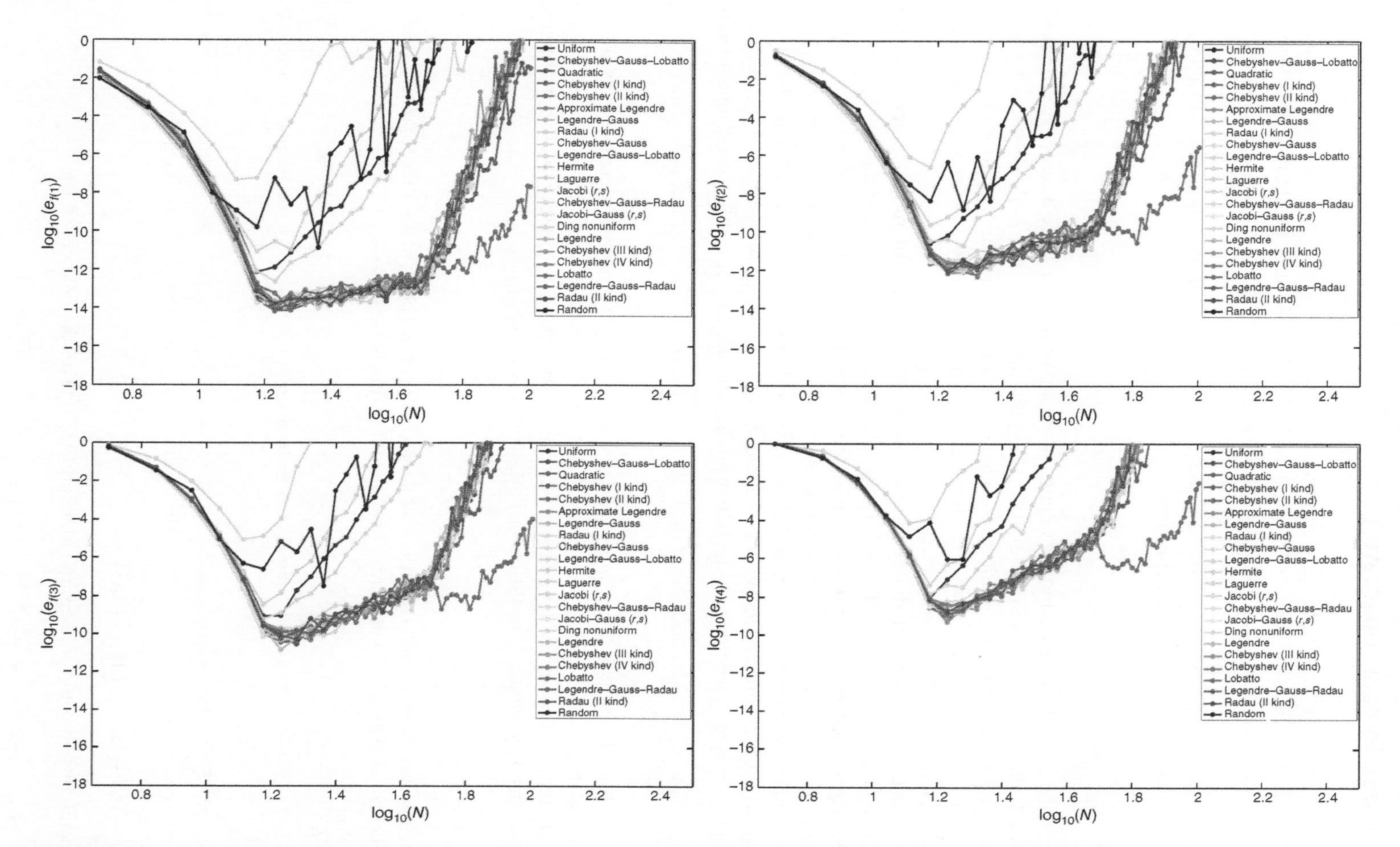

Figure 14.14 Accuracy of the first four derivatives of the function $\cos(x)$ by the use of the Chebyshev of the first kind polynomials as basis functions.

14.4 Generalized Integral Quadrature

In the present section, we illustrate an integration procedure based on the weighting coefficients of the GDQ method. According to a weak formulation-based element method, such as finite elements, it is required to evaluate the integral of some unknown quantities, which makes the numerical integration a key point in computational mechanics.

The GIQ has been introduced by Shu [2], starting from the idea that the numerical integration of a function $f(x)$ in a domain $[a, b]$ can be written in the following form

$$\int_a^b f(x)dx = \sum_{k=1}^{N} w_k f(x_k) \tag{14.110}$$

w_k being the weighing coefficients that can be evaluated by means of the GDQ-based recursive formulae under fixed assumptions. Unlike the Gaussian integration schemes, this approach can handle the coefficients without any restriction on the point distribution defining the domain. The GIQ method has been developed following the same concepts of the GDQ technique, whereby the fundamental idea of the PDQ has been applied to determine the value of the integral.

A continuous function $f(x)$ is considered over a domain $[a, b]$, which is divided into $N-1$ intervals with some grid points, such as $a = x_1, x_2, \ldots, b = x_N$. Because of the continuity of $f(x)$, and under the assumption of N grid points discretizing the domain, the given function can be approximated by a polynomial of order $N-1$. Moreover, when the functional values are known at the whole grid points, the function $f(x)$ can be approximated by the Lagrange interpolation polynomials. This means that the integral of this approximated polynomial in the interval $[x_i, x_j]$ could involve the functional values outside the integral domain. Thus, the integral of $f(x)$ over a certain domain can be approximated by a linear combination of all the functional values in the overall domain as follows

$$\int_{x_i}^{x_j} f(x)dx = \sum_{k=1}^{N} w_k^{ij} f(x_k) \tag{14.111}$$

The limits x_i, x_j are interchangeable. When $a = x_i$ and $b = x_j$, Eq. (14.111) becomes a conventional integral. As already stated, the polynomial of order $N-1$ is a linear N-dimensional vector space, and if all the basis polynomials satisfy Eq. (14.111), the same is satisfied for all the polynomials of the space. As a result, under the assumption of Lagrange interpolating polynomials $l_k(x)$, the coefficients w_k^{ij} can be evaluated as

$$w_k^{ij} = \int_{x_i}^{x_j} l_k(x)dx \tag{14.112}$$

This expression, however, has a complex solution, which justifies the adoption of another approach. The continuous function $f(x)$ can be defined as a derivative of an auxiliary function $F(x)$ as

$$f(x) = \frac{dF(x)}{dx} \tag{14.113}$$

Hence, if $f(x)$ is a $N-1$ order polynomial, then $F(x)$ must have order N. In fact, the polynomial approximation of $f(x)$ can be written as

$$f(x) = a_1 + a_2 x + \cdots + a_N x^{N-1} \tag{14.114}$$

where a_i are arbitrary constants, for $i = 1, 2, \ldots, N$. The function $F(x)$ can be computed by integration of Eq. (14.114)

$$F(x) = \int_{\varepsilon}^{x} f(t)dt + u(\varepsilon) = I_f(x, \varepsilon) + u(\varepsilon) \tag{14.115}$$

where ε is a general constant and the integrand function $I_f(x, \varepsilon)$ takes the following form

$$I_f(x, \varepsilon) = x\left(a_1 + \frac{a_2}{2}x + \cdots + \frac{a_N}{N}x^{N-1}\right) - \varepsilon\left(a_1 + \frac{a_2}{2}\varepsilon + \cdots + \frac{a_N}{N}\varepsilon^{N-1}\right) \tag{14.116}$$

It is obvious that $I_f(x, \varepsilon)$ depends on N constants and it is a N-dimensional linear polynomial vector space. The following set can be selected as its basis polynomials

$$p_k(x) = (x - \varepsilon)l_k(x) \quad \text{for } k = 1, 2, \ldots, N \tag{14.117}$$

According to the Weierstrass theorem and the properties of the linear vector space, the function $I_f(x, \varepsilon)$ can be approximated as

$$I_f(x, \varepsilon) = \sum_{j=1}^{N} p_j(x)d_j = \sum_{j=1}^{N} (x - \varepsilon)l_j(x)d_j \tag{14.118}$$

Thus, the derivative of the function $I_f(x, \varepsilon)$ assumes the following form

$$\begin{aligned} \left.\frac{dI_f(x, \varepsilon)}{dx}\right|_{x_i} &= \left.\frac{d}{dx}\left(\sum_{j=1}^{N} p_j(x)d_j\right)\right|_{x_i} = \left.\left(\sum_{j=1}^{N} d_j(l_j(x) + (x - \varepsilon)l_j^{(1)}(x))\right)\right|_{x_i} \\ &= \sum_{j=1}^{N} d_j(l_j(x_i) + (x_i - \varepsilon)l_j^{(1)}(x_i)) \end{aligned} \tag{14.119}$$

Furthermore, the derivative of the function $I_f(x, \varepsilon)$ can be evaluated using the DQ law (14.3) as follows

$$\left.\frac{dI_f(x, \varepsilon)}{dx}\right|_{x_i} = I_f^{(1)}(x, \varepsilon)\Big|_{x_i} = \sum_{j=1}^{N} \bar{\varsigma}_{ij}^{(1)} I_f(x_j, \varepsilon) = \sum_{j=1}^{N} \bar{\varsigma}_{ij}^{(1)} \sum_{k=1}^{N} p_k(x_j)d_k = \sum_{j=1}^{N} \bar{\varsigma}_{ij}^{(1)} \sum_{k=1}^{N} d_k(x_j - \varepsilon)l_k(x_j) \tag{14.120}$$

It is well-known that $l_j^{(1)}(x_i) = \varsigma_{ij}^{(1)}$ are the weighting coefficients of the first-order derivative and that $l_k(x_j) = \delta_{kj}$ stands for the Kronecker delta function. Thus, by comparing the relations (14.119) and (14.120), we get the following equation

$$\begin{aligned} \left.\frac{dI_f(x, \varepsilon)}{dx}\right|_{x_i} &= \sum_{j=1}^{N} d_j(l_j(x_i) + (x_i - \varepsilon)\varsigma_{ij}^{(1)}) = \sum_{j=1}^{N} \bar{\varsigma}_{ij}^{(1)} \sum_{k=1}^{N} d_k(x_j - \varepsilon)l_k(x_j) \\ &= \sum_{j=1}^{N} \bar{\varsigma}_{ij}^{(1)}(x_j - \varepsilon)l_j(x_j)d_j = \sum_{j=1}^{N} \bar{\varsigma}_{ij}^{(1)}(x_j - \varepsilon)d_j \end{aligned} \tag{14.121}$$

Using Eq. (14.121), the following coefficients are found

$$\begin{aligned} \bar{\varsigma}_{ij}^{(1)} &= \frac{x_i - \varepsilon}{x_j - \varepsilon}\varsigma_{ij}^{(1)} \quad \text{for } i \neq j \\ \bar{\varsigma}_{ij}^{(1)} &= \varsigma_{ii}^{(1)} + \frac{1}{x_i - \varepsilon} \quad \text{for } i = j \end{aligned} \tag{14.122}$$

Note that ε is an arbitrary constant and can be set to $\varepsilon = x_N + 10^{-10}$ in order to obtain stable and accurate results. Writing Eq. (14.120) in matrix form, the following relation is obtained

$$\mathbf{I}_f^{(1)} = \bar{\varsigma}^{(1)}\mathbf{I}_f \tag{14.123}$$

where

$$\begin{aligned} \mathbf{I}_f^{(1)} &= \mathbf{f} = [I_f^{(1)}(x_1, \varepsilon), I_f^{(1)}(x_2, \varepsilon), \ldots, I_f^{(1)}(x_N, \varepsilon)]^T \\ \mathbf{I}_f &= \mathbf{F} = [I_f(x_1, \varepsilon), I_f(x_2, \varepsilon), \ldots, I_f(x_N, \varepsilon)]^T \end{aligned} \tag{14.124}$$

By substituting Eqs. (14.113) and (14.115) into (14.123), we get

$$\mathbf{f} = \bar{\varsigma}^{(1)}\mathbf{I}_f \tag{14.125}$$

where $\mathbf{I}_f$ can be defined as

$$\mathbf{I}_f = \left[\int_{\varepsilon}^{x_1} f(x)dx, \int_{\varepsilon}^{x_2} f(x)dx, \dots, \int_{\varepsilon}^{x_N} f(x)dx \right]^T \tag{14.126}$$

In conclusion, the integral (14.115) can be evaluated by invertion of the matrix $(\overline{\varsigma}^{(1)})^{-1} = \mathbf{W}$ as

$$\int_{\varepsilon}^{x_i} f(x)dx = \sum_{k=1}^{N} w_{ik} f(x_k) \quad i = 1, 2, \dots, N \tag{14.127}$$

In order to evaluate Eq. (14.111), using the properties of integrals, we get the final form of the integral weighting coefficients

$$\int_{\varepsilon}^{x_j} f(x)dx + \int_{x_i}^{\varepsilon} f(x)dx = \int_{\varepsilon}^{x_j} f(x)dx - \int_{\varepsilon}^{x_i} f(x)dx = \int_{x_i}^{x_j} f(x)dx = \sum_{k=1}^{N} (w_{jk} - w_{ik}) f(x_k) \tag{14.128}$$

Finally, the GIQ weighting coefficients can be computed as

$$w_k^{ij} = w_{jk} - w_{ik} \tag{14.129}$$

Relation (14.78) is useful to map from a domain $\zeta_k \in [c = \zeta_1, d = \zeta_N]$ to another one $x_k \in [a = x_1, b = x_N]$. In this case, the following transformation should be adopted

$$\begin{aligned} \int_a^b f(x)dx &= \frac{b-a}{d-c} \int_c^d f\left(\frac{b-a}{d-c}(\zeta - c) + a \right) d\zeta \\ &= \frac{b-a}{d-c} \sum_{k=1}^{N} (\tilde{w}_{Nk} - \tilde{w}_{1k}) f\left(\frac{b-a}{d-c}(\zeta_k - c) + a \right) \\ &= \frac{b-a}{d-c} \sum_{k=1}^{N} \tilde{w}_k^{1N} f\left(\frac{b-a}{d-c}(\zeta_k - c) + a \right) = \frac{b-a}{d-c} \sum_{k=1}^{N} \tilde{w}_k^{1N} f(x_k) \end{aligned} \tag{14.130}$$

Thus, the transformation of the GIQ weighting coefficients w_k^{1N} can be deduced to map from the interval $[c, d]$ to a general one $[a, b]$ using the following relation

$$\int_a^b f(x)dx = \sum_{k=1}^{N} w_k^{1N} f(x_k) \tag{14.131}$$

where

$$w_k^{1N} = \frac{b-a}{d-c} \tilde{w}_k^{1N} \tag{14.132}$$

For the sake of clarity, note that the coefficients $\tilde{w}_k^{1N}$ are related to the set $[c, d]$, whereas the coefficients w_k^{1N} are related to the general interval $[a, b]$. Differently from the Gaussian integration schemes, the present approach determines the weighting coefficients for the numerical integration without any restriction on the grid distributions used to discretize the domain.

14.4.1 Generalized Taylor-Based Integral Quadrature

An alternative approach based on the Taylor series is now presented to approximate integrals. This approach is named generalized Taylor-based integral quadrature (GTIQ). Let $f(x)$ be a one-dimensional function, written in the neighborhood of a generic point x_i as follows

$$f(x) \cong \sum_{r=0}^{m-1} \frac{(x - x_i)^r}{r!} \left. \frac{d^r f}{dx^r} \right|_{x_i} \quad m \leq N \tag{14.133}$$

The integral of $f(x)$ in the closed interval $[x_i, x_j]$ can be evaluated as

$$\int_{x_i}^{x_j} f(x)dx = \sum_{r=0}^{m-1} \frac{(x_j - x_i)^{r+1}}{(r+1)!} \left.\frac{d^r f}{dx^r}\right|_{x_i} \tag{14.134}$$

The same integration can be alternatively performed by splitting the reference interval in two parts, i.e.

$$\int_{x_i}^{x_j} f(x)dx = \int_{x_i}^{\frac{x_j+x_i}{2}} f(x)dx + \int_{\frac{x_j+x_i}{2}}^{x_j} f(x)dx = \int_{x_i}^{\frac{x_j+x_i}{2}} f(x)dx - \int_{x_j}^{\frac{x_j+x_i}{2}} f(x)dx \tag{14.135}$$

By means of Eq. (14.134), the integral in (14.135) assumes the following form

$$\begin{aligned}\int_{x_i}^{x_j} f(x)dx &= \sum_{r=0}^{m-1} \frac{(x_j - x_i)^{r+1}}{2^{r+1}(r+1)!} \left.\frac{d^r f}{dx^r}\right|_{x_i} - \sum_{r=0}^{m-1} \frac{(x_i - x_j)^{r+1}}{2^{r+1}(r+1)!} \left.\frac{d^r f}{dx^r}\right|_{x_j} \\ &= \sum_{r=0}^{m-1} \frac{(x_j - x_i)^{r+1}}{2^{r+1}(r+1)!} \left(\left.\frac{d^r f}{dx^r}\right|_{x_i} + (-1)^{r+2} \left.\frac{d^r f}{dx^r}\right|_{x_j} \right)\end{aligned} \tag{14.136}$$

Thus, the basics of the DQ (14.14) are now recalled as follows

$$\left.\frac{d^r f}{dx^r}\right|_{x_i} = \sum_{k=1}^{N} \varsigma_{ik}^{(r)} f(x_k) \tag{14.137}$$

whose expression is combined with Eq. (14.136) to yield

$$\int_{x_i}^{x_j} f(x)dx = \sum_{r=0}^{m-1} \frac{(x_j - x_i)^{r+1}}{(r+1)!} \sum_{k=1}^{N} \varsigma_{ik}^{(r)} f(x_k) = \sum_{k=1}^{N} \left(\sum_{r=0}^{m-1} \frac{(x_j - x_i)^{r+1}}{(r+1)!} \varsigma_{ik}^{(r)} \right) f(x_k) = \sum_{k=1}^{N} w_k^{ij} f(x_k) \tag{14.138}$$

Similarly, the (14.136) becomes

$$\begin{aligned}\int_{x_i}^{x_j} f(x)dx &= \sum_{r=0}^{m-1} \frac{(x_j - x_i)^{r+1}}{2^{r+1}(r+1)!} \sum_{k=1}^{N} \varsigma_{ik}^{(r)} f(x_k) - \sum_{r=0}^{m-1} \frac{(x_i - x_j)^{r+1}}{2^{r+1}(r+1)!} \sum_{k=1}^{N} \varsigma_{jk}^{(r)} f(x_k) \\ &= \sum_{k=1}^{N} \left(\sum_{r=0}^{m-1} \left(\frac{(x_j - x_i)^{r+1}}{2^{r+1}(r+1)!} \varsigma_{ik}^{(r)} - \frac{(x_i - x_j)^{r+1}}{2^{r+1}(r+1)!} \varsigma_{jk}^{(r)} \right) \right) f(x_k) \\ &= \sum_{k=1}^{N} \left(\sum_{r=0}^{m-1} \frac{(x_j - x_i)^{r+1}}{2^{r+1}(r+1)!} (\varsigma_{ik}^{(r)} + (-1)^{r+2} \varsigma_{jk}^{(r)}) \right) f(x_k) = \sum_{k=1}^{N} w_k^{ij} f(x_k)\end{aligned} \tag{14.139}$$

w_k^{ij} being the weighting coefficients, here summarized for both the approaches

$$w_k^{ij} = \sum_{r=0}^{m-1} \frac{(x_j - x_i)^{r+1}}{(r+1)!} \varsigma_{ik}^{(r)} \tag{14.140}$$

$$w_k^{ij} = \sum_{r=0}^{m-1} \frac{(x_j - x_i)^{r+1}}{2^{r+1}(r+1)!} (\varsigma_{ik}^{(r)} + (-1)^{r+2} \varsigma_{jk}^{(r)}) \tag{14.141}$$

Note also that

$$\left.\frac{d^0 f}{dx^0}\right|_{x_i} = \sum_{k=1}^{N} \varsigma_{ik}^{(0)} f(x_k) = f(x_i) \tag{14.142}$$

since $\varsigma_{ii}^{(0)} = 1$, for $i = j$, and $\varsigma_{ij}^{(0)} = 0$, for $i \neq j$. In other words, the matrix of the weighting coefficients for 0th order derivatives corresponds to the identity matrix ($\varsigma^{(0)} = \mathbf{I}$). It should be recalled that the recursive formulae provided by Shu [2] can be applied to evaluate the weighting coefficients for the derivatives. Analogously, the matrix multiplication approach can be adopted.

Compared to the GIQ method, it can be easily observed that the inversion of the coefficient matrix is not required because the coefficients for integration can be computed recursively. Nevertheless, this approach needs the computation and the summation of the weighting coefficients for the derivatives up to higher orders. On the contrary, the GIQ method requires only the coefficients for the first-order derivatives, although it needs the inversion of the coefficient matrix, which could be ill-conditioned. This drawback can be overcome by this Taylor-based approach.

14.4.2 Classic Integral Quadrature Methods

Some classic integration schemes are briefly described here, together with the pertaining discretizations and expressions of the weighting coefficients needed for the integration.

14.4.2.1 Trapezoidal Rule with Uniform Discretization

The points are located according to the following rule

$$x_i = h(i-1) + a, \quad \text{with } h = \frac{b-a}{N-1}, \quad \text{for } i = 1, 2, \dots, N \tag{14.143}$$

and the weighting coefficients are defined as

$$w_1 = w_N = \frac{h}{2}, \quad w_i = h, \quad \text{for } i = 2, 3, \dots, N-1 \tag{14.144}$$

14.4.2.2 Simpson's Method (One-third Rule) with Uniform Discretization

The points are located according to the following rule

$$x_i = h(i-1) + a, \quad \text{with } h = \frac{b-a}{N-1}, \quad \text{for } i = 1, 2, \dots, N \text{ and } N \text{ odd} \tag{14.145}$$

and the weighting coefficients are defined as

$$\begin{aligned} & w_1 = w_N = \frac{h}{3}, \quad w_i = \frac{4h}{3}, \quad \text{for } i = 2, 4, 6, 8, \dots, N-1 \\ & w_i = \frac{2h}{3}, \quad \text{for } i = 3, 5, 7, 9, \dots, N-2 \end{aligned} \tag{14.146}$$

14.4.2.3 Chebyshev–Gauss Method (Chebyshev of the First Kind)

The points are located according to the following rule

$$x_{N-i+1} = \cos\left(\frac{2i-1}{2N}\pi\right), \quad \text{with } w_{0i} = \sqrt{1-x_i^2}, \quad \text{for } i = 1, 2, \dots, N \tag{14.147}$$

and the weighting coefficients are defined as

$$w_i = \frac{\pi w_{0i}}{N}, \quad \text{for } i = 1, 2, \dots, N \tag{14.148}$$

14.4.2.4 Chebyshev–Gauss Method (Chebyshev of the Second Kind)

The points are located according to the following rule

$$x_{N-i+1} = \cos\left(\frac{i}{N+1}\pi\right), \quad \text{with } w_{0i} = \frac{1}{\sqrt{1-x_i^2}}, \quad \text{for } i = 1, 2, \dots, N \tag{14.149}$$

and the weighting coefficients are defined as

$$w_i = \frac{\pi w_{0i}}{N+1}\left(\sin\left(\frac{i\pi}{N+1}\right)\right)^2, \quad \text{for } i = 1, 2, \dots, N \tag{14.150}$$

14.4.2.5 Chebyshev–Gauss Method (Chebyshev of the Third Kind)

The points are located according to the following rule

$$x_{N-i+1} = \cos\left(\frac{2i-1}{2N+1}\pi\right), \quad \text{with } w_{0i} = \frac{\sqrt{1+x_i}}{\sqrt{1-x_i}}, \quad \text{for } i = 1, 2, \dots, N \tag{14.151}$$

and the weighting coefficients are defined as

$$w_i = \frac{4\pi w_{0i}}{2N+1}\left(\sin\left(\frac{(N-i+1)\pi}{2N+1}\right)\right)^2, \quad \text{for } i = 1, 2, \dots, N \tag{14.152}$$

14.4.2.6 Chebyshev–Gauss Method (Chebyshev of the Fourth Kind)

The points are located according to the following rule

$$x_{N-i+1} = \cos\left(\frac{2i}{2N+1}\pi\right), \quad \text{with } w_{0i} = \frac{\sqrt{1-x_i}}{\sqrt{1+x_i}}, \quad \text{for } i = 1, 2, \dots, N \tag{14.153}$$

and the weighting coefficients are defined as

$$w_i = \frac{4\pi w_{0i}}{2N+1}\left(\sin\left(\frac{i\pi}{2N+1}\right)\right)^2, \quad \text{for } i = 1, 2, \dots, N \tag{14.154}$$

14.4.2.7 Chebyshev–Gauss–Radau Method (Chebyshev of the First Kind)

The points are located according to the following rule

$$x_{N-i+1} = \cos\left(\frac{2(i-1)}{2N-1}\pi\right), \quad \text{with } w_{0i} = \sqrt{1-x_i^2}, \quad \text{for } i = 1, 2, \dots, N \tag{14.155}$$

and the weighting coefficients are defined as

$$w_N = \frac{\pi w_{01}}{2N-1}, \quad w_i = \frac{2\pi w_{0i}}{2N-1}, \quad \text{for } i = 1, 2, \dots, N-1 \tag{14.156}$$

14.4.2.8 Chebyshev–Gauss–Lobatto Method (Chebyshev of the First Kind)

The points are located according to the following rule

$$x_{N-i+1} = \cos\left(\frac{i-1}{N-1}\pi\right), \quad \text{with } w_{0i} = \sqrt{1-x_i^2}, \quad \text{for } i = 1, 2, \dots, N \tag{14.157}$$

and the weighting coefficients are defined as

$$w_1 = \frac{\pi w_{01}}{2(N-1)}, \quad w_N = \frac{\pi w_{0N}}{2(N-1)}, \quad w_i = \frac{\pi w_{0i}}{N-1}, \quad \text{for } i = 2, 3, \dots, N-1 \tag{14.158}$$

14.4.2.9 Gauss–Legendre or Legendre–Gauss Method

The discretization is determined finding the roots x_i of the Legendre polynomial $L_N(x)$ of degree N, for $i = 1, 2, \dots, N$, whereby the weighting coefficients are defined by the following expression

$$w_i = \frac{2}{(1-x_i^2)(L_N^{(1)}(x_i))^2}, \quad \text{for } i = 1, 2, \dots, N \tag{14.159}$$

where $L_N^{(1)}(x)$ represents the first-order derivative of the Legendre polynomial $L_N(x)$ of degree N.

14.4.2.10 Gauss–Legendre–Radau or Legendre–Gauss–Radau Method

The discretization is determined finding the roots x_i of the polynomial $(1+x)(L_N(x)+L_{N-1}(x))$, for $i = 1, 2, \dots, N$. The weighting coefficients are defined by

$$w_1 = \frac{2}{N^2}, \quad w_i = \frac{1-x_i}{N^2(L_{N-1}(x_i))^2}, \quad \text{for } i = 2, 3, \dots, N \tag{14.160}$$

14.4.2.11 Gauss–Legendre–Lobatto or Legendre–Gauss–Lobatto Method

The discretization is related to the roots x_i of the polynomial $(1-x)^2 L_{N-1}^{(1)}(x)$, for $i = 1, 2, \ldots, N$, whereby the weighting coefficients read

$$w_1 = w_N = \frac{2}{N(N-1)}, \quad w_i = \frac{2}{N(N-1)(L_{N-1}(x_i))^2}, \quad \text{for } i = 2, 3, \ldots, N-1 \tag{14.161}$$

14.4.3 Numerical Applications: Integral Quadrature

Here, we report some numerical applications, in order to demonstrate the accuracy of the GIQ method, when computing the integral of some functions in the interval $[\pi/4, \pi]$. In detail, we account for the integration of the same functions previously analyzed for derivation purposes, i.e.

$$f(x) = x^5, \quad f(x) = \sqrt{x}, \quad f(x) = \cos(x) \tag{14.162}$$

The integral is numerically computed using the same basis polynomials considered in the derivative cases, while using different types of discretizations. For each approach analyzed here, the accuracy of the integral quadrature law is checked for an increasing number of grid points N (up to 101 points), for the whole discretizations. The relative error is considered as a parameter for the integral accuracy, namely

$$e_I = \left| \frac{I_{\langle \mathrm{GIQ} \rangle} - I_{\langle \mathrm{exact} \rangle}}{I_{\langle \mathrm{exact} \rangle}} \right| \tag{14.163}$$

where $I_{\langle \mathrm{GIQ} \rangle}$ is the integral numerical value, whereas $I_{\langle \mathrm{exact} \rangle}$ is its corresponding exact value. The integrals of the functions (14.163) are performed considering the following basis functions: Lagrange polynomials (PDQ), monomial functions, Hermite polynomials, and Chebyshev polynomials of the first kind.

Figure 14.15 depicts the results in terms of accuracy (in bilogarithmic scale) related to the integral of the polynomial $f(x) = x^5$. The maximum accuracy is reached for a very small number of points ($N = 5$), similarly to the derivation case, because of the polynomial integrand function. The solution obtained via Lagrange polynomials is more stable compared to the ones obtained through the other interpolating functions. Hermite polynomials and power functions feature an oscillating behavior for a large number of grid points ($N > 51$). Moreover, the solution based on the Chebyshev polynomials of the first kind diverges completely from the exact prediction, for a large number of points. This behavior is caused by the ill-conditioning of the coefficient matrix that yields numerical errors during the inversion of the same matrix. This effect is avoided by the PDQ approach because the weighting coefficients are evaluated using exact analytical functions. It must be highlighted that the ill-conditioning effect is amplified in a mathematical integration because the ill-conditioned matrix has to be inverted twice with respect to the derivation. During a mathematical integration, the weighting coefficients collected in the matrix $\mathbf{W}$ are obtained by an inversion of the first-order derivatives $\bar{\varsigma}^{(1)}$ of the weighting coefficients matrix, as follows

$$\mathbf{W} = (\bar{\varsigma}^{(1)})^{-1} \tag{14.164}$$

Similar considerations can be written for the functions $f(x) = \sqrt{x}$ and $f(x) = \cos(x)$, whose results are evaluated in terms of accuracy in the Figures 14.16 and 14.17. In these last two cases, the highest accuracy is obtained using more points with respect to the first case because of the non-polynomial nature of both functions.

Finally, it is worth noticing that, for a few grid points, the behavior of the solutions maintains almost the same, independently of the adopted basis. This aspect is fundamental for GIQ-based applications related to the domain decomposition technique. Based on the above-mentioned examples, we suggest to adopt meshes with more elements and approximating polynomials of low order.

A comparative evaluation of the numerical integration techniques presented above is represented in the plots of Figure 14.18, for the function $f(x) = \sqrt{x}$ in the interval $[\pi/4, \pi]$. From Figure 14.18, the approaches based on Chebyshev polynomials have clearly a lower accuracy and require a large number of points to resemble the exact solution. This application shows, once again, the excellent accuracy and stability features of the GIQ method.

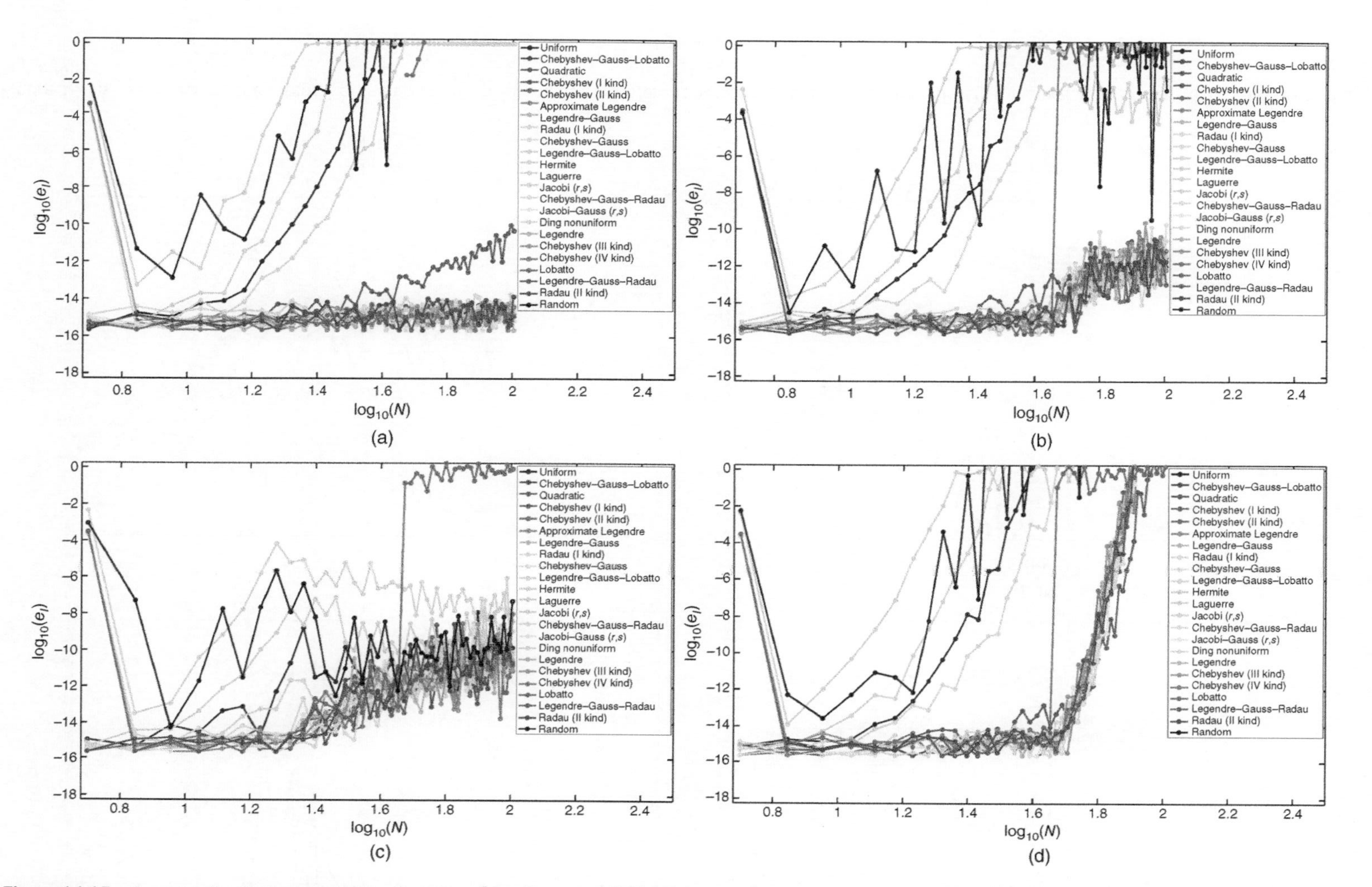

Figure 14.15 Accuracy for the integral of the function x^5 by the use of different basis functions. (a) Lagrange (PDQ). (b) Power. (c) Hermite. (d) Chebyshev of the first kind.

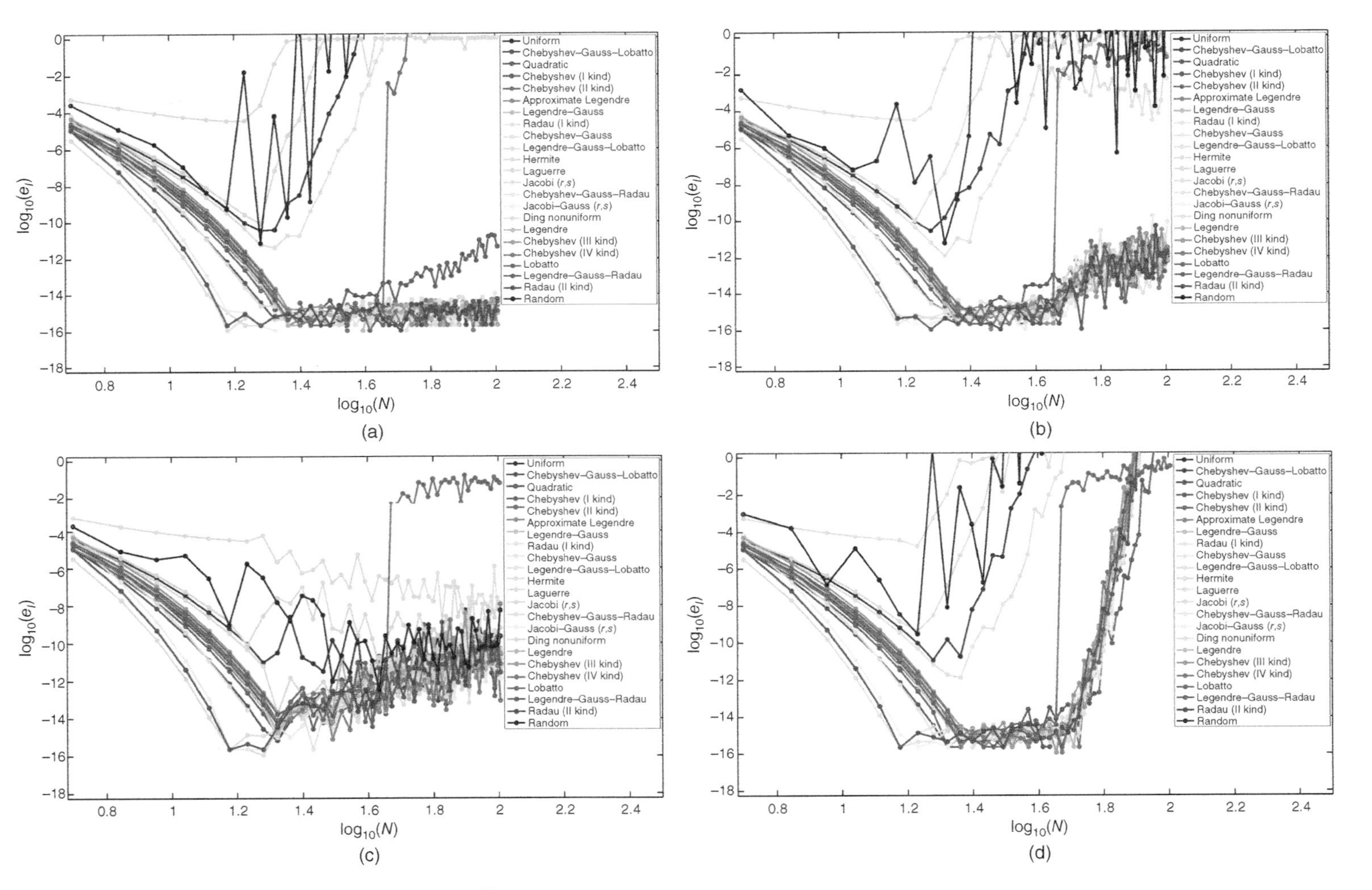

Figure 14.16 Accuracy for the integral of the function $\sqrt{x}$ by the use of different basis functions. (a) Lagrange (PDQ). (b) Power. (c) Hermite. (d) Chebyshev of the first kind.

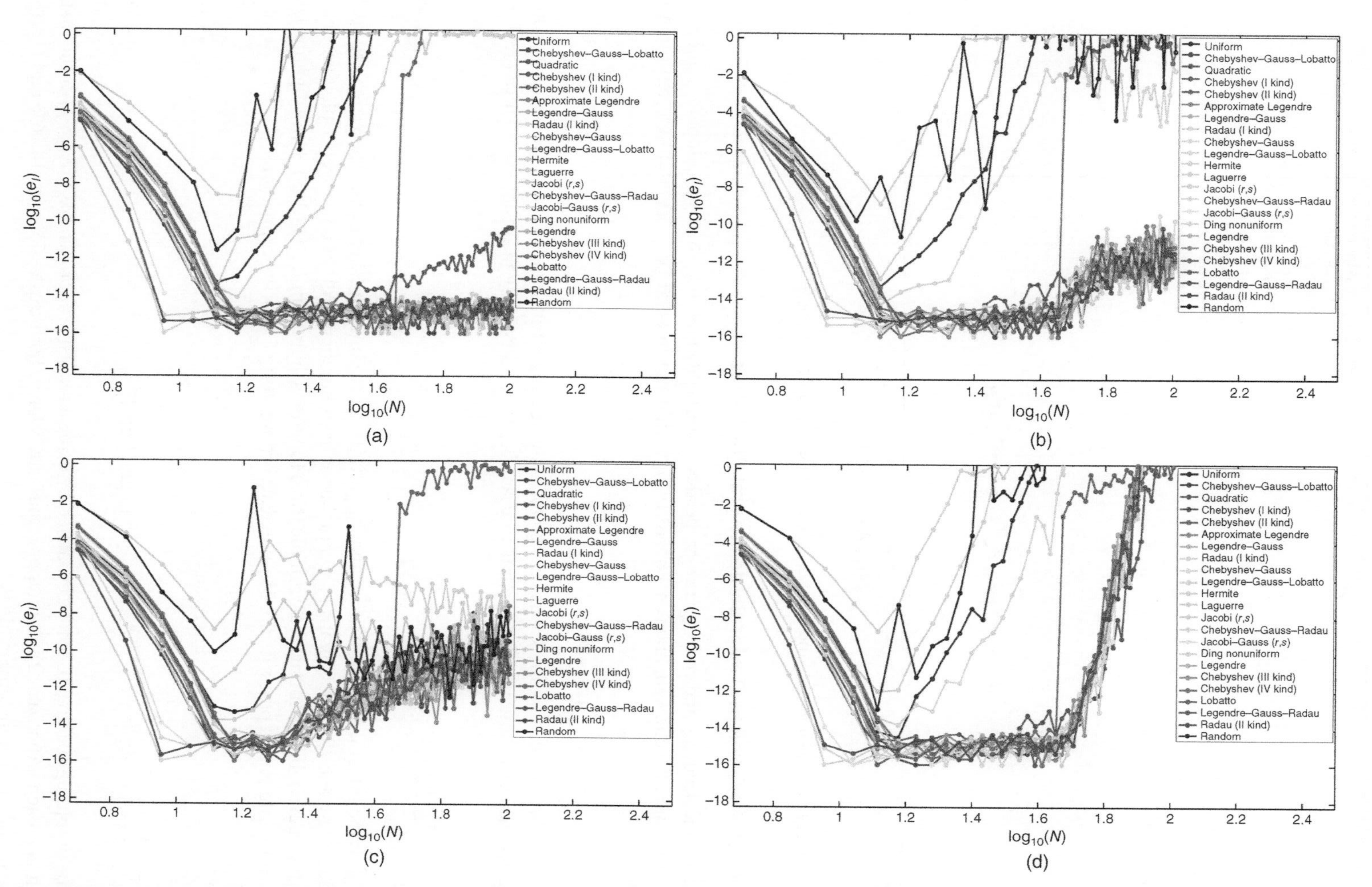

Figure 14.17 Accuracy for the integral of the function cos(x) by the use of different basis functions. (a) Lagrange (PDQ). (b) Power. (c) Hermite. (d) Chebyshev of the first kind.

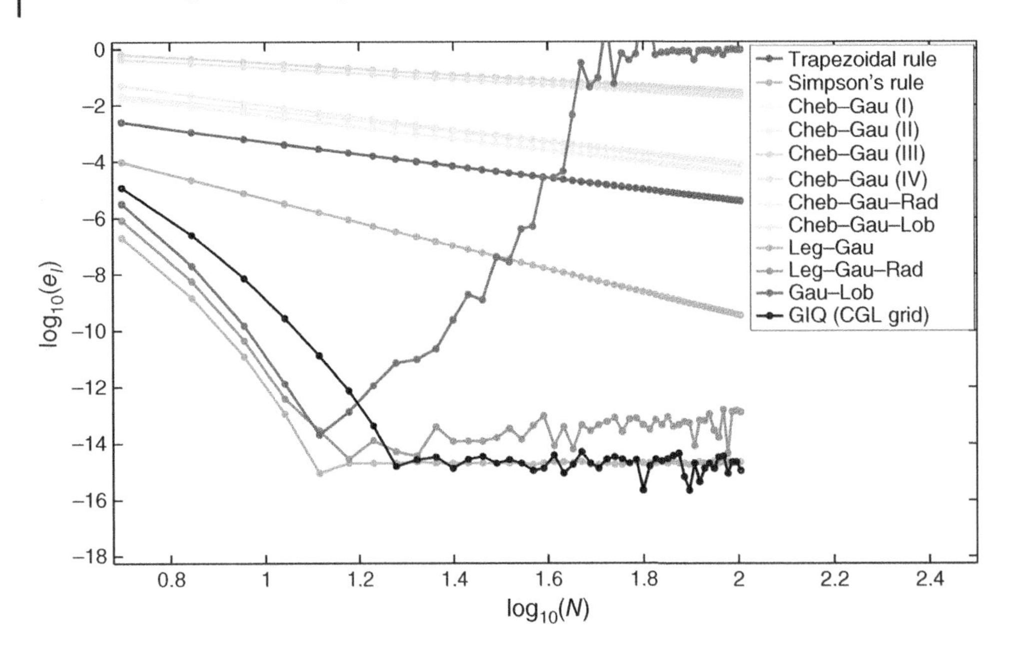

Figure 14.18 Accuracy for the integral of the function $\sqrt{x}$ by the use of different numerical integration schemes.

14.4.4 Numerical Applications: Taylor-Based Integral Quadrature

Now, we check for the accuracy and convergence properties of the GTIQ method. Thus, we compute numerically the integral of the same functions defined in Eq. (14.162), within the interval $[\pi/4, \pi]$, and here rewritten for simplicity as follows

$$f(x) = x^5, \quad f(x) = \sqrt{x}, \quad f(x) = \cos(x) \tag{14.165}$$

The accuracy of the integral quadrature law is verified for an increasing number of grid points N up to 101 points, for all the selected grid distributions. The integral accuracy is here evaluated by means of the relative error estimation, namely

$$e_I = \left| \frac{I_{\langle \mathrm{GTIQ} \rangle} - I_{\langle \mathrm{exact} \rangle}}{I_{\langle \mathrm{exact} \rangle}} \right| \tag{14.166}$$

in which $I_{\langle \mathrm{GTIQ} \rangle}$ is the integral numerical value and $I_{\langle \mathrm{exact} \rangle}$ is its corresponding exact value. The weighting coefficients are evaluated by selecting the Lagrange polynomials as basis functions. All the convergence plots are shown in Figures 14.19–14.21, in bilogarithmic scale, for the three functions defined in Eq. (14.165). The weighting coefficients adopted for integration purposes are computed according to the following expressions

$$w_k^{ij} = \sum_{r=0}^{m-1} \frac{(x_j - x_i)^{r+1}}{(r+1)!} \varsigma_{ik}^{(r)} \quad \rightarrow \quad \text{Expression 1} \tag{14.167}$$

$$w_k^{ij} = \sum_{r=0}^{m-1} \frac{(x_j - x_i)^{r+1}}{2^{r+1}(r+1)!} (\varsigma_{ik}^{(r)} + (-1)^{r+2} \varsigma_{jk}^{(r)}) \quad \rightarrow \quad \text{Expression 2} \tag{14.168}$$

In addition, the weighting coefficients can be evaluated by using a recursive approach (Approach 1) or the matrix multiplication approach (Approach 2). Both approaches can refer to both expressions in (14.167) and (14.168). This provides the following four combinations

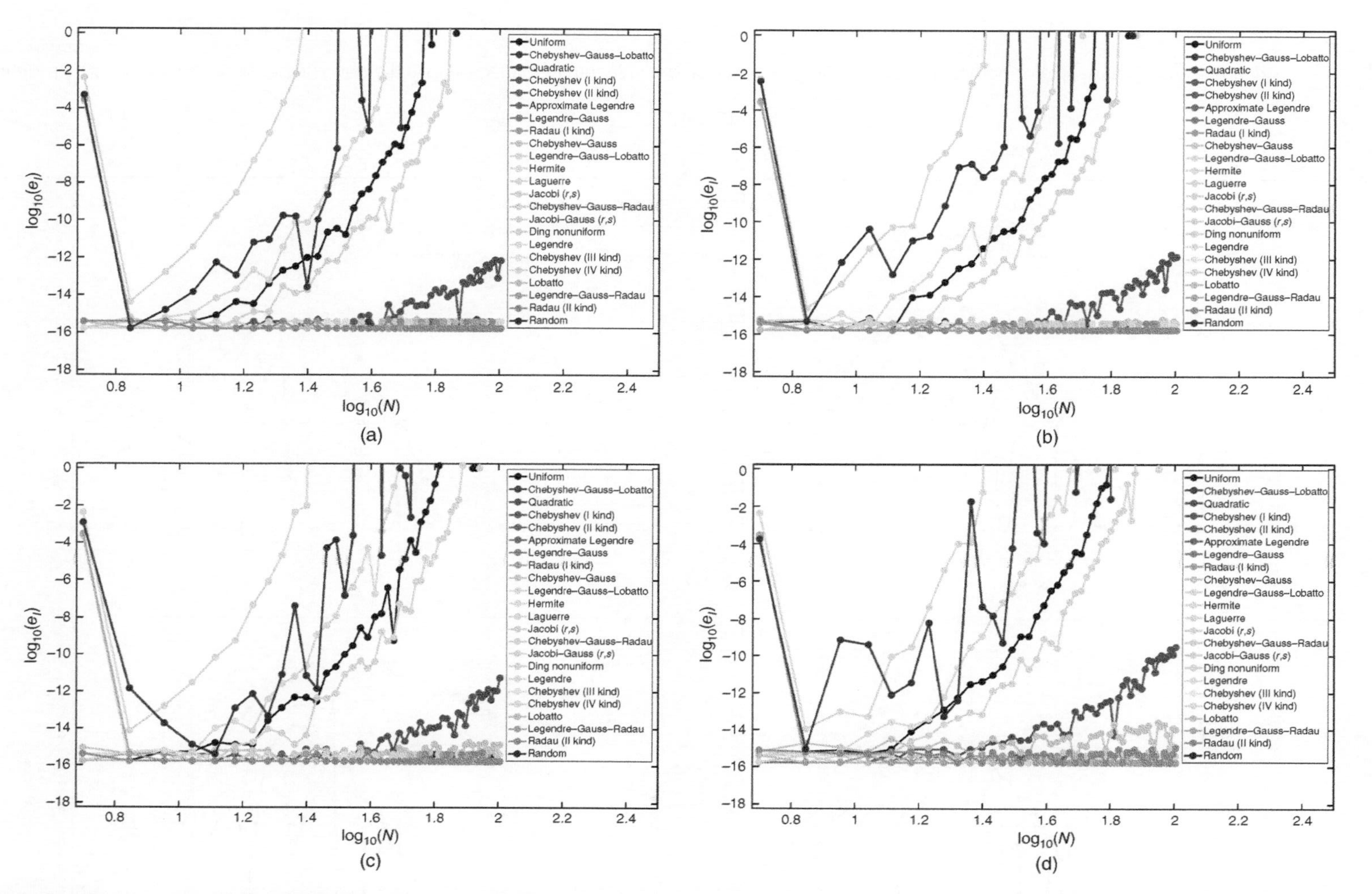

Figure 14.19 Accuracy for the integral of the function x^5 for different approaches based on GTIQ method. (a) Method 1. (b) Method 2. (c) Method 3. (d) Method 4.

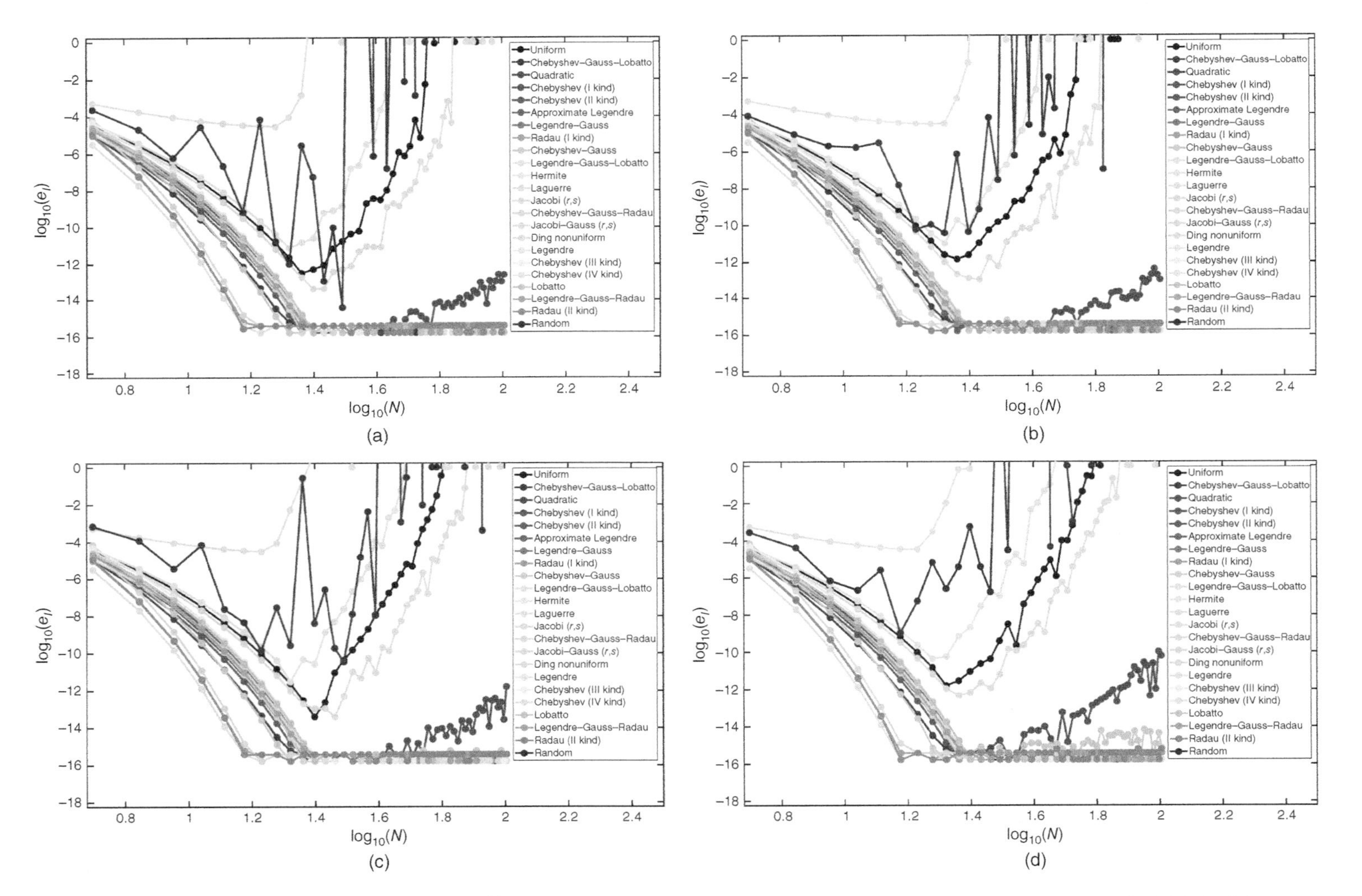

Figure 14.20 Accuracy for the integral of the function $\sqrt{x}$ for different approaches based on GTIQ method. (a) Method 1. (b) Method 2. (c) Method 3. (d) Method 4.

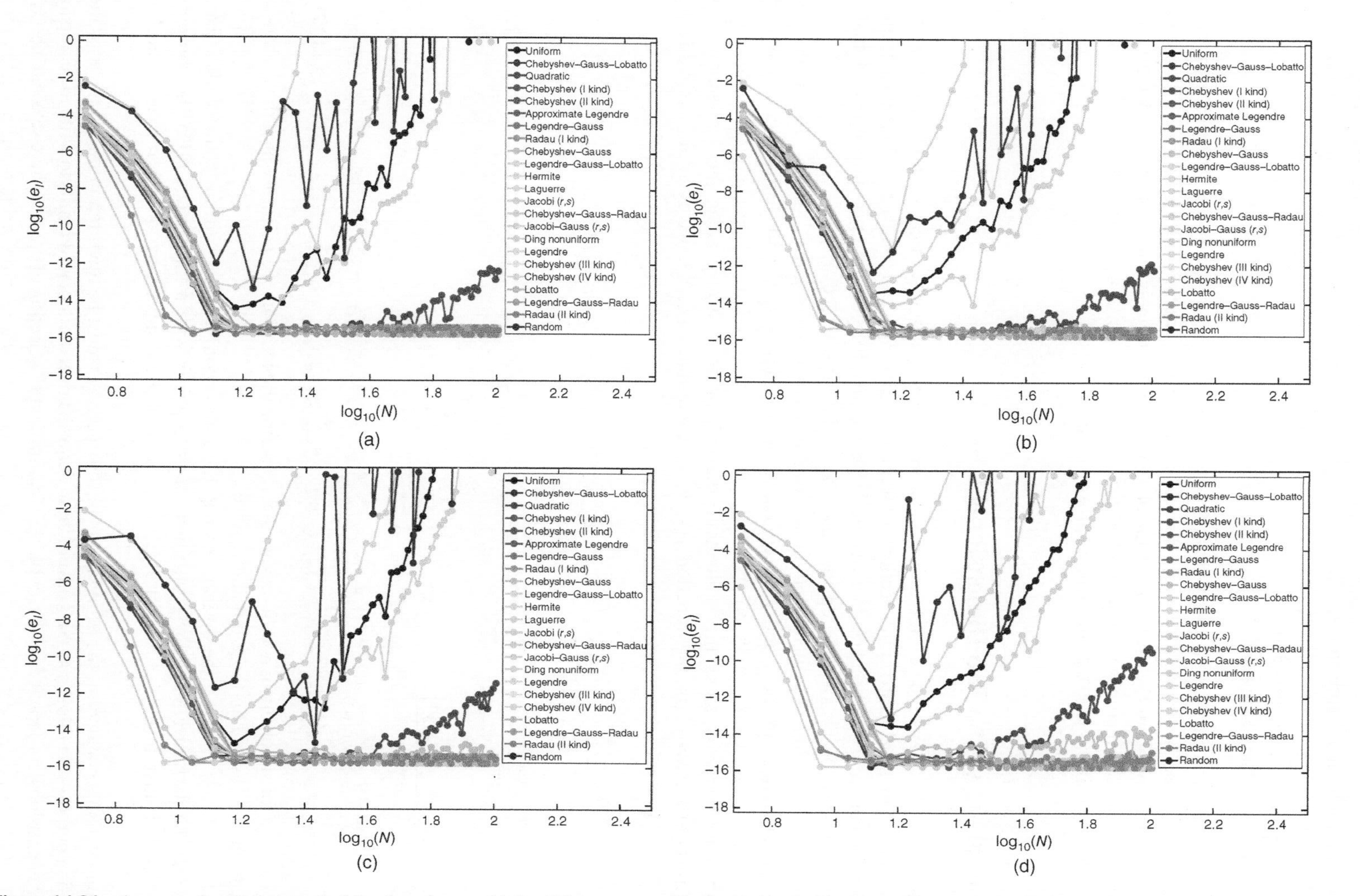

Figure 14.21 Accuracy for the integral of the function cos(x) for different approaches based on GTIQ method. (a) Method 1. (b) Method 2. (c) Method 3. (d) Method 4.

(a) *Method 1*: Approach 1 and Expression 2;
(b) *Method 2*: Approach 1 and Expression 1;
(c) *Method 3*: Approach 2 and Expression 2;
(d) *Method 4*: Approach 2 and Expression 1.

A similar convergence behavior can be observed with respect to the GIQ method, although the numerical solution in the present case becomes more stable once the maximum level of accuracy is reached ($\approx 10^{-16}$). In other words, possible oscillations about the maximum value of accuracy are extremely small for the selected grid distributions, whereby the four methods provide similar plots. Finally, it is worth mentioning that such a level of accuracy is reached by using $m = 30$ for the Taylor approximation.

14.5 General View: The Two-Dimensional Case

A general extension of the DQ method to multidimensional applications is now discussed, at least for a regular discretization domain. More specifically, in this section, we want to give a simple and compact mathematical framework for the computation of weighting coefficients for derivatives and integrals in a two-dimensional domain. Thus, some preliminary concepts are repeated once again below, for the sake of clarity.

Let us consider a function $f(x, y)$ defined in a two-dimensional rectangular domain. This definition domain must be discretized in the x and y directions, while selecting a fixed grid distribution. Let N and M be the number of the discretization points along the x and y directions, respectively. Thus, to determine the nth order derivative along the x direction, the mth order derivative along the y direction, and the mixed $(n+m)$th derivative of the function $f(x, y)$ at coordinates (x_i, y_j), we refer to the following expressions

$$f_x^{(n)}(x_i, y_j) = \left.\frac{\partial^{(n)} f(x,y)}{\partial x^{(n)}}\right|_{x=x_i, y=y_j} = \sum_{k=1}^{N} \varsigma_{x(ik)}^{(n)} f_{(kj)} = \varsigma_{x(i1)}^{(n)} f_{(1j)} + \varsigma_{x(i2)}^{(n)} f_{(2j)} + \cdots + \varsigma_{x(iN)}^{(n)} f_{(Nj)} \tag{14.169}$$

$$f_y^{(m)}(x_i, y_j) = \left.\frac{\partial^{(m)} f(x,y)}{\partial y^{(m)}}\right|_{x=x_i, y=y_j} = \sum_{l=1}^{M} \varsigma_{y(jl)}^{(m)} f_{(il)} = \varsigma_{(j1)}^{(m)} f_{(i1)} + \varsigma_{(j2)}^{(m)} f_{(i2)} + \cdots + \varsigma_{(jM)}^{(m)} f_{(iM)} \tag{14.170}$$

$$\begin{aligned} f_{xy}^{(n+m)}(x_i, y_j) &= \left.\frac{\partial^{(n+m)} f(x,y)}{\partial x^{(n)} \partial y^{(m)}}\right|_{x=x_i, y=y_j} = \sum_{k=1}^{N} \varsigma_{x(ik)}^{(n)} \left(\sum_{l=1}^{M} \varsigma_{y(jl)}^{(m)} f_{(kl)} \right) \\ &= \varsigma_{x(i1)}^{(n)} (\varsigma_{y(j1)}^{(m)} f_{(11)} + \varsigma_{y(j2)}^{(m)} f_{(12)} + \cdots + \varsigma_{y(jM)}^{(m)} f_{(1M)}) + \varsigma_{x(i2)}^{(n)} (\varsigma_{y(j1)}^{(m)} f_{(21)} + \varsigma_{y(j2)}^{(m)} f_{(22)} + \cdots + \varsigma_{y(jM)}^{(m)} f_{(2M)}) \\ &\quad + \cdots + \varsigma_{x(iN)}^{(n)} (\varsigma_{y(j1)}^{(m)} f_{(N1)} + \varsigma_{y(j2)}^{(m)} f_{(N2)} + \cdots + \varsigma_{y(jM)}^{(m)} f_{(NM)}) \end{aligned} \tag{14.171}$$

where, $f(x_i, y_j) = f_{(ij)}$, for simplicity of notation. Note that i, j define the location of a point at which the derivative is computed, whereas k, l stand for the summation indices.

Based on Eqs. (14.169)–(14.171), the derivative along the x direction considers all the points with a fixed index j, whereas the derivative along the y direction refers to points with a fixed index i. On the contrary, the mixed order derivative includes all the grid points because of the compresence of both indices k and l in Eq. (14.171).

From a computational standpoint, the expressions (14.169)–(14.171) must be rewritten in matrix form for the implementation purposes. In this sense, it is convenient to define a novel order of grid points to write the compact form of these expressions. As depicted in Figure 14.22, a "column-type" selection of points is here selected, i.e. $(x_1, y_1), (x_2, y_1), \ldots, (x_N, y_1), (x_1, y_2), \ldots, (x_N, y_2), \ldots, (x_1, y_M), \ldots\ldots, (x_N, y_M)$ whose arrow refers to the selected order.

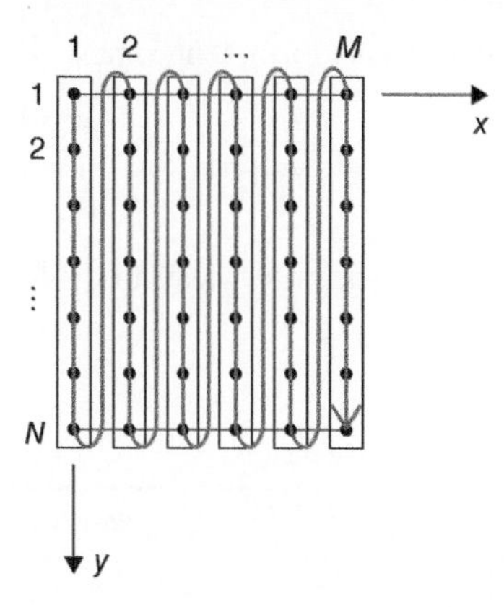

Figure 14.22 Grid point order adopted to define the vector $\boldsymbol{\pi}$.

These points are collected in a vector $\boldsymbol{\pi}$ as follows

$$\boldsymbol{\pi} = \left[\underbrace{(x_1,y_1)_1\ (x_2,y_1)_2\ \dots\ (x_N,y_1)_N}_{\text{first column}}\ \underbrace{(x_1,y_2)_{N+1}\ \dots\ (x_N,y_2)_{2N}}_{\text{second column}} \cdots \underbrace{(x_1,y_M)_{N\cdot M-N+1}\ \cdots\ (x_N,y_M)_{N\cdot M}}_{\text{last column}}\right]^T \tag{14.172}$$

The whole vector can be considered as M times a vector of length N, whose general element is expressed in the following compact form:

$$\pi_k = (x_i, y_j)_k \quad \text{for} \quad \begin{array}{l} i = 1, 2, \dots, N \\ j = 1, 2, \dots, M \\ k = i + (j-1)N \end{array} \tag{14.173}$$

Note that the sequence of points is not unique, but it can be defined arbitrarily in order to satisfy the computational necessities of the user. For example, a "row-type" order could be alternatively chosen, such that $\boldsymbol{\pi}$ would be defined as

$$\boldsymbol{\pi} = \left[\underbrace{(x_1,y_1)_1\ (x_1,y_2)_2\ \dots\ (x_1,y_M)_M}_{\text{first row}}\ \underbrace{(x_2,y_1)_{M+1}\ \dots\ (x_2,y_M)_{2M}}_{\text{second row}} \cdots \underbrace{(x_N,y_1)_{N\cdot M-M+1}\ \cdots\ (x_N,y_M)_{N\cdot M}}_{\text{last row}}\right]^T \tag{14.174}$$

In compact notation, it reads

$$\pi_k = (x_i, y_j)_k \quad \text{for} \quad \begin{array}{l} i = 1, 2, \dots, N \\ j = 1, 2, \dots, M \\ k = j + (i-1)M \end{array} \tag{14.175}$$

The same procedure could be applied for function $f(x, y)$. Since expressions (14.169)–(14.171) require that a two-dimensional function must be evaluated in all the points of the discretized domain of coordinates (x_i, y_j), the definition criterion of vector $\boldsymbol{\pi}$ can be used to order the functional values of the grid, i.e.

$$\mathbf{f} = \left[\underbrace{f(x_1,y_1)_1\ f(x_2,y_1)_2\ \dots\ f(x_N,y_1)_N}_{\text{first column}}\ \underbrace{f(x_1,y_2)_{N+1}\ \dots\ f(x_N,y_2)_{2N}}_{\text{second column}} \cdots \underbrace{f(x_1,y_M)_{N\cdot M-N+1}\ \cdots\ f(x_N,y_M)_{N\cdot M}}_{\text{last column}}\right]^T \tag{14.176}$$

In compact form, it becomes

$$f_k = f(x_i, y_j)_k \quad \text{for} \quad \begin{aligned} i &= 1, 2, \ldots, N \\ j &= 1, 2, \ldots, M \\ k &= i + (j-1)N \end{aligned} \tag{14.177}$$

A row-type order yields to the following definitions

$$\mathbf{f} = \left[\underbrace{f(x_1,y_1)_1 \; f(x_1,y_2)_2 \; \ldots \; f(x_1,y_M)_M}_{\text{first row}} \; \underbrace{f(x_2,y_1)_{M+1} \; \ldots \; f(x_2,y_M)_{2M}}_{\text{second row}} \ldots \underbrace{f(x_N,y_1)_{N\cdot M-M+1} \; \ldots \; f(x_N,y_M)_{N\cdot M}}_{\text{last row}} \right]^T \tag{14.178}$$

$$f_k = f(x_i, y_j)_k \quad \text{for} \quad \begin{aligned} i &= 1, 2, \ldots, N \\ j &= 1, 2, \ldots, M \\ k &= j + (i-1)M \end{aligned} \tag{14.179}$$

Note that vectors $\boldsymbol{\pi}$ and $\mathbf{f}$ have dimension $NM \times 1$. Once vector $\mathbf{f}$ is defined, a criterion for reorganizing the weighting coefficients of (14.169)–(14.171) must be found. The easiest and more compact way to tackle the problem is to use the Kronecker product $\otimes$, i.e. an operation between two square matrices, whose result is a block matrix.

If $\varsigma_x^{(n)}$ and $\varsigma_y^{(m)}$ indicate the weighting coefficient matrices for the derivatives along the x and y directions, respectively, for all the grid points, the block matrices read as follows

$$\underset{NM\times NM}{\mathbf{C}_x^{(n)}} = \underset{M\times M}{\mathbf{I}} \otimes \underset{N\times N}{\varsigma_x^{(n)}} \tag{14.180}$$

$$\underset{NM\times NM}{\mathbf{C}_y^{(m)}} = \underset{M\times M}{\varsigma_y^{(m)}} \otimes \underset{N\times N}{\mathbf{I}} \tag{14.181}$$

$$\underset{NM\times NM}{\mathbf{C}_{xy}^{(n+m)}} = \underset{M\times M}{\varsigma_y^{(m)}} \otimes \underset{N\times N}{\varsigma_x^{(n)}} \tag{14.182}$$

where $\mathbf{I}$ is the identity matrix. Derivatives of any order can be evaluated easily through the expressions (14.180)–(14.182). Thus, each row of matrices $\mathbf{C}_x^{(n)}$, $\mathbf{C}_y^{(m)}$, and $\mathbf{C}_{xy}^{(n+m)}$ represents the derivative approximation at an arbitrary point of coordinates (x_i, y_j).

The components of these matrices are labeled $C_{x(kl)}^{(n)}$, $C_{y(kl)}^{(m)}$, and $C_{xy(kl)}^{(n+m)}$, for $k, l = i + (j-1)N$ with $i = 1, 2, \ldots, N$ and $j = 1, 2, \ldots, M$. In view of a row-type order approach of points, the coefficient matrices defined in (14.180)–(14.182) assume the following form

$$\underset{NM\times NM}{\mathbf{C}_x^{(n)}} = \underset{N\times N}{\varsigma_x^{(n)}} \otimes \underset{M\times M}{\mathbf{I}} \tag{14.183}$$

$$\underset{NM\times NM}{\mathbf{C}_y^{(m)}} = \underset{N\times N}{\mathbf{I}} \otimes \underset{M\times M}{\varsigma_y^{(m)}} \tag{14.184}$$

$$\underset{NM\times NM}{\mathbf{C}_{xy}^{(n+m)}} = \underset{N\times N}{\varsigma_x^{(n)}} \otimes \underset{M\times M}{\varsigma_y^{(m)}} \tag{14.185}$$

Definitions (14.180)–(14.182), or the alternative expressions (14.183)–(14.185), allow us to rewrite Eqs. (14.169)–(14.171) as simple matrix products, i.e.

$$\mathbf{f}_x^{(n)} = \mathbf{C}_x^{(n)}\mathbf{f} \tag{14.186}$$

$$\mathbf{f}_y^{(m)} = \mathbf{C}_y^{(m)}\mathbf{f} \tag{14.187}$$

$$\mathbf{f}_{xy}^{(n+m)} = \mathbf{C}_{xy}^{(n+m)}\mathbf{f} \tag{14.188}$$

where $\mathbf{f}_x^{(n)}$, $\mathbf{f}_y^{(m)}$, and $\mathbf{f}_{xy}^{(n+m)}$ denote the vectors of size $NM \times 1$ that collect the derivative with respect to x, y, and both of them, for a function $f(x, y)$ at the selected grid points. The same approach can be followed for the integration of a function $f(x, y)$ in a two-dimensional regular domain. In this case, Eq. (14.131) referred to the integration of a one-dimensional function is easily extended to two-dimensional case, namely

$$\int_c^d \int_a^b f(x,y)dxdy = \sum_{i=1}^{N}\sum_{j=1}^{M} w_i^{1N} w_j^{1M} f(x_i, y_j) \tag{14.189}$$

where $x \in [a, b]$ and $y \in [c, d]$. The coefficients w_i^{1N} and w_j^{1M} denote the weighting coefficients for the mathematical integration along x and y, respectively.

By collecting the values of function $f(x_i, y_j)$ in any discrete grid point in the vector $\mathbf{f}$ defined by Eq. (14.176), it is possible to redefine the weighting coefficients in compact form as follows

$$\underset{1\times NM}{\mathbf{W}_{xy}} = \underset{1\times M}{\mathbf{W}_y} \otimes \underset{1\times N}{\mathbf{W}_x} \tag{14.190}$$

where $\mathbf{W}_x$ and $\mathbf{W}_y$ include the coefficients w_i^{1N} and w_j^{1M} for the integration of a function along the x and y directions, respectively. The definitions (14.176) and (14.190) can be used to define the double integral in (14.189) as a simple matrix product, i.e.

$$I = \mathbf{W}_{xy}\mathbf{f} \tag{14.191}$$

where I is the numerical value of the integral. The relevant advantage of expressions (14.186)–(14.188) and (14.191) from a computational standpoint is almost clear and will be discussed in the following paragraphs.

Some numerical applications are reported to demonstrate the accuracy of this method for the evaluation of derivatives and integrals of some two-dimensional functions in a regular domain. In detail, we consider the following functions

$$\begin{aligned} &f_1(x,y) = x^3 + x^2y^2 + y^3, \quad f_2(x,y) = \frac{1}{1-xy}, \quad f_3(x,y) = \cos(x)\sin(y), \\ &f_4(x,y) = \cos(x)\sinh(y), \quad f_5(x,y) = \tan(x)\tanh(y) \end{aligned} \tag{14.192}$$

defined in the domain $[0, \pi/4] \times [0, \pi/4]$, whose derivatives and double integrals are computed numerically by the use of several types of discretization. The accuracy of the differential and integral quadrature laws is discussed for all the approaches, increasing the number of grid points along N and M (up to 101 points along each direction). Since the best results are obtained for square domains with the same number of grid point along x and y, henceforth, we set $N = M$.

Once again, we check for the accuracy of derivatives by quantifying the following relative errors

$$\begin{aligned} e_{f_x^{(n)}} &= \left\| \frac{\mathbf{f}_{x\langle DQ\rangle}^{(n)} - \mathbf{f}_{x\langle exact\rangle}^{(n)}}{\left|\mathbf{f}_{x\langle exact\rangle}^{(n)}\right|} \right\|_\infty = \max_{\substack{i=1,2,\dots,N\\ j=1,2,\dots,M}} \left| \frac{\mathbf{f}_{x\langle DQ\rangle}^{(n)} - \mathbf{f}_{x\langle exact\rangle}^{(n)}}{\left|\mathbf{f}_{x\langle exact\rangle}^{(n)}\right|} \right| \\ e_{f_y^{(m)}} &= \left\| \frac{\mathbf{f}_{y\langle DQ\rangle}^{(m)} - \mathbf{f}_{y\langle exact\rangle}^{(m)}}{\left|\mathbf{f}_{y\langle exact\rangle}^{(m)}\right|} \right\|_\infty = \max_{\substack{i=1,2,\dots,N\\ j=1,2,\dots,M}} \left| \frac{\mathbf{f}_{y\langle DQ\rangle}^{(m)} - \mathbf{f}_{y\langle exact\rangle}^{(m)}}{\left|\mathbf{f}_{y\langle exact\rangle}^{(m)}\right|} \right| \quad \text{for } n,m = 1,2 \\ e_{f_{xy}^{(n+m)}} &= \left\| \frac{\mathbf{f}_{xy\langle DQ\rangle}^{(n+m)} - \mathbf{f}_{xy\langle exact\rangle}^{(n+m)}}{\left|\mathbf{f}_{xy\langle exact\rangle}^{(n+m)}\right|} \right\|_\infty = \max_{\substack{i=1,2,\dots,N\\ j=1,2,\dots,M}} \left| \frac{\mathbf{f}_{xy\langle DQ\rangle}^{(n+m)} - \mathbf{f}_{xy\langle exact\rangle}^{(n+m)}}{\left|\mathbf{f}_{xy\langle exact\rangle}^{(n+m)}\right|} \right| \end{aligned} \tag{14.193}$$

where $\mathbf{f}_{x\langle DQ\rangle}^{(n)}$, $\mathbf{f}_{y\langle DQ\rangle}^{(m)}$, and $\mathbf{f}_{xy\langle DQ\rangle}^{(n+m)}$ are the vectors of the numerical derivatives, whereas $\mathbf{f}_{x\langle exact\rangle}^{(n)}$, $\mathbf{f}_{y\langle exact\rangle}^{(m)}$, and $\mathbf{f}_{xy\langle exact\rangle}^{(n+m)}$ refer to the vector of the exact solutions in the same domain.

The derivatives of functions are computed under the selection of the Lagrange polynomials as basis functions. The weighting coefficients and the matrices $\mathbf{C}_x^{(n)}$, $\mathbf{C}_y^{(m)}$, and $\mathbf{C}_{xy}^{(n+m)}$ are determined for all the discretizations previously defined. Figures 14.23–14.27 describe the accuracy of the derivatives $f_x^{(1)}, f_y^{(1)}, f_x^{(2)}, f_y^{(2)}, f_{xy}^{(2)}$, and $f_{xy}^{(4)}$ of the functions in (14.192), while providing an estimate of the rate of convergence.

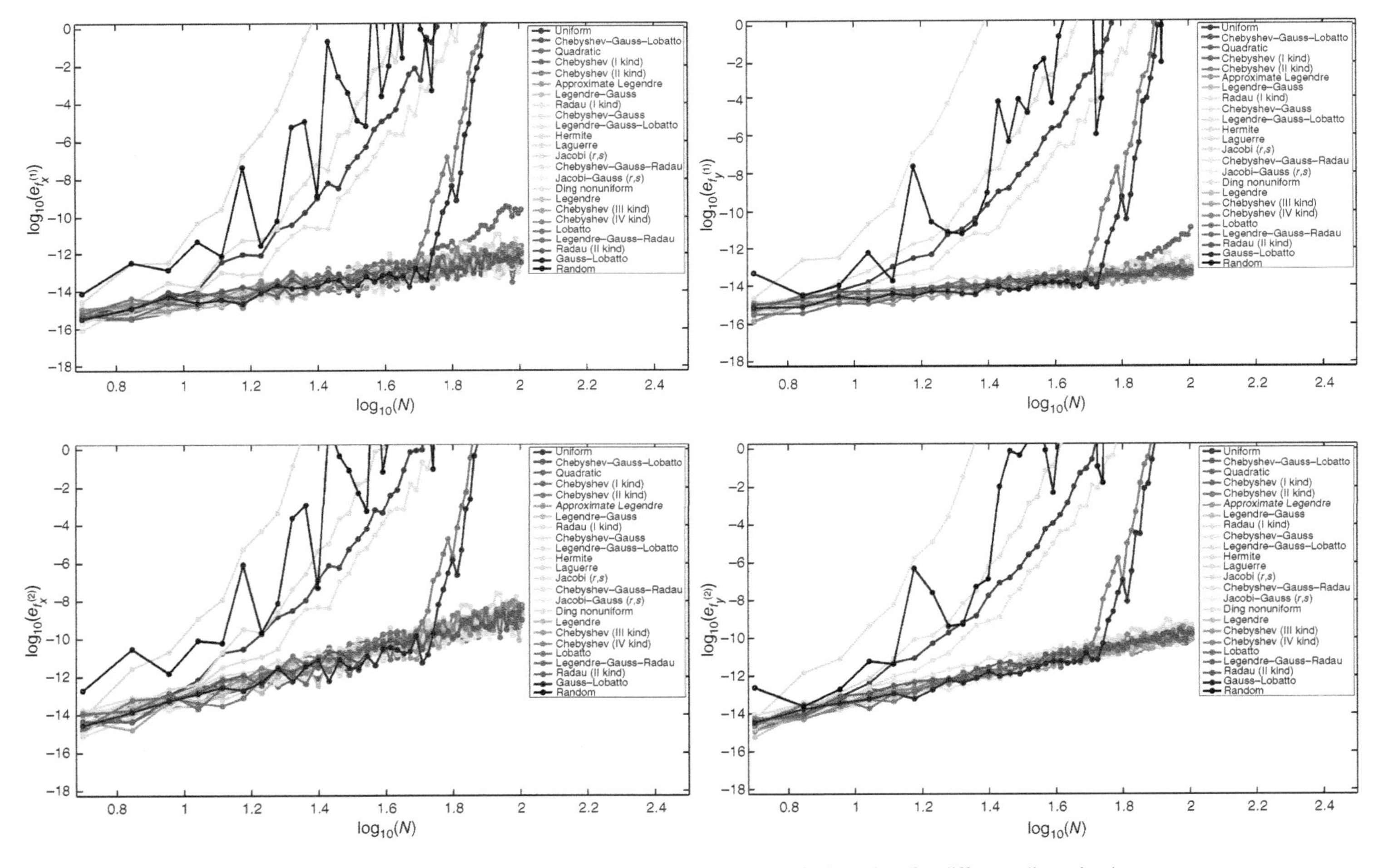

Figure 14.23 Accuracy of the numerical derivatives of the function $f_1(x, y)$ vs. the number of grid points for different discretizations.

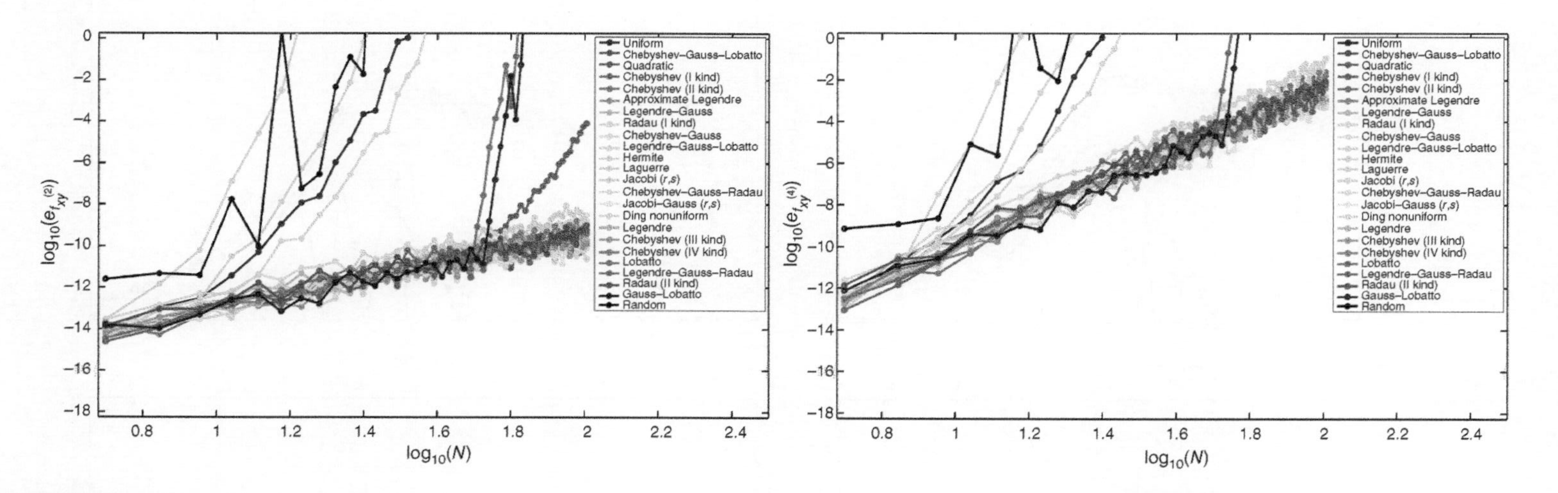

Figure 14.23 (*Continued*)

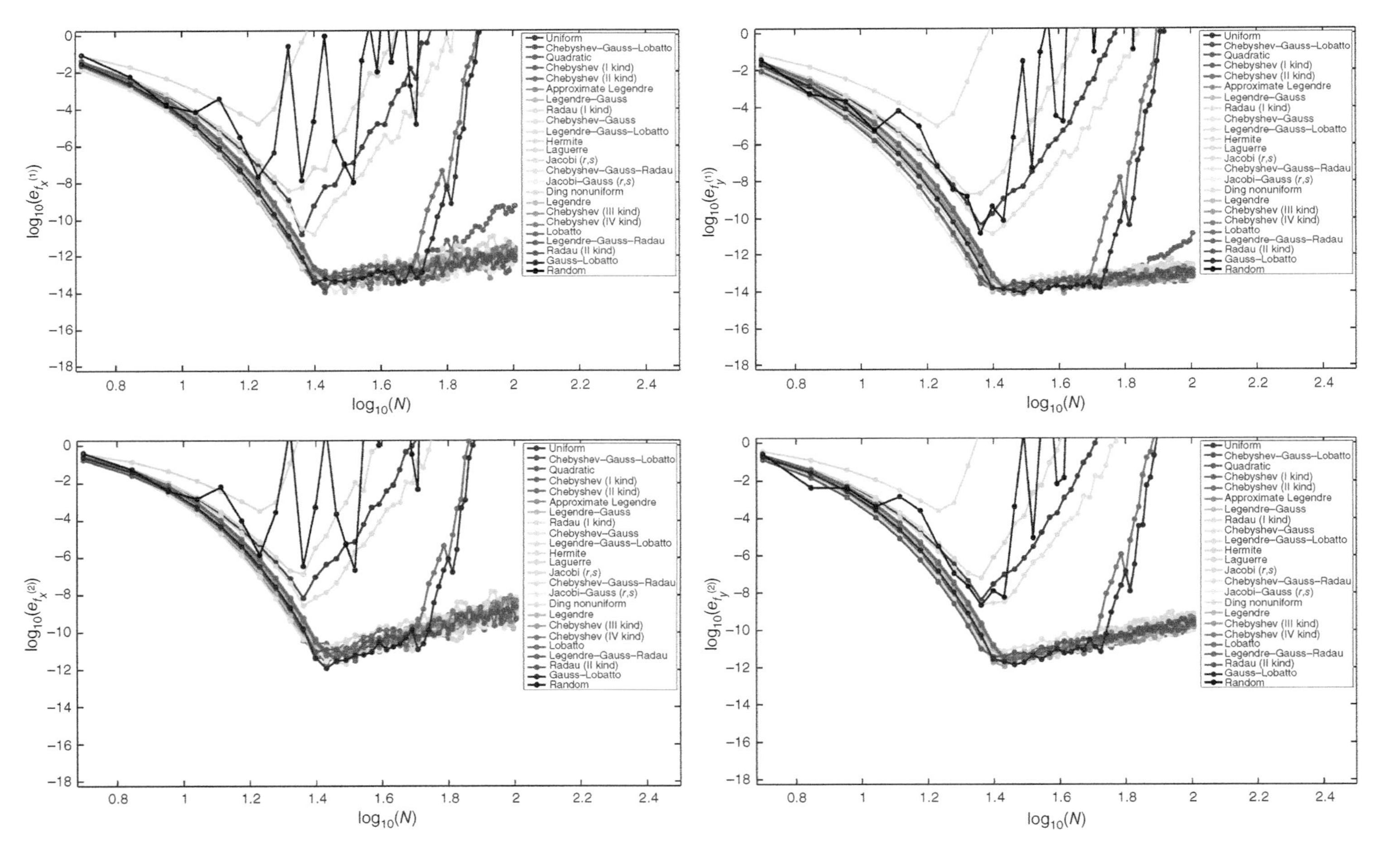

Figure 14.24 Accuracy of the numerical derivatives of the function $f_2(x, y)$ vs. the number of grid points for different discretizations.

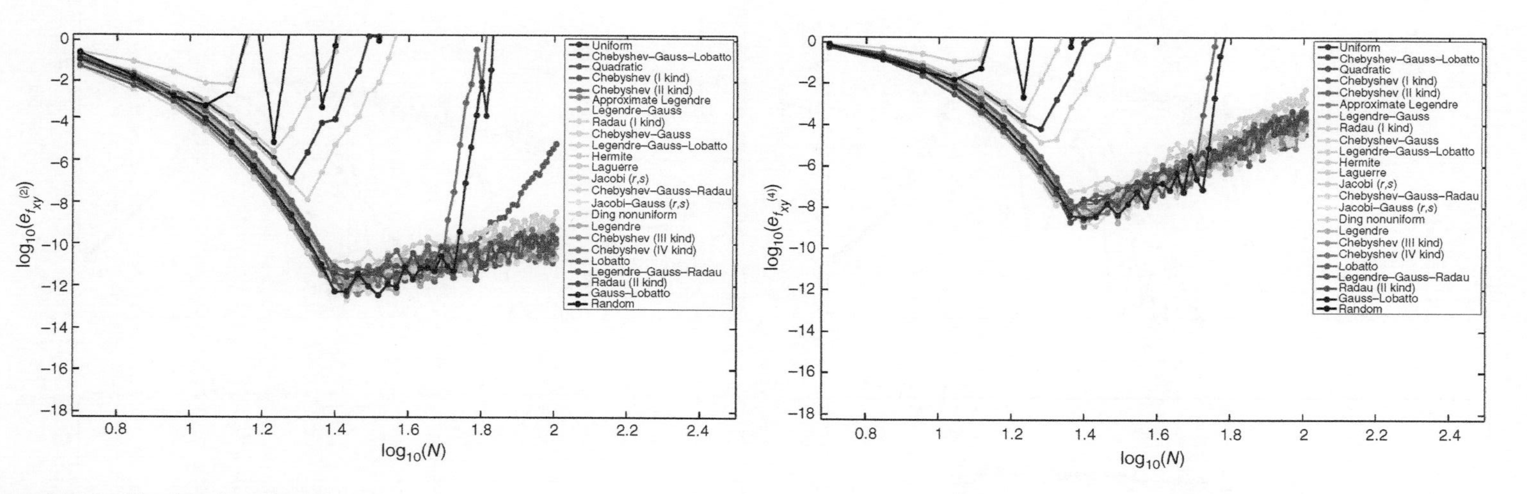

Figure 14.24 *(Continued)*

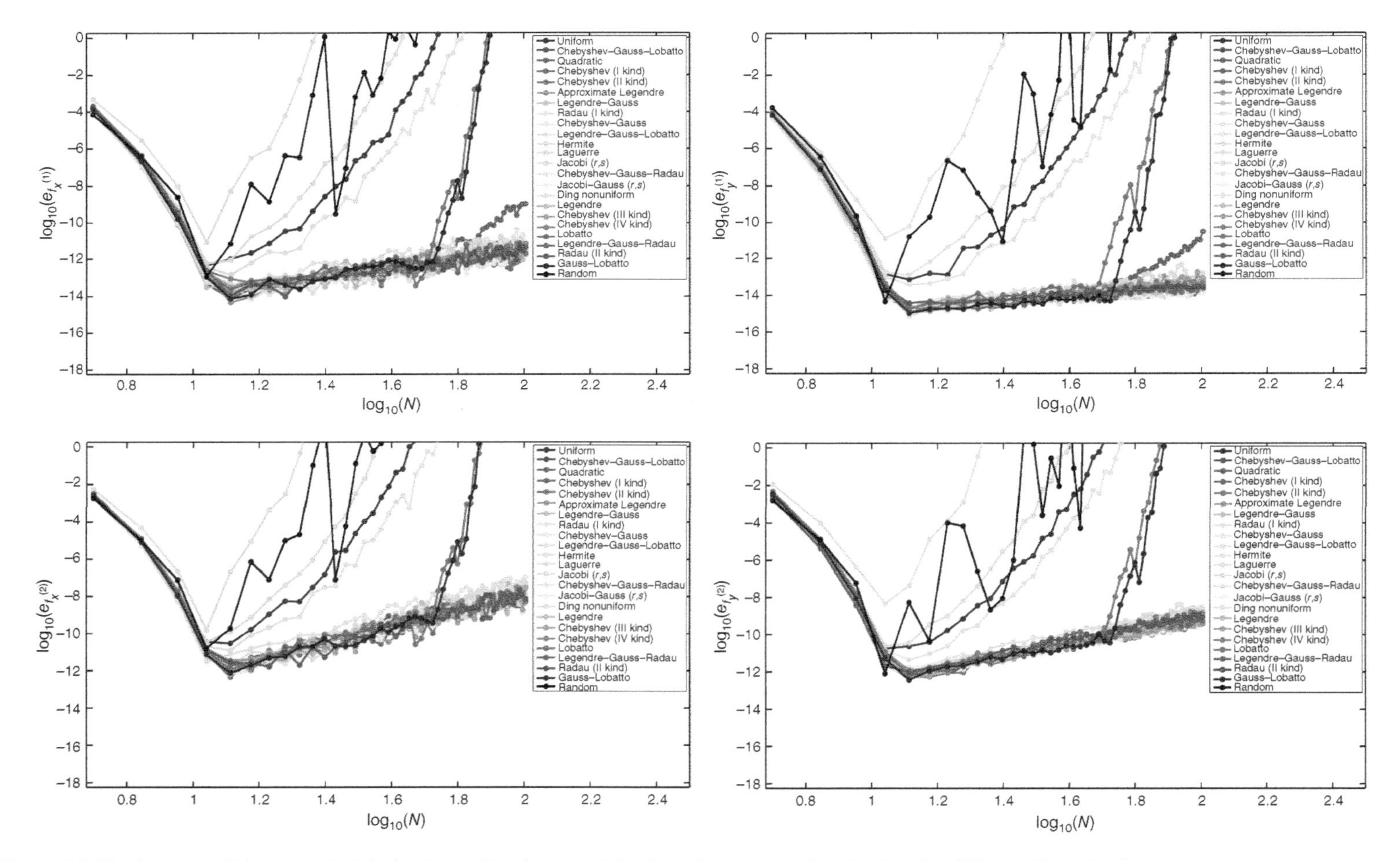

Figure 14.25 Accuracy of the numerical derivatives of the function $f_3(x, y)$ vs. the number of grid points for different discretizations.

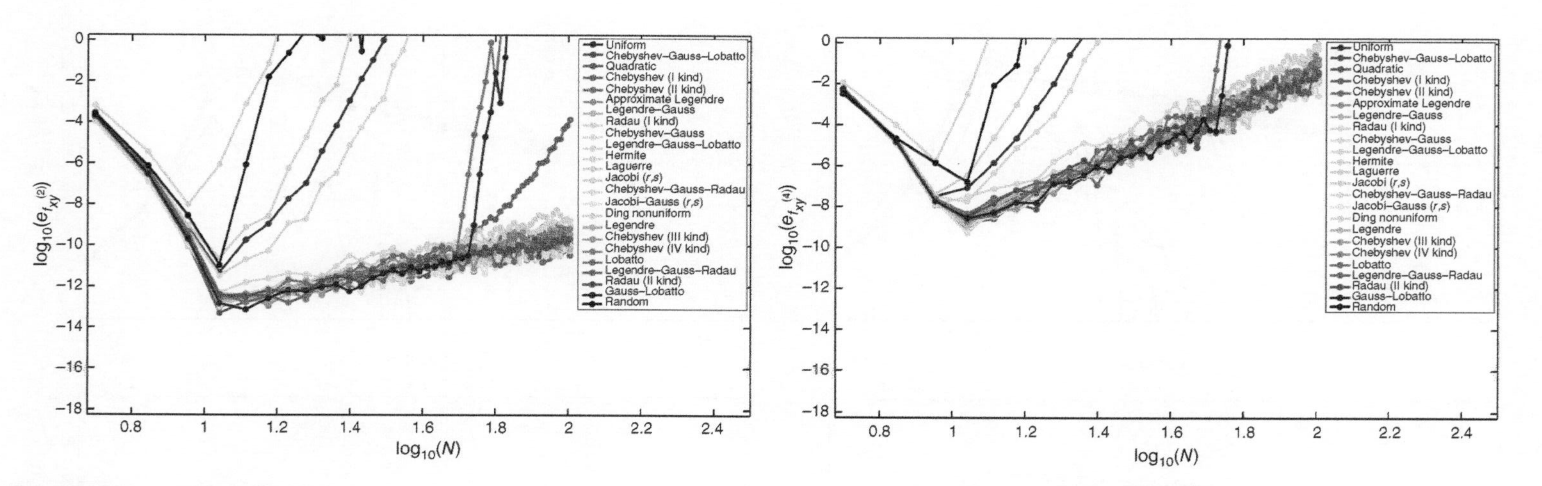

Figure 14.25 *(Continued)*

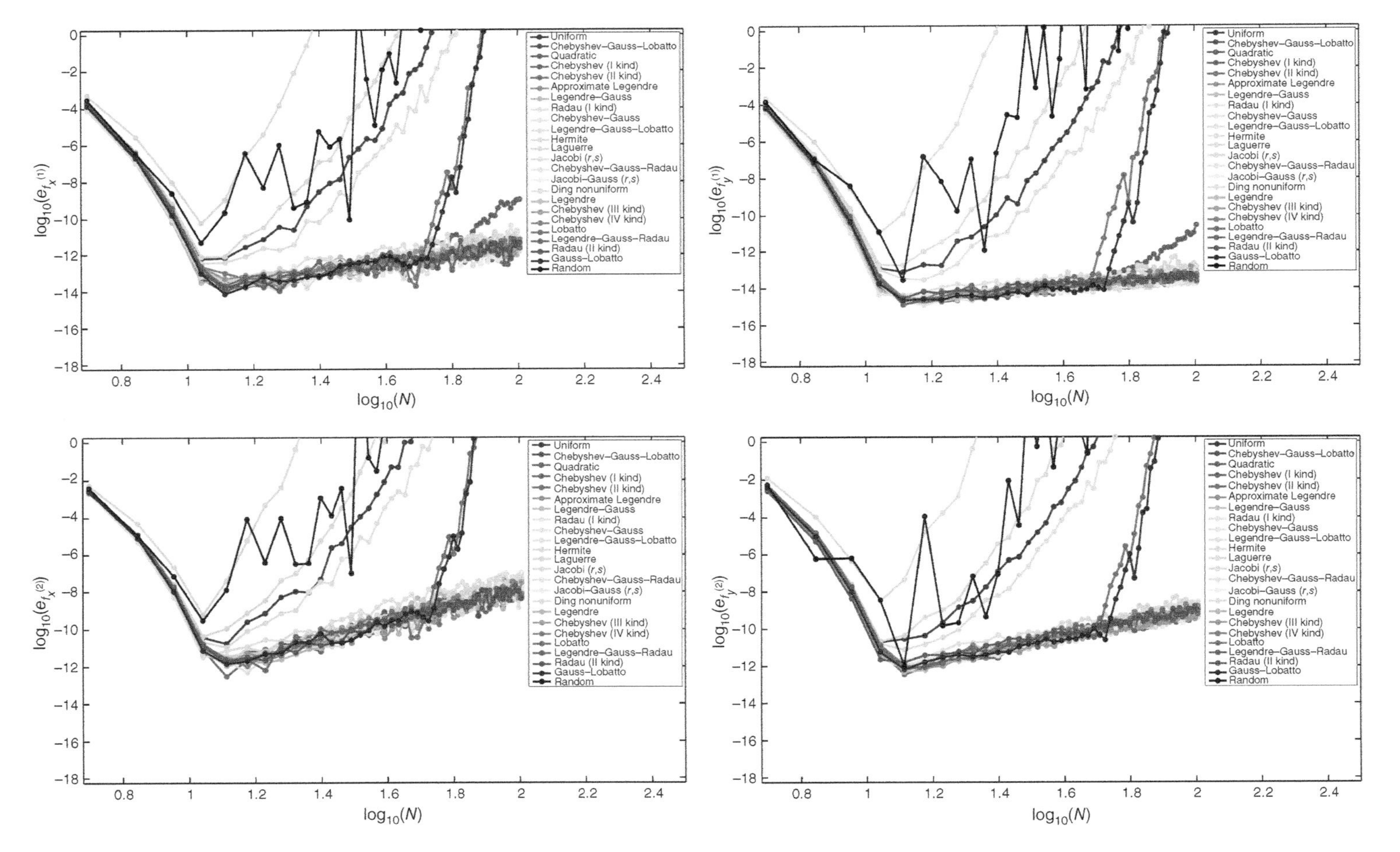

Figure 14.26 Accuracy of the numerical derivatives of the function $f_4(x, y)$ vs. the number of grid points for different discretizations.

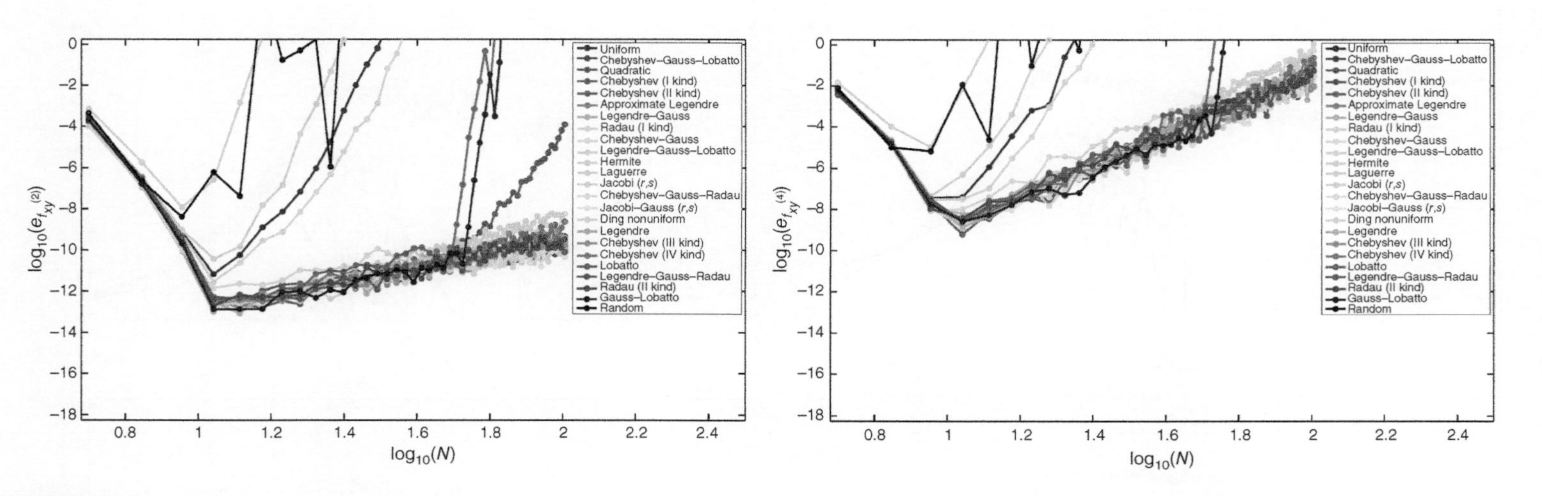

Figure 14.26 *(Continued)*

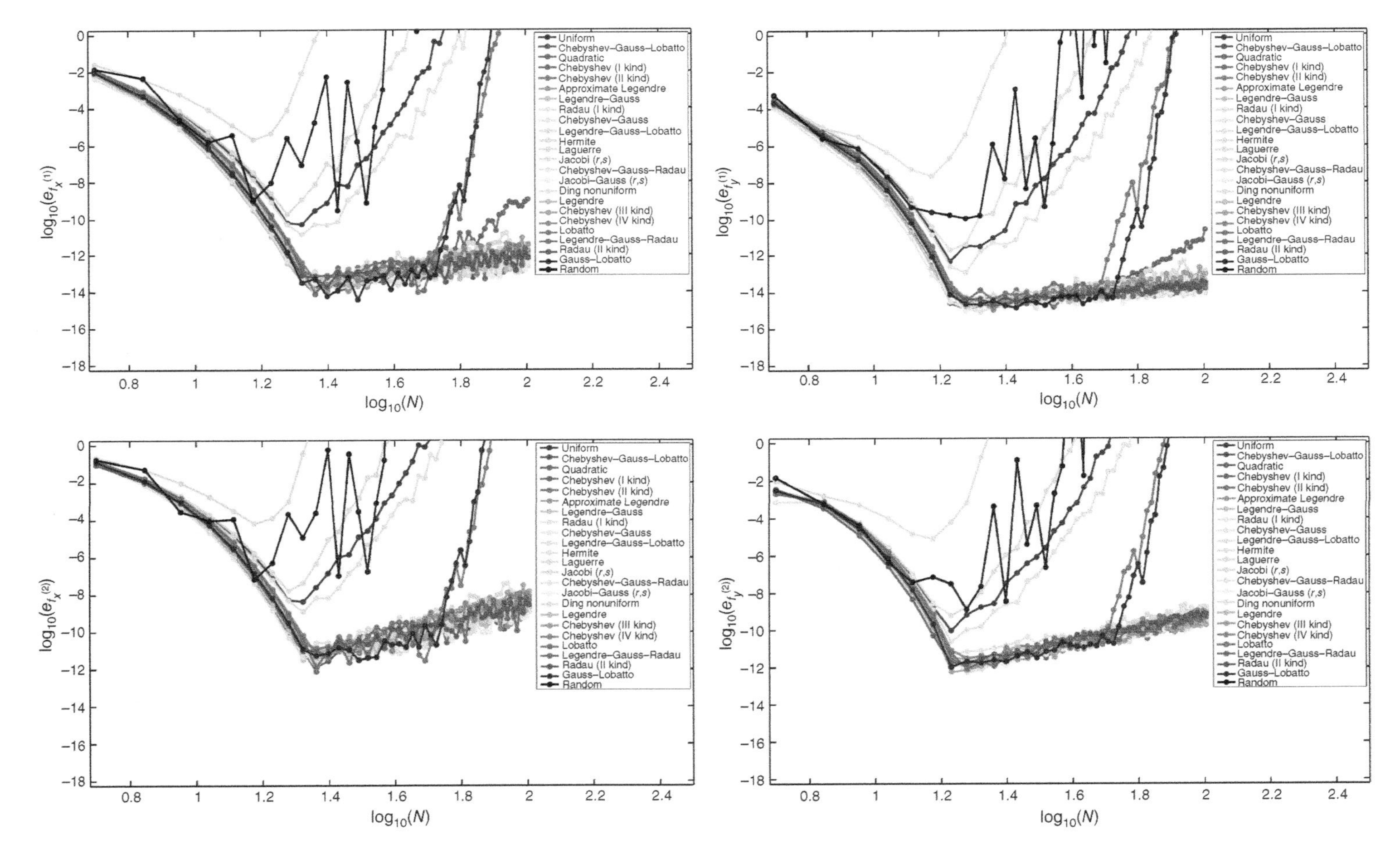

Figure 14.27 Accuracy of the numerical derivatives of the function $f_5(x, y)$ vs. the number of grid points for different discretizations.

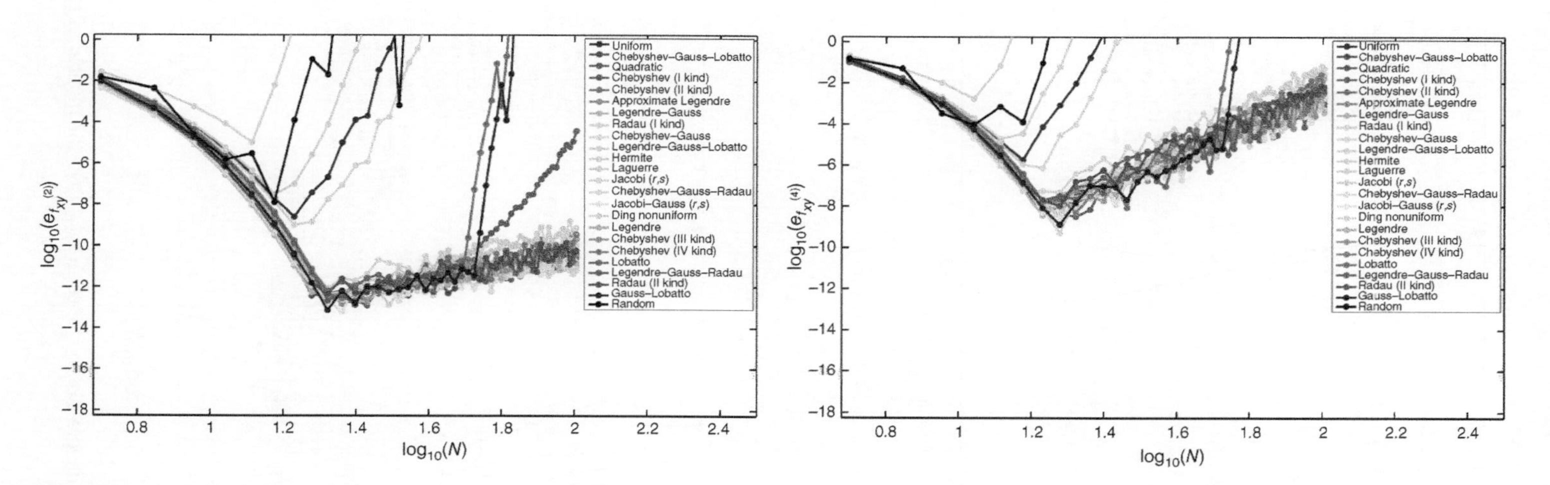

Figure 14.27 *(Continued)*

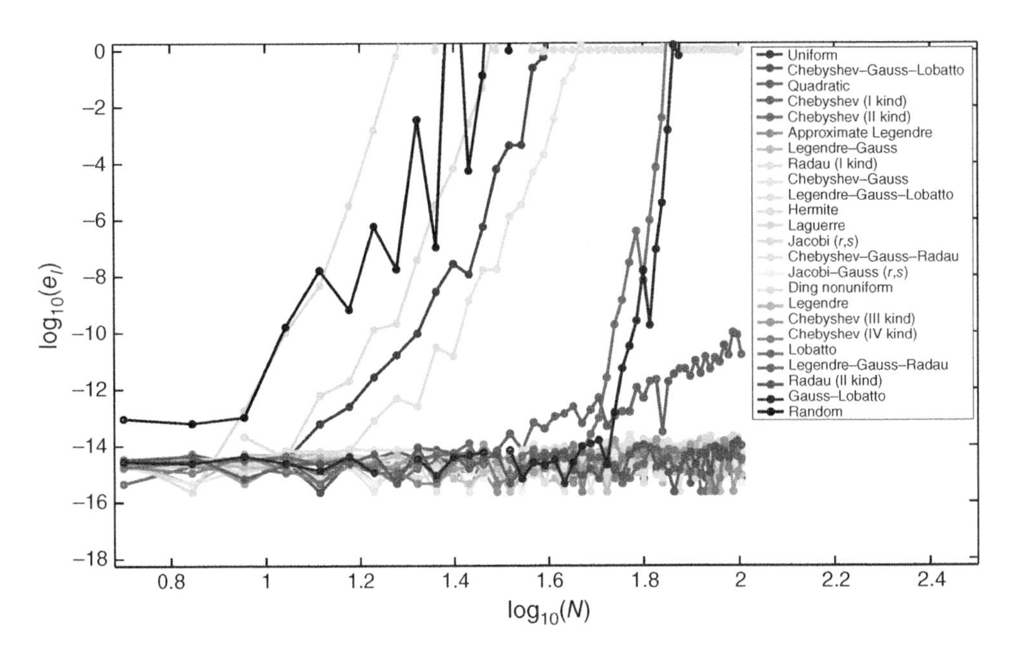

Figure 14.28 Accuracy of the numerical integrals of the function $f_1(x, y)$ vs. the number of grid points for different discretizations.

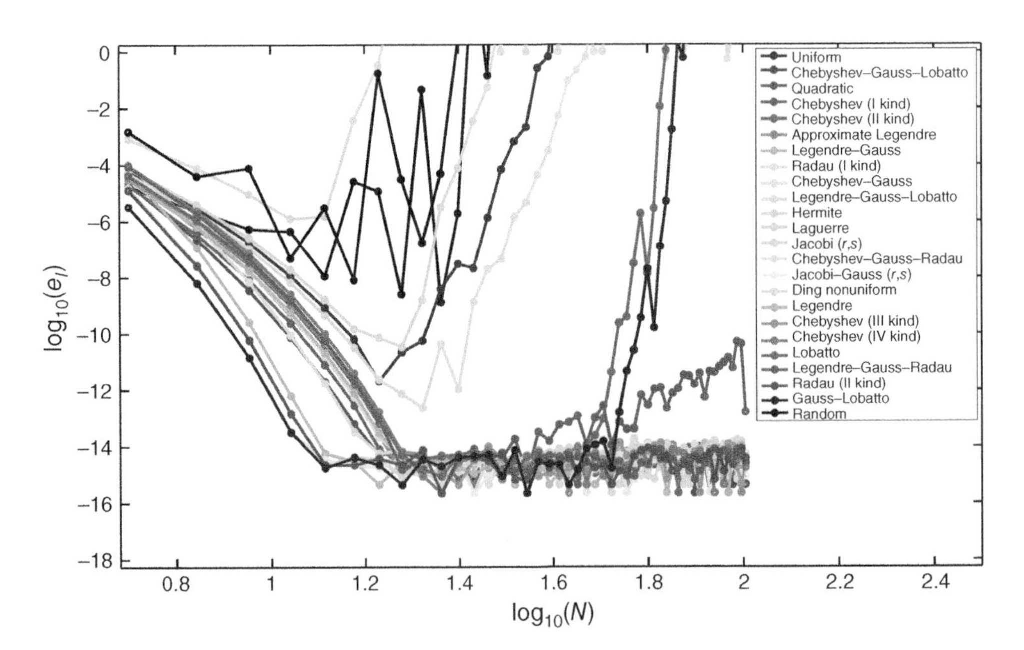

Figure 14.29 Accuracy of the numerical integrals of the function $f_2(x, y)$ vs. the number of grid points for different discretizations.

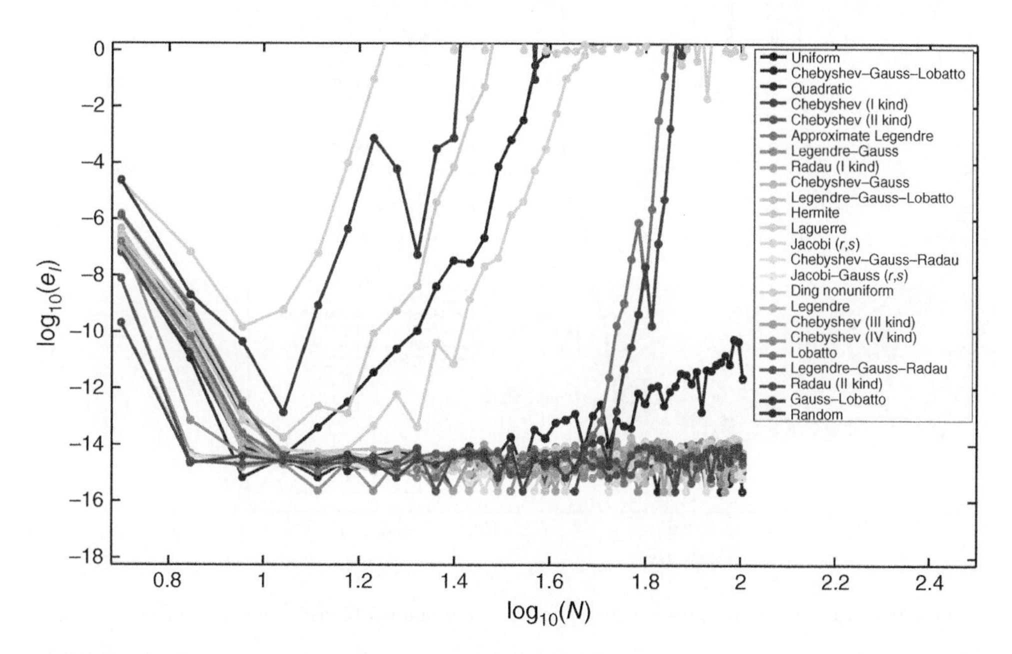

Figure 14.30 Accuracy of the numerical integrals of the function $f_3(x, y)$ vs. the number of grid points for different discretizations.

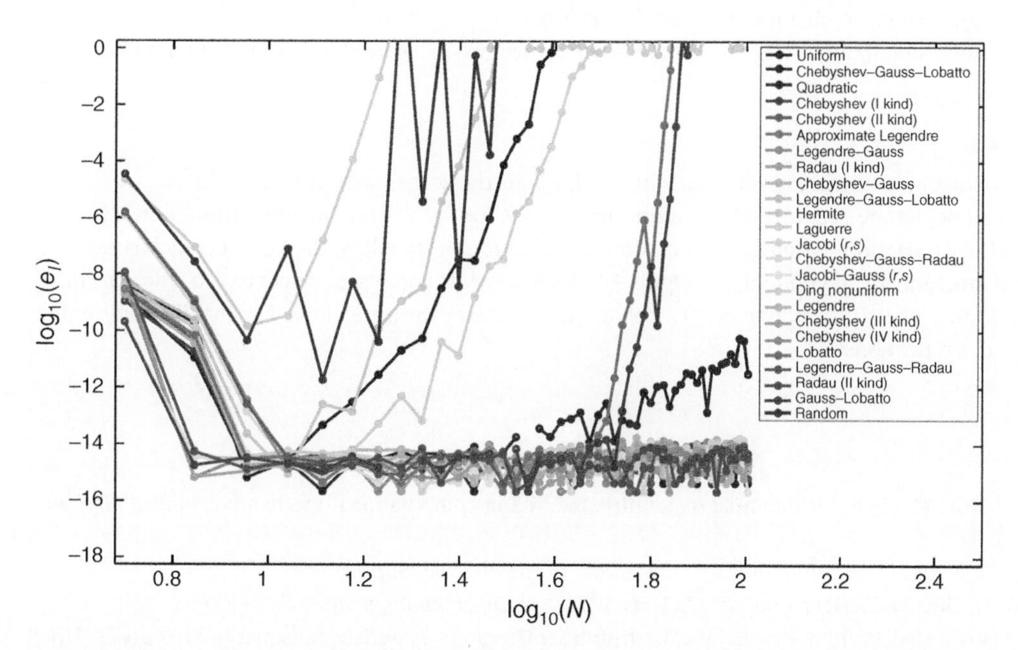

Figure 14.31 Accuracy of the numerical integrals of the function $f_4(x, y)$ vs. the number of grid points for different discretizations.

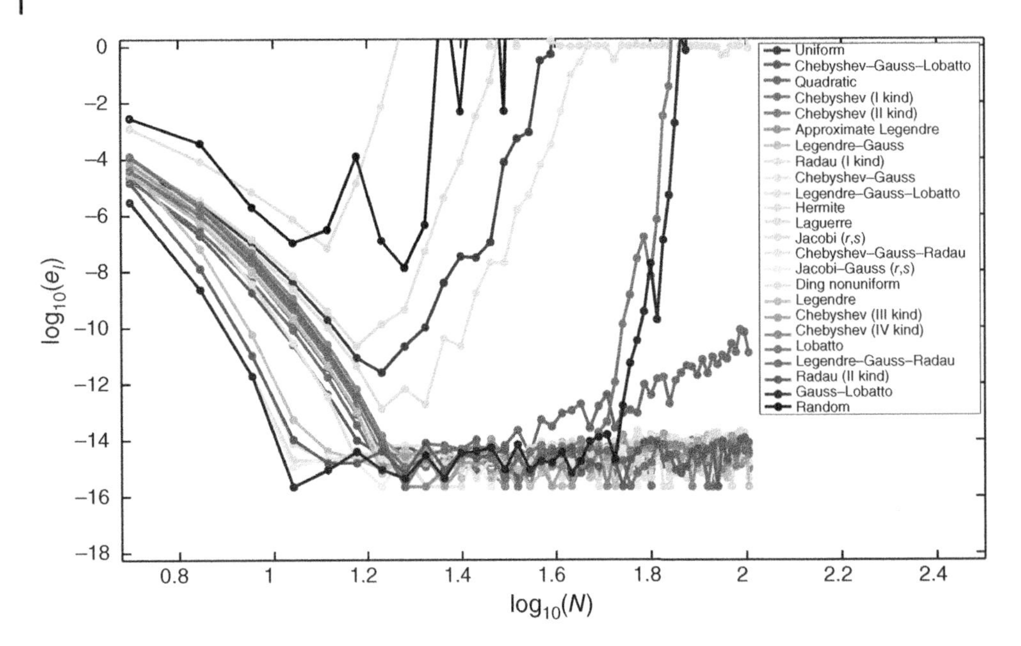

Figure 14.32 Accuracy of the numerical integrals of the function $f_5(x, y)$ vs. the number of grid points for different discretizations.

It is observed that the maximum accuracy for any derivative is reached with few grid points and a reduced computational effort. Moreover, it is easily noticed that the accuracy decreases for increasing derivative orders. The numerical solution features a high stability for all the discretizations considered. Nonpolynomial functions would require many grid points, to obtain the maximum accuracy.

As far as the two-dimensional numerical integration is concerned, we measure the accuracy as follows

$$e_I = \left| \frac{I_{\langle \mathrm{GIQ} \rangle} - I_{\langle \mathrm{exact} \rangle}}{I_{\langle \mathrm{exact} \rangle}} \right| \tag{14.194}$$

where $I_{\langle \mathrm{GIQ} \rangle}$ is the value of the numerical integral, whereas $I_{\langle \mathrm{exact} \rangle}$ is the corresponding exact value.

Figures 14.28–14.32 depict the accuracy of the numerical results for the double integral of functions (14.192) and several types of discretization. Based on a comparative evaluation of the plots, the best accuracy is reached when the integrand function is a polynomial (see Figure 14.28). A lot of grid points are necessary to reach the maximum accuracy of results, when the function is not a polynomial, whereby the problem suffers the ill-conditioning drawback for some discretizations.

References

1 Bert, C.W. and Malik, M. (1996). Differential quadrature method in computational mechanics. *Applied Mechanics Reviews* 49: 1–27.

2 Shu, C. (2000). *Differential Quadrature and Its Application in Engineering*. Springer.

3 Chen, C.N. (2006). *Discrete Element Analysis Methods of Generic Differential Quadratures*. Springer.

4 Tornabene, F. (2007). Modellazione e Soluzione di Strutture a Guscio in Materiale Anisotropo. PhD thesis. University of Bologna.

5 Zong, Z. and Zhang, Y.Y. (2009). *Advanced Differential Quadrature Methods*. CRC Press.

6 Tornabene, F. (2012). *Meccanica delle Strutture a Guscio in Materiale Composito*. Esculapio.

7 Tornabene, F. and Fantuzzi, N. (2014). *Mechanics of Laminated Composite Doubly-Curved Shell Structures. The Generalized Differential Quadrature Method and the Strong Formulation Finite Element Method*. Esculapio.

8 Tornabene, F., Fantuzzi, N., Ubertini, F., and Viola, E. (2015). Strong formulation finite element method based on differential quadrature: a survey. *Applied Mechanics Reviews* 67: 020801 (55 pages).

9 Tornabene, F., Fantuzzi, N., Bacciocchi, M., and Viola, E. (2015). *Strutture a Guscio in Materiale Composito I. Geometria Differenziale. Teorie di Ordine Superiore*. Esculapio.

10 Tornabene, F., Fantuzzi, N., Bacciocchi, M., and Viola, E. (2015). *Strutture a Guscio in Materiale Composito II. Quadratura Differenziale e Integrale. Elementi Finiti in Forma Forte*. Esculapio.

11 Wang, X. (2015). *Differential Quadrature and Differential Quadrature Based Element Methods Theory and Applications*. Elsevier.

12 Tornabene, F., Fantuzzi, N., Bacciocchi, M., and Viola, E. (2016). *Laminated Composite Doubly-Curved Shell Structures I. Differential Geometry. Higher-Order Structural Theories*. Esculapio.

13 Tornabene, F., Fantuzzi, N., Bacciocchi, M., and Viola, E. (2016). *Laminated Composite Doubly-Curved Shell Structures II. Differential and Integral Quadrature. Strong Formulation Finite Element Method*. Esculapio.

14 Tornabene, F. and Bacciocchi, M. (2018). *Anisotropic Doubly-Curved Shells. Higher-Order Strong and Weak Formulations for Arbitrary Shaped Shell Structures*. Esculapio.

15 Bellman, R. and Casti, J. (1971). Differential quadrature and long-term integration. *Journal of Mathematical Analysis and Applications* 34: 235–238.

16 Civan F. (1978). Solution of Transport Phenomena type models by the method of differential quadratures as illustrated on the LNG vapor dispersion process modeling. PhD thesis. University of Oklahoma.

17 Björck, A. and Pereyra, V. (1970). Solution of Vandermonde systems of equations. *Mathematics of Computation* 24: 112.

18 Quan, J.R. and Chang, C.T. (1989). New insights in solving distributed system equations by the quadrature method—I. Analysis. *Computers and Chemical Engineering* 13 (7): 779–788.

19 Shu C. (1991). Generalized differential-integral quadrature and application to the simulation of incompressible viscous flows including parallel computation. PhD thesis. University of Glasgow, Glasgow, Scotland.

20 Shu, C. and Richards, B. (1992). Application of generalized differential quadrature to solve two-dimensional incompressible Navier-Stokes equations. *International Journal for Numerical Methods in Fluids* 15: 791–798.

21 Shu, C. and Chew, Y.T. (1998). On the equivalence of generalized differential quadrature and highest order finite difference scheme. *Computer Methods in Applied Mechanics and Engineering* 155 (3–4): 249–260.

22 Shu, C. and Xue, H. (1997). Explicit computation of weighting coefficients in the harmonic differential quadrature. *Journal of Sound and Vibration* 204 (3): 549–555.

15

Brain Activity Reconstruction by Finding a Source Parameter in an Inverse Problem

Amir H. Hadian-Rasanan[1] *and Jamal Amani Rad*[2]

[1] *Institute for Cognitive and Brain Sciences, Shahid Beheshti University, Evin, Tehran, Iran*
[2] *Department of Cognitive Modeling, Institute for Cognitive and Brain Sciences, Shahid Beheshti University, Tehran, Iran*

15.1 Introduction

Brain activity reconstruction is the field of studying and reconstructing human brain activities from imaging data such as functional magnetic resonance imaging (fMRI), electroencephalography (EEG), magnetoencephalography (MEG), and so on. This field of neuroscience has attracted a lot of attention recently and different tools and algorithms have been developed for this purpose. On the other hand, because deep learning methods obtained outperform results in different problems of science, some researchers employed these methods for their problems and found out that a good tool for reconstruction of the brain activities are deep learning methods. Despite the deep learning methods have good accuracy and are so useful, all of them have a major problem, which is considering the learning algorithm as a black box. Moreover, there is no meaningful way that can explain "Why do the employed deep learning methods get good results?". Fortunately, this major problem can be solved by using a suitable mathematical model. There are various mathematical tools for modeling this phenomenon. One of the most useful mathematical models for this purpose is inverse model. As a definition of an inverse problem, we can remark that this is a mathematical framework that is used to extract information about a (unknown) system from the observed treatments of the system. In fact, the problem of discovering information about source of the electrical currents from EEG data can be treated as inverse problem. Figure 15.1 shows that how the brain activities can be considered as an inverse model.

The procedure of measuring electrical currents and applying some preprocessing algorithm on this raw data to prepare these data to monitoring is called forward problem. On the other hand, the procedure of estimating neural activity sources corresponding to a set of measured data is called inverse problem. It is worth to mention that in the field of brain activity reconstruction of solving both forward and inverse problem is very related to each other because if the forward problem does not solve accurately, the measured data does not have enough accuracy and we are not able to solve the inverse problem accurately by using this noisy data.

In addition to the brain activity reconstruction, inverse problems and especially determining a source parameter in an inverse problem from additional information have many applications in different fields of science and engineering. Generally, source parameter evaluating in a parabolic partial differential equations from additional information has been employed to specify the unknown attributes of a special region by evaluating the properties of its boundary or specific area of the domain. These unknown attributes contain valuable information about the (unknown) system, but measuring them directly is expensive and impossible [1].

Mathematical Methods in Interdisciplinary Sciences, First Edition. Edited by Snehashish Chakraverty.

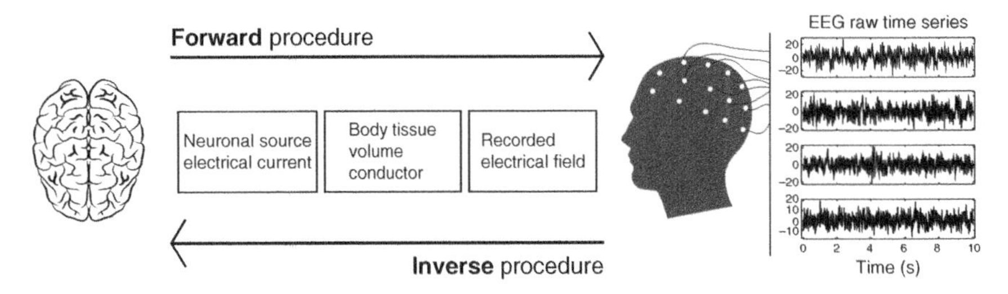

Figure 15.1 The graph of problem modeling procedure for the brain activities.

Despite of the applicability of the inverse problems and source parameter evaluation, these problems have some challenging aspects including solution existence, solution uniqueness, and instability of the solution [2]. In the following sentences we explain each of these major problems briefly:

1) *Existence of the solution*: Since the measured data contain noise and moreover in the process of modeling of the phenomena, some simplifying and approximation is added to the problem, so it is possible that the model does not fit to the data.
2) *Uniqueness of the solution*: If the model fits to the data, the existence of the solution is satisfied, but the second major is uniqueness of the solution, which means there may be more than one solution for the model. This problem usually occurs in the potential field models.
3) *Ill-posedness of the solution*: The nature of the inverse problems is ill-posed, which means that small changes in the measured data can cause big changes in the results. In order to overcome this major problem, some regularization algorithms can be employed to prevent obtaining an ill-conditioned system of linear algebraic equations from the inverse problem.

In Section 15.1.1, we illustrate the problem that is considered to be solved in this chapter.

15.1.1 Statement of the Problem

In this chapter, we study the following nonlinear inverse problem of simultaneously finding unknown function $w(\mathbf{x}, t)$ and unknown source parameter $p(t)$ from the following parabolic equation:

$$\frac{\partial w(\mathbf{x}, t)}{\partial t} = \Delta w(\mathbf{x}, t) + p(t)w(\mathbf{x}, t) + \phi(\mathbf{x}, t), \quad t \in [0, T], \quad \mathbf{x} \in \Omega \subset R^d, d = 1, 2, 3 \tag{15.1}$$

with initial condition

$$w(\mathbf{x}, 0) = f(\mathbf{x}), \quad \mathbf{x} \in \Omega \cup \partial\Omega \tag{15.2}$$

and boundary condition

$$w(\mathbf{x}, t) = h(\mathbf{x}, t), \quad \mathbf{x} \in \partial\Omega, 0 \leq t \tag{15.3}$$

subject to an over specialization at a point in the spatial domain

$$w(\mathbf{x}^*, t) = E(t), \quad 0 \leq t \tag{15.4}$$

In order to overcome to the problem of nonlinearity of the problem and finding two unknown functions simultaneously, a transformation is used. By employing a pair of transformations $r(t) = e^{-\int_0^t p(s)ds}$ and $u(\mathbf{x}, t) = r(t)w(\mathbf{x}, t)$, Eqs. (15.1)–(15.4) are written in the following form:

$$\frac{\partial u(\mathbf{x}, t)}{\partial t} = \Delta u(\mathbf{x}, t) + r(t)\phi(\mathbf{x}, t), \quad t \in [0, T], \quad \mathbf{x} \in \Omega \subset R^d, d = 1, 2, 3 \tag{15.5}$$

with initial condition

$$u(\mathbf{x}, 0) = f(\mathbf{x}), \quad \mathbf{x} \in \Omega \cup \partial\Omega \tag{15.6}$$

and boundary condition

$$u(\mathbf{x}, t) = r(t)h(\mathbf{x}, t), \quad \mathbf{x} \in \partial\Omega, 0 \leq t \tag{15.7}$$

subject to

$$u(\mathbf{x}^*, t) = r(t)E(t), \quad 0 \leq t \tag{15.8}$$

which yields

$$r(t) = \frac{u(\mathbf{x}^*, t)}{E(t)} \tag{15.9}$$

As you can see with this transformation, the nonlinear equation (15.1) is transformed to linear equation (15.5). Moreover, the role of $p(t)$ is transferred to $r(t)$. As mentioned in Section 15.1, , this problem has special importance in different fields of science. Therefore, many researchers developed different numerical algorithms to solve this problem. Some of these efforts are presented in Section 15.1.2.

15.1.2 Brief Review of Other Methods Existing in the Literature

Since finding a source parameter has a significant role in different fields of science, this problem has been attracted over the past decade. Many researchers analyze different aspects of this problem. For example, the existence and uniqueness of the solutions to these problems and also some more applications are discussed in [3–9]. Some authors developed different numerical algorithms for simulating this problem. These numerical methods include various classes of basic numerical algorithms such as finite difference, meshless methods, or spectral method. We summarized some numerical schemes that are used to solve this problem as follows:

1) *Finite difference methods*: Dehghan is a pioneer in using various finite difference schemes to solve this problem. About 2000, he developed different finite difference schemes and discussed about the stability and convergence of these schemes in [10–13]. He used these schemes for solving one or multidimensional problems. After one decade, Mohebbi and Dehghan developed some high-order schemes. They presented an algorithm based on hybrid of boundary value method and a high-order finite difference scheme in [14]. After that, in 2015, Mohebbi and Abbasi developed a compact finite difference scheme for this problem [1]. In addition, some other finite difference methods were presented by some other researchers [15–17]. Recently, the authors have developed a multigrid compact finite difference method to solve such a problem from this family in [18].
2) *Meshless methods*: Meshless methods are a powerful class of numerical algorithms that are used to solve different problems. One of the advantages of these algorithms is their ability of handling nonrectangular domains easily. Dehghan and Tatari used radial basis function method to solve one-dimensional version of this problem [19, 20]. Cheng used a moving least-square method for solving this problem [21]. Parzlivand and Shahrezaee developed a meshless algorithm based on inverse quadratic radial basis functions to solve this problem [22]. Recently, some new improvements on this problem have accomplished using meshless algorithms. For example: Shivanian and Jafarabadi developed a spectral meshless radial point algorithm to solve the multidimensional version of this problem. They also applied their algorithm for solving the problem on nonrectangular domains [23]. Also, a greedy version of meshless local Petrov-Galerkin (MLPG) method is employed to solve this problem by Takhtabnoos and Shirzadi [24].
3) *Spectral and pseudo-spectral methods*: The spectral methods are very famous for their stability and high convergence rate. These methods have been used frequently for solving different problems in science and engineering. In [25], a Tau method based on shifted Legendre polynomials has been employed to solve one-dimensional

version of this problem. Moreover, some other pseudo-spectral methods based on Legendre polynomials are used to solve one-dimensional [26] and multidimensional [27] version of the problem. Ritz Galerkin method [28] and Ritz least-squares method [29] are used for this problem by some other researchers. Additionally, Chebyshev cardinal functions [30], Cardinal Hermite collocation method [31], sinc-collocation method [32, 33], Bernstein Galerkin method [34], Cubic B-spline scaling functions [35], and some other spectral methods were developed by mathematicians. Recently, in 2019, Gholampoor and Kajani have developed a direct numerical method for solving this problem based on block-pulse/Legendre hybrid functions [36].

4) *Analytical and semianalytical methods*: These methods have used for solving different problems. Tatari and Dehghan developed Adomian decomposition method [37] for solving this problem in 2007. They also presented variational iteration method [38] for the same purpose in that year. Moreover, some other analytical and semianalytical methods have been employed for this problem including reproducing kernel space [39, 40], homotopy perturbation method, and homotopy analysis method [41, 42].

In addition to the above-mentioned methods, some other researchers used different methods for solving the inverse control parameter such as method of lines [43] and predictor-corrector method [44].

Weighted residual methods have attracted much attention in recent years. There are various types of the weighted residual methods such as Tau, Galerkin, Petrov–Galerkin, and collocation. Collocation algorithm is a powerful method based on the expansion of an unknown function by basis functions. Using different basis functions can help us to solve many different problems that have different dynamics [45–48]. In addition, Chebyshev polynomials and their various types such as rational and fractional are a suitable choice as basis functions. Many researchers have been using these polynomials in their approaches [49–56]. In this chapter, for spatial approximation, the Chebyshev collocation method is used, which yields a linear algebraic system of equations. On the other hand, in order to temporal approximation, θ-weighted finite difference scheme is used. The rest of this chapter is organized as follows. In Section 15.2, we introduce different parts of proposed method, which includes weighted residual method, collocation method, and function approximation using Chebyshev polynomials. In Section 15.3, we apply the proposed method on Eqs. (15.5)–(15.8), and in Section 15.4, we report the numerical experiments of solving Eqs. (15.1)–(15.4). Finally, a conclusion is given in Section 15.5.

15.2 Methodology

Before we explain how to solve the model in the form of Eqs. (15.5)–(15.8), we focus on the numerical tools that are used in this chapter. At first, the weighted residual method and collocation algorithm as a special case of the weighted residual method are introduced; afterward, the properties of Chebyshev polynomials and function approximation by using these polynomials are expressed.

15.2.1 Weighted Residual Methods and Collocation Algorithm

The spectral and pseudo-spectral methods are used frequently to solve many problems in different fields such as computational neuroscience [57], cognitive science [58], biology [59], fluid dynamics [60], engineering [61], and so on. There is no obvious historical start point for these methods but there may be the main idea that these methods are inspired by Fourier analysis method. In 1820, Navier employed an approach based on expanding the unknown function by sine functions for solving the Laplace equation. He expands the two value unknown function in the following form:

$$f(x, y) = \sum_{m=0}^{\infty} \sum_{n=0}^{\infty} f_{m,n} \sin(mx) \sin(ny) \tag{15.10}$$

in which $f_{m,n}$ can be obtained using the orthogonality of the sine functions. Because the computational cost of this method was high and does not have acceptable accuracy, this method was not used frequently in that time. By growing the available computational resources, these methods have been attracted very much and some researchers developed these methods in different aspects such as error estimation analysis, convergence and stability analysis, parallelism, etc. The weighted residual methods are powerful approach to approximate the solution of differential equations, which are the general form of the spectral methods. Before we explain the procedure and the basics of the weighted residual methods, we should remark the definition of the weight function. The definition of the weight function is presented in Definition 15.1.

Definition 15.1 The function $w(x)$ is called weight function over the interval $[0, \eta]$ whenever it has the following properties:

- $w(x)$ is integrable, which means $\int_0^\eta w(x)dx < \infty$.
- $w(x)$ is a positive function, which means for each $x \in [0, \eta]$, $0 \leq w(x)$.
- For each subinterval of $[0, \eta]$, $w(x) \neq 0$

In order to illustrate the weighted residual method, consider a multidimensional (nonlinear) partial differential equation as follows:

$$H[u(\mathbf{x})] = f(\mathbf{x}), \quad \mathbf{x} \in \Omega \subset R^d \tag{15.11}$$

with the boundary condition

$$K[u(\mathbf{x})] = g(\mathbf{x}), \quad \mathbf{x} \in \partial\Omega \subset R^d \tag{15.12}$$

where H is a (nonlinear) partial differential operator of the equation, K operates on the boundary conditions, Ω is the domain of the equation, and $\partial\Omega$ is the boundary of the domain. In order to approximate the solution of the above equation, we select $\{\phi^1_{i_1}(x_1)\}_{i_1=0}^{N_1-1}, \{\phi^2_{i_2}(x_2)\}_{i_2=0}^{N_2-1}, \ldots, \{\phi^d_{i_d}(x_d)\}_{i_d=0}^{N_d-1}$ as basis functions on the space of $\mathbf{x} = (x_1, x_2, \ldots, x_d)$ and assumed the solution as a finite summation of multiplication of basis functions. Therefore, we obtain $\overline{u}(\mathbf{x})$ as follows:

$$\overline{u}(\mathbf{x}) = \sum_{i_1=0}^{N_1-1} \sum_{i_2=0}^{N_2-1} \cdots \sum_{i_d=0}^{N_d-1} a_{i_1,i_2,\ldots,i_d} \phi^1_{i_1}(x_1)\phi^2_{i_2}(x_2) \ldots \phi^d_{i_d}(x_d) \tag{15.13}$$

By substituting Eq. (15.13) in Eqs. (15.11) and (15.12), the following equations are obtained, which are named residual functions:

$$R_h(\mathbf{x}) = H[\overline{u}(\mathbf{x})] - f(\mathbf{x}) \neq 0, \quad \mathbf{x} \in \Omega \subset R^d \tag{15.14}$$

$$R_k(\mathbf{x}) = K[\overline{u}(\mathbf{x})] - g(\mathbf{x}) \neq 0, \quad \mathbf{x} \in \partial\Omega \subset R^d \tag{15.15}$$

because the approximated solution and the exact solution are not equal, so the residuals are not equal to zero, but the weighted residual method will force the residual to minimize as much as possibly closer to zero.

To obtain unknown coefficients $\{a_{i_1,i_2,\ldots,i_d}\}_{i_1=0,i_2=0,\ldots,i_d=0}^{N_1-1,N_2-1,\ldots,N_d-1}$ weighted inner product of Eqs. (15.15) and (15.16) which is used by a set of test functions $\{\psi^1_{i_1}(x_1)\}_{i_1=0}^{N_1-1}, \{\psi^2_{i_2}(x_2)\}_{i_2=0}^{N_2-1}, \ldots, \{\psi^d_{i_d}(x_d)\}_{i_d=0}^{N_d-1}$. So we have

$$\langle \ldots \langle \langle R_h, \psi^1_{i_1} \rangle_{w_1}, \psi^2_{i_2} \rangle_{w_2} \ldots, \psi^d_{i_d} \rangle_{w_d} = \int_\Omega \cdots \int_\Omega \int_\Omega R_h \psi^1_{i_1} w_1 \psi^2_{i_2} w_2 \cdots \psi^d_{i_d} w_d \, dx_1 \, dx_2 \cdots dx_d \tag{15.16}$$

$$\langle \ldots \langle \langle R_k, \psi^1_{i_1} \rangle_{w_1}, \psi^2_{i_2} \rangle_{w_2} \ldots, \psi^d_{i_d} \rangle_{w_d} = \int_{\partial\Omega} \cdots \int_{\partial\Omega} \int_{\partial\Omega} R_k \psi^1_{i_1} w_1 \psi^2_{i_2} w_2 \cdots \psi^d_{i_d} w_d \, dx_1 \, dx_2 \cdots dx_d \tag{15.17}$$

where w_i are positive weight functions. The choice of test functions determines the type of spectral methods. Here, a brief review on the different types of the weighted residual method is presented by considering $R(\mathbf{x})$ as a residual function:

1) *Collocation method*: In the collocation method, the test function is Dirac delta function, and because of some properties in this test function, it yields that the residual function should be forced to zero at some points of the domain.

$$R(\mathbf{x}_i) = 0, \quad \text{for some} \quad \mathbf{x}_i \in \Omega \tag{15.18}$$

2) *Subdomain method*: In this type of weighted residual method, the domain is split into some subdomain parts and the test function in the each subdomain is considered equal to 1.

$$\int_{\Omega} R(\mathbf{x})\psi_i(x_i)d\mathbf{x} = \sum \int_{\Omega_i} R(\mathbf{x})d\mathbf{x} \tag{15.19}$$

Therefore, this method forces the residual function to be zero over the all subdomains.

$$\int_{\Omega_i} R(\mathbf{x})d\mathbf{x} = 0 \tag{15.20}$$

3) *Least-squares method*: The least-squares method is the oldest of all the methods of weighted residuals. The history of using least-squares method returns to 1775 when Gauss introduced this method. In this method, the residual function is considered as test function, so the problem is reduced to minimizing the following expression:

$$\min J = \int_{\Omega} R^2(\mathbf{x})d\mathbf{x} \tag{15.21}$$

4) *Moment method*: In the moment method, the test functions are considered equal to the family of single-term polynomials. Therefore, in this method, $1, x, x^2, \ldots$ are considered as test function. This method has some major problems, for example, it is ill-conditioned.
5) *Galerkin method*: The Galerkin method is a modification of the least-squares method. In this method, the basis functions are considered as test function. In 1915, Galerkin introduced this method for the first time [62].

In the above-mentioned methods recently, some new pseudo-spectral methods based on a new family of Lagrange functions have been introduced [63, 64]. In order to solve our model, we employed the collocation algorithm. In the following, this algorithm is explained as a special case of weighted residual method in details.

The collocation algorithm is the simplest and most common weighted residual method that is used to solve many problems in different areas of science. In 1934, Salter and Kantrovic employed this method for the first time. Afterward, about three years later, Frazer, Jones, and Skan used this algorithm for approximating the solution of ordinary differential equations. Lanczos illustrated the significance of choosing basis functions and collocation points in 1938. From 1957 till 1964, some studies were done by Clenshaw, Norton, and Wright, which cased to revive this method. As mentioned, the collocation method is obtained by choosing Dirac delta function as the test function. Dirac delta function has some special properties, which should be remarked. The definition and some essential properties of Dirac delta function are presented.

Remark 15.1 The well-known Dirac delta function is defined as follows:

$$\delta(x - x_i) = \begin{cases} 1, & x = x_i \\ 0, & \text{otherwise} \end{cases}.$$

Moreover, this function has two useful properties listed here:

1) The integration of Dirac delta function over the interval $[x_i - c, x_i + c]$ is equal to 1,

$$\int_{x_i-c}^{x_i+c} \delta(x - x_i)dx = 1, \quad c \to 0$$

2) The integration of multiplication Dirac delta function and another function over the domain is equal to the value of that function in the x_i,

$$\int_{\Omega} f(x)\delta(x - x_i)dx = f(x_i)$$

This method is a member of weighted residual methods, which force the residual to zero at some points of the domain. As a result of choosing $(x_{j_1}, x_{j_2}, \dots, x_{j_d})$ in the role of collocation points and Dirac delta function as test functions, it yields that $\psi^1_{i_1}(x_{j_1}) = \delta_{i_1 j_1}, \psi^2_{i_2}(x_{j_2}) = \delta_{i_2 j_2}, \dots, \psi^d_{i_d}(x_{j_d}) = \delta_{i_d j_d}$. Therefore, if we take $w_1 = w_2 = \dots = w_d = 1$, Eqs. (15.16) and (15.17) become as follows:

$$R_h(x_{j_1}, x_{j_2}, \dots, x_{j_d}) = 0 \tag{15.22}$$

$$R_k(x_{j_1}, x_{j_2}, \dots, x_{j_d}) = 0 \tag{15.23}$$

Now, if we select $N_1 \times N_2 \times \dots \times N_d$ equations from Eqs. (15.22) and (15.23), a $(N_1 \times N_2 \times \dots \times N_d) \times (N_1 \times N_2 \times \dots \times N_d)$ linear or nonlinear system of equations is obtained. The unknown coefficients are specified from solving of this system. The pseudo-code of collocation algorithm is presented in Algorithm 15.1

Algorithm 15.1 Collocation Algorithm

- Input $N_1, N_2, \dots, N_d$.
- Consider $\{a_{i_1,i_2,\dots,i_d}\}_{i_1=0,i_2=0,\dots,i_d=0}^{N_1-1,N_2-1,\dots,N_d-1}$ as unknown coefficients.
- Consider $(x_{j_1}, x_{j_2}, \dots, x_{j_d})$ as collocation points such that it covers all the domain.
- $\overline{u}(\mathbf{x}) \leftarrow \sum_{i_1=0}^{N_1-1} \sum_{i_2=0}^{N_2-1} \cdots \sum_{i_d=0}^{N_d-1} a_{i_1,i_2,\dots,i_d} \phi^1_{i_1}(x_1)\phi^2_{i_2}(x_2) \cdots \phi^d_{i_d}(x_d)$.
- Substitute $\overline{u}(\mathbf{x})$ into Eqs. (15.11) and (15.12).
- Construct the residual functions in the following form:

$$R_h(\mathbf{x}) = H[\overline{u}(\mathbf{x})] - f(\mathbf{x})$$

$$R_k(\mathbf{x}) = K[\overline{u}(\mathbf{x})] - g(\mathbf{x})$$

- Construct a system of linear or nonlinear algebraic equations by collocating the collocation points in the obtained residual functions and selecting $N_1 \times N_2 \times \dots \times N_d$ equation.
- $\{a_{i_1,i_2,\dots,i_d}\}_{i_1=0,i_2=0,\dots,i_d=0}^{N_1-1,N_2-1,\dots,N_d-1} \leftarrow$ the solution of obtained system in the previous step.
- Return $\{a_{i_1,i_2,\dots,i_d}\}_{i_1=0,i_2=0,\dots,i_d=0}^{N_1-1,N_2-1,\dots,N_d-1}$.

15.2.2 Function Approximation Using Chebyshev Polynomials

Orthogonal polynomials have significant role in function approximation, spectral methods, and specially in the collocation algorithm. These functions have many applications in different fields of science and engineering.

For example, in pattern recognition, many support vector machine algorithms have been developed based on orthogonal functions [65–69]. Moreover, some algorithms for signal and image representation [70] use orthogonal functions. In the field of complex network, computing the reliability of the network has a special importance. Recently, some approaches based on orthogonal functions have employed to obtain the reliability of the network [71]. In the spectral methods, choosing suitable basis function has significant effect on convergence and accuracy of the algorithm. These polynomials have suitable properties to choose as a basis function. The general properties that the basis functions should have are as follows:

- convenient in computations
- completeness
- high convergence rate

Some classes of the orthogonal functions have the above-mentioned properties simultaneously. One of these suitable properties is completeness over the arbitrary close finite interval. In Theorem 15.1, we explain this property.

Theorem 15.1 *Let $\{\phi_n(x)\}$ be the sequence of orthogonal polynomials determined by the positive integrable function $w(x)$. These polynomials are assumed to be orthogonal over the close finite interval $[0, \eta]$, that is*

$$\int_0^\eta \phi_n(x)\phi_m(x)w(x)dx = c_n\delta_{m,n}$$

where c_n is a constant and $\delta_{m,n}$ is the Kronecker delta function. Then, the set $\{\phi_n(x)\}$ is complete with respect to continues functions over $[0, \eta]$.

It is worth to mention that all classes of orthogonal functions are not complete and the mentioned theorem is only about the orthogonal functions that are the solution of a Sturm–Liouville differential equation. One of these orthogonal polynomials is Chebyshev polynomials. The Chebyshev polynomials have been used frequently in many areas of numerical analysis and scientific computing. The Chebyshev polynomials are a special case of Gegenbauer and Jacobi polynomials and have four kinds. The first kind of Chebyshev polynomials is solutions of the following Sturm–Liouville differential equation:

$$(1-x^2)\frac{d^2y}{dx^2} - x\frac{dy}{dx} + n^2y = 0 \tag{15.24}$$

The solutions of this differential equation depends on n. The nth solution is $\cos(n\theta)$, where $\theta = \cos^{-1}(x)$ and the range of x is $[-1, 1]$. Therefore, by denoting the nth solution by $T_n(x)$, we have:

$$T_n(x) = \cos(n\cos^{-1}(x)) \tag{15.25}$$

where $T_n(x)$ is the nth member of the first kind of Chebyshev polynomials. All classical orthogonal polynomials can be obtained by a recursive formula. The recursive formula for these polynomials is as follows:

$$\begin{gathered} T_0(x) = 1, \quad T_1(x) = x \\ T_{n+1}(x) = 2xT_n(x) - T_{n-1}(x), \quad n = 1, 2, 3, \ldots \end{gathered} \tag{15.26}$$

In addition to the recursive formula, there is some direct formula to obtain the Chebyshev polynomials. For example, the following series give the nth Chebyshev polynomial:

$$T_n(x) = \sum_{k=0}^{\lfloor \frac{n}{2} \rfloor} (-1)^k \frac{n!}{(2k)!(n-2k)!}(1-x^2)^k x^{n-2k} \tag{15.27}$$

Or

$$T_n(x) = (-1)^n \frac{2^n n!}{(2n)!} \sqrt{1-x^2} \frac{d^n}{dx^n} \sqrt{(1-x^2)^{2n-1}} \tag{15.28}$$

As mentioned above, the Chebyshev polynomials have the orthogonality property. These functions are orthogonal over the interval $[-1, 1]$ by $w_k(x)$ weight functions as follows:

$$\int_{-1}^{1} T_n(x) T_m(x) w(x) dx = c\delta_{m,n} \tag{15.29}$$

where $\delta_{m,n}$ is the Kronecker delta function, and

$$w(x) = \frac{1}{\sqrt{1-x^2}}, \quad c = \begin{cases} \pi, & m = n = 0 \\ \frac{\pi}{2}, & \text{otherwise} \end{cases} \tag{15.30}$$

Another important property of the Chebyshev polynomials is that the nth member of Chebyshev series has n real roots. The following theorem explain this fact formally.

Theorem 15.2 *The Chebyshev polynomial $T_n(x)$ has precisely n real roots over interval $[-1, 1]$ in the following form:*

$$x_k = \cos\left(\frac{(2k-1)\pi}{2n}\right), \quad k = 1, 2, \ldots, n$$

and has precisely n real extremums in the following form:

$$x_k^{(1)} = \cos\left(\frac{k\pi}{n}\right), \quad k = 1, 2, \ldots, n$$

Because these polynomials are orthogonal over the interval $[-1, 1]$ and sometimes the problem is defined over the another interval, so we should use a transformation that shifts the problem to the interval $[-1, 1]$ or the Chebyshev polynomials to the interval of the problem. If we want to shift these functions from $[-1, 1]$ to $[a, b]$, we can use the following transformation:

$$x = \frac{2\bar{x} - b - a}{b - a} \tag{15.31}$$

where $\bar{x} \in [a, b]$ and $x \in [-1, 1]$. In this paper, we use shifted first kind of Chebyshev polynomials on $[0, 1]$. The remaining thing, which is essential for the spectral method, is calculating the differentiation of these polynomials. In order to calculate the differentiation of the Chebyshev polynomials, we can use an operational matrix of differentiation. By denoting the operational matrix of first derivative of shifted Chebyshev polynomials by D, its elements can be obtained as follows:

$$D_{i,j} = \frac{2}{b-a} \times \begin{cases} \frac{2i}{p_j}, & i > j, i+j \text{ is odd} \\ 0, & \text{otherwise} \end{cases} \tag{15.32}$$

in which

$$p_j = \begin{cases} 1, & j = 1, 2, \ldots, n \\ 2, & j = 0 \end{cases} \tag{15.33}$$

It is worth to mention that for obtaining a higher order of derivative of Chebyshev polynomials by using this operational matrix, we can use the powers of this operational matrix. For example, in order to obtain the second order of derivative of Chebyshev polynomials, it is just needed to compute D^2.

In order to approximate functions using shifted Chebyshev polynomials, some definition and theorem are needed, which expressed in the rest of this section.

Definition 15.2 For any real function $f(x)$, $x > 0$, if there exists a real number $p > \mu$, such that $f(x) = x^p f_1(x) \in C(0, \infty)$, is said to be in space C_μ, $\mu \in R$, and it is in the space C_μ^n, if and only if $f^{(n)} \in C_\mu$, $n \in N$.

Theorem 15.3 *Suppose that* $f^{(k)}(x) \in C(0, \eta)$, *where* $k = 0, 1, \dots, m$ *and* $\eta > 0$. *Then, we have*

$$f(x) = \sum_{i=0}^{m-1} \frac{x^i f^{(i)}(0^+)}{i!} + \frac{x^m}{m!} f^{(m)}(\xi)$$

with $0 < \xi \leq \eta$, $\forall x \in (0, \eta]$. *And thus for* $M > |f^{(m)}(\xi)|$:

$$|f(x) - \sum_{i=0}^{m-1} \frac{x^i f^{(i)}(0^+)}{i!}| < M \frac{x^m}{m!}$$

Definition 15.3 Consider $\Gamma = \{x | 0 \leq x \leq \eta\}$ and $L_w^2(\Gamma) = \{f : \Gamma \to \mathbb{R} | f \text{ is measurable and} ||f||_w < \infty\}$, where

$$w(x) = \frac{1}{\sqrt{1 - x^2}}$$

and

$$||f(x)||_w = \left(\int_0^\eta f^2(x) w(x) dx \right)^{\frac{1}{2}}$$

is the norm induced by the inner product of the space

$$\langle f(x), g(x) \rangle_w = \int_0^\eta f(x) g(x) w(x) dx$$

It is known that any function $f(x) \in C(0, \eta)$ can be expanded as follows:

$$f(x) = \sum_{n=0}^{\infty} a_n \overline{T}_n(x) \tag{15.34}$$

where $\overline{T}(x)$ denotes shifted Chebyshev polynomials to $[0, \eta]$ and

$$a_i = \langle f(x), \overline{T}_i(x) \rangle = \langle \sum_{n=0}^{\infty} a_n \overline{T}_n(x), \overline{T}_i(x) \rangle \tag{15.35}$$

that is,

$$a_i = \frac{1}{c_1} \int_0^\eta \overline{T}_i(x) f(x) w(x) dx \tag{15.36}$$

Now, let us assume that

$$V_m = \text{span}\{\overline{T}_0(x), \overline{T}_1(x), \dots, \overline{T}_m(x)\}$$

is a finite dimensional subspace, as mentioned in Theorem 15.1 V_m is a complete subspace of $L_w^2(\Gamma)$ [47, 51]. Let us define the $L_w^2(\Gamma)$-orthogonal projection $\Pi_{N,w} : L_w^2(\Gamma) \to V_m$, that for any function $f \in L_w^2(\Gamma)$:

$$\langle \Pi_{N,w} f - f, v \rangle = 0, \quad \forall v \in V_m \tag{15.37}$$

It is clear that $\Pi_{N,w} f$ is the best approximation of $f(x)$ in V_m and can be expanded as [72]:

$$\Pi_{N,w} f = f_m(x) = \sum_{i=0}^{m} a_i \overline{T}_i(x) \tag{15.38}$$

15.3 Implementation

In this part, we demonstrate that how to apply numerical methods that are explained in Section 15.2. First, we discretize the temporal dimension of the problem using a finite difference scheme and then we apply the collocation algorithm for spatial dimensions. Many researchers have developed different finite difference schemes to approximate the fractional or integer order of derivatives of a (unknown) function. Moreover, some other researchers have used and combined these schemes with some other methods to solve various problems. The basic idea of the finite difference method is inspired from Taylor series. In this chapter, we use θ-weighted finite difference scheme for temporal approximation. The stability of the used scheme is a challenging problem in the numerical methods. Crank–Nicolson is the most famous θ-weighted finite difference scheme that is unconditionally stable.

The domain of problem is $[0,1]^d \times [0,T]$. In order to discretize temporal dimension, the interval $[0,T]$ is divided into N steps with time step size $\delta t = \frac{T}{N}$. Note that $t_n = n \times \delta t$ and we denote $u(\mathbf{x}, t_n)$ by $u|^n$ and $r(t_n)$ by $r(t)|^n$. By applying θ-weighted finite difference scheme on Eq. (15.5), we obtain the following equation:

$$\frac{u|^{n+1} - u|^n}{\delta t} = \theta\left[\Delta u|^{n+1} + r(t)|^{n+1}\phi(\mathbf{x},t)|^{n+1}\right] + (1-\theta)\left[\Delta u|^n + r(t)|^n\phi(\mathbf{x},t)|^n\right] \tag{15.39}$$

in which $0 \le \theta \le 1$. If we choose $\theta = 0$, the mentioned scheme is called explicit; if $\theta = 1$, the scheme is called implicit; and if $\theta = \frac{1}{2}$, the Crank–Nicolson scheme is obtained. In the above equation, the goal is obtaining $u|^{n+1}$, so by reordering it, we have:

$$u|^{n+1} - \delta t\theta\left[\Delta u|^{n+1} + r(t)|^{n+1}\phi(\mathbf{x},t)|^{n+1}\right] = \delta t(1-\theta)\left[\Delta u|^n + r(t)|^n\phi(\mathbf{x},t)|^n\right] + u|^n \tag{15.40}$$

In order to approximate u in different time steps, we expand it as follows:

$$u|^n = \sum_{i_1=0}^{N_1-1}\sum_{i_2=0}^{N_2-1}\cdots\sum_{i_d=0}^{N_d-1} a^n_{i_1,i_2,\dots,i_d} T^*_{i_1}(x_1)T^*_{i_2}(x_2)\cdots T^*_{i_d}(x_d) \tag{15.41}$$

or equivalently in the matrix form:

$$u|^n = A_n^T \times \mathcal{T}_1^* \otimes \mathcal{T}_2^* \otimes \mathcal{T}_3^* \otimes \cdots \otimes \mathcal{T}_d^* \tag{15.42}$$

where

$$A_n = (a^n_{0,0,\dots,0}, a^n_{1,0,\dots,0}, \dots, a^n_{N_1-1,0,\dots,0}, \dots, a^n_{0,N_2-1,\dots,N_d-1}, a^n_{1,N_2-1,\dots,N_d-1}, \dots, a^n_{N_1-1,N_2-1,\dots,N_d-1})^T \tag{15.43}$$

and

$$\mathcal{T}_i^* = (T_0^*(x_i), T_1^*(x_i), \dots, T^*_{N_i-1}(x_i))^T \tag{15.44}$$

It is worth to mention that $T^*(x)$ is the shifted Chebyshev polynomial to interval $[0,1]$. Therefore, if we consider $u|^n$ as Eq. (15.41) for $\Delta u|^n$, it is obtained:

$$\begin{aligned}\Delta u|^n = &\sum_{i_1=0}^{N_1-1}\sum_{i_2=0}^{N_2-1}\cdots\sum_{i_d=0}^{N_d-1} a^n_{i_1,i_2,\dots,i_d} T_i^{*\prime\prime}(x_1)T_i^*(x_2)\dots T_i^*(x_d)\\ &+\sum_{i_1=0}^{N_1-1}\sum_{i_2=0}^{N_2-1}\cdots\sum_{i_d=0}^{N_d-1} a^n_{i_1,i_2,\dots,i_d} T_i^*(x_1)T_i^{*\prime\prime}(x_2)\dots T_i^*(x_d)\\ &\qquad\vdots\\ &+\sum_{i_1=0}^{N_1-1}\sum_{i_2=0}^{N_2-1}\cdots\sum_{i_d=0}^{N_d-1} a^n_{i_1,i_2,\dots,i_d} T_i^*(x_1)T_i^*(x_2)\dots T_i^{*\prime\prime}(x_d)\end{aligned} \tag{15.45}$$

or equivalently in the matrix form:

$$\begin{aligned}\Delta u|^n = A_n^T \times (D^2 \times \mathcal{T}_1^* \otimes \mathcal{T}_2^* \otimes \mathcal{T}_3^* \cdots \otimes \mathcal{T}_d^* \\ + \mathcal{T}_1^* \otimes D^2 \times \mathcal{T}_2^* \otimes \mathcal{T}_3^* \cdots \otimes \mathcal{T}_d^* \\ \vdots \\ + \mathcal{T}_1^* \otimes \mathcal{T}_2^* \otimes \mathcal{T}_3^* \cdots \otimes D^2 \times \mathcal{T}_d^*)\end{aligned} \tag{15.46}$$

We construct the residual functions as follows:

$$\begin{aligned}R_h|^{n+1}(\mathbf{x}) = u|^{n+1} - \delta t\theta\Delta u|^{n+1} - \delta t\theta r(t)|^{n+1}\phi(\mathbf{x},t)|^{n+1} \\ -\delta t(1-\theta)\Delta u|^n - \delta t(1-\theta)r(t)|^n\phi(\mathbf{x},t)|^n - u|^n\end{aligned} \tag{15.47}$$

$$R_k|^{n+1}(\mathbf{x}) = u|^{n+1} - r(t)|^{n+1}h(\mathbf{x},t)|^{n+1} \tag{15.48}$$

By choosing $\{(x_1, x_2, \ldots, x_d)|x_i \in \{z|T_{N_i}^*(z) = 0\}\}$ as collocation point and substituting Eqs. (15.41)–(15.45) in residual functions and equal them to zero, then selecting $N_1 \times N_2 \times \cdots \times N_d$ equations from them, we obtain $N_1 \times N_2 \times \cdots \times N_d$ linear algebraic equations, which can be solved for the unknown coefficients $\{a_{i_1,i_2,\ldots,i_d}^{n+1}\}_{i_1=0,i_2=0,\ldots,i_d=0}^{N_1-1,N_2-1,\ldots,N_d-1}$ using any linear solver such as LU decomposition or successive over-relaxation (SOR) in each time step.

The aim of this problem is finding unknown source parameter. For obtaining unknown source parameter, $p(t)$ from Eqs. (15.1) and (15.4) we have

$$p(t)|^n = \frac{1}{E(t)|^n}(E'(t)|^n - \Delta w(\mathbf{x}^*,t)|^n - \phi(\mathbf{x}^*,t)|^n) \tag{15.49}$$

Note that for $n = 0$, we can write

$$p(0) = \frac{1}{f(\mathbf{x}^*)}(E'(0) - \Delta f(\mathbf{x}^*) - \phi(\mathbf{x}^*,0)) \tag{15.50}$$

15.4 Numerical Results and Discussion

In this section, we are going to test the presented algorithm in different aspects. In order to this purpose, five examples with fixed values of their parameters are provided to test the accuracy, efficiency, stability, and convergence of the proposed algorithm. Because usually the inverse problems are ill-posed, the stability of the algorithm should be checked. In order to check the stability of the proposed algorithm, some noises are added to overspecified condition. We add 1%, 3%, and 5% of noise to $E(t)$ as follows:

$$\overline{E}(t) = (1 + \text{noise}) \times E(t) \tag{15.51}$$

in which *noise* is a random number in interval $[-r \times 10^{-2},\ r \times 10^{-2}]$, where $r = 1, 3, 5$, and $\overline{E}(t)$ denotes the noisy condition. On the other hand, the convergence and order of convergence of the algorithm are demonstrated by increasing the number of collocation points, which means it is demonstrated that by increasing the number of collocation points, the error of approximated solutions decreased. In order to compute the error of approximated solutions and also compare our results to each other and with other methods in literature, the following norms are used:

$$||e||_\infty(u) = \max_{\mathbf{x}\in\Omega}\{|u(\mathbf{x}) - \overline{u}(\mathbf{x})|\} \tag{15.52}$$

$$||e||_2 = \left[\int_\Omega (u(\mathbf{x}) - \overline{u}(\mathbf{x}))^2 d\,\mathbf{x}\right]^{\frac{1}{2}} \tag{15.53}$$

It is worth to mention that we performed all computations using Maple software. Moreover, in our computations, it is considered that $N_1 = N_2 = \cdots = N_d = M$ and $T = 1$.

15.4.1 Test Problem 1

Let us consider the following inverse problem as the first test example

$$w_t = w_{xx} + p(t)w + [\pi^2 - (t+1)^2]e^{-t^2}[\sin(\pi x) + \cos(\pi x)] \tag{15.54}$$

where $w(0.25, t) = \sqrt{2}e^{-t^2}$, and the exact solution is given as follows:

$$w(x,t) = e^{-t^2}[\sin(\pi x) + \cos(\pi x)] \tag{15.55}$$

$$p(t) = 1 + t^2 \tag{15.56}$$

This equation is solved in [1] by Mohebbi and Abbasi using a compact finite difference scheme. Here, we solve this equation by considering $\theta = \frac{1}{2}$ in our computations for this example. In Table 15.1, the obtained absolute error of presented numerical algorithm for unknown function is compared with the compact finite difference scheme, which is presented in [1]. Additionally, the comparison between the absolute errors for source parameter is presented in Table 15.2. It is concluded from Tables 15.1 and 15.2 that the proposed method is more accurate than finite difference type solvers.

As can be seen in Tables 15.1 and 15.2, the obtained results for the proposed method are extremely more accurate than the finite difference method, which is presented in [1]. Another aspect of the proposed method, which should be checked, is stability. Figure 15.2 shows the stability of the presented algorithm. In Figure 15.2, the absolute errors of proposed method by using different numbers of collocation points for the first test problem with various percentage of noisy condition are presented, which show the stability of the mentioned method. As shown in Figure 15.2, the accuracy of the proposed method is about 10^{-2} for noisy data, which is acceptable accuracy for this problem. We can conclude from Figure 15.2 that the presented approach is stable for different numbers of collocation points and different temporal steps.

Moreover, convergence of the algorithm is very significant. Figure 15.3 is presented to show the convergence of the suggested numerical method. In Figure 15.3, we plot the residual error for this example by using different numbers of collocation points in Figure 15.3a, and on the other hand, we plot the 2 norm of error function for different values of collocation points in Figure 15.3b. As can be seen from Figure 15.3, the order of convergence of the algorithm is exponential.

Table 15.1 A comparison between absolute error obtained by the methods of [1] and the presented method for $w(x, T)$ in test problem 1.

x	**Compact [1]** $N = 100, M = 40$	**Present method** $N = 10, M = 18$	**Present method** $N = 10, M = 25$
0.1	1.3323×10^{-15}	7.0×10^{-17}	9.0×10^{-27}
0.2	3.3307×10^{-16}	4.9×10^{-17}	6.2×10^{-27}
0.3	2.1094×10^{-15}	7.2×10^{-17}	9.5×10^{-27}
0.4	1.7764×10^{-15}	2.9×10^{-16}	3.7×10^{-26}
0.5	1.1102×10^{-16}	5.7×10^{-16}	7.4×10^{-26}
0.6	8.8818×10^{-16}	8.9×10^{-16}	1.1×10^{-25}
0.7	1.0547×10^{-15}	1.2×10^{-15}	1.5×10^{-25}
0.8	1.2490×10^{-16}	1.5×10^{-15}	1.9×10^{-25}
0.9	3.6082×10^{-16}	1.9×10^{-15}	2.4×10^{-25}

Table 15.2 A comparison between absolute error obtained by the methods of [1] and the presented method for $p(t)$ in test problem 1.

t	Exact p	Compact [1] $N = 100, M = 40$	Present method $N = 10, M = 18$	Present method $N = 10, M = 25$
0.1	1.01	$4.170\,395 \times 10^{-6}$	2.4×10^{-15}	6.5×10^{-25}
0.2	1.04	$4.170\,405 \times 10^{-6}$	1.2×10^{-14}	1.9×10^{-24}
0.3	1.09	$4.170\,393 \times 10^{-6}$	1.5×10^{-14}	2.3×10^{-24}
0.4	1.16	$4.170\,401 \times 10^{-6}$	1.3×10^{-14}	2.0×10^{-24}
0.5	1.25	$4.170\,394 \times 10^{-6}$	1.5×10^{-14}	2.3×10^{-24}
0.6	1.36	$4.170\,405 \times 10^{-6}$	1.5×10^{-14}	2.2×10^{-24}
0.7	1.49	$4.170\,395 \times 10^{-6}$	1.5×10^{-14}	2.3×10^{-24}
0.8	1.64	$4.170\,405 \times 10^{-6}$	1.6×10^{-14}	2.4×10^{-24}
0.9	1.81	$4.170\,388 \times 10^{-6}$	1.6×10^{-14}	2.3×10^{-24}
1.0	2.00	$4.170\,416 \times 10^{-6}$	1.7×10^{-14}	2.6×10^{-24}

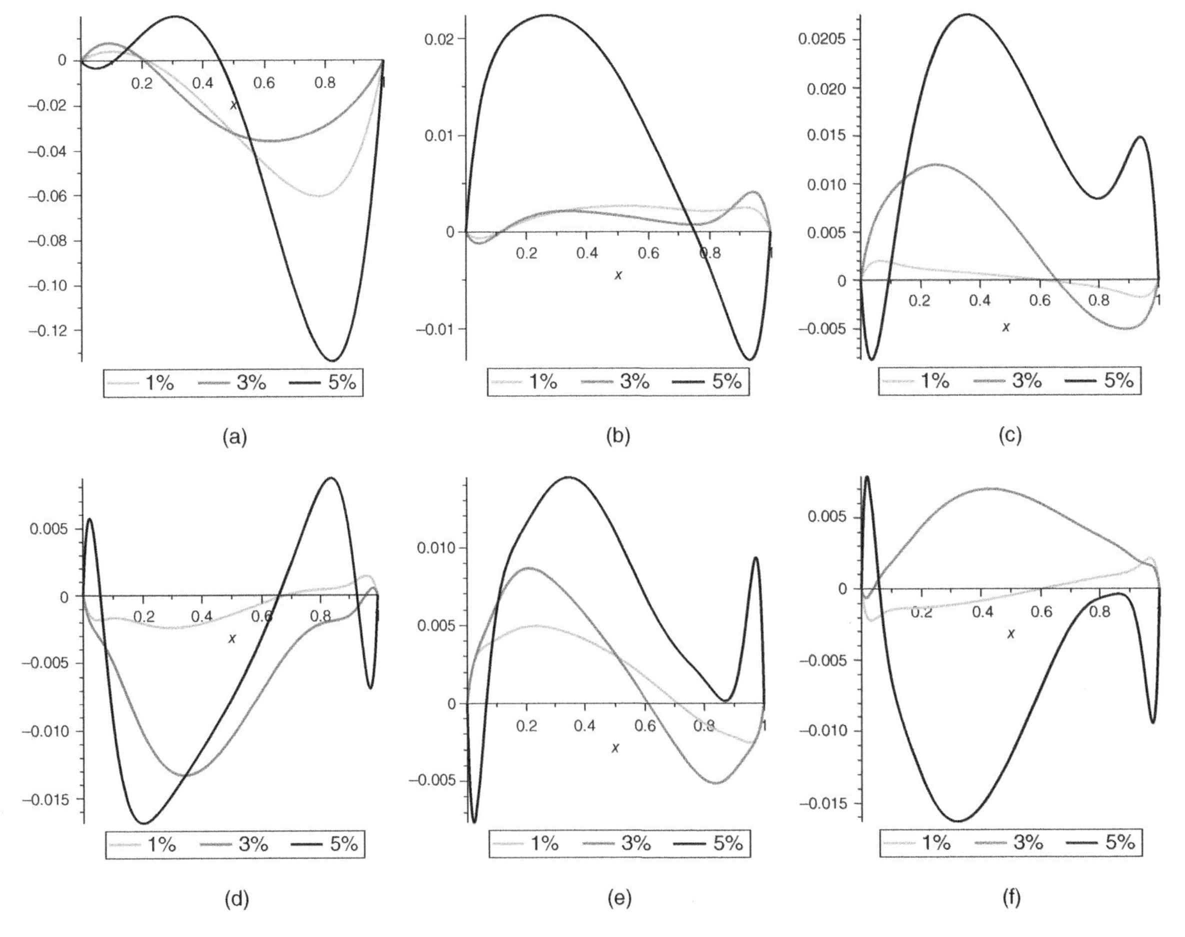

Figure 15.2 Obtaining errors in approximating $w(x, T)$, for test problem 1 with different percents of noise when (a) $N = M = 5$, (b) $N = M = 10$, (c) $N = M = 15$, (d) $N = M = 20$, (e) $N = M = 25$, and (f) $N = M = 30$.

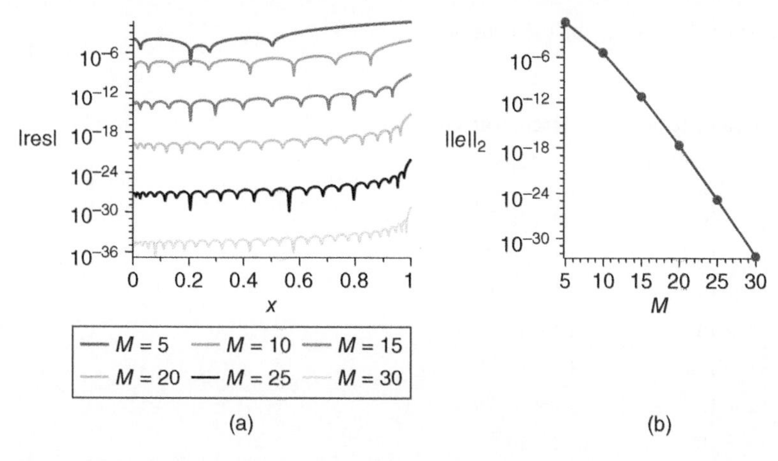

Figure 15.3 The graph of (a) $|res|$ and (b) $||e||_2$ for the various numbers of collocation points for test problem 1.

15.4.2 Test Problem 2

In the second example, consider the following one-dimensional inverse problem:

$$w_t = w_{xx} + p(t)w + e^{-t^3-t^2-x^2}[t\sin(x) - 2\sin(x) - 2x\cos(x) + 2x^2\sin(x)] \tag{15.57}$$

where $w(0.5,t) = -\frac{1}{2}\sin\left(\frac{1}{2}\right)e^{-t^3-t^2-\frac{1}{4}}$, and the exact solution is given as follows:

$$w(x,t) = -\frac{1}{2}\sin(x)e^{-t^3-t^2-x^2} \tag{15.58}$$

$$p(t) = -3t^2 - 1 \tag{15.59}$$

Similar to the previous example, this problem is solved in [1] by compact finite difference method. Here, by applying the presented numerical algorithm with $\theta = \frac{1}{2}$ on this example, we obtain high-accuracy results, which are presented in Tables 15.3 and 15.4. In Table 15.3, the obtained absolute error of the proposed method for an unknown function is compared with a compact finite difference results [1]; in addition, the comparison between

Table 15.3 A comparison between absolute error obtained by the methods of [1] and the presented method for $w(x,T)$ in test problem 2.

x	Compact [1] $N = 1000, M = 50$	Present method $N = 10, M = 15$	Present method $N = 10, M = 20$
0.1	2.8836×10^{-10}	1.8×10^{-12}	8.2×10^{-23}
0.2	4.7905×10^{-10}	3.1×10^{-12}	1.4×10^{-22}
0.3	5.0098×10^{-10}	3.5×10^{-12}	1.6×10^{-22}
0.4	3.3019×10^{-10}	2.5×10^{-12}	1.1×10^{-22}
0.5	0	0	0
0.6	4.0129×10^{-10}	4.2×10^{-12}	1.9×10^{-22}
0.7	7.4437×10^{-10}	1.0×10^{-11}	4.7×10^{-22}
0.8	8.8099×10^{-10}	1.8×10^{-11}	8.3×10^{-22}
0.9	6.6972×10^{-10}	2.8×10^{-11}	1.3×10^{-21}

Table 15.4 A comparison between absolute error obtained by the methods of [1] and the presented method for $p(t)$ in test problem 2.

t	Exact p	Compact [1] $N = 1000, M = 50$	Present method $N = 10, M = 15$	Present method $N = 10, M = 25$
0.1	−1.03	7.3556×10^{-7}	4.5×10^{-9}	2.0×10^{-19}
0.2	−1.12	7.2429×10^{-7}	7.9×10^{-9}	3.6×10^{-19}
0.3	−1.27	6.9548×10^{-7}	6.3×10^{-9}	2.9×10^{-19}
0.4	−1.48	6.7260×10^{-7}	7.0×10^{-9}	3.2×10^{-19}
0.5	−1.75	6.7260×10^{-7}	6.9×10^{-9}	3.1×10^{-19}
0.6	−2.08	6.7260×10^{-7}	6.8×10^{-9}	3.1×10^{-19}
0.7	−2.47	6.6262×10^{-7}	7.1×10^{-9}	3.2×10^{-19}
0.8	−2.92	6.6771×10^{-7}	6.8×10^{-9}	3.1×10^{-19}
0.9	−3.43	6.8867×10^{-7}	7.1×10^{-9}	3.3×10^{-19}

the obtained absolute error of presented method and the method of [1] for evaluating the source parameter is provided in Table 15.4. As can be seen in Tables 15.3 and 15.4, we can obtain a more accurate approximated solution by using less spatial points and time steps.

Similar to test problem 1, we test the stability of the proposed algorithm by adding some noise to the overspecified condition. For a different percent of noise, we apply the proposed method by various numbers of collocation points. The obtained results are presented in Figure 15.4, which shows the stability of the provided approach.

Moreover, the convergence of provided method is shown in Figure 15.5. The process of decreasing the value of error and residual function is shown in Figure 15.5, which means, by increasing the number of collocation points, the residual and error of approximated solution decrease in the exponential sense.

15.4.3 Test Problem 3

In the third example, we consider the following two-dimensional inverse problem:

$$w_t = w_{xx} + w_{yy} + p(t)w + \left(\frac{5\pi^2}{16} - 5t\right) e^t \sin\left(\frac{\pi}{4}(x + 2y)\right) \tag{15.60}$$

in which $w(0.4, 0.2, t) = e^t \sin(0.2\pi)$ and its exact solution is :

$$w(x, y, t) = e^t \sin\left(\frac{\pi}{4}(x + 2y)\right) \tag{15.61}$$

$$p(t) = 1 + 5t \tag{15.62}$$

Shivanian and Jafarabadi developed a spectral meshless method for solving this problem in [23]. In Table 15.5, the obtained results of the proposed method by using $\theta = \frac{1}{2}$ are compared with the meshless method, which is presented in [23], in terms of $||e||_\infty$ for unknown function and source parameter, in different numbers of collocation points and time steps. Table 15.5 shows that the presented algorithm is more accurate than meshless method, which is introduced in [23].

As can be seen from Table 15.5, the presented method has the higher convergence rate in comparison with the method of [23]. The convergence of the presented method is illustrated in Figure 15.6. In addition to Table 15.5, for convergence examination of the suggested algorithm, the values of $||\text{res}||_2^2$ and $||e||_2$, which are obtained by using different number collocation points, is shown in Figure 15.6.

On the other hand, in order to show that this method is stable, the provided algorithm is tested by noisy data. The obtained results by adding different percent of noise to $E(t)$ is presented in Figure 15.7, in which the number

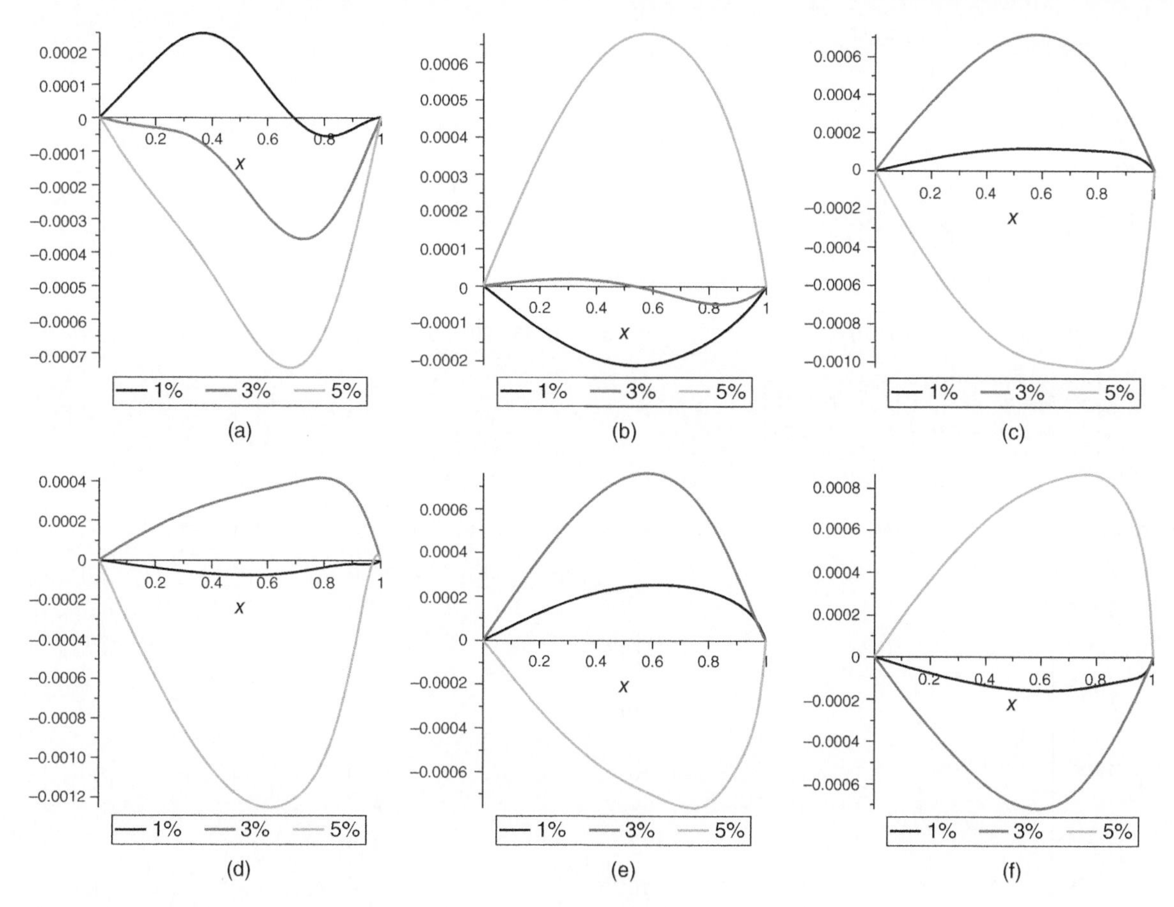

Figure 15.4 Obtaining errors in approximating $w(x, T)$, for test problem 2 with a different percent of noise when (a) $N = M = 5$, (b) $N = M = 10$, (c) $N = M = 15$, (d) $N = M = 20$, (e) $N = M = 25$, and (f) $N = M = 30$.

Table 15.5 A comparison between the presented method and the method of [23] in the sense of $||e||_\infty$ for test problem 3.

	Meshless [23] $\|\|e\|\|_\infty(w)$	Present method $\|\|e\|\|_\infty(w)$	Meshless [23] $\|\|e\|\|_\infty(p)$	Present method $\|\|e\|\|_\infty(p)$
$N = M = 8$	1.2363×10^{-4}	1.1×10^{-7}	6.8035×10^{-1}	7.6×10^{-6}
$N = M = 10$	5.7001×10^{-5}	3.2×10^{-10}	4.3220×10^{-1}	3.6×10^{-8}
$N = 100, M = 16$	4.4928×10^{-6}	1.9×10^{-19}	3.8632×10^{-3}	4.2×10^{-17}

of collocation points is $M = 12$. Moreover, the obtained results by using 20 collocation points is presented in Figure 15.8. In both cases, the number of time steps is equal to 10.

15.4.4 Test Problem 4

In the fourth example, we consider another two-dimensional inverse problem as follows:

$$w_t = w_{xx} + w_{yy} + p(t)w + \left(\frac{5\pi^2}{16} + e^t - 3\right) e^t \sin\left(\frac{\pi}{4}(x + 2y)\right) \tag{15.63}$$

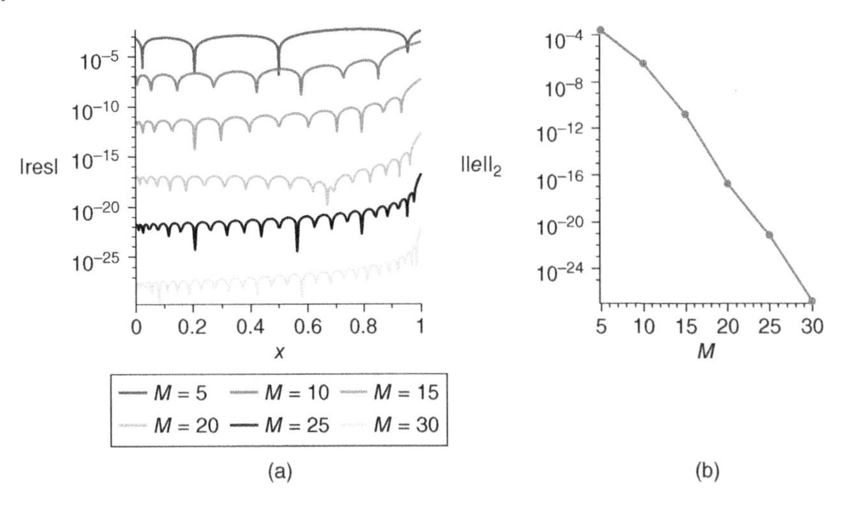

Figure 15.5 The graph of (a) $|res|$ and (b) $||e||_2$ for the various numbers of collocation points for test problem 2.

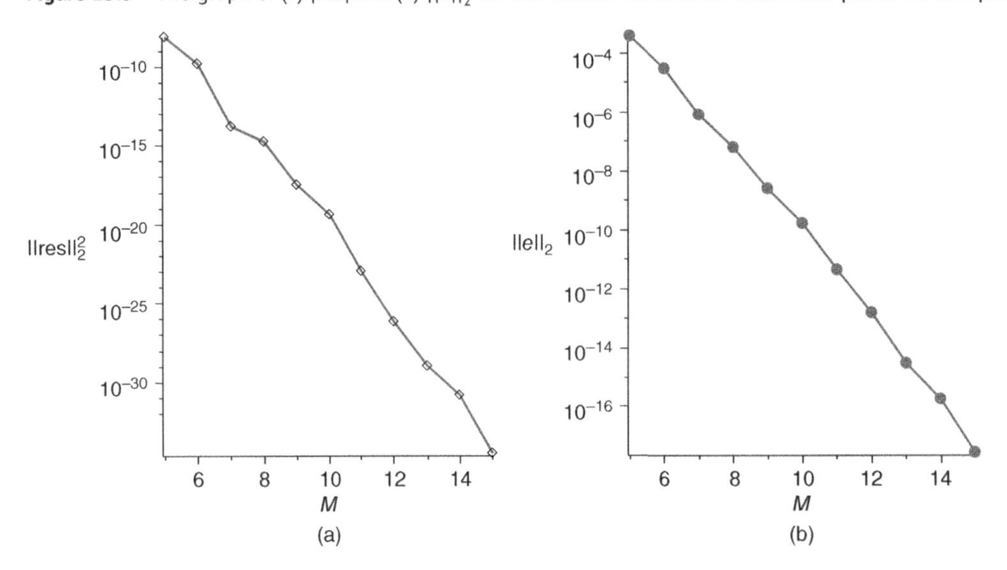

Figure 15.6 The graph of (a) $||res||_2^2$ and (b) $||e||_2$ for the various numbers of collocation points for test problem 3.

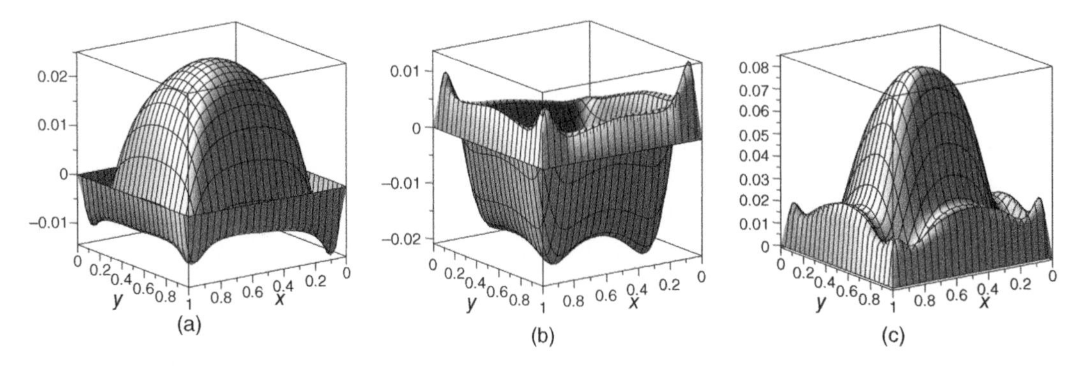

Figure 15.7 Obtaining errors in approximating $w(x, y, T)$, for test problem 3 with $M = 12$ and $N = 10$ by: (a) 1% noise, (b) 3% noise, and (c) 5% noise.

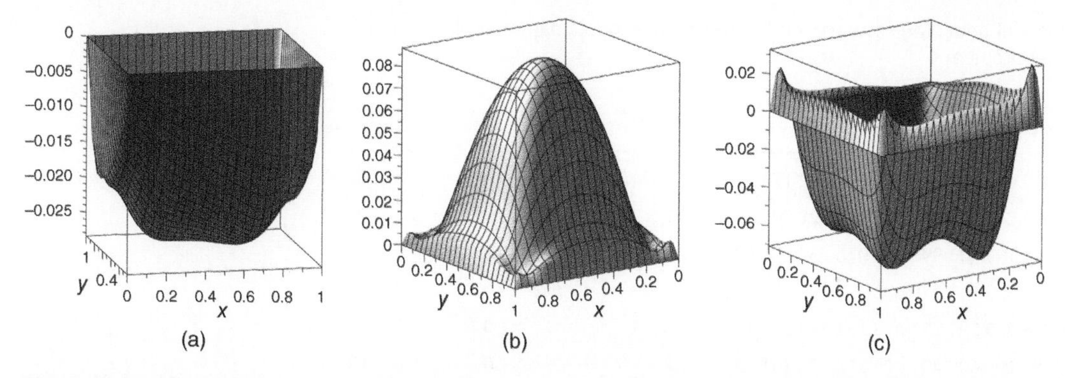

Figure 15.8 Obtaining errors in approximating $w(x, y, T)$ for test problem 3 with $M = 20$ and $N = 10$ by: (a) 1% noise, (b) 3% noise, and (c) 5% noise.

Table 15.6 Table of obtaining accuracy for unknown function by using 10 time steps in test problem 4.

x	y	z	$M = 5$	$M = 10$	$M = 15$
0.1	0.1	0.1	3.6×10^{-4}	1.5×10^{-11}	2.8×10^{-18}
0.2	0.2	0.2	3.4×10^{-4}	1.5×10^{-11}	3.9×10^{-18}
0.3	0.3	0.3	1.1×10^{-3}	1.1×10^{-10}	7.9×10^{-18}
0.4	0.4	0.4	1.6×10^{-3}	1.7×10^{-10}	1.1×10^{-17}
0.5	0.5	0.5	1.7×10^{-3}	4.8×10^{-11}	1.3×10^{-17}
0.6	0.6	0.6	1.7×10^{-3}	1.9×10^{-10}	1.3×10^{-17}
0.7	0.7	0.7	1.3×10^{-3}	1.5×10^{-10}	1.1×10^{-17}
0.8	0.8	0.8	8.7×10^{-4}	9.2×10^{-11}	7.5×10^{-18}
0.9	0.9	0.9	2.6×10^{-4}	4.9×10^{-11}	3.0×10^{-18}

in which $w(0.4, 0.1, t) = e^t \sin(0.15\pi)$ and its exact solution is

$$w(x, y, t) = e^t \sin\left(\frac{\pi}{4}(x + 2y)\right) \tag{15.64}$$

$$p(t) = 4 - e^t \tag{15.65}$$

This equation is solved by a splitting finite difference method in [16] and also by a greedy MLPG meshless method in [24]. The best obtained solution by the method of [16] is about 10^{-5} by using 10 000 time steps and 50 nodes for each spatial dimension. In Tables 15.6 and 15.7, the obtained results of proposed method by considering $\theta = \frac{1}{2}$ is reported.

In addition to high accuracy of the algorithm, the convergence of the presented method can be conclude from Tables 15.6 and 15.7. Moreover, in order to show the convergence of the presented algorithm, the values of $||\text{res}||_2^2$ and $||e||_2$, which are obtained by using different numbers of collocation points, are shown in Figure 15.9.

On the other hand, in order to show that the proposed method is stable, the provided algorithm is tested by noisy data. The obtained results by adding various percents of noise to $E(t)$ are presented in Figure 15.10, in which the number of collocation points is $M = 12$. Moreover, the obtained results by using 20 collocation points are presented in Figure 15.11.

Table 15.7 Table of obtaining accuracy for discovering the source parameter by using 10 time steps in test problem 4.

t	Exact p	$M = 5$	$M = 10$	$M = 15$
0.1	2.894 829 081 924 352 375 188 292 173 51	4.3×10^{-2}	9.8×10^{-8}	3.0×10^{-15}
0.2	2.778 597 241 839 830 166 078 928 005 36	5.1×10^{-2}	9.5×10^{-8}	3.1×10^{-15}
0.3	2.650 141 192 423 996 896 016 255 686 67	4.7×10^{-2}	9.7×10^{-8}	3.0×10^{-15}
0.4	2.508 175 302 358 729 682 175 147 047 16	4.9×10^{-2}	9.5×10^{-8}	3.0×10^{-15}
0.5	2.351 278 729 299 871 853 151 349 212 19	4.8×10^{-2}	9.6×10^{-8}	3.0×10^{-15}
0.6	2.177 881 199 609 491 025 124 632 331 84	4.8×10^{-2}	9.6×10^{-8}	3.0×10^{-15}
0.7	1.986 247 292 529 523 478 375 450 611 42	4.8×10^{-2}	9.6×10^{-8}	3.0×10^{-15}
0.8	1.774 459 071 507 532 395 420 462 468 60	4.8×10^{-2}	9.6×10^{-8}	3.0×10^{-15}
0.9	1.540 396 888 843 050 336 199 873 436 40	4.8×10^{-2}	9.6×10^{-8}	3.0×10^{-15}

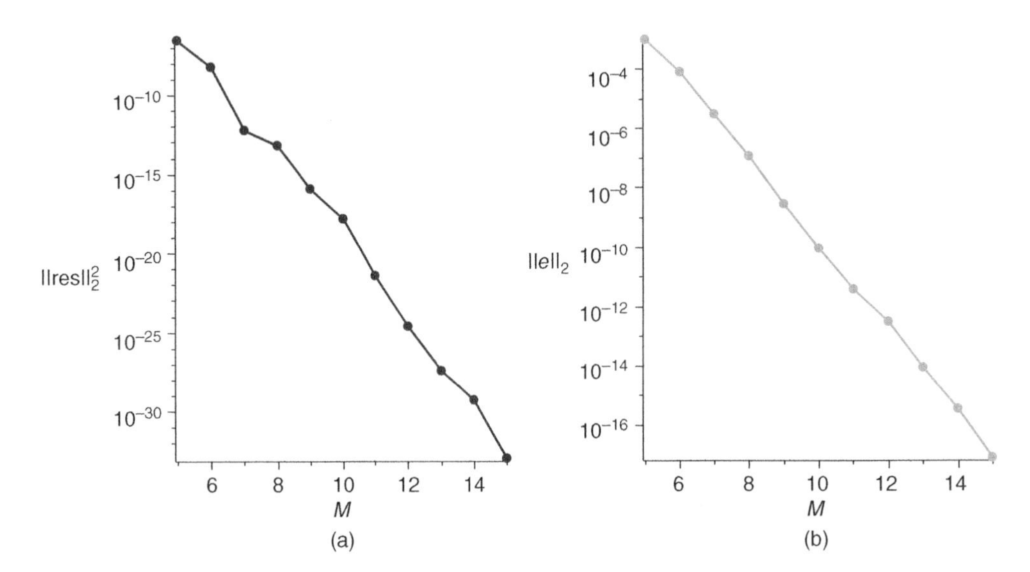

Figure 15.9 The graph of (a) $||\text{res}||_2^2$ and (b) $||e||_2$ for the various numbers of collocation points for test problem 4.

15.4.5 Test Problem 5

As the final example, the following three-dimensional inverse problem is considered as fifth test example:

$$w_t = w_{xx} + w_{yy} + w_{zz} + p(t)w + \left(\frac{7\pi^2}{8} - 10t\right) e^t \sin\left(\frac{\pi}{4}(x + 2y + 3z)\right) \tag{15.66}$$

where $w(0.6, 0.4, 0.2, t) = e^t$, and it has the following exact solution:

$$w(x, y, z, t) = e^t \sin\left(\frac{\pi}{4}(x + 2y + 3z)\right) \tag{15.67}$$

$$p(t) = 1 + 10t \tag{15.68}$$

Dehgan solved this equation in [13] by a seven-point finite difference scheme. The best accuracy, which is reported in [13], is 4.3×10^{-4} for $w(x, y, z, 1)$ and 8.9×10^{-3} for $p(t)$, while it is using 15 000 time steps and 50

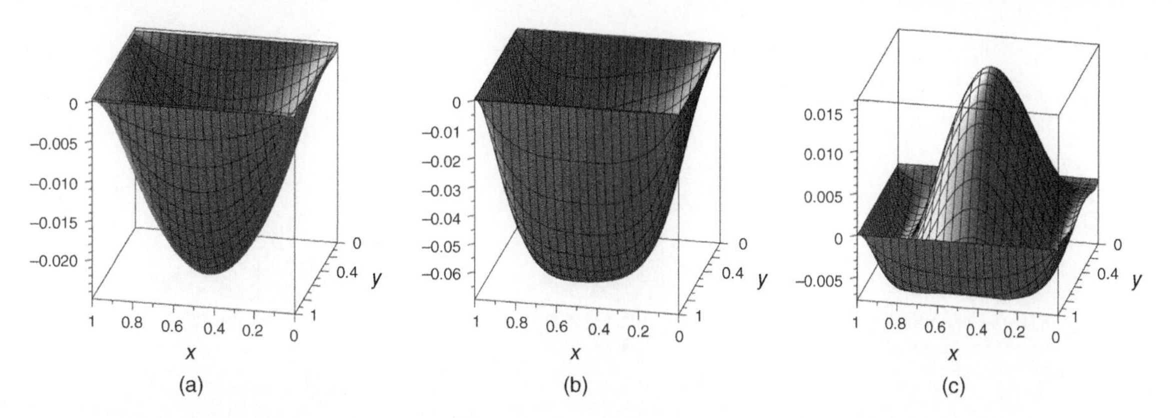

Figure 15.10 Obtaining errors in approximating $w(x, y, T)$ for test problem 4 with $M = 5$ and $N = 18$ by: (a) 1% noise, (b) 3% noise, and (c) 5% noise.

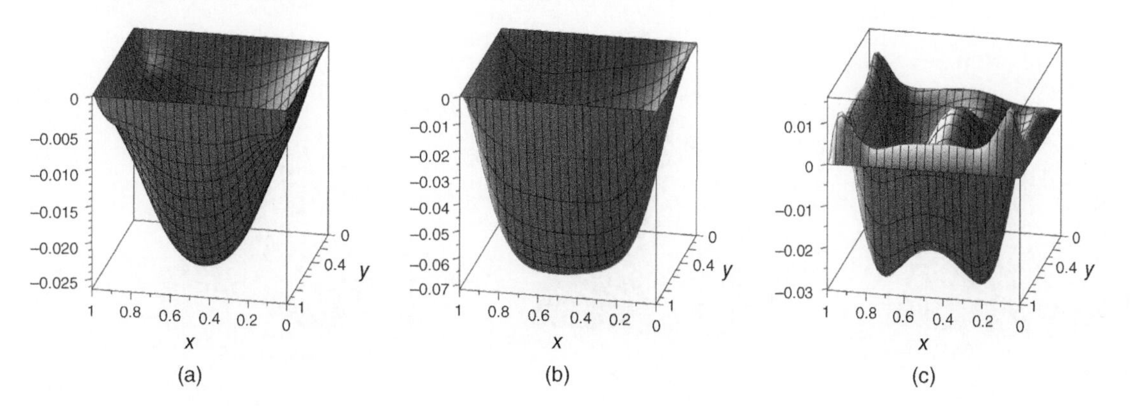

Figure 15.11 Obtaining errors in approximating $w(x, y, T)$ for test problem 4 with $M = 10$ and $N = 18$ by: (a) 1% noise, (b) 3% noise, and (c) 5% noise.

Table 15.8 Table of obtaining results for different time steps and different numbers of collocation points for test problem 5 in the sense of $||e||_\infty$.

	$\|\|e\|\|_\infty(w)$	$\|\|e\|\|_\infty(p)$
$N = 10, M = 4$	0.3375	0.5656
$N = 50, M = 6$	4.7×10^{-3}	1.4×10^{-2}
$N = 100, M = 8$	3.0×10^{-5}	4.8×10^{-4}

nodes for each spatial dimension. In this example, the computations are done by considering $\theta = 1$. The $||e||_\infty$ is a measure of comparison between the presented algorithm and method of [13], which is discussed in Table 15.8.

As can be seen in Table 15.8, the presented method can obtain more accurate results by much less than time steps and collocation pints for spatial dimensions in comparison with finite difference method of [13]. Moreover, Table 15.8 yields the convergence of the algorithm because by increasing the time steps and collocation points, the infinite norm of error decreased. On the other hand, the stability of presented algorithm is tested by adding some percent of noise to overspecified condition. The results of adding the noise to overspecified condition are illustrated in Figures 15.12–15.14.

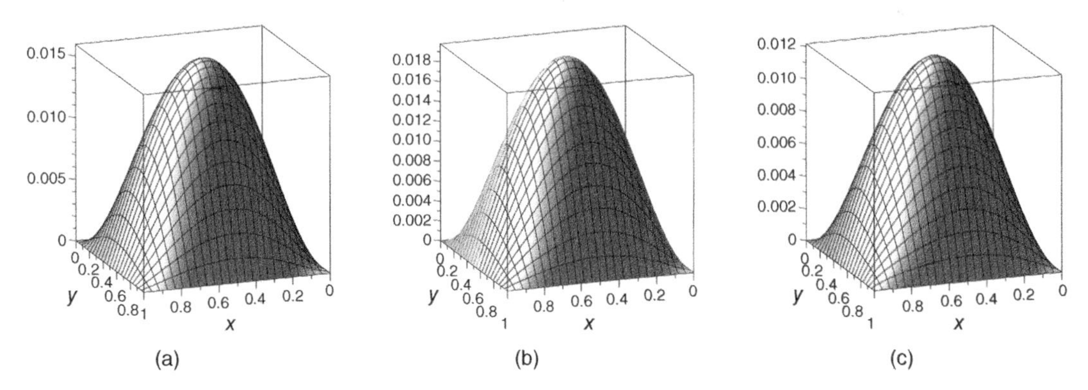

Figure 15.12 Obtaining errors in approximating $w(x, y, z, T)$ for test problem 5 with $M = 10$, $N = 10$, and 5% noise when (a) $z = 0.25$, (b) $z = 0.5$, and (c) $z = 0.75$.

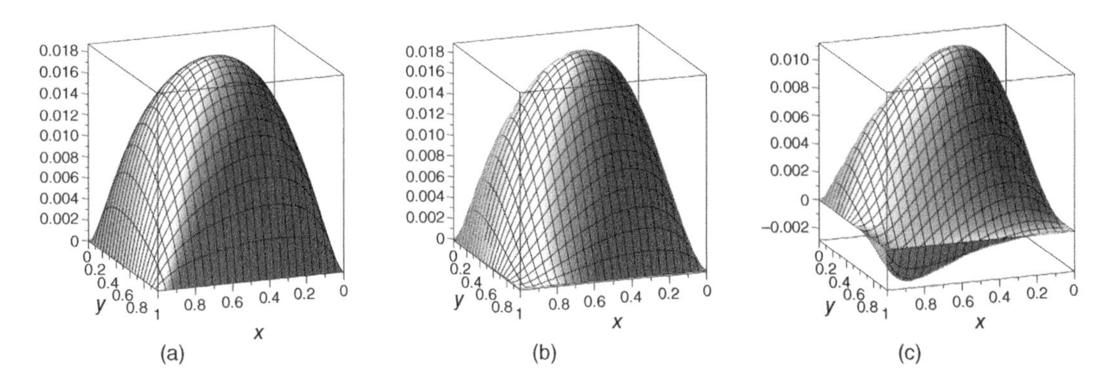

Figure 15.13 Obtaining errors in approximating $w(x, y, z, T)$, for test problem 5 with $M = 10$, $N = 10$, and 3% noise when (a) $z = 0.25$, (b) $z = 0.5$, and (c) $z = 0.75$.

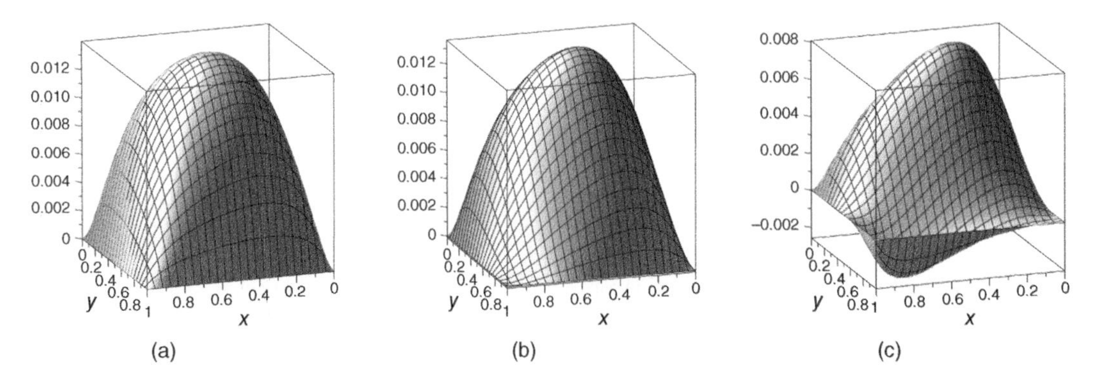

Figure 15.14 Obtaining errors in approximating $w(x, y, z, T)$ for test problem 5 with $M = 10$, $N = 10$, and 1% noise when (a) $z = 0.25$, (b) $z = 0.5$, and (c) $z = 0.75$.

15.5 Conclusion

In this chapter, a multidimensional inverse problem of finding a source parameter is considered, and the connection between the brain activity reconstruction and evaluating a source parameter in an inverse problem is illustrated. On the other hand, in order to approximate the solution of this inverse problem, a finite difference/pseudo-spectral method was developed. In the presented method, a finite difference scheme is used to discretize the temporal dimension, and a pseudo-spectral scheme based on collocation algorithm is used to approximate the spatial dimensions. We choose a θ scheme weighted for the finite different method. Moreover, in order to function approximation, the first kind of Chebyshev polynomials are selected as basis functions. The presented algorithm is tested for convergence and stability. In order to test the convergence of the algorithm, we show that by increasing the amount of basis functions and collocation points, the value of residual is decreased. On the other hand, in order to show the stability of the presented method, some noises are added to the overspecified condition and the obtained numerical results are reported. In addition, computational experiments confirmed the high accuracy of the proposed method in comparison with other method such as meshless methods, finite difference methods, and other relevant methods in the literature. Another important point of the presented algorithm is discovering both linear and nonlinear source parameters strongly.

References

1 Mohebbi, A. and Abbasi, M. (2015). A fourth-order compact difference scheme for the parabolic inverse problem with an overspecification at a point. *Inverse Problems in Science and Engineering* 23 (3): 457–478.

2 Aster, R.C., Borchers, B., and Thurber, C.H. (2018). *Parameter Estimation and Inverse Problems*. Elsevier.

3 Cannon, J.R. and Lin, Y. (1990). An inverse problem of finding a parameter in a semi-linear heat equation. *Journal of Mathematical Analysis and Applications* 145 (2): 470–484.

4 MacBain, J.A. and Bednar, J.B. (1986). Existence and uniqueness properties for the one-dimensional magnetotellurics inversion problem. *Journal of Mathematical Physics* 27 (2): 645–649.

5 Deng, Z.C., Yang, L., Yu, J.N., and Luo, G.W. (2009). An inverse problem of identifying the coefficient in a nonlinear parabolic equation. *Nonlinear Analysis: Theory Methods & Applications* 71 (12): 6212–6221.

6 Cannon, J.R., Lin, Y., and Wang, S. (1992). Determination of source parameter in parabolic equations. *Meccanica* 27 (2): 85–94.

7 Cannon, J.R., Lin, Y., and Xu, S. (1994). Numerical procedures for the determination of an unknown coefficient in semi-linear parabolic differential equations. *Inverse Problems* 10 (2): 227.

8 Liu, S. and Triggiani, R. (2011). Global uniqueness and stability in determining the damping and potential coefficients of an inverse hyperbolic problem. *Nonlinear Analysis: Real World Applications* 12 (3): 1562–1590.

9 Prilepko, A.I. and Solovev, V.V. (1987). Solvability of the inverse boundary-value problem of finding a coefficient of a lower-order derivative in a parabolic equation. *Differential Equations* 23 (1): 101–107.

10 Dehghan, M. (2000). Finite difference schemes for two-dimensional parabolic inverse problem with temperature overspecification. *International Journal of Computer Mathematics* 75 (3): 339–349.

11 Dehghan, M. (2001). Implicit solution of a two-dimensional parabolic inverse problem with temperature overspecification. *Journal of Computational Analysis and Applications* 3 (4): 383–398.

12 Dehghan, M. (2003). Identifying a control function in two-dimensional parabolic inverse problems. *Applied Mathematics and Computation* 143 (2–3): 375–391.

13 Dehghan, M. (2003). Determination of a control function in three-dimensional parabolic equations. *Mathematics and Computers in Simulation* 61 (2): 89–100.

14 Mohebbi, A. and Dehghan, M. (2010). High-order scheme for determination of a control parameter in an inverse problem from the over-specified data. *Computer Physics Communications* 181 (12): 1947–1954.

15 Dehghan, M. (2005). Parameter determination in a partial differential equation from the overspecified data. *Mathematical and Computer Modelling* 41 (2–3): 196–213.

16 Daoud, D.S. and Subasi, D. (2005). A splitting up algorithm for the determination of the control parameter in multi dimensional parabolic problem. *Applied Mathematics and Computation* 166 (3): 584–595.

17 Fatullayev, A.G. and Cula, S. (2009). An iterative procedure for determining an unknown spacewise-dependent coefficient in a parabolic equation. *Applied Mathematics Letters* 22 (7): 1033–1037.

18 Hadian Rasanan, A.H. and Rad, J.A. (2019). A computational source modeling of brain activity: an inverse problem. *Neurodevelopmental Cognition* 1 (1): 69–78.

19 Dehghan, M. and Tatari, M. (2006). Determination of a control parameter in a one-dimensional parabolic equation using the method of radial basis functions. *Mathematical and Computer Modelling* 44 (11–12): 1160–1168.

20 Dehghan, M. and Tatari, M. (2007). The radial basis functions method for identifying an unknown parameter in a parabolic equation with overspecified data. *Numerical Methods for Partial Differential Equations: An International Journal* 23 (5): 984–997.

21 Cheng, R. (2008). Determination of a control parameter in a one-dimensional parabolic equation using the moving least-square approximation. *International Journal of Computer Mathematics* 85 (9): 1363–1373.

22 Parzlivand, F. and Shahrezaee, A. (2016). The use of inverse quadratic radial basis functions for the solution of an inverse heat problem. *Bulletin of the Iranian Mathematical Society* 42 (5): 1127–1142.

23 Shivanian, E. and Jafarabadi, A. (2018). An inverse problem of identifying the control function in two and three-dimensional parabolic equations through the spectral meshless radial point interpolation. *Applied Mathematics and Computation* 325: 82–101.

24 Takhtabnoos, F. and Shirzadi, A. (2018). A greedy MLPG method for identifying a control parameter in 2D parabolic PDEs. *Inverse Problems in Science and Engineering* 26 (11): 1676–1694.

25 Dehghan, M. and Saadatmandi, A. (2006). A tau method for the one-dimensional parabolic inverse problem subject to temperature overspecification. *Computers & Mathematics with Applications* 52 (6–7): 933–940.

26 Abdelkawy, M.A., Ahmed, E.A., and Sanchez, P. (2015). A method based on legendre pseudo-spectral approximations for solving inverse problems of parabolic types equations. *Mathematical Sciences Letters* 4 (1): 81–90.

27 Shamsi, M. and Dehghan, M. (2012). Determination of a control function in three-dimensional parabolic equations by legendre pseudospectral method. *Numerical Methods for Partial Differential Equations* 28 (1): 74–93.

28 Dehghan, M., Yousefi, S.A., and Rashedi, K. (2013). Ritz–Galerkin method for solving an inverse heat conduction problem with a nonlinear source term via bernstein multi-scaling functions and cubic B-spline functions. *Inverse Problems in Science and Engineering* 21 (3): 500–523.

29 Khorshidi, M.N. and Yousefi, S.A. (2016). Ritz-least squares method for finding a control parameter in a one-dimensional parabolic inverse problem. *Journal of Applied Analysis* 22 (2): 169–179.

30 Lakestani, M. and Dehghan, M. (2010). The use of Chebyshev cardinal functions for the solution of a partial differential equation with an unknown time-dependent coefficient subject to an extra measurement. *Journal of Computational and Applied Mathematics* 235 (3): 669–678.

31 Ashpazzadeh, E., Lakestani, M., and Razzaghi, M. (2017). Cardinal Hermite interpolant multiscaling functions for solving a parabolic inverse problem. *Turkish Journal of Mathematics* 41 (4): 1009–1026.

32 Molai, A.A. and Houlari, T. (2013). Resolution of an inverse parabolic problem using Sinc-Galërkin method. *TWMS Journal of Applied and Engineering Mathematics* 3 (2): 160–181.

33 Zolfaghari, R. (2013). Parameter determination in a parabolic inverse problem in general dimensions. *Computational Methods for Differential Equations* 1 (1): 55–70.

34 Yousefi, S.A. (2009). Finding a control parameter in a one-dimensional parabolic inverse problem by using the Bernstein Galerkin method. *Inverse Problems in Science and Engineering* 17 (6): 821–828.

35 Lakestani, M. and Dehghan, M. (2008). A new technique for solution of a parabolic inverse problem. *Kybernetes* 37 (2): 352–364.

36 Gholampoor, I. and Kajani, M.T. (2019). A direct numerical method for approximate solution of inverse reaction diffusion equation via two-dimensional Legendre hybrid functions. *Numerical Algorithms* 1–18. https://doi.org/10.1007/s11075-019-00691-0.

37 Tatari, M. and Dehghan, M. (2007). Identifying a control function in parabolic partial differential equations from overspecified boundary data. *Computers & Mathematics with Applications* 53 (12): 1933–1942.

38 Tatari, M. and Dehghan, M. (2007). He's variational iteration method for computing a control parameter in a semi-linear inverse parabolic equation. *Chaos, Solitons & Fractals* 33 (2): 671–677.

39 Wang, W., Han, B., and Yamamoto, M. (2013). Inverse heat problem of determining time-dependent source parameter in reproducing kernel space. *Nonlinear Analysis: Real World Applications* 14 (1): 875–887.

40 Mohammadi, M., Mokhtari, R., and Isfahani, F.T. (2014). Solving an inverse problem for a parabolic equation with a nonlocal boundary condition in the reproducing kernel space. *Iranian Journal of Numerical Analysis and Optimization* 4 (1): 57–76.

41 Khader, M.M. (2017). Analytical solution for determination the control parameter in the inverse parabolic equation using HAM. *Applications and Applied Mathematics: An International Journal* 12 (2): 1072–1087.

42 Rostamian, M. and Shahrezaee, A. (2016). Numerical solution of a semi-linear inverse parabolic problem via HPM. *International Journal of Computing Science and Mathematics* 7 (4): 330–339.

43 Dehghan, M. and Shakeri, F. (2009). Method of lines solutions of the parabolic inverse problem with an overspecification at a point. *Numerical Algorithms* 50 (4): 417–437.

44 Deng, Z.C. and Yang, L. (2011). An inverse problem of identifying the coefficient of first-order in a degenerate parabolic equation. *Journal of Computational and Applied Mathematics* 235 (15): 4404–4417.

45 Parand, K., Rezaei, A.R., and Taghavi, A. (2010). Numerical approximations for population growth model by rational Chebyshev and Hermite functions collocation approach: a comparison. *Mathematical Methods in the Applied Sciences* 33 (17): 2076–2086.

46 Parand, K. and Rad, J.A. (2012). Numerical solution of nonlinear Volterra–Fredholm–Hammerstein integral equations via collocation method based on radial basis functions. *Applied Mathematics and Computation* 218 (9): 5292–5309.

47 Parand, K. and Delkhosh, M. (2017). Accurate solution of the Thomas–Fermi equation using the fractional order of rational Cchebyshev functions. *Journal of Computational and Applied Mathematics* 317: 624–642.

48 Parand, K., Lotfi, Y., and Rad, J.A. (2017). An accurate numerical analysis of the laminar two-dimensional flow of an incompressible Eyring-Powell fluid over a linear stretching sheet. *The European Physical Journal Plus* 132 (9): 397.

49 Baseri, A., Abbasbandy, S., and Babolian, E. (2018). A collocation method for fractional diffusion equation in a long time with Chebyshev functions. *Applied Mathematics and Computation* 322: 55–65.

50 Agarwal, P. and El-Sayed, A.A. (2018). Non-standard finite difference and Chebyshev collocation methods for solving fractional diffusion equation. *Physica A: Statistical Mechanics and Its Applications* 500: 40–49.

51 Boyd, J.P. (2001). *Chebyshev and Fourier Spectral Methods.* Courier Corporation.

52 Boyd, J.P. (2011). Chebyshev spectral methods and the Lane-Emden problem. *Numerical Mathematics: Theory Methods and Applications* 4 (2): 142–157.

53 Babolian, E., Fattahzadeh, F., and Raboky, E.G. (2007). A Chebyshev approximation for solving nonlinear integral equations of Hammerstein type. *Applied Mathematics and Computation* 189 (1): 641–646.

54 Parand, K. and Razzaghi, M. (2004). Rational Chebyshev tau method for solving Volterra's population model. *Applied Mathematics and Computation* 149 (3): 893–900.

55 Parand, K., Shahini, M., and Taghavi, A. (2009). Generalized Laguerre polynomials and rational Chebyshev collocation method for solving unsteady gas equation. *International Journal of Contemporary Mathematical Sciences* 4 (21): 1005–1011.

56 Parand, K. and Shahini, M. (2010). Rational Chebyshev collocation method for solving nonlinear ordinary differential equations of Lane-Emden type. *International Journal of Information and Systems Sciences* 6: 72–83.

57 Olmos, D. and Shizgal, B.D. (2009). Pseudospectral method of solution of the Fitzhugh–Nagumo equation. *Mathematics and Computers in Simulation* 79 (7): 2258–2278.

58 Parand, K., Moayeri, M.M., Latifi, S., and Rad, J.A. (2019). Numerical study of a multidimensional dynamic quantum model arising in cognitive psychology especially in decision making. *The European Physical Journal Plus* 134 (3): 109–126.

59 Yüzbaşı, Ş. (2012). A numerical approach to solve the model for HIV infection of $CD4^{+}T$ cells. *Applied Mathematical Modelling* 36 (12): 5876–5890.

60 Parand, K., Moayeri, M.M., Latifi, S., and Delkhosh, M. (2017). A numerical investigation of the boundary layer flow of an eyring-powell fluid over a stretching sheet via rational Chebyshev functions. *The European Physical Journal Plus* 132 (7): 325–336.

61 Parand, K., Latifi, S., Moayeri, M.M., and Delkhosh, M. (2018). Generalized lagrange Jacobi Gauss-Lobatto (GLJGL) collocation method for solving linear and nonlinear Fokker-Planck equations. *Communications in Theoretical Physics* 69 (5): 519–531.

62 Galerkin, B.G. (1915). Vestnik inzhenerov i tekhnikovi. *Tech* 19: 897–908.

63 Delkhosh, M. and Parand, K. (2019). Generalized pseudospectral method: theory and applications. *Journal of Computational Science* 34: 11–32.

64 Delkhosh, M., Parand, K., and Hadian-Rasanan, A.H. (2019). A development of lagrange interpolation, Part I: Theory. *arXiv preprint arXiv:1904.12145.*

65 Ozer, S., Chen, C.H., and Cirpan, H.A. (2011). A set of new Chebyshev kernel functions for support vector machine pattern classification. *Pattern Recognition* 44 (7): 1435–1447.

66 Parodi, M. and Gómez, J.C. (2014). Legendre polynomials based feature extraction for online signature verification. Consistency analysis of feature combinations. *Pattern Recognition* 47 (1): 128–140.

67 Benouini, R., Batioua, I., Zenkouar, K. et al. (2019). New set of generalized legendre moment invariants for pattern recognition. *Pattern Recognition Letters* 123: 39–46.

68 Moghaddam, V.H. and Hamidzadeh, J. (2016). New Hermite orthogonal polynomial kernel and combined kernels in support vector machine classifier. *Pattern Recognition* 60: 921–935.

69 Padierna, L.C., Carpio, M., Rojas-Domínguez, A. et al. (2018). A novel formulation of orthogonal polynomial kernel functions for SVM classifiers: the Gegenbauer family. *Pattern Recognition* 84: 211–225.

70 Benouini, R., Batioua, I., Zenkouar, K. et al. (2019). Fractional-order orthogonal Chebyshev moments and moment invariants for image representation and pattern recognition. *Pattern Recognition* 86: 332–343.

71 Robledo, F., Romero, P., and Sartor, P. (2013). A novel interpolation technique to address the edge-reliability problem. In: *2013 5th International Congress on Ultra Modern Telecommunications and Control Systems and Workshops (ICUMT)*, 187–192. IEEE.

72 Ben-Yu, G. (1998). *Spectral Methods and Their Applications*. World Scientific.

16

Optimal Resource Allocation in Controlling Infectious Diseases

A.C. Mahasinghe, S.S.N. Perera, and K.K.W.H. Erandi

Research and Development Center for Mathematical Modeling, Department of Mathematics, University of Colombo, Colombo 03, Sri Lanka

16.1 Introduction

Infectious diseases continue to burden the world as the second leading cause of death [1]. According to World Health Organization, infectious diseases cause 63% of all childhood deaths and 48% of premature deaths over the world [2]. That being said, regions affected most seriously by infectious disease are developing countries [3–5]. One major hindrance to the control of infectious diseases in such countries is the limitation of resources [6]. Therefore, health-planning during an outbreak of an infectious disease in a developing country becomes highly nontrivial and challenging. Let alone the straightforward factors such as transmission dynamics, it is necessary to take into account environmental, geographical, and behavioral factors when designing control strategies. Further, explicit objectives must be identified in order to review and assess the control process. Eventually, a *resource schedule* in such a control process must prescribe *how the resources must be distributed*, *what control strategies must be implemented*, and *what exactly is to be achieved* by utilizing the resources.

Considering the pragmatic situation in many disease-prone developing countries, control strategies with the above-mentioned characteristics are too far from reality. For instance, it is customary to see abrupt and random fumigation and cleaning projects in order to control dengue epidemic in Sri Lanka, although utilizing the same resources more systematically would yield much benefits. Further, there is no substructure in Sri Lanka to rely on, when deciding the amounts of resources allocated to regions, except for some trivial indicators such as dengue incidence or infectious percentage of that region, although inter-regional connections add an entirely different complexion on the matter. In addition, it is not clear if the objectives are explicitly identified before the implementation of control strategies. There could be different objectives such as *minimizing the infectious persons in the whole country*, *minimizing the highly susceptible population*, and *reducing the overall disease transmission rate*, which could possibly contradict with each other. Once the resources are allocated and control strategies are implemented, it is impossible even to *assess* the process, unless the objectives were made explicit before resource allocation. Further, confining to conventional control strategies might yield only a little benefit during an outbreak; thus, it is very helpful if appropriate novel strategies could be introduced, explaining how the resources must be allocated to implement them.

This chapter is aimed at developing resource allocation models addressing these concerns as much as possible, ultimately providing a complete guide to resource allocation in the control process during a disease outbreak. Our approach consists of several optimization models and relevant solution methods as well. Although these models are more appropriate for infectious diseases in a country with ample human mobility and a few highly connected local regions, moderated and extended versions might work in different contexts. In order to generate experimental

Mathematical Methods in Interdisciplinary Sciences, First Edition. Edited by Snehashish Chakraverty.

results, we used the statistics relevant to dengue transmission in Sri Lanka. Because of the scope of the book and our concern on developing resource allocation models, relevant algorithms will not be described in detail in this chapter, although we provide necessary references and discuss the feasibility of proposed solutions and methods in different contexts.

Under the proposed resource allocation criteria, strategies are supposed to be implemented at three different levels. First, the country's national budget is proposed to be distributed to provinces, based on interprovincial human mobility. The relevant optimization problem is discussed in detail in Section 16.2 with computational results. Secondly, it is proposed to utilize the provincial resources to reduce the connection strength of the epidemiological network in that region. This is discussed in Section 16.3 with experimental results. Both optimization problems are highly nonlinear; hence, solving them for optimum is a challenging task. This computational hardness is overcome by replacing respective nonlinear functions by their piecewise linear approximations. Finally, novel control strategies are proposed at the individual level, which can be found in Section 16.4, applicable to very small regions. The aim of these control strategies is utilization of the resources in very small health administrative divisions, in order to minimize the disease infection risk of susceptible inhabitants in those regions.

16.2 Mobility-Based Resource Distribution

16.2.1 Distribution of National Resources

When a disease outbreak takes place, the government allocates a certain amount of resources for the control process. These resources are first distributed to large administrative divisions such as provinces or districts. Recall the allocation to each province is decided in most cases upon reported incidence in that region, and factors such as interprovincial human movement are often neglected. Therefore, it is highly probable that some provinces do not get sufficient resources, eventually being unable to contribute optimally to the national fight against the disease.

Several previous works have explored the problem of allocating national resources to regions during an epidemic [7–13]. A number of works including Refs. [8, 10, 11] have considered resource allocation with simple epidemic models, aimed at minimizing the number of infected population, thus minimizing the burden of infection during the epidemic. One may find in Ref. [14] a survey of models with similar goals together with other health optimization models. It is noteworthy that most analyses including Refs. [13, 15] have been proposed to allocate the regional budget proportional to infectious population. However, it was proposed in Ref. [12] to prioritize the regions with relatively lower infectious levels. In Ref. [8], a combination of higher and lower level of infectious population reduction based on landscape scale has been proposed. Although with different goals, most of these models assume in general that regions are independent, or regional benefits of allocation may only depend on neighboring total or infected populations.

Undoubtedly, these models would yield fruitful results in the contexts they were supposed to serve. In our view, every country should consider its own environmental, geographical, and behavioral factors when designing resource allocation models. We consider a small country with ample human mobility, such as Sri Lanka. In this context, inter-regional connections are mainly defined by human movement. Therefore, our optimization model in this section serves a special purpose and is significantly different from what is available in the literature.

In order to incorporate inter-regional mobility into the model, we consider the general transmission dynamics of the disease. Most often, the transmission dynamics are captured by compartment models and respective differential equations.

16.2.2 Transmission Dynamics

16.2.2.1 Compartment Models

Mathematical modeling has been a successful tool in understanding the spreading mechanism of a disease and also in predicting the future course of the epidemic, which is highly significant to control the spread of the epidemic. In the widely used compartment model approach, the population is divided into several classes called compartments, and the timely variation of their numbers is formulated. The choice of compartments to use in the model depends on the characteristics of the disease and objectives. In literature, three main deterministic models are commonly used based on acquired immunity [16–18]. Recall that we implement our resource allocation techniques during a short-term epidemic outbreak period, and the susceptible-infectious (SI) model is the most appropriate selection [19].

16.2.2.2 SI Model

This is the simplest among compartment epidemic models. In this model, a constant population of size N is divided into *susceptible* (S) and *infected* (I) components. Let β denote the rate of virus transmission from infected human to susceptible human through vectors. Let us suppose the vector population to be constant and the infected vector population to be proportional to the initially infected human population. It is realistic to assume a constant human population for a short period of time; thus, every individual is either in compartment S or I. New infections occur as a result of contact between infectious and susceptible individuals. Once the susceptible individuals become infected with the disease, they move to the infectious class. Therefore, the model is described by the following differential equations.

$$\frac{dS}{dt} = -\beta IS \tag{16.1a}$$

$$\frac{dI}{dt} = \beta IS \tag{16.1b}$$

16.2.2.3 Exact Solution

One advantage in the SI model is that it can be solved analytically by standard methods [20]. Because total population is constant for a considered time period, we can substitute $S = N - I$ into Eq. (16.1b). Thus, the rate of change of I is given by,

$$\frac{dI}{dt} = \beta I(N - I) \tag{16.2}$$

Equation (16.2) can be restated as follows

$$\frac{dI}{dt} - \beta NI = -\beta I^2 \tag{16.3}$$

Equation (16.3) is a standard Bernoulli equation, hence we let $u = I^{-1}$. Then,

$$\frac{du}{dt} = -I^{-2}\frac{dI}{dt} \tag{16.4}$$

Multiplying Eq. (16.3) by $-I^{-2}$ and substituting u and $\frac{du}{dI}$ gives the linear equation,

$$\frac{du}{dt} + \beta Nu = \beta \tag{16.5}$$

Observe that Eq. (16.5) is a linear equation of the form

$$\frac{dx}{dt} + A(t)x = B(t) \tag{16.6}$$

with $A(t) = \beta N$ and $B(t) = \beta$. Then, the integrating factor $\mu(t)$ is given by,

$$\mu(t) = \exp\left(\int A(t)dt\right) = e^{\beta Nt} \tag{16.7}$$

Multiplying Eq. (16.5) by $\mu(t)$, it follows that,

$$\frac{d(\mu(t)u)}{dt} = \beta e^{\beta Nt} \tag{16.8}$$

Integrating Eq. (16.8) and substituting the initial condition $I(0) = I_0$ (then, $u(0) = \frac{1}{I_0}$), the analytical solution to the SI model or the infected human population at time t according to SI can be expressed as follows (Figure 16.1):

$$I(t) = \frac{I_0 N}{I_0 + (N - I_0)e^{(-\beta Nt)}} \tag{16.9}$$

16.2.2.4 Transmission Rate and Potential

The transmission rate symbolized by β is interpreted similarly in all compartment models and usually supposed to be constant. However, a few recent works have explored the possibility of a time-varying transmission rate [21–23]. Indeed, this transmission rate represents the potential of virus transmission from infected human to susceptible human. Also, such a potential must depend on regional and inter-regional human interaction, hence upon the respective densities of infected human population. A control strategy implemented in some region reduces the potential of that region, contributing to reduce the potentials of other regions as well, depending on the level of inter-regional human interactions.

Considering a region i, which is connected to another region j through human mobility, β_i, or the epidemic potential of region i depends on the degree of its connection with j. Also, it depends on the number of infectious persons and favorable conditions for the disease spread in the jth region, which is inversely proportional to the resources allocated to that region. Therefore, the potential of disease transmission in the ith region is expressible as follows:

$$\beta_i = \frac{\zeta_i I_{0i}}{x_i} + \eta_i \sum_{j \neq i} \frac{w_{ij} I_{0j}}{x_j} \tag{16.10}$$

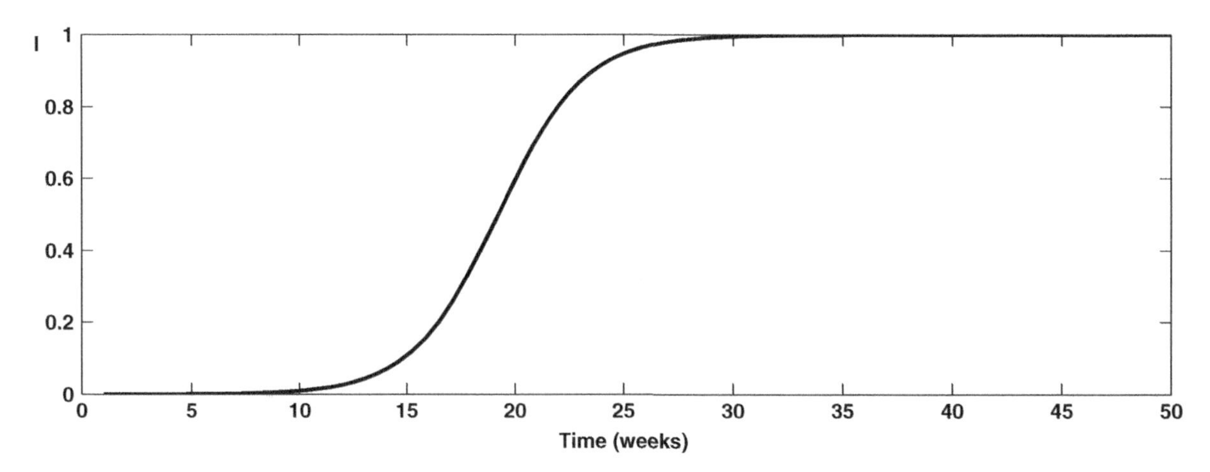

Figure 16.1 Solution for infected population with respect to t.

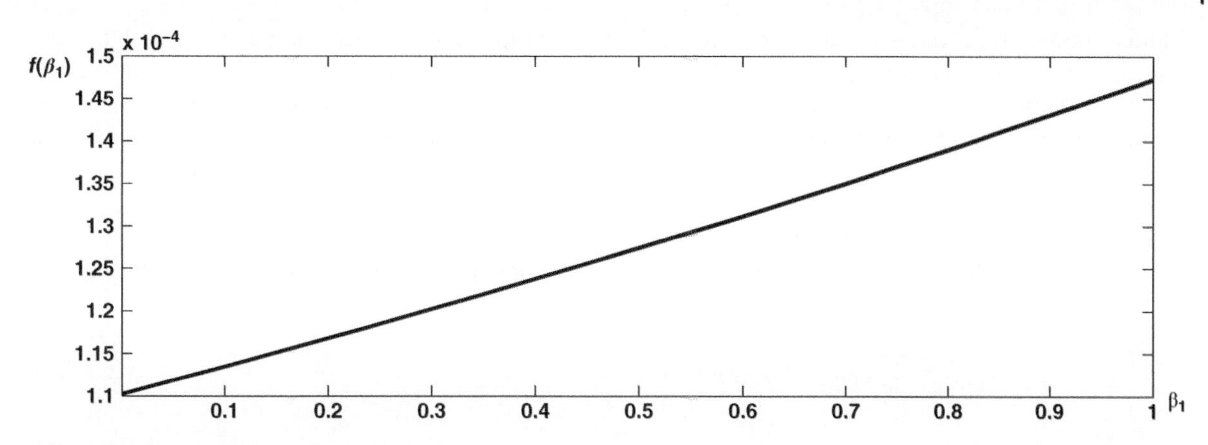

Figure 16.2 Function f_1 with respect to β_1.

where x_i denotes the amount of resources allocated to the ith region. The terms ζ_i and η_i are constants such that $\zeta_i : \eta_i$ is the ratio of internal to outward human movement of the ith region. The term w_{ij} is a mobility-based interaction measure between ith and jth regions. I_{0i} is the number of infectious people in the ith region just before the control strategy implementation. Using this Eq. (16.10) together with Eq. (16.9), we formulate the problem of best regional resource allocation as a nonlinear optimization problem (Figure 16.2).

16.2.3 Nonlinear Problem Formulation

We wish to minimize the total number of infectious persons in the country throughout the period of control strategy implementation. Let us denote by $I_i(t)$ the infectious persons in the ith region at time t and suppose that there are n regions. Then, the objective function to be minimized is,

$$Z = \sum_{i=1}^{n} I_i(\beta_i)dt \tag{16.11}$$

where

$$I_i(\beta_i) = \frac{I_{0i}N_i}{I_{0i} + (N_i - I_{0i})e^{-\beta_i N_i t}} \tag{16.12}$$

which will be followed by relevant constraints. It follows straightaway that n constraints must appear due to Eq. (16.10), relating the two types of decision variables x_i's and β_i's nonlinearly. That is,

$$\forall i \in \{1, 2, \dots, n\}, \quad \beta_i = \frac{\zeta_i I_{0i}}{x_i} + \eta_i \sum_{j \neq i} \frac{w_{ij} I_{0j}}{x_j} \tag{16.13}$$

Let the total (maximum possible) budget for resources be denoted by C. Then, the budgetary constraint can be expressed as,

$$\sum_{i=1}^{n} x_i \leq C \tag{16.14}$$

Finally, the nonnegativity of x_i, β_i must be specified by a set of $2n$ constraints.

$$\forall i \in \{1, 2, \dots, n\}, \quad \beta_i, x_i \geq 0 \tag{16.15}$$

Hence, our first optimization problem is minimizing (16.11), subject to constraints (16.13), (16.14), and (16.15). Clearly, this is a nonlinear optimization problem. In addition, it is nonconvex, adding more complexity to the solution process. However, our nonlinear objective function is expressible as a summation of f_i's, each f_i being a function of the single decision variable β_i. This property is identified as *separability* in optimization literature. Nonlinear optimization problems involving functions with this specific property can be efficiently solved under certain conditions, and such techniques have long been studied under the title *Separable Programming* [24–26]. This motivates us to restate our problem in the form of a linearly approximated problem as in Ref. [27].

16.2.3.1 Piecewise Linear Reformulation

In order to convert the problem into standard separable programming format, we restate the constraints in Eqs. (16.13) and (16.14) as summations of single-variabled functions as follows.

$$\forall i \in \{1, 2, \ldots, n\}, \quad \sum_{j=1}^{n} g_{ij}(x_j) + g_{i(n+1)}(\beta_i) = 0 \tag{16.16a}$$

$$\sum_{j=1}^{n} g_{(n+1)j}(x_j) \leq C \tag{16.16b}$$

where

$$g_{ii}(x_i) = -\frac{\zeta_i I_{0i}}{x_i} \tag{16.17a}$$

$$g_{ij}(x_j) = -\frac{\eta_i w_{ij} I_{0j}}{x_j} \quad \text{when } i \neq j \text{ and } i,j \leq n \tag{16.17b}$$

$$g_{i(n+1)}(\beta_i) = \beta_i \tag{16.17c}$$

$$g_{(n+1)j}(x_j) = x_j \tag{16.17d}$$

$$x_i, \beta_i \geq 0 \quad \forall i \in \{1, 2, \ldots, n\} \tag{16.17e}$$

Instead of solving Eqs. (16.11) and (16.13) directly, it is possible to approximate each nonlinear term by piecewise linear functions [27]. Although this provides a strategy of overcoming nonlinearity, nonconvexity is still an obstacle when searching for the global optimum. In order to circumvent this issue, it is possible to adopt the method of reducing down to mixed integer programming formulation. Our experimental results were generated using this method, and the readers are encouraged to refer [26] for a detailed description of the relevant reduction and algorithm.

16.2.3.2 Computational Experience

For our implementation, we used reported dengue incidence in January 2012, and relevant transportation statistics in Sri Lanka. We experimented by solving the piecewise linear approximation to our problem by implementing the algorithm in Ref. [26]. We used mixed integer programming model in *SAGEmath* [28] as the computational platform and our results were generated efficiently. These results indicate that several provinces experience a significant shortage of resources, if the infectious percentage in provinces is regarded the only indicator. This is shown graphically in Figures 16.3 and 16.4.

Applying our method in a different context might not incur too much extra computational cost. There are only $4(nk + nl)$ variables in the final mixed integer linear program, whenever n regions are considered. It is not realistic to assume that n would grow indefinitely, as regional resource allocations would require regions with sizable geographical area. Hence, the final mixed integer linear programing (MILP) will still be solvable in a reasonable time, in spite of the fact that generic MILP optimization is Non-deterministic Polynomial-time hard (NP-hard).

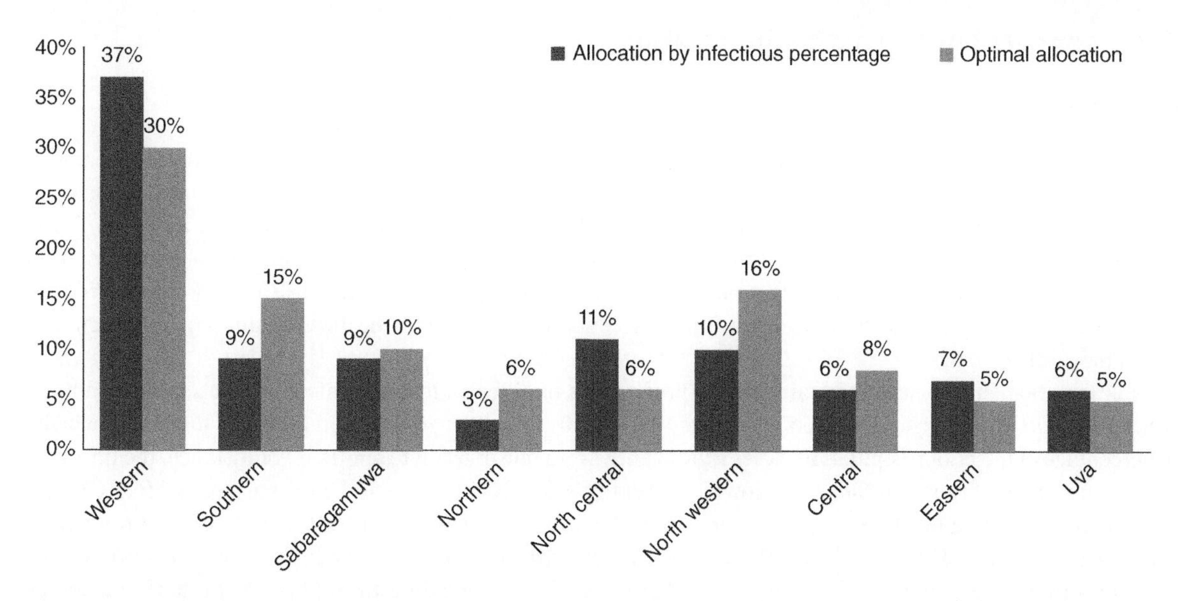

Figure 16.3 Comparison of resource allocation percentages (allocation by infectious percentage and optimal allocation).

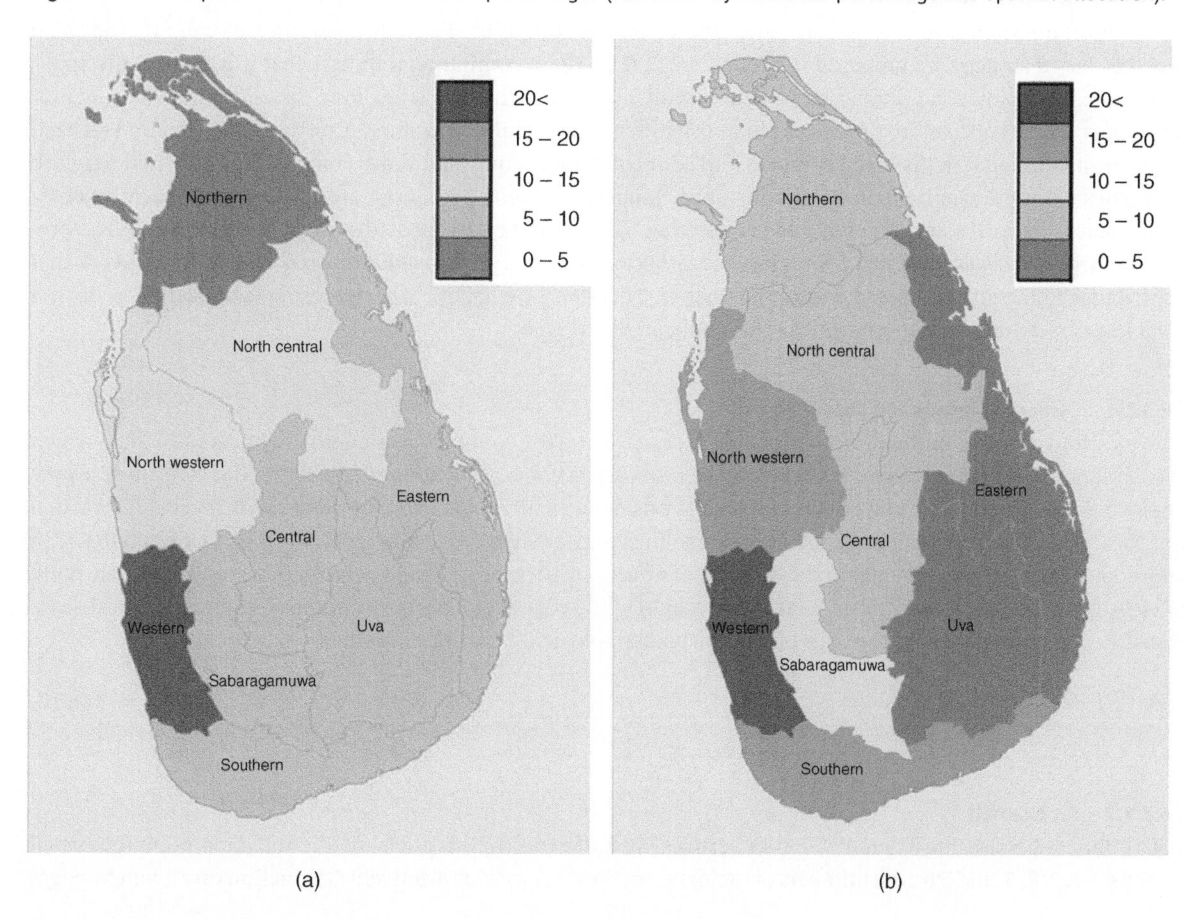

Figure 16.4 (a) Allocation by infectious percentage and (b) optimal allocation.

16.3 Connection–Strength Minimization

16.3.1 Network Model

Once a province receives its portion of resources, the next immediate decision problem is distribution of these resources within the province. In Sri Lanka, provincial resources are provided to small administrative divisions, first to districts and then to municipalities and medical officers of health areas, usually known as MOH. Each such smaller division receives its portion of resources for local implementation of dengue control activities. It is noteworthy that these regions too are interconnected through mobility. However, we have already considered the mobility factor in resource distribution; hence, our main concern here is a more localized factor than inter-regional human movement.

In a pragmatic point of view, there are disease-prone areas of different levels of infection in each local administrative division. Considering Sri Lanka, there are 13 local administrative divisions in Colombo district, of which the percentage of infectious population varies significantly from one division to another. For instance, the infected percentage in nine of these divisions is below 4%, although it is 43% in *Colombo Municipal Council* (CMC), the most urbanized geographical area. Many other divisions are highly connected to CMC and it acts as a transmission hub for the entire district [29]. In addition, Kolonnawa and Dehiwala too have high infectious ratio, which contributes to transmit the dengue to the district. Thus, interdivisional connections play a key role in spreading dengue in Colombo district.

A natural question to ask now is whether the resources should be distributed within Colombo by the same mobility-based strategy we proposed in Section 16.2? A careful examination indicates that it is not equally applicable here. Interprovincial human movement mainly takes place through a few long-distance bus routes; hence, *network properties* relevant to provinces are not significant in resource allocation. Contrastingly, a more localized division such as a district is made compact and connected by complicated inner connections, of which network properties cannot be ignored at all. For instance, suburbs in Colombo district are connected to each other by roads, lanes, channels, and different other mediums. This could be regarded a network in which vertices represent suburbs and edges represent connection levels between them. It is the *connection strength* of this network that contributes to the disease spread within the district. Therefore, we believe that the resource allocation at district level must be done aimed at minimizing this connection strength.

16.3.1.1 Disease Transmission Potential

One may find different methods from literature for measuring the connection strength of graphs [30, 31], of which *Wiener index* [32] is the conventional choice. We introduce a connection strength measure that suits our purpose and could be regarded a variant of Wiener index. Consider a small region i connected to j with level of interaction w_{ij}. Now, the disease transmission potential due to the connectedness of i and j is inversely proportional to w_{ij}. In addition, it must be proportional to the number of infectious persons in i and j as well. In a graph theoretic point of view, i and j are vertices, while w_{ij} acts as the weight of the edge (i, j). This motivates us to express H_{ij} the disease transmission potential along the edge (i, j) of the epidemiological network as follows:

$$H_{ij} = \frac{I_i I_j}{w_{ij}} \tag{16.18}$$

16.3.1.2 An Example

Let us illustrate this using a graph with four vertices in Figure 16.5. In this graph, a, b, c, and d represent four small regions. Let 2, 8, 5, and 5 be the numbers of incidence in these regions, respectively. Interactions are given by edges

Figure 16.5 Example graph on four subregions.

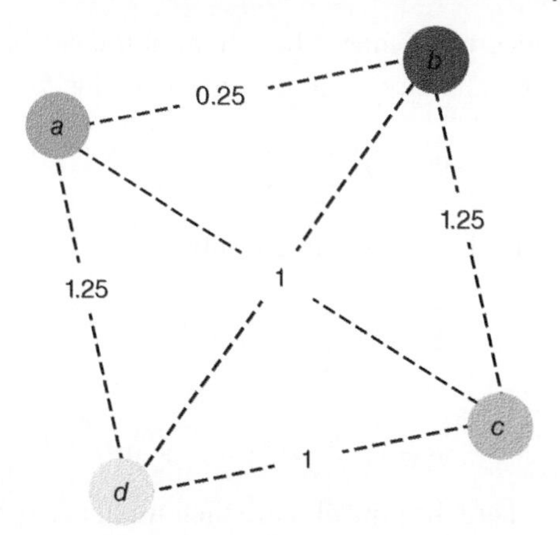

$(a, b), (a, c), (a, d), (b, c), (b, d), and(c, d)$ and weights of interaction are represented by matrix W, in which rows and columns are indexed by a, b, c, d.

$$W = \begin{bmatrix} 0 & 0.25 & 1 & 1.25 \\ 0.25 & 0 & 2 & 1 \\ 1 & 2 & 0 & 1 \\ 1.25 & 1 & 1 & 0 \end{bmatrix} \tag{16.19}$$

Consider the vertices c and d. Notice that $I_c = I_d$ and $w_{bd} < w_{bc}$. It is possible to calculate the disease transmission potential along (b, c) and (b, d).

$$H_{bc} = \frac{8 \times 5}{1.25} = 32 \tag{16.20a}$$

$$H_{bd} = \frac{8 \times 5}{1} = 40 \tag{16.20b}$$

It can be observed that $H_{bc} < H_{bd}$. That is, the transmission potential is inversely proportional to the interaction measure. Now, consider H_{bd} and H_{cd}. Notice that $w_{bd} = w_{cd}$ and $I_b > I_c$. It can be seen that $H_{bd} = 40 > H_{cd} = 25$. That is, the transmission potential increases because of the number of infected individuals. On the other hand, the potential of disease transmission can be reduced by increasing the per infectious resource allocation. Incorporating the per infectious resource allocation, the potential of disease transmission along the edge (i,j) can be restated as follows:

$$H_{ij} = \frac{I_i I_j}{x_i x_j w_{ij}} \tag{16.21}$$

where x_i denotes the resource allocation for i and I_i its incidence.

16.3.2 Nonlinear Problem Formulation

16.3.2.1 Connection Strength Measure

Once the transmission potential is defined for an edge, the next task would be defining the connection strength of the network. As mentioned before, our measure is motivated by Wiener index of graphs, which is one of the early attempts of measuring the connection strength of a graph, defined as

$$\overline{W} = \sum_{i=1}^{\hat{n}} \sum_{j \neq i} \sigma_i \sigma_j D_{ij} \tag{16.22}$$

where σ_i denotes the weight of the vertex i and D_{ij} is the distance between the vertices i and j [33]. The close resemblance of our measure $\hat{W}$ to the Wiener index is explicit from Eq. (16.23).

$$\hat{W} = \sum_{i=1}^{\hat{n}} \sum_{j \neq i} \frac{I_i I_j}{x_i x_j w_{ij}} \tag{16.23}$$

This is to be minimized subject to

$$\sum_{i=1}^{\hat{n}} x_i \leq C \tag{16.24}$$

and

$$\forall i \in \{1, 2, \dots, \hat{n}\}, \quad x_i \geq 0 \tag{16.25}$$

where the symbols have their usual meaning as in Section 16.2.

16.3.2.2 Piecewise Linear Approximation

It is straightforward to see that we are supposed to solve a nonlinear optimization problem again, somewhat similar to the national budget allocation in Section 16.2.3. In contrast to that, functions in this optimization problem do not have the remarkable property that they can be expressed as summations of single-variabled functions, thus making it impossible to be approximated by piecewise linear functions directly. However, some substitution will help, letting us to restate the original problem using new variables. For instance, it is possible to use

$$x_i x_j = y_k, \quad k \in \{1, 2, \dots, \hat{n}(\hat{n} - 1)/2\} \tag{16.26}$$

to transform the objective function to the separable form. Then, the each term $x_i x_j$ in the objective function is replaced by y_k and the following constraints in the separable form,

$$\log x_i + \log x_j = \log y_k \tag{16.27}$$

are introduced. Then, the objective function in Eq. (16.23) can be restated as,

$$\hat{W} = \sum_{k=1}^{\frac{\hat{n}(\hat{n}-1)}{2}} \frac{\alpha_k}{y_k} \tag{16.28}$$

where $\alpha_k = \frac{I_i I_j}{w_{ij}}$, for some $i \in \{1, 2, \dots, \hat{n}\}, j > i$, with separable constraints

$$\forall i \in \{1, 2, \dots, \hat{n}\}, \forall j > i, \quad \log x_i + \log x_j = \log y_k \tag{16.29}$$

Also, the non-negativity of y_k is specified by

$$\forall k \in \{1, 2, \dots, \hat{n}(\hat{n} - 1)/2\}, \quad y_k \geq 0 \tag{16.30}$$

Hence, our second nonlinear separable optimization problem is minimizing (16.28) subject to

$$\forall i \in \{1, 2, \dots, \hat{n}\}, \forall j > i, \quad \log x_i + \log x_j = \log y_k \tag{16.31a}$$

$$\sum_{i=1}^{\hat{n}} x_i \leq C \tag{16.31b}$$

$$\forall k \in \{1, 2, \dots, \hat{n}(\hat{n} - 1)/2\}, \quad y_k \geq 0 \tag{16.31c}$$

$$\forall i \in \{1, 2, \dots, \hat{n}\}, \quad x_i \geq 0 \tag{16.31d}$$

for which the same separable programming techniques in Section 16.2 are applicable. Once the optimal assignment of variable is made for the approximated formula, the optimal resource allocation for each small region could be determined.

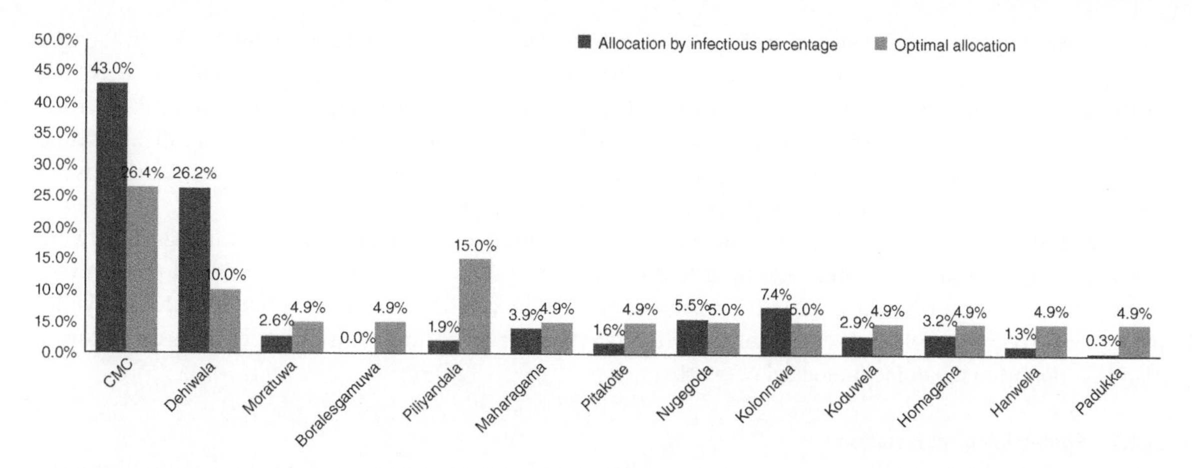

Figure 16.6 Comparison of resource allocation percentages (allocation by infectious percentage and optimal allocation).

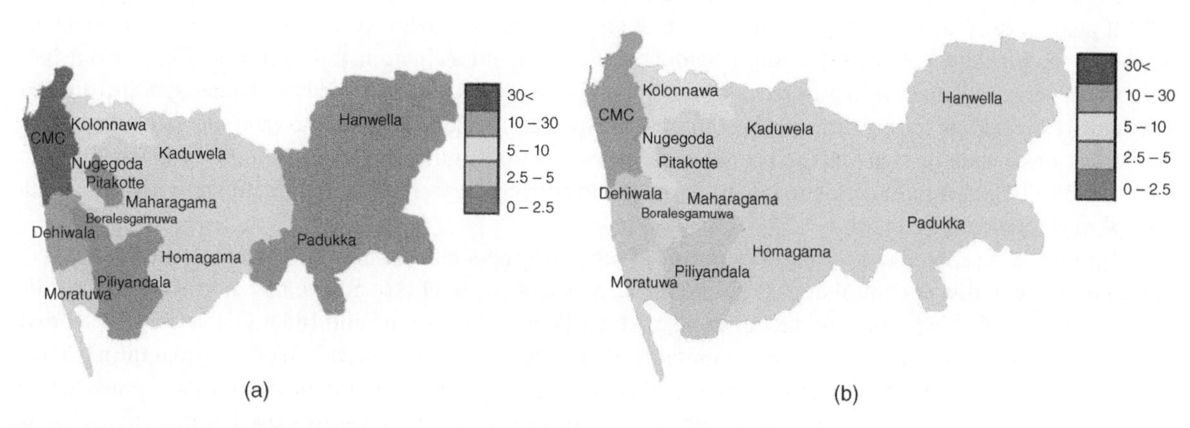

Figure 16.7 (a) Allocation by infectious percentage. (b) Optimal allocation.

16.3.2.3 Computational Experience

We experimented this using real data related to dengue transmission in Colombo district, Sri Lanka. For our implementation, we used reported dengue incidences in January 2011. Further, we defined the w_{ij} as a measure of human mobility between the ith and the jth small regions. If there is a direct public transport between the ith and the jth, then w_{ij} was set to one, otherwise $w_{ij} = 2$. In order to solve the approximated linear programming formulation, we used mixed integer programming model in Ref. *SAGEmath* [28]. Figure 16.6 represents a comparison of the proposed resource allocation percentage and infected percentage of each division in Colombo district (Figure 16.7).

16.4 Risk Minimization

16.4.1 Novel Strategies for Individuals

Once the resources are distributed to small administrative divisions as in Section 16.3, it is worth to consider how these resources must be utilized to achieve best results. Recall our proposal is on a more systematic and well-planned expenditure of the usual budget; it is helpful if novel strategies that serve our purpose could be defined, by slightly modifying the currently used strategies, rather than introducing totally dissimilar ones.

It is important in this point of view to review very localized disease control strategies in Sri Lanka. In the case of dengue, once an incidence is reported from a particular house, a public health inspector (*widely known as* PHI) visits the household of the patient to scan the mosquito breeding sites and to implement relevant control activities. Likewise, equal attention is paid to every infected individual and household. It is not difficult to see that this is far from the optimum, in achieving whatever particular objectives of resource allocation. On the other hand, it is not feasible for the inspectors to look into every household in the local area.

In contrast to this, we propose to identify the most influential individuals in the local epidemiological network and to spend more resources on them and their households. For instance, if a person is identified to have a very high influence in transmitting the disease, the resources must be allocated to clean the household of this particular individual, also recommending essential homestay for a certain period of time. Thus, this person is undergone a certain *isolation* from the epidemiological network.

16.4.1.1 Epidemiological Isolation

The notion of isolation existed for more than half a millennium in human society, and it helped controlling communicable disease outbreaks [34–37]. Examples include quarantine, special clothing as warning signs [35, 36], bills of health issued to ships [34], and regulation of groups of persons who were believed to be responsible for spreading the infection [34, 35, 37]. Besides, before the invention of vaccination, most successful infection avoidance was having no contact with infected persons. During the middle ages, individuals with leprosy were expelled from cities to avoid contact with them and communities expected people with leprosy to wear special clothing with yellow color as a warning sign [35]. Red paper and cloth were hung around the beds of children with smallpox in China, India, Turkey, and Asian Georgia [36]. From the twelfth century onwards, red treatment was practiced in European countries as well [36].

Quarantine was adopted as an obligatory means of separating persons that may have been exposed to a contagious disease and used to control disease outbreaks such as leprosy, smallpox, and yellow fever. Later, in middle ages, hospitals called *lazar* or *leper houses* were used, particularly in Europe and India to colonize the leprosy infectious. These houses were located on mountains or outside town limits in order to ensure quarantine. Some of these hospitals had private gardens, chapels, and cemeteries [35]. Another form of isolation was practiced on passengers arriving on ships or flights from infected countries [34, 37]. This was exercised during the outbreak of plague in India [37]. Hospitals called *lazaretto* were located in Mediterranean ports to quarantine passengers come from infected countries, particularly through India, and were given special attention to passengers with fever, cough, or chills. When a ship landed, passengers who displayed symptoms that might indicate pneumonic plague were placed under the surveillance of the local health department in lazarettos for 40 days [34].

The isolation we propose has a similar purpose but different strategy. In the case of dengue and other vector-borne diseases, the resources can be utilized to provide mosquito repellents to individuals identified as very influential in spreading the disease, thus *isolating* them from the relevant epidemiological network. Also, the health inspectors may pay special attention to these individuals, contrary to their previous method of treating every case uniformly. We emphasize that the isolation must be done only within an acceptable ethical ground, and it must be strictly confined to an *epidemiological* isolation, unlike its medieval era version.

16.4.1.2 Identifying Objectives

It is important to identify the objectives of implementing the epidemiological isolation. This enables the health planners to evaluate the control process and to analyze if the resources were utilized optimally. Among several possible objectives, we select two particular objectives, which seem to have much practical significance over the others. The first is reducing the high-risk population. That is, once the susceptible population is considered, an epidemiological risk value for each person can be calculated, classifying this population into two disjoint classes as *high risk* and *low risk*. It is possible to utilize the resources aimed at minimizing the number of individuals in the high-risk category. On the other hand, they can be utilized to reduce the total risk of the susceptible population.

These are the two major objectives (in particular, the earlier objective) that we see as most appropriate in our context.

16.4.2 Minimizing the High-Risk Population

The risk level of each individual can be categorized as *external* and *internal* risk of being infected [38, 39]. The external risk of being infected depends on the level of interaction with infectious individuals [38]. We define the level of interaction between ith susceptible and jth infectious by v_{ij}. If the ith and jth individuals have no interaction, then, $v_{ij} = 0$ and $v_{ij} = 1$ represent the maximum level of interaction. The internal risk of being infected of ith susceptible person depends on the person's immunity level [39, 40]. For example, someone having diabetes has a higher risk of being infected with dengue than a nondiabetic person [41]. Hence, the immunity level is inversely related to the risk level. We denote the internal possibility of not being infected or the immunity level of the person by p_i. Considering the internal and external risk factors together, the risk level of ith susceptible person can be defined as follows:

$$r_i = (1 - p_i) \sum_{j \in I} v_{ij} \tag{16.32}$$

Notice that the upper limit of the risk value changes with the number of interactions per individuals by Eq. (16.32). To overcome this, we divide the risk function by number of interactions per individual (v).

$$r_i = \frac{1}{v}(1 - p_i) \sum_{j \in I} v_{ij} \tag{16.33}$$

Then, we define two risk levels for susceptible persons as *high* and *low* with respect to a threshold risk. Notice that all the values of $(1 - p_i)$ and $\frac{1}{v}\sum_{j \in I} v_{ij}$ are in [0, 1]. Hence, we take 0.5 as the *threshold value* (η) for both internal and external risk factors. Consequently, the threshold value for the risk function is 0.25, and we classify all $r_i \geq 0.25$ as high and all $r_i < 0.25$ as low.

16.4.2.1 An Example

Let us illustrate the concept of isolation using the example graph on eight vertices in Figure 16.8. In this graph, $I = \{0, 1, 2\}$ represent infected individuals and $S = \{3, 4, 5, 6, 7\}$ represent susceptible individuals. The edge set $\{(0, 3), (0, 4), (0, 5), (0, 6), (0, 7), (1, 3), (1, 5), (1, 6), (2, 4), \textit{ and } (2, 6)\}$ represent the interaction between susceptible and infected individuals and the weight of each interaction is represented by V.

$$V = \begin{bmatrix} 0 & 0 & 0 & 0.8 & 0.3 & 0.2 & 0.2 & 0.5 \\ 0 & 0 & 0 & 1 & 0 & 0.2 & 0.7 & 0 \\ 0 & 0 & 0 & 0 & 0.8 & 0 & 0.4 & 0 \\ 0.8 & 1 & 0 & 0 & 0 & 0 & 0 & 0 \\ 0.3 & 0 & 0.8 & 0 & 0 & 0 & 0 & 0 \\ 0.2 & 0.2 & 0 & 0 & 0 & 0 & 0 & 0 \\ 0.2 & 0.7 & 0.4 & 0 & 0 & 0 & 0 & 0 \\ 0.5 & 0 & 0 & 0 & 0 & 0 & 0 & 0 \end{bmatrix} \tag{16.34}$$

The vector p represents the immunity levels of 3rd, 4th, 5th, 6th, and 7th individuals

$$p^T = \begin{bmatrix} 0.4 & 0.2 & 0.6 & 0.3 & 0.7 \end{bmatrix} \tag{16.35}$$

According to Eq. (16.33), $r_3 = 0.36, r_4 = 0.29, r_5 = 0.05, r_6 = 0.303$, and $r_7 = 0.05$. Notice that there are three high-risk individuals in our example (r_3, r_4, r_6). Now, we shall isolate vertex 0; that is, we nullify the weights of edges incident upon vertex 0. Then, by calculating the risk values for susceptible individuals again, we can

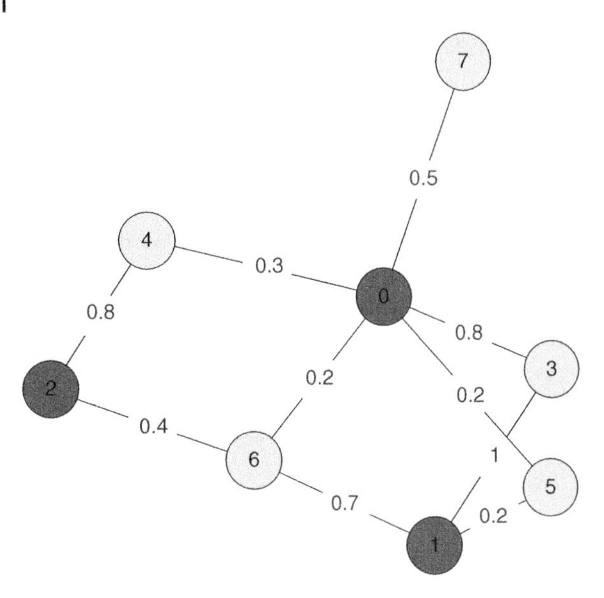

Figure 16.8 Example graph on eight individuals.

see that all susceptibles are in low-risk category. Moreover, notice that edges incident upon vertex 0 have a total weight 2. Because the isolation cost is proportional to the weight of interaction, the cost of isolation strategies is proportional to 2 units. It can be seen that the results are quite different if all cases were treated uniformly. That is, if all v_{ij}'s in our example were reduced by 0.2 and calculated the new risk value $\hat{r_i}$ for all susceptible individuals, then $\hat{r_3} = 0.28, \hat{r_4} = 0.187, \hat{r_5} = 0, \hat{r_6} = 0.16$, and $\hat{r_7} = 0.03$, which is proven not to nullify the high-risk individuals.

16.4.2.2 Model Formulation

Our method is motivated by the discrete optimization model in [42]. Define a Boolean decision variable γ_j for each infected individual based on isolation as follows:

$$\gamma_j = \begin{cases} 1 & \text{if } j \in I \text{ is isolated} \\ 0 & \text{otherwise} \end{cases} \tag{16.36}$$

Recall that the allocated amount for each local area is determined before this utilization; thus, isolation is to be limited by a predefined number K. Hence, the resource constraint can be expressed as follows:

$$\sum_{j \in I} \gamma_j \leq K \tag{16.37}$$

Now, we define another Boolean decision variable h_i for each susceptible individual based on the risk level.

$$h_i = \begin{cases} 1 & \text{if } r_i > \eta \\ 0 & \text{otherwise} \end{cases} \tag{16.38}$$

Because h_i and r_i are related nonlinearly, we choose a sufficiently large number M and define the following two constraints to relate h_i to r_i through linear constraints.

$$r_i - Mh_i \leq \eta \tag{16.39}$$

$$r_i + M(1 - h_i) > \eta \tag{16.40}$$

Including the notion of isolation, the risk function in Eq. (16.33) can be reformulated as follows:

$$r_i = \frac{1}{v}(1 - p_i) \sum_{j \in I} v_{ij}(1 - \gamma_j) \tag{16.41}$$

16.4.2.3 Linear Integer Program

Now, the entire problem is expressible as a linear binary integer program. The objective function to be minimized is,

$$Z_1 = \sum_{i \in S} h_i \tag{16.42}$$

subject to the following constraints:

$$\frac{1}{v}((1 - p_i) \sum_{j \in I} v_{ij}(1 - \gamma_j)) - Mh_i \leq \eta \tag{16.43a}$$

$$\frac{1}{v}((1 - p_i) \sum_{j \in I} v_{ij}(1 - \gamma_j)) + M(1 - h_i) > \eta \tag{16.43b}$$

$$\sum_{j \in I} \gamma_j \leq K \tag{16.43c}$$

$$\gamma_j, h_i \in \{0, 1\} \tag{16.43d}$$

16.4.2.4 Computational Experience

In order to solve the binary integer program given by Eqs. (16.42) and (16.43), mixed integer programming model in *SAGEmath* [28] has been used. We experimented the proposed control strategies on a network with 500 individuals. To determine the $S:I$ ratio, we considered the infected percentage of population for each MOH area in Colombo district for several years.

From Table 16.1, it can be observed that the yearly infected percentage of population is less than 1% for each area. Hence, we present our results by defining initial infected ratio as 5%.

Figure 16.9 represents the variation of high-risk population against the number of isolated individuals. Notice that the gain by isolation shows an exponential behavior; that is, we can reduce the number of high-risk individuals rapidly in the first phase of isolation.

Table 16.1 Yearly dengue-infected ratio–Colombo MOH areas.

Area	Infected percentage – 2009	Infected percentage – 2010	Infected percentage – 2011
Dehiwala	0.246 337 075	0.568 717 334	0.766 322 509
Piliyandala	0.148 101 75	0.204 369 798	0.199 061 491
Homagama	0.164 900 273	0.117 253 209	0.153 299 248
Kaduwela	0.101 706 495	0.124 998 059	0.183 615 161
Kolonnawa	0.180 473 949	0.177 397 689	0.919 801 889
Kotte	0.162 278 573	0.185 027 906	0.582 382 917
Maharagama	0.209 395 251	0.188 903 973	0.410 465 92
CMC	0.280 465 103	0.339 167 101	0.584 904 596
Moratuwa	0.144 523 89	0.179 630 908	0.285 537 079
Nugegoda	0.345 236 314	0.369 402 856	0.620 274 578
Padukka	0.147 242 213	0.289 977 012	0.166 774 344
Boralesgamuwa	0.191 973 825	0.277 110 043	0.325 520 833
Hanwella	0.243 155 853	0.181 935 762	0.206 941 151

Sources: Data from MOH Areas Report, Epidemiology Unit, Ministry of Health, Sri Lanka.

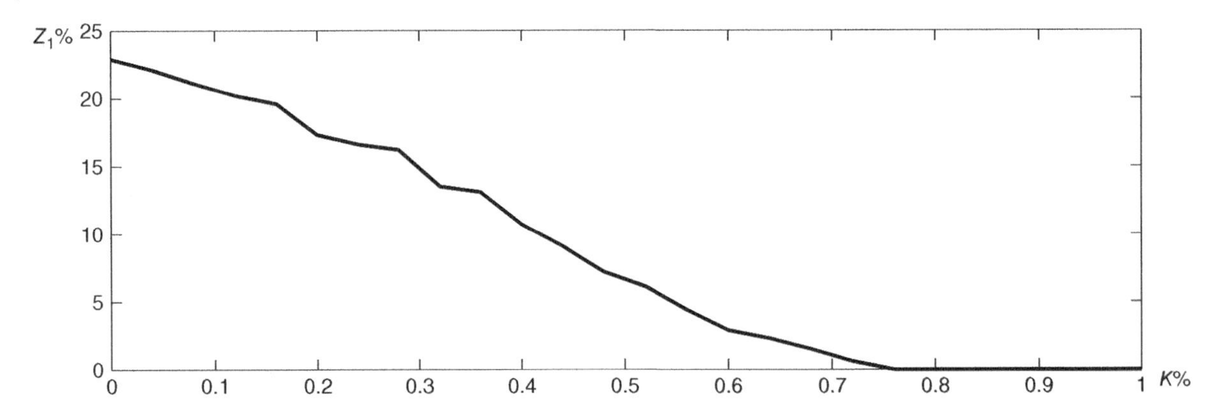

Figure 16.9 Variation of high-risk population percentage of susceptible population with isolation percentage of infected population. ($K\%$= isolation per infected population, $Z_1\%$= high-risk population per susceptible population).

We suggest to apply this model only for small regions, as solving a binary integer programming for a large number of variables is a challenging task. It is well-known that generic binary integer programming is NP-hard. Having said that, recent developments in scientific computing such as *tabu search* [43, 44], *simulated annealing* [45], and *quantum annealing* [46, 47] seem to have produced promising solutions for several binary integer programming instances. It would be an interesting research task to experiment with these methods and to extend the applicability of our discrete model.

16.4.3 Minimizing the Total Risk

Instead of minimizing the high-risk individuals, one may be interested in minimizing total risk of that population. Then, our integer programming formulation can be easily modified to meet the new objective. The new objective function to be minimized is

$$Z_2 = \sum_{i \in S} \frac{1}{v}((1 - p_i) \sum_{j \in I} v_{ij}(1 - \gamma_j)) \tag{16.44}$$

subject to similar constraints:

$$\frac{1}{v}((1 - p_i) \sum_{j \in I} v_{ij}(1 - \gamma_j)) - Mh_i \leq \eta \tag{16.45a}$$

$$\frac{1}{v}((1 - p_i) \sum_{j \in I} v_{ij}(1 - \gamma_j)) + M(1 - h_i) > \eta \tag{16.45b}$$

$$\sum_{j \in I} \gamma_j \leq K \tag{16.45c}$$

$$\gamma_j, h_i \in \{0, 1\} \tag{16.45d}$$

Figure 16.10 illustrates the variation of total risk function against the number of isolated individuals, and it can be observed that total risk reduction is almost linear.

16.4.4 Goal Programming Approach

We believe that minimizing the high-risk individuals is of highest significance of all possible objective functions. Also, some conditions might insist selecting total risk minimization. Although it is natural to select one of these objectives, in certain cases, the health planners might be interested in achieving both objectives instead of one. In order to achieve this, it is possible to make use of *goal programming* [48, 49].

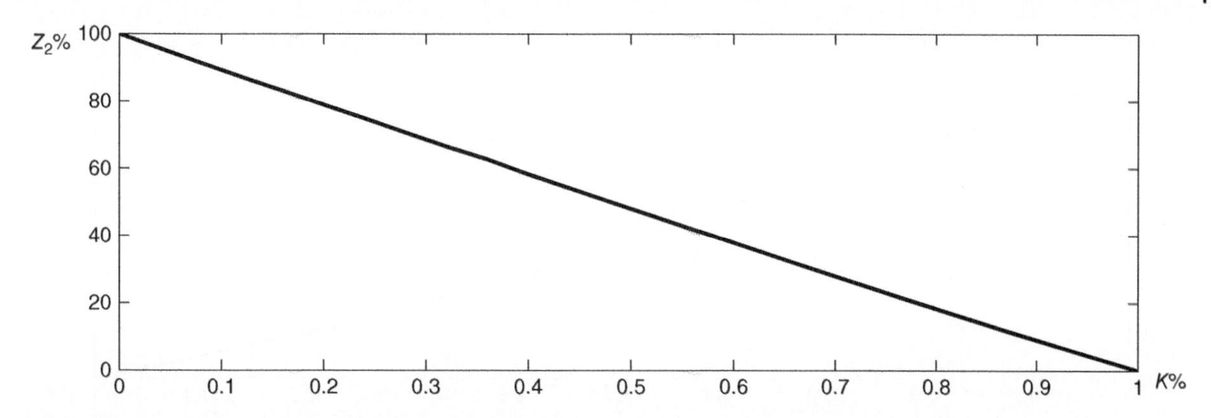

Figure 16.10 Variation of total risk percentage with isolation percentage of infected population. (K%= isolation per infected population, Z_2%= total risk percentage.)

The goal programming model has a multidimensional objective function, and we specify a goal δ_i for each objective function (Z_i). Then, we define deviations above and below for ith goal by U_i and L_i. Total deviation of the ith goal is given by $U_i + L_i$. The aim of goal programming is to minimize the total deviation of goals subject to a given set of constraints. Hence, the objective function of goal programming can be expressed as,

$$\text{Minimize} \quad G = \sum_{i\in 1}^{m} (U_i + L_i) \tag{16.46}$$

where m is the number of goals in the model. Then, each goal value could be incorporated as a goal constraint.

$$\forall i \in \{1, 2, \dots, m\}, \quad Z_i - U_i + L_i = \delta_i \tag{16.47}$$

Finally, the non-negativity of U_i and L_i are specified by,

$$\forall i \in \{1, 2, \dots, m\}, \quad U_i, L_i \geq 0 \tag{16.48}$$

Recall we have two goals to be achieved; first, we minimize the total number of high-risk individuals to zero ($\delta_1 = 0$). Secondly, we minimize the total risk of the population to zero ($\delta_2 = 0$). Hence, the objective of two-dimensional goal programming can be expressed as,

$$\text{Minimize} \quad G = U_1 + L_1 + U_2 + L_2 \tag{16.49}$$

subject to,

$$Z_1 - U_1 + L_1 = 0 \tag{16.50a}$$

$$Z_2 - U_2 + L_2 = 0 \tag{16.50b}$$

$$\frac{1}{v}\left((1 - p_i) \sum_{j\in I} v_{ij}(1 - \gamma_j)\right) - Mh_i \leq \eta \tag{16.50c}$$

$$\frac{1}{v}\left((1 - p_i) \sum_{j\in I} v_{ij}(1 - \gamma_j)\right) + M(1 - h_i) > \eta \tag{16.50d}$$

$$\sum_{j\in I} \gamma_j \leq K \tag{16.50e}$$

$$\gamma_j, h_i \in \{0, 1\} \tag{16.50f}$$

$$U_1, L_1, U_2, L_2 \geq 0 \tag{16.50g}$$

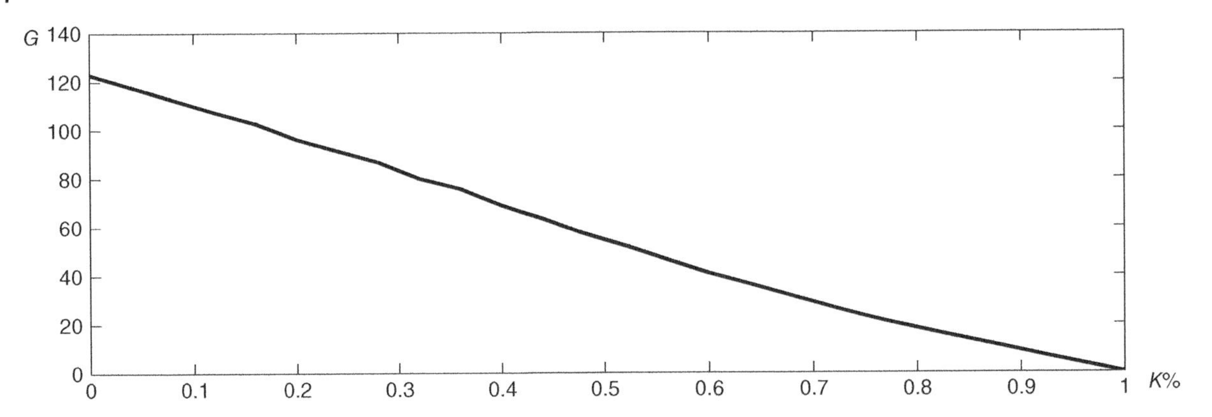

Figure 16.11 Total deviation of goals with isolation. ($K\%$= isolation per infected population, G= total deviation of goals.)

Moreover, the goal programming model allows us to rank goals according to order of importance to the decision makers [48, 49]. For instance, if we want to achieve goal 1, minimize the number of high-risk individuals first, and we can reformulate the objective function with respect to goal 1 without regard to goal 2. Then, the objective function in Eq. 16.49 can be modified as (Figure 16.11):

$$\text{Minimize} \quad G = U_1 + L_1 \tag{16.51}$$

subject to the similar constraints:

$$Z_1 - U_1 + L_1 = 0 \tag{16.52a}$$

$$Z_2 - U_2 + L_2 = 0 \tag{16.52b}$$

$$\frac{1}{v}\left((1-p_i)\sum_{j\in I} v_{ij}(1-\gamma_j)\right) - Mh_i \leq \eta \tag{16.52c}$$

$$\frac{1}{v}\left((1-p_i)\sum_{j\in I} v_{ij}(1-\gamma_j)\right) + M(1-h_i) > \eta \tag{16.52d}$$

$$\sum_{j\in I} \gamma_j \leq K \tag{16.52e}$$

$$\gamma_j, h_i \in \{0,1\} \tag{16.52f}$$

$$U_1, L_1, U_2, L_2 \geq 0 \tag{16.52g}$$

Once we solve Eqs. (16.51) and (16.52), we get exactly the same solution as the original discrete formulation represented by Figure 16.9. Similarly, when the objective function of goal problem is set to achieve goal 2, the system also gives the exact same solution as the original discrete formulation represented by Figure 16.10.

16.5 Conclusion

Systematic health planning would yield many benefits during a disease outbreak, in particular, to developing countries with limited resources. This chapter provides a complete guide to resource allocation, which is broken down to three levels. The national and provincial resource distributions were determined by two different continuous optimization models, which were eventually boiled down into approximate linear programs. Further, discrete models were developed for risk minimization of susceptible individuals, applicable to very small administrative divisions. Experimental results were presented for dengue-related data in Sri Lanka, Colombo district, and CMC area. These results indicate that systematic resource allocation would be much helpful for health planning in developing countries like Sri Lanka during a disease outbreak.

References

1 Fauci, A.S. (2001). Infectious diseases: considerations for the 21st century. *Clinical Infectious Diseases* 32 (5): 675–685.

2 World Health Organization (2000). WHO Report on Global Surveillance of Epidemic-Prone Infectious Diseases. Tech. rep. 892Geneva: World Health Organization.

3 Boutayeb, A. (2010). The burden of communicable and non-communicable diseases in developing countries. In: *Handbook of Disease Burdens and Quality of Life Measures* (ed. V.R. Preedy and R.R. Watson), 531–546. New York: Springer.

4 Jamison, D.T., Breman, J.G., Measham, A.R. et al. (2006). *Disease Control Priorities in Developing Countries*. The World Bank.

5 Lopez, A.D. and Murray, C.C. (1998). The global burden of disease, 1990–2020. *Nature Medicine* 4 (11): 1241.

6 Vilar-Compte, D., Camacho-Ortiz, A., and Ponce-de León, S. (2017). Infection control in limited resources countries: challenges and priorities. *Current Infectious Disease Reports* 19 (5): 20.

7 Dye, C. and Gay, N. (2003). Modeling the SARS epidemic. *Science* 300 (5627): 1884–1885.

8 Mbah, M.L.N. and Gilligan, C.A. (2011). Resource allocation for epidemic control in metapopulations. *PLoS One* 6 (9): e24577.

9 Nowzari, C., Preciado, V.M., and Pappas, G.J. (2015). Optimal resource allocation for control of networked epidemic models. *IEEE Transactions on Control of Network Systems* 4 (2): 159–169.

10 Preciado, V.M., Zargham, M., Enyioha, C. et al. (2013). Optimal vaccine allocation to control epidemic outbreaks in arbitrary networks. In: *52nd IEEE Conference on Decision and Control*, 7486–7491. IEEE.

11 Preciado, V.M., Zargham, M., Enyioha, C. et al. (2014). Optimal resource allocation for network protection against spreading processes. *IEEE Transactions on Control of Network Systems* 1 (1): 99–108.

12 Rowthorn, R.E., Laxminarayan, R., and Gilligan, C.A. (2009). Optimal control of epidemics in metapopulations. *Journal of the Royal Society Interface* 6 (41): 1135–1144.

13 Zaric, G.S. and Brandeau, M.L. (2002). Dynamic resource allocation for epidemic control in multiple populations. *Mathematical Medicine and Biology* 19 (4): 235–255.

14 Rais, A. and Viana, A. (2011). Operations research in healthcare: a survey. *International Transactions in Operational Research* 18 (1): 1–31.

15 Dushoff, J., Plotkin, J.B., Viboud, C. et al. (2007). Vaccinating to protect a vulnerable subpopulation. *PLoS Medicine* 4 (5): e174.

16 Allen, L.J., Brauer, F., Van den Driessche, P., and Wu, J. (2008). *Mathematical Epidemiology*, vol. 1945. Springer.

17 Medzhitov, R. (2007). Recognition of microorganisms and activation of the immune response. *Nature* 449 (7164): 819.

18 Wolfe, N.D., Dunavan, C.P., and Diamond, J. (2007). Origins of major human infectious diseases. *Nature* 447 (7142): 279.

19 Zaric, G.S. and Brandeau, M.L. (2001). Resource allocation for epidemic control over short time horizons. *Mathematical Biosciences* 171 (1): 33–58.

20 Kermack, W.O. and McKendrick, A.G. (1927). A contribution to the mathematical theory of epidemics. *Proceedings of the Royal Society of London. Series A, containing papers of a Mathematical and Physical character* 115 (772): 700–721.

21 Boatto, S., Bonnet, C., Cazelles, B., and Mazenc, F. (2018). *SIR model with time dependent infectivity parameter: approximating the epidemic attractor and the importance of the initial phase.*

22 Mummert, A. (2013). Studying the recovery procedure for the time–dependent transmission rate (s) in epidemic models. *Journal of Mathematical Biology* 67 (3): 483–507.

23 Pollicott, M., Wang, H., and Weiss, H. (2012). Extracting the time-dependent transmission rate from infection data via solution of an inverse ODE problem. *Journal of Biological Dynamics* 6 (2): 509–523.

24 Charnes, A. and Lemke, C.E. (1954). Minimization of non-linear separable convex functionals. *Naval Research Logistics Quarterly* 1 (4): 301–312.

25 Li, H.L. and Yu, C.S. (1999). A global optimization method for nonconvex separable programming problems. *European Journal of Operational Research* 117 (2): 275–292.

26 Markowitz, H.M. and Manne, A.S. (1957). On the solution of discrete programming problems. *Econometrica: Journal of the Econometric Society* 25 (1): 84–110.

27 Sinha, S. (2005). *Mathematical Programming: Theory and Methods*. Elsevier.

28 Developers, T.S. (2016). SageMath, the Sage Mathematics Software System (Version 7.1). http://www.sagemath.org.

29 Wickramaarachchi, W., Perera, S., and Jayasinghe, S. (2015). Modelling and analysis of dengue disease transmission in urban Colombo: a wavelets and cross wavelets approach. *Journal of the National Science Foundation of Sri Lanka* 43 (4): 337–345.

30 Harris, C.D. (1954). The, market as a factor in the localization of industry in the United States. *Annals of the Association of American Geographers* 44 (4): 315–348.

31 Marchiori, M. and Latora, V. (2000). Harmony in the small-world. *Physica A: Statistical Mechanics and its Applications* 285 (3–4): 539–546.

32 Wiener, H. (1947). Structural determination of paraffin boiling points. *Journal of the American Chemical Society* 69 (1): 17–20.

33 Iyer, V.K. (2012). *Wiener Index of Graphs: Some Graph-Theoretic and Computational Aspects*. LAP Lambert Academic Publishing.

34 Cassar, P. (1987). A tour of the lazzaretto buildings. *Melita Historica: Journal of the Malta Historical Society* 9 (4): 369–380.

35 Covey, H.C. (2001). People with leprosy (Hansen's disease) during the middle ages. *The Social Science Journal* 38 (2): 315–321.

36 Fenner, F. (1982). Global eradication of smallpox. *Reviews of Infectious Diseases* 4 (5): 916–930.

37 Titball, R.W. and Leary, S.E. (1998). Plague. *British Medical Bulletin* 54 (3): 625–633.

38 Bouzid, M., Colón-González, F.J., Lung, T. et al. (2014). Climate change and the emergence of vector-borne diseases in Europe: case study of dengue fever. *BMC Public Health* 14 (1): 781.

39 Whitehorn, J. and Simmons, C.P. (2011). The pathogenesis of dengue. *Vaccine* 29 (42): 7221–7228.

40 Premaratne, M., Perera, S., Malavige, G.N., and Jayasinghe, S. (2017). Mathematical modelling of immune parameters in the evolution of severe dengue. *Computational and Mathematical Methods in Medicine* 2017 Article ID 2187390, 9 pages.

41 Chen, L.C., Lei, H.Y., Liu, C.C. et al. (2006). Correlation of serum levels of macrophage migration inhibitory factor with disease severity and clinical outcome in dengue patients. *The American Journal of Tropical Medicine and Hygiene* 74 (1): 142–147.

42 Charkhgard, H., Subramanian, V., Silva, W., and Das, T.K. (2018). An integer linear programming formulation for removing nodes in a network to minimize the spread of influenza virus infections. *Discrete Optimization* 30: 144–167.

43 Battiti, R. and Tecchiolli, G. (1994). The reactive tabu search. *ORSA Journal on Computing* 6 (2): 126–140.

44 Glover, F. and Laguna, M. (1998). Tabu search. In: *Handbook of Combinatorial Optimization* (ed. D.Z. Du and P.M. Pardalos), 2093–2229. Boston, MA: Springer.

45 Kirkpatrick, S., Gelatt, C.D., and Vecchi, M.P. (1983). Optimization by simulated annealing. *Science* 220 (4598): 671–680.

46 Mahasinghe, A., Hua, R., Dinneen, M.J., and Goyal, R. (2019). Solving the Hamiltonian Cycle Problem using a quantum computer. In: *Proceedings of the Australasian Computer Science Week Multiconference*, 8. ACM.

47 McGeoch, C.C. (2014). Adiabatic quantum computation and quantum annealing: theory and practice. *Synthesis Lectures on Quantum Computing* 5 (2): 1–93.

48 Coello, C.A.C., Lamont, G.B., and Van Veldhuizen, D.A. (2007). *Evolutionary Algorithms for Solving Multi-Objective Problems*, vol. 5. Springer.

49 Lee, Sang.M.. (1972). *Goal Programming for Decision Analysis*. Philadelphia, PA: Auerbach Publishers.

17

Artificial Intelligence and Autonomous Car

Merve Antürk[1,2], *Sırma Yavuz*[1], *and Tofigh Allahviranloo*[2]

[1] *Faculty of Electrical and Electronics, Department of Computer Engineering, Yıldız Technical University, Istanbul, Turkey*
[2] *Faculty of Engineering and Natural Sciences, Bahcesehir University, Istanbul, Turkey*

17.1 Introduction

With the advancement of technology, the use of computers and smart devices is increasing day by day. Increased usage and desire trigger the next step by making life easier. At this point, artificial intelligence, which aims to solve problems and eliminate requests, comes into play. In this section, first of all, artificial intelligence is explained and general usage areas are explained with examples. The objectives and levels of natural language processing (NLP), which is one of the subfields of artificial intelligence, are mentioned. The classification of robots is given in detail by making the most important definitions about the very popular robotic field that attracts everyone's attention. Image preprocessing, image enhancement, image separation, feature extraction, and image classification techniques are explained by explaining the logic of image processing, which is one of the most important subfields of artificial intelligence. Problem solving, which is the basic task of artificial intelligence, is explained in detail. Optimization related to problem solving is defined and applications related to this field are given with examples. Finally, today's technology and autonomous systems that we will be more closely integrated in the future are explained in detail.

17.2 What Is Artificial Intelligence?

Artificial intelligence is a field that provides the most accurate solutions for any problem by combining with the computer commands or software of the human mind. It aims not only to solve problems in computer science but also to solve problems in other sciences. Figure 17.1 shows the areas where artificial intelligence is used.

In essence, artificial intelligence means solving problems. There are many subfields of artificial intelligence that allow machines to think and work like humans: NLP, robotics, image processing, machine learning, problem solving, optimization, autonomous systems, etc. Let us examine the subfields of artificial intelligence in more detail.

17.3 Natural Language Processing

NLP is the science and engineering branch that deals with the design and realization of computer systems whose main function is to analyze, understand, interpret, and produce a natural language. NLP is a subcategory of

Mathematical Methods in Interdisciplinary Sciences, First Edition. Edited by Snehashish Chakraverty.

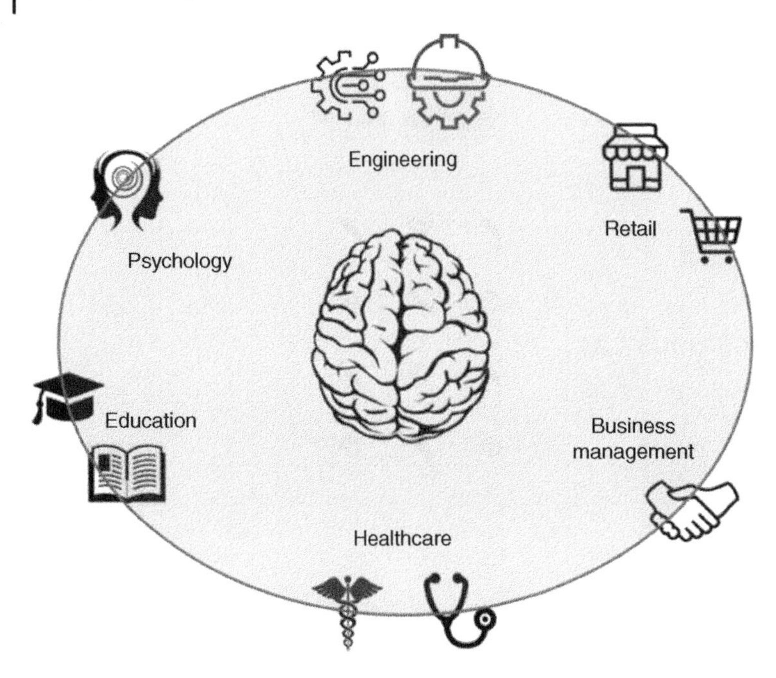

Figure 17.1 General usage of artificial intelligence.

artificial intelligence and linguistics, which examines the processing of texts, words, syllables, and sounds in the computer language, aiming at the processing and use of natural languages.

NLP combines theories, methods, and technologies developed in a wide range of fields such as artificial intelligence (knowledge representation, planning, reasoning, etc.), formal language theory (language analysis), theoretical linguistics, and cognitive psychology. This subject, which is seen as a small subfield of artificial intelligence in the 1950s and 1960s, is now recognized as a fundamental discipline of computer science after achieving the achievements of researchers and practices.

NLP is one of the candidate technologies to bring fundamental changes in people's interaction with computers in the years to come. Because of the fact that NLP is based on cheap computing power, the emergence of powerful computers for the NLP applications, which are very expensive compared to the expected result, has made major changes.

The main objectives in research in the field of NLP have generally been the following:

- Better understand the function and structure of natural languages
- Use natural language as an interface between computers and people and thus facilitate communication between people and the computer
- Language translation with computer

NLP systems are examined at five main levels. These levels are as follows:

- *Phonology*: Phonology examines the sounds of letters and how they are used in the language. All languages have an alphabet. The sound of each letter is different from the others, and these units are called phonemes. The main problem in phonology is that the phonemes are discrete, although the sound waves are continuous. To understand speech, a continuous-flowing speech should be divided into discrete sounds. Then, sounds and phonemes should be classified. This complex problem is handled within the scope of pattern recognition.

- *Morphology*: Morphology is the science that deals with word formation. At this stage, the main aim is to correctly identify and classify the root and the suffixes of the word. In a word, the order and structure of the inserts, the construction, and shooting of the process such as the determination of attachments are made at this level.
- *Syntax*: Syntaxology examines how words should be sorted in order to create a sentence. It defines the structural tasks of words in sentences, phrases, and subphrases that make up them. Combining words and creating sentences or parsing of a sentence is performed at this stage.
- *Semantics*: Semantics is the level in which language communicates with the real world. It examines the meanings of words and how they come together to form the meaning of the sentence.
- *Pragmatics*: Pragmatics examines situations in which sentences are used in a different sense from the normal meaning of sentences. For example, the phrase "can you open the window" is not a question. It is considered by some as a subset of semantics.

17.4 Robotics

Reprogrammable mechanical components are called robots. The concept of robotics defines the working area of mechanical engineering, computer engineering, electrical engineering, and control engineering. With the development of microchip technology over time, the concept of robotics has become widespread [1].

Applications of robot technology are listed below.

- Mechanical production: Parts selection, sorting, and placement;
- In the assembly of parts;
- Tooling and workpiece attachment, disassembly, and replacement;
- Deburring and polishing;
- Loading and unloading of hot parts (such as forged casting) to the counter (heat treatment);
- Measurement and control of finished parts; and
- Used for loading, transfer, and packaging of parts, machine tools, plastic parts manufacturing, presses, die casting, precision casting, forging, filling, and unloading of furnaces in storage processes.

If we classify the robot technology, we can collect it under two main classes.

17.4.1 Classification by Axes

The capacity of a robot movement is determined by the movements in the axes that are possible to be controlled. It is very similar to movements in numerical control. Industrial robots are made in different types and sizes. They can make various arm movements and have different motion systems.

17.4.1.1 Axis Concept in Robot Manipulators

If a manipulator is turning from a single joint around its axis or doing one of its forward/backward linear movements, this robot (1 degree of freedom) is called a uniaxial robot. If the manipulator has two connection points and each of these nodes has linear or rotational movements, this robot (2 degrees of freedom) is called a biaxial robot. If the manipulator has three connection points and each of these nodes has linear or rotational motions, this robot (3 degrees of freedom) is called a three-axis robot. Industrial robots have at least three axes. These movements can be summarized as the rotation around its axis and its ability to move up and down and back and forth.

17.4.2 Classification of Robots by Coordinate Systems

a. *Cartesian coordinate system*: All robot movements; they are at right angles to each other. This is a robot design form with the most limited freedom of movement. The process for assembling some parts is done by Cartesian configuration robots. This robot shape has parts moving in three axes perpendicular to each other. Moving parts X, Y, and Z move parallel to the Cartesian coordinate system axes. The robot can move the lever to points within the three-dimensional rectangular prism volume.
b. *Cylindrical coordinate system*: It can rotate around a basic bed and has a main body with other limbs. The movement is provided vertically and the main body is provided radially when the axis is accepted. Therefore, within the working volume, a zone is formed up to the volume of the main body, which the robot cannot reach. In addition, because of mechanical properties, the trunk cannot fully rotate 360°.
c. *Spherical coordinate system*: Mathematically, the spherical coordinate system has three axes, two circular axes and one linear axis. There are basically two movements. These are horizontal and vertical rotation. A third movement is the linear movement of the elongation arm. The linear motion behaves just like the movement of any coordinate from the Cartesian coordinates.
d. *Rotating coordinate system*: If a robot does the work in a circular motion, this type of robot is called a robot with a rotating coordinate system. The connections of the robot arm are mounted on the body so that it rotates around it and carries two separate sections with similar abutments. Rotating parts can be mounted horizontally and vertically.
e. *Multiaxis (articulated) robots*: Arm-jointed robots, according to their abilities, were made to undertake the tasks that the human arm could fulfill. They move freely in six axes. Three are for arm movement and the other three are for wrist movement. The biggest advantage of this connection is the ability to reach every point in the work area.
f. *SCARA Robotlar*: The Selective Compliance Assembly stands for the Robotic Arm, i.e. the assembly robot arm that matches the selected. SCARA-type robots are a type of robot with very high speed and best repetition capability. There are three general features: accuracy, high speed, and easy installation.

17.4.3 Other Robotic Classifications

There are many other classification methods such as classification by skill level, classification by control type, classification by technology level, classification by energy source, classification according to their work, holders, and limiters in robots.

- *Robot holders*: It is used very easily in the assembly process, welding operations, and painting operations. However, in an assembly line, it is seen that the same holder can do more than one job, or carry parts of various features, which can be seen as functional difficulties. In this case, it is a fact that a general purpose robot is needed. Robot hands have actuators based on various technologies that produce the power required for the movement of joints. The most widely used actuator technologies are electric motors, hydraulic actuators, and pneumatic actuators.
- *Mechanical holders*: They hold the parts between the mechanical holders and the fingers move mechanically.
- *Vacuum holders*: It is used to hold flat objects such as glass. The workpiece is held with the help of the vacuum formed between the holder and the holder.
- *Magnetic holders*: They are used to hold metal materials.
- *Adhesive holders*: It is used for carrying flexible materials such as adhesive materials and fabric.

The specifications of the programming languages of robots are listed below:

- The programming language is specific to the robot. (Example: Melfa Basic IV for Mitsubishi robot arms.)
- The language sets the electrical traffic for the robot components, processing the sensor data.
- It is simple to use and is limited by the capacity of the robot.

17.5 Image Processing

Basically, image processing is a name for analyzing a picture or video and identifying its characteristics. The characteristic feature of detection is, of course, based on distinctive character [2]. All of the facial recognition systems used today are performed with image processing.

The image can be modeled in a two-dimensional (2D) and $n \times m$ matrix. Each element of this matrix determines the magnitude of that pixel between black and white. The values of the pixels in bits and the size of the image in horizontal and vertical pixels affect the quality and size of the image. In an 8-bit black and white image, if a pixel is 0, it is black, and if a pixel is 255, it is white. The values are also proportional to black-and-white values.

Let us give another example. If we have an image whose size is 200×200 and each of the pixels in the image is 8-bits, the total size of the image is 40 000 byte (40 kB). For the calculation, the formula is used $\left(\text{size} = \frac{\text{total number of bits}}{\text{bits per byte}}\right)$. This information is correct if the image is stored as a bitmap. There are different compressed image formats that will reduce the size of these files.

The display format of color images is slightly different. The most common is the RGB (Red, Green, and Blue) representation, although there are different display formats. This impression assumes that each color can be created with a mixture of these three colors. Mathematically, these three colors can be considered for individual matrices.

17.5.1 Artificial Intelligence in Image Processing

The information expected to be obtained by visual processing has gone far beyond the labeling of a cat or dog on the photograph. The factors leading to this increase in expectation are both the increase in the number of data that can be processed and the development of the algorithms used. We are now able to work in important areas such as visual processing techniques to diagnose cancer cells or to develop sensors to be used in military fields. This advanced technology in image processing is made possible by the use of side heads such as machine learning, deep learning, and artificial neural networks.

Face recognition technologies, driverless tools, or Google's translation mode are among the simplest examples of machine learning-based image processing systems. Google Translate creates the image processing system in the translation application by machine learning and artificial neural network algorithms.

Google Translate's camera-to-camera mode allows you to save letters of any language with image processing and machine learning. When we scan the word with the camera, it combines the letters in the system database to reveal the words. After the emergence of the word or sentence as in the classic Google Translate application with its own language processing algorithm translates words. The important point for this mode to work is that the words you read with the camera are recognized by the system, which is thanks to the machine learning.

Artificial intelligence, in addition to providing advanced technology in image processing, also reduces processing time. The electro-optical sensor systems used in the military field have been used for the determination of mobile or fixed targets for many years. Nowadays, these systems are strengthened by artificial intelligence to provide both faster and more precise target detection.

17.5.2 Image Processing Techniques

Recently, new development techniques have been developed to support classical image processing methods or to walk through completely new algorithms. Most of the techniques are designed to develop images from unmanned spacecraft, space probes, and military reconnaissance flights. However, the developments have shown the effect in all areas where image processing is used.

A digital image consists of the numbers that gives information about the image or color. The main advantages of digital image processing methods are multidirectional scanning capability, reproducibility, and original data accuracy. Major image processing techniques are image preprocessing, image enhancement, image separation, feature extraction, and image classification.

17.5.2.1 Image Preprocessing and Enhancement

Images from satellites or digital cameras produce contrast and brightness errors due to limitations in imaging subsystems and lighting conditions when capturing images. Images can have different types of noise. Image preprocessing prevents the geometry and brightness values of the pixels. These errors are corrected using empirical or theoretical mathematical models.

Image enhancement improves the appearance of the image, allowing the image to be transformed into a more suitable form for human or machine interpretation, or emphasizes specific image properties for subsequent analysis. Visual enhancement is mainly provided by fluctuations on pixel brightness values. Examples of applications include contrast and edge enhancement, pseudo-coloration, noise filtering, sharpening, and magnification.

The enhancement process does not increase the content of internal information in the data. It emphasizes only certain image properties. Enhancement algorithms are usually a preparation for subsequent image processing steps and are directly related to the application.

17.5.2.2 Image Segmentation

Image segmentation is the process of dividing the image into self-constituent components and objects. It is determined when image segmentation will be stopped according to the application subject. Segmentation is stopped as soon as the searched object or component is obtained from the image. For example, airborne targets are firstly divided into photographs, roads, and vehicles as a whole. Each tool and path alone forms the split parts of this image. The image segmentation process is stopped when the appropriate subpicture is reached according to the target criteria.

Image threshold generation methods are used for image segmentation. In these limiting methods, the object pixels are set to a gray level and the background to different pixels. Usually, object pixels are black and the background is white. These binary images resulting from pixel difference are evaluated according to gray scale and image separation is made. Segmentation of images includes not only the distinction between objects and background but also the distinction between different regions.

17.5.2.3 Feature Extraction

Feature extraction can be considered as a top version of image segmentation. The goal is not only to separate objects from the background; the object is to define the object according to features such as size, shape, composition, and location. By mathematical expression, it is the process of extracting from raw data information to increase the variability of the class pattern while minimizing the in-class pattern variability. In this way, quantitative property measurements, classification, and identification of the object are facilitated. Because the recognition system has an observable effect on the efficiency, the feature extraction phase is an important step in the visual processing processes.

17.5.2.4 Image Classification

Classification is one of the most frequently used information extraction methods. In its simplest form, a pixel or pixel group is labeled based on its gray value. It uses multiple properties of an object during labeling, which makes it easier to classify it as having many images of the same object in the database. In vehicles using remote sensing technology, most of the information extraction techniques analyze the spectral reflection properties of images. Spectral analyses are performed by a number of specialized algorithms and are generally carried out by two different systems, supervised and uncontrolled multispectral classification.

17.5.3 Artificial Intelligence Support in Digital Image Processing

All of these methods are the basic principles of image processing, which can help us to define the definition of image processing. The widespread use of image processing in the technological field and the expectation of increase in image processing applications necessitated the introduction of artificial intelligence into image processing.

17.5.3.1 Creating a Cancer Treatment Plan

Some cancer centers are developing artificial intelligence technologies to automate their radiation treatment processes. The researchers at the center produced a technology combining artificial intelligence and image processing to develop a customized individual radiotherapy treatment plan for each patient. AutoPlanning establishes quantitative relationships between patients using machine learning and image processing techniques. An appropriate treatment plan is created based on the visual data associated with the disease information, diagnosis, and treatment details of each patient. Learn to model the shape and intensity of radiation therapy in relation to the images of the patient. Thus, the planning times are reduced from hours to minutes. In particular, it uses a database of thousands of high-quality plans to produce the most appropriate solutions to the problems experienced in the planning phase.

17.5.3.2 Skin Cancer Diagnosis

Computer scientists at the Standford Artificial Intelligence Laboratory have created an algorithm for skin cancer detection by combining deep learning with image processing. This algorithm can visually diagnose potential cancer with a high accuracy rate over a database containing approximately 130 000 skin disease images. The algorithm is fed as raw pixels associated with each image. This method is more useful than classical training algorithms because it requires very little processing of images before classification. The major contribution of deep learning to the algorithm is that it can be trained to solve a problem rather than adapting programmed responses to the current situation.

17.6 Problem Solving

Problem solving is mostly related to the performance outputs of reasonable creature. Problem solving is the activity of the mind when searching for a solution to a problem [3]. The term problem is a bit misleading. We often think of the problem as sadness and danger. Although this is true in some cases, it is not true in every case. For example, analyzing a potential merger is an opportunity research and can be considered as problem solving. Likewise, a new technology research is a problem-solving process. The term problem solving was first used by mathematicians. Decision making is used in the sense of problem solving in business world.

17.6.1 Problem-solving Process

The definition of the problem-solving process depends on the training and experience of the researchers. For example, Bel et al. has proposed several approaches to problem solving and decision-making that vary quantitatively according to people's intuition. Generally, six basic steps can be observed in the process: identifying and defining the problem, setting criteria for finding a solution, creating alternatives, searching for and evaluating solutions, making selections, and making recommendations and implementing them.

Some scientists use different classifications. Simon's classical approach, for example, has three phases: intelligence, design, and choice. Although the process in Figure 17.2 is shown as linear, it is rarely linear. In real life, some of these steps can be combined, some steps can be the basic steps, or revisions can be made in the initial steps. In short, this process is iterative. A brief description of each step is given below.

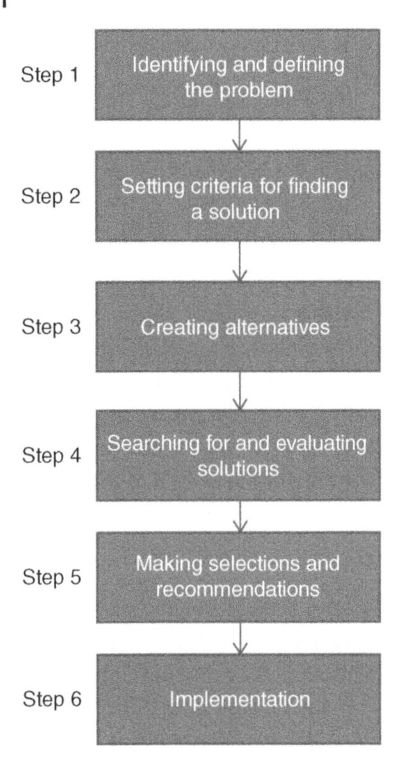

Figure 17.2 Steps of problem-solving process.

- *Step 1 – Identifying and defining the problem*: A problem (or opportunity) must be recognized first. The magnitude and significance of the problem (or opportunity) is specified and identified.
- *Step 2 – Determining criteria*: The solution of a problem depends on the criterion used to compare possible alternatives. For example, a good investment depends on criteria such as security, liquidity, and rejection rate. In this step, we determine the criteria and their relative importance to each other.
- *Step 3 – Creating alternatives*: According to the definition, there must be two or more chances to make a decision. Creating potential solutions requires creativity and intelligence.
- *Step 4 – Finding and evaluating solutions*: In this step, solution options are examined in the light of predetermined criteria. This step is basically a search process because we are trying to find the best or "good enough" solutions. Several step-by-step methodologies can be used in this step.
- *Step 5 – Making choices and making recommendations*: The result of the search is choosing a solution to recommend as a remedy to a problem.
- *Step 6 – Implementation*: To solve the problem, the recommended solution must be successfully implemented. In fact, this process is more complex because at each step, there may be several interim decisions, each following a similar process.

Applied Artificial Intelligence technologies can be used to support all of these six steps. However, most of the artificial intelligence movements take place in steps 4 and 5. In particular, expert systems are used to find solutions from the presumed alternatives. The role of artificial intelligence is mainly to manage search and evaluation by using some inference capabilities. Today, despite the limited role of artificial intelligence, it is hoped that after a certain time, technologies will play a greater role in the steps of the process. Nevertheless, artificial intelligence technologies have another great advantage. Although artificial intelligence effectively uses only two steps of the problem-solving process, artificial intelligence is used in many other tasks that are not classified as problem solving and decision-making. For example, expert systems help to develop computer commands. Expert systems are also used to simulate people's help centers, which provide information in catalogs and manuals, and for planning, complex scheduling, and information interpretation.

17.7 Optimization

Optimization tries to find the best possible solution using mathematical formulas that model a particular situation. The problem area must be structured in accordance with the rules, and optimization is managed by a one-step formula or an algorithm. Remember that the algorithm is a step-by-step search process in which solutions are produced and tested for possible improvements. Wherever possible, improvements are made and the new solution is subjected to an improvement test. This process continues until it becomes impossible to improve. Optimization is widely used in nonintelligence technologies such as business research (management science) and mathematics. In artificial intelligence, blind search and heuristic search are widely used [4].

Optimization is the job of finding the best solution under the constraints given by a basic definition. The solution of a problem is trying to find an optimal option that is listed in the solution set. These options are based on mathematical expressions. The optimal option of any problem is not the efficient solution for all problems. The solution or optimal option can be changed by the problem scope and the definition. The methods used do not guarantee the real solution. However, they are generally successful in finding the best possible solution at an acceptable speed. Various classifications of optimization problems have been made. However, according to the generally accepted classification [5];

- Minimization or maximization of objective function without any limitation of parameters, *unrestricted optimization*,
- Optimization problem with constraints or constraints on parameters is *constrained optimization*,
- According to whether the objective function and parameters are linear, an optimization problem is *linear optimization*,
- Nonlinear optimization if objective function or parameters are *nonlinear*,
- The problem of optimal regulation, grouping, or selection of discrete quantities *discrete optimization*,
- If the values of design variables are continuous values, such problems are called *continuous optimization* problems.

Different techniques are used to solve optimization problems. There are three main categories that are mathematical programming, stochastic process techniques, and artificial intelligence optimization techniques [5]. There are many examples of mathematical programming techniques: classical analysis, nonlinear programming, linear programming, dynamic programming, game theory, squared programming, etc. Statistical decision theory, Markov processes, regeneration theory, simulation method, and reliability theory are the examples of stochastic process techniques. In the last category, which is an artificial intelligence optimization technique, there are many examples such as genetic algorithm, simulated annealing, taboo search, ant colony algorithm, differential development algorithm, and artificial immune algorithm.

17.7.1 Optimization Techniques in Artificial Intelligence

Optimization techniques are called metaintuitive and research techniques. The aim of these algorithms is solving the problem. Some of the algorithms about the solution of the problem can guarantee the exact solution, but an appropriate solution can guarantee it.

Application results show that artificial intelligence optimization algorithms are problem-dependent algorithms. In other words, they may be successful in one problem and not in the same way for another problem. The most important artificial intelligence optimization algorithms are artificial heat treatment algorithms, genetic algorithms, and taboo search algorithms [6].

Artificial heat treatment algorithm: Artificial heat treatment algorithms, which are also mentioned in the literature as an annealing simulation method, are basically a different form of application of the best search methods by accepting the better solution encountered first. Unlike the search method with gradient, which aims to go to a better point at each step, the main disadvantage of this algorithm is that it may lead to a worse solution with a decreasing probability in the process.

Genetic algorithm: Genetic algorithms were proposed by Holland based on the biological evolution process. Unlike the other two methods described here, the genetic algorithm evaluates multiple solutions at the same time instead of a single solution. This feature is described as the parallel search feature of the genetic algorithm. Problem solving is expressed by a chromosome consisting of genes.

Taboo search: Taboo search is basically an adaptation of the best fit strategy around a solution that seeks the best local solution. The last steps taken by the algorithm are declared taboo so that the algorithm does not return to the same local best solution immediately after it leaves a local best solution. The taboo list is dynamic. Each time a new element enters the taboo list, the most remaining element in the taboo list is removed from the list. In this way, the algorithm is given a memory.

Applications of artificial intelligence optimization techniques are listed below:

Scheduling: Production schedules, workforce planning, class schedules, machine scheduling, and business schedules

Telecommunication: Search route problems, path assignment problems, network design for service, and customer account planning

Design: Computer-aided design, transport network design, and architectural area planning

Production and finance: Manufacturing problems, material requirements planning (MRP) capacities, selection of production department, production planning, cost calculations, and stock market forecasts

Location determination and distribution: Multiple trading problems, trade distribution planning, oil, and mineral exploration

Route: Vehicle route problems, transportation routes problems, mobile salesman problems, computer networks, and airline route problems

Logic and artificial intelligence: Maximum satisfaction problems, logic of probability, clustering, model classification and determination, and data storage problems

Graphics optimization: Graphics partitioning problems, graphic color problems, graphics selection problems, P-median problems, and image processing problems

Technology: Seismic technology problems, electrical power distribution problems, engineering structure design, and robot movements

General combinational optimization problems: 0–1 Programming problems, partial and multiple optimization, nonlinear programming, “all-or-none” networks, and general optimization

17.8 Autonomous Systems

17.8.1 History of Autonomous System

After Turing, the father of computer science, John McCarthy is the father of artificial intelligence. He is also the one who indirectly brings us together with Siri. He established the Artificial Intelligence Laboratory at Stanford University and developed the list processing (LISP), which is considered the programming language of artificial intelligence based on lambda calculations. He first described the concept as science and engineering for the construction of intelligent machines, especially intelligent computer programs, at a conference in Dartmouth in 1956. The term reason is, of course, the subject of scientific and philosophical discussions. Some researchers believe that strong artificial intelligence can be done, and that we can develop an artificial intelligence that matches or exceeds human intelligence, and that it is possible to accomplish any intellectual task that a human can accomplish. Other researchers believe that weak artificial intelligence applications may be possible and that software can be used to perform certain tasks much better than humans, but can never cover all of human cognitive abilities. No one is yet sure which one is right.

When we look at the history of autonomous vehicles in 1920–1930, some prominent systems already gave the gospel of autonomous vehicles. The first models that could travel on their own were introduced in the 1980s. The first vehicle was built in 1984 by Carnegie Mellon University with Navlab and autonomous land vehicle (ALV) projects. This was followed in 1987 by the Eureka Prometheus project, jointly implemented by Mercedes-Benz and the Bundeswehr University. After these two vehicles, countless companies have built thousands of autonomous cars, some of which are currently in traffic in several countries.

Autonomous automobiles seem to lead to a revolution in the passenger car sector, leading to a major transformation in the automotive sector. The rapid development of driver vehicle technology will soon change the rules of Henry Ford's game dramatically. Especially, the automotive industry will be one of the most affected units from this development. It is thought that driverless or autonomous cars will lead to an unprecedented economic, social, and environmental change in the near future. It is also clear that it will lead to an excellent equality of social status on the basis of citizens. Autonomous tools ensure that young people, the elderly, or physically disabled people have the same level of freedom to travel.

In addition to a few fundamental changes in the technological field, changes in the demands of consumers are being caused by the production of more powerful batteries, the provision of more environmentally friendly fuels, and the production of autonomous vehicles. These changes in the demands of consumers also lead to drastic changes in the planning of the manufacturing companies. In addition, the sharing economy has reached the automotive industry with BMW's diverse services, such as DriveNow in Paris or Autolib. Tesla has taken advantage of such developments so far, offering products that can be offered to the market among the most successful candidates. However, manufacturers in other companies such as BMW, Toyota, and General Motors in Silicon Valley are also making ambitious developments in this field. These developments seem to be one of the biggest steps toward the fourth industrial revolution in the automotive sector. Roland Berger, one of the largest strategic consulting firms based in Europe, says that the automotive, technology, and telecommunications sectors will unite at some point. Nowadays, considering the vehicles that have gained value in the market, we see that approximately 30% of the vehicles are based on electronics, and this ratio is expected to increase to 80% with the future innovations of the sector.

17.8.2 What Is an Autonomous Car?

Autonomous vehicles are automobiles that can go on the road without the intervention of the driver by detecting the road, traffic flow, and environment without the need of a driver thanks to the automatic control systems. Autonomous vehicles can detect objects around them by using technologies and techniques such as radar, light detection and ranging (LIDAR), GPS, odometry, and computer vision. Let us see detailed information about the sensors.

LIDAR, which stands for light detection and ranging, was developed in 1960s for the detection of airborne submarines and started to be used in 1970s. In the following years, there has been an increase in the usage area and types of sensors using LIDAR technique, both from air and land. LIDAR technology is currently used in architecture, archeology, urban planning, oil and gas exploration, mapping, and forest and water research. An LIDAR instrument principally consists of a laser, scanner, and specialized GPS receiver [7]. You can see an example of LIDAR view in Figure 17.3.

Both natural and manmade environments can be displayed by LIDAR. In a study, it is said to be very advantageous to use LIDAR for the emergency response operations [7].

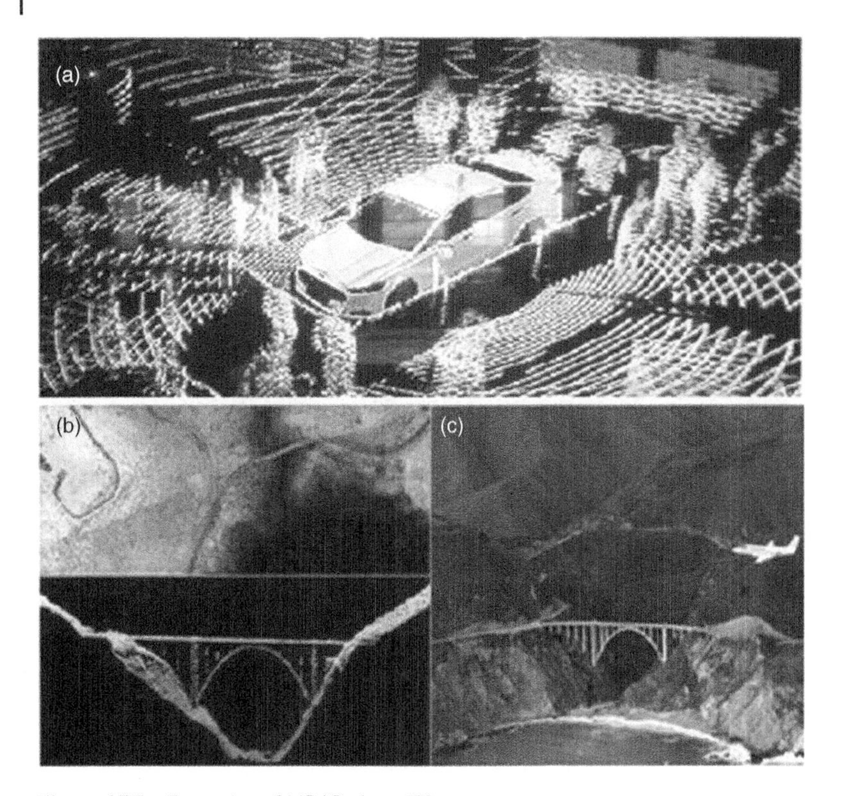

Figure 17.3 Examples of LIDAR views [7].

GPS, which stands for global positioning system, is a system that helps to provide a location with satellites, receivers, and algorithms. It is generally used for traveling such as air, sea, and land locations. GPS works at all times and in almost all weather conditions [8].

In general, there are five important usage of GPS:

- *Location*: Detects the location and position.
- *Navigation*: Gets from a location to another.
- *Tracking*: Monitors objects or personal movement.
- *Mapping*: Creates maps of the world.
- *Timing*: Calculates the precise time measurements.

17.8.3 Literature of Autonomous Car

There are several studies describing the level of autonomy in the literature. However, the most common of these is suggested by Thomas B. Sheridan [9]. According to this scale, the order of increasing autonomy level from 1 to 10 is given below.

1. Computer will not help; human makes the entire task.
2. The computer provides a set of motion options that includes all options.
3. The computer reduces the options up to a few.
4. The computer recommends a single action.
5. The computer executes the movement if the person approves.
6. The computer gives the person the right to refuse for a limited time before automatic execution.

7. The computer automatically executes the task and informs the person necessarily.
8. The computer informs the person after the automatic execution if he wishes.
9. Computer automatically informs people if they want after the automatic execution.
10. The computer decides and executes everything and does not care about people.

The first developments in the field of advanced vehicle control systems were discovered by the General Motors Research Group in the early 1960s. This group developed and exemplified the automatic control of automobile steering, speed, and brakes on test tracks [10]. This research triggered other research groups. In the late 1960s, Ohio State University and Massachusetts University of Technology began working on the application of these techniques to address urban transport problems.

From the first days of intelligent transportation systems, we can see that the results of the studies carried out in the laboratory environment can be seen in advanced driving support systems in the vehicles on sale. Large vehicle manufacturers are constantly researching new and improved driving assistance systems [11]. The purpose of these systems is to help drivers rather than replace them. The most important developments in this field are cruise control [12], dynamic stability control [9], antilock brakes (ABs) [13], and pedestrian detection with night vision systems [14], collision avoidance [15], and semiautonomous parking system [16]. In these applications, longitudinal control (acceleration and braking actions) is mostly emphasized.

One of the leaders of lateral control studies on the steering wheel movement is Ackermann [17]. Ackermann's approach is used to combine active steering with feedback of the swing rate to strongly distinguish between skidding and lateral movements. It is possible to control nonlinear dynamic systems such as steering control thanks to techniques that allow fast, trouble-free, and high-quality control. In order for the vehicle to continue to follow the reference trajectory; fuzzy logic [18], optimization of linear matrix inequalities [19], and skew rate control [20].

Another important change in the steering system has been the replacement of hydraulic power steering system (HPS) with electric power steering system (EPS) in new-generation vehicles. Such a system has advanced control simulations [17–21]. Sugeno and Murakami published a landmark publication in 1984. They have shown that any industrial process can be controlled by a simple process of human process or human experience [18]. The problem is reduced to finding the right control rules and fine-tuning them based on a drive experience. Several studies have compared fuzzy control and classical control techniques in autonomous vehicles, and fuzzy controllers have shown good results [22].

The first serious step in fuzzy logic was demonstrated in an article written by Azerbaijani scientist Lütfi Aliasker Zade in 1965 [23]. In this study, Zade stated that a large part of human thought is blurred and not precise. Therefore, human reasoning system cannot be expressed with dual logic. Human logic uses expressions such as hot and cold, as well as expressions such as warm and cool.

After the introduction of fuzzy logic, it was not accepted for a long time in the Western world. The idea found its place in practice only nine years after its production, when Mamdani carried out the inspection of a steam engine [24]. As the 1990s approached, fuzzy logic attracted attention in Japan and began to be used in many applications. In 1987, the Japanese Sendai subway designed by Hitachi managed to reduce energy consumption by 10% while using fuzzy logic to provide more comfortable travel, smoother slowdown, and acceleration in the subway [25]. The products of fuzzy logic applications were introduced to consumers in Japan in the early 1990s. According to the degree of pollution in dishwashers and washing machines, washing technology has yielded fruitful results [26]. The Western world could not remain indifferent to these developments, and in 1993, IEEE Transactions on Fuzzy Systems magazine began to publish.

In order to develop autonomous vehicles, there are models developed on ATVs (all-terrain vehicles). These models have a structure built on the acceleration and steering systems of the vehicle, as in almost all autonomous vehicle studies. Rather than developing a new vehicle from the start, it seems reasonable to make an already existing vehicle autonomous. One of the reasons for this is the opportunity to work with vehicles that have already been tested to work in all harsh conditions. Another reason is that the product that will be produced as a result

Figure 17.4 (a) Ensco, (b) Spirit of Las Vegas, (c) Overbot, and (d) Cajunbot.

of the studies will be independent of the vehicle to be integrated, and this will maximize the applicability. In this context, four teams whose names are Ensco [27], Spirit of Las Vegas [28], Overbot [29], and Cajunbot [30] that are built on the ATVs shown in Figure 17.4 produced for the DARPA Grand Challenge in 2004 and 2005 were examined. All the four teams controlled acceleration using a servo motor on the accelerator pedal. For steering, Cajunbot and Spirit of Las Vegas teams preferred DC Motors, while Ensco and Overbot teams preferred servo motors.

Researchers use DGPS (differential global positioning system) for positioning. In this way, the vehicle is able to determine its position with an average error of 10 cm instead of locating with an average of 15 m error with normal GPS [31]. The operating principle of DGPS is that in addition to the normal GPS system, an antenna with a location on the ground is taken as a reference. The satellites are detected with the reference antenna after the connection with the user and the reference antenna, and the location of the user is detected with very small errors. DGPS is the only solution yet available for positioning in outdoor applications.

In order to determine the exact movement of the vehicle, the researchers used a three-axis gyroscope and an accelerometer. The three-axis gyroscope is a key instrument in marine and aviation and is now being used in land vehicles. It works according to the basic physics rules and is based on the conservation of angular momentum. The gyroscope, which is also used in smartphones today, can easily be determined in which direction an object is inclined.

The accelerometer measures the acceleration to which a mass is exposed. The principle of operation is to measure the forces applied to the test mass from the mass in the reference axis. With this device, the forces applied to the mass such as acceleration, deceleration, and skidding can be measured.

Researchers have chosen a wide range of power sources for their projects. Power is an important issue because it is a major factor determining all dynamics in a drone. Although it is first considered that it meets the needs when supplying power to the system, the preferred power source should not place excessive volume and weight on the vehicle. In this context, internal combustion engines, high-current batteries, generators, and uninterruptible power supplies are used as power supplies. When required, the required voltages are produced with DC/DC converters.

Processors, random access memory (RAM), and hard drives used to process data and control devices can be catered for a wide variety of projects. Because the vehicles will move, it is generally taken care not to use moving parts. In this context, it is seen that solid-state disks are used for storage purposes. According to the needs of the project, wireless technologies are used to send the data to the computer.

Environment detection is the most important need of unmanned vehicles. The vehicles have to perform all their movements in the light of the data coming from the sensors. LIDAR, cameras, and ultrasonic sensors are the most commonly used sensors in unmanned land vehicle projects [32]. In the aforementioned studies, all teams used both LIDAR and camera. LIDAR is an optical distance sensing technology that can detect and measure objects by direction and distance, usually by laser-transmitted rays. LIDAR systems used in unmanned land vehicles have laser range finders of the type, which are reflected by rotating mirrors [33]. The device is very high-priced and helps to identify objects around the vehicle. Google's driverless vehicle can detect the perimeter of the vehicle.

The cameras used in the studies are two of types, mono and stereo. Mono cameras are generally used to monitor the vehicle's surrounding information, while stereo cameras are used to understand depth information. The images taken with all these sensors are combined, interpreted, and interpreted with advanced image processing methods. Open source software is widely used in the studies. Developed according to project-specific requirements, these softwares are shaped according to the requirements of the hardware used.

17.8.4 How Does an Autonomous Car Work?

The autopilot drive of autonomous vehicles starts briefly with the ultrasonic sensors on its wheels detecting the positions of vehicles that are braking or parked, and data from a wide range of sensors are analyzed with a central computer system, and events such as steering control, braking, and acceleration are performed. This event can only be described as a beginning. With the technology becoming cheaper, the future of driverless vehicles becomes increasingly realizable.

In autonomous vehicles, systems such as sensor data, extensive data analysis, machine learning, and the M2M (machine-to-machine) communication system are essential for the successful implementation of the Internet of Things philosophy. The autonomous vehicle projects currently on the market, evaluates the sensor data, analyzes and provides machine movements as well as machine learning. At present, there are not enough autonomous vehicles in today's traffic to design a vehicle suitable for M2M communication system. However, when a certain threshold is reached in this issue, interesting developments such as autonomous traffic control systems and autonomous traffic control can also be allowed.

Such a system would have a very extraordinary impact on industry and society as a whole. At the level of community and living standards, the biggest impacts are the decrease in the number of accidents, the increase in the use of passenger cars that easily accompany, and the decrease in crowds in public transportation centers. Together with the increase in the use of vehicles, the increase in the amount of fuel consumption because of the increase in oil trade seems to cause a new activity in the country's economy.

In the next eight years, Korea plans to produce autonomous vehicles that will be developed entirely with its own technology. Currently in the monopoly of American- and European-based companies in this sector in the future, market is expected to expand further into Asia. Until 2019, the Korean Government set aside a budget of approximately US$ 2 billion to produce the eight main components required for driverless vehicles, including visual sensors and radars. However, in 2024, 100% of Korean-made autonomous vehicles are planned to hit the road. It is clear that autonomous vehicles have a very bright future around the world. Korean car manufacturers and technology giants such as Samsung have already been thrown into developments in the autonomous vehicle industry. The structure of an autonomous car is given in Figure 17.5.

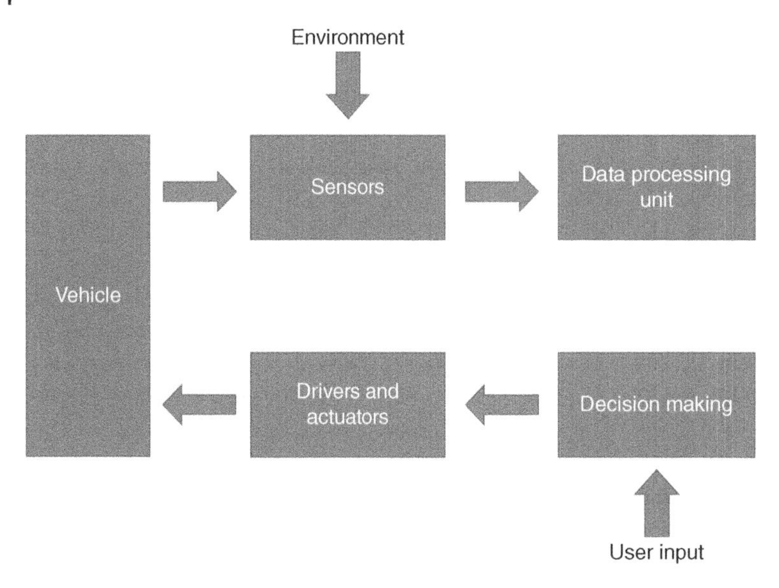

Figure 17.5 The structure of autonomous car.

There are two different types of autonomous car systems. The traditional approach divides the task into some basic sections, which is lane detection [34, 35], path planning [36, 37], and control logic [35, 38], and is often investigated separately. Image processing techniques such as Hough Transform, edge detection algorithms are usually used for detecting lane marks. After detecting lane marks, path planning can be done, and control logic helps to check conditions. In this approach, feature extraction and interpretation of image data are done. On the other hand, the end-to-end learning approach to self-driving tools is using convolutional neural networks (CNNs) [39]. End-to-end learning takes the raw image as the input and outputs the control signal automatically. The model optimizes itself based on training data, and there are no manually defined rules. These become two important advantages of end-to-end learning: better performance and less manual effort. Because the model is self-optimizing based on data to give maximum overall performance, the intermediate parameters are automatically adjusted to be optimal. In addition, there is no need to design control logic based on identifying certain predefined objects, labeling these objects during training, or observing them. As a result, less manual effort is required.

17.8.5 Concept of Self-driving Car

The concept of self-driving car has a relationship between computer vision, sensor fusion, deep learning, path planning, actuator, and localization [40]. These are the main parts of the self-driving car.

Computer vision: Computer vision is an interdisciplinary field that can understand high-level form of digital images or videos. Computer can see like from human's eye with the advanced computer vision algorithms [41, 42].

Sensor fusion: Sensor fusion is basically the combination of the data that is derived from the source [43, 44].

Deep learning: Deep learning is part of machine learning methods based on learning data representations. There are three types of methods, which are supervised, partially supervised, or unsupervised, for the learning [45, 46].

Path planning: Path planning is the most important field in the concept of robotics. In this field, shortest path can be calculated by the algorithm, and the calculated path helps autonomous mobile robots to move individually. Otherwise optimal paths could be paths that minimize the amount of turning, the amount of braking, or whatever a specific application requires [40].

Actuator system: An actuator is the main component of the machine for controlling a machine or system. It is the heart of the system.

Localization: Localization has a close relationship with navigation. For the navigation, robot or autonomous system needs a local or real-time map. For generating a map of the robot or autonomous system, the system needs to know its position in the frame and plans a path toward the target. In order to navigate in its environment, the robot or any other mobility device requires representation, i.e. a map of the environment and the ability to interpret that representation [40].

An autonomous vehicle needs data that come from the different types of sensors such as camera, radar, laser, etc., to see the perspective of the car. This is used for creating a digital map that is very important for object detection [47].

17.8.5.1 Image Classification

Image classification is the main action of self-driving car in the concept of object detection. The image is taken by the camera, and then computer vision algorithms are run on the image, and finally, items or objects are detected and labeled. There is an example of object detection in Figure 17.6.

Object detection consists of two processes, image classification and image localization. Image classification is used for determining the objects in the image, such as an animal or a plane, while image localization is used for determining the specific location of these objects. In the localization part, the objects are seen by the bounding boxes.

To perform image classification, a CNN is trained to recognize various objects, such as traffic lights and pedestrians. A CNN performs many operations on images in order to classify them.

17.8.5.2 Object Tracking

Object tracking algorithms can be divided into two categories that are tracking-by-detection methods and generative methods [48]. In tracking-by-detection methods, object is detected in every time. For the detection, a pretrained detector and the positions are used [49]. In this approach, all object categories must be trained previously. Generative object tracking methods are especially used for autonomous car. In these approaches, the most important thing is the appearance of the object. The object model is often updated online, adapting to appearance variations.

Figure 17.6 A simple example of object detection.

17.8.5.3 Lane Detection

Lane detection is one of the main actions for the autonomous vehicles. In that part, computer vision algorithms help to detect and find the lanes. In that point, a large amount of data is needed. A huge amount of data and complex algorithms mean deep learning. Object and lane detection for self-driving car is done by deep learning. There are many kinds of datasets such as ImageNet [50], ActivityNet [51], MS COCO, Open Images [52], YouTube-8M, and YouTube-BB. Recently, in the field of autonomous car or driving, various datasets have been published including CamVid [53], KITTI [54], TORCS [55], GTA5 [56], Cityscapes [57], and SYNTHIA [58]. In these datasets, user can find high-level scenes. Semantic segmentation is required for analyzing complex scenes. Segmenting objects means to detect all objects that are static or dynamic. After detecting, classification can be done according to the analysis of object [59].

At present, 90% of traffic accidents are caused by human error, but autonomous vehicles designed and developed today will be presented as programmed not to crash in the near future. Research is being carried out to solve some of the ongoing technology problems to ensure this technology. It will also be difficult to coordinate traffic with mixed drivers and driverless vehicles. In this transition period, the biggest problem that may arise in the traffic consisting of autonomous and nonautonomous vehicles is that both vehicles will share the road while the driverless automobiles will try to dominate the road.

In one company, which is located at Turkey, a self-driving robot has been developed for cleaning the streets of the big cities. This robot can also analyze and report the problems that are located in the streets.

17.8.5.4 Introduction to Deep Learning

The software that is designed to make human life easier can create excitement in the world of science, sometimes giving results beyond human intelligence. Deep learning is one of the most current and popular examples of this situation. You can think of deep learning as a more advanced version of the concept of machine learning. Deep learning is an approach used in machines to perceive and understand the world. The myth of artificial intelligence is only possible with deep learning. Everything from the automatic scanning of unnecessary emails to a mailbox to the driving cars without drivers or cancer research done without a doctor is the work of this chain of algorithms. In addition to top-ranking research companies, universities, social media companies, and technology giants are increasingly investing in this field.

Rather than constructing software step-by-step, lead up to make this fiction by computer is the goal of deep learning. In this way, it can produce computer model solutions against alternative scenarios. Although the scenarios that programmers can produce in the software phase are limited, there are innumerable solutions that deep learning machines can offer. Therefore, what kind of objects a computer needs to recognize must first be loaded into the computer as a training set. Because this set consists of layers, it can also be likened to nerve cells interacting in the human brain. Each object to be introduced must be uploaded to the system as labeled/tagged data as the first job. By referring to the deep neural networks in the human body, deep learning algorithms begin to establish their own causal relationships over time. Although useless codes are subject to elimination, useful codes are being used more frequently and efficiently.

Some basic applications, which are used in real life, of deep learning are listed below.

- *Face recognition*: The ability to open the lock screen with face recognition, which is used in smart phones or to suggest the person to be tagged in a photo uploaded to social media by the program, can be given as an example of face recognition.
- *Voice recognition*: In telephone banking, it can be considered as an example of voice recognition that voice can be detected and directed to the necessary services. Voice assistants on the phones are also inspired by deep learning. Likewise, there is no mistake to predict that all devices and people will stay in constant contact with voice recognition in the long run.

- *Use in vehicles*: In addition to autopilot feature, even when the field of view is limited, making the right decision and creating a safe area during lane change continue. Among the work of automobile giants, there is also a driverless vehicle.
- *Use in defense and security fields*: Deep learning facilitates the work of security and defense companies through video recognition. Rather than constantly monitoring camera recordings, technologies such as the receipt of a written record of the recording or the activation of the alarm system in unusual movements are possible through in-depth learning.
- *Health care*: One of the most exciting uses of deep learning is definitely health. Especially in cancer research, there are studies that eliminate time loss. Deep learning algorithms introducing cancer cell samples are both faster and more successful in diagnosing whether new cells are cancerous.

17.8.6 Evaluation

Until now, all studies on autonomous vehicles have been tested mostly during daytime or in well-lit environments. Uber accident that occurred in the night in recent years has frightened and upset everyone. After the accident, Uber interrupted the test drive on its driverless vehicles. For this reason, driverless vehicles continue to work in night/dark environments. Although there is a certain amount of illumination of the road and environment without a driver at night, the biggest problem of this issue is the lack of illumination. Apart from that, the current traffic is seen as a pattern of bright lights on a black background. To solve this problem, the fusion of the sensors is also required. Multiple sensors (LIDAR, GPS, advanced driver assistance systems – ADAS, etc.) and data from these sensors must be processed at the same time. At this point, testing this subject with deep learning will facilitate the procedures. When the previous studies are examined, the algorithm is generally followed as follows:

1. First, the data from the sensors used are collected and the collected data are simplified/extracted.
2. With a preferred algorithm, the object is detected and the objects around the vehicle are detected and labeled.
3. Lane detection is done by using deep learning architecture and image processing algorithms.
4. After all these determinations are made, decision-making is carried out. (Gas/brake adjustment, updating steering angle, stop/start the car, etc.)

The above steps continue until the autonomous vehicle reaches the destination.

In another study, a leading vehicle was placed in front of the autonomous vehicle and the vehicle in front was followed on a dark and curvy road [60]. A real-time autonomous car chaser operating optimally at night (RACCOON) solves these problems by creating an intermediate map structure that records the trajectory of the leading vehicle. The path is represented by points in a global reference frame and the computer-controlled vehicle is routed from point to point. The autonomous vehicle follows this track at any speed and takes into account the taillights of the leading vehicle. Using this approach, the autonomous vehicle travels around corners and obstacles, not inside the vehicle. RACCOON was tested successfully applied on Carnegie Mellon Navlab I1. RACCOON consists of the following subsystems:

- Start
- Image acquisition
- Rear lamp (taillight) tracking
- Leader vehicle position calculation
- Consistency check and error recovery
- Output

In another study, object classification was successfully performed by using LIDAR, which is one of the most useful sensors for autonomous vehicles in night vision and image processing techniques. The collected samples were processed with CNN. The system works as follows:

1. Autonomous vehicle receives RGB image and LIDAR data from the camera.
2. These data are clipped by forming color and depth pairs.
3. The edited data are trained in the CNN architecture and the result is recorded on the driving cognitive map.

17.9 Conclusion

As a result, artificial intelligence takes place in our lives with all fields. The most important point here is to use this field correctly and in accordance with ethical rules. Five years from now, machines can discuss and discuss new areas. Those who want to work in this field should first of all have good mathematics and be a good researcher. Do not hesitate to share your work so that many hands make light work. The more important it is to train an individual well, the more important it is to code a machine well.

References

1 Özdemir, S. (2018). Robotik Nedir. https://www.muhendisbeyinler.net/robotik-nedir/ (accessed 1 June 2019).

2 Information about computer vision. http://www.elektrik.gen.tr/2015/08/bilgisayar-gorusu-ve-imge-isleme/360 (accessed 1 June 2019).

3 Lecture Notes of an instructor: Information about Artificial Intelligence. https://web.itu.edu.tr/~sonmez/lisans/ai/yapay_zeka_problem_cozme.pd (accessed 1 June 2019).

4 Coşkun, A. *Yapay Zeka Optimizasyon Teknikleri: Literatür Değerlendirmesi*. Ankara: Gazi Üniversitesi Endüstriyel Sanatlar Eğitim Fakültesi Bilgisayar Eğitimi Bölümü.

5 Karaboğa, D. (2004). *Yapay Zeka Optimizasyon Algo-ritmaları*, 75–112. İstanbul: Atlas Yayınevi.

6 Ulusoy, G. (2002). Proje Planlamada Kaynak Kısıtlı Çizelgeleme. *Sabancı Üniversitesi Mühendislik ve Doğa Bilimleri Fakültesi Dergisi* 8: 23–29.

7 LIDAR. https://oceanservice.noaa.gov/facts/lidar.html (accessed 2 June 2019).

8 GPS. https://www.geotab.com/blog/what-is-gps/ (accessed 2 June 2019).

9 Sheridan, T.B. and Verplank, W.L. (1978). *Human and Computer Control of Undersea Teleoperators*. Cambridge, MA: Massachusetts Institute of Technology Man-Machine Systems Laboratory.

10 Gardels, K. (1960). *Automatic Car Controls for Electronic Highways*. MI: General Motors Research Laboratories.

11 Cheng, H., Zheng, N., Zhang, X. et al. (2007). Interactive road situation analysis for driver assistance and safety warning systems: framework and algorithms. *IEEE Transactions on Intelligent Transportation Systems* 8 (1): 157–167.

12 van Arem, B., van Driel, C.J.G., and Visser, R. (2006). The impact of cooperative adaptive cruise control on traffic-flow characteristics. *IEEE Transactions on Intelligent Transportation Systems* 7 (4): 429–436.

13 Lin, J.S. and Ting, W.E. (2007). Nonlinear control design of anti-lock braking systems with assistance of active suspension. *IET Control Theory and Applications* 1 (1): 343–348.

14 Bi, L., Tsimhoni, O., and Liu, Y. (2009). Using image-based metrics to model pedestrian detection performance with night-vision systems. *IEEE Transactions on Intelligent Transportation Systems* 10 (1): 155–164.

15 Gehrig, S.K. and Stein, F.J. (2007). Collision avoidance for vehicle-following systems. *IEEE Transactions on Intelligent Transportation Systems* 8 (2): 233–244.

16 Li, T.H.S. and Chang, S.J. (2003). Autonomous fuzzy parking control of a car-like mobile robot. *IEEE Transactions on Systems, Man, and Cybernetics Part A: Systems and Humans* 33 (4): 451–465.

17 Ackermann, J., Bünte, T., and Odenthal, D. (1999). *Advantages of Active Steering for Vehicle Dynamics Control*. Cologne: German Aerospace Center.

18 Sugeno, M. and Murakami, K. (1984). Fuzzy parking control of model car. *The 23rd IEEE Conference on Decision and Control*, Las Vegas, Nevada, USA (12–14 December 1984). IEEE.

19 Tan, H.S., Bu, F., and Bougler, B. (2007). A real-world application of lane-guidance technologies – automated snowblower. *IEEE Transactions on Intelligent Transportation Systems* 8 (3): 538–548.

20 White, R. and Tomizuka, M. (2001). Autonomous following lateral control of heavy vehicles using laser scanning radar. *Proceedings of the 2001 American Control Conference*, Arlington, VA, USA (25–27 June 2001). IEEE.

21 Chen, X., Yang, T., and Zhou, X.C.K. (2008). A generic model-based advanced control of electric power-assisted steering systems. *IEEE Transactions on Control Systems Technology* 16 (6): 1289–1300.

22 Chaib, S., Netto, M.S., and Mammar, S. (2004). H^{∞}, adaptive, PID and fuzzy control: a comparison of controllers for vehicle lane keeping. *IEEE Intelligent Vehicles Symposium*, Parma, Italy (14–17 June 2004). IEEE.

23 Zade, L.A. (1965). Fuzzy sets and information granularity. *Information and Control* 8: 338–353.

24 Mamdani, E.H. (1974). Application of fuzzy algorithms for control of simple dynamic plant. *Proceedings of the Institution of Electrical Engineers* 121 (12): 1585–1588.

25 Alçı M., Karatepe E. (2002). Bulanık Mantık ve MATLAB Uygulamaları In: *Chapter 6, MATLAB Fuzzy Logic Toolbox Kullanarak Bulanık Mantık Tabanlı Sistem Tasarımı, İzmir*. 8–10.

26 Elmas, Ç. (2011). Yapay Zeka Uygulamaları. In: *Bulanık Mantık Denetleyici Sistemler*, 185–187. Ankara: Seçkin Yayıncılık.

27 Team Ensco (2004). 5th IFAC/EURON Symposium on Intelligent Autonomous Vehicles Instituto Superior Técnico, Lisboa, Portugal July 5–7, 2004. DARPA Grand Challenge Technical Paper.

28 Team Spirit of Las Vegas (2004). Defense Advanced Research Projects Agency (DARPA) and DARPA Grand Challenge Technical Paper.

29 Overboat Team (2005). Defense Advanced Research Projects Agency (DARPA) and Team Banzai Technical Paper DARPA Grand Challenge.

30 Cajunbot Team (2005). Defense Advanced Research Projects Agency (DARPA) and Technical Overview of CajunBot.

31 Differential GPS. Wikipedia. http://www.webcitation.org/query?url=http%3A%2F%2Fen.wikipedia.org%2Fwiki%2FDifferential_GPS&date=2015-07-20 (accessed 20 July 2015).

32 Cremean, L.B., Foote, T.B., Gillula, J.H. et al. (2006). Alice: an information-rich autonomous vehicle for high-speed desert navigation. *Journal of Field Robotics* 23 (9): 777–810.

33 Hachman, M. 'Eyes' of Google's self-driving car may bust crooks. PCMAG. https://uk.pcmag.com/cars/64978/eyes-of-googles-self-driving-car-may-bust-crooks (accessed 20 July 2015).

34 Zhao, J., Xie, B., and Huang, X. (2014). Real-time lane departure and front collision warning system on an FPGA. In: *2014 IEEE High Performance Extreme Computing Conference (HPEC)*, 1–5. IEEE.

35 Humaidi, A.J. and Fadhel, M.A. (2016). Performance comparison for lane detection and tracking with two different techniques. In: *2016 Al-Sadeq International Conference on Multidisciplinary in IT and Communication Science and Applications (AIC-MITCSA)*, 1–6. IEEE.

36 Li, C., Wang, J., Wang, X., and Zhang, Y. (2015). A model based path planning algorithm for self-driving cars in dynamic environment. In: *2015 Chinese Automation Congress (CAC)*, 1123–1128. IEEE.

37 Yoon, S., Yoon, S.E., Lee, U. et al. (2015). Recursive path planning using reduced states for car-like vehicles on grid maps. *IEEE Transactions on Intelligent Transportation Systems* 16 (5): 2797–2813.

38 Wang, D. and Qi, F. (2001). Trajectory planning for a four-wheel-steering vehicle. In: *Proceedings 2001 ICRA. IEEE International Conference on Robotics and Automation*, vol. 4, 3320–3325. IEEE.

39 Bojarski, M., Testa, D.D., Dworakowski, D. et al. (2016). End to end learning for selfdriving cars. CoRR, vol. abs/1604.07316. http://arxiv.org/abs/1604.07316.

40 Aziz, M.V.G., Prihatmanto, A.S., and Hindersah, H. (2017). Implementation of lane detection algorithm for self-driving car on Toll Road Cipularang using Python language. *International Conference on Electric Vehicular Technology (ICEVT)*, Bali, Indonesia (2–5 October 2017).

41 Ballard, D.H. and Brown, C.M. (1982). *Computer Vision*. Prentice Hall. ISBN: 0-13-165316-4.

42 Huang, T. (1996). Computer vision: evolution and promise. In: *19th CERN School of Computing* (ed. C.E. Vandoni), 21–25. Geneva: CERN https://doi.org/10.5170/CERN-1996-008.21. ISBN: 978-9290830955.

43 Elmenreich, W. (2002). Sensor fusion in time-triggered systems. PhD thesis (PDF). Vienna University of Technology, Vienna, Austria. p. 173.

44 Haghighat, M.B.A., Aghagolzadeh, A., and Seyedarabi, H. (2011). Multi-focus image fusion for visual sensor networks in DCT domain. *Computers and Electrical Engineering* 37 (5): 789–797.

45 Bengio, Y., Courville, A., and Vincent, P. (2013). Representation learning: a review and new perspectives. *IEEE Transactions on Pattern Analysis and Machine Intelligence* 35 (8): 1798–1828. arXiv:1206.5538doi:https://doi.org/10.1109/tpami.2013.50.

46 Schmidhuber, J. (2015). Deep learning in neural networks: an overview. *Neural Networks* 61: 85–117. PMID: 25462637. arXiv:1404.7828doi:https://doi.org/10.1016/j.neunet.2014.09.003.

47 About an Information of self-driving car. https://towardsdatascience.com/how-do-self-driving-cars-see-13054aee2503 (accessed 1 June 2019).

48 Yang, H., Shao, L., Zheng, F. et al. (2011). Recent advances and trends in visual tracking: a review. *Neurocomputing* 74 (18): 3823–3831.

49 Geiger, A., Lauer, M., Wojek, C. et al. (2014). 3d traffic scene understanding from movable platforms. *IEEE Transactions on Pattern Analysis and Machine Intelligence* 36: 1012–1025.

50 Russakovsky, O., Deng, J., Su, H. et al. (2015). ImageNet large scale visual recognition challenge. *International Journal of Computer Vision* 115 (3): 211–252.

51 Heilbron, F.C., Escorcia, V., Ghanem, B., and Niebles, J.C. (2015). ActivityNet: a large-scale video benchmark for human activity understanding. *2015 IEEE Conference on Computer Vision and Pattern Recognition (CVPR)*, Boston, MA, USA (7–12 June 2015). IEEE.

52 Krasin, I., Duerig, T., Alldrin, N. et al. (2016). *Openimages: A Public Dataset for Large-Scale Multi-label and Multiclass Image Classification*. Springer. https://github.com/openimages.

53 Brostow, G.J., Fauqueur, J., and Cipolla, R. (2008). Semantic object classes in video: a high-definition ground truth database. *Pattern Recognition Letters* 30: 88–97.

54 Geiger, A., Lenz, P., and Urtasun, R. (2012). Are we ready for autonomous driving? The KITTI vision benchmark suite. *2012 IEEE Conference on Computer Vision and Pattern Recognition*, Providence, RI, USA (16–21 June 2012). IEEE.

55 Chen C., Seff A., Kornhauser A., and Xiao J. (2015). Deepdriving: learning affordance for direct perception in autonomous driving. In *Proceedings of 15th IEEE International Conference on Computer Vision* (ICCV2015),

56 Richter S. R., Vineet V., Roth S., and Koltun V. (2016). Playing for data: ground truth from computer games. In *14th European Conference on Computer Vision (ECCV 2016), Computer Vision and Pattern Recognition (cs.CV)*.

57 Cordts, M., Omran, M., Ramos, S. et al. (2016). The cityscapes dataset for semantic urban scene understanding. *2016 IEEE Conference on Computer Vision and Pattern Recognition (CVPR)*, Las Vegas, NV, USA (27–30 June 2016). IEEE.

58 Ros, G., Sellart, L., Materzynska, J. et al. (2016). The synthia dataset: a large collection of synthetic images for semantic segmentation of urban scenes. *2016 IEEE Conference on Computer Vision and Pattern Recognition (CVPR)*, Las Vegas, NV, USA (27–30 June 2016). IEEE.

59 Kim, J. and Park, C. (2017). End-to-end ego lane estimation based on sequential transfer learning for self-driving cars. *IEEE Conference on Computer Vision and Pattern Recognition Workshops*, Honolulu, HI, USA (21–26 July 2017). IEEE.

60 Bolc, L. and Cytowski, J. (1992). *Search Methods for Artificial Intelligence*. London: Academic Press.

18

Different Techniques to Solve Monotone Inclusion Problems

Tanmoy Som[1], Pankaj Gautam[1], Avinash Dixit[1], and D. R. Sahu[2]

[1] *Department of Mathematical Sciences, Indian Institute of Technology (Banaras Hindu University), Varanasi, Uttar Pradesh, 221005, India*
[2] *Department of Mathematics, Banaras Hindu University, Varanasi, Uttar Pradesh, 221005, India*

18.1 Introduction

Consider a monotone operator $T: H \to 2^H$. A point $x \in H$ is zero of T if $0 \in Tx$. The set of all zeros of T is denoted by "zer(T)." A wide class of problems can be reduced to a fundamental problem of nonlinear analysis is to find a zero of a maximal monotone operator T in a real Hilbert space H:

$$\text{find } x \in H: \ 0 \in T(x) \tag{18.1}$$

A more general problem we can consider here is to find $x \in H$ such that $d \in T(x)$, for some $d \in H$. This problem finds many important application in scientific fields such as signal and image processing [1], inverse problems [2, 3], convex optimization [4], and machine learning [5].

Suppose T is the gradient of a differentiable convex function f, i.e. $T = \nabla f$, the most simple approach to solve Problem (18.1) is via gradient projection method, given by

$$x_{n+1} = (Id - \lambda_n \nabla f)(x_n)$$

where $\lambda_n > 0$ are stepsizes. The operator $(Id - \lambda_n T)$ is called as forward–backward and the above scheme is nothing else than classical method of steepest descent.

In case f is nondifferentiable function, gradient method generalizes to subgradient method, given by

$$x_{n+1} = x_n - \gamma_n y_n, \quad \text{for some } y_n \in \partial f(x_n)$$

For a general monotone operator T, the classical approach to solve Problem (18.1) is proximal point method, which converts the maximal monotone operator inclusion problem into a fixed point problem of a firmly nonexpansive mapping via resolvent operator. The proximal point algorithm [6, 7] is one of the most influential method to solve Problem (18.1) and has been studied extensively both in theory and practice [8–10]. Proximal point algorithm is given by

$$x_{n+1} = (Id + \lambda_n T)^{-1}(x_n)$$

where $\lambda_n > 0$ is regularization parameter.

The operator $(Id + \lambda_n T)^{-1}$ is called resolvent operator, introduced by Moreau [11]. The resolvent operator is also known as a backward operator. This method was further extended to monotone operator by Rockafellar [10]. Rockafeller also studied the weak convergence behavior of proximal point algorithm under some mild assumptions. The aim of this chapter is to study the convergence behavior of proximal point algorithm and its different modified form.

Mathematical Methods in Interdisciplinary Sciences, First Edition. Edited by Snehashish Chakraverty.

In the Section 18.2, we have discussed some definitions and results used to prove the convergence analysis of algorithms. In Section 18.3, we discussed the convergence behavior of proximal point algorithm. In Section 18.4, we have discussed about the splitting algorithms to solve monotone inclusion problems. Section 18.5 deals with the inertial methods to solve monotone inclusion problems. Numerical example is also shown to compare the convergence speed of different algorithms based on inertial technique.

18.2 Preliminaries

Through this chapter, we will denote the Hilbert space by H.

Definition 18.1 A set-valued operator $A : H \to 2^H$ is said to be

(i) monotone if

$$\langle x - y, u - v\rangle \geq 0, \ \ (\forall(x,u) \in graA)(\forall(y,v) \in graA)$$

where $graA \ : \ = \{(x,u) \in H \times 2^H : u \in Ax\}$.

(ii) maximal monotone if there does not exist any monotone operator $B : H \to 2^H$ such that $graB$ properly contains $graA$.

Definition 18.2 Let $f : H \to (-\infty, +\infty)$ be a proper function. The subdifferentiable of f is the set-valued operator $\partial f : H \to 2^H$ is defined by

$$\partial f(x) : \ = \{u \in H : (\forall u \in H)\langle y - x, u\rangle + f(x) \leq f(y)\}$$

Let $x \in H$. Then, f is subdifferentiable at x if $\partial f(x) \neq \emptyset$, and the elements of $\partial f(x)$ are the subgradients of f at x.

An operator $T : H \to 2^H$ is said to be uniformly monotone on a subset $C \subseteq H$ if there exists an increasing function $\phi : R_+ \to [0, +\infty)$ vanishing only at 0 such that $\forall x \in C$ $(\forall y \in C)$ $(\forall u \in Ax)$ $(\forall v \in Ay)$

$$\langle x - y, u - v\rangle \geq \phi(\|x - y\|)$$

The set of fixed points of function T is denoted by Fix(T).

Lemma 18.1 *[12] Let $T : H \to 2^H$ be monotone and let $\gamma \in \mathbb{R}_+$. Then* $\mathrm{Fix}(J_{\gamma T}) = \mathrm{zer}(T)$.

Lemma 18.2 *[12] Let C be a nonempty, closed, convex subset of a real Hilbert space H, $T : C \to H$ be a nonexpansive and $\{x_n\}$ be a sequence in H and $x \in H$ such that $x_n \rightharpoonup x$ and $Tx_n - x_n \to 0$ as $n \to \infty$. Then, $x \in$* Fix(T).

Lemma 18.3 *[13, 14] Let $\{\phi_n\}$, $\{\delta_n\}$, and $\{\alpha_n\}$ be sequences in $[0, \infty)$ such that $\phi_{n+1} \leq \phi_n + \alpha_n(\phi_n - \phi_{n-1}) + \delta_n$ for all $n \geq 1$, $\sum_{n\in\mathbb{N}}\delta_n < \infty$ and there exist a real number α such that $0 \leq \alpha_n \leq \alpha < 1$ for all $n \in \mathbb{N}$. Then, the following holds:*

(i) $\sum_{n\geq 1}[\phi_n - \phi_{n-1}]_+ < \infty$, where $[t]_+ = \max\{t, 0\}$;
(ii) there exists $\phi^* \in [0, \infty)$ such that $\lim\limits_{n\to\infty} \phi_n = \phi^*$.

Lemma 18.4 ***(Opial Lemma)*** *Let H be a Hilbert space and $\{x_n\}$ a sequence such that there exists a nonempty set $S \subset H$ verifying*

(a) For every $z \in S$, $\lim\limits_{n\to\infty} \|x_n - z\|$ exists;
(b) If $x_{n_k} \rightharpoonup x$ weakly in H for a subsequence $n_k \to \infty$, then $x \in S$.

Then, there exists $\overline{x} \in S$ such that $x_n \rightharpoonup \overline{x}$ weakly in H as $n \to \infty$.

18.3 Proximal Point Algorithm

Theorem 18.1 *[12] Let $T : H \to 2^H$ be a maximal monotone operator such that* zer$(T) \neq \emptyset$*, let $\{\gamma_n\}$ be a sequence in $\mathbb{R}_+$ such that $\sum_{n\in\mathbb{N}}\gamma_n^2 = +\infty$ and let $x_0 \in H$. Then, the sequence generated by proximal point algorithm*

$$x_{n+1} = J_{\gamma_n T}x_n \tag{18.2}$$

satisfies the following:

(i) $\{x_n\}$ converges weakly to a point in zer(T).
(ii) Suppose that T is uniformly monotone on every bounded subset of H. Then, $\{x_n\}$ converges strongly to the unique point in zer(T).

Proof: Set $(\forall n \in \mathbb{N})\ u_n = \frac{x_n - x_{n+1}}{\gamma_n}$.
Then $(\forall n \in \mathbb{N})\ u_n \in Ax_{n+1}$ and $x_{n+1} - x_{n+2} = \gamma_{n+1}u_{n+1}$. Hence, by monotonicity and Cauchy–Schwarz inequality,

$$\begin{aligned}(\forall n \in \mathbb{N})\ 0 &\leq \langle x_{n+1} - x_{n+2}, u_n - u_{n+1}\rangle/\gamma_{n+1}\\ &= \langle u_{n+1}, u_n - u_{n+1}\rangle\\ &= \langle u_{n+1}, u_n\rangle - \|u_{n+1}\|^2\\ &= \|u_{n+1}\|(\|u_n\| - \|u_{n+1}\|)\end{aligned} \tag{18.3}$$

which implies that $\{\|u_n\|\}$ converges. Now, let $z \in \text{zer}(A)$. Because Lemma 18.1 asserts that $z \in \bigcap_{n\in\mathbb{N}}\text{Fix}(J_{\gamma_n A})$, we deduce from (18.2) and the fact that resolvent of a maximal monotone operator is firmly nonexpansive that

$$\begin{aligned}(\forall n \in \mathbb{N})\ \|x_{n+1} - z\|^2 &= \|J_{\gamma_n T}x_n - J_{\gamma_n T}z\|^2\\ &\leq \|x_n - z\|^2 - \|x_n - J_{\gamma_n T}x_n\|^2\\ &= \|x_n - z\|^2 - \gamma_n^2\|u_n\|^2\end{aligned}$$

Thus, $\{x_n\}$ is bounded and Fejér monotone with respect to zer(T) and $\sum_{n\in\mathbb{N}}\gamma_n^2\|u_n\|^2 \leq +\infty$. In turn, since $\sum_{n\in\mathbb{N}}\gamma_n^2 \leq +\infty$, we have *liminf* $\|u_n\| = 0$ and therefore $u_n \to 0$ since $\{\|u_n\|\}$ converges.

(i) Let x be a weak sequential cluster point of $\{x_n\}$, say $x_{n_k+1} \rightharpoonup x$. Since $0 \leftarrow u_{n_k} \in Tx_{n_k+1}$, yields $0 \in Tx$ and hence we have the result. (ii) The assumption imply that T is strictly monotone. Hence, zer(T) reduces to a singleton. Now, let x be the weak limit in (i). Then, $0 \in Tx$ and $u_n \in Tx_{n+1}$, $\forall n \in \mathbb{N}$. Hence, since $\{x_n\}$ lies in a bounded set, there exists an increasing function $\phi : \mathbb{R}_+ \to [0, \infty)$ that vanishes only at 0 such that

$$\langle x_{n+1} - x, u_n\rangle \geq \phi(\|x_{n+1} - x\|),\ \forall n \in \mathbb{N} \tag{18.4}$$

Since $u_n \to 0$ and $x_{n+1} \rightharpoonup x$, we have $\|x_{n+1} - x\| \to 0$.

18.4 Splitting Algorithms

In many cases, it is difficult to evaluate the resolvent of operator T. Interestingly, sometimes, it is as difficult as the original problem itself. To overcome this problem, the operator T splits into sum of two maximal monotone operators A and B such that resolvent of A and B are easier to compute than the full resolvent $(Id + \lambda T)^{-1}$. Based on splitting techniques, many iterative methods has been designed to solve Problem (18.1). Some popular methods are Peaceman–Rachford splitting algorithm [15], Douglas–Rachford splitting algorithm [16], and forward–backward algorithm [17].

18.4.1 Douglas–Rachford Splitting Algorithm

Douglas–Rachford splitting algorithm is proposed to solve a general class of monotone inclusion problems, when A and B are maximally monotone. Let $x_0 \in H$ and $\{\lambda_n\}$ be a sequence in [0,2]. Then, algorithm is as given as follows:

$$\begin{aligned} y_n &= J_{\gamma B} x_n \\ z_n &= J_{\gamma A}(2y_n - x_n) \\ x_{n+1} &= x_n + \lambda_n(z_n - y_n) \end{aligned} \tag{18.5}$$

The convergence behavior of Douglas–Rachford splitting algorithm can be summarized as follows:

Theorem 18.2 *[12] Let A and B be maximally monotone operators from H to 2^H such that* $\mathrm{zer}(A+B) \neq \emptyset$. *Let $\gamma \in \mathbb{R}$ is strictly positive, $x_0 \in H$, and $\{\lambda_n\}$ be a sequence in [0,2] such that $\sum_{n\in\mathbb{N}}^{\infty} \lambda_n(2-\lambda_n) = \infty$. Consider $\{x_n\}$ is the sequence generated by the Algorithm (18.5). Then, there exists $x \in \mathrm{Fix}(R_{\gamma A}R_{\gamma A})$ such that the following hold:*

(i) $J_{\gamma B}x \in \mathrm{zer}(A+B)$;
(ii) $\{y_n - z_n\}$ converges strongly to 0;
(iii) $\{x_n\}$ converges weakly to x;
(iv) $\{y_n\}$ converges weakly to $J_{\gamma B}x$;
(v) $\{z_n\}$ converges weakly to $J_{\gamma B}x$;
(vi) Suppose that A is normal cone operator of a closed affine subspace C of H. Then, $\{P_C x_n\}$ converges weakly to $J_{\gamma B}x$;
(vii) Suppose that one of the following holds:
  (a) A is uniformly monotone on every nonempty bounded subset of dom A;
  (b) B is uniformly monotone on every nonempty bounded subset of dom B.

Then, sequences $\{y_n\}$ and $\{z_n\}$ converges weakly to the unique point in $\mathrm{zer}(A+B)$.

18.4.2 Forward–Backward Algorithm

The splitting method discussed above is complicated, and the simplest one is forward–backward method. The forward–backward algorithm, the composition of forward step with respect to B followed by backward step with respect to A, is given by

$$x_{n+1} = (Id + \lambda_n A)^{-1}(Id - \lambda_n B)x_n \tag{18.6}$$

The forward–backward splitting method is studied extensively in [18–20]. For the general monotone operators A and B, the weak convergence of Algorithm (18.6) has been studied under some restriction on the step-size.

Passty studied the convergence of forward–backward, which can be summarized as follows.

Theorem 18.3 *[17] Let H be a real Hilbert space and T, A, and B be maximal monotone operators such that $T = A + B$. Further, let $\{x_n\}_{n=0}^{\infty}$, $\{y_n\}_{n=0}^{\infty}$, $\{z_n\}_{n=0}^{\infty} \subseteq H$, and $\{\lambda_n\}_{n=0}^{\infty} \subseteq (0,\infty)$ be sequences satisfying the conditions, for all $n \geq 0$*

$$\begin{aligned} x_{n+1} &= (Id + \lambda_n B)^{-1}(x_n - \lambda_n y_n) \\ y_n &= Ax_n \\ z_n &= \frac{\sum_{j=0}^{n} \lambda_j x_{j+1}}{\sum_{j=0}^{n} \lambda_j} \end{aligned}$$

where $\{y_n\}$ is bounded and $\sum_{n=0}^{\infty} \lambda_n = \infty$ and $\sum_{n=0}^{\infty} \lambda_n^2 \leq \infty$. Then, if $\mathrm{zer}(T) = \emptyset$, $\|z_n\| \to \infty$, *while if* $\mathrm{zer}(T) \neq \emptyset$, *$\{z_n\}$ converges weakly to some zero of T.*

Theorem 18.4 *Let A and B be maximal monotone operators on $\mathbb{R}^n$ such that* $\text{zer}(A+B) \neq \emptyset$ *and A^{-1} is strongly monotone with modulus α. Let* $\{\lambda_n\}_{n=0}^{\infty} \subseteq (0,\infty)$ *and* $\{x_n\}_{n=0}^{\infty} \subseteq \mathbb{R}^n$ *be sequences such that*

$$x_{n+1} = (Id + \lambda_n B)^{-1}(Id - \lambda_n A)x_n \ \forall n \geq 0$$
$$0 < \inf_{n\geq 0} \lambda_n \leq \sup_{n\geq 0} \lambda_n < 2\alpha$$

(Note that A must be single-valued.) Then $\{x_n\}$ converges to zero of $A+B$.

Proof: Choose $x^* \in \text{zer}(A+B)$ and let $y^* = Ax^*$. Then, $-y^* \in Bx^*$. For all $n \geq 0$, define

$$y_n = Ax_n$$
$$z_n = \frac{1}{\lambda_n}(x_n - x_{n+1}) - y_n$$

Note that

$$\begin{aligned} x_{n+1} = (Id + \lambda_n B)^{-1}(x_n - \lambda_n y_n) &\Leftrightarrow x_n - \lambda_n y_n \in (Id + \lambda_n B)x_{n+1} \\ &\Leftrightarrow x_n - x_{n+1} - \lambda_n y_n \in \lambda_n B x_{n+1} \\ &\Leftrightarrow z_n \in Bx_{n+1} \end{aligned}$$

Also, by rearranging the definition of z_n, one has

$$z_{n+1} = x_n - \lambda_n(y_n + z_n)$$

Using this identity, we have for all $n \geq 0$ that

$$\begin{aligned} &\langle x_{n+1} - x^*, y_n - y^*\rangle + \lambda_n\langle z_n + y^*, y_n - y^*\rangle \\ &= \langle x_n - x^*, y_n - y^*\rangle - \lambda_n\langle y_n + z_n, y_n - y^*\rangle + \lambda_n\langle z_n + y^*, y_n - y^*\rangle \\ &= \langle x_n - x^*, y_n - y^*\rangle - \lambda_n y_n - y^8, y_n - y^*\rangle \\ &\geq \alpha\|y_n - y^*\|^2 - \lambda_n\|y_n - y^*\|^2 \\ &= (\alpha - \lambda_n)\|y_n - y^*\|^2 \end{aligned}$$

where the inequality follows from the strong monotonicity of A^{-1}. Since $z_n \in Bx_{n+1}$ and $-y^* \in Bx^*$

$$\langle x_{n+1} - x^*, z_n + y^*\rangle \geq 0$$

Therefore, adding the previous inequality, one obtains

$$\begin{aligned} &\langle x_{n+1} - x^*, y_n + z_n\rangle + \lambda_n\langle z_n + y^*, y_n - y^*\rangle \geq (\alpha - \lambda_n)\|y_n - y^*\|^2 \\ &\Leftrightarrow \langle x_{n+1} - x^*, \frac{1}{\lambda_n}(x_n - x_{n+1})\rangle + \lambda_n\langle z_n + y^*, y_n - y^*\rangle \geq (\alpha - \lambda_n)\|y_n - y^*\|^2 \\ &\Leftrightarrow \frac{1}{\lambda_n}\langle x_{n+1} - x^*, (x_n - x_{n+1})\rangle \geq (\alpha - \lambda_n)\|y_n - y^*\|^2 - \lambda_n\langle z_n + y^*, y_n - y^*\rangle \\ &\Leftrightarrow \frac{1}{2\lambda_n}[\|x_n - x^*\|^2 - \|x_{n+1} - x^*\|^2 - \|x_n - x_{n-1}\|^2] \geq (\alpha - \lambda_n)\|y_n - y^*\|^2 - \lambda_n\langle z_n + y^*, y_n - y^*\rangle \\ &\Leftrightarrow \|x_{n+1} - x^*\|^2 \leq \|x_n - x^*\|^2 - \|x_n - x_{n-1}\|^2 + 2\lambda_n^2\langle z_n + y^*, y_n - y^*\rangle + (2\lambda_n^2 - 2\lambda_n\alpha)\|y_n - y^*\|^2 \end{aligned}$$

Also,

$$\begin{aligned} \|x_n - x_{n+1}\|^2 &= \|\lambda_n(y_n + z_n)\|^2 \\ &= \lambda_n^2\|y_n - y^* + z_n + y^*\|^2 \\ &= \lambda_n^2\|y_n - y^*\|^2 + 2\lambda_n\langle z_n + y^*, y_n - y^*\rangle + \lambda_n^2\|z_n + y^*\|^2 \end{aligned}$$

Therefore, substituting for $\|x_n - x_{n+1}\|^2$ gives

$$\|x_{n+1} - x^*\|^2 \leq \|x_n - x^*\|^2 - \lambda_n^2\|z_n + y^*\|^2 - (\lambda_n - 2\alpha\lambda_n)\|y_n - y^*\|^2$$

Let

$$\epsilon = \min\left\{\inf_{n\geq 0}\{\lambda_n\}, 2\alpha - \sup_{n\geq 0}\{\lambda_n\} \geq 0\right\}$$

Then, for all $n \geq 0$,

$$\|x_{n+1} - x^*\|^2 \leq \|x_n - x^*\|^2 - \epsilon^2\|z_n + y^*\|^2 - \epsilon^2\|y_n - y^*\|^2 \tag{18.7}$$

Hence $\|x_{n+1} - x^*\|^2 \leq \|x_n - x^*\|^2$ for all $n \geq 0$, and $\{x_n\}$ is bounded. Adding the inequality (18.7) over $n = 0, 1, \ldots, N-1$, we obtain

$$\|x_N - x^*\|^2 \leq \|x_0 - x^*\|^2 - \epsilon^2\left[\sum_{n=0}^{N-1}\|z_n + y^*\|^2 + \sum_{n=0}^{N-1}\|y_n - y^*\|^2\right]$$

and letting $N \to \infty$, one concludes that

$$\lim_{n\to\infty} z_n = -y^* \qquad \lim_{n\to\infty} y_n = y^*$$

Since $\{x_n\}$ is bounded, it possesses a limit point x_∞. Let $\{x_{n_k}\}$ be a subsequence of $\{x_n\}$ converging to x_∞. Then,

$$\lim_{k\to\infty}\{x_{n_k}, y_{n_k}\} = (x_\infty, y^*)$$

$$\{x_{n_k}, y_{n_k}\} \in A \ \forall k \geq 0$$

$$\lim_{k\to\infty}\{x_{n_k}, z_{n_k} - 1\} = (x_\infty, -y^*)$$

$$\{x_{n_k}, z_{n_k} - 1\} \in A \ \forall k \geq 0$$

Therefore, by the maximal monotonicity of A and B, we have $(x_\infty, y^*) \in A$ and $(x_\infty, -y^*) \in B$, hence $x_\infty \in \text{zer}(A+B)$ was chosen arbitrary, we can substitute x_∞ for x^*, obtaining $\|x_{n+1} - x_\infty\|^2 \leq \|x_n - x_\infty\|^2$ for all $n \geq 0$. It follows that x^* is the only limit point of $\{x_n\}$ and $\{x_n\}$ converges to x^*.

18.5 Inertial Methods

The classical proximal point algorithm converts the maximal monotone operator inclusion problem into a fixed point problem of firmly nonexpansive mapping via resolvent operators. Consider $f : \mathbb{R}^n \to \mathbb{R}$ and $\omega : \mathbb{R} \to \mathbb{R}^n$ are differentiable functions. The proximal point algorithm can be interpreted as one step discretization method for the ordinary differential equations

$$\omega' + \nabla f(w) = 0 \tag{18.8}$$

where ω' represents the derivative of ω and ∇f is the gradient of f. The iterative method mentioned above are one step, i.e. the new iteration term depends on only its previous iterate. Multistep methods have been proposed to accelerate the convergence speed of the algorithm, which are discretization of second-order ordinary differential equation

$$w'' + \gamma w' + \nabla f(w) = 0 \tag{18.9}$$

where $\gamma > 0$. System (18.9) roughly shows the motion of heavy ball rolling under its own inertia over the graph of f until friction stops it at a stationary point of f. Inertial force, friction force, and graving force are the three parameters in System (18.9), that is why the system is named heavy-ball with friction (HBF) system. The energy function is given by

$$E(t) = \frac{1}{2}\|w'(t)\|^2 + f(w(t)) \tag{18.10}$$

which is always decreasing with t unless w' vanishes. Convexity of f insures that trajectory will attain minimum point. The convergence speed of the solution trajectories of the System (18.9) is greater than the System (18.8). Polyak [21] first introduced the multistep method using the property of heavy ball System (18.9) to minimize a convex smooth function f. Polyak's multistep algorithm was given by

$$y_n = x_n + \alpha_n(x_n - x_{n-1})$$
$$x_{n+1} = y_n - \lambda_n \nabla f(x_n)$$

where $\alpha_n \in [0, 1)$ is a momentum parameter and λ_n is a step-size parameter. It significantly improves the performance of the scheme. Nestrov [22] improves the idea of Polyak by evaluating the gradient at the inertial rather than the previous term. The extrapolation parameter α_n is chosen to obtain the optimal convergence of the scheme. The scheme is given by

$$y_n = x_n + \alpha_n(x_n - x_{n-1})$$
$$x_{n+1} = y_n - \lambda_n \nabla f(x_n)$$

where $\lambda_n = \frac{1}{L}$.

18.5.1 Inertial Proximal Point Algorithm

In 2001, by combining the idea of heavy ball method and proximal point method, Alvarez and Attouch proposed the inertial proximal point algorithm, which can be written as

$$y_n = x_n + \alpha_n(x_n - x_{n-1})$$
$$x_{n+1} = (Id + \lambda_n T)^{-1}(y_n) \tag{18.11}$$

where $T : H \to 2^H$ is a set-valued operator and $\{\alpha_n\}$ and $\{\gamma_n\}$ are the sequences satisfying some conditions.

Theorem 18.5 *[13] Let $\{x_n\}$ be the sequence generated by the inertial proximal point algorithm, for some maximal monotone operator $T : H \to 2^H$ such that $S\ :\ = T^{-1}(0) \neq \emptyset$ with the parameters α_n and λ_n satisfying*

(i) $\exists \lambda > 0$ such that $\forall n \in \mathbb{N}$, $\lambda_n \geq \lambda$;
(ii) $\exists \alpha \in [0, 1)$ such that $\forall n \in \mathbb{N}$, $0 \leq \alpha_n \leq \alpha$;
(iii) $\sum_{n=1}^{\infty} \alpha_n \|x_n - x_{n-1}\|^2 < \infty$.

Then, there exists $x^* \in S$ such that $\{x_n\}$ converges weakly to a point in H as $n \to \infty$.

Proof: Consider the case $\alpha_n = 0$, which reduces the inertial proximal point algorithm into proximal point algorithm, whose convergence is already known. Now, let $x^* \in T^{-1}(0)$. Define the sequence

$$\phi_n = \frac{1}{2}\|x_n - x^*\|^2 \tag{18.12}$$

Using Eqs. (18.11) and (18.12), we have

$$\phi_{n+1} = \phi_n + \langle x_{n+1} - x_n, x_{n+1} - x^* \rangle - \frac{1}{2}\|x_{n+1} - x_n\|^2 \tag{18.13}$$

If $\alpha_n = 0$, then $0 \in x_{n+1} - x_n + \lambda T(x_{n+1})$ and from the monotonicity of T, we have $\langle x_{n+1} - x_n, x_{n+1} - x^* \rangle \leq 0$, which implies that

$$\phi_{n+1} - \phi_n \leq -\frac{1}{2}\|x_{n+1} - x_n\|^2 \tag{18.14}$$

Therefore, $\{\phi_n\}$ is nonincreasing and hence convergence. Since x^* is an arbitrary point in $T^{-1}(0)$, condition (a) of Opial Lemma is satisfied. On the other hand from (18.14), we obtain

$$\sum_{n=1}^{\infty} \|x_{n+1} - x_n\|^2 \leq 2\phi_1 = \|x_1 - z\|^2 \tag{18.15}$$

and consequently $\|x_{n+1} - x_n\| \to 0$ as $n \to \infty$. As λ_n is bounded away from zero, we deduce that $d(0, T(x_n)) \to 0$ as $n \to \infty$. Let x be a weak cluster point of $\{x_n\}$. Since the graph of the maximal monotone operator T is closed in $H \times H$ for the weak–strong topology, we have that $0 \in T(x)$ i.e. $x \in T^{-1}(0)$. Thus, the condition (b) of Opial's Lemma is also satisfied, which proves that theorem when $\alpha_n = 0$.

We now turn to the case $\alpha_n \geq 0$ for some $n \in \mathbb{N}$. We have

$$\langle x_{n+1} - x_n - \alpha_n(x_n - x_{n-1}), x_{n+1} - z\rangle + \lambda_n\langle T(x_{n+1}), x_{n+1} - z\rangle = 0$$

and the monotonicity of T yields

$$\langle x_{n+1} - x_n - \alpha_n(x_n - x_{n-1}), x_{n+1} - z\rangle \leq 0$$

Let us rewrite this inequality in terms of $\phi_{n-1}, \phi_n, \phi_{n+1}$. Observe that

$$\begin{aligned}&\langle x_{n+1} - x_n - \alpha_n(x_n - x_{n-1}), x_{n+1} - z\rangle \\ &\quad = \phi_{n+1} - \phi_n + \frac{1}{2}\|x_{n+1} - x_n\|^2 - \alpha_n\langle x_n - x_{n-1}, x_{n+1} - z\rangle\end{aligned}$$

and since

$$\begin{aligned}&\langle x_n - x_{n-1}, x_{n+1} - z\rangle \\ &\quad = \langle x_n - x_{n-1}, x_n - z\rangle + \langle x_n - x_{n-1}, x_{n+1} - x_n\rangle \\ &\quad = \phi_n - \phi_{n-1} + \frac{1}{2}\|x_n - x_{n-1}\|^2 + \langle x_n - x_{n-1}, x_{n+1} - x_n\rangle\end{aligned}$$

it follows that

$$\begin{aligned}&\phi_{n+1} - \phi_n - \alpha_n(x_n - x_{n-1}) \\ &\quad \leq -\frac{1}{2}\|x_{n+1} - x_n\|^2 + \alpha_n\langle x_n - x_{n-1}, x_{n+1} - x_n\rangle + \frac{\alpha_n}{2}\|x_n - x_{n-1}\|^2 \\ &\quad = -\frac{1}{2}\|x_{n+1} - x_n - \alpha_n(x_n - x_{n-1})\|^2 + \frac{1}{2}(\alpha_n + \alpha_n^2)\|x_n - x_{n-1}\|^2\end{aligned}$$

Hence,

$$\phi_{n+1} - \phi_n - \alpha_n(x_n - x_{n-1}) \leq -\frac{1}{2}\|v_{n+1}\|^2 + \alpha_n\|x_n - x_{n-1}\|^2 \tag{18.16}$$

where $v_{n+1} := x_{n+1} - x_n - \alpha_n(x_n - x_{n-1})$.

Setting, $\theta_n := \phi_n - \phi_{n-1}$ and $\delta_n := \alpha_n\|x_n - x_{n-1}\|^2$, we obtain $\theta_{n+1} \leq \alpha_n\theta_n + \delta_n \leq \alpha_n[\theta_n]_+ + \delta_n$, where $[t]_+ := \max\{t, 0\}$ and consequently, $[\theta_{n+1}]_+ \leq \alpha[\theta_n]_+ + \delta_n$, with $\alpha \in [0, 1)$ given by (ii). This yields

$$[\theta_{n+1}]_+ \leq \alpha^n[\theta_n]_+ + \sum_{j=0}^{n-1} \alpha^j\delta_{n-j}$$

and therefore

$$\sum_{n=0}^{\infty} [\theta_{n+1}]_+ \leq \frac{1}{1-\alpha}\left([\theta_1]_+ + \sum_{n=1}^{\infty} \delta_n\right)$$

which is finite because of (iii). Consider the sequence defined by $w_n := \phi_n - \sum_{j=1}^{n} [\theta_j]_+$. Since $\phi_n \geq 0$ and $\sum_{j=1}^{n} [\theta_j]_+ \leq \infty$, it follows that $\{w_n\}$ is bounded from below.

$$w_{n+1} := \phi_{n+1} - [\theta_{n+1}]_+ - \sum_{j=1}^{n} [\theta_j]_+ \leq \phi_{n+1} - \phi_{n+1} + \phi_n - \sum_{j=1}^{n} [\theta_j]_+ = w_n$$

so that $\{w_n\}$ is nonincreasing. We thus deduce that $\{w_n\}$ is convergent and hence

$$\lim_{n\to\infty} \phi_n = \lim_{n\to\infty} w_n + \sum_{j=1}^{\infty} [\theta_j]_+$$

Consequently, for every $z \in S$, $\lim_{n\to\infty} \|x_n - z\|$ exists. On the other hand, from (18.16), we obtain the estimate

$$\frac{1}{2}\|v_{n+1}\|^2 \leq \phi_n - \phi_{n+1} + \alpha[\theta_n]_+ + \delta_n$$

and it follows that

$$\frac{1}{2}\sum_{n=1}^{\infty} \|v_{n+1}\|^2 \leq \phi_1 + \sum_{n=1}^{\infty}(\alpha[\theta_n]_+ + \delta_n) \leq \infty$$

Consequently, for every $v_{n+1} \to 0$ as $n \to \infty$ and since $v_{n+1} + \lambda_n A(x_{n+1})$, by (i), we have $d(0, A(x_n)) \to 0$ as $n \to \infty$. The rest of the proof runs as in the case $\alpha_n = 0$ by applying Opial's lemma.

18.5.2 Splitting Inertial Proximal Point Algorithm

In 2003, Moudafi and Oliny [23] modified the inertial proximal point algorithm by splitting into maximal monotone operators A and B with B is single valued, Lipschitz continuous operator and γ-cocoercive. The proposed algorithm can be given as follows:

$$\begin{aligned} y_n &= x_n + \alpha_n(x_n - x_{n-1}) \\ x_{n+1} &= (Id + \lambda_n A)^{-1}(y_n - \lambda_n B(x_n)) \end{aligned} \tag{18.17}$$

Theorem 18.6 *[23] Let $\{x_n\} \subset H$ be a sequence generated by (18.17), where A, B are maximal monotone operators with B γ-cocoercive and suppose that parameters α_n, λ_n, and ϵ_n satisfied*

(i) $\exists \epsilon\ \exists \lambda > 0$ such that $\forall n \in \mathbb{N}$, $\lambda \leq \lambda_n \leq 2\gamma - \epsilon$.
(ii) $\exists \alpha \in [0, 1)$ such that $\forall n \in \mathbb{N}\ 0 \leq \alpha_n \leq \alpha$.
(iii) $\sum_{n=1}^{\infty} \epsilon_n < +\infty$.

If the following condition holds:

$$\sum_{n=1}^{\infty} \alpha_n \|x_n - x_{n-1}\|^2 < \infty$$

Then, there exists $\overline{x} \in S$ such that $\{x_n\}$ weakly converges to $\overline{x}$ as $n \to +\infty$.

18.5.3 Inertial Douglas–Rachford Splitting Algorithm

This section is dedicated to the formulation of an inertial Douglas–Rachford splitting algorithm, which approaches the set of zeros of the sum of two maximally monotone operators and to the investigation of its convergence properties.

Theorem 18.7 *[24] Let $A, B : H \to 2^H$ be maximally monotone operators such that $\text{zer}(A + B) \neq \emptyset$. Consider the following iterative scheme:*

$$\begin{aligned} y_n &= J_{\gamma B}[x_n + \alpha_n(x_n - x_{n-1})] \\ z_n &= J_{\gamma A}[2y_n - x_n - \alpha_n(x_n - x_{n-1})] \\ x_{n+1} &= x_n + \alpha_n(x_n - x_{n-1}) + \lambda_n(z_n - y_n) \end{aligned}$$

where $\gamma > 0, x_0, x_1$ are arbitrarily chosen in H, $\{\alpha_n\}$ is nondecreasing with $\alpha_1 = 0$ and $0 \leq \alpha_n \leq \alpha < 1$ for every $n \leq 1$ and γ, σ, δ are such that

$$\delta > \frac{\alpha^2(1+\alpha) + \alpha\sigma}{1 - \alpha^2} \text{ and } 0 \leq \lambda \leq \lambda_n \leq 2\frac{\delta - \alpha[\alpha(1+\alpha) + \alpha\delta + \sigma]}{\delta[1 + \alpha(1+\alpha) + \alpha\delta + \sigma]}$$

Then, there exists $x \in \text{Fix}(R_{\gamma A}R_{\gamma B})$ such that the following statements are true:

(i) $J_{\gamma B} \in \text{zer}(A + B)$;
(ii) $\sum_{n\in\mathbb{N}} \|x_{n+1} - x_n\|^2 < +\infty$;
(iii) $\{x_n\}$ *converges weakly to* x;
(iv) $\{y_n - z_n\} \to 0$ *as* $n \to \infty$;
(v) $\{y_n\}$ *converges weakly to* $J_{\gamma B}x$;
(vi) $\{z_n\}$ *converges weakly to* $J_{\gamma B}x$;
(vii) if A or B is uniformly monotone, then $\{y_n\}$ and $\{z_n\}$ converge strongly to the unique point in $\text{zer}(A + B)$.

18.5.4 Pock and Lorenz's Variable Metric Forward–Backward Algorithm

Recently, with the inspiration of Nestrov's accelerated gradient method, Lorenz and Pock [24] proposed a modification of the forward–backward splitting to solve monotone inclusion problems by the sum of a monotone operator whose resolvent is easy to compute and another monotone operator is cocoercive. They have introduced a symmetric, positive definite map M, which is considered as a preconditioner or variable metric. The algorithm can be given as follows:

$$\begin{aligned} y_n &= x_n + \alpha_n(x_n - x_{n-1}) \\ x_{n+1} &= (Id + \lambda_n M^{-1}A)^{-1}(Id - \lambda_n M^{-1}B)^{-1}(y_n) \end{aligned} \tag{18.18}$$

where $\alpha_n \in [0, 1)$ is an extrapolation factor, λ_n is a step-size parameter.

Definition 18.3 The operator B is said to be cocoerceive with respect to the solution set $S = (A + B)^{-1}(0)$ and a linear, self-adjoint, and positive definite map L, if for all $x \in H, y \in S$

$$\langle B(x) - B(y), x - y\rangle \leq \|B(x) - B(y)\|^2_{L^{-1}}$$

where $\|x\|_M$ is the norm corresponding to M-norm defined as $\langle x, y\rangle_M = \langle x, My\rangle$.

Theorem 18.8 *[24] Let H be a real Hilbert space and $A, B : H \to 2^H$ be maximally monotone operators. Further assume that $M, L : H \to H$ are linear, bounded, self-adjoint, and positive definite maps and that B is single-valued and cocoerceive w. r. t L^{-1}. Moreover, let $\lambda_n > 0$, $\alpha < 1$, $\alpha_n \in [0, \alpha]$, $x_0 = x_{-1} \in X$ and let the sequence $\{x_n\}$ and $\{y_n\}$ be defined by (18.18). If*

(i) $S_n = M - \frac{\lambda_n}{2}L$ *is positive definite for all n and*
(ii) $\sum_{n=1}^{\infty} \|x_n - x_{n-1}\|^2_M \leq \infty$,

then, $\{x_n\}$ converges weakly to a solution of the inclusion $0 \in (A + B)(x)$.

Proof: Let $x^* \in \text{zer}(A+B)$, so we have

$$-B(x^*) \in A(x^*)$$

Furthermore, the second line in (18.18) can be equivalently expressed as

$$M(y_n - x_{n+1}) - \lambda_n B(y_n) \in \lambda_n A(x_{n+1})$$

For convenience, we define for any symmetric positive definite matrix M,

$$\begin{aligned}
\phi_M^n &= \frac{1}{2}\|x_n - x^*\|_M^2 = \frac{1}{2}\langle M(x_n - x^*), x_n - x^*\rangle \\
\Delta_M^n &= \frac{1}{2}\|x_n - x^*\|_M^2 = \frac{1}{2}\langle M(x_n - x_{n-1}), x_n - x_{n-1}\rangle \\
\Gamma_M^n &= \frac{1}{2}\|x_{n+1} - y_n\|_M^2 = \frac{1}{2}\langle M(x_{n+1} - y_n), x_{n+1} - y_n\rangle
\end{aligned}$$

From the well-known inequality

$$\langle a-b, a-c\rangle_M = \frac{1}{2}\|a-b\|_M^2 + \frac{1}{2}\|a-c\|_M^2 - \frac{1}{2}\|b-c\|_M^2 \tag{18.19}$$

we have by using the definition of the inertial extrapolate y_n that

$$\begin{aligned}
\phi_M^n - \phi_M^{n+1} = \Delta_M^{n+1} &+ \langle y_n - x_{n+1}, x_{n+1} - x^*\rangle_M \\
&- \alpha_n\langle x_n - x_{n-1}, x_{n+1} - x^*\rangle_M
\end{aligned} \tag{18.20}$$

Then, by using the monotonicity of T, we deduce that

$$\begin{aligned}
\langle \lambda_n A(x_{n+1} - \lambda_n A(x^*)), x_{n+1} - x^*\rangle \geq 0 \\
\langle M(y_n - x_{n+1}) - \lambda_n B(y_n) + \lambda_n B(x^*), x_{n+1} - x^*\rangle \geq 0
\end{aligned}$$

and

$$\langle y_n - x_{n+1}, x_{n+1} - x^*\rangle_M + \lambda\langle B(x^*) - B(y_n), x_{n+1} - x^*\rangle \geq 0$$

Combining with (18.20), we obtain

$$\phi_M^n - \phi_M^{n+1} \geq \Delta_M^{n+1} + \lambda_n\langle B(y_n) - bx^*, x_{n+1} - x^*\rangle - \alpha_n\langle x_n - x_{n+1}, x_{n+1} - x^*\rangle_M \tag{18.21}$$

From the cocoercivity property of B, we have that

$$\begin{aligned}
&\langle B(y_n) - B(x^*), x_{n+1} - x^*\rangle \\
&\quad = \langle B(y_n) - B(x^*), x_{n+1} - y_n + y_n - x^*\rangle \\
&\quad \geq \|B(y_n) - B(x^*)\|_{L^{-1}}^2 + \langle B(y_n) - B(x^*), x_{n+1} - y_n\rangle \\
&\quad \geq \|B(y_n) - B(x^*)\|_{L^{-1}}^2 - \|B(y_n) - B(x^*)\|_{L^{-1}}^2 - \frac{1}{2}\Gamma_L^n \\
&\quad = -\frac{1}{2}\Gamma_L^n
\end{aligned}$$

Substituting back into (18.21), we arrive at

$$\phi_M^n - \phi_M^{n+1} \geq \Delta_M^{n+1} - \frac{\lambda_n}{2}\Gamma_L^n - \alpha_n\langle x_n - x_{n+1}, x_{n+1} - x^*\rangle_M$$

Invoking again (18.19), it follows that

$$\begin{aligned}
\phi_M^{n+1} - \phi_M^n - \alpha_n(\phi_M^n - \phi_M^{n-1}) &\leq -\Delta_M^{n+1} + \frac{\lambda_n}{2}\Gamma_L^n + \alpha_n(\Delta_M^{n+1} + \langle x_n - x_{n+1}, x_{n+1} - x^*\rangle_M) \\
&= -\Gamma_M^n + \frac{\lambda_n}{2}\Gamma_L^n + (\alpha_n + \alpha_n^2)\Delta_M^n
\end{aligned} \tag{18.22}$$

The rest of the proof closely follows the proof of theorem 2.1 in [25]. By the definition of S_n and using $(\alpha_n + \alpha_n^2)/2 \leq \alpha_n$, we have

$$\phi_M^{n+1} - \phi_M^n \alpha_n(\phi_M^n - \phi_M^{n-1}) \leq -\Gamma_{S_n}^n + 2\alpha_n \Delta_M^n \tag{18.23}$$

By assumption (i), the first term is nonpositive and since $\alpha_n \geq 0$, the second term is nonnegative. Now, define $\theta_n = \max\{0, \phi_M^n - \phi_M^{n-1}\}$ and setting $\delta_n = 2\alpha_n \Delta_M^n = \alpha_n \|x_n - x_{n-1}\|_M^2$, we obtain

$$\theta_{n+1} \leq \alpha_n \theta_n + \delta_n \leq \alpha_n \theta_n + \delta_n \tag{18.24}$$

Applying this inequality recursively, one obtains a geometric series of the form

$$\theta_{n+1} \leq \alpha_n \theta_1 + \sum_{i=0}^{n-1} \alpha_i \delta_{n-i} \tag{18.25}$$

Summing this inequality from $n = 0, \ldots, \infty$, one has

$$\sum_{n=0}^{\infty} \theta_{n+1} \leq \frac{1}{1-\alpha}\left(\theta_1 + \sum_{n=1}^{\infty} \delta_n\right) \tag{18.26}$$

Note that the series on the right-hand side converges by assumption (ii).

Now, we set $t_n = \phi_M^n - \sum_{i=1}^{n} \theta_i$ and since $\phi_M^n \geq 0$ and $\sum_{i=1}^{n} \theta_i$ is bounded independently of n, we see that t_n is bounded from below. On the other hand,

$$\begin{aligned} t_{n+1} &= \phi_M^{n+1} - \theta_{n+1} - \sum_{i=1}^{n} \theta_i \\ &\leq \phi_M^{n+1} - \phi_M^{n+1} + \phi_M^n - \sum_{i=1}^{n} \theta_i = t_n \end{aligned}$$

and hence t_n is also nondecreasing, thus convergent. This implies that ϕ_M^{n+1} is convergent and especially that $\theta_n \rightarrow 0$. From (18.23), we get

$$\begin{aligned} \frac{1}{2}\|x_{n+1} - y_n\|_{S_n}^2 &\leq -\theta_{n+1} - \alpha\theta_n + \delta_n \\ \frac{1}{2}\|x_{n+1} - x_n - \alpha_n(x_n - x_{n-1})\|_{S_n}^2 &\leq -\theta_{n+1} - \alpha\theta_n + \delta_n \end{aligned}$$

Since δ_n is summable, it follows that $\|x_n - x_{n-1}\|_{S_n} \rightarrow 0$ and hence

$$\lim_{n\rightarrow\infty} \|x_{n+1} - x_n - \alpha_n(x_n - x_{n-1})\|_{S_n} = 0$$

We already know that x_n is bounded; hence, there is a convergent subsequence $x_{n_k} \rightharpoonup \overline{x}$. Then, we also get that $y_{n_k} = (1 + \alpha_{n_k})x_{n_k} - \alpha_{n_k} x_{n_k - 1} \rightharpoonup \overline{x}$. Now, we get from (18.18) that

$$x_{n_k} = (Id + \lambda_{n_k} M^{-1}A)^{-1}(y_{n_k} - \lambda_{n_k} M^{-1}B(y_{n_k}))$$

and pass to the limit to obtain

$$\overline{x} = (Id + \overline{\lambda} M^{-1}A)^{-1}(\overline{x} - \overline{\lambda} M^{-1}B(\overline{x}))$$

which is equivalent to

$$-B(\overline{x}) \in A(\overline{x})$$

which in turn shows that $\overline{x}$ is a solution. Opial's Lemma concludes the proof.

Recently, some inertial iterative algorithms have been proposed, which replaced the condition (iii) of Theorem 18.5 with some mild conditions, which makes the algorithms easy to use for real-world problems. In

2015, Bot et al. [26] proposed the inertial Mann algorithm for nonexpansive operators and studied the weak convergence in real Hilbert space framework. The inertial Mann algorithm for some initial points $x_0, x_1 \in C$:

$$\begin{aligned} y_n &= x_n + \alpha_n(x_n - x_{n-1}) \\ x_{n+1} &= (1-\beta_n)y_n + \beta_n T(y_n)\ \forall n \geq 1 \end{aligned} \tag{18.27}$$

where $\alpha_n, \beta_n \in [0,1)$, and C are nonempty, closed, and affine subsets of H.

Theorem 18.9 *[26] Let C be a nonempty, closed, and affine subsets of H and* $T : C \to C$ *be a nonexpansive mapping such that* $\text{Fix}(T) \neq \emptyset$. *Suppose that* $\{\alpha_n\}$ *is nondecreasing with* $\alpha_1 = 0$ *and* $0 \leq \alpha_n \leq \alpha < 1$ *for every* $n \geq 1$ *and* $\lambda, \sigma, \delta > 0$ *are such that*

$$\delta > \frac{\alpha^2(1+\alpha)+\alpha\sigma}{1-\alpha^2} \text{ and } 0 < \lambda \leq \lambda_n \leq \frac{\delta - \alpha[\alpha(1+\alpha)+\alpha\delta+\sigma]}{\delta[1+\alpha(1+\alpha)+\alpha\delta+\sigma]}\ \forall n \geq 1 \tag{18.28}$$

Then, the sequence generated by algorithm 18.28 satisfies the following:

(i) $\sum_{n\in\mathbb{N}} \|x_{n+1} - x_n\|^2 < \infty$;
(ii) $\{x_n\}$ *converges weakly to a point in* $\text{Fix}(T)$.

Proof: We notice that because of the choice of δ, $\lambda_n \in (0,1)$ for every $n \geq 1$. Since C is affine, the iterative scheme provides a well-defined sequence in C.

(i) We denote

$$w_n := x_n + \alpha_n(x_n - x_{n-1})\ \forall n \geq 1$$

Then, the iterative scheme reads for every $n \geq 1$:

$$x_{n+1} = w_n + \lambda_n(Tw_n - w_n) \tag{18.29}$$

Let us fix an element $y \in \text{Fix}(T)$ and $n \geq 1$. Since T is nonexpansive, so we have

$$\begin{aligned} \|x_{n+1} - y\|^2 &= (1-\lambda_n)\|w_n - y\|^2 + \lambda_n\|Tw_n - Ty\|^2 - \lambda_n(1-\lambda_n)\|Tw_n - w_n\|^2 \\ &\leq \|w_n - y\|^2 - \lambda_n(1-\lambda_n)\|Tw_n - w_n\|^2 \end{aligned} \tag{18.30}$$

Therefore, we have

$$\begin{aligned} \|w_n - y\|^2 &= \|(1+\alpha_n)(x_n - y) - \alpha_n(x_{n-1} - y)\|^2 \\ &= (1+\alpha_n)\|x_n - y\|^2 - \alpha_n\|x_{n-1} - y\|^2 + \alpha_n(1+\alpha_n)\|x_n - x_{n-1}\|^2 \end{aligned}$$

hence by (18.30) we obtain

$$\begin{aligned} &\|x_{n+1} - y\|^2 - (1+\alpha_n)\|x_n - y\|^2 + \alpha_n\|x_{n-1} - y\|^2 \leq -\lambda_n(1-\lambda_n)\|Tw_n - w_n\|^2 \\ &\quad + \alpha_n(1+\alpha_n)\|x_n - x_{n-1}\|^2 \end{aligned} \tag{18.31}$$

Further, we have

$$\begin{aligned} \|Tw_n - w_n\|^2 &= \left\|\frac{1}{\lambda_n}(x_{n+1} - x_n) + \frac{\alpha_n}{\lambda_n}(x_{n-1} - x_n)\right\|^2 \\ &= \frac{1}{\lambda_n^2}\|x_{n+1} - x_n\|^2 + \frac{\alpha_n^2}{\lambda_n^2}\|x_{n-1} - x_n\|^2 + 2\frac{\alpha_n}{\lambda_n^2}\langle x_{n+1} - x_n, x_{n-1} - x_n\rangle \\ &\geq \frac{1}{\lambda_n^2}\|x_{n+1} - x_n\|^2 + \frac{\alpha_n^2}{\lambda_n^2}\|x_n - x_{n-1}\|^2 \\ &\quad + \frac{\alpha_n}{\lambda_n^2}\left(-\rho_n\|x_{n+1} - x_n\|^2 - \frac{1}{\rho_n}\|x_n - x_{n-1}\|^2\right) \end{aligned} \tag{18.32}$$

where we denote $\rho_n := \frac{1}{\alpha_n + \delta\lambda_n}$.

We derive from (18.31) and (18.32) the inequality

$$\|x_{n+1}-y\|^2-(1+\alpha_n)\|x_n-y\|^2+\alpha_n\|x_{n-1}-y\|^2 \le \frac{(1-\lambda_n)(\alpha_n\rho_n-1)}{\lambda_n}\|x_{n+1}-x_n\|^2 + \gamma_n\|x_n-x_{n-1}\|^2 \tag{18.33}$$

where

$$\gamma_n : = \alpha_n(1+\alpha_n)+\alpha_n(1-\lambda_n)\frac{(1-\rho_n\alpha_n)}{\rho_n\lambda_n} > 0 \tag{18.34}$$

Taking again into account the choice of ρ_n, we have

$$\delta = \frac{1-\rho_n\alpha_n}{\rho_n\lambda_n}$$

and by (18.34) it follows

$$\gamma_n = \alpha_n(1+\alpha_n)+\alpha_n(1-\lambda_n)\delta \le \alpha(1+\alpha)+\alpha\delta \ \forall n \ge 1 \tag{18.35}$$

In the following, we use some techniques from [25] adapted to our setting. We define the sequence $\phi_n : = \|x_n-y\|^2$ for all $n \in \mathbb{N}$ and $\mu_n : = \phi_n-\alpha_n\phi_{n-1}+\gamma_n\|x_n-x_{n-1}\|^2$ for all $n \ge 1$. Using the monotonicity of $\{\alpha_n\}$ and the fact that $\phi_n \ge 0$ for all $n \in \mathbb{N}$, we get

$$\mu_{n+1}-\mu_n = \phi_{n+1}-(1+\alpha_n)\phi_n+\alpha_n\phi_n+\gamma_{n+1}\|x_{n+1}-x_n\|^2-\gamma_n\|x_n-x_{n-1}\|^2 \tag{18.36}$$

which gives by (18.33)

$$\mu_{n+1}-\mu_n \le \left(\frac{(1-\lambda_n)(\alpha_n\rho_n-1)}{\lambda_n}+\gamma_{n+1}\right)\|x_{n+1}-x_n\|^2 \ \forall n \ge 1 \tag{18.37}$$

We claim that

$$\frac{(1-\lambda_n)(\alpha_n\rho_n-1)}{\lambda_n}+\gamma_{n+1} \le -\sigma \ \forall n \ge 1 \tag{18.38}$$

Let $n \ge 1$. Indeed, by the choice of ρ_n, we obtain

$$\frac{(1-\lambda_n)(\alpha_n\rho_n-1)}{\lambda_n}+\gamma_{n+1} \le -\sigma$$

$$\Leftrightarrow \lambda_n(\gamma_n+\sigma)+(\alpha_n\rho_n-1)(1-\lambda_n) \le 0$$

$$\Leftrightarrow \lambda_n(\gamma_n+\sigma)-\frac{\delta\lambda_n(1-\lambda_n)}{\alpha_n+\delta\lambda_n} \le 0$$

$$\Leftrightarrow (\alpha_n+\delta\lambda_n)(\gamma_{n+1}+\delta)+\delta\lambda_n \le \delta$$

Thus, by using (18.35), we have

$$(\alpha_n+\delta\lambda_n)(\gamma_{n+1}+\sigma)+\delta\lambda_n \le (\alpha+\delta\lambda_n)(\alpha(1+\alpha_n)+\alpha\delta+\sigma)+\delta\lambda_n \le \delta$$

where the last inequality follows by taking into account the upper bound considered for $\{\lambda_n\}$ in (18.28). Hence, the claim in (18.28) is true.

We obtain from (18.37) and (18.38) that

$$\mu_{n+1}-\mu_n \le -\sigma\|x_{n+1}-x_n\|^2 \ \forall n \ge 1 \tag{18.39}$$

The sequence $\{\mu_n\}$ is nonincreasing and the bound for $\{\alpha_n\}$ delivers

$$-\alpha\phi_{n-1} \le \phi_n-\alpha\phi_{n-1} \le \mu_n \le \mu_1 \tag{18.40}$$

We obtain

$$\phi_n \le \alpha^n\phi_0+\mu_1\sum_{k=0}^{n-1}\alpha^k \le \alpha^n\phi_0+\frac{\mu_1}{1-\alpha} \ \forall n \ge 1$$

where we notice that $\mu_1 = \phi_1 \geq 0$. Combining (18.39) and (18.40), we get

$$\sigma \sum_{k=1}^{n} \|x_{k+1} - x_k\|^2 \leq \mu_1 - \mu_{n+1}\mu_1 + \alpha\phi_n \leq \alpha^{n+1}\phi_0 + \frac{\mu_1}{1-\alpha} \quad \forall n \geq 1 \tag{18.41}$$

which shows that $\sum_{n\in\mathbb{N}} \|x_{n+1} - x_n\|^2 < \infty$.

(ii) We prove this by using the result of Opial Lemma. We have proven above that for an arbitrary $y \in \text{Fix}(T)$ the inequality (18.33) is true. By part (i), (18.35) and Lemma 18.3, we derive that $\lim\limits_{n\to\infty} \|x_n - y\|$ exists. On the other hand, let x be a sequential weak cluster point of $\{x_n\}$, that is, the letter has a subsequence $\{x_{n_k}\}$ fulfilling $x_{n_k} \rightharpoonup x$ as $k \to \infty$. By part (i), the definition of w_n and the upper bound requested for $\{\alpha_n\}$, we get $w_{n_k} \rightharpoonup x$ as $k \to \infty$. Further, by (18.29) we have

$$\|Tw_n - w_n\| = \frac{1}{\lambda_n}\|x_{n+1} - w_n\| \leq \frac{1}{\lambda}\|x_{n+1} - w_n\| \leq \frac{1}{\lambda}(\|x_{n+1} - x_n\| + \alpha\|x_n - x_{n-1}\|) \tag{18.42}$$

thus by (i), we obtain $Tw_{n_k} - w_{n_k} \to 0$ as $k \to \infty$. Applying now Lemma 18.2 for the sequence $\{w_{n_k}\}$, we conclude that $x \in \text{Fix}(T)$. Since the two assertions of Opial Lemma are verified, it follows that $\{x_n\}$ converges weakly to a point in $\text{Fix}(T)$.

Remark 18.1 For the choice $\alpha_n = 0$, the inertial Mann algorithm reduces to classical Mann algorithm.

In 2018, Dong et al. [27] proposed a generalized form of inertial Mann algorithm. They included an extra inertial term in inertial Mann algorithm. The algorithm is given as follows:

$$\begin{aligned} y_n &= x_n + \alpha_n(x_n - x_{n-1}) \\ z_n &= x_n + \beta_n(x_n - x_{n-1}) \\ x_{n+1} &= (1-\lambda_n)y_n + \lambda_n T(z_n) \ \forall n \geq 1 \end{aligned} \tag{18.43}$$

where $\alpha_n, \beta_n, \lambda_n \in [0, 1)$.

Convergence analysis can be summarized as follows:

Theorem 18.10 *[27] Suppose $T : H \to H$ is nonexpansive with $\text{Fix}(T) \neq \emptyset$. Consider the sequence $\{\alpha_n\}, \{\beta_n\}, \{\lambda_n\}$ satisfies the following:*

(B1) $\{\alpha_n\} \subset [0, \alpha]$ and $\{\beta_n\} \subset [0, \beta]$ are nondecreasing with $\alpha_1 = \beta_1 = 0$ and $\alpha, \beta \in [0, 1)$;
(B2) for any $\lambda, \sigma, \delta > 0$ are such that

$$\delta > \frac{\alpha\psi(1+\psi) + \alpha\sigma}{1-\alpha^2} \text{ and } 0 < \lambda \leq \lambda_n \leq \frac{\delta - \alpha[\psi(1+\psi) + \alpha\delta + \sigma]}{\delta[1 + \psi(1+\psi) + \alpha\delta + \sigma]} \ \forall n \geq 1$$

where $\psi = \max\{\alpha, \beta\}$.

Then, the sequence $\{x_n\}$ generated by the general inertial Mann algorithm converges weakly to a point of $\text{Fix}(T)$.

Remark 18.2 The following choices of α_n, β_n reduces to generalized inertial Mann algorithm to classical inertial Mann algorithm in any of the following conditions
(i) $\alpha_n = \beta_n$, (ii) $\alpha_n = 0$, (iii) $\beta_n = 0$.

In 2019, Dixit et al. [28] proposed the inertial normal algorithm and studied the weak convergence of the proposed algorithm. The algorithm is given as follows:

$$\begin{aligned} y_n &= x_n + \alpha_n(x_n - x_{n-1}) \\ x_{n+1} &= T[(1-\lambda_n)y_n + \lambda_n T(y_n)], \quad \forall n \in \mathbb{N} \end{aligned} \tag{18.44}$$

Theorem 18.11 *[28] Let C be a nonempty closed affine subset of a real Hilbert space H and $T : C \to C$ be a nonexpansive mapping with $\text{Fix}(T) \neq \emptyset$. Let $\{\alpha_n\}, \{\lambda_n\}$ satisfies the following conditions:*

(C1) $\{\alpha_n\} \subset [0, \alpha]$ *is a nondecreasing sequence with* $\alpha \in [0, 1)$ *and* $\alpha_1 = 0$;

(C2) for any constants $\lambda, \sigma, \delta > 0$ *satisfying*

$$\delta > \frac{2q\alpha(1+\alpha)\alpha + \sigma}{1 - \alpha^2(1-\lambda)} \text{ and } 0 < \lambda \leq \lambda_n \leq \frac{\delta - \alpha[2q\alpha(1+\alpha) + \alpha\delta(1-\lambda) + 2q\sigma]}{\delta[1 + 2q\alpha(1+\alpha) + 2\delta(1-\lambda) + 2q\sigma]}$$

where $q = 1 + \frac{1}{\lambda^2}$.

Then, the sequence $\{x_n\}$ *generated by Algorithm (18.44) converges weakly to a point in* Fix(T).

18.5.5 Numerical Example

In order to compare the convergence speed of Algorithm (18.27), (18.43), and (18.44) in real Hilbert space $H = \mathbb{R}^2$ with Euclidean norm, we consider the nonexpansive mapping

$$T(u_1, u_2) = (\sin u_1, \sin u_2) \quad \text{for } (u_1, u_2) \in \mathbb{R}^2$$

To show T is nonexpansive mapping, we have for $u = (u_1, u_2)$ and $v = (v_1, v_2)$

$$\begin{aligned} \|T(u) - T(v)\|^2 &= \|(\sin u_1, \sin u_2) - (\sin v_1, \sin v_2)\|^2 \\ &\leq |\sin u_1 - \sin v_1|^2 + |\sin u_2 - \sin v_2|^2 \\ &\leq |u_1 - v_1|^2 + |u_2 - v_2|^2 \\ &= \|u - v\|^2 \end{aligned} \tag{18.45}$$

This implies that $\|T(u) - T(v)\| \leq \|u - v\|$, thus T is nonexpansive. For comparison, we choose the sequences $\alpha_n = \frac{1}{20}$, $\beta_n = \frac{1}{14}$, and $\lambda_n = \frac{1}{2}$. The convergence behaviors of the algorithms are shown in figure.

Figure 18.1 shows that the Algorithms (18.27) and (18.43) have nearly same convergence speed but greater than the Mann algorithm. The convergence speed of Algorithm (18.44) is highest. This figure shows the importance of Algorithm (18.44) to calculate the zeros of a nonexpansive mapping.

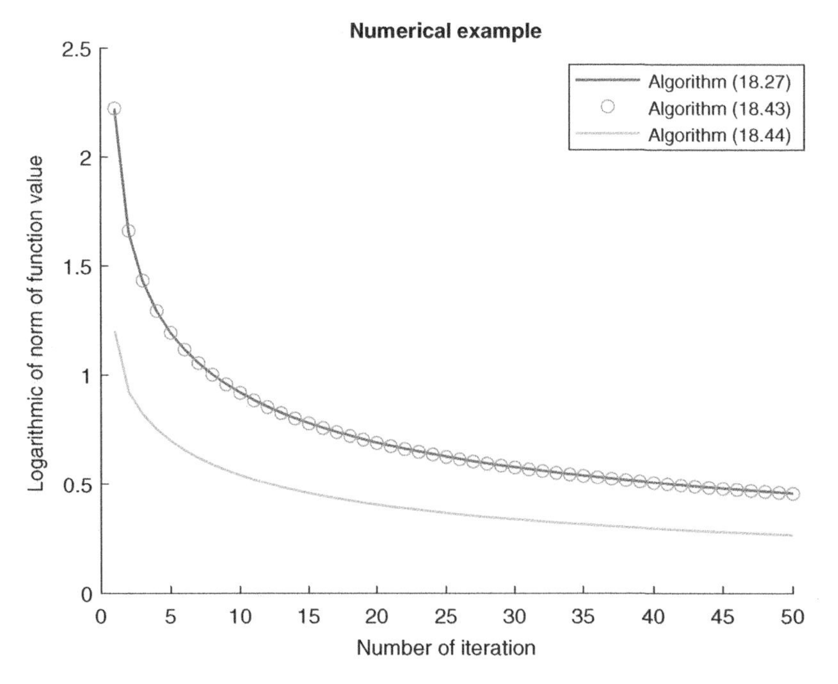

Figure 18.1 Semilog graph between number of iterations and iteration value.

18.6 Numerical Experiments

Further to compare the convergence speed and accuracy of iterative algorithms (18.27), (18.43), and (18.44) on real-world problems, we conducted numerical experiments for **regression problem** on high dimensional datasets that are publicly available. Consider the convex minimization problem given by

$$\min_{x\in\mathbb{R}^n} F(x) = \frac{1}{2m}\|Ax - b\|_2^2 + \rho\|x\|_1 \tag{18.46}$$

where $A : \mathbb{R}^n \to \mathbb{R}^m$ is a linear map, $b \in \mathbb{R}^m$ and ρ is sparsity controlling parameter. According to the Karush–Kuhn–Tucker (KKT) condition a point $x^* \in \mathbb{R}^n$ solves (18.46) if and only if

$$0 \in \frac{1}{2m}\nabla\|Ax^* - b\|_2^2 + \partial\rho\|x^*\|_1 \tag{18.47}$$

For any $\lambda > 0$, Eq. (18.47) becomes

$$\begin{aligned}
&0 \in \frac{\lambda}{2m}\nabla\|Ax - b\|_2^2 + \lambda\partial\rho\|x\|_1\\
&\Leftrightarrow 0 \in \frac{\lambda}{2m}\nabla\|Ax - b\|_2^2 - x^* + x^* + \lambda\partial\rho\|x\|_1\\
&\Leftrightarrow x^* - \frac{\lambda}{2m}\nabla\|Ax^* - b\|_2^2 \in x^* + \lambda\partial\rho\|x\|_1\\
&\Leftrightarrow x^* = \left(I - \frac{\lambda}{2m}\nabla\|A\cdot -b\|_2^2\right)^{-1}(I + \lambda\partial\rho\|\cdot\|_1)(x^*)\\
&\Leftrightarrow x^* = \mathrm{Prox}_{\lambda\partial\rho\|\cdot\|_1}\left(I - \frac{\lambda}{2m}\nabla\|A\cdot -b\|_2^2\right)(x^*)
\end{aligned} \tag{18.48}$$

Thus, x^* minimizes $F(x)$ iff x^* is the fixed point of the operator $\mathrm{Prox}_{\lambda\partial\rho\|\cdot\|_1}\left(I - \frac{\lambda}{2m}\nabla\|A\cdot -b\|^2\right)$.

Note that here operator $\mathrm{Prox}_{\lambda\partial\rho\|\cdot\|_1}\left(I - \frac{\lambda}{2m}\nabla\|A\cdot -b\|_2^2\right)$ is nonexpansive as l_1-norm is proper lower semicontinuous convex function. Thus, we can apply the Algorithms (18.27), (18.43), and (18.44) to solve the convex minimization problem (18.46).

For numerical experiments, we consider the operator $A := [A_1, A_2, \ldots, A_n] \in \mathbb{R}^{m\times n}$ data matrix having n-features and m-samples, where each A_i is a m-dimensional vector $i = 1, 2, \ldots, n$ contains m responses. We have taken here datasets from two types of cancer patients: Colon-cancer dataset and Carcinom dataset.

(i) *Colon-cancer dataset*: Colon cancer is a cancer of large intestine (colon). This dataset is collected from 62 patients having 2000 gene expressions. It contains 40 tumor biopsies from tumors and 22 normal biopsies from healthy parts of the large intestine of the patient.

(i) *Carcinom dataset*: Carcinoma cancer occurs because of uncontrolled growth of a cell. This dataset contains 174 samples with 9182 features.

For the numerical experiment, we selected $\{\alpha_n\} = \{\frac{n-1}{14n+2.5}\}$, $\{\beta_n\} = \{\frac{n-1}{20n+2.5}\}$, and $\{\lambda_n\} = \{0.5 + \frac{1}{200n}\}$. For this choice of $\{\alpha_n\}$, $\{\beta_n\}$, and $\{\lambda_n\}$, algorithms satisfies their respective convergence criteria. The sparsity controlling parameter ρ is taken as $\theta\|A^T b\|_\infty$, where θ is tuned in the range $[1, 10^{-10}]$ in the multiple of 0.1. The maximum number of iteration is set to 1000 and to stop the procedure, difference between consecutive iteration should be less than 0.001.

In first experiment, we compare the convergence speed of the Algorithms (18.27), (18.43), and (18.44). We plotted the graphs between number of iteration and corresponding function value for each algorithm.

From Figure 18.2, we can observe that for both the datasets, convergence speed of Algorithm (18.44) is highest and convergence speed of Algorithms (18.27) and (18.43) are nearly the same.

In second experiment, we compare the Algorithms (18.27), (18.43), and (18.44) on the basis of their accuracy for colon dataset and carcinom dataset. We plotted the graphs between number of iterations and corresponding root mean square error (RMSE) value.

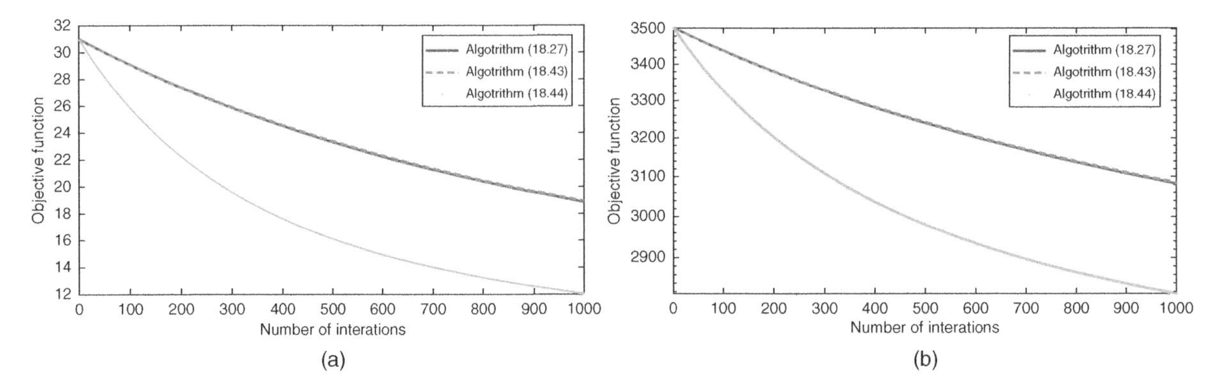

Figure 18.2 The semilog graphs are plotted between number of iteration vs. corresponding objective function value for different datasets. (a) Colon and (b) carcinom.

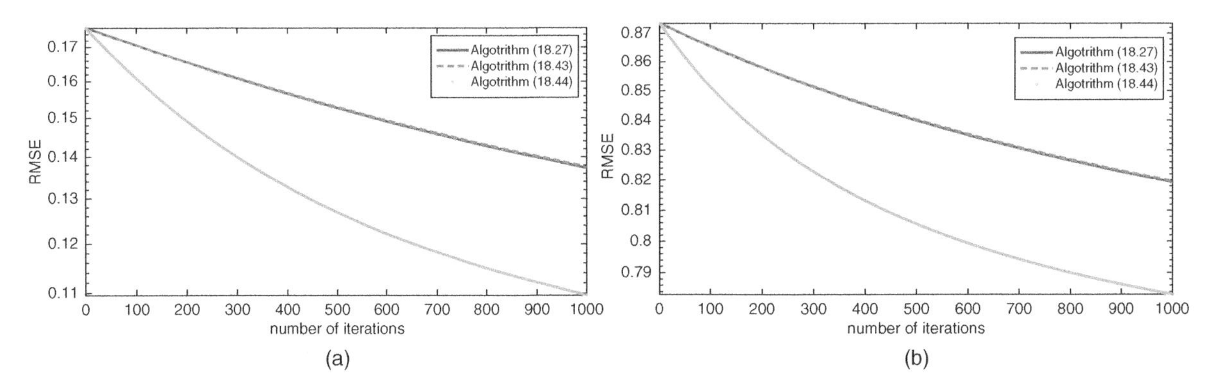

Figure 18.3 The semilog graphs are plotted between number of iterations and corresponding root mean square error of the function. (a) Colon and (b) carcinom.

From Figure 18.3, we observe that the RMSE value for Algorithm (18.44) is least while the RMSE values for Algorithms (18.27) and (18.43) are nearly the same.

Thus, we conclude from the observations of Figures 18.2 and 18.3 that Algorithm (18.44) outperforms over Algorithms (18.27) and (18.43).

References

1 Byrne, C. (2003). A unified treatment of some iterative algorithms in signal processing and image reconstruction. *Inverse Problems* 20 (1): 103.

2 Combettes, P.L. and Wajs, V.R. (2005). Signal recovery by proximal forward-backward splitting. *Multiscale Modeling and Simulation* 4 (4): 1168–1200.

3 Daubechies, I., Defrise, M., and De Mol, C. (2004). An iterative thresholding algorithm for linear inverse problems with a sparsity constraint. *Communications on Pure and Applied Mathematics: A Journal Issued by the Courant Institute of Mathematical Sciences* 57 (11): 1413–1457.

4 Bauschke, H.H. and Borwein, J.M. (1996). On projection algorithms for solving convex feasibility problems. *SIAM Review* 38 (3): 367–426.

5 Koh, K., Kim, S.-J., and Boyd, S. (2007). An interior-point method for large-scale l1-regularized logistic regression. *Journal of Machine Learning Research* 8: 1519–1555.

6 Minty, G.J. (1962). Monotone (nonlinear) operators in Hilbert space. *Duke Mathematical Journal* 29 (3): 341–346.

7 Martinet, B. (1970). Brève communication. Régularisation d'inéquations variationnelles par approximations successives. *ESAIM: Mathematical Modelling and Numerical Analysis-Modélisation Mathématique et Analyse Numérique* 4 (R3): 154–158.

8 Eckstein, J. and Bertsekas, D.P. (1992). On the Douglas–Rachford splitting method and the proximal point algorithm for maximal monotone operators. *Mathematical Programming* 55 (1–3): 293–318.

9 Güler, O. (1992). New proximal point algorithms for convex minimization. *SIAM Journal on Optimization* 2 (4): 649–664.

10 Rockafellar, R.T. (1976). Monotone operators and the proximal point algorithm. *SIAM Journal on Control and Optimization* 14 (5): 877–898.

11 Moreau, J.-J. (1965). Proximité et dualité dans un espace Hilbertien. *Bulletin de la Société mathématique de France* 93: 273–299.

12 Bauschke, H.H. and Patrick, L.C. (2011). *Convex Analysis and Monotone Operator Theory in Hilbert Spaces*, vol. 408. New York: Springer.

13 Alvarez, F. and Attouch, H. (2001). An inertial proximal method for maximal monotone operators via discretization of a nonlinear oscillator with damping. *Set-Valued Analysis* 9 (1–2): 3–11.

14 Alvarez, F (2004). Weak convergence of a relaxed and inertial hybrid projection-proximal point algorithm for maximal monotone operators in Hilbert space. *SIAM Journal on Optimization* 14 (3): 773–782.

15 Peaceman, D.W. and Rachford, H.H. Jr. (1955). The numerical solution of parabolic and elliptic differential equations. *Journal of the Society for Industrial and Applied Mathematics* 3 (1): 28–41.

16 Douglas, J. and Rachford, H.H. (1956). On the numerical solution of heat conduction problems in two and three space variables. *Transactions of the American Mathematical Society* 82 (2): 421–439.

17 Passty, G.B. (1979). Ergodic convergence to a zero of the sum of monotone operators in Hilbert space. *Journal of Mathematical Analysis and Applications* 72 (2): 383–390.

18 Combettes, P.L. and Wajs V.R. (2005). Signal recovery by proximal forward-backward splitting. *Multiscale Modeling and Simulation* 4 (4): 1168–1200.

19 Lions, P.-L. and Mercier, B. (1979). Splitting algorithms for the sum of two nonlinear operators. *SIAM Journal on Numerical Analysis* 16 (6): 964–979.

20 Bauschke, H.H., Matoušková, E., and Reich, S. (2004). Projection and proximal point methods: convergence results and counterexamples. *Nonlinear Analysis: Theory Methods & Applications* 56 (5): 715–738.

21 Polyak, B.T. (1964). Some methods of speeding up the convergence of iteration methods. *USSR Computational Mathematics and Mathematical Physics* 4 (5): 1–17.

22 Nesterov, Y.E. (1983). A method for solving the convex programming problem with convergence rate $o(1/k^2)$. *Doklady Akademii Nauk SSSR* 269: 543–547.

23 Moudafi, A. and Oliny, M. (2003). Convergence of a splitting inertial proximal method for monotone operators. *Journal of Computational and Applied Mathematics* 155 (2): 447–454.

24 Lorenz, D.A. and Pock, T. (2015). An inertial forward-backward algorithm for monotone inclusions. *Journal of Mathematical Imaging and Vision* 51 (2): 311–325.

25 Alvarez, F. and Attouch, H. (2001). An inertial proximal method for maximal monotone operators via discretization of a nonlinear oscillator with damping. *Set-Valued Analysis* 9 (1–2): 3–11.

26 Bot, R.I., Csetnek, E.R., and Hendrich, C. (2015). Inertial Douglas–Rachford splitting for monotone inclusion problems. *Applied Mathematics and Computation* 256: 472–487.

27 Dong, Q.-L., Cho, Y.J., and Rassias, T.M. (2018). General inertial Mann algorithms and their convergence analysis for nonexpansive mappings. In: *Applications of Nonlinear Analysis* (ed. T.M. Rassias), 175–191. Cham: Springer.

28 Dixit, A., Sahu, D.R., Singh, A.K., and Som, T. (2019). Application of a new accelerated algorithm to regression problems. *Soft Computing* 24 (2): 1–14.

Index

a
Activation function 2, 52
Affine arithmetic 220, 222, 229
Affine-contractor 215–217
Algebraic system of aggregates (ASA) 63, 64
Artificial intelligence 1, 33, 50, 186, 253, 391, 392, 395, 397–401, 408
Artificial neural network 1, 50
automl 27
Autonomous car 402, 405
Autonomous systems 400

b
Backpropagation algorithm 56
Basis functions 278
Bernstein polynomials 291
Bessel polynomials 292
Boubaker polynomials 292
Boundary value problem 7
Brain activity reconstruction 343

c
Carcinom dataset 429
Canonical coordinates 254
Cartesian power 171
Chebyshev polynomials 289, 290, 349
Classic integral quadrature methods 314
Classical interval 174, 217
Classic orthogonal polynomials 282
Colon-cancer dataset 429
Contractor 223, 229–231, 235

d
Data synchronization 75
Decision systems 154
Deep learning 23, 39, 49, 408
Dependency problem 174
Differential equation 4
Differential quadrature 273–275, 277, 297
Differential transform method (DTM) 241
Diffusion problems 99
Digital intervals 72
Discernibility matrix-based approaches 154
Disease transmission potential 376
Douglas–Rachford splitting algorithm 416
Dynamic behavior 239, 242

e
Electrostatic field 258, 260
Emotional speech databases 33
Epidemiological isolation 380
Euler–Bernoulli nanobeam 240

f
Finite difference 102
Finite element 102
Fluid dynamic problems 125
Forward–backward algorithm 416
Fractional order differential equations 115
Fredholm integral equation 5, 6, 9
Functional dependence 184
Function link neural network 1, 3
Fuzzy arithmetic 100

Mathematical Methods in Interdisciplinary Sciences, First Edition. Edited by Snehashish Chakraverty.

Fuzzy implicator 148
Fuzzy Laplace transform 118
Fuzzy number 100, 126
Fuzzy rough set 145, 146, 149, 155–157, 159
Fuzzy set 126
Fuzzy synchronization 92
Fuzzy triangular norm 148
Fuzzy-valued functions 116

g
Gaussian mixture model 37
Gegenbauer polynomials 288
Generalized integral quadrature 310
Generalized Taylor-based integral quadrature 312
Genetic algorithm 1
Genetics 23
Grid distributions 293
Goal programming 384

h
Heat transfer problem 104
High-risk population 381
Homotopy perturbation method 125, 126, 129, 130
Hukuhara derivative 116
Hyperbolic tangent activation function 53

i
Image and preimage of arelation 171
Inertial Douglas–Rachford splitting algorithm 421
Inertial methods 418
Inertial proximal point algorithm 419
Infectious diseases 369
Infinitesimal transformation 254
Initial value problem 6
Integral equation 5, 16
Integral quadrature 273, 316
Interval arithmetic 99
Interval computations 169
Interval dependency 176, 179, 181
Interval dependency problem 176, 219, 220, 222
Interval enclosures 184
Interval subdivision 185

j
Jacobi polynomials 283
Jeffery–Hamel problem 127, 128, 131, 137, 142

l
Laguerre neural network 7, 8
Lagrange polynomials 281
Lane-Emden equation 12
Legendre polynomials 289
Lie symmetry 255
Lobatto polynomials 289
Logistic sigmoid activation function 52

m
Markov model 36
Microcantilever beam 258, 260, 266, 268
Monotone inclusion problems 413

n
Nanobeam 238, 240, 242
Natural language processing 391
Nonlinear differential equation 215, 225, 228, 230, 232

o
One-parameter groups 254
Ordinal tuple 170

p
Parametric approach 127
Parametric intervals 186–188
Parametric representation of fuzzy number 101
Partial and total operations 172
Pock and Lorenz's variable metric forward–backward algorithm 422
Proximal point algorithm 415

q
Quantification dependence 177

r
Radon diffusion 105, 107, 108, 111
Rectified linear unit activation function 54
Regression problem 429
Resource allocation 369
Risk minimization 379
Robotics 393

s
Set inversion via interval analysis (SIVIA) 216, 223–225, 229, 230
S-field algebra 201
Sigmoidal function 3

Small-scale parameter 245–248
Speech emotion recognition 33
Splitting algorithms 415
Splitting inertial proximal point algorithm 421
Support vector machine 1, 38

t

Taylor-based integral quadrature 320
Transmission dynamics 371
Trigonometric Lagrange polynomials 282

u

Universal intervals 192, 193, 196, 197

v

Vibration 239, 241, 242, 244, 247, 253, 256, 258, 260
Volterra integral equation 5

w

Weighted residual methods 346
Whitham–Broer–Kaup equations 134

Printed and bound by CPI Group (UK) Ltd, Croydon, CR0 4YY

05/01/2023

03177925-0001